T0406870

Artificial Intelligence and Advanced Analytics for Food Security

Chandrasekar Vuppalapati

Lecturer, San Jose State University, San Jose, California, USA
Founder, Hanumayamma Innovations and Technologies, Inc., Fremont, USA

CRC Press
Taylor & Francis Group
Boca Raton London New York

CRC Press is an imprint of the
Taylor & Francis Group, an **informa** business
A SCIENCE PUBLISHERS BOOK

First edition published 2023
by CRC Press
6000 Broken Sound Parkway NW, Suite 300, Boca Raton, FL 33487-2742

and by CRC Press
4 Park Square, Milton Park, Abingdon, Oxon, OX14 4RN

CRC Press is an imprint of Taylor & Francis Group, LLC

Library of Congress Cataloging-in-Publication Data (applied for)

ISBN: 978-1-032-34618-2 (hbk)
ISBN: 978-1-032-34619-9 (pbk)
ISBN: 978-1-003-32309-9 (ebk)

DOI: 10.1201/9781003323099

Typeset in Times New Roman
by Radiant Productions

Preface

The World is at a critical juncture. The challenges to ending hunger, food insecurity, and all forms of malnutrition keep growing, causing detrimental effects. In 2021, almost 40 million people were facing Food Security Emergency or worse (IPC/CH Phase 4 or above) conditions, across 36 countries. The number of people who faced these dire consequences was four times that observed in 2020, and accounted for nearly one in three people who did not have access to adequate food, i.e., between 720 and 811 million people faced hunger. The COVID-19 pandemic has further highlighted the fragilities in our agrifood systems and the inequalities in our societies, driving further increases in world hunger and severe food insecurity. The most recent evidence available suggests that the number of people unable to afford a healthy diet around the world rose by 112 million to almost 3.1 billion, showing the impact of rising consumer food prices during the pandemic. The state of Food Security and Nutrition in the World 2022 report makes it easy to understand any lingering doubts that the world is moving backwards in its efforts to end hunger, food insecurity and malnutrition in all its forms.

Damage from the COVID-19 pandemic as well as the Russian invasion of Ukraine magnified the slowdown in the global economy and has increased an unprecedented threat to food security. Climate change and extreme weather have repetitively caused droughts, water deficits, and monsoon & crop failures and have inverted agricultural yields in many farm-focused countries across the world. In essence, humanity is at a critical juncture and this unprecedented movement has thrusted upon us—the dependency on agriculture professionals and technologists of the world—to develop climate smart food security enhanced innovative products to address the multi-pronged food security challenges.

Agricultural innovation is key to overcoming concerns of food insecurity; the infusion of data science, artificial intelligence, advanced analytics, satellites, sensor technologies, and climate modeling with traditional agricultural practices such as soil engineering, fertilizers, phenological stages, and agronomy is one of the best ways to achieve this. Data science helps farmers to unravel patterns in equipment usage and transportation and storage costs. As well as the impact of drought & heat waves on yield per hectare, and weather trends to better plan and spend resources. As part of the book, I have proposed macroeconomic pricing models that data mines financial signals, climatology, and the influence of global economic trends on small farm sustainability to provide actionable insights for farmers to avert any financial disasters due to recurrent economic & climate crises.

The mission of the book, Artificial Intelligence and Advanced Analytics for Food Security, is to prepare current and future data science, data engineering, software engineering teams with the skills and tools to fully utilize advanced data science. Through artificial intelligence, time series techniques, climatology, and economic models in order to develop software capabilities that help to achieve sustained food security for the future of the planet and generations to come! Ensuring food security is my personal and professional obligation, and the sole theme of the works of the book is to help achieve that goal, making the world a better place than what I have inherited. The book also covers:

- Advanced Time Series Techniques for creating forecasting food security models.
- Time Series advanced modeling techniques—Multivariate Vector Auto Regressive (VAR), Models, and Prophet model techniques.
- Food Security Drivers & Key Signal Pattern Analysis
- Energy Shocks and Macroeconomic Linkage Analytics

Contents

Preface iii

Section I: Advanced Analytics

1. Time Series and Advanced Analytics **3**
- Milk Pricing Linkage & Exploratory Data Analysis (EDA) 7
- Auto Correlation 15
- Partial Autocorrelation Function (PACF) 18
- Stationarity Check: Augmented Dickey Fuller (ADF) & Kwiatkowski-Phillips-Schmidt-Shin (KPSS) Tests 19
- Non-stationary Stochastic Time Series 27
- Granger's Causality Test 33
- Transformation & Detrending by Differencing 34
- Models 36
- VAR Linkage Model: CPI Average Milk Prices, Imported Oil Prices, and All Dairy Products (milk-fat milk-equivalent basis): Supply and Use 37
- Regressive Linkage Model: CPI Average Milk Prices, Imported Oil Prices, and All Dairy Products (milk-fat milk-equivalent basis): Supply and Use 62
- Prophet Time Series Model: CPI Average Milk Prices, Imported Oil Prices, and All Dairy Products (milk-fat milk-equivalent basis): Supply and Use 67

References 72

2. Data Engineering Techniques for Artificial Intelligence and Advanced Analytics **76**
- Food Security Data 77
- Data Encoding—Categorical to Numeric 78
- Data Enrichment 86
- Data Resampling 103
- Synthetic Minority Oversampling Technique (SMOTE) & Adaptive Synthetic Sampling (ADASYN) 113
- Machine Learning Model: Kansas Wheat Yield SMOTE Model 114
- Machine Learning Model: Kansas Wheat Yield with Adaptive Synthetic (ADASYN) Sampling 124

References 129

Section II: Food Security & Machine Learning

3. Food Security **135**
Domino Effect 138
Food Security is National Security! 138
Food Security Frameworks 142
The Food Security Bell Curve—Machine Learning (FS-BCML) Framework 149
Machine Learning Model: Who In the World is Food Insecure? Prevalence of Moderate or Severe Food Insecurity in the Population 177
Machine Learning Model: Who In the World is Food Insecure? Prevalence of Moderate or Severe Food Insecurity in the Population—Ordinary Least Squares (OLS) Model 191
Machine Learning Model: Who In the World is Food Insecure? Prevalence of Severe Food Insecurity in the Population (%) 194
Machine Learning Model: Who In the World is Food Insecure? Prevalence of Severe Food Insecurity in the Population—Ordinary Least Squares (OLS) Model 199
Guidance to Policy Makers 202
References 203

4. Food Security Drivers and Key Signal Pattern Analysis **210**
Food Security Drivers & Signal Analysis 213
Afghanistan 213
Afghanistan Macroeconomic Key Drivers & Linkage Model - Prevalence of Undernourishment 229
Afghanistan Macroeconomic Key Drivers, Food Security Parameters, & Linkage Model - Prevalence of Undernourishment 239
Sri Lanka 246
Sri Lanka Macroeconomic Key Drivers & Linkage Model - Prevalence of Undernourishment 257
Guidance to Policy Makers 263
References 264

Section III: Prevalence of Undernourishment and Severe Food Insecurity in the Population Models

5. Commodity Terms of Trade and Food Security **271**
Mechanics of Food Inflation 275
Machine Learning Model: Food Grain Producer Prices & Consumer Prices 283
Machine Learning Model: Prophet—Food Grain Producer Prices & Consumer Prices 303
FAO in Emergencies—CTOT & Food Inflation 306
Trade and Food Security 306
Food Security is National Security! 309
References 310

6. Climate Change and Agricultural Yield Analytics **313**
Wheat Phenological Stages 316
Climate Change & Wheat Yield 321
Wheat Futures 323

NOAA Star Global Vegetation Health (VH) 324
Coupled Model Intercomparison Project Climate Projections (CMIP) 326
Shared Socioeconomic Pathway (SSP) Projection Models 329
Kansas & Wheat Production 331
Mathematical Modeling 338
Machine Learning Model: Drought & Wheat Yield Production Linkage in Kansas 339
India & Wheat Production 357
Machine Learning Model: Heat Waves & Wheat Yield Production Linkage in India 371
Machine Learning Forecasting Model: Heat Waves & Wheat Yield Production in India 380
Machine Learning (Mid-century 2050) Projection Model: Heat Waves with Increased Frequencies and Intensities & Wheat Yield Production in India 384
Food Security and Climate Resilient Economy: Heatwaves and Dairy Productivity Signal Mining to create a Smart Climate Sensor for Enhanced Food Security 388
Machine Learning Model: Drought & Heatwave Signature Mining through the Application of Sensor, Satellite Data to reduce overall Food Insecurity 390
References 397

7. Energy Shocks and Macroeconomic Linkage Analytics **402**
Climate Change and Energy Shocks Linkage 403
Unexpected Price Shocks—Gasoline, Natural Gas, and Electricity 405
Impact of Higher Gas prices on Macroeconomic Level 406
The Channels of Transmission & Behavioral Economy 411
Fertilizer Use and Price 411
The U.S. Dollar Index 415
U.S. Dollar and Global Commodity Prices Linkage 416
Farm Inputs & Fertilizer Linkage Model 417
Machine Learning Model Energy Prices & Fertilizer Costs—Urea 423
Machine Learning Model Energy Prices and Fertilizer Costs—Phosphate 444
Machine Learning Model—Commodities Demand and Energy Shocks on the Phosphate Model 454
Machine Learning Model—Prophet Time Series Commodities Demand and Energy Shocks on Phosphate Model 460
References 465

Section IV: Conclusion

8. Future **473**

Appendices **474**
Appendix A—Food Security & Nutrition **474**
The 17 Sustainable Development Goals (SDGs) 474
Food Security Monitoring System (FSMS) 474
NHANES 2019–2020 Questionnaire Instruments—Food Security 476
Macroeconomic Signals that Could help Predict Economic Cycles 476
List of World Development Indicators (WDI) 477
Food Aids 478
The United Nations—17 Sustainable Development Goals (SDGs) 479

The Statistical Distributions of Commodity Prices in Both Real and Nominal Terms 480
Poverty Thresholds for 2019 by Size of Family and Number of Related Children Under 18 Years 483

Appendix B—Agriculture **484**
Agricultural Data Surveys 484
USDA—The Foreign Agricultural Service (FAS) Reports and Databases 484
USDA Data Products 486
Data Sources 487
Conversion Factors 488
National Dairy Development Board (NDDB) India 488
Worldwide—Artificial Intelligence (AI) Readiness 490

Appendix C—Data World **492**
Global Historical Climatology Network monthly (GHCNm) 492
Labor Force Statistics from the Current Population Survey 502

Appendix D—Data U.S. **508**
U.S. Bureau of Labor Statistics 508
Dollars/Bushel : Dollars/Tonne Converter 510
NOAA - Storm Events Database 511
Consumer Price Index, 1913 512

Appendix E—Economic Frameworks & Macroeconomics **516**
Macroeconomic Signals that could help Predict Economic Cycles 516
The U.S. Recessions 516
Labor Force Statistics from the Current Population Survey 521
United Nations Statistics Department (UNSD) 522
Reserve Bank of India—HANDBOOK OF STATISTICS ON INDIAN ECONOMY 524
Department of Commerce 525
United Nations Data Sources 526
Poverty Thresholds for 2019 by Size of Family and Number of Related Children Under 18 Years 533
Rice Production Manual 534

Index **537**

Section I: Advanced Analytics

Chapter 1

Time Series and Advanced Analytics

This chapter introduces Time Series (TS) modeling techniques. In accordance with classical data science KDD, the chapter starts with exploratory data analysis. Fundamental time series validation techniques such as autocorrelation and partial autocorrelation are introduced. The chapter provides an in-depth view on detecting non-stationary time series and fixing them to stationary before modeling the TS. Time Series modeling techniques are detailed next. Multivariate Vector Auto Regressive (VAR), Regressive Models, and Prophet model techniques have been studied. Finally, a time series analysis has been conducted on all USDA dairy products (milk-fat milk-equivalent basis): Supply, CPI of Average prices, and Imported Oil Energy prices to develop predictive models. Finally, the chapter concludes with the evaluation of time series mode models.

A time series is the universal data signature. Almost all meaningful measured quantities have time associated with them. A time series (TS) represents the value of a particular characteristic with respect to time. A generalized version of a time series consists of three components: trend, seasonality, and randomness. Mathematically a time series is represented as (please see Equation 1),

$$Y_t = m_t + s_t + \varepsilon_t \qquad \text{EQ. 1}$$

where: m_t is trend, s_t is seasonality, and ε_t is a random variable.[1]

The mentioned characteristics lead to the following behaviors.

- No trend or seasonality (white noise)—Figure 1

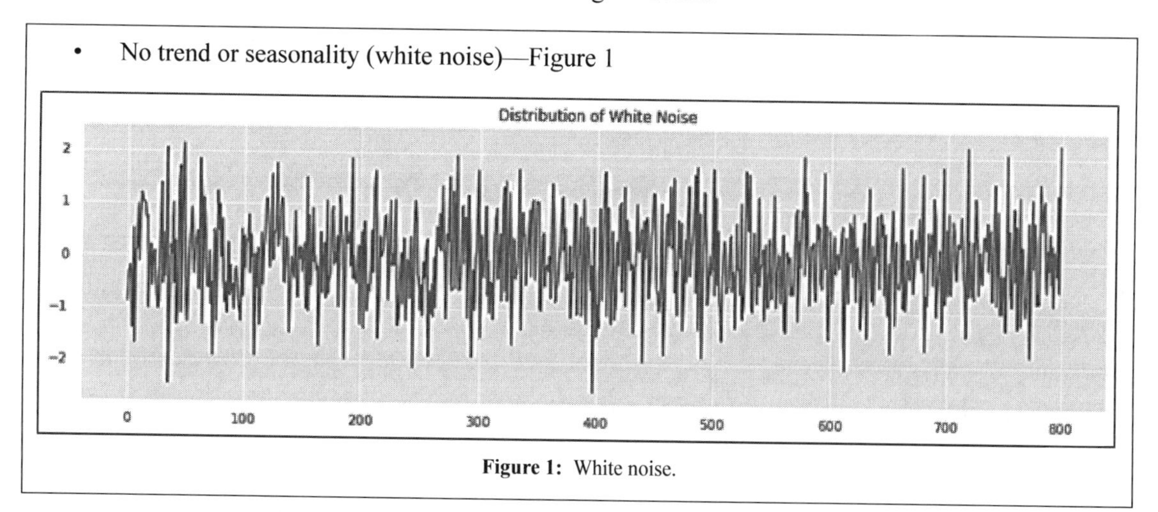

Figure 1: White noise.

- Has trend but no seasonality (a good example is yearly U.S. Dairy Shipments to states and territories)—Figure 2.

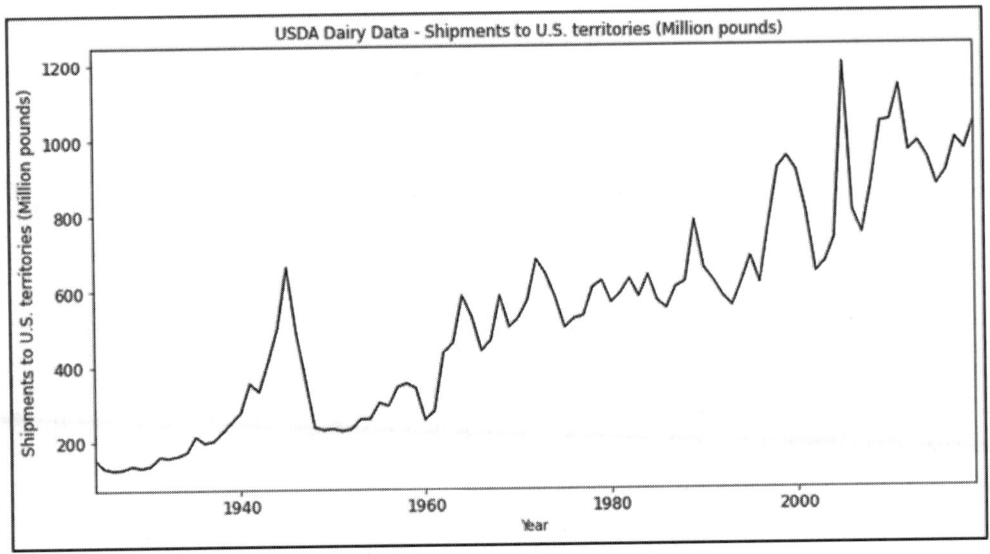

Figure 2: U.S. Dairy Shipments.

One can see an increase in shipments (million pounds) on a yearly basis but not on a seasonal basis.

- Has seasonality, but no trend

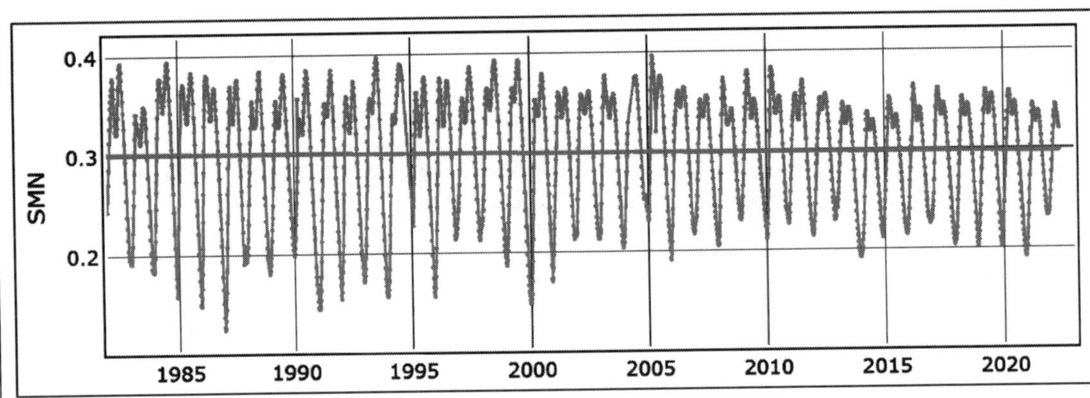

Figure 3: SMN.

A common characteristic of a time series that has seasonality without trend would be a repetitive pattern that occurs periodically. In a similar tone, capturing the reading of a particular scalar quantity on a regular basis—take the example of a No noise (smoothed) Normalized Difference Vegetation Index[2] (SMN)—one that is captured by weather satellites at a Global resolution of 4 km, a validated 7-day composite. The data in the above Figure is from STAR—Global Vegetation Health Products:[3] Province-Averaged VH—for California Crop land (please see Figure 3).

- Has both seasonality and trend (Figure 4).

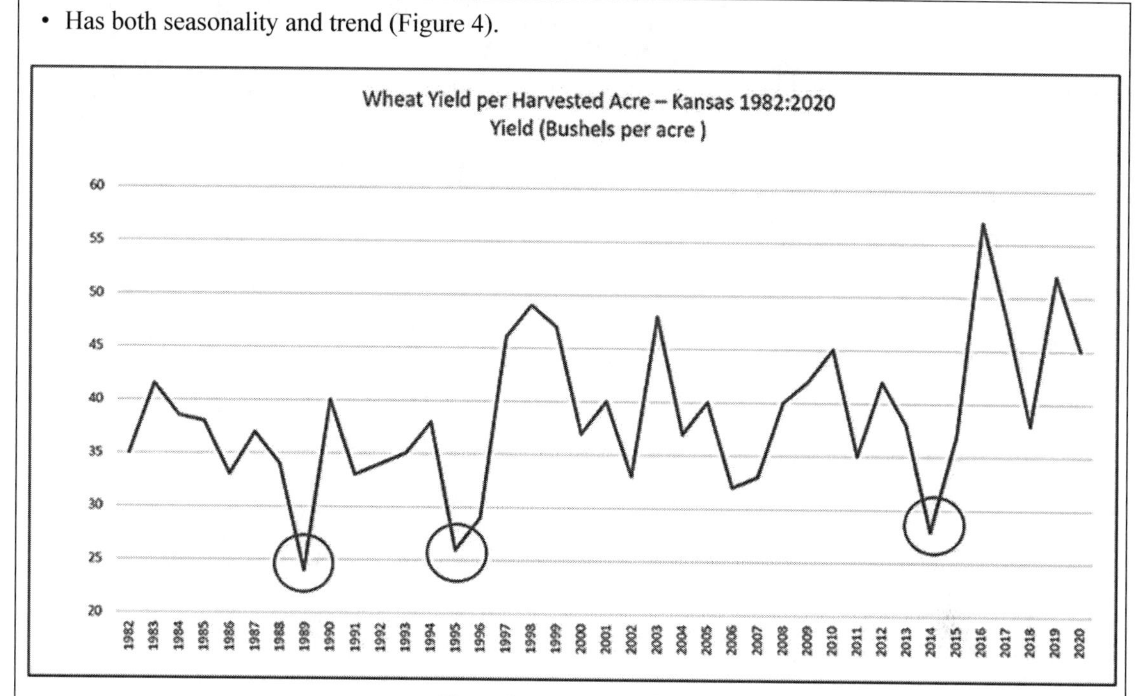

Figure 4: Kansas Wheat Yield.

Kansas state wheat yield per harvested acre (1982:2020) exhibits both trend and seasonality, especially during a drought.

The data points of a Time series can be stationary or non-stationary—the data points are often non-stationary or have statistical means, variances, and covariances that change over time. Non-stationary behaviors can be *trends, cycles, random walks,* or combinations of all three. To extract maximum information from the time series and to forecast it, the *data points need to be stationary*. Non-stationary data, as a rule, is *unpredictable* and cannot be modeled or forecasted. The results obtained by using a non-stationary time series may be *spurious* in that it may indicate a relationship between two variables that does not exist. In essence, we cannot construct a meaningful predictive model using non-stationary datasets. We need to transform a non-stationary data point into one that is stationary before applying it to time series models. To receive consistent, reliable results, the non-stationary nature of the data needs to be transformed to a stationary form. The following diagram shows the processing of time series. Exploratory data analysis (EDA) follows standard data processing checks (please see Figure 5): inspect data, check for nulls, imputation strategy, and dataset creation for processing the time series.

Let us understand how a multivariate time series is formulated. The simple K-equations of a multivariate time series where each equation is a lag (legacy) of the other series are given below. X is the exogenous series here (please see Equations 2 & 3). The objective is to see if the series is affected by its own past and the past of the other series.[4]

$$y_{1,t} = f(y_{1,t-1} \ldots, y_{k,t-1}, \ldots y_{1,t-p} \ldots y_{k,t-p}, \ldots X_{t-1}\, X_{t-2} \ldots\ldots\ldots) \qquad \text{EQ. 2}$$

$$y_{k,t} = f(y_{1,t-1} \ldots, y_{k,t-1}, \ldots y_{1,t-p} \ldots y_{k,t-p}, \ldots X_{t-1}\, X_{t-2} \ldots\ldots\ldots) \qquad \text{EQ. 3}$$

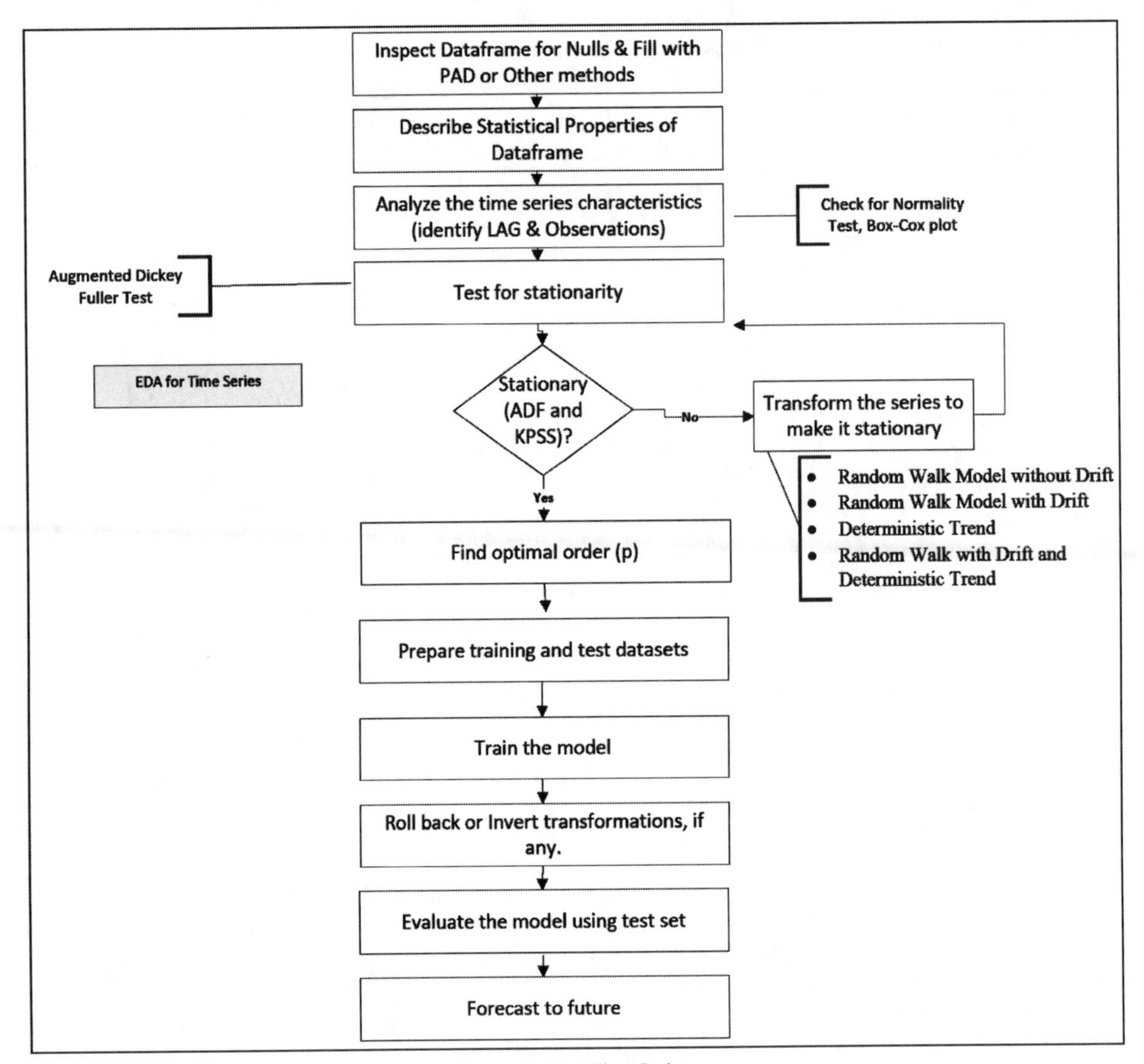

Figure 5: EDA Time Series.

This kind of series allows us to model the dynamics of the series itself and the interdependence of other series [1]. A good example of multiplicative time series interactions includes food nourishment (dependent variable) and macroeconomic, agriculture, climate change, and governmental policies (independent variables). A recent governmental action, for instance, that was instrumental when the price of cheese bottomed out in April 2020 was the Department of Agriculture announcing plans to spend $3 billion to buy food from farmers, including $100 million per month on a variety of dairy products. The announcement helped put an underprice floor that helped to restore food security.[5]

$$y_{\text{food security}} = f(X_{\text{macroeconomic key drivers}}, X_{\text{Food-Fuel shocks}}, X_{\text{Agriculture Co-movement}}, X_{\text{weather \& climate linkage}}) \quad \text{EQ. 4}$$

A schematic of linkages can be seen in the figure below. Mathematically, food security (Y) is a function of demand, macroeconomic conditions, agricultural inputs, farm supplies, governmental policies, energy markets (food-to-fuel), climatology (please see Figure 6), and shocks (wars, conflicts, energy crisis) (please see Equation 4).

The order of variables (M1 - M20 & A1-A20) is arranged to engineer the Machine Learning (ML) models. Each driver has its own sphere of influence, for instance, domestic food price level index a good *proxy* for food consumer price inflation (CPI), some a mere reflection of the whole, for example Gross

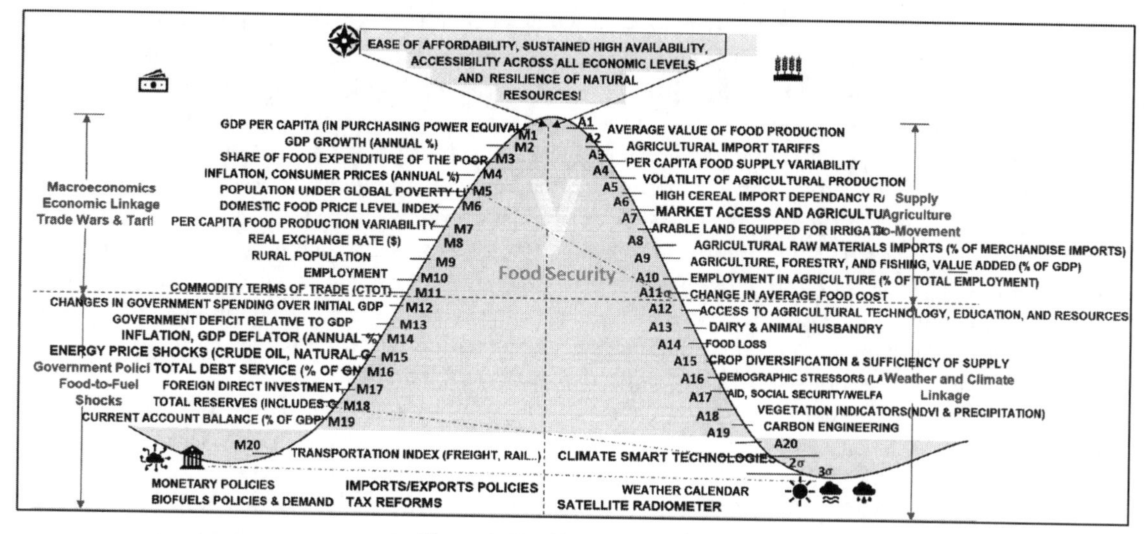

Figure 6: Food Security ML Model.

Domestic Product (GDP) summarizing the economic output of a nation, and some *exogenous* variables, for instance, Satellite Radiometer data or Normalized Difference Vegetation Index (NDVI) & geospatial Precipitation data. As such, there is no precedence among these individual drivers given that the causality of a variable with others normalizes the spread of influence between or among other co-moving, *orthogonal*, or substitute drivers.

Milk Pricing Linkage & Exploratory Data Analysis (EDA)

Exploratory data analysis is quite extensive for a multivariate time series. It is advisable to conduct all statistical tests to ensure a clear understanding of data distribution.

Common techniques include:

- Inspect the data for NULLs and fill them with imputation strategies.
- Analyze statistical properties.
- Inspect and understand data distribution.
- Inspect and fix representation under a variable class (if any) using SMOTE and ADASYN among others.

Common imputation strategies replace the nulls of a column with the mean, median or mode value with forward or backward filling. Some other techniques used are kNN (clustering) and Multivariate Imputation by Chained Equation (MICE). Many of these imputation strategies shift the statistical properties of a dataset. It is vital to recognize that when applying them to macroeconomic datasets, instead of applying classical techniques of filling nulls with the means, it is recommended to consider the *macroeconomic* key driver's country income group or band. I will cover these techniques in the next chapter.

Before we begin with time series analysis, let's see the importance of milk to food security. Milk is a nutrient-dense food product, contributing importantly to healthy diets, especially for children and women, but also to those for adolescents and the elderly. The composition properties can vary during the production and processing procedures, but high nutritional value remains. Milk supply[6] in low-and middle-income countries has been increasing over the past two decades in response to milk demand and consumption and offers numerous pathways to enhance food and nutrition security for most vulnerable populations [2].

Let us see the historical relationship among Milk Prices, Chedder Cheese, Yogurt, and Imported Oil. The time series datasets for these items are collected from the U.S. Bureau of Labor Statistics (BLS)—CPI Average Data[7] (please see Appendix D - U.S. Bureau of Labor Statistics). Specifically, average prices are retrieved from the CPI Data retrieval tool—One-*Screen Data Search*.[8] The average price of Fresh whole milk, fortified, sold per gallon regardless of packaging type. Includes organic and non-organic milk—*(Milk,*

Fresh, Whole, Fortified (Cost per Gallon/3.8 Liters)) (APU0000709112).[9] *The data used for the following analysis is not monthly data, but a 12-month average, resampled using time series techniques*. At the U.S. national level, the BLS publishes approximately 70 average prices for food items, 6 for utility gas (natural gas) and electricity, and 5 for automotive fuels. If the sample size is sufficiently big, all average prices are also published at the four major geographic regions—Northeast, Midwest, South, and West. In addition to national and regional levels, utility gas, electricity, and automotive fuel average prices are published for population class sizes, city size by region (regional/size cross classifications), divisions, and published areas.[10]

It is important to keep in mind that the Consumer Price Index (CPI) sample is optimized for the calculation of price indexes, not for the calculation of average prices. Also, due to the narrow definition of many average prices, the number of prices used to calculate some average prices can be substantially smaller than the number of prices used in index calculation. Average prices are estimated from a subset of CPI data and are subject to the same set of sampling and collection errors as the index.

The USDA ERS analyzed dairy price movements to understand how quickly and how completely swings in farm-level prices reach consumers.[11] The research focused on whole milk and Cheddar cheese. Findings confirm that *fluctuations* in retail prices for whole milk and cheddar cheese are generally less pronounced than swings in prices received by dairy farmers, and price changes experienced by farmers take time to reach consumers. The same behavior can be seen in other commodities.[12] Moreover, *farm-level and retail prices do not track as closely for Cheddar cheese* as they do for whole milk [3]. Over the past decade, milk prices have been volatile. In November 2014, the milk prices peaked at \$3.858 per gallon. By August 2016, the prices fell to \$3.062 per gallon. By April 2018, the average price (please see Figure 7): milk, fresh, whole, fortified price was down to \$2.839 per gallon. The milk prices exhibit co-movement with other dairy products.

Now let's analyze cheddar cheese prices. Before we go on to cheddar cheese numbers, let's see the importance of the cheddar cheese market to dairy farmers? It is pure economics—as the cheese market goes, so go farmer milk prices.[13] The relationship is determined by the USDA and market demand. The supply and demand combined with federal and state dairy policies are used to establish prices farmers receive for their raw milk [4].

Over the past 50 years, a complex system of public and private pricing institutions has developed to manage milk production, assembly, and distribution.[14] Milk pricing in the US is part market-determined, and part publicly administered through a wide variety of pricing regulations. Some of the regulations include milk price supports, Federal milk marketing orders, import restrictions, export subsidies, domestic and international food aid programmes, state-level milk marketing programmes, and a multi-state milk pricing organisation. Non-government pricing institutions such as the dairy cooperative are also important. All government and

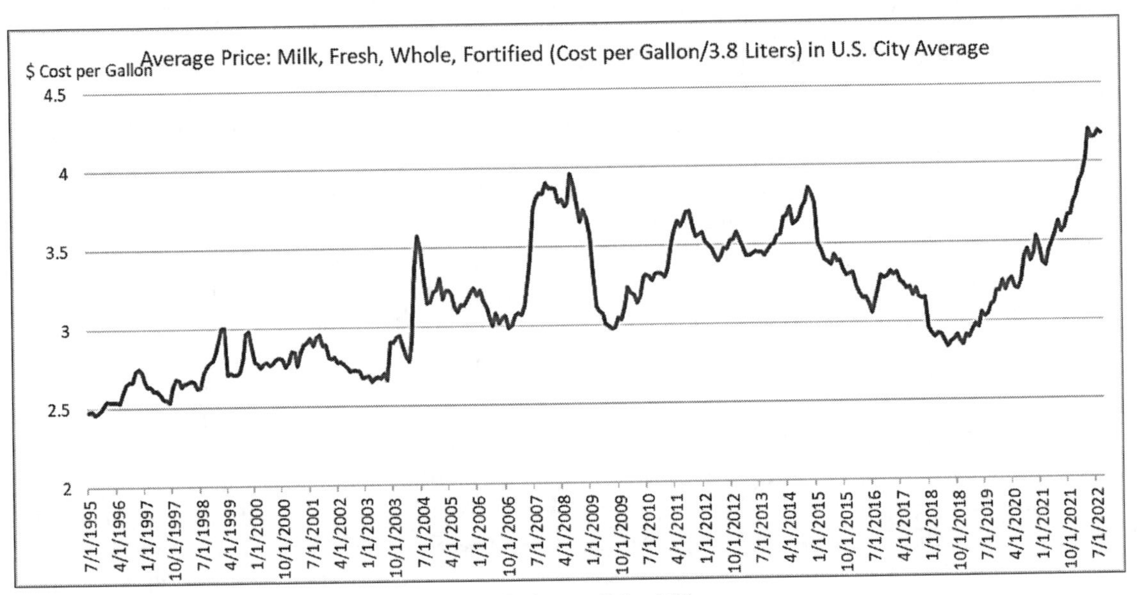

Figure 7: Average Price: Milk.

non-government institutions work together to ensure that the public gets the needed milk, while the dairy industry gets the economic returns required to provide the milk [5].

Most milk is priced according to its end use, with products grouped into four classes:

- Class I is milk used for fluid, or beverage products;
- Class II is milk going into 'soft' manufactured products such as sour cream, cottage cheese, ice cream, and yogurt;
- Class III is milk used for making hard cheeses and cream cheese;
- And Class IV milk is used to make butter and dry products.

In order to calculate the base Class I skim milk price, both the Class III and IV advanced skim milk pricing factors must be calculated. In essence, the average of the advanced Class III and IV skim milk pricing factors, plus $0.74, determines the advanced base class I skim (please see Table 1) milk price.[15]

Table 1: Milk Price Calculation.

Step	Advanced Component Price		Differential		
1	Average of Advanced Class III and Class IV Skim Milk Pricing Factor ($/cwt)	+	$0.74 ($/cwt)	=	**Base Class 1 Skim Milk Price ($/cwt)**
	Advanced Commodity Price Make Allowance		Yield		Advanced Component Price
2	Advanced Butter Price ($/lb) - $ 0.1715 ($/lb)	**X**	1.211 (lb butter/lb butterfat)	=	Advanced Butterfat Pricing Factor ($/lb)
	Calculated in Step 1 Yield		Calculated in Step 2		
3	Base Class 1 Skim Price ($/cwt) X 0.965 (cwt skim/cwt milk)	+	Advanced Butterfat Pricing Factor ($/lb) X 3.5 (lb butterfat/cwt milk)	=	Base Class 1 Price($/cwt)

In summary, for most, farm milk price[16] is based on federal minimums, plus (or minus) various premiums, hauling charges or other pricing factors not regulated under Federal Orders.[17] The federal order "blend price", which represents most of the farm price, is a weighted average of the four class prices.

- Class III prices are one of the movers of the Class I price, which impacts every area, but Class III prices are especially important in the Upper Midwest, and to a lesser extent in California, where Class III utilization is so high.
- Class III prices are driven by the *wholesale prices of certain cheeses, bulk butter, and dry whey powder*. The cheese price is a weighted average of prices for bulk cheddar cheese made in blocks, and barrel cheese, used primarily to make processed cheese products.
- The Class III price changes are based on the combination of price movements in these four dairy product categories.

For the year 2018–2019, in the last few months since April 2019, the cheese price has been mostly pulling up but is on the low side, the butter price has been high and stable, and the dried whey price has been declining. Since last September 2018, the Class III price has been dropping but looks to be stabilizing in 2019. It is hoped that cheese prices will continue to recover, assuming cheese stocks return to more normal levels [6].

In summary, the cheese price pretty much determines the farmers milk price. And, that the price of milk in the store has little to do with the dairy farmer price. In essence, cheese prices are the key market indicators for the entire dairy industry. Retailers usually give cheese the highest profit margin off all dairy products; since the 1980s (Figure 8), retail cheese prices have risen more rapidly than wholesale prices.[18] Dairy farmers receive about 48 cents for the milk used in $1 worth of cheese at the retail level; processing, packaging, and marketing take the other 52 cents [7]. The sustainability of small farmers is highly influenced by milk prices that are based on cheese prices. The basic association principle dictates that small farmer sustainability is

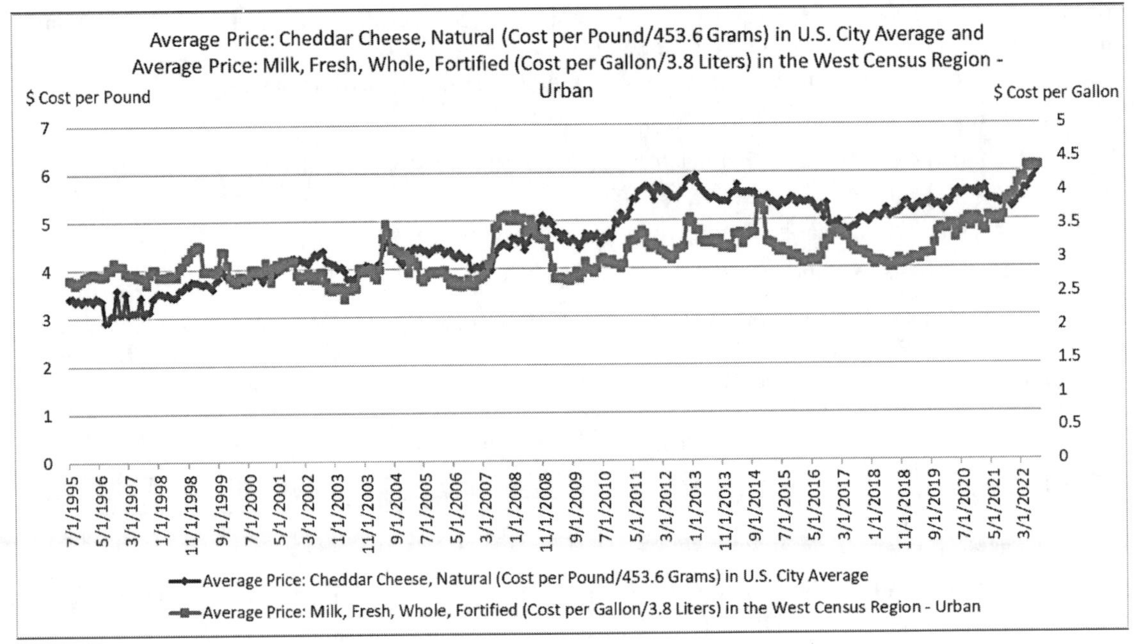

Figure 8: Average Price Cheddar Cheese and Milk.

dependent on cheese. Finally, Cheese production tends to follow seasonal milk production patterns and price equilibrium has been maintained thus far [6]!

	Cheese pretty much determines the farmers milk price.[19]

The price dynamics of cheese has changed recently. Cheddar Cheese Pricing Rules the Dairy Industry and the price elasticity of demand for cheese is like that for any other commodity: when cheese is cheaper, more is consumed. When cheese is expensive, less is consumed.[20] When cheddar production has taken a major dive during 2018 and 2019, the price (classical demand and supply) of *chedder cheese has increased and driven the price of milk* [8].

The cheese market was not volatile at all. Shifts in pandemic demand wreaked havoc on a market that's rarely so volatile [9]. During the peak of pandemic-related fears in March and April 2020, consumers rushed to grocery stores to stockpile cheese for their coming quarantines. Retail sales surged more than 70 percent from a year earlier. However, that wasn't enough to offset the drop in demand from shuttered restaurants and educational institutions, which together account for at least half the sales of bulk commodity cheese, according to industry estimates. The fall in cheese demand spilled over into the dairy market, contributing to a *plunge* in milk prices (Figure 9).

Like the price of oil, silver and hogs, cheese prices are set, in part, by traders in commodity markets. Each trading day at 11 a.m. Chicago time, the Chicago Mercantile Exchange operates a 10-minute session in which buyers and sellers[21]—typically large dairy food cooperatives, cheese producers or other companies active in the industry—electronically trade roughly 40,000-pound truckloads of young, mild Cheddar. All the techniques that analyze commodity markets apply to cheese! The price correlation [10] between Milk and

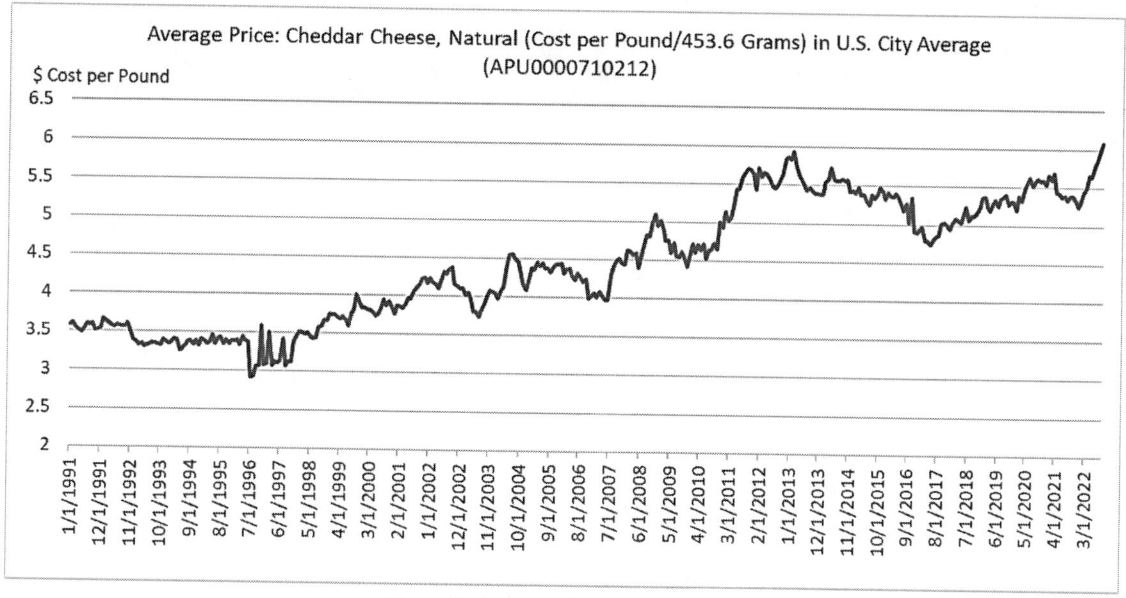

Figure 9: Average Price Cheddar Cheese.

Chedder Cheese is 74% (Figure 10). The sources for prices are from ST. Louis FED (APU0000710212[22] and APU0000709112[23]).

	Average Price: Cheddar Cheese, Natural (Cost per Pound/453.6 Grams) in U.S. City Average	Average Price: Milk, Fresh, Whole, Fortified (Cost per Gallon/ 3.8 Liters) in U.S. City Average
Average Price: Cheddar Cheese, Natural (Cost per Pound/453.6 Grams) in U.S. City Average	**1**	
Average Price: Milk, Fresh, Whole, Fortified (Cost per Gallon/3.8 Liters) in U.S. City Average	**0.755909543**	**1**

Next, let's see price elasticity of Milk and Yogurt. The data is very limited (1995:2004). It is (series id: APU0000710122) Yogurt, natural, fruit flavored, per 8 oz. (226.8 gm) in U.S. city average, average price, not seasonally adjusted (available from January, 1980 to March 2003). And a new series (APU0000FJ4101 "Yogurt, per 8 oz. (226.8 gm) in U.S. city average, average price, not seasonally adjusted") available from April 2018 to April 2022, the time at which this book was being written. The new series covers only Yogurt whereas the old series covered diverse flavors.

The price relationship of Yogurt to milk is negative (intuition is derived from a limited dataset). Though not a substitute, the price relationship is based on the manufacturing process. Firstly, the Class II family of dairy milk goes into 'soft' manufactured products such as sour cream, cottage cheese, ice cream, and yogurt. The primary reason for less than perfect Class II and Class III products is the utilization of Nonfat Dry Milk (NDM) in them. As will be shown in the VAR ML Model section, significant quantities of NDM are used to make cheese (excluding cottage cheese) and soft products (including cottage cheese, sour cream, ice cream

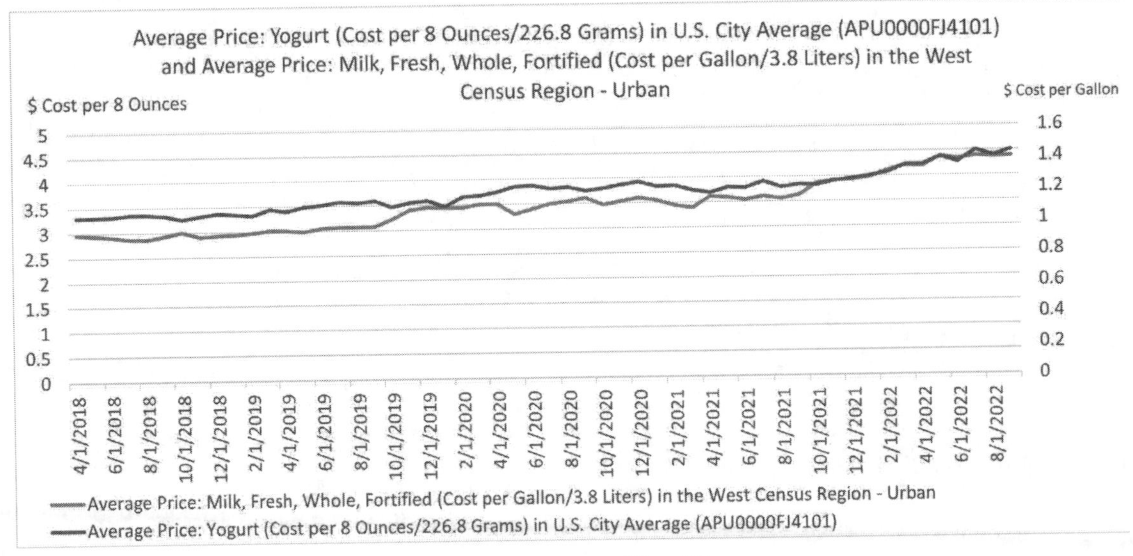

Figure 10: Average Price Milk and Yogurt.

and yogurt). The quantity of NDM used to make soft products increases when milk supplies are tight as was the case in the late 1970s and the early 1990s. However, since the establishment of Class IIIA there has been a sharp increase in the utilization of NDM to make soft products [11]. The quantity of NDM to make cheese likewise increased sharply after Class IIIA pricing began. It suggests that as much as half of the nonfat solids contained in soft products came from NDM, historically, and surplus milk is converted into NDM for storage [11—Chapter 4]. Another chief reason for the negative correlation with Yogurt prices is justified (Figure 11) (excessive milk implies depressed milk prices (classic demand and supply))!

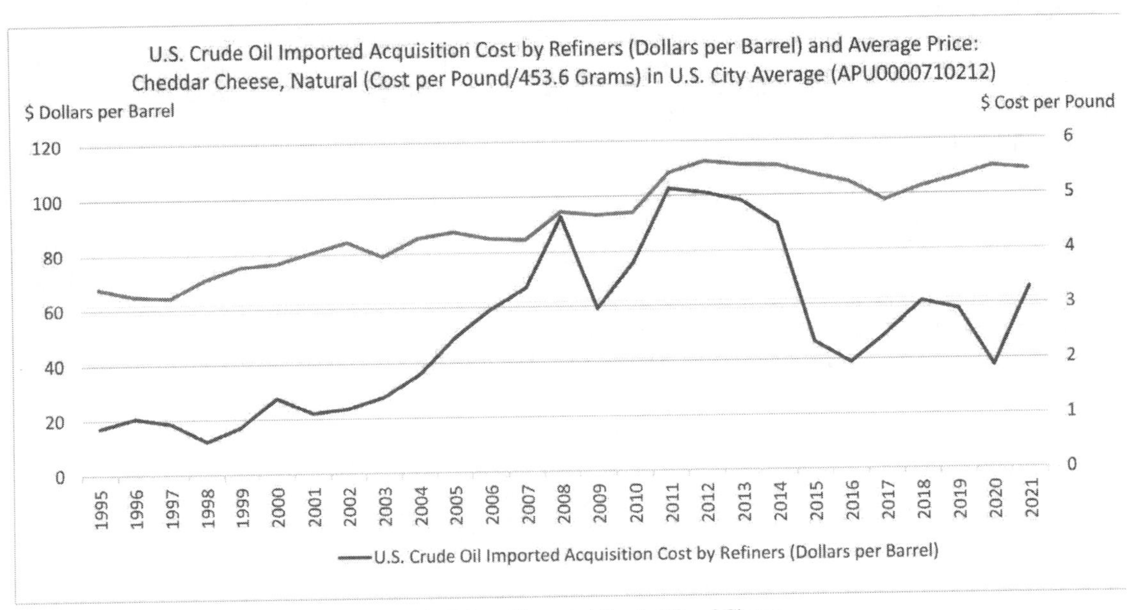

Figure 11: Price of Imported Crude Oil and Cheese.

Historically, oil prices have served as an indicator of global demand for products [12]. As shown in the ML model, the relationship is significantly weaker for cheese (as shown below) and nonexistant for butter.

Except for the peak curves in 2008 (due to the great recession) and 2011, Oil and Milk prices exhibit a weaker relationship (Figure 12).

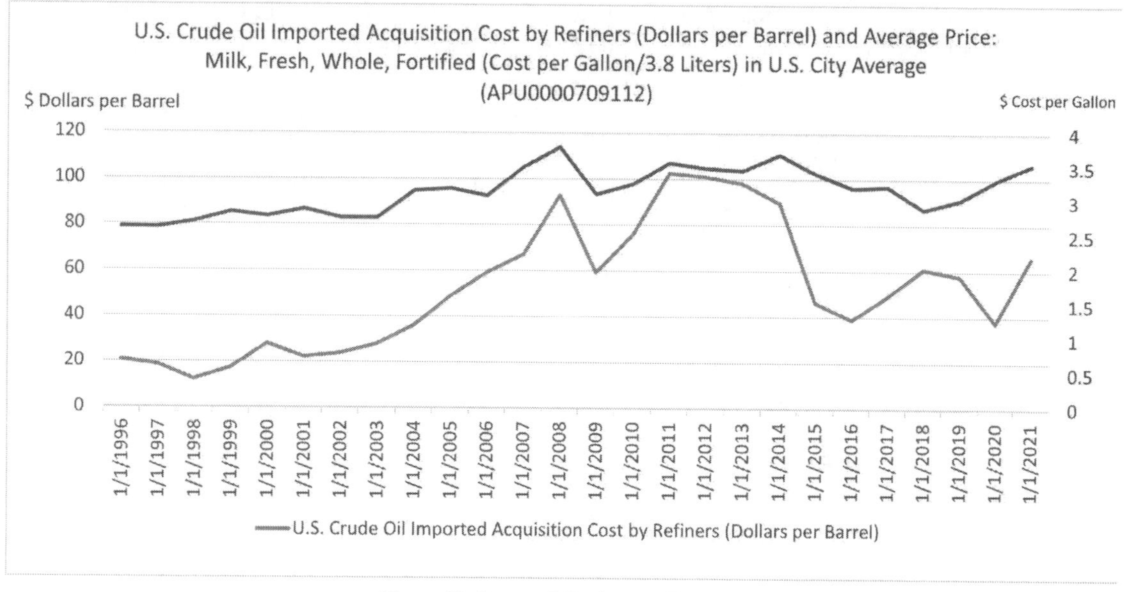

Figure 12: Imported Crude vs. Milk Prices.

The predictive models developed below have shown the relationship between Milk Prices, Cheddar Cheese, Yogurt Ice Cream, Imported Oil Gas, and Stocks to Use ratio. The Stocks to Use ratio was computed based on Ending stocks (Million pounds) divided by Total Disappearance. The Total Disappearance is computed as, the Total supply (Million pounds) less Ending stocks (Million pounds).

```
dfAlldairyproductsSupplyandUse['TotalDisappearance'] = dfAlldairyproductsSupplyandUse['Total supply (Million pounds)'] - dfAlldairyproductsSupplyandUse['Ending stocks (Million pounds)']
dfAlldairyproductsSupplyandUse['StockstoUse'] = 100 * (dfAlldairyproductsSupplyandUse['Ending stocks (Million pounds)']/dfAlldairyproductsSupplyandUse['TotalDisappearance'])
```

The milk prices could be forecasted with a higher accuracy (104.943%) when the time series model includes Price of Chedder Cheese, Imported Crude Oil Price ($/barrel)—Real, Stocks to Use, and yougurt Ice Cream prices (Figure 13).

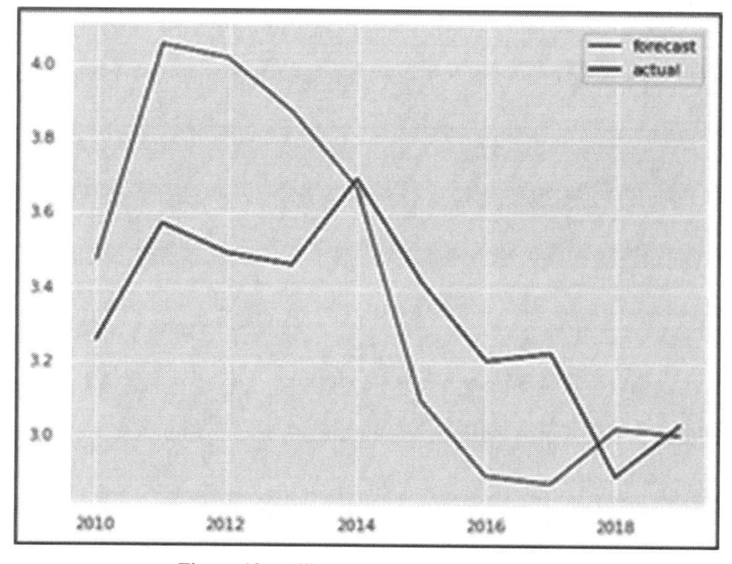

Figure 13: Milk process, actual vs. forecast.

A lack of key added parameters resulted in the model exhibiting a lower predictability and accuracy. The following Milk price prediction included only Oil and Stocks to Use (Figure 14).

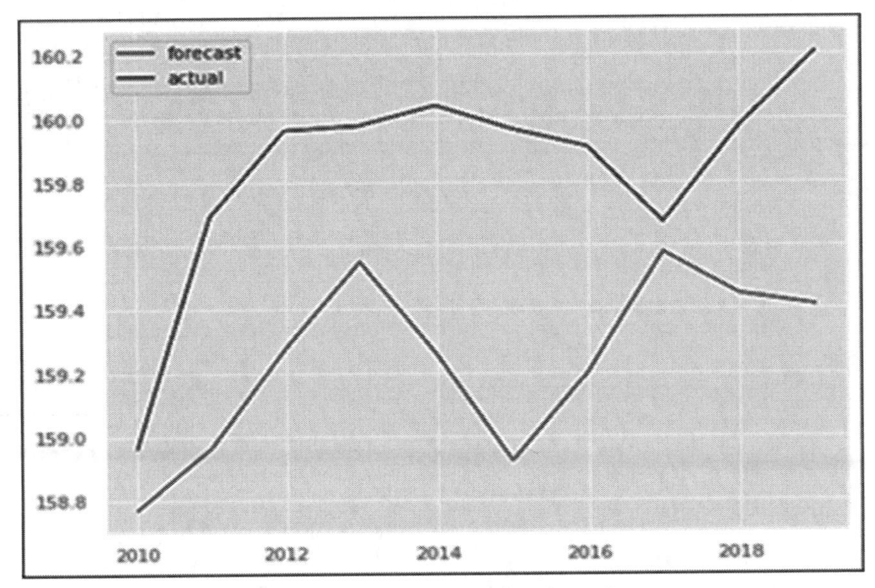

Figure 14: Milk Prices (actual vs. forecast).

The same was observed with milk price predictions with only Oil and Cheese (Figure 15).

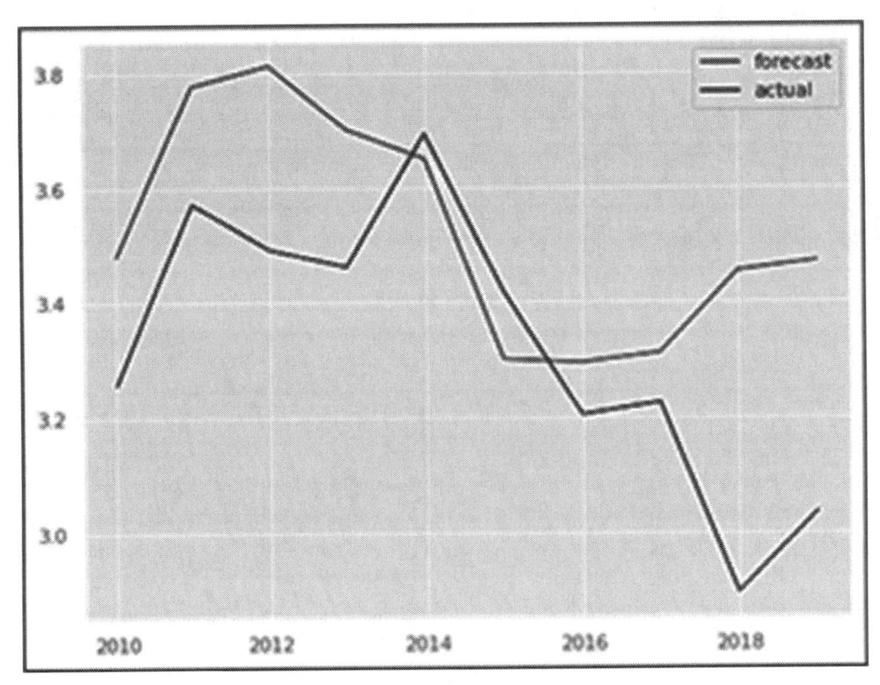

Figure 15: Cheese Prices actual vs. forecast.

Seasonality is less obvious but remains a distinct element of US price patterns.[24]

- Farm milk prices have a distinct *seasonal pattern*, with lows in the early spring and highs in the fall.
- The seasonal nadir and zenith of milk prices has shifted as milk production has moved to the Southwest [13].
- New cycles obscure the seasonal component.

Trend is no longer particularly interesting from a practical standpoint, although it remains economically important:

- Milk price trends are driven by linear trends in productivity and population (y = 1119x + 87969. & R^2 = 0.8225) (please see figure 16 shown below).
- Productivity has grown more quickly than population.
- Thus, farm milk prices increase at a lower rate than inflation or declining "real" farm milk prices.

Only when feed prices are very high relative to those for milk, do farmers make management decisions about yield that are short term (Figure 16 and Figure 17). Otherwise, the key decision is "how many cows?" are required [13].

Cow Necklace[25] sensors are essential to improve productivity of dairy farms. In this era of labor shortage, climate change, and unprecedented inflation, Climate-Smart sensors are instrumental to small farmer sustainability.

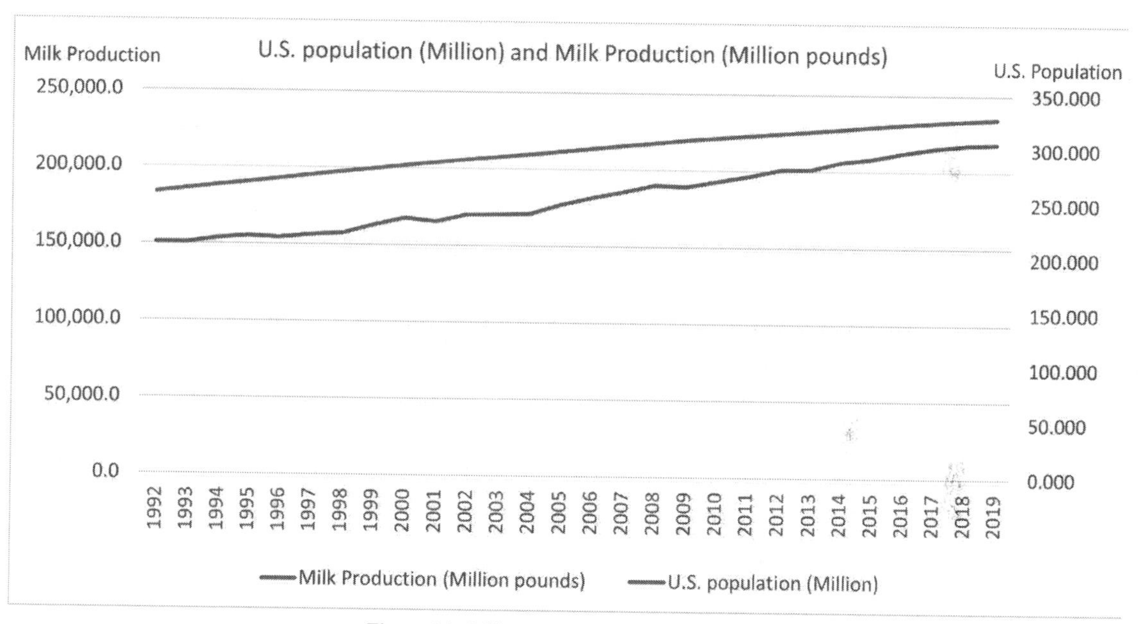

Figure 16: Milk Production & U.S. Population.

Auto Correlation

The term autocorrelation refers to the degree of similarity between (A) a given time series, and (B) a lagged version of itself, over (C) successive time intervals. In other words, autocorrelation is intended to measure the relationship between a variable's *present value and past values* that you may have access to. We can calculate the correlation for time series observations with observations with previous time steps, called lags or legacies. Because the correlation of the time series observations is calculated with values of the same series at previous times, this is called a serial correlation,[26] or an autocorrelation [14]. Auto-correlation or serial correlation can be a significant problem in analyzing historical data if we do not know how to look out for it [15].

To analyze auto correlation, one of the important items to be considered is "LAG". A time series (y) with k^{th} lag is its version that is 't-k' periods behind in time, i.e., y(t-k). A time series with lag (k=1) is a version of the original time series that is 1 period behind in time, i.e., y(t-1). For instance, consider the autocorrelation of U.S. Energy Information and Administration[27] (EIA) short term energy outlook (please see Appendix D—EIA Short-Term Energy Outlook—Monthly Average Imported Crude Oil Price (please see Figure 18)) of Imported Gasoline yearly prices (based on Base CPI (5/2022) and EIA Short-Term Energy Outlook,

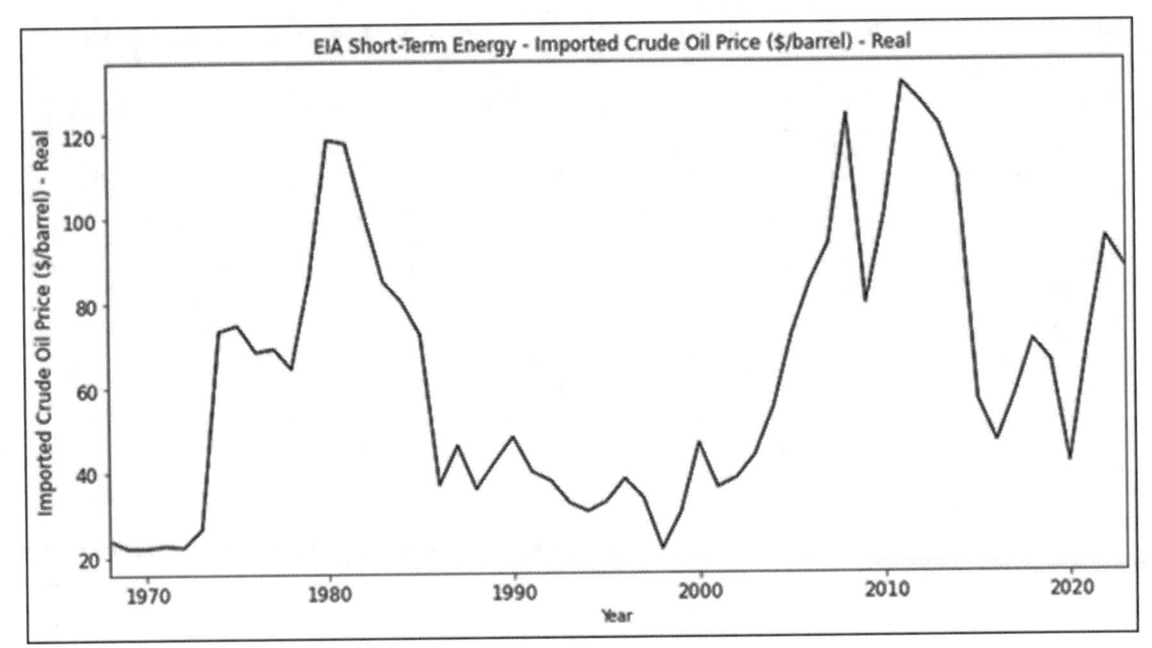

Figure 17: Crude Oil Prices real.

May 2022)—current year price with the price 'k' periods behind in time. So, the autocorrelation with lag (k=1) is the correlation with current year price y(t) and last year's price y(t-1). Similarly, for k=2, the autocorrelation is computed between y(t) and y(t-2). Of course, given the dynamics of a market, a year is too long to average the trend of a price. As seen in the figure below (imported oil gas real prices data)[28], the auto correlation: lag value along the x-axis and the correlation on the y-axis between –1 and 1.

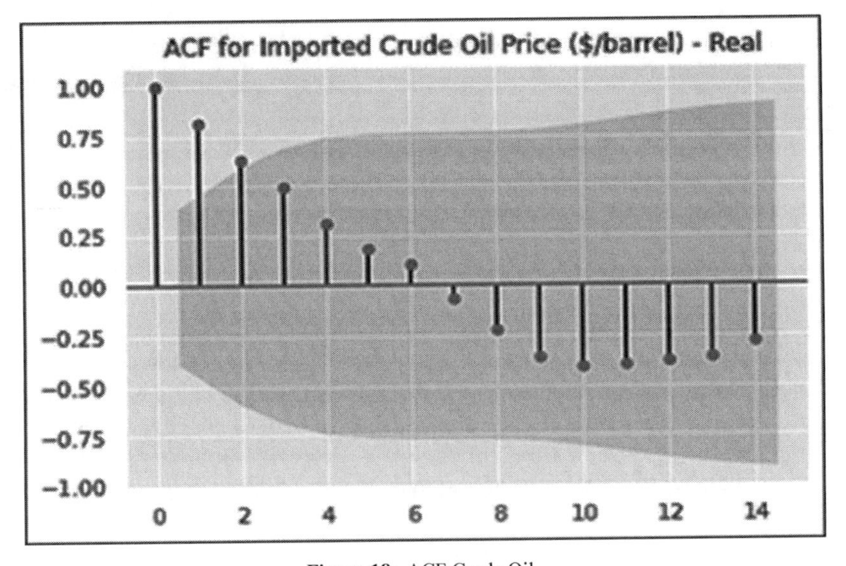

Figure 18: ACF Crude Oil.

Confidence intervals are drawn as a cone. By default, this is set to a 95% confidence interval, suggesting that correlation values outside of this code are very likely a correlation and not a statistical fluke [14]. We see here from the above plots, the autocorrelation of +1 which represents a perfect positive correlation which means, an increase seen in one time series leads to a proportionate increase in the other time series. We need to apply a transformation and neutralize this to make the series stationary. It measures linear relationships; even if the autocorrelation is minuscule, there may still be a nonlinear relationship between a time series and a lagged version of itself [15].

In Python, the following code provides auto correlation:

```
import pandas as pd
import matplotlib.pyplot as plt
import statsmodels.api as sm
MAX_LAGS=3
# plots the autocorrelation plots for each stock's price at 150 lags
for i in dataset:
    sm.graphics.tsa.plot_acf(dataset[i])
    plt.title('ACF for %s' % i)
    plt.show()
```

"plot_acf()" provides auto correlation with the option of specifying a lag.
You can compute the auto correlation by using the following (please see Equation 5):

$$r_k = \frac{\sum_{t=k+1}^{T} (Yt - \bar{Y})\,(Yt - k - \bar{Y})}{\sum_{t=1}^{T} (Yt - \bar{Y})2} \qquad \text{EQ. 5}$$

A closer examination of the two components of the numerator shows that the mean of the original time series, mean (y) or ($\bar{Y}$), is being subtracted from Y_t and lagged version of Y_{t-k}, not mean (Y_t) and mean ($Y_{(t-k)}$), respectively. The denominator is squared version of Y_t and subtract mean $\bar{Y}$.

Year	Imported Crude Oil Price ($/barrel) - Real
1990	47.96674111
1991	39.65629366
1992	37.42208821
1993	32.20197041
1994	30.22895147
1995	32.43889193
1996	37.9057184
1997	33.21346386
1998	21.34635503
1999	29.89817912
2000	46.42507575
2001	35.82253037
2002	38.01608146
2003	43.45465626
2004	54.79021153
2005	72.19574439
2006	84.47988287
2007	93.44000011
2008	124.0172316
2009	79.3436134
2010	100.2661133
2011	131.5153802
2012	126.9678128
2013	121.4626899
2014	109.1936852
2015	56.38668638
2016	46.50146254
2017	57.62407155
2018	70.44533058
2019	65.37012557
2020	41.46524904

Lag(k=1)

Year	Imported Crude Oil Price ($/barrel) - Real
1990	47.96674111
1991	39.65629366
1992	37.42208821
1993	32.20197041
1994	30.22895147
1995	32.43889193
1996	37.9057184
1997	33.21346386
1998	21.34635503
1999	29.89817912
2000	46.42507575
2001	35.82253037
2002	38.01608146
2003	43.45465626
2004	54.79021153
2005	72.19574439
2006	84.47988287
2007	93.44000011
2008	124.0172316
2009	79.3436134
2010	100.2661133
2011	131.5153802
2012	126.9678128
2013	121.4626899
2014	109.1936852
2015	56.38668638
2016	46.50146254
2017	57.62407155
2018	70.44533058
2019	65.37012557
2020	41.46524904

Lag(k=2)

Year	Imported Crude Oil Price ($/barrel) - Real
1990	47.96674111
1991	39.65629366
1992	37.42208821
1993	32.20197041
1994	30.22895147
1995	32.43889193
1996	37.9057184
1997	33.21346386
1998	21.34635503
1999	29.89817912
2000	46.42507575
2001	35.82253037
2002	38.01608146
2003	43.45465626
2004	54.79021153
2005	72.19574439
2006	84.47988287
2007	93.44000011
2008	124.0172316
2009	79.3436134
2010	100.2661133
2011	131.5153802
2012	126.9678128
2013	121.4626899
2014	109.1936852
2015	56.38668638
2016	46.50146254
2017	57.62407155
2018	70.44533058
2019	65.37012557
2020	41.46524904

Lag(k=5)

Year	Imported Crude Oil Price ($/barrel) - Real
1990	47.96674111
1991	39.65629366
1992	37.42208821
1993	32.20197041
1994	30.22895147
1995	32.43889193
1996	37.9057184
1997	33.21346386
1998	21.34635503
1999	29.89817912
2000	46.42507575
2001	35.82253037
2002	38.01608146
2003	43.45465626
2004	54.79021153
2005	72.19574439
2006	84.47988287
2007	93.44000011
2008	124.0172316
2009	79.3436134
2010	100.2661133
2011	131.5153802
2012	126.9678128
2013	121.4626899
2014	109.1936852
2015	56.38668638
2016	46.50146254
2017	57.62407155
2018	70.44533058
2019	65.37012557
2020	41.46524904

Lag(k=15)

Figure 19: Lag Order.

For K=1 Lag order, the Auto Correlation computes the difference between the current and Y_{t-1} order. For K = 2, it is Y_{t-2}. The following code outputs autocorrelation for specified lags (K=15) (please see Figure 19).

```
def compute_auto_corr(df, nlags=2):

   def autocorr(y, lag=2):

   y = np.array(y).copy()
   y_bar = np.mean(y) #y_bar = mean of the time series y
   denominator = sum((y - y_bar) ** 2) #sum of squared differences between y(t) and y_bar
   numerator_p1 = y[lag:] - y_bar #y(t)-y_bar: difference between time series (from 'lag' till the end) and
y_bar
   numerator_p2 = y[:-lag] - y_bar #y(t-k)-y_bar: difference between time series (from the start till lag)
and y_bar
   numerator = sum(numerator_p1 * numerator_p2) #sum of y(t)-y_bar and y(t-k)-y_bar
   return (numerator / denominator)

 acf = [1] #initializing list with autocorrelation coefficient for lag k=0 which is always 1
 for i in range(1, (nlags + 1)):
   acf.append(autocorr(df.iloc[:, 0].values, lag=i)) #calling autocorr function for each lag 'i'
 return np.array(acf)
```

Output:

```
Imported Crude Oil Price ($/barrel) - Real
statsmodels acf:
[1. 0.81555769 0.63224325 0.48934334 0.30691418 0.18474574
 0.1102879 -0.07219308 -0.22628682 -0.36034683 -0.4097954 -0.39654324
 -0.3777347 -0.36002538 -0.28309978]
compute_auto_corr:
[1. 0.81555769 0.63224325 0.48934334 0.30691418 0.18474574
 0.1102879 -0.07219308 -0.22628682 -0.36034683 -0.4097954 -0.39654324
 -0.3777347 -0.36002538 -0.28309978]
```

We can calculate the correlation for time series observations with observations with previous time steps, called lags. Because the correlation of the time series observations is calculated with values of the same series at previous times, this is called a serial correlation, or an autocorrelation.

Partial Autocorrelation Function (PACF)

A partial autocorrelation is a summary of the relationship between an observation in a time series with observations at prior time steps with the relationships of intervening observations removed [14]. The partial autocorrelation at lag k is the correlation that results after removing the effect of any correlations due to the terms at shorter lags [16]. The autocorrelation for an observation and an observation at a prior time step is comprised of both the direct correlation and indirect correlations. These indirect correlations are a linear function of the correlation of the observation, with observations at intervening time steps.

It is these indirect correlations that the partial autocorrelation function seeks to remove (please see Figure 20). This is the intuition for the partial autocorrelation.

In the above diagram Partial Auto Correlation (PACF) for imported Crude Oil is shown. Compared to ACF, plots of the autocorrelation function and the partial autocorrelation function for a time series tell a *very different story*.

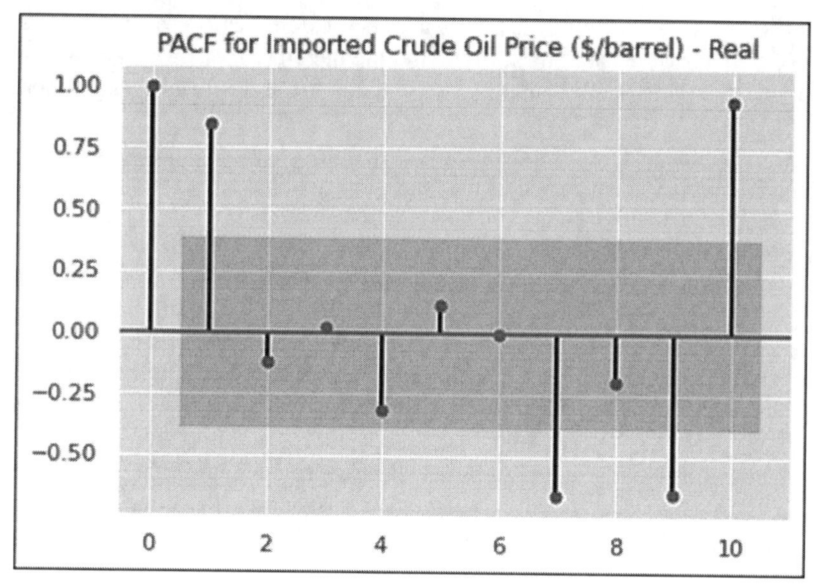

Figure 20: PACF Crude Oil.

Real Imported Crude Oil Prices – ACF & PACF Intuition

We know that the ACF describes the autocorrelation between an observation and another observation at a prior time step that includes direct and indirect dependence information. This means we would expect the ACF for the Imported Oil time series to be strong for a lag of k and the inertia of that relationship would carry on to subsequent lag values, trailing off at some point as the effect is weakened.

We know that the PACF only describes the direct relationship between an observation and its lag. This would suggest that there would be no correlation for lag values beyond k. PACF can only compute partial correlations for lags up to 50% of the sample size. So, for imported oil, a sample size of 12 can have PACF computed for lags < 12. For ACF, we can give the entire sample size as k. This is exactly the expectation of the ACF and PACF plots of imported real Oil prices (1995:2020).

In general, using k=10 is recommended (assuming a sufficient sample size - tests of the null hypothesis[29] of stationarity cannot be recommended for applied work, especially economics, unless the sample size is very large [17]) for non-seasonal data and for seasonal data, k=2 m where m is the period of seasonality. These suggestions were based on power considerations. We want to ensure that k is large enough to capture any meaningful and troublesome correlations.[30] For seasonal data, it is common to have correlations at multiples of the seasonal lag remaining in the residuals, so we wanted to include at least two seasonal lags [18]. For Oil markets, generally, two seasons (m=2) can be observed—summer is the peak demand season due to increased travel and the normal season is during the remainder of the year. Seasonal peaks usually appear in April, May, and November; troughs in January and September.[31]

Stationarity Check: Augmented Dickey Fuller (ADF) & Kwiatkowski-Phillips-Schmidt-Shin (KPSS) Tests

A time series is said to be stationary if it has the following three properties:

- It has a constant statistical mean, i.e., mean remains constant over time.
- It has a constant variance, i.e., variance remains constant over time.
- It has a constant covariance. The value of the covariance between the two time periods depends on *the lag between the two time periods* and not on the time at which the covariance is computed. This means that the covariance between series at time t=2 and t=4 should be roughly the same as the covariance between series at time t=7 and time t=9.

A stationary time series will have the same statistical properties, i.e., same mean, variance, and covariance no matter at what point we measure them, these properties are therefore *time invariant*. The series will have no predictable patterns in the long-term. Time plots will show the series to be roughly horizontal with a constant mean and variance.

Why do we need a stationary time series? Because if the time series is not stationary, we can study its behavior only for a single period. Each period of the time series will have its own distinct behavior and it is not possible to predict or generalize for future time periods if the series is not stationary.

A stationary time series will tend to return to its mean value and the fluctuations around this mean will have a constant magnitude. Thus, *a stationary time series will not drift too much from its mean value because of the finite variance* [19].

Consider, for example, the USDA ERS Food Availability (Per Capita) Data System (FADS) that includes three distinct but related data series on food and nutrient availability for consumption: food availability data, loss-adjusted food availability data, and nutrient availability data. The *data serves as proxies for actual consumption at the national level*, a good source for food security. Within these data systems, let us look at the Dairy Production—the case of Annual commercial disappearance, milk fat, 1909–2019 (millions of pounds), USDA Dairy Products:[32] dymfg.xls[33] from the Food Availability (Per Capita) Data System.[34] Please see Appendix D—All dairy products (milk-fat milk-equivalent basis): Supply and use (Figure 21).

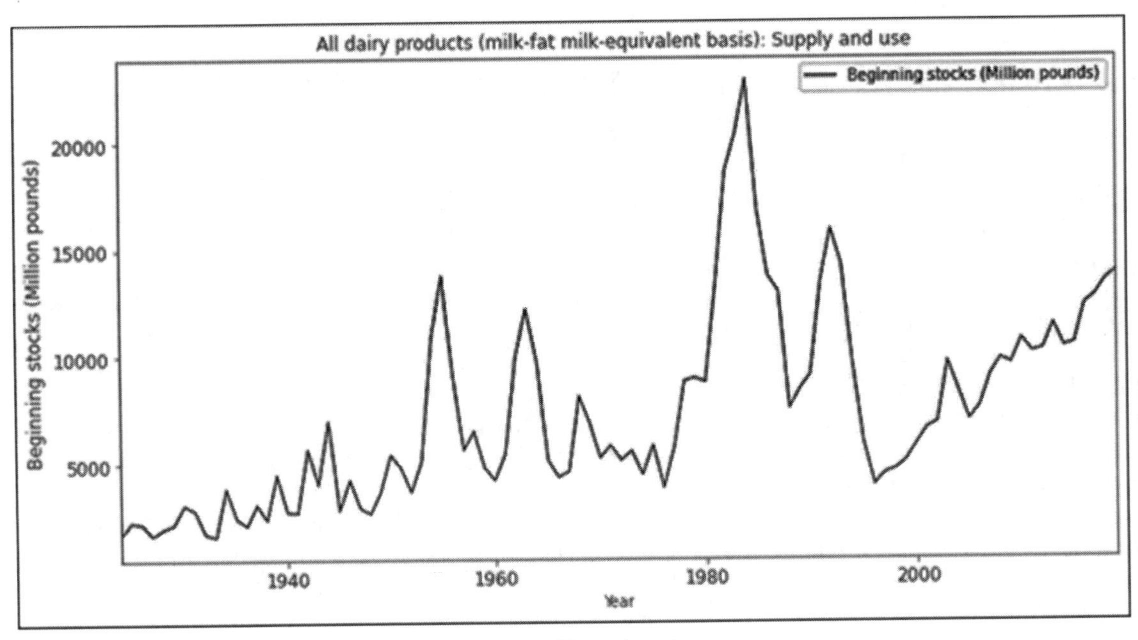

Figure 21: Milk supply and use.

The data distribution exhibits statistical properties, all units are in million pounds, minimum:1560.00 (million pounds), mean:7284.27 (million pounds), median:5789.95, mode:1560.00, and maximum:22851.00 (million pounds) (Figure 22).

The Augmented Dickey Fuller test (ADF Test) is a common statistical test used to test whether a given Time series is stationary or not. It is one of the most used statistical tests when it comes to analyzing the stationarity of a series. Another point to remember is that the ADF test is fundamentally a statistical significance test. That means, there is hypothesis testing involved with a null and alternate hypothesis and as a result a test statistic is computed, and p-values get reported. It is from the test statistic and the p-value, you can make an inference as to whether a given series is stationary or not.

ADF Test belongs to a category of tests called "Unit Root Test", which is the proper method for testing the stationarity of a time series. Unit root is a characteristic of a time series that makes it non-stationary.

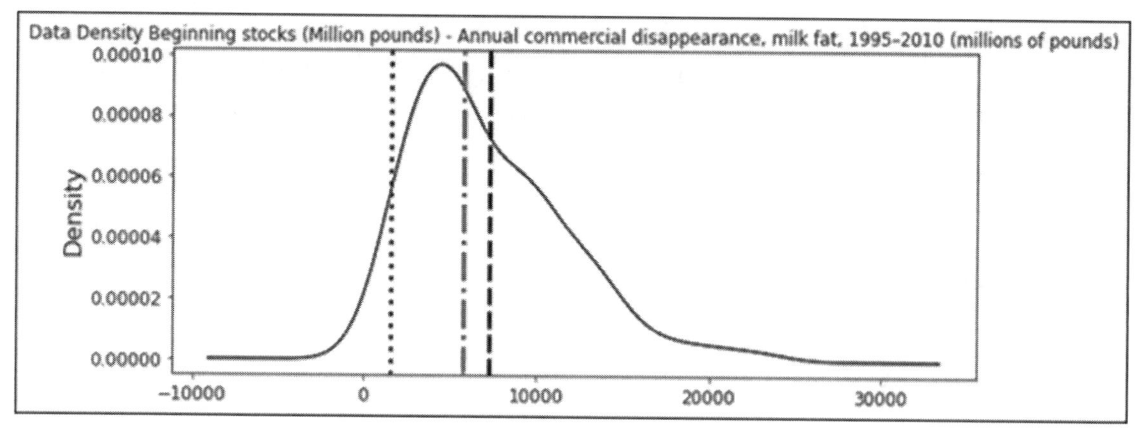

Figure 22: Data Density - Beginning stocks.

Technically speaking, a unit root is said to exist in a time series, alpha = 1 (coefficient of the first lag on Y in the below equation (a lower[35] value of alpha for stationary time series) (see Equation 6):

$$Y_t = \alpha\, Y_{(t-1)} + \beta\, X_e + \varepsilon \qquad \text{EQ. 6}$$

where Y_t is the value of the time series at time 't' and X_e is an exogenous variable (a separate explanatory variable, which is also a time series) (see Equation 7). So, what it signifies is that the value of a time series Y_t at time t is influenced by its historical value at (t-1) plus other time series values and an error function. The trouble with this is (a) it masks out the true influencer of Y_t and (b) creates a causation that may not be true.

$$y_t = c + \beta\, t + \alpha\, y_{(t-1)} + \phi\, \Delta\, Y_{t-1} + e_t \qquad \text{EQ. 7}$$

where,

- C is constant after differencing
- $y_{(t-1)}$ is the first lag
- Δ Y t-1 first difference of the series at time (t-1)

The presence of a unit root means the time series is non-stationary. Besides, the number of unit roots contained in the series corresponds to the number of differencing operations (i.e., a shifting time series) required to make the series stationary. The only caveat is that each unit root consumes one time series observation period (that could be in seconds, days, quarters, or years) as differencing need to shift a time series.

Fundamentally, it has a similar null hypothesis as the unit root test (please see Equation 8). That is, the coefficient of Y(t-1) is 1, implying the presence of a unit root. If not rejected, the series is taken to be non-stationary [21].

$$y_t = c + \beta\, t + \alpha\, y_{(t-1)} + \phi_1\, \Delta\, Y_{t-1} + \phi_2\, \Delta\, Y_{t-1} + \ldots + \phi_p\, \Delta\, Y_{t-p} + et \qquad \text{EQ. 8}$$

If you notice, we have only added more differencing terms, while the rest of the equation remains the same. This adds more thoroughness to the test. The null hypothesis however is still the same as the Dickey Fuller test. The general way of testing the stationarity of a time series is the application of the augmented Dickey Fuller test. When we conduct this test to check for stationarity, we check for the presence of a unit root.

- Null Hypothesis Ho: There is a unit root i.e., the time series is nonstationary
- Alternate Hypothesis Ha: There is no unit root i.e., the time series is stationary

A key point to remember here is, since the null hypothesis assumes the presence of unit root, which is α=1, the p-value obtained should be less than the significance level (say 0.05) to reject the null hypothesis, therefore, inferring that the series is stationary.

In Python, Augmented Dickey Fuller (ADF) Test is as follows:

```
import statsmodels.tsa.stattools as sm
def augmented_dickey_fuller_statistics(time_series):
  result = sm.adfuller(time_series.values, autolag='AIC')
  print('ADF Statistic: %f' % result[0])
  print('p-value: %f' % result[1])
  print('Critical Values:')
  for key, value in result[4].items():
  print('\t%s: %.3f' % (key, value))
```

Method to use when automatically determining[36] the lag length among the values 0, 1, …, maxlag.

- If "Akaike information criterion[37] (AIC)" (default) or Bayesian Information Criterion ("BIC"), then the number of lags is chosen to minimize the corresponding information criterion.
  - The Akaike information criterion (AIC) is a mathematical method for evaluating how well a model fits the data it was generated from. In statistics, AIC is used to compare different models and determine which one is the best fit for the data.
  - The Bayesian Information Criterion, or BIC for short, is a method for scoring and selecting a model. It is named for the field of study from which it was derived: Bayesian probability and inference. Like AIC, it is appropriate for fitting models under the maximum likelihood estimation framework.
- "t-stat" based choice of maxlag. Starts with maxlag and drops a lag until the t-statistic on the last lag length is significant using a 5%-sized test.
- If None, then the number of included lags is set to maxlag.

The stats model package provides a reliable implementation of the ADF test via the adfuller () function. It returns the following outputs:

- statsmodels.tsa. stattools the p-value
- The value of the test statistics
- Number of lags considered for the test
- The critical value cutoffs.

The KPSS test

KPSS is another test for checking the *stationarity of a time series*. The null and alternate hypothesis for the KPSS test is opposite that of the ADF test.[38] The KPSS figures out if a time series is stationary around a mean or linear trend or is non-stationary due to a unit root [15, 20].

- Null hypothesis: The time series is stationary.
- Alternative hypothesis: The series has a unit root (series is not stationary).

```
from statsmodels.tsa.stattools import kpss
def kpss_test(df):
  statistic, p_value, n_lags, critical_values = kpss(df.values)

  print(f'KPSS Statistic: {statistic}')
  print(f'p-value: {p_value}')
  print(f'num lags: {n_lags}')
  print('Critial Values:')
  for key, value in critical_values.items():
     print(f' {key} : {value}')
kpss_test(df_train[AverageMilkPrices])
```

KPSS[39] takes the time series as input and computes and returns KPSS statistics, the p-value of the test, the truncation lag parameter, and the critical values at 10%, 5%, 2.5% and 1%.

To estimate sigma^2 the Newey-West estimator is used. If lags are "legacy", the truncation lag parameter is set to int (12 * (n/100) ** (1/4)). If the computed statistic is outside the table of critical values, then a warning message is generated. If the call parameters to KPSS is "legacy", then observations are used to compute the lags. If not, AUTOLAG is used.

```
if nlags == "legacy":
    nlags = int(np.ceil(12.0 * np.power(nobs / 100.0, 1 / 4.0)))
    nlags = min(nlags, nobs - 1)
  elif nlags == "auto" or nlags is None:
    if nlags is None:
        # TODO: Remove before 0.14 is released
        warnings.warn(
            "None is not a valid value for nlags. It must be an integer, "
            "'auto' or 'legacy'. None will raise starting in 0.14",
            FutureWarning,
            stacklevel=2,
        )
     # autolag method of Hobijn et al. (1998)
     nlags = _kpss_autolag(resids, nobs)
     nlags = min(nlags, nobs - 1)
```

KPSS AUTO LAG is computed as follows: compute the number of lags for covariance matrix estimation in the KPSS test.[40]

```
def _kpss_autolag(resids, nobs):
    """
    Computes the number of lags for covariance matrix estimation in KPSS test
    using method of Hobijn et al (1998). See also Andrews (1991), Newey & West
    (1994), and Schwert (1989). Assumes Bartlett / Newey-West kernel.
    """
    covlags = int(np.power(nobs, 2.0 / 9.0))
    s0 = np.sum(resids ** 2) / nobs
    s1 = 0
    for i in range(1, covlags + 1):
        resids_prod = np.dot(resids[i:], resids[: nobs - i])
        resids_prod /= nobs / 2.0
        s0 += resids_prod
        s1 += i * resids_prod
    s_hat = s1 / s0
    pwr = 1.0 / 3.0
    gamma_hat = 1.1447 * np.power(s_hat * s_hat, pwr)
    autolags = int(gamma_hat * np.power(nobs, pwr))
    return autolags
```

When the test statistic is lower than the critical value (0.05), you reject the null hypothesis and infer that the time series is stationary [21]. Augmented Dickey Fuller test for "Beginning stocks (Million pounds) time series" confirms the data is stationary (p-value less the 0.05):

AverageMilkPrices
KPSS Statistic: 0.4431932220716194
p-value: 0.05853740427947442
num lags: 8
Critial Values:
 10% : 0.347
 5% : 0.463
 2.5% : 0.574
 1% : 0.739

For the KPSS test, the p-values are all greater than 0.05, the alpha level therefore, we cannot reject the null hypothesis and conclude that the time series is stationary. KPSS test using 8 lags order.

Stationary is a critical assumption in time series models. Stationary implies homogeneity in the sense that the series behaves in a similar way regardless of time, which means that its statistical properties do not change over time.[41] More precisely, stationary means that the data fluctuates about a mean or a constant level. Many time series models assume equilibrium processes. Many statistical models require the series to be stationary to make effective and precise predictions. However, in practice, a much weaker definition of stationary, often referred to as weak stationary, is used. The condition of weak stationary requires that the elements of the time series should have a common finite expected value and that the auto covariance of two elements should depend only on their temporal separation. Generally, a stationary time series is one whose statistical properties such as mean, variance and autocorrelation are all constant over time. Mathematically, these conditions are Mean = μ and variance = σ^2 are constant for all values of t [22].

It is always better to apply both the tests, so that it can be ensured that the series is truly stationary. The outcomes of applying these stationary tests are as follows:

- Case 1: Both tests conclude that the series is not stationary—The series is not stationary.
- Case 2: Both tests conclude that the series is stationary—The series is stationary.
- Case 3: KPSS indicates stationarity and ADF indicates non-stationarity—The series is trend stationary. Trend needs to be removed to make the series strict stationery. The detrended series is checked for stationarity.
- Case 4: KPSS indicates non-stationarity and ADF indicates stationarity—The series is different stationary. Differencing is to be used to make the series stationary. The difference in series is checked for stationarity.

Another time series, ending stock, exhibits stationarity within the Annual commercial disappearance, milk fat, 1909–2019 (millions of pounds) data series—source: USDA Dairy Products:[42] dymfg.xls[43] from the Food Availability (Per Capita) Data System,[44]

The data distribution exhibits statistical properties, data units million pounds, minimum:1560.00 (million pounds), mean:7408.61 (million pounds), median:5829.50, mode:1560.00, and maximum:22851.00 (million pounds) (please see Figure 23).

The ADF Test on "Ending Stock" data series results in a p-value of 0.026078, confirming that the series is stationary (Figure 24).

Ending stocks (Million pounds)
ADF Statistic: –3.106467
p-value: **0.026078**
Critical Values:
 1%: –3.502
 5%: –2.893
 10%: –2.583

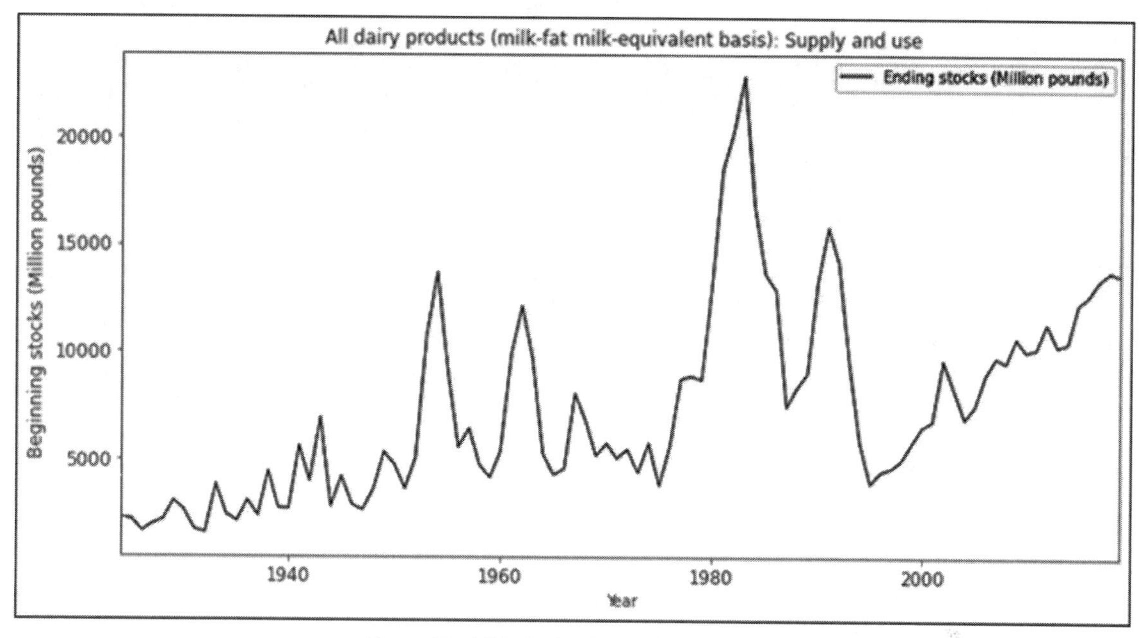

Figure 23: All Dairy Products (Supply and use).

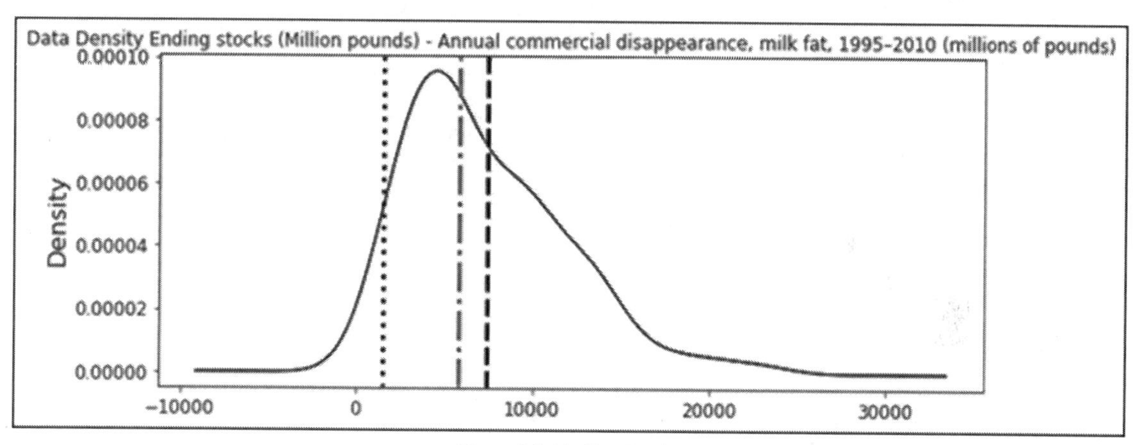

Figure 24: Ending Stocks.

The Beginning and End stocks stationarities confirm that the millions of pounds of milk from 1909–2019 maintain a gradual growth with mean & variance remaining constant.

$$y_t = c + \beta t + \alpha y_{(t-1)} + \phi \Delta Y_{t-1} + e_t$$

Finally, 'White Noise' is a classical stationary time series. Due to its inherent stationarity, it has no predictable pattern in the long term. It is thus memory less.

```
# Plot for White Noise with Mean 0 and standard deviation as 0.8
wnoise= np.random.normal(loc=0, scale=0.8, size=1800)
plt.figure(figsize=(15, 4)),
plt.plot(wnoise);
plt.title('Distribution of White Noise');
```

To plot them:

```
import pandas as pd
import numpy as np
wnoiseDF = pd.DataFrame(columns=['wNoise'])
for value in np.nditer(wnoise):
    wnoiseDF = wnoiseDF.append({'wNoise':value},ignore_index = True)
wnoiseDF
```

```
wnoiseDF['wNoise']=wnoiseDF['wNoise'].astype(float)
```

```
def show_density(var_name,var_data):
    from matplotlib import pyplot as plt

    print("\n" + var_name + "\n")
    rng = var_data.max() - var_data.min()
    var = var_data.var()
    std = var_data.std()
    print('\n{}:\n - Range: {:.2f}\n - Variance: {:.2f}\n - Std.Dev: {:.2f}\n'.format(var_name, rng, var, std))
    # Get statistics
    min_val = var_data.min()
    max_val = var_data.max()
    mean_val = var_data.mean()
    med_val = var_data.median()
    mod_val = var_data.mode()[0]
print('Minimum:{:.2f}\nMean:{:.2f}\nMedian:{:.2f}\nMode:{:.2f}\nMaximum:{:.2f}\n'.format(min_
val, mean_val, med_val, mod_val, max_val))
    fig = plt.figure(figsize=(10,4))
    # Plot density
    var_data.plot.density()
    # Add titles and labels
    plt.title('Data Density ' + var_name, fontsize=12)
    plt.ylabel('Density', fontsize=16)
    plt.tick_params(axis = 'both', which = 'major', labelsize = 12)
    plt.tick_params(axis = 'both', which = 'minor', labelsize = 12)
    # Show the mean, median, and mode
    plt.axvline(x=var_data.mean(), color = 'black', linestyle='dashed', linewidth = 3)
    plt.axvline(x=var_data.median(), color = 'darkgray', linestyle='dashdot', linewidth = 3)
    plt.axvline(x=var_data.mode()[0], color = 'gray', linestyle='dotted', linewidth = 3)
    # Show the figure
    plt.show()
```

```
show_density ('wNoise',wnoiseDF['wNoise'])
```

Here is how a distribution of White Noise will look like: (Figure 25)

Distribution properties: Minimum: –2.55, Mean:0.01, Median:0.01, Mode: –2.55, and Maximum: 2.47 with, Range: 5.02, Variance: 0.62, and Std. Dev: 0.79.

As you can see from the plot above, the distribution is constant across the mean and is completely random. It is difficult to predict the next movement of the time series. If we plot the autocorrelation of this series, one will observe complete zero autocorrelation. This means that the correlation between the series at any time t and its lagged values is zero.

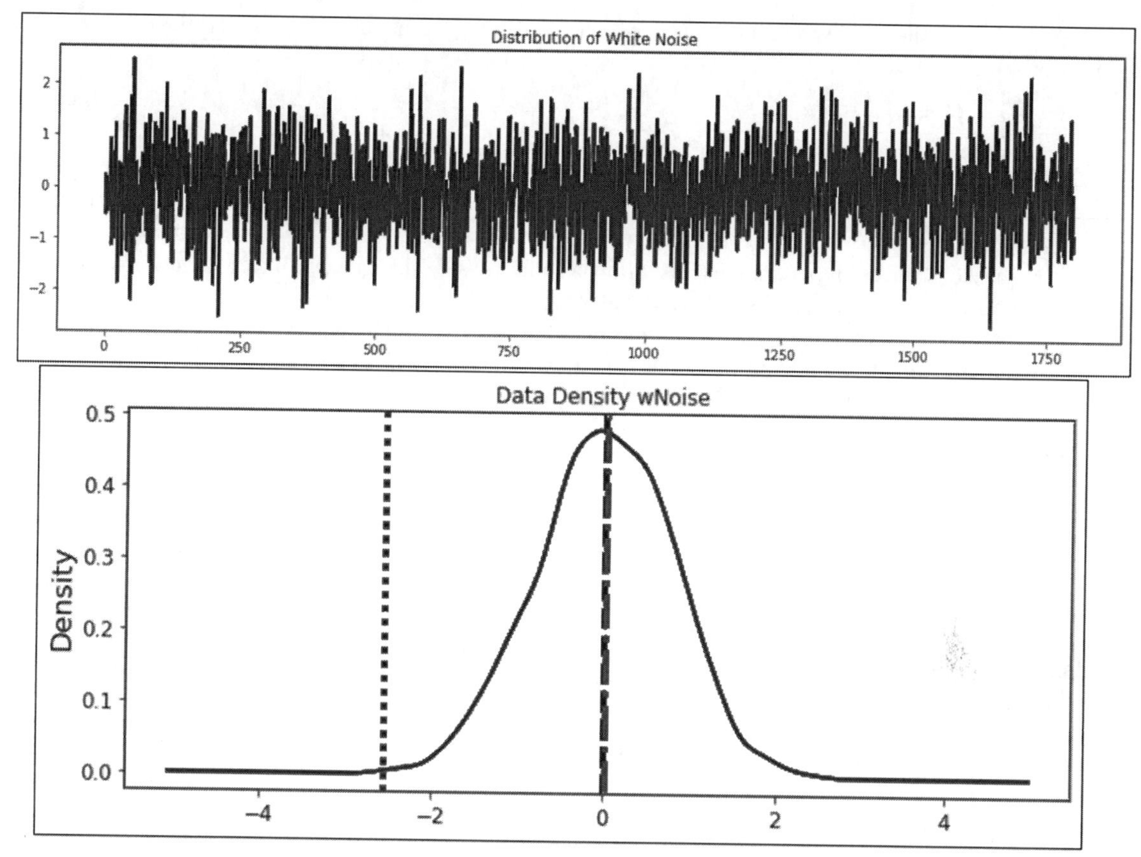

Figure 25: White Noise (density and distribution).

Non-stationary Stochastic Time Series

The presence of a trend component, in most observed time series, results in the series being nonstationary. Furthermore,[45] the trend can be either deterministic or stochastic, depending on which appropriate transformations must be applied to achieve stationarity. For example, a stochastic trend, or commonly known as a unit root, is eliminated by differencing the series. However, differencing a series that in fact contains a deterministic trend result in a unit root in the moving-average process. Similarly, subtracting a deterministic trend from a series that in fact contains a stochastic trend does not render a stationary series [15]. Hence, it is important to identify whether nonstationarity is due to a deterministic or a stochastic trend before applying the proper transformations. There are two kinds of non-stationary stochastic time series that are observed [23]:

- Random Walk Model without Drift
- Random Walk Model with Drift
- Deterministic Trend
- Random Walk with Drift and Deterministic Trend

Random Walk without Drift

In Random Walk[46] Without Drift, it assumes that, at each point in time, the series merely takes a random step away from its last recorded position, with steps whose mean value is zero. The mean value of the series is constant and is equal to its initial or starting value, but the variance increases with time. Thus, the series is nonstationary.

The value of the time series Y at time t (please see Equation 9) is equal to its value at time t-1 plus a random shock, i.e.,

$$Y_{(t)} = Y_{(t-1)} + \varepsilon_t \qquad \text{EQ. 9}$$

Random walk predicts that the value at time "t" will be equal to the last period value plus a stochastic (non-systematic) component that is a white noise, which means ε_t is independent and identically distributed with mean "0" and variance "σ^2".

Usually, the stock prices and exchange rates follow Random Walk. The data exhibiting Random Walk has long periods of up or down trends and sudden and unpredictable changes in direction. Consider for example, the data distribution of Nominal Advanced Foreign Economies U.S. Dollar Index[47] (TWEXAFEGSMTH). One can see the random change in the time series (please see Figure 26). The Dickey Fuller Augmented test would result in a p-value greater than 0.05 and hence the series is non-stationary.

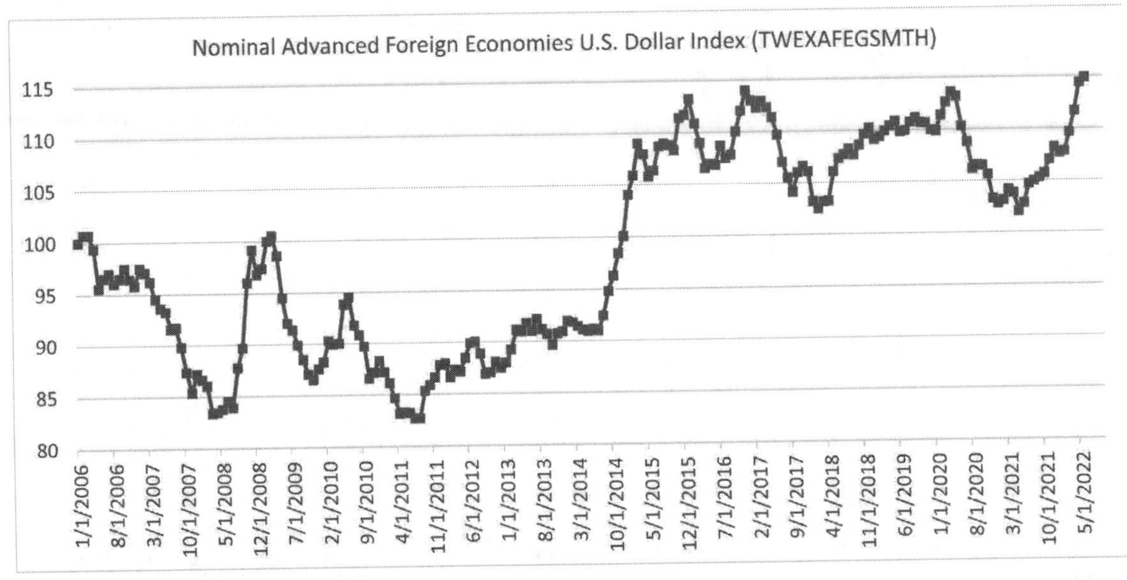

Figure 26: U.S. Dollar Index.

Nominal Advanced Foreign Economies U.S. Dollar Index (TWEXAFEGSMTH)
ADF Statistic: –1.059046
p-value: 0.731125
Critical Values:
1%: –3.466
5%: –2.877
10%: –2.575

The data distribution has a bi-modal distribution indicating two modes of data peaks or distribution.

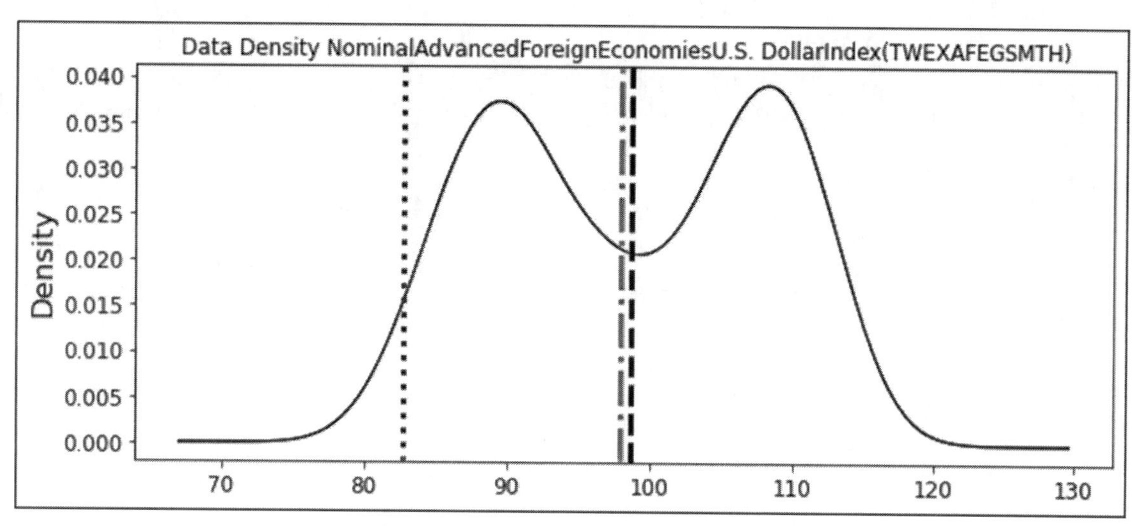

Figure 27: Distribution U.S. Dollar Index (2006 : 2020).

The distribution of the (please see figure 27) data includes Minimum:82.68, Mean:98.66, Median:97.97, Mode:82.68, and Maximum:114.01. Range:31.34, Variance: 90.65 and Std. Dev:9.52.

Random Walk with Drift

$$Y_t = \alpha + Y_{(t-1)} + \varepsilon_t \qquad \text{EQ. 10}$$

If the random walk model predicts that the value at time "t" will equal the last period's value plus a constant, $Y_{(t-1)}$, or drift (α), and a white noise term (ε_t), then the process is a random walk with a drift (please see Equation 10). It also does not revert to a long run mean and has a variance dependent on time.[48]

For example, consider the Yearly cheddar cheese price from USDA. The average cheddar cheese price from 1995 is increasing with random drift: $Y_t = \alpha + Y_{(t-1)} + \varepsilon_t$.
This is obviously non-stationary (Figure 28). To prove it, let's apply the Augmented Dickey-Fuller Test.

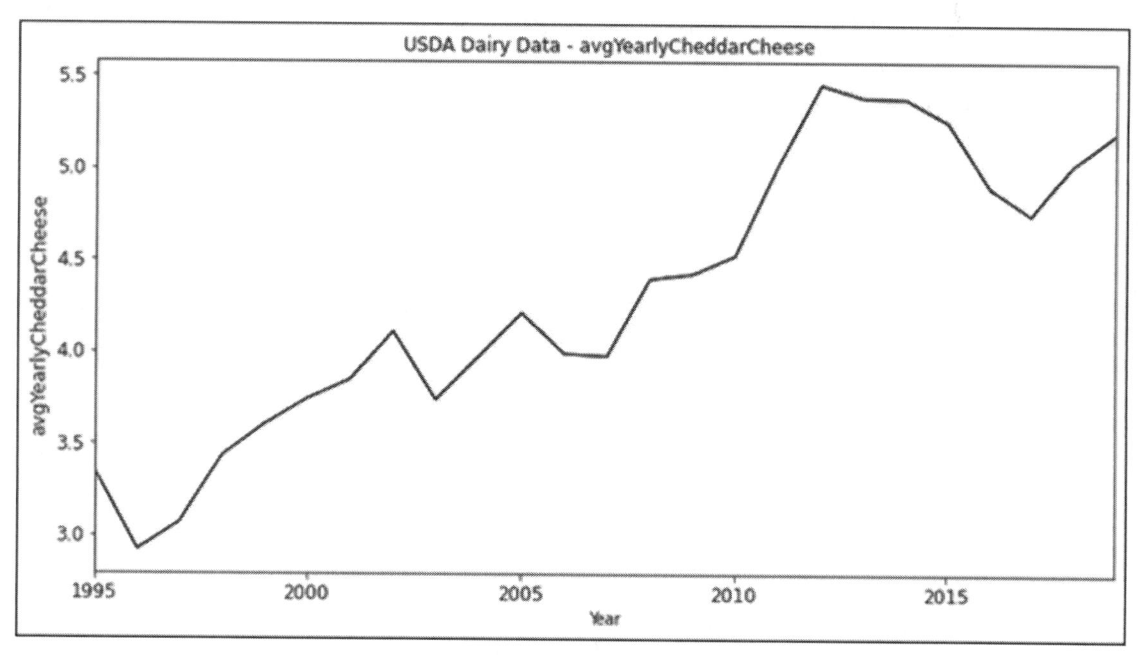

Figure 28: Cheddar Cheese Prices (1995:2020).

```
import statsmodels.tsa.stattools as sm
def augmented_dickey_fuller_statistics(time_series):
    result = sm.adfuller(time_series.values, autolag='AIC')
    print('ADF Statistic: %f' % result[0])
    print('p-value: %f' % result[1])
    print('Critical Values:')
    for key, value in result[4].items():
        print('\t%s: %.3f' % (key, value))
```

Output:

```
Augmented Dickey-Fuller Test: avgYearlyCheddarCheese
ADF Statistic: 0.802302
p-value: 0.991673
Critical Values:
    1%: -3.964
    5%: -3.085
    10%: -2.682
```

Since the p-value is greater than 0.05, it rejects the null hypothesis. That is the time series is not stationary.

Deterministic Trend

Often a random walk with a drift is confused for a deterministic trend. Both include a drift and a white noise component, but the value at time "t" in the case of a random walk is regressed on the last period's value (Y_{t-1}), while in the case of a *deterministic trend it is regressed on a time trend (βt).* A non-stationary process with a deterministic trend has a mean that grows around a fixed trend, which is constant and independent of time (please see Equation 11).

$$Y_t = \alpha + \beta t + \varepsilon_t \qquad \text{EQ. 11}$$

For example, consider the U.S. population growth since the 1940s. As it can be seen, the population growth is pretty deterministic and trending upwards (Figure 29).

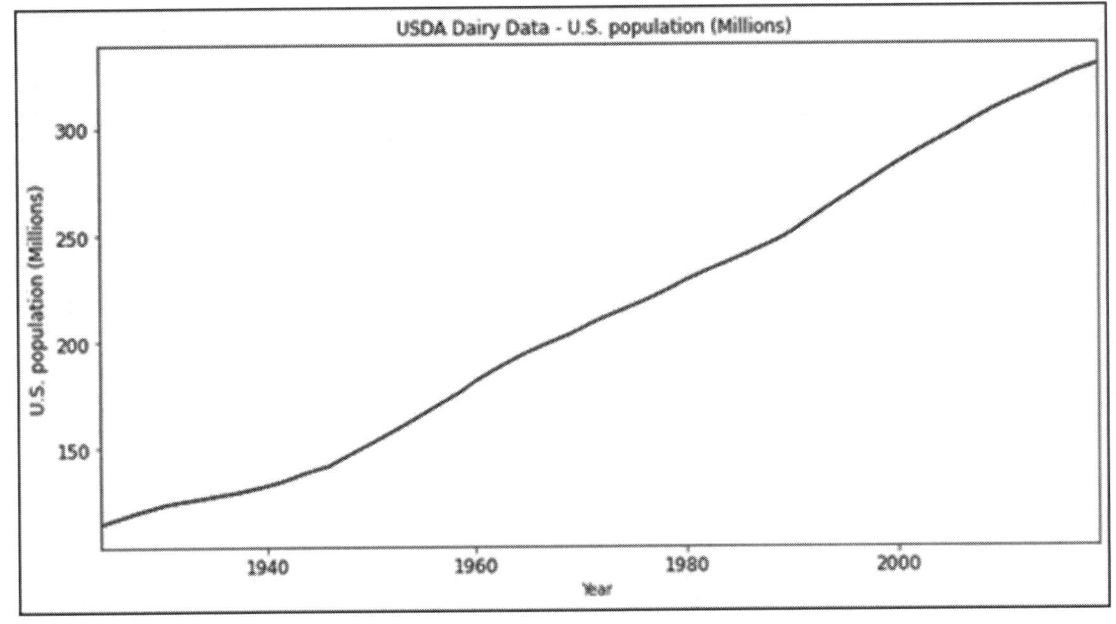

Figure 29: U.S. Population (millions) 1936 : 2020.

Trend + Cycle

Consider U.S. Dairy Shipments to U.S. territories (million pounds) (Figure 30).

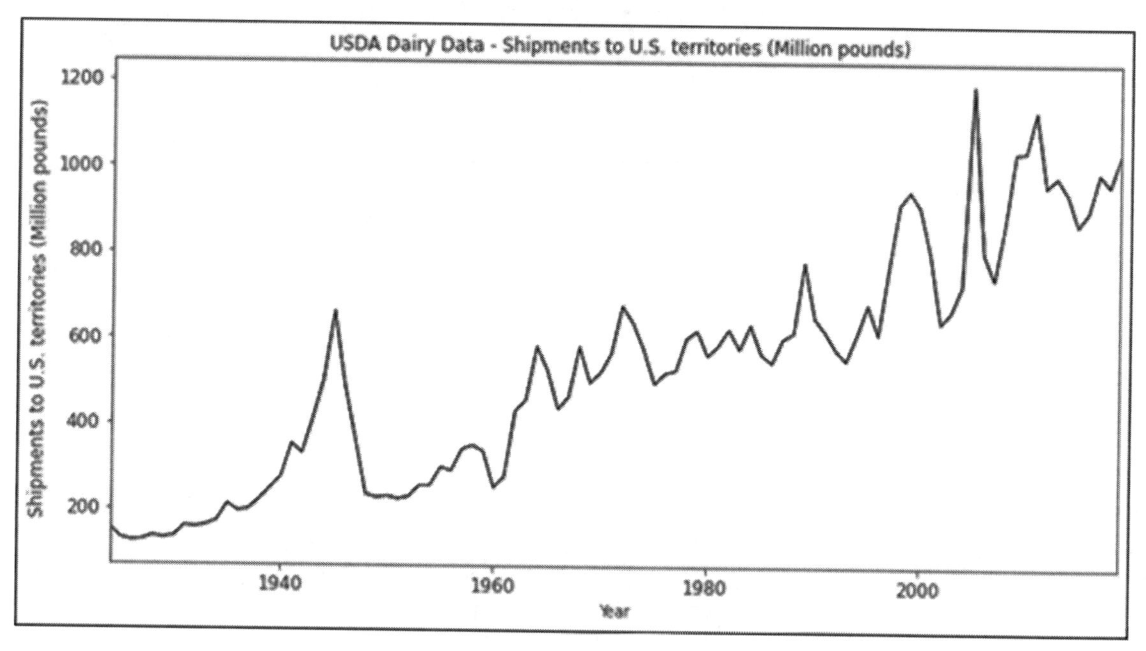

Figure 30: Dairy Data shipments (1936 : 2020).

The U.S. Dairy shipments data has a constant increasing trend plus a cycle within the data—for instance, every 15 years there is spiel in shipments. So, we have trend plus cycle.

Random Walk with Drift and Deterministic Trend

$$Y_t = \alpha + Y_{(t-1)} + \varepsilon_t + \beta t \qquad \text{EQ. 12}$$

Another example is a non-stationary process that combines a random walk with a drift component (α) and a deterministic trend (βt) (please see Equation 12). It specifies the value at time "t" from the last period's value, a drift, a trend, and a stochastic component. A good example is the price movements of imported crude oil ($/barrel).[49] It shows a deterministic downward trend, especially during recessive periods due to a drop in demand. In the figure below, recessionary periods are shaded grey areas,[50] marked as deterministic price[51] movement behavior. Random walk of the imported Oil prices is difficult to predict (please see Figure 31).

The distribution (please see Figure 32) of the data includes Minimum: 21.35, Mean: 62.55, Median: 57.01, Mode: 21.35, and Maximum: 131.52. Range: 110.17, Variance: 1006.57 and Std. Dev: 31.73. Finally, the Augmented Dickey Fuller test confirms non-stationarity of the series.

```
ADF Statistic: –2.239740
p-value: 0.192140
Critical Values:
   1%: –3.555
   5%: –2.916
   10%: –2.596
```

The p-value greater than 0.05 results in the rejection of the null-hypothesis, confirming a non-stationary time series!

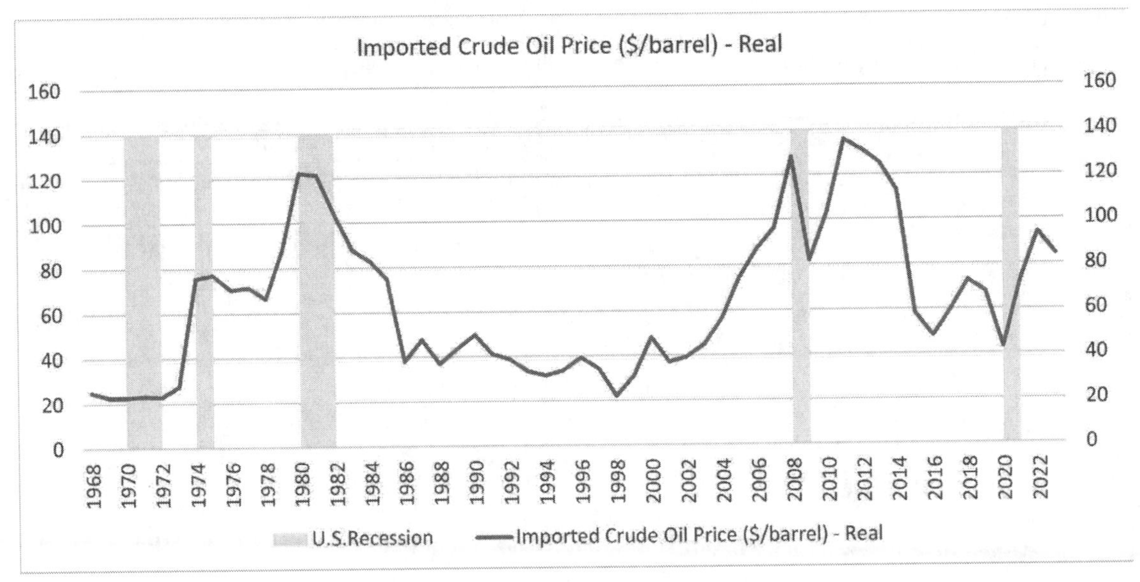

Figure 31: Imported Crude Oil (1968:2022).

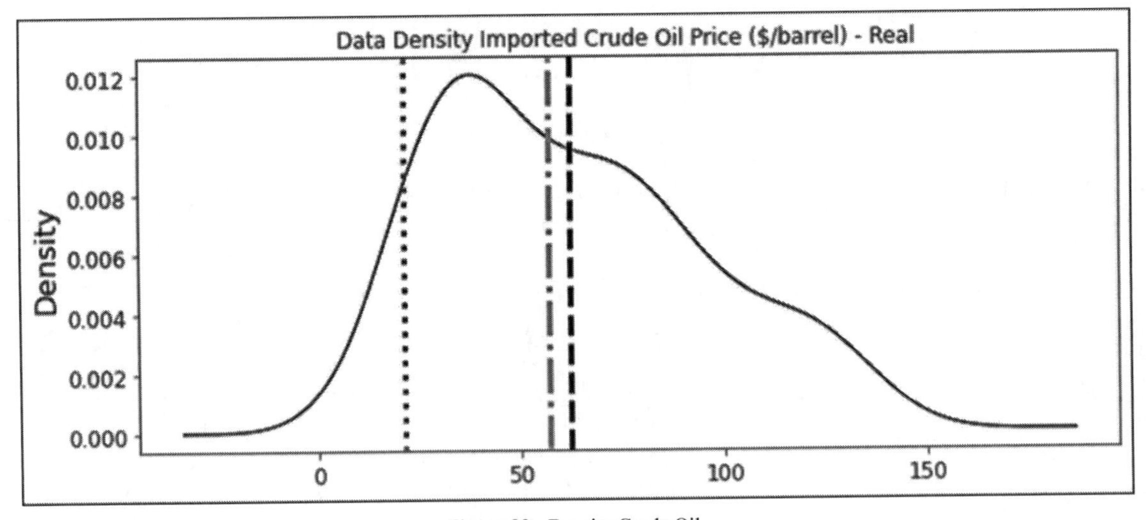

Figure 32: Density Crude Oil.

The first difference of the Random Walk Without Drift is a stationary time series. For the above equation, $Y_{(t)} = Y_{(t-1)} + \varepsilon_t$

The error term is a white noise or stationary series with mean 0 and constant variance Thus, the forecast for this model is the last observation as future movements are unpredictable and are equally likely to go up or down.

In contrast to the non-stationary process that has a variable variance and a mean that does not remain near or returns to a long-run mean over time, the stationary process reverts around a constant long-term mean and has a constant variance independent of time.

Granger's Causality Test

The Granger causality test is a statistical hypothesis test for determining whether one time series is useful for forecasting another.[52] If the probability value is less than any α level, then the hypothesis would be rejected at that level [24]. The formal definition of Granger causality can be explained as whether past values of x aid in the prediction of y(t), conditional on having already accounted for the effects on y(t) of past values of y (and perhaps of past values of other variables). If they do, then x is said to "Granger cause" y.

```
import statsmodels.tsa.stattools as sm
maxlag=MAX_LAGS
test = 'ssr-chi2test'
def grangers_causality_matrix(X_train, variables, test = 'ssr_chi2test', verbose=False):
    dataset = pd.DataFrame(np.zeros((len(variables), len(variables))), columns=variables, index=variables)
    for c in dataset.columns:
      for r in dataset.index:
            test_result = sm.grangercausalitytests(X_train[[r,c]], maxlag=maxlag, verbose=False)
            p_values = [round(test_result[i+1][0][test][1],4) for i in range(maxlag)]
            if verbose: print(f'Y = {r}, X = {c}, P Values = {p_values}')
            min_p_value = np.min(p_values)
            dataset.loc[r,c] = min_p_value
    dataset.columns = [var + '_x' for var in variables]
    dataset.index = [var + '_y' for var in variables]
    return dataset
grangers_causality_matrix(dataset, variables = dataset.columns)
```

Output:

	AverageMilkPrice_x	AverageCheddarCheesePrice_x	AverageYogartPrice_x	Imported Crude Oil Price ($/barrel) - Real_x	StockstoUse_x
AverageMilkPrice_y	1.0	0.0	0.0000	0.0013	0.0000
AverageCheddarCheesePrice_y	0.0	1.0	0.0070	0.0000	0.0006
AverageYogartPrice_y	0.0	0.0	1.0000	0.0000	0.0000
Imported Crude Oil Price ($/barrel) - Real_y	0.0	0.0	0.0000	1.0000	0.0000
StockstoUse_y	0.0	0.0	0.0002	0.0000	1.0000

For example, given a question: could we use today's Imported Crude Oil Price ($/barrel) price to predict tomorrow's Milk price? If this is true, our statement will be Crude Oil Price ($/barrel) price[53] Granger causes (please see Figure 33) Milk price. If this is not true, we say Crude Oil Price ($/barrel) price does not Granger cause Milk price [20].

The rows are the responses (y), and the columns are the predictor series (x).

- If we take the value 0.00 in (row 1, column 2), we can reject the null hypothesis and conclude that Imported Crude Oil Price ($/barrel) - Real_x Granger causes AverageMilkPrices_y. Likewise, the 0.00 in (row 2, column 1) refers to Imported Crude Oil Price ($/barrel) - Real_y Granger causes AverageMilkPrices_x [20] and so on.
- Similarly, Cheddar Cheese Price & Crude Oil Price, Yogurt Cream Price, and Crude Oil Prices all are Granger causes as seen in the following figure.[54]

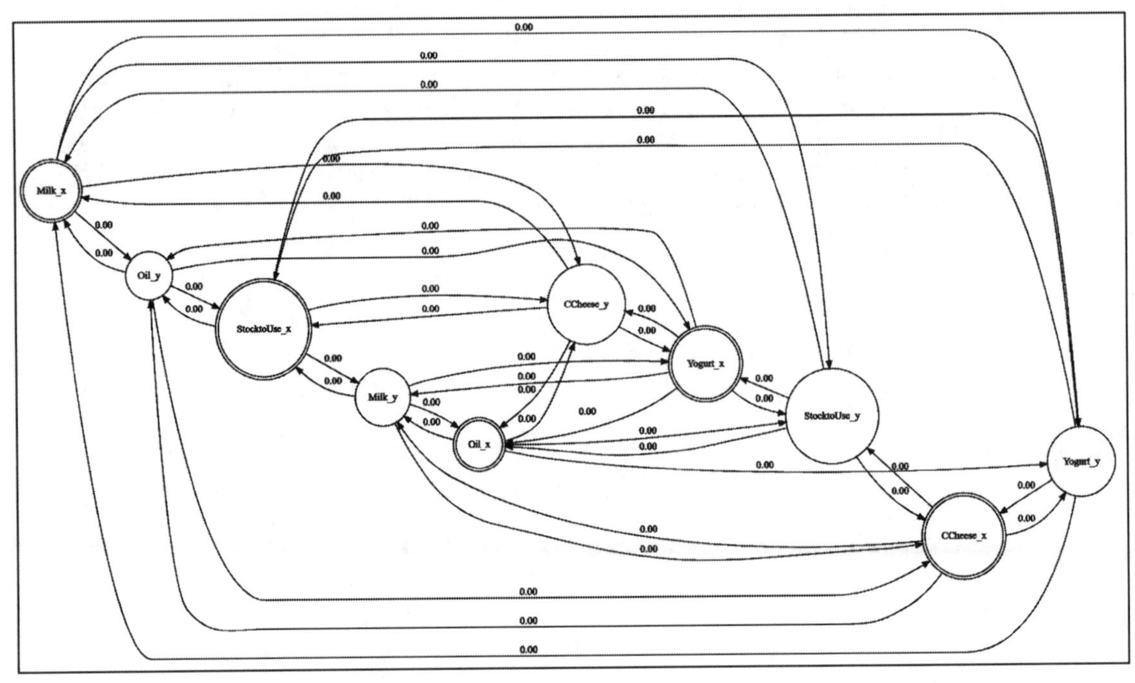

Figure 33: Granger Casuality.

Transformation & Detrending by Differencing

After cross-checking ADF and KPSS tests, if the time series is proven nonstationary, we need to transform the time series to stationary by the difference method. It is one of the simplest methods for detrending a time series. A new series is constructed where the value at the current time step is calculated as the difference between the original observation and the observation at the previous time step. First differencing is applied on the training set to make all the series stationary. However, this is an iterative process where after the first differencing, the series may still be non-stationary. We shall have to apply a second difference or log transformation to standardize the series in such cases [1].

```
DF["TS_diff"] = DF["TS"] - DF["TS"].shift(1)
DF["TS_diff"].dropna().plot(figsize=(12, 8))
```

Pandas[55] Shift index by desired number of periods with an optional time frequency.
In the above code, we have detrended by shifting 1. Other ways to achieve this are:

```
transform_data.diff().dropna()
```

Panda's diff:[56] Calculates the difference of a Dataframe element compared with another element in the Dataframe (the default element is in the previous row) (Figure 34).

A time series data is the data on a response variable Y(t) observed at different points in time t. Data on the variable is collected at regular intervals and in a chronological order. Anything that is observed sequentially over time is a time series. For example, prevalence of food insecurity collected over a one-year or three-year average for a nation. Another example is the price of milk for a marketing year. Here data collection could be on a monthly or yearly basis.

U.S. 2018 milk production totaled 217.6 billion pounds,[57] an increase of 2.0 billion pounds over 2017. The 2018 total is a record high for the ninth consecutive year and is an increase of 1.0% over the 215.5 billion pounds produced in 2017. Milk production has exceeded earlier year totals during 20 of the last 22 years, with exceptions occurring in 2001 and 2009. Production has increased by 16.3 billion pounds over the past five years (2018 versus 2013), a gain of 8.1%.

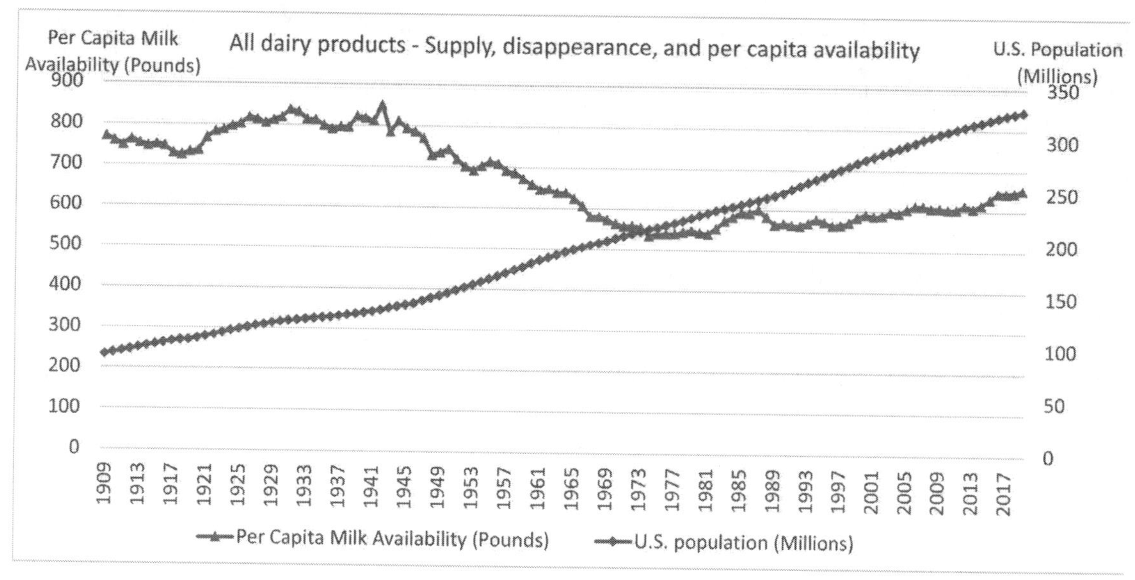

Figure 34: All dairy products vs. U.S. Population.

Sixteen states (please see Figure 35) registered annual milk production increases during 2018, the combined increase being 2.3%. Colorado recorded the largest percentage gain, with milk production increasing by 8.8%. Two other states recorded increases of five percent or more: Texas (+6.6%); and Kansas (+6.0%). Thirty-four states posted production decreases during 2018, with combined production falling by 1.8% compared with 2017.

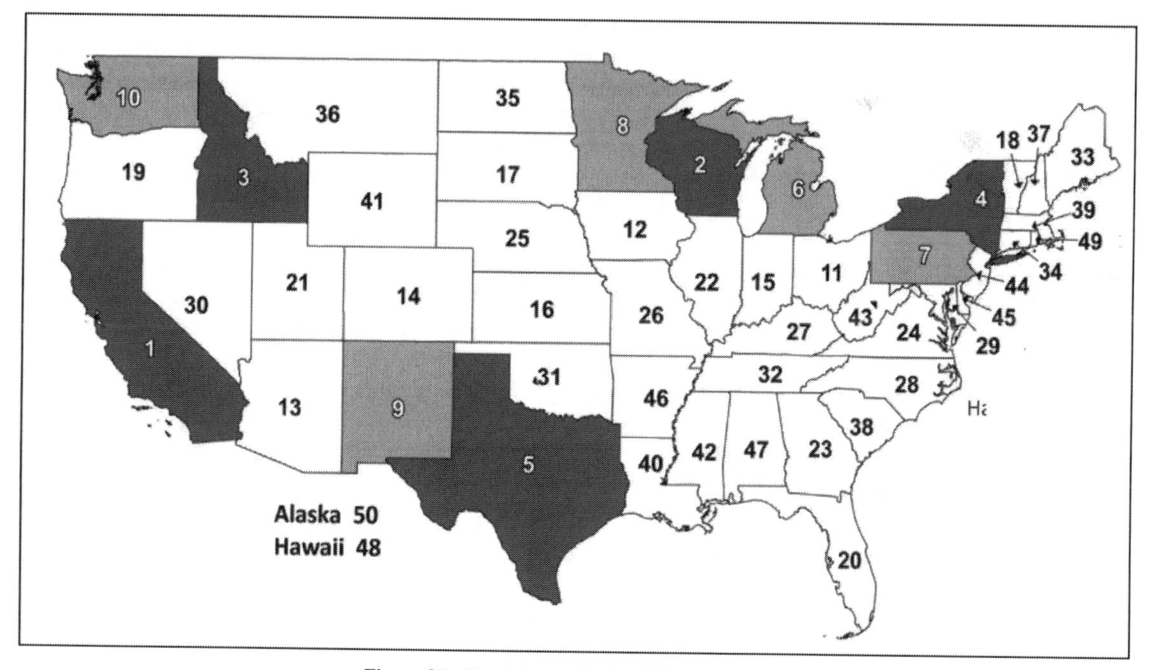

Figure 35: Top dairy producing U.S states (2017).

The aim of forecasting time series data is to understand how the sequence of observations will continue in the future.

A time series data will have one or more than one of the following components:

1. Trend Component—It is the consistent fluctuating data over the entire time span. The trend can be both linear and nonlinear.
2. Seasonality Component—It is the repetitive upward or downward fluctuation from the trend that occurs within a calendar year at fixed intervals. It always has a fixed and known frequency.
3. Cyclical Component—Cyclical fluctuations occur due to macro-economic factors like recession. Here, the interval between the repetition is more than a few years. Periodicity of cyclical fluctuation is not fixed. The average length of a cyclical pattern is longer than the average length of the seasonal pattern.
4. Irregular Fluctuation (Also called White Noise)—It is the uncorrelated random component of the time series data. If the time series data only has White Noise as a component, it cannot be used for prediction. This is because the series has observations which are identically and independently distributed with a mean of 0 and a constant variance.

Models

The choice of the forecasting model will depend on one or several components present in the time series. The time series forecasting models can be broadly classified into Simple Models (Mean Model, Linear Trend Model, Random Walk Model), Average and Smoothing Models (Moving Average, Exponential Smoothing), Vector Auto Regressive (VAR), Linear Regression Models, and ARIMA Models. This chapter enables us to construct both Regression and Time series models. The purpose of the regressive models is to establish the coefficient of regression to see the impact of macroeconomic, co-movement, and other significant variables on the prevalence of under nourishment. On the other hand, time series models are constructed to see future predictions as time series models differing from regression models in that time series models depend on the order of the observations. In regression, the order of the observations does not affect the model [25, 26]. Time series models use historic values of a series to predict future values. Some time series models try to use seasonal and cyclical patterns. Let's understand All dairy products (milk-fat milk-equivalent basis): Supply and use of Milk, i.e., fresh, whole, fortified, per gal. (3.8 lit) in U.S. city averages, average prices, seasonally unadjusted. The Milk Prices data is collected from the CPI FRED data.

Accuracy

Evaluation statistics used are the mean error, mean absolute percent error (MAPE), root mean square error (RMSE), and maximum absolute error (MAE). The mean error measures a procedure's bias. The RMSE, which reflects both bias and variability of the estimate, measures a procedure's accuracy. RMSE is the square root of the averaged squared deviations of a forecast from the truth. MAPE is a relative measure of the absolute size of the forecast errors. Forecast errors are transformed to truth percentages. The absolute value of these percentages is averaged over the number of time periods. MAPE does not penalize large deviations as severely as the RMSE does when the truth is also large. Also, the MAE will be used because an alternative procedure must have a smaller maximum error than the current procedure. Recall that larger than expected errors in the last quarter of 1988 prompted this analysis. The official Final two-state estimate is considered as the truth for computing evaluation statistics [25].

For the equations describing the statistics, i indexes, the procedures and k indexes the time periods are in months. Mean error for i^{th} procedure is defined as (please see Equations 13–16):

$$ME_i = [(1/t) \sum_{k=1}^{t} (f_{ik} - Y_k)] \qquad \text{EQ. 13}$$

The RMSE is defined as,

$$RMSE_i = [(1/t) \sum_{k=1}^{t} (f_{ik} - Y_k)^2]^{1/2} \qquad \text{EQ. 14}$$

The MAPE for the procedure is defined as,

$$MAPE_i = [(1/t) \sum_{k=1}^{t} |(f_{ik} - Y_k)/Y_k|.\ 100 \qquad \text{EQ. 15}$$

The MAE for the ith procedure is defined as:

$$MAE_i = MAX\ (|(f_{ik} - Y_k)|) \qquad \text{EQ. 16}$$

where,

- f_{ik} = the forecast from procedure i for the kth time, and
- Y k = the truth for the kth time.

 	Software code for this model: DairyData_CPI_TimeSeries ipynb (Jupyter Notebook Code)

VAR Linkage Model: CPI Average Milk Prices, Imported Oil Prices, and All Dairy Products (milk-fat milk-equivalent basis): Supply and Use

Data Sources

To model the influence of the CPI on milk prices and to create a forecasting model, the following data sources have been used. Despite having all dairy products (milk-fat milk-equivalent basis): Supply and use from 1909: 2019, due to the price short time series of CPI Milk Prices from 1995, the resolution of all other datasets must be limited. Average food prices are among the oldest data series published by BLS and date back to 1890. The average price data from 1980 has been accessed from our online database.[58] Secondly, BLS CPI monthly data was resampled to yearly to meet the requirements of the time series of other exogenous variables.

Parameter	**Reference**	**Source**
		U.S. BUREAU OF LABOR STATISTICS
CPI: Average price data	U.S. Bureau of Labor Statistics	

Series Title[59]	**Series ID**	**Description**
Milk, fresh, whole, fortified, per gal (3.8 lit)	**709112**	Fresh whole milk, fortified, sold per gallon regardless of packaging type. Includes organic and non-organic milk.
Milk, fresh, low fat, per gal (3.8 lit)	**709213**	Fresh, low fat or light milk, sold per gallon, regardless of packaging. Includes organic and non-organic milk.
Butter, in sticks, per pound	**FS1101**	Regular butter, sold in sticks, per pound (16 oz.). Includes organic and non-organic butter.
Butter, salted, grade AA, stick, per lb. (453.6 gm)	710111	Regular butter, salted, grade AA, sold in sticks, per pound (16 oz.). Includes organic and non-organic butter.
Cheddar cheese, natural, per lb. (453.6 gm)	**710212**	Natural cheddar cheese sold at the store, regardless of variety. Includes organic and non-organic cheese.
Ice cream, prepackaged, bulk, regular, per ½ gal (1.9 lit)	**710411**	Pre-packaged, bulk ice cream, sold in cardboard container regardless of package size, flavor, or dietary features. Includes organic and non-organic ice cream.
Yogurt, per 8 ounces	FJ4101	All yogurts sold in a plastic container or in a carton of between 1 and 8.999 oz. Includes organic and non-organic.

Parameter	Reference	Source
		eia
Energy Information Administration[60]	Real Gas prices[61]	
FRED Economic Data	• Average Price: Cheddar Cheese[62] • Average Price: Milk[63] • Average Price: Yogurt[64]	FRED ECONOMIC DATA | ST. LOUIS FED
USDA Dairy Products[65]	All dairy products (milk-fat milk-equivalent basis): Supply and use: dymfg.xls[66]	USDA

Dataset:[67] https://data.bls.gov/PDQWeb/ap

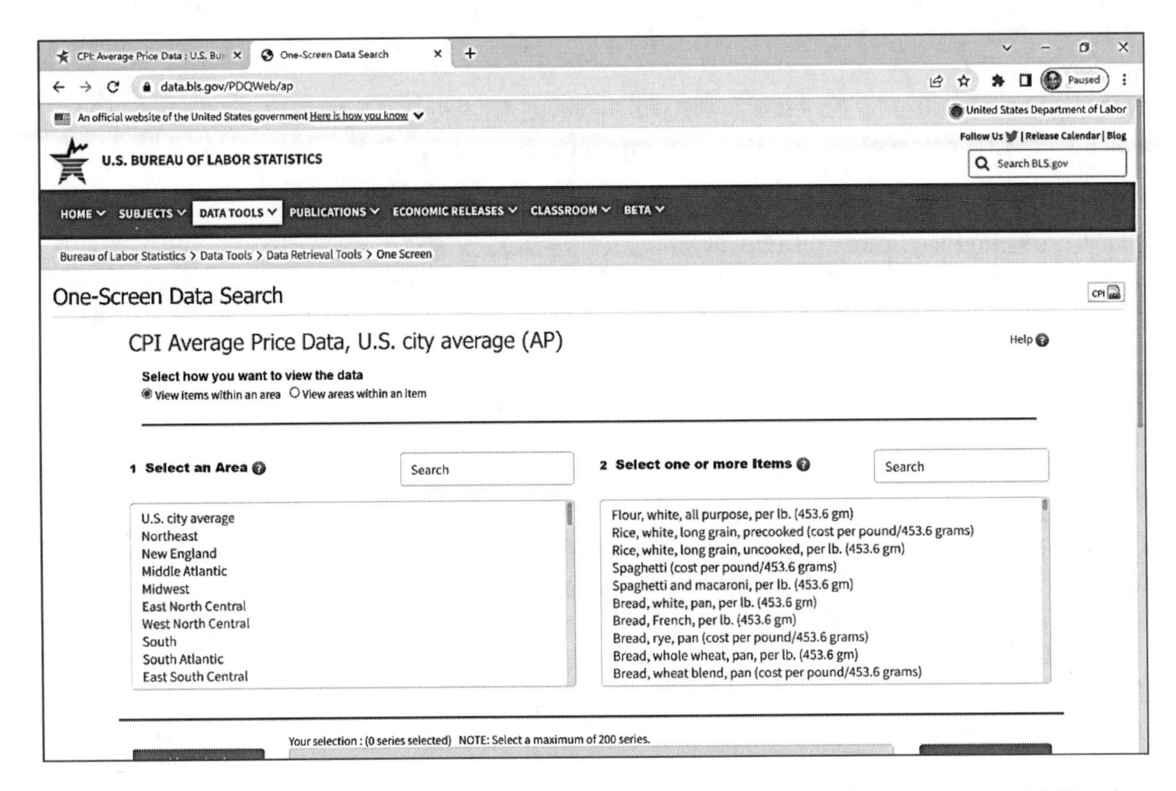

Dairy Data: dymfg.xls[68] from Food Availability (Per Capita) Data System.[69] The food availability data include estimates for over 200 commodities, including individual fruits, vegetables, grains, added sugars & sweeteners, dairy products, nuts, meat, poultry, and seafood. Due to the discontinuation of the Census Bureau's Current Industrial Reports (CIR) in 2011, data for added fats & oils (except butter), durum flour, and candy & other confectionery products are no longer available [27].

This Dairy Data page [28] includes data files covering domestic supply, demand, and international trade of various dairy products. Data on the U.S. dairy situation and commercial disappearance are updated monthly, whereas the U.S. milk production and related data are updated quarterly. All other dairy data files are updated annually [28]. These files include data on supply and allocation of milk fat and skim solids; dairy per capita consumption; fluid milk sales; milk supply by State and region; milk production and factors affecting supply and utilization of milk in all dairy products; and numbers and size of milk bottling plants.

USDA Milk production

For price spreads, the ending stocks play an important role [28]. This is generally true for the commodities market. For instance, wheat ending stocks have an influence on the price set for a new marketing year. Coming to milk-based marketing products, due to the limited shelf nature, the intuition is, ending stocks play an important role for relatively long shelf-life products, Cheddar Cheese, and will have a negative price influence on liquid-based products such as milk and yogurt. The model provides concrete & conclusive evidence.

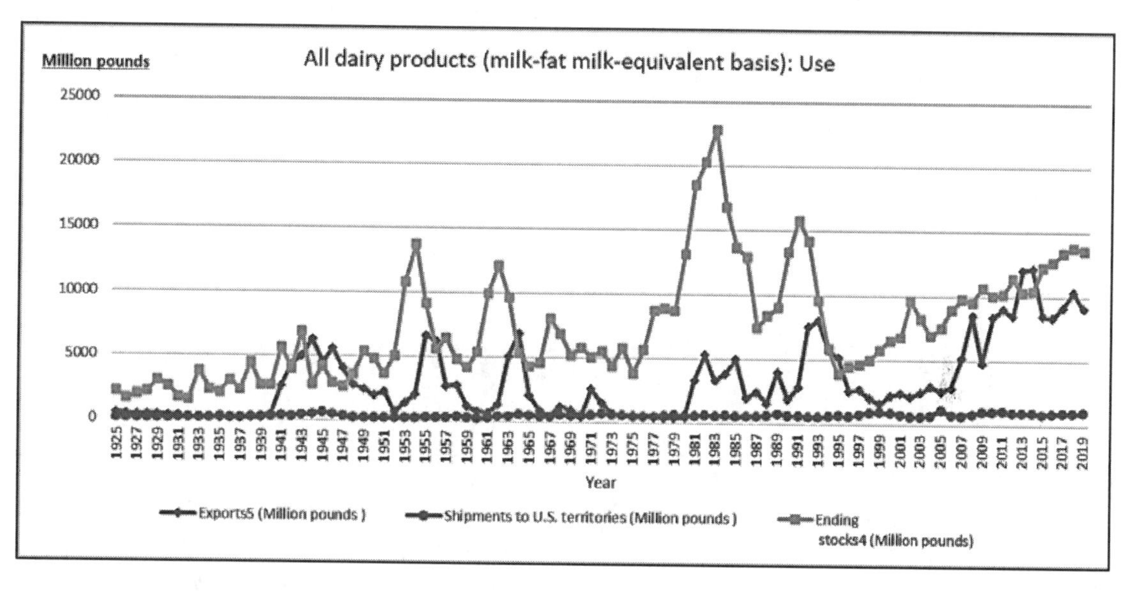

Milk Availability (Per Capita)

The information from the USDA's Economic Research Service (ERS) adds data from 2020 to account for per capita dairy consumption dating back to 1975 when the average American consumed just 539 pounds of dairy foods per year. Last year, the average American consumed 655 pounds of dairy products in milk, cheese, yogurt, ice cream, butter, and other wholesome and nutritious dairy foods, demonstrating a resilient and growing love for all things dairy. The 2020 figure represents an increase of 3 pounds per person over the previous year.[70]

Sources:

- Consumer Price Index for All Urban Consumers: Dairy and Related Products in U.S. City Averages[71]
- Dairy Data: dymfg.xls[72] from Food Availability (Per Capita) Data System. The food availability data include estimates for over 200 commodities, including individual fruits, vegetables, grains, added sugars & sweeteners, dairy products, nuts
- FAO[73] Dairy Prices' indices, Historical Time Series[74]

What 2020 shows us is that Americans are choosing to include dairy in all parts of their day because it's delicious, nutritious, and fits almost any occasion, despite challenges posed by the pandemic to all parts of the supply chain in 2020 - including the near-overnight loss of the foodservice sector—per capita dairy consumption continued to surge upward thanks to growth in ice cream, butter and yogurt. Last year's consumption figures are nearly 70 percentage points above the annual average, showing America's growing appreciation for their favorite dairy products.

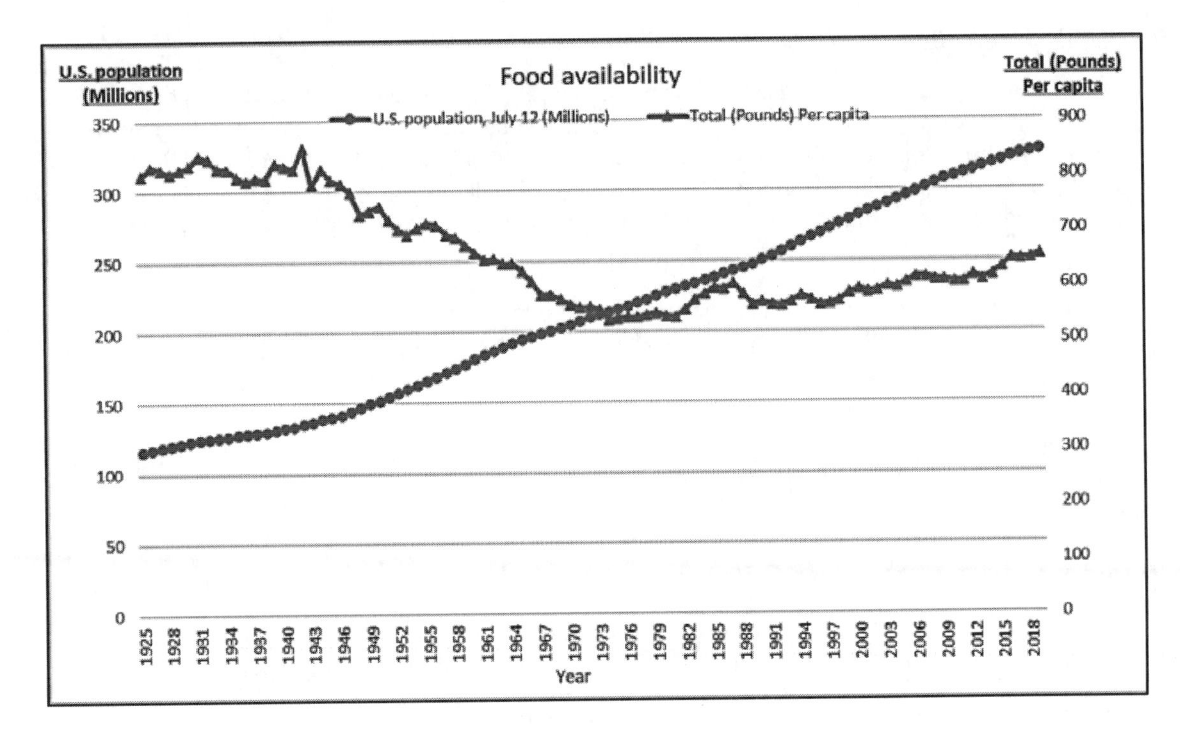

Step 1: Load Python libraries to process the data

Load Machine Learning and Python libraries to process data and create a mathematical model.

```
import pandas as pd
import numpy as np
import matplotlib.pyplot as plt
import seaborn as sns
import warnings
warnings.filterwarnings('ignore')
%matplotlib inline
from statsmodels.tsa.seasonal import seasonal_decompose
from numpy import mean
from sklearn.metrics import mean_squared_error
import math
from statsmodels.graphics.tsaplots import plot_acf, plot_pacf
from statsmodels.tsa.stattools import adfuller
from statsmodels.tsa.arima_model import ARIMA
import statsmodels.api as sm
##import pmdarima as pm
from statsmodels.tsa.api import ExponentialSmoothing
from matplotlib import pyplot
import warnings
import itertools
```

Step 2: Load CPI Average—Yogurt, natural, fruit flavored, per 8 oz. (226.8 gm) Data

Load Yogurt data. This dataset holds CPI averages for all yogurt sold in a plastic containers or in cartons of 1 to 8.999 oz. capacities. It includes organic and non-organic yogurt.

```
import pandas as pd

dfAveragePriceYogurt = pd.read_csv('APU0000710122 - Average Price Yogurt, Natural, Fruit Flavored
(Cost per 8 Ounces226.8 Grams).csv')
dfAveragePriceYogurt.head()
MAX_LAGS= 7 #3
dfAveragePriceYogurt
```

```
dfAveragePriceYogurt= dfAveragePriceYogurt.astype({'DATE': 'datetime64[ns]'})
```

```
dfAveragePriceYogurt=dfAveragePriceYogurt.set_index('DATE')
```

Output:

In order to merge the data set with other CPI time series data, let's convert the Date field and set the index on a date.

	DATE	Average Price: Yogurt, Natural, Fruit Flavored (Cost per 8 Ounces/226.8 Grams)
0	1/1/1980	0.449
1	2/1/1980	0.448
2	3/1/1980	0.454
3	4/1/1980	0.459
4	5/1/1980	0.479
...	...	...
503	11/1/2017	NaN
504	12/1/2017	NaN
505	1/1/2018	NaN
506	2/1/2018	NaN
507	3/1/2018	NaN

508 rows × 2 columns

The dataset contains nulls for the missing data for the period, 11/1/2017 to 3/1/2018. Later, we will input it using a forward fill strategy. To plot the data,

```
import matplotlib.pyplot as plt

#plot weekly sales data
plt.plot(dfAveragePriceYogurt.index, dfAveragePriceYogurt['Average Price: Yogurt, Natural, Fruit
Flavored (Cost per 8 Ounces/226.8 Grams) '], linewidth=3)
```

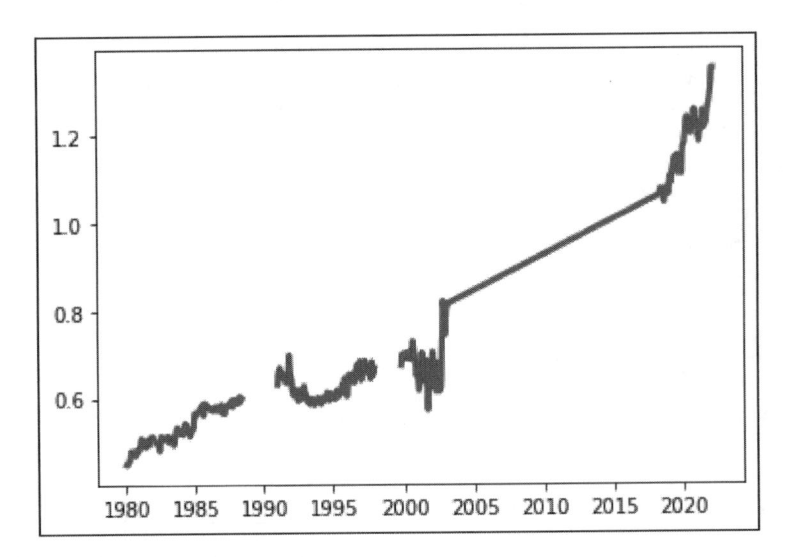

Step 3: Load CPI Average Price: Cheddar Cheese, Natural (Cost per Pound/453.6 Grams)

Load natural, cheddar cheese sold at stores, regardless of variety. Includes organic and non-organic cheese. To merge the data set with other CPI time series data, let's convert the Date field and set the index on the date.

```
dfCheddarcheese = pd.read_csv('APU0000710212 - Average Price Cheddar Cheese, Natural (Cost per Pound453.6 Grams).csv')
dfCheddarcheese.head()

dfCheddarcheese= dfCheddarcheese.astype({'DATE': 'datetime64[ns]'})
dfCheddarcheese=dfCheddarcheese.set_index('DATE')
                    dfCheddarcheese
```

```
dfCheddarcheese= dfCheddarcheese.astype({'DATE': 'datetime64[ns]'})
dfCheddarcheese=dfCheddarcheese.set_index('DATE')
dfCheddarcheese
```

Output:

DATE	Average Price: Cheddar Cheese, Natural (Cost per Pound/453.6 Grams)
1984-01-01	3.048
1984-02-01	3.036
1984-03-01	3.044
1984-04-01	3.03
1984-05-01	3.026
...	...
2021-12-01	5.257
2022-01-01	5.315
2022-02-01	5.447
2022-03-01	5.486
2022-04-01	5.656

460 rows × 1 columns

Step 4: Load CPI Average Price Milk, Fresh, Whole, Fortified (Cost per Gallon3.8 Liters)

Load Fresh whole milk, fortified, sold per gallon regardless of packaging type. Includes organic and non-organic milk data. This is the milk that is sold in the retail space in the U.S.,

```
dfMilkPrices = pd.read_csv('APU0000709112 -Average Price Milk, Fresh, Whole, Fortified (Cost per Gallon3.8 Liters).csv')
dfMilkPrices.head()
```

```
dfMilkPrices= dfMilkPrices.astype({'DATE': 'datetime64[ns]'})
dfMilkPrices=dfMilkPrices.set_index('DATE')
```

Output:

DATE	Average Price: Milk, Fresh, Whole, Fortified (Cost per Gallon/3.8 Liters)
1995-07-01	2.477
1995-08-01	2.482
1995-09-01	2.459
1995-10-01	2.473
1995-11-01	2.493
...	...
2021-12-01	3.743
2022-01-01	3.787
2022-02-01	3.875
2022-03-01	3.917
2022-04-01	4.012

322 rows × 1 columns

Step 5: Merge CPI Average Price Data for Dairy Products (Milk, Cheddar Cheese, and Yogurt) Time Series
Merge the time series data that includes milk, cheddar cheese, and yogurt.

```
dfCPIUSAveragePricesCMY = dfMilkPrices.merge(dfCheddarcheese , on = "DATE")
dfCPIUSAveragePricesCMY = dfCPIUSAveragePricesCMY.merge(dfAveragePriceYogurt , on = "DATE")
dfCPIUSAveragePricesCMY
```

Output:
Merge on Date index. The data contains nulls and needs a fix.

DATE	Average Price: Milk, Fresh, Whole, Fortified (Cost per Gallon/3.8 Liters)	Average Price: Cheddar Cheese, Natural (Cost per Pound/453.6 Grams)	Average Price: Yogurt, Natural, Fruit Flavored (Cost per 8 Ounces/226.8 Grams)
1995-07-01	2.477	3.418	0.611
1995-08-01	2.482	3.434	0.609
1995-09-01	2.459	3.347	0.628
1995-10-01	2.473	3.402	0.620
1995-11-01	2.493	3.34	0.644
...	...	...	...
2021-12-01	3.743	5.257	1.265
2022-01-01	3.787	5.315	1.281
2022-02-01	3.875	5.447	1.303
2022-03-01	3.917	5.486	1.350
2022-04-01	4.012	5.656	1.350

322 rows × 3 columns

Step 6: Check for the missing data and Nulls.

Using IsNulll SUM to identify all columns that have NULLS.

```
dfCPIUSAveragePricesCMY.isnull().sum()
```

Output:

```
Average Price: Milk, Fresh, Whole, Fortified (Cost per Gallon/3.8 Liters)        0
Average Price: Cheddar Cheese, Natural (Cost per Pound/453.6 Grams)              0
Average Price: Yogurt, Natural, Fruit Flavored (Cost per 8 Ounces/226.8 Grams)  203
dtype: int64
```

As can be seen Yogurt has 203 Nulls that can be fixed with imputation strategies. For now, just apply forward filling and we will revisit other imputation strategies in the next chapter.

Step 7: Time Series missing data imputation—Last observation carried forward (LOCF).

It is a common statistical approach for the analysis of longitudinal repeated measures data when some follow-up observations are missing. Another approach—Forward filling and backward filling are two approaches to fill missing values. Forward filling means filling missing values with previous or historical data. Backward filling means filling missing values with the following or future time series data points.

```
dfCPIUSAveragePricesCMY = dfCPIUSAveragePricesCMY.fillna(method = 'ffill')
dfCPIUSAveragePricesCMY.isnull().sum()
```

Output:

```
Average Price: Milk, Fresh, Whole, Fortified (Cost per Gallon/3.8 Liters) 0
Average Price: Cheddar Cheese, Natural (Cost per Pound/453.6 Grams)       0
Average Price: Yogurt, Natural, Fruit Flavored (Cost per 8 Ounces/226.8 Grams) 0
dtype: int64
```

Step 8: Resample Time Series - From Monthly to Annual Frequency

The Imported Oil Prices, All Dairy Supply and Use time series are in yearly frequency. So, it is required to convert CPI Average data to Yearly.

1	dfCPIUSAveragePricesCMY['Average Price: Cheddar Cheese, Natural (Cost per Pound/453.6 Grams) ']=dfCPIUSAveragePricesCMY['Average Price: Cheddar Cheese, Natural (Cost per Pound/453.6 Grams) '].astype(float) dfCPIUSAveragePricesCMY['Average Price: Yogurt, Natural, Fruit Flavored (Cost per 8 Ounces/226.8 Grams) ']=dfCPIUSAveragePricesCMY['Average Price: Yogurt, Natural, Fruit Flavored (Cost per 8 Ounces/226.8 Grams) '].astype(float)
2	dfCPIUSAveragePricesCMYYearly = dfCPIUSAveragePricesCMY.resample('Y').mean() dfCPIUSAveragePricesCMYYearly
3	dfCPIUSAveragePricesCMYYearly['Year'] = pd.DatetimeIndex(dfCPIUSAveragePricesCMYYearly.index).year dfCPIUSAveragePricesCMYYearly
4	dfCPIUSAveragePricesCMYYearly=dfCPIUSAveragePricesCMYYearly.set_index('Year') dfCPIUSAveragePricesCMYYearly

Here are the sub-steps:

Firstly, convert "Average Price: Cheddar Cheese, Natural (Cost per Pound/453.6 Grams)" and "Average Price: Yogurt, Natural, Fruit Flavored (Cost per 8 Ounces/226.8 Grams)" from object to float.

Next, resample Data Frame CPI US Average Prices of Cheddar Cheese, Milk, and Yogurt from monthly re to yearly resolution.

Get yearly mean values into the data frame as the third step. Finally, re-index on year.

Year	Average Price: Milk, Fresh, Whole, Fortified (Cost per Gallon/3.8 Liters)	Average Price: Cheddar Cheese, Natural (Cost per Pound/453.6 Grams)	Average Price: Yogurt, Natural, Fruit Flavored (Cost per 8 Ounces/226.8 Grams)
1995	2.483667	3.388500	0.623500
1996	2.623167	3.248083	0.650333
1997	2.614000	3.219667	0.661333
1998	2.703750	3.547500	0.666000
1999	2.842750	3.769750	0.669250
2000	2.780667	3.829917	0.694417
2001	2.884250	4.026917	0.645000
2002	2.757250	4.218083	0.679333
2003	2.761083	3.948417	0.806167
2004	3.155917	4.272917	0.813000
2005	3.186833	4.381500	0.813000
2006	3.081333	4.254833	0.813000
2007	3.503250	4.229917	0.813000
2008	3.795333	4.725333	0.813000
2009	3.109000	4.668750	0.813000
2010	3.258500	4.709750	0.813000
2011	3.571583	5.418750	0.813000
2012	3.492500	5.625417	0.813000
2013	3.461917	5.557500	0.813000
2014	3.693667	5.542667	0.813000

Output: Plot the U.S. Bureau of Labor Statistics (BLS) CPI Average data:

Average Price of Milk:

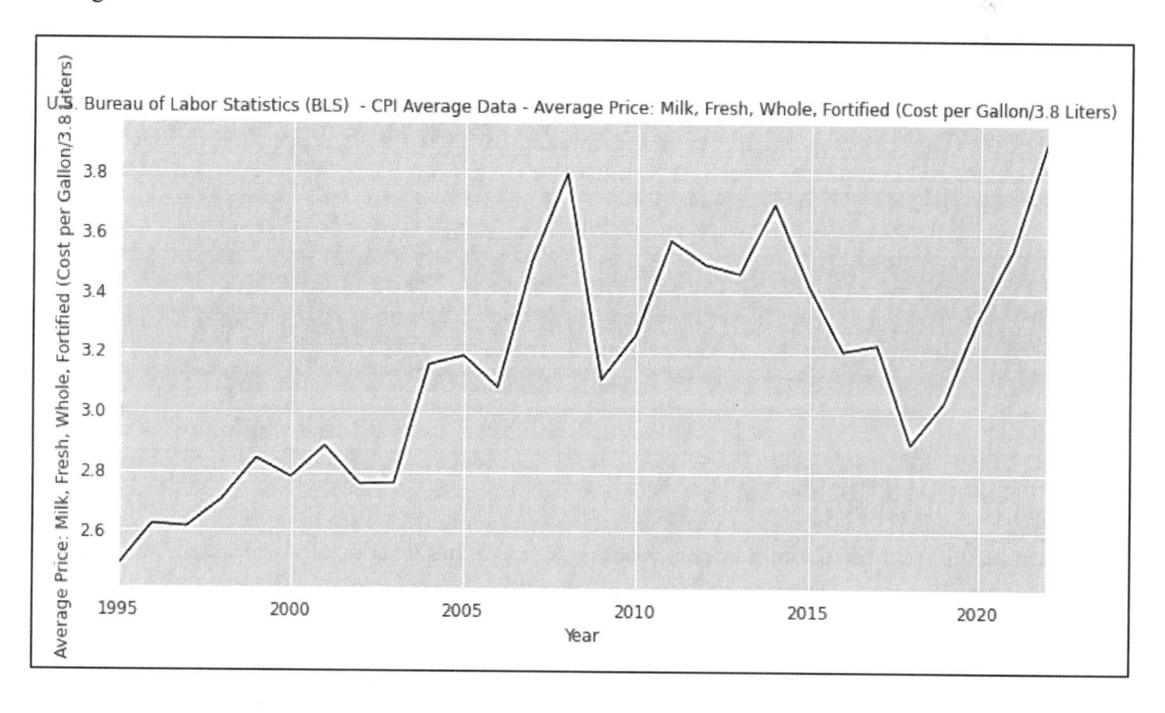

Average Price of Cheddar Cheese:

Average Price of Yogurt:

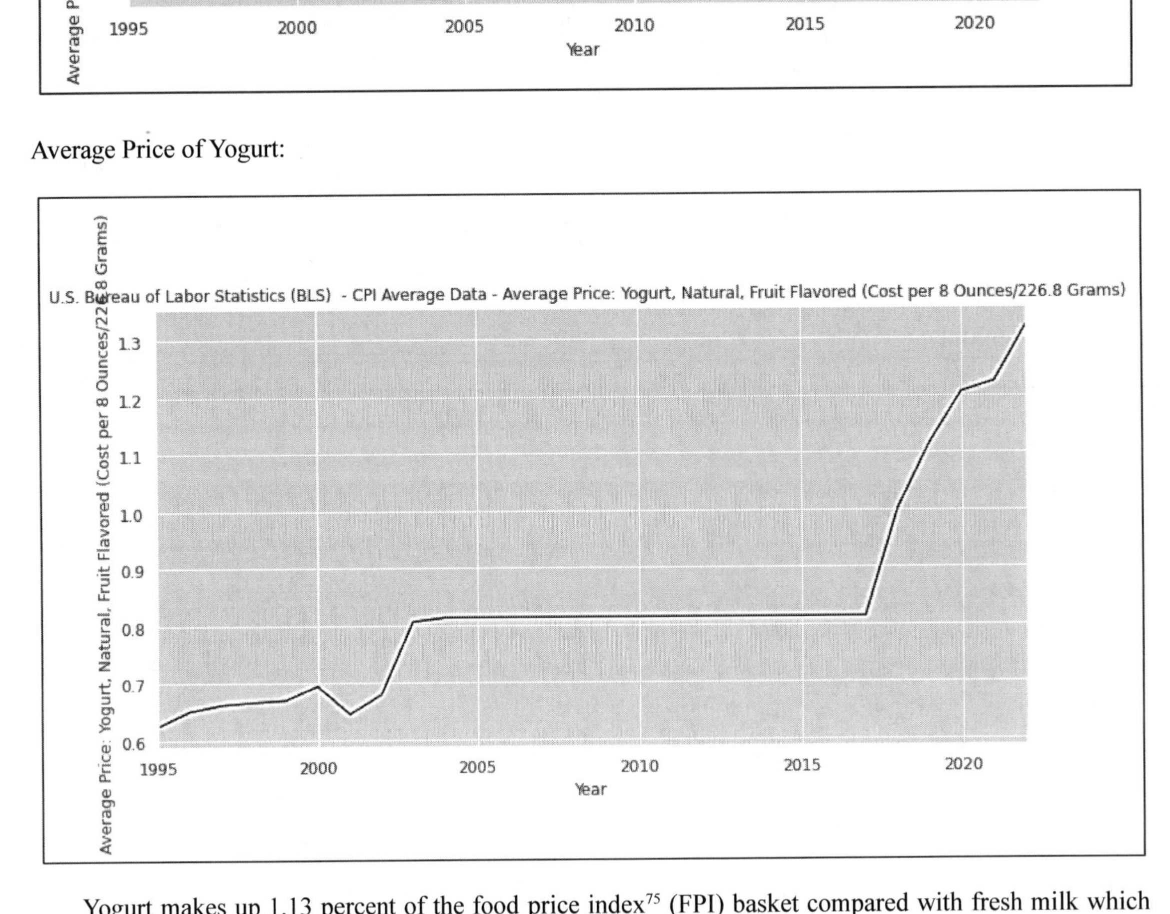

Yogurt makes up 1.13 percent of the food price index[75] (FPI) basket compared with fresh milk which makes up 3.60 percent. Due to the increase in CPI 2017, the prices jump for yogurt [29].

Step 9: Load U.S. Energy Information Administration Short-Term Energy Prices

Load U.S. Energy Information Administration[76] all real and nominal energy prices.[77] Real Petroleum Prices are computed by dividing the nominal price for each month by the ratio of the Consumer Price Index (CPI) in that month to the CPI in some "base" period.

1	dfCrudeOilPrices = pd.read_csv('EIAShort-TermEnergyOutlookAnnualAverageImportedCrudeOilPrice.csv') dfCrudeOilPrices.head()
2	dfCrudeOilPrices=dfCrudeOilPrices.set_index('Year')
3	mask = (dfCrudeOilPrices.index > 1980) & (dfCrudeOilPrices.index <= 2021) dfCrudeOilPrices=dfCrudeOilPrices.loc[mask] dfCrudeOilPrices

Output:

Year	Consumer Price Index (1982-84=1)	Imported Crude Oil Price ($/barrel) - Nominal	Imported Crude Oil Price ($/barrel) - Real
1981	0.909333	37.099725	117.650532
1982	0.965333	33.568900	100.278106
1983	0.995833	29.314416	84.886945
1984	1.039333	28.876824	80.119988
1985	1.076000	26.991317	72.336604
1986	1.096917	13.934332	36.631848
1987	1.136167	18.138013	46.035618
1988	1.182750	14.602182	35.601731
1989	1.239417	18.071613	42.046117
1990	1.306583	21.733567	47.966741
1991	1.361667	18.725638	39.656294
1992	1.403083	18.208123	37.422088
1993	1.444750	16.133509	32.201970
1994	1.482250	15.538111	30.228951
1995	1.523833	17.141829	32.438892
1996	1.568583	20.618925	37.905718
1997	1.605250	18.488877	33.213464
1998	1.630083	12.066664	21.346355
1999	1.665833	17.271497	29.898179
2000	1.721917	27.721609	46.425076

Plot the energy prices.

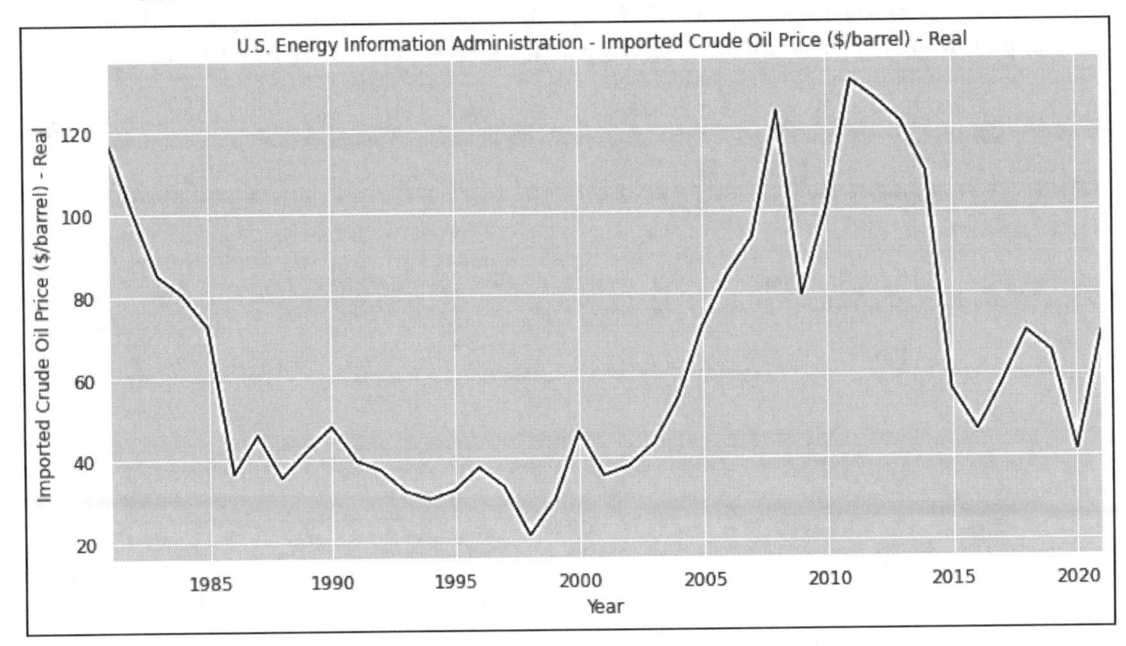

Since 2012, the U.S. import and export price indexes have experienced a downward trend. However, in 2016, these indexes reversed the trend as both recorded over-the-year increases. Crude oil prices and exchange rates tell much of the story. This article discusses how changes in these two fundamental economic characteristics contributed to 2016 trends and examines the price movements found in detailed indexes for U.S. imports and exports.[78]

Step 10: Load USDA Food Availability (Per Capita) Data System—Dairy products (Milk Fat milk-equivalent basis) Supply and Use

Load the Food Availability (Per Capita) Data System. The ERS Food Availability (Per Capita) Data System (FADS) includes three distinct but related data series on food and nutrient availability for consumption: food availability data, loss-adjusted food availability data, and nutrient availability data. The datasets serve as proxies for actual consumption at the national level. The food availability data series serves as the foundation for the other two series. Loss-adjusted food availability data (LAFA) is derived from food availability data by adjusting for food spoilage, plate waste, and other losses to more closely approximate actual consumption. The second data series is considered preliminary because the underlying food loss assumptions and estimates require further improvement. USDA Dairy Products:[79] dymfg.xls[80] from the Food Availability (Per Capita) Data System,[81]

```
dfAlldairyproductsSupplyandUse = pd.read_csv('Alldairyproducts(milk-fat milk-equivalent basis)_SupplyandUse.csv')
```

```
dfAlldairyproductsSupplyandUse=dfAlldairyproductsSupplyandUse.set_index('Year')
```

```
mask = (dfAlldairyproductsSupplyandUse.index > 1980) & (dfAlldairyproductsSupplyandUse.index <= 2021)
dfAlldairyproductsSupplyandUse=dfAlldairyproductsSupplyandUse.loc[mask]
dfAlldairyproductsSupplyandUse
```

Output:

Year	U.S. population (Millions)	MilkProduction(Million pounds)	Farm milk fed to calves (Million pounds)	For human use (Million pounds)	Imports(Million pounds)	Beginning stocks (Million pounds)	Total supply (Million pounds)	Exports (Million pounds)	Shipments to U.S. territories (Million pounds)	Ending stocks (Million pounds)	Total Domestic availability, not including USDA donations
1981	229.966	132,770.00	1,429.00	131,330.00	2,329.00	13,126.00	146,785.00	3,343.00	586	18,552.00	120,068.00
1982	232.188	135,505.00	1,534.00	133,958.00	2,477.00	18,552.00	154,987.00	5,320.00	624	20,296.00	121,449.00
1983	234.307	139,588.00	1,520.00	138,051.00	2,617.00	20,296.00	160,964.00	3,313.00	577	22,851.00	122,331.00
1984	236.348	135,351.00	2,129.00	133,202.00	2,741.00	22,851.00	158,794.00	3,851.00	634	16,784.00	126,587.00
1985	238.466	143,012.00	1,745.00	141,246.00	2,776.00	16,784.00	160,806.00	4,986.00	566	13,682.00	130,257.00
1986	240.651	143,124.00	1,714.00	141,389.00	2,732.00	13,682.00	157,803.00	2,001.00	546	12,922.00	132,693.00
1987	242.804	142,709.00	1,599.00	141,091.00	2,490.00	12,922.00	156,503.00	2,446.00	602	7,473.00	135,265.00
1988	245.021	145,034.00	1,589.00	143,437.00	2,394.00	7,473.00	153,304.00	1,582.00	615	8,378.00	136,040.00
1989	247.342	143,893.00	1,496.00	142,393.00	2,498.00	8,378.00	153,269.00	3,995.00	779	9,036.00	134,114.00
1990	250.132	147,721.00	1,484.00	146,235.00	2,690.00	9,036.00	157,961.00	1,886.00	651	13,359.00	137,835.00
1991	253.493	147,697.00	1,480.00	146,216.00	2,625.00	13,359.00	162,200.00	2,845.00	619	15,840.00	138,012.00
1992	256.894	150,847.00	1,436.00	148,481.00	2,521.00	15,840.00	166,842.00	7,569.00	578	14,214.00	140,693.00
1993	260.255	150,636.00	1,330.00	149,305.00	2,806.00	14,214.00	166,325.00	8,049.00	552	9,570.00	144,292.00
1994	263.436	153,602.00	1,267.00	152,334.00	2,880.00	9,570.00	164,784.00	5,725.00	613	5,867.10	149,071.90
1995	266.557	155,292.00	1,216.00	154,074.70	2,294.00	5,867.10	162,235.70	5,153.50	682	3,904.30	150,965.90

Step 11: Merge Dairy Prices, Dairy Manufacturing, and Oil Prices

Merge USDA Dairy Supply and Use, Dairy Prices, and EIA Oil Prices into one single dataset.

```
dfAlldairyproductsSupplyandUse = dfAlldairyproductsSupplyandUse.
merge(dfCPIUSAveragePricesCMYYearly , on = "Year")
dfAlldairyproductsSupplyandUse
```

```
dfAlldairyproductsSupplyandUse = dfAlldairyproductsSupplyandUse.merge(dfCrudeOilPrices , on =
"Year")
dfAlldairyproductsSupplyandUse
```

```
dfAlldairyproductsSupplyandUse.rename({'Average Price: Milk, Fresh, Whole, Fortified (Cost per
Gallon/3.8 Liters) ': 'AverageMilkPrice',
                'Average Price: Cheddar Cheese, Natural (Cost per Pound/453.6 Grams)
':'AverageCheddarCheesePrice',
                'Average Price: Yogurt, Natural, Fruit Flavored (Cost per 8 Ounces/226.8 Grams)
':'AverageYogartPrice'}, axis=1, inplace=True)
dfAlldairyproductsSupplyandUse
```

Output:

Merge Dairy Use and Supply, CPI US Average, and All Dairy Products Use datasets.

Step 12: Perform Exploratory Analysis

Lets visualize the dataset to perform exploratory analysis.

```
fig, axes = plt.subplots(nrows=4, ncols=4, dpi=120, figsize=(10,6))
for i, ax in enumerate(axes.flatten()):
    d = dataset[dataset.columns[i]]
    ax.plot(d, color='red', linewidth=1)
    # Decorations
    ax.set_title(dataset.columns[i])
    ax.xaxis.set_ticks_position('none')
    ax.yaxis.set_ticks_position('none')
    ax.spines['top'].set_alpha(0)
    ax.tick_params(labelsize=6)
plt.tight_layout();
```

Output:

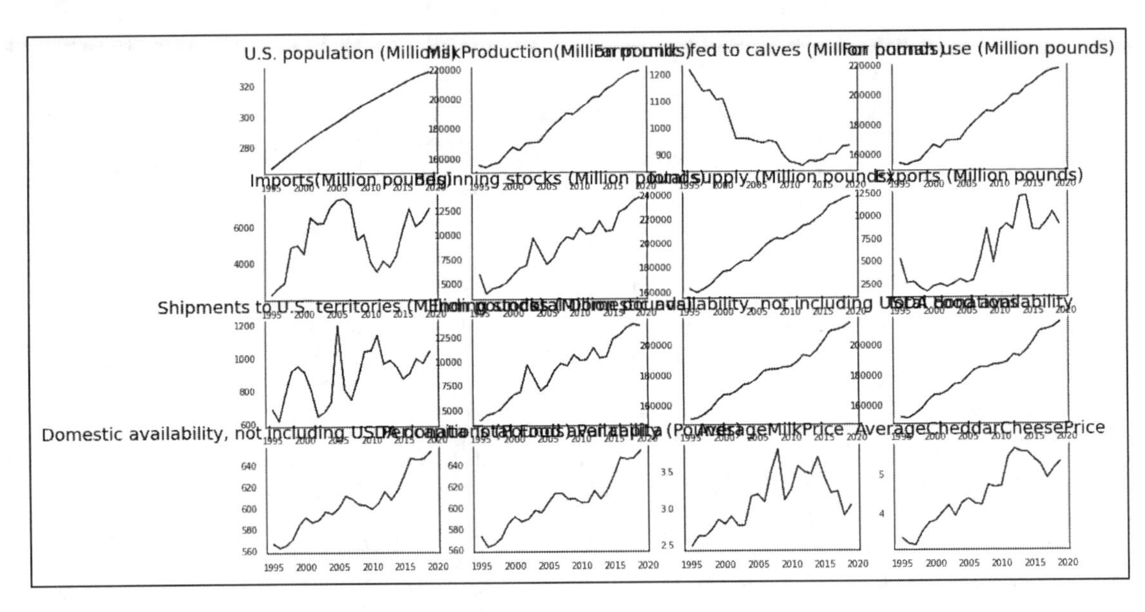

Step 13: Model Dependent and Independent variables Time Series

Drop all time series variables that have no direct influence on price forecast modeling.

```
dataset.drop(['U.S. population (Millions)', 'MilkProduction(Million pounds)',
    'Farm milk fed to calves (Million pounds)',
    'For human use (Million pounds)', 'Imports(Million pounds)',
    'Beginning stocks (Million pounds)', 'Total supply (Million pounds)',
    'Exports (Million pounds)',
    'Shipments to U.S. territories (Million pounds)',
    'Ending stocks (Million pounds)',
    'Total Domestic availability, not including USDA donations',
    'Total Food availability',
    'Domestic availability, not including USDA donations (Pounds) Per capita',
    'Per capita Total Food availability (Pounds)','Consumer Price Index (1982-84=1)',
        'Imported Crude Oil Price ($/barrel) - Nominal', 'TotalDisappearance','Consumer Price Index
    (1982-84=1)'],axis=1, inplace=True)
```

```
dataset.columns
```

Output:

```
Index(['AverageMilkPrice', 'AverageCheddarCheesePrice', 'AverageYogartPrice',
    'Imported Crude Oil Price ($/barrel) - Real', 'StockstoUse'],
    dtype='object')
```

Step 14: Perform Normality Test

To extract maximum information from our data, it is important to have a normal or Gaussian data distribution. To check for that, we have done a normality test based on the Null and Alternate Hypothesis intuition.

```
from statsmodels.tsa.stattools import adfuller
def stationary_test(df,col):
    print("++++++++++++++++++++++++++++++++")
    print("Observations of Dickey-fuller test")
    dftest = adfuller(df[col],autolag='AIC')
    dfoutput=pd.Series(dftest[0:4],index=['Test Statistic','p-value','#lags used','number of observations
   used'])
    for key,value in dftest[4].items():
        dfoutput['critical value (%s)'%key]= value
    print(dfoutput)
    if dfoutput[0] < dfoutput[4] or dfoutput[0] < dfoutput[5] or dfoutput[0] < dfoutput[6]:
        print(f"{col}: The series is stationary ")
    else :
        print(f"{col}: The series is not stationary ")
```

```
for i in dataset.columns:
    stationary_test(dataset,i )
```

Output:

Except for Imported Crude Oil Price ($/barrel)—Real, all other columns (Average Milk Price, Average Cheddar Cheese Price, AverageYogurt Price, and Stocks to Use) are not stationary.

So, we needed to convert it into a stationary time series.

Step 15: Perform Auto Correlation (AUF) and Partial Auto Correlation (PACF) Tests

Auto-correlation or serial correlation can be a significant problem in analyzing historical data if we do not know how to look out for it.

```
import pandas as pd
import matplotlib.pyplot as plt
import statsmodels.api as sm
# plots the autocorrelation plots for each stock's price at 150 lags
for i in dataset:
    sm.graphics.tsa.plot_acf(dataset[i], lags=15)
    plt.title('ACF for %s' % i)
plt.show()
```

Output:

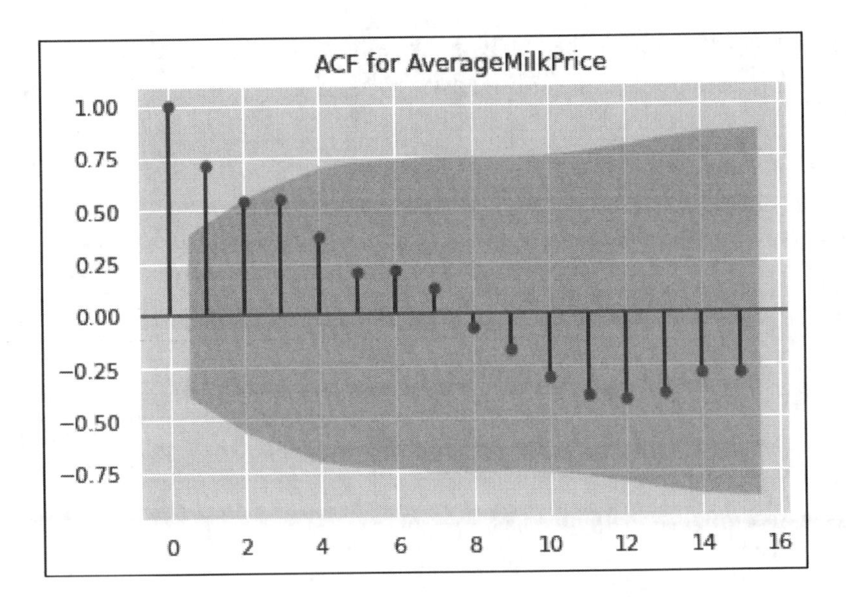

Auto Correlation of 1 could be an issue.
Perform Partial Auto Correlation.

```
import pandas as pd
import matplotlib.pyplot as plt
import statsmodels.api as sm
# plots the autocorrelation plots
for i in dataset:
    sm.graphics.tsa.plot_pacf(dataset[i], lags=11)
    plt.title('PACF for %s' % i)
plt.show()
```

Output:

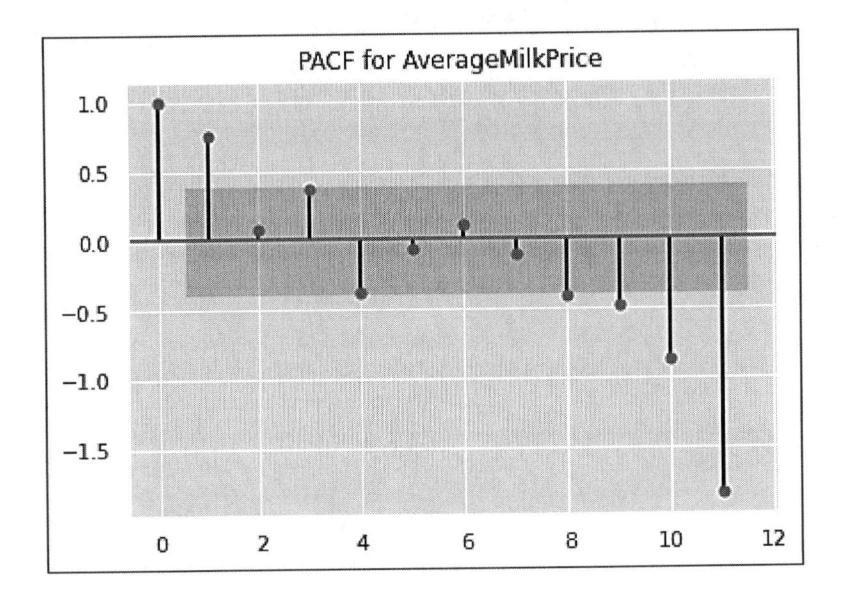

Step 16: Train and Test Data

Before transforming the data set to stationary, let's split it.

```
n_obs=6
X_train, X_test = dataset[0:-n_obs], dataset[-n_obs:]
print(X_train.shape, X_test.shape)
```

Step 17: Transformation

Apply first differencing on the training set to make all the series stationary. However, this is an iterative process where after the first differencing, the series may still be non-stationary. We shall have to apply a second difference or log transformation to standardize the series in such cases.

```
transform_data = X_train.diff().dropna()
transform_data.head()
```

We difference the data 4 times or shift it 4 times to make it stationary.

Re-Perform AUF and KPSS Tests

Applying first differencing Re-perform

```
or col in transform_data.columns:
    print('************')
    print(col)
    augmented_dickey_fuller_statistics(transform_data[col] )
```

Output:

```
************
AverageMilkPrice
ADF Statistic: –7.481422
p-value: 0.000000
Critical Values:
  1%: –4.665
  5%: –3.367
  10%: –2.803
************
AverageCheddarCheesePrice
ADF Statistic: –8.644678
p-value: 0.000000
Critical Values:
  1%: –4.473
  5%: –3.290
  10%: –2.772
```

```
************
AverageYogurtPrice
ADF Statistic: –5.264031
p-value: 0.000006
Critical Values:
  1%: –4.665
  5%: –3.367
  10%: –2.803
************
Imported Crude Oil Price ($/barrel) - Real
ADF Statistic: -4.604236
p-value: 0.000127
Critical Values:
  1%: –4.665
  5%: –3.367
  10%: –2.803
************
StockstoUse
ADF Statistic: –7.957003
p-value: 0.000000
Critical Values:
  1%: –4.473
  5%: –3.290
  10%: –2.772
```

```
for col in transform_data.columns:
    print('************')
    print(col)
    kpss_test(transform_data[col] )
```

Output:

```
************
AverageMilkPrice
KPSS Statistic: 0.4434199121420594
p-value: 0.05843969304221579
num lags: 8
Critical Values:
    10% : 0.347
    5% : 0.463
    2.5% : 0.574
    1% : 0.739
************
AverageCheddarCheesePrice
KPSS Statistic: 0.4058687437901143
p-value: 0.07462554146977833
num lags: 8
Critical Values:
    10% : 0.347
    5% : 0.463
    2.5% : 0.574
    1% : 0.739
************
AverageYogartPrice
KPSS Statistic: 0.346294750404284
p-value: 0.1
num lags: 8
Critical Values:
    10% : 0.347
    5% : 0.463
    2.5% : 0.574
    1% : 0.739
************
Imported Crude Oil Price ($/barrel) - Real
KPSS Statistic: 0.3044135047306327
p-value: 0.1
num lags: 8
Critical Values:
    10% : 0.347
    5% : 0.463
    2.5% : 0.574
    1% : 0.739
************
StockstoUse
KPSS Statistic: 0.39606784360785496
p-value: 0.07885006741040734
num lags: 8
Critial Values:
    10% : 0.347
    5% : 0.463
    2.5% : 0.574
    1% : 0.739
```

We determine which AUF and KPSS tests confirm that the time series is stationary.
Plot the transformed dataset.

```
fig, axes = plt.subplots(nrows=2, ncols=2, dpi=120, figsize=(10,6))
for i, ax in enumerate(axes.flatten()):
    d = transform_data[transform_data.columns[i]]
    ax.plot(d, color='red', linewidth=1)
    # Decorations
    ax.set_title(dataset.columns[i])
    ax.xaxis.set_ticks_position('none')
    ax.yaxis.set_ticks_position('none')
    ax.spines['top'].set_alpha(0)
    ax.tick_params(labelsize=6)
plt.tight_layout();
```

Output:

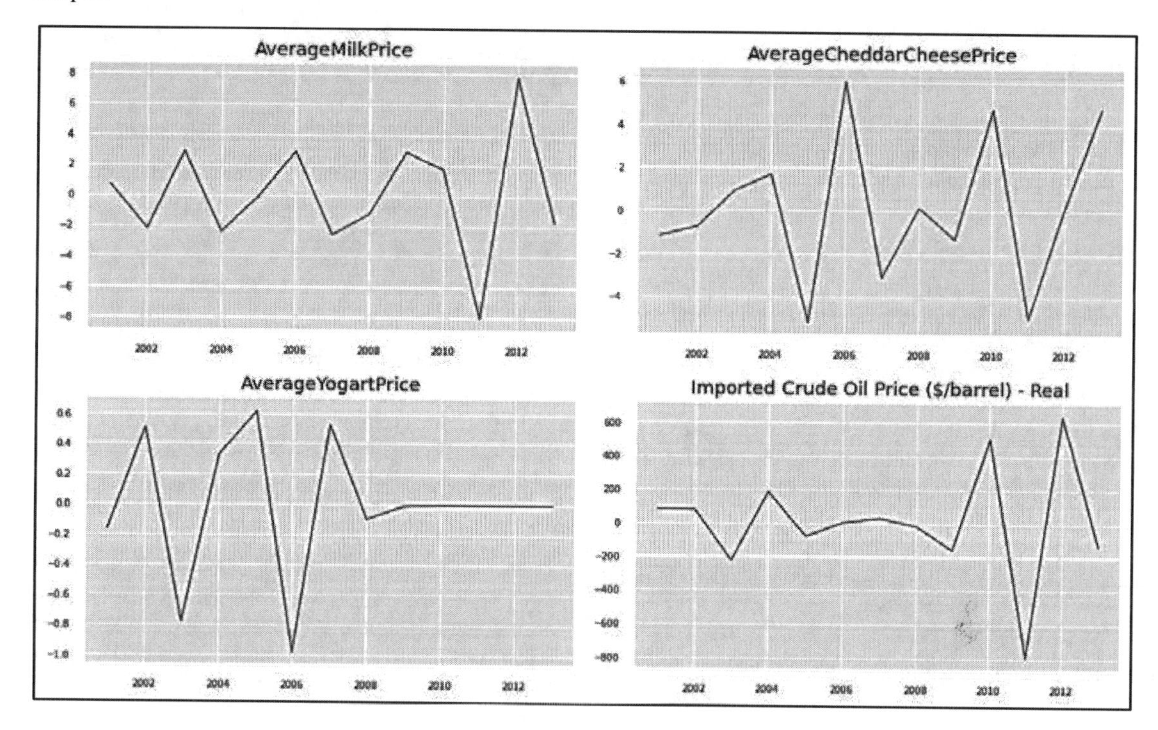

Step 18: Perform Grangers Causality Test

To validate causation of the model time series, it is required to perform a Granger causation test.

```
import statsmodels.tsa.stattools as sm
maxlag=MAX_LAGS
test = 'ssr-chi2test'
def grangers_causality_matrix(X_train, variables, test = 'ssr_chi2test', verbose=False):
    dataset = pd.DataFrame(np.zeros((len(variables), len(variables))), columns=variables, index=variables)
    for c in dataset.columns:
        for r in dataset.index:
            test_result = sm.grangercausalitytests(X_train[[r,c]], maxlag=maxlag, verbose=False)
            p_values = [round(test_result[i+1][0][test][1],4) for i in range(maxlag)]
            if verbose: print(f'Y = {r}, X = {c}, P Values = {p_values}')
            min_p_value = np.min(p_values)
            dataset.loc[r,c] = min_p_value
    dataset.columns = [var + '_x' for var in variables]
    dataset.index = [var + '_y' for var in variables]
    return dataset
```

```
grangers_causality_matrix(dataset, variables = dataset.columns)
```

Output:
All variables are in granger causation.

	AverageMilkPrice_x	AverageCheddarCheesePrice_x	AverageYogurtPrice_x	Imported Crude Oil Price ($/barrel) - Real_x	StockstoUse_x
AverageMilkPrice_y	1.0	0.0	0.0000	0.0013	0.0000
AverageCheddarCheesePrice_y	0.0	1.0	0.0070	0.0000	0.0006
AverageYogurtPrice_y	0.0	0.0	1.0000	0.0000	0.0000
Imported Crude Oil Price ($/barrel) - Real_y	0.0	0.0	0.0000	1.0000	0.0000
StockstoUse_y	0.0	0.0	0.0002	0.0000	1.0000

So, looking at the p-Values, we can assume that all the other variables (time series) in the system are interchangeably causing each other. This justifies the VAR modeling approach for this system of multi time-series to forecast. The granger causation interaction node diagram depicts relationships. In a way, it forms an ideal basis to run VAR and other time series algorithms.

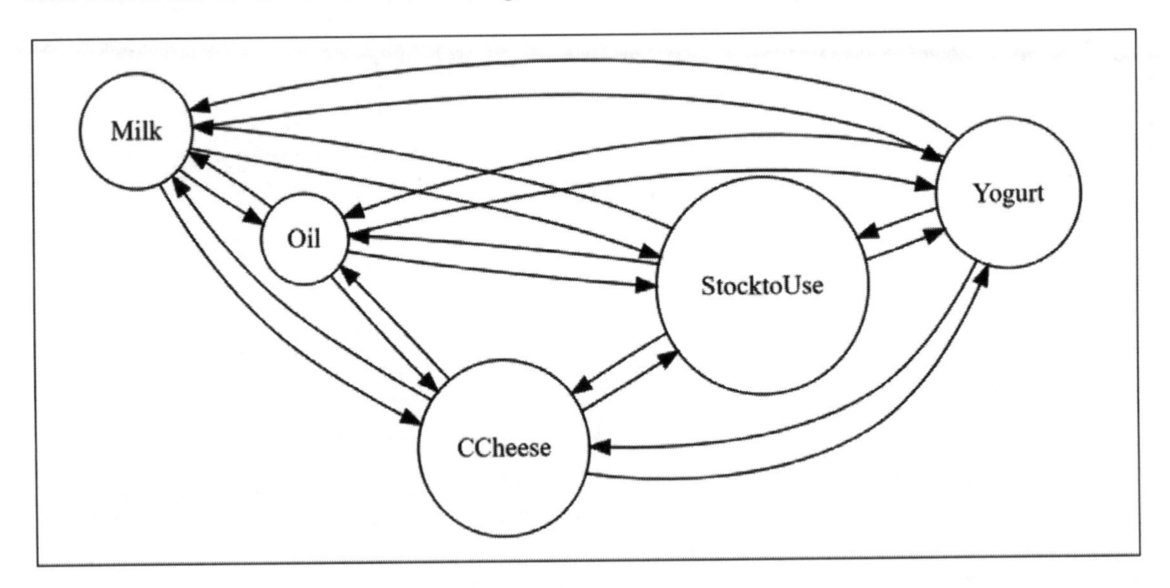

Step 19: Vector Auto Regressive Model (VAR)

VAR requires stationarity of the series which means, the means to the series do not change over time (we can find this out from the plot drawn next to the Augmented Dickey-Fuller Test). Perform the Vector Auto Regressive (VAR) model. Here each series is modeled by its own lag and other series' lag. $y_{\text{average milk prices}}$ {1, t-1, $y_{\text{cheddar cheese price}}$ {2, t-1}, $y_{\text{Yogurt price}}$ {3, t-1}, $y_{\text{Stocks to Use}}$ {4, t-1}, and $y_{\text{imported oil}}$ {5, t-1} …. are the lags of the time series of the respective commodities. Since the y terms in the equations are interrelated, the y's are considered as endogenous variables, rather than as exogenous predictors. To thwart the issue of structural instability, the VAR framework was utilized choosing the lag length according to AIC.

Next, fit the VAR model on the training set and then use the fitted model to forecast the following observations. These forecasts will be compared against the actual ones present in the test data. I have taken the maximum lag (AIC decided) to identify the required lags for the VAR model.

Historically, several trends can be observed in U.S. Milk prices: several cycles in the period, 1988–2007[82]

- Triennial, 36-month cycle - large and exploding (Lag = 3)
- Biennial, 26-month cycle - quite large and exploding (Lag = 2)
- **Annual, 12-month cycle - smaller, erratic (Lag = 1) – VAR model suggests Lag = 1**
- 9-month cycle—small and stable (Lag = 1)

```
import numpy as np
import pandas
import statsmodels.api as sm
from statsmodels.tsa.api import VAR
mod = VAR (transform_data)
results = mod.fit(ic='aic')
print(results.summary())
```

The AIC criterion asymptotically overestimates the order with positive probability, whereas the BIC and HQ criteria estimate the order consistently under general conditions if the true order p is less than or equal to p_{max}. For more information on the use of model selection criteria in VAR models[83] refer to [26].

Output:

```
Summary of Regression Results
==================================
Model:                         VAR
Method:                        OLS
Date:            Sun, 05, Jun, 2022
Time:                     08:13:15
--------------------------------------------------------------------
No. of Equations:          5.00000    BIC:                   4.43937
Nobs:                      12.0000    HQIC:                  2.77828
Log likelihood:           -74.4989    FPE:                   41.2711
AIC:                       3.22710    Det(Omega_mle):        5.43488
--------------------------------------------------------------------
```

The biggest correlations are 0.7310 (Milk Price & Cheddar Cheese) and –1.0429 (Milk Price& Imported Crude Oil); the following impactful correlations are 0.6692 (Yogurt Price and Stocks to Use) and –0.9699 (Cheddar Cheese and Yogurt prices); Finally, –0.8955 (Milk Price and Stocks to Use); however, these are small enough to be ignored in this case.

The price relationship between Milk and Stocks to Use: Wholesale prices for cheese [30], butter, and most dried dairy products are determined by market supply and demand factors. Wholesale prices rise when supply is short relative to demand. On the other hand, prices fall when supply grows and/or demand falls.[84] Classical demand-supply infused price movements are presented in the following table:

	Average MilkPrice	Average Cheddar CheesePrice	Average YogurtPrice	Imported Crude Oil Price ($/barrel) - Real	StockstoUse
AverageMilkPrice	1	**0.731043**	–0.710348	–1.042969	**–0.895542**
AverageCheddarCheese Price	**0.731043**	1	–0.969997	–0.137611	–0.595156
AverageYogurtPrice	–0.710348	–0.969997	1	0.046058	**0.669221**
Imported Crude Oil Price ($/barrel) – Real	–0.147427	–0.137611	0.046058	1	–0.069933
StockstoUse	**–0.895542**	–0.595156	**0.669221**	–0.069933	1

Stocks to Use: When milk production is long relative to profitable sales, the dairy industry, primarily through operating cooperatives, looks to bank the surplus in a form that is easy to make and cheap to store. Typically, *that means nonfat dry milk (NDM) and butter* [6].

The price of milk and surplus (stocks to use) milk production & stocks have a negative relationship. As surplus milk production eases,[85] the stocks, prices for butter, cheese, and nonfat dry milk increase, to reduce stocks through commercial market sales. And hence, Milk Price to Stock To Use (**–0.8955)** & Cheddar Cheese Price to Stocks **(–0.5951**) are justified [31].

The second factor is cheese stocks. Cheese stocks exist for three major reasons. First, there is always pipeline inventory, product held for a short time for a near-term future sale. Second, some cheeses are aged as part of their product development. Three to nine months, maybe a year, is common. There is probably less aged cheese today than there used to be. In either the first or second case, these cheese stocks are not in "surplus". This kind of storage is really part of the normal marketing process. The last category is unplanned or unbudgeted cheese stocks, stuff we made but couldn't sell. Often, I would say this happens because we were surprised by demand (it's less) than production (it's more) [6]. Hence, cheese stocks to use negatively correlated **(–0.5951**), a reinforcement of the model outcome.

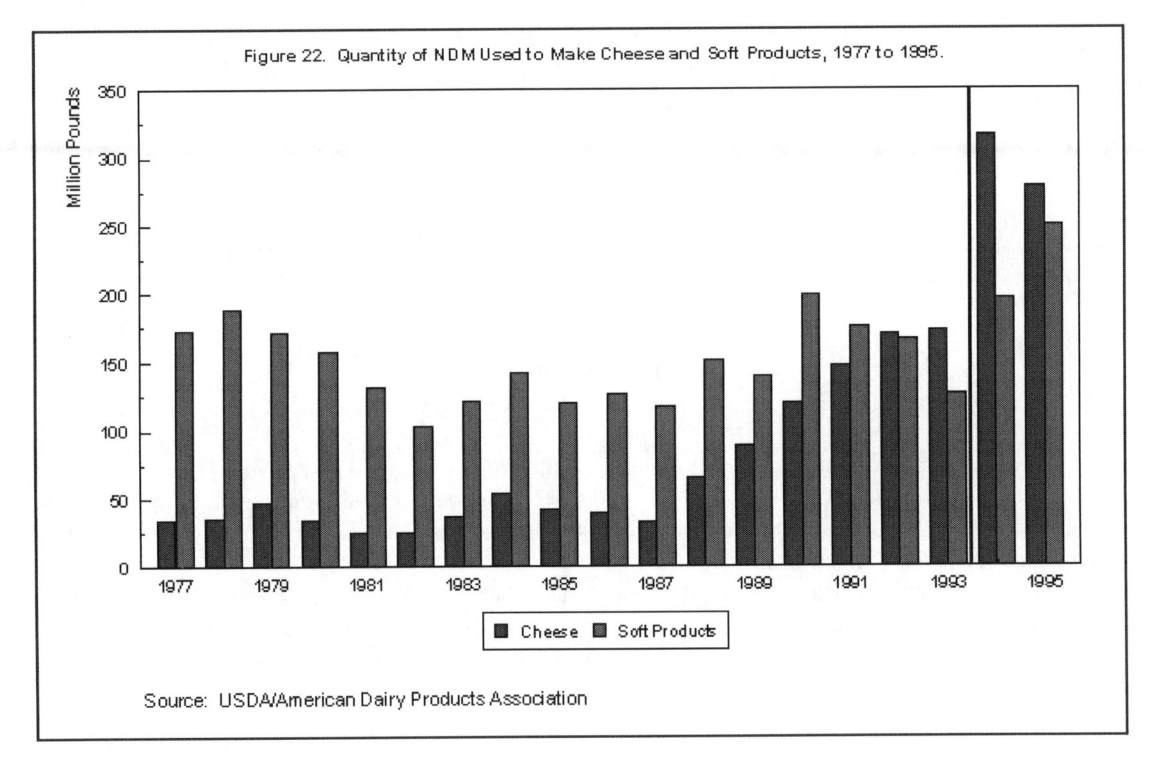

Source: Figure 22[86] & Chapter 4[87]

Finally, Yogurt positively correlated with Stock to Use as it belonged to the Class II family of dairy classifications for milk derivatives going into 'soft' manufactured products such as sour cream,[88] cottage cheese, ice cream, and yogurt. The primary reason is that NDM[89] is being utilized in Class II and Class III products which are less than perfect. The above figure shows the quantities of NDM used to make cheese (excluding cottage cheese) and soft products (including cottage cheese, sour cream, ice cream and yogurt). The quantity of NDM used to make soft products increases when milk supplies are tight as was the case in the late 1970s and the early 1990s. However, since the establishment of Class IIIA there has been a sharp increase in the utilization of NDM to make soft products [11]. The quantity of NDM to make cheese likewise increased sharply after Class IIIA pricing began. It suggests that as much as half of the nonfat solids contained in soft products came from NDM, historically, and surplus milk is converted into NDM for storage [11—Chapter 4]. Another chief reason for the positive correlation with Stocks to Use, 0.669221, is justified!

Step 20: Residual plot

The residual plot looks normal with a constant mean throughout apart from some large fluctuations during 2003, 2004, 2010, 2011, and 2012.

```
y_fitted = results.fittedvalues
residuals = results.resid
plt.figure(figsize = (15,5))
plt.plot(residuals, label='resid')
plt.plot(y_fitted, label='VAR prediction')
plt.xlabel('Date')
plt.xticks(rotation=45)
plt.ylabel('Residuals')
plt.grid(True)
```

Output:

The price spikes are due to adjacent commodity prices and global dynamics.

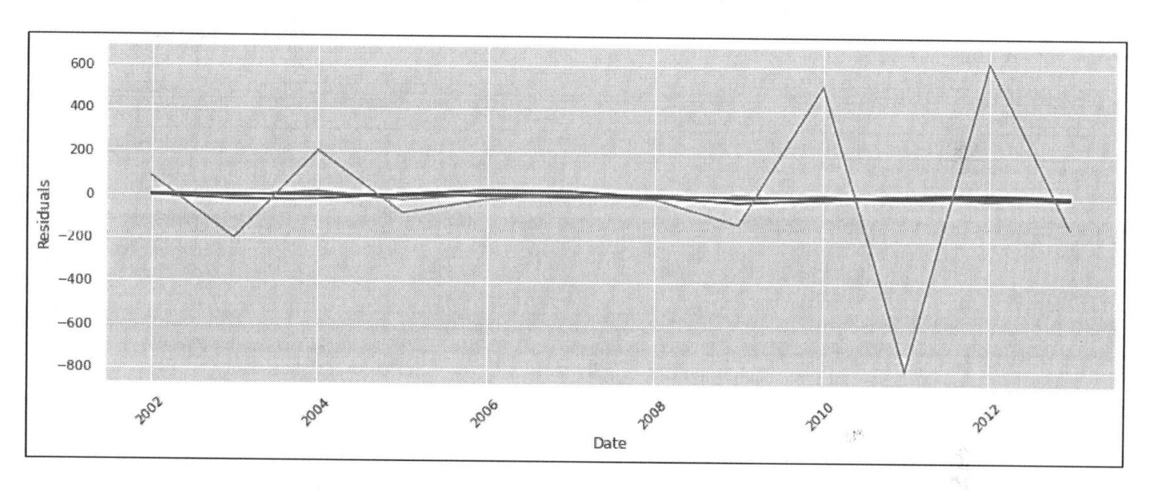

Step 21: Durbin-Watson Statistic

The Durbin-Watson Statistic is related to auto correlation. The Durbin-Watson statistic will always have a value between 0 and 4. A value of 2.0 means that there is no autocorrelation detected in the sample. Values from 0 to less than 2 indicate positive autocorrelation and values from 2 to 4 indicate negative autocorrelation. A rule of thumb is that test statistic values in the range of 1.5 to 2.5 are relatively normal. Any value outside this range (Milk Prices, Imported Oil, and Stock to Use) could be a cause for concern.

```
from statsmodels.stats.stattools import durbin_watson
out = durbin_watson(results.resid)
for col,val in zip(transform_data.columns, out):
  print((col), ":", round(val,2))
```

Output:

```
AverageMilkPrice : 2.83
AverageCheddarCheesePrice : 2.25
AverageYogurtPrice : 2.32
Imported Crude Oil Price ($/barrel) - Real : 1.04
StockstoUse : 2.83
```

There is no autocorrelation, majority, (2.0) exists; so, we can proceed with the forecast.

Step 22: Prediction

To forecast, the VAR model expects to lag up the ordered number of observations from the past data. This is because the terms in the VAR model are essentially the lags of the various time series in the dataset, so we need to provide as many of the previous values as indicated by the lag order used by the model.

```
# Get the lag order
lag_order = results.k_ar
lag_order
```

```
print(lag_order)
# Input data for forecasting
input_data = transform_data.values[-lag_order:]
print(input_data)
# forecasting
pred = results.forecast(y=input_data, steps=n_obs)
pred = (pd.DataFrame(pred, index=X_test.index, columns=X_test.columns + '_pred'))
print(pred)
```

Output:

```
      AverageMilkPrice_pred  AverageCheddarCheesePrice_pred    AverageYogurtPrice_pred
Year
2014              -0.883868                       -3.202134                  -0.505489
2015               1.669666                        1.240682                   0.463309
2016              -0.610692                       -0.192130                   0.023862
2017              -0.834274                        0.954508                  -0.697580
2018               1.518351                       -2.035140                   0.913773
2019              -0.847967                        2.443385                  -0.593475

      Imported Crude Oil Price ($/barrel) - Real_pred  StockstoUse_pred
Year
2014                                      -296.127914         -2.974771
2015                                       369.369323          8.308542
2016                                       -94.459054        -15.203828
2017                                      -309.556048         15.778190
2018                                       506.288901         -8.205272
2019                                      -373.734892         -3.561183
```

Step 23: Invert the transformation

The forecasts are generated but it is on the scale of the training data used by the model. So, to bring it back up to its original scale, we need to de-difference it. The way to convert the differencing is to add these differences consecutively to the base number. An easy way to do it is to first determine the cumulative sum at index and then add it to the base number. This process can be reversed by adding the observation at the prior time step to the difference value.

inverted(ts) = differenced(ts) + observation(ts-1)

```
# inverting transformation
def invert_transformation(X_train, pred):
    forecast12 = pred.copy()
    columns = X_train.columns
    #col='CPI_AllUrban_FoodandBeverages'
    #forecast[str(col)+'_pred'] = dataset[col].iloc[-1] + forecast[str(col)+'_pred'].cumsum()
    for col in columns:
        print('++++++++++++++++')
        print(col)
        forecast12[str(col)+'_pred'] = X_train[col].iloc[-5] + forecast12[str(col)+'_pred'].cumsum()
        print('++++++++++++++++')
    return forecast12

output = invert_transformation(X_train, pred)
output
```

Output:

Year	AverageMilkPrice_pred	AverageCheddarCheesePrice_pred	AverageYogurtPrice_pred	Imported Crude Oil Price ($/barrel) - Real_pred	StockstoUse_pred
2014	2.225132	1.466616	0.307511	-216.784300	2.566905
2015	3.894797	2.707298	0.770820	152.585022	10.875447
2016	3.284106	2.515168	0.794682	58.125969	-4.328380
2017	2.449831	3.469676	0.097102	-251.430080	11.449810
2018	3.968182	1.434536	1.010875	254.858821	3.244537
2019	3.120215	3.877921	0.417400	-118.876071	-0.316646

```
#combining predicted and real data set
combine = pd.concat([output['AverageCheddarCheesePrice_pred'], X_test['AverageCheddarCheesePrice']], axis=1)
combine['accuracy'] = round(combine.apply(lambda row: row['AverageCheddarCheesePrice_pred'] / row['AverageCheddarCheesePrice'] * 100, axis = 1),2)
combine['accuracy'] = pd.Series(["{0:.2f}%".format(val) for val in combine['accuracy']],index = combine.index)
combine = combine.round(decimals=2)
```

Output:

Year	AverageCheddarCheesePrice_pred	AverageCheddarCheesePrice	accuracy
2014	1.47	5.54	26.46%
2015	2.71	5.38	50.33%
2016	2.52	5.23	48.05%
2017	3.47	4.90	70.88%
2018	1.43	5.14	27.94%
2019	3.88	5.31	73.06%

Step 24: Evaluation

To evaluate the forecasts, a comprehensive set of metrics, such as MAPE, ME, MAE, MPE and RMSE can be computed. We have computed some of these below.

```
#combining predicted and real data set
combine = pd.concat([output['AverageMilkPrice_pred'], X_test['AverageMilkPrice']], axis=1)
combine['accuracy'] = round(combine.apply(lambda row: row['AverageMilkPrice_pred'] / row['AverageMilkPrice'] * 100, axis = 1),2)
combine['accuracy'] = pd.Series(["{0:.2f}%".format(val) for val in combine['accuracy']],index = combine.index)
combine = combine.round(decimals=2)
```

```
from sklearn.metrics import mean_absolute_error
from sklearn.metrics import mean_squared_error
import math
#Forecast bias
print('Mean absolute error:', mean_absolute_error(combine['AverageMilkPrice'].values, combine['AverageMilkPrice_pred'].values))
print('Mean squared error:', mean_squared_error(combine['AverageMilkPrice'].values, combine['AverageMilkPrice_pred'].values))
print('Root mean squared error:', math.sqrt(mean_squared_error(combine['AverageMilkPrice'].values, combine['AverageMilkPrice_pred'].values)))
```

Output:

```
Mean absolute error: 0.6566666666666666
Mean squared error: 0.6864333333333333
Root mean squared error: 0.8285127237003264
```

Mean absolute error tells us how big of an error we can expect from the forecast on average. Our error rates are quite low here, indicating we have the right fit for the model.

The VAR model is a popular tool for the purpose of predicting joint dynamics of multiple time series based on linear functions of past observations. More analysis, e.g., impulse response (IRF) and forecast error variance decomposition (FEVD) can also be done along-with VAR for assessing the impacts of shock from one asset on another.

Regressive Linkage Model: CPI Average Milk Prices, Imported Oil Prices, and All Dairy Products (milk-fat milk-equivalent basis): Supply and Use

The VAR model is known for modeling multivariate economic parameters' time series. The time series models are constructed to see future prediction as time series models differing from regression models in that time series models depend on the order of the observations. The regressive models, on the other hand, establish coefficients of regression to see the impact of macroeconomic, co-movement, and other significant variables on milk prices. In regression, the order of the observations does not affect the model [25, 26]. Time series models use historic values of a series to predict future values. Regressive models predict based on the feature scalar values. Let's apply a regressive model for Dairy milk predictions. The exploratory data analysis and data frame construction can be utilized from the above VAR model. So, we can jump directly to model construction (Step 16 onwards):

Step 1: Train and Split

```
X = dataset[['AverageCheddarCheesePrice', 'Imported Crude Oil Price ($/barrel) - Real',
    'AverageYogurtPrice', 'StockstoUse'
    ]]
y = dataset['AverageMilkPrice']
from sklearn.model_selection import train_test_split
X_train, X_test, y_train, y_test = train_test_split(X, y, test_size = 0.3, random_state = 0)
```

Split the dataset with 70% for training and 30% for testing. The sample size is pretty low and hence these training percentages.

Step 2: Construct Linear Regressive Model

Construct Linear regressive model.

```
# Train the model
from sklearn.linear_model import LinearRegression
# Fit a linear regression model on the training set
model = LinearRegression(normalize=False).fit(X_train, y_train)
print (model)
```

Output:

LinearRegression()

Step 3: Predict the model with test data

Using test dataset, predict the model.

```
import numpy as np
predictions = model.predict(X_test)
np.set_printoptions(suppress=True)
print('Predicted labels: ', np.round(predictions))
print('Actual labels : ' ,y_test)
```

Output:

```
Predicted labels: [3. 3. 4. 4. 3. 3. 4. 3.]
Actual labels : Year
2000    2.780667
1997    2.614000
2014    3.693667
2011    3.571583
2006    3.081333
2017    3.225500
2012    3.492500
2019    3.035667
Name: AverageMilkPrice, dtype: float64
```

Step 4: Plot Model Accuracy

Now, plot the model accuracy:

```
import matplotlib.pyplot as plt
%matplotlib inline
plt.scatter(y_test, predictions)
plt.xlabel('Actual Labels')
plt.ylabel('Predicted Labels')
plt.title('Predictions')
# overlay the regression line
z = np.polyfit(y_test, predictions, 1)
p = np.poly1d(z)
plt.plot(y_test,p(y_test), color='magenta')
plt.show()
```

Output: the model linear slope line connecting feature variables and predictor.

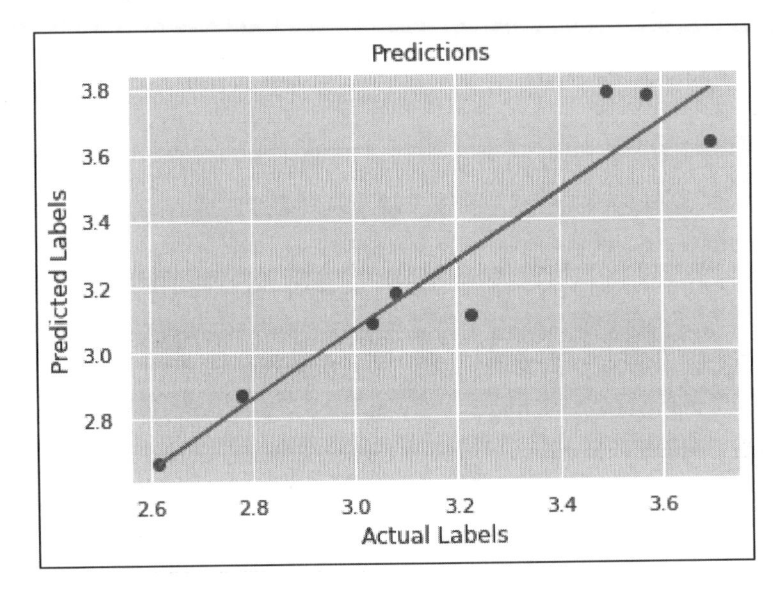

Step 5: Evaluate the Model

Evaluate model accuracy parameters.

```
from sklearn.metrics import mean_squared_error, r2_score
mse = mean_squared_error(y_test, predictions)
print("MSE:", mse)
rmse = np.sqrt(mse)
print("RMSE:", rmse)
r2 = r2_score(y_test, predictions)
print("R2:", r2)
```

Output:

```
MSE: 0.020391074099694505
RMSE: 0.14279731825105996
R2: 0.8411493821658248
```

The model exhibiting very low MSE, a similar output seen as part of the VAR model.

Step 6: Develop Model Equation (y=mx+c)

Construct slope (mx) part:

```
mx=""
for ifeature in range(len(X.columns)):
    if model.coef_[ifeature] <0:
        # format & beautify the equation
        mx += " - " + "{:.2f}".format(abs(model.coef_[ifeature])) + " * " + X.columns[ifeature]
    else:
        if ifeature == 0:
            mx += "{:.2f}".format(model.coef_[ifeature]) + " * " + X.columns[ifeature]
        else:
            mx += " + " + "{:.2f}".format(model.coef_[ifeature]) + " * " + X.columns[ifeature]

print(mx)
```

Output:

```
0.20 * AverageCheddarCheesePrice + 0.01 * Imported Crude Oil Price ($/barrel) - Real - 0.54 * AverageYogurtPrice + 0.00 * StockstoUse
```

Step 7: Construct governing equation

Generate the model governing equation:

```
# y=mx+c
if(model.intercept_ <0):
    print("The formula for the " + y.name + " linear regression line (y=mx+c) is = " + " - {:.2f}".format(abs(model.intercept_)) + " + " + mx )
else:
    print("The formula for the " + y.name + " linear regression line (y=mx+c) is = " + "{:.2f}".format(model.intercept_) + " + " + mx )
```

Output:

```
The formula for the AverageMilkPrice linear regression line (y=mx+c) is = 2.12 + 0.20 * AverageCheddarCheesePrice + 0.01 * Imported Crude Oil Price ($/barrel) - Real - 0.54 * AverageYogurtPrice + 0.00 * StockstoUse
```

Step 8: Model Explainability

Using model Explainability, construct important features.

```
import eli5
eli5.show_weights(model,feature_names = model_features)
```

Output:

y top features

Weight?	Feature
+2.118	<BIAS>
+0.202	Imported Crude Oil Price ($/barrel) - Real
+0.008	AverageCheddarCheesePrice
+0.002	StockstoUse
-0.544	AverageYogurtPrice

Step 9: Explain Model Predictions

```
from eli5 import show_prediction
show_prediction(model, X_test[5:6], show_feature_values=True, feature_names=model_features)
```

Output:

y (score **3.109**) top features

Contribution?	Feature	Value
+2.118	<BIAS>	1.000
+0.987	Imported Crude Oil Price ($/barrel) - Real	4.895
+0.435	AverageCheddarCheesePrice	57.624
+0.011	StockstoUse	6.093
-0.442	AverageYogurtPrice	0.813

Step 10: Model summary

The following table has VAR and regression coefficients significant with respect to Milk Price predictor or class variables: (although we cannot compare regression coefficients with VAR, the table provides a comparative view)

	VAR	Regression
AverageCheddarCheesePrice	0.731043	0.20
Imported Crude Oil Price ($/barrel) – Real	–1.042969	0.01
AverageYogurtPrice	–0.710348[†]	**-0.54**[‡] – It falls into Class II 'soft' manufactured products & only surplus milk is converted into NDM that makes Yogurt. The relationship with milk price is negative as excessive milk implies dried prices!
StockstoUse	–0.895542	0.0001 [11]

[† & ‡] Yogurt positively correlated with Stock to Use. Here is the reason: firstly, Class II family of dairy classification to milk going into 'soft' manufactured products such as sour cream,[90] cottage cheese, ice cream, and yogurt. The primary reason is NDM[91] is being utilized in Class II and Class III products which are less than perfect. As shown in the VAR section, the quantities of NDM are used to make cheese (excluding cottage cheese) and soft products (including cottage cheese, sour cream, ice cream and yogurt). The quantity of NDM used to make soft products increases when milk supplies are tight as was the case in the late 1970s and the early 1990s. However, since the establishment of Class IIIA there has been a sharp increase in the utilization of NDM to make soft products [11]. The quantity of NDM to make cheese likewise increased sharply after Class IIIA pricing began. It suggests that as much as half of the nonfat solids contained in soft products came from NDM, historically, and surplus milk is converted into NDM for storage [11—Chapter 4]. Another chief reason for the negative correlation with Yogurt prices, **–0.7103**, is justified (excessive milk implies depressed milk prices (classic demand and supply)!

Prophet Time Series Model: CPI Average Milk Prices, Imported Oil Prices, and All Dairy Products (milk-fat milk-equivalent basis): Supply and Use

Prophet is a procedure for forecasting time series data based on an additive model where non-linear trends are fit with yearly, weekly, and daily seasonality with holiday effects. It works best with time series' that have strong seasonal effects and several seasons of historical data. Prophet is robust to missing data and shifts in the trend, and typically handles outliers well.

Step 1: Load Dataset to model Prophet Model

We can re-use data & exploratory analysis – Step 16 of the VAR model.

```
dfAlldairyproductsSupplyandUse
dfAlldairyproductsSupplyandUse.info()
dfProphetMultiVarDataset=dfAlldairyproductsSupplyandUse.copy()
```

Output:

```
<class 'pandas.core.frame.DataFrame'>
Int64Index: 25 entries, 1995 to 2019
Data columns (total 22 columns):
U.S. population (Millions) 25 non-null float64
MilkProduction(Million pounds) 25 non-null float64
Farm milk fed to calves (Million pounds) 25 non-null float64
For human use (Million pounds) 25 non-null float64
Imports(Million pounds) 25 non-null float64
Beginning stocks (Million pounds) 25 non-null float64
Total supply (Million pounds) 25 non-null float64
Exports (Million pounds) 25 non-null float64
Shipments to U.S. territories (Million pounds) 25 non-null float64
Ending stocks (Million pounds) 25 non-null float64
Total Domestic availability, not including USDA donations 25 non-null float64
Total Food availability 25 non-null float64
Domestic availability, not including USDA donations (Pounds) Per capita 25 non-null float64
Per capita Total Food availability (Pounds) 25 non-null float64
AverageMilkPrice 25 non-null float64
AverageCheddarCheesePrice 25 non-null float64
AverageYogurtPrice 25 non-null float64
Consumer Price Index (1982-84=1) 25 non-null float64
Imported Crude Oil Price ($/barrel) - Nominal 25 non-null float64
Imported Crude Oil Price ($/barrel) - Real 25 non-null float64
TotalDisappearance 25 non-null float64
StockstoUse 25 non-null float64
```

Step 2: Split the data (train and test)

The Prophet Time Series model needs both trained and test datasets to validate the model accuracies. For both, we need to define "ds" and "y" columns.

```
n_obs=15
X_train, X_test = dfProphetMultiVarDataset[0:-n_obs], dfProphetMultiVarDataset[-n_obs:]
print(X_train.shape, X_test.shape)
```

Output:
(10, 22) (15, 22)

Step 3: Prepare Data Frame to add Y column

We fit the model by instantiating a new Prophet object. Next, prepare a train data frame with a predictor column. Any settings to the forecasting procedure are passed into the constructor.

```
from fbprophet import Prophet
model = Prophet()
train_df = X_train.rename(columns={"AverageMilkPrice":'y'})
#train_df = X_train.rename(columns={"AverageCheddarCheesePrice":'y'})
train_df["ds"] = train_df.index
```

Step 4: Add additional Regressors

Additional regressors can be added to the linear part of the model using the add_regressor method. A column with the regressor value will need to be present in both the fitting and prediction data frames. In the following, Milk Production (million pounds), Imported Crude Oil Price ($/barrel) – Real, Averaged Cheddar Cheese, Average Yogurt Price, and Stocks to Use are the additional regressors.

```
from fbprophet import Prophet
model = Prophet()
model.add_regressor('AverageYogurtPrice')
model.add_regressor('StockstoUse')
model.add_regressor('Imported Crude Oil Price ($/barrel) - Real')
model.add_regressor('AverageCheddarCheesePrice')
model.fit(train_df)
```

Output:
<fbprophet.forecaster.Prophet at 0x7f6e6bee0e10>

```
model.params
```

Output:

```
{'k': array([[0.32084222]]),
'm': array([[0.67907221]]),
'delta': array([[-0. , -0.00000425, -0.00006804, -0.13785798, -0.00001128,
        0. , -0. ]]),
'sigma_obs': array([[0.01401531]]),
'beta': array([[ 0.00708289, 2.64587702, 0.0133649 , 2.43459889, 0.01804522,
        2.08246153, 0.02032321, 1.58945424, 0.01939842, 0.95556204,
        0.01447072, 0.18076567, 0.00474035, -0.73495842, -0.01059203,
        -1.79163803, -0.03232522, -2.98930523, -0.06125743, -4.32799637,
        -0.10087136, 0.01336009, 0.05780734]])}
```

Step 5: Prepare Test Dataset

Test Dataset is used to assess model accuracy. We need to define the Data Time column and the Dependent variable (Y").

```
fProphetMultiVarTestDataset=X_test
dfProphetMultiVarTestDataset['Date']=dfProphetMultiVarTestDataset.index
```

```
dfProphetMultiVarTestDataset = dfProphetMultiVarTestDataset.set_index('Date')
dfProphetMultiVarTestDataset["ds"] = dfProphetMultiVarTestDataset.index
```

```
dfProphetMultiVarTestDataset
```

Output:

	U.S. population (Millions)	MilkProduction(Million pounds)	Farm milk fed to calves (Million pounds)	For human use (Million pounds)	Imports(Million pounds)	Beginning stocks (Million pounds)	Total supply (Million pounds)	Exports (Million pounds)	Shipments to U.S. territories (Million pounds)	Ending stocks (Million pounds)	...	Per capita Total Food availability (Pounds)
Date												
2005	296.186	176931.0	949.0	175978.3	7421.0	6914.3	190313.6	2592.3	1194.3	7548.0	...	604.3
2006	298.996	181782.0	943.0	180834.1	7484.0	7548.0	195866.1	2860.0	802.6	8986.1	...	612.8
2007	302.004	185654.0	952.0	184698.8	7178.0	8986.1	200862.9	5210.0	741.8	9768.0	...	613.0
2008	304.798	189978.0	944.0	189026.3	5260.0	9768.0	204054.3	8527.0	872.8	9533.6	...	607.4
2009	307.439	189202.0	901.0	188285.9	5559.0	9533.6	203378.6	4809.7	1036.7	10678.8	...	607.8
2010	309.741	192877.0	873.0	191860.9	4043.0	10678.8	206582.7	8366.2	1040.4	10053.0	...	604.1
2011	311.974	196255.0	867.0	195373.0	3504.0	10053.0	208930.0	9056.0	1134.2	10186.4	...	604.4
2012	314.168	200642.0	858.0	199776.0	4110.0	10186.4	214072.4	8506.0	960.0	11371.7	...	615.1
2013	316.295	201260.0	877.0	200369.9	3772.0	11371.7	215513.6	12074.0	983.3	10305.9	...	607.5
2014	318.577	206048.0	874.0	205114.7	4372.0	10305.9	219792.6	12174.0	943.3	10443.4	...	616.0
2015	320.871	208508.0	880.0	207555.3	5759.0	10443.4	223757.7	8501.0	869.1	12287.5	...	629.8
2016	323.161	212451.0	900.0	211527.1	6936.0	12287.5	230750.5	8388.0	904.4	12673.9	...	646.1
2017	325.206	215527.0	901.0	214604.6	5987.8	12673.9	233266.2	9212.2	991.3	13397.3	...	644.7
2018	326.924	217568.0	928.0	216556.5	6300.0	13397.3	236253.8	10375.4	963.1	13790.9	...	645.8
2019	328.476	218382.0	932.0	217408.4	6934.8	13790.9	238134.1	9102.4	1035.1	13638.1	...	652.6

Step 6: Build Forecast

Call Predict to build forecast output.

```
future=dfProphetMultiVarTestDataset
future
```

```
forecast = model.predict(future)
model.plot(forecast)
```

Output:

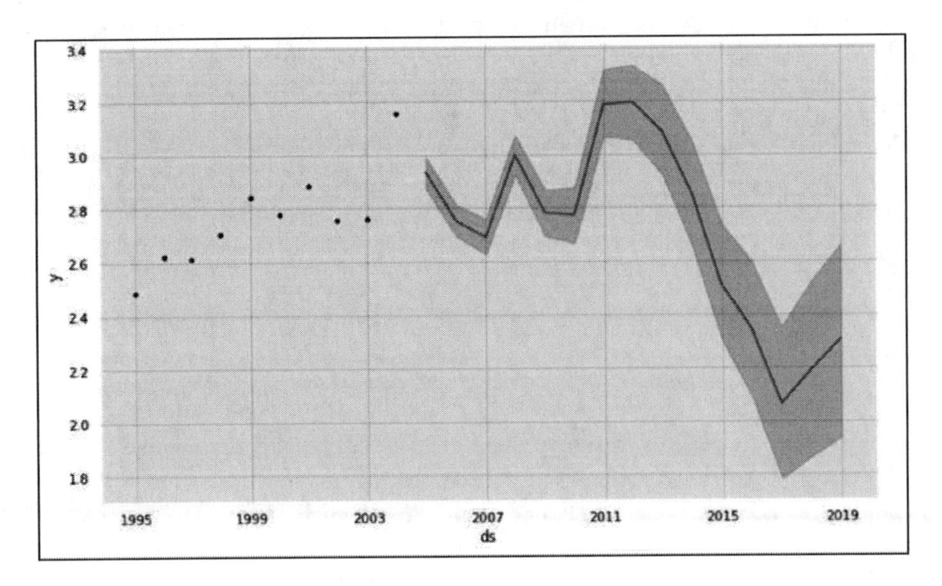

Then validate forecast patterns of regressors.

```
fig = model.plot_components(forecast)
```

Output:

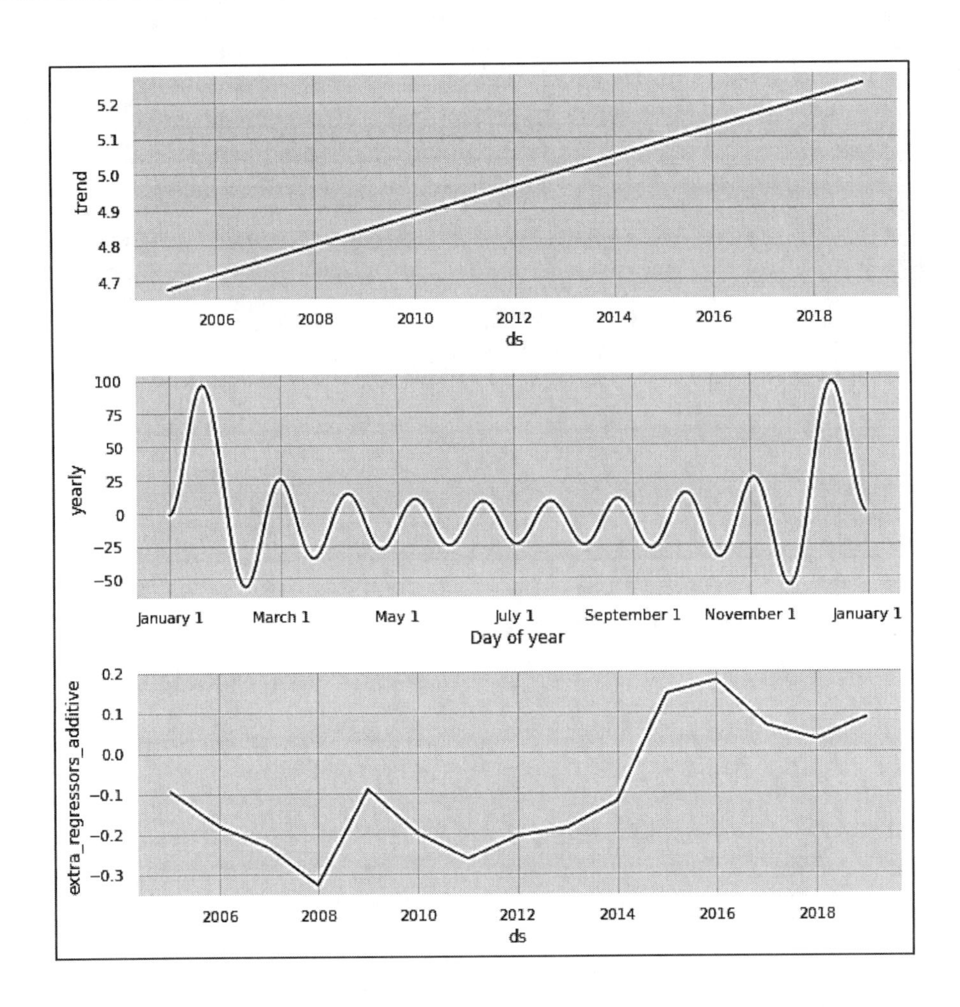

Step 7: Evaluate Model Accuracy
Compare model accuracies.

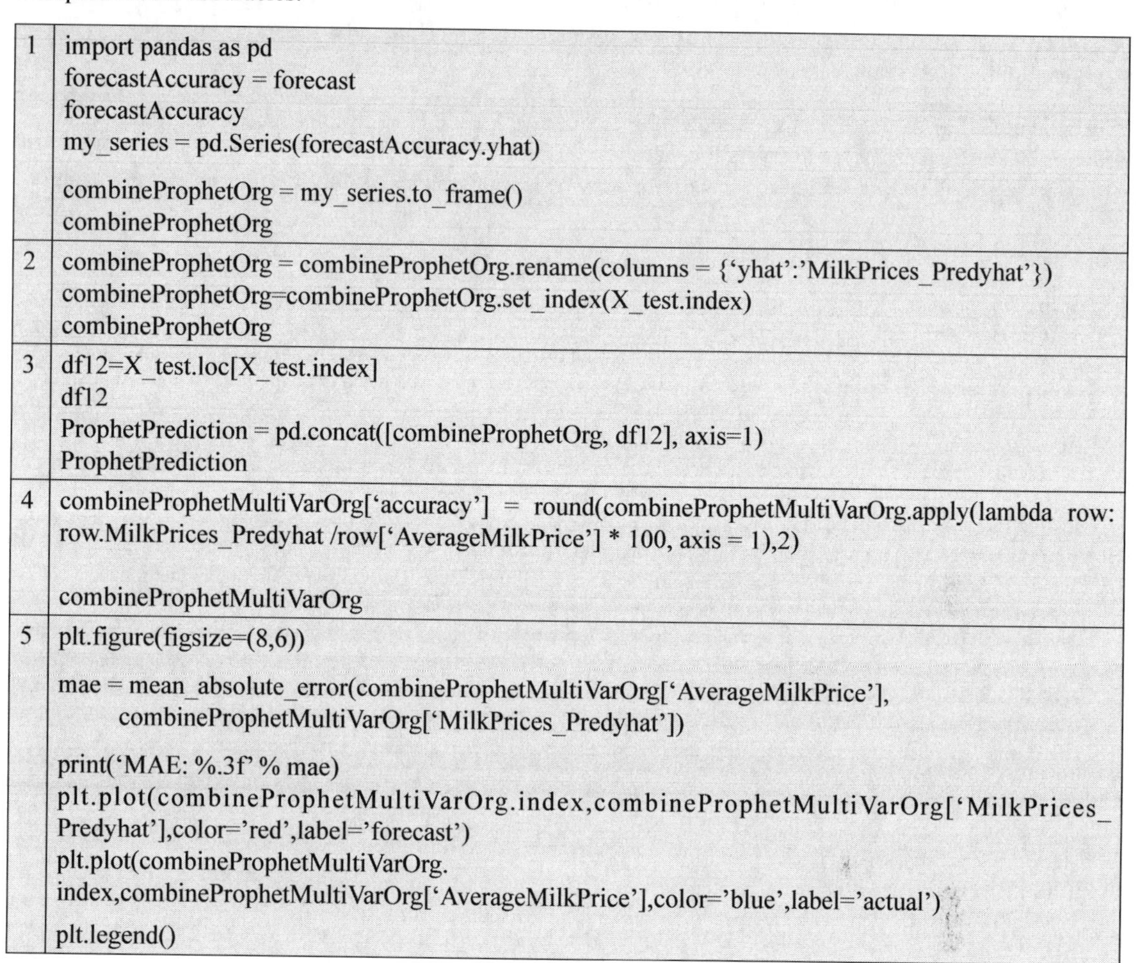

1	import pandas as pd forecastAccuracy = forecast forecastAccuracy my_series = pd.Series(forecastAccuracy.yhat) combineProphetOrg = my_series.to_frame() combineProphetOrg
2	combineProphetOrg = combineProphetOrg.rename(columns = {'yhat':'MilkPrices_Predyhat'}) combineProphetOrg=combineProphetOrg.set_index(X_test.index) combineProphetOrg
3	df12=X_test.loc[X_test.index] df12 ProphetPrediction = pd.concat([combineProphetOrg, df12], axis=1) ProphetPrediction
4	combineProphetMultiVarOrg['accuracy'] = round(combineProphetMultiVarOrg.apply(lambda row: row.MilkPrices_Predyhat /row['AverageMilkPrice'] * 100, axis = 1),2) combineProphetMultiVarOrg
5	plt.figure(figsize=(8,6)) mae = mean_absolute_error(combineProphetMultiVarOrg['AverageMilkPrice'], combineProphetMultiVarOrg['MilkPrices_Predyhat']) print('MAE: %.3f' % mae) plt.plot(combineProphetMultiVarOrg.index,combineProphetMultiVarOrg['MilkPrices_ Predyhat'],color='red',label='forecast') plt.plot(combineProphetMultiVarOrg. index,combineProphetMultiVarOrg['AverageMilkPrice'],color='blue',label='actual') plt.legend()

Output:

```
MAE = 0.438
```

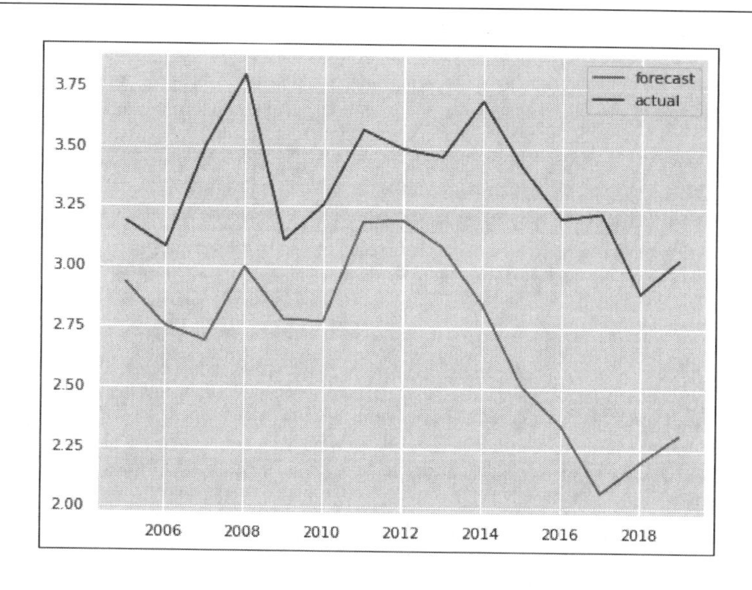

Step 8: Model summary

The following table has Prophet model predictions based on regressors Imported Crude Oil Price ($/barrel) and Real Average Cheddar Cheese Price. The mean absolute error is 0.438. The time series model can predict the trends of milk price changes – peaks & troughs.

After reading this chapter, you should be able to apply Time Series techniques—stationarity tests, autocorrelation tests, causation analysis for real world use cases across the industries. You should be able to answer questions on time series advanced modeling techniques such as VAR, Regression, and Prophet.

References

[1] Ashish Rajbhandari, Unit-root tests in Stata, 21 June 2016, https://blog.stata.com/2016/06/21/unit-root-tests-in-stata/, Access Date: June 1, 2022.

[2] Dominguez-Salas, P., Galiè, A., Omore, A., Omosa, E. and Ouma, E. 2019. Contribution of milk production to food and nutrition security. *In*: Ferranti, P., Berry, E.M. and Anderson, J.R. (eds.). Encyclopedia of Food Security and Sustainability 3: 278–291.

[3] Hayden, Stewart, William Hahn, Jerry Cessna and Christopher G. Davis, Processing and Marketing Blunt the Impact of Volatile Farm Prices on Retail Dairy Prices, August 01, 2016, https://www.ers.usda.gov/amber-waves/2016/august/processing-and-marketing-blunt-the-impact-of-volatile-farm-prices-on-retail-dairy-prices/, Access Date: June 3, 2022.

[4] John, F. Oncken. The ups and downs of a farmer's milk price, June 24, 2020, https://www.wisfarmer.com/story/opinion/columnists/2020/06/24/ups-and-downs-farmers-milk-price/3247088001/, Access Date: June 3, 2022.

[5] Samaneh Azarpajouh, United States: Milk pricing and its future, 2009, https://www.dairyglobal.net/industry-and-markets/market-trends/united-states-milk-pricing-and-its-future/, Access Date: June 5, 2022.

[6] Christopher A. Sims, Macroeconomics and reality, 1980, https://www.pauldeng.com/pdf/Sims%20macroeconomics%20and%20reality.pdf,Access Date: June 5, 2022.

[7] Harold W. Lough. Cheese Pricing, December 1980, https://ageconsearch.umn.edu/record/307889/files/aer462.pdf, Access Date: June 3, 2022.

[8] John Geuss, Cheddar Cheese Pricing Rules the Dairy Industry. – Here are the Details and Dynamics, 13 December 2019, https://www.dairybusiness.com/cheddar-cheese-pricing-rules-the-dairy-industry-here-are-the-details-and-dynamics/, Access Date: June 6, 2022.

[9] John F. Oncken, What's Gotten Into the Price of Cheese? Matt Phillips, https://www.nytimes.com/2020/06/22/business/cheese-cheddar-prices.html, Access Date: June 3, 2022.

[10] John, F. Oncken, The ups and downs of a farmer's milk price, June 24, 2020, https://www.wisfarmer.com/story/opinion/columnists/2020/06/24/ups-and-downs-farmers-milk-price/3247088001/, Access Date: June 3, 2022.

[11] BFP University Study Committee, AN ECONOMIC EVALUATION OF BASIC FORMULA PRICE (BFP) ALTERNATIVES, October 1996, https://www.afpc.tamu.edu/research/publications/89/bfp1.htm,Access Date: June 5, 2022.

[12] Matt, Gould. The link between oil and milk prices, 10 June 2016, https://www.progressivedairy.com/news/industry-news/the-link-between-oil-and-milk-prices, Access Date: June 6, 2022.

[13] Andrew, M. Novakovic and Mark W. Stephenson. Price Volatility in US Dairy Markets, https://www.fsa.usda.gov/Internet/FSA_File/8_andy_price_volat_diac_jun.pdf, Access Date: June 5, 2022.

[14] Jason, Brownlee. A Gentle Introduction to Autocorrelation and Partial Autocorrelation, February 6, 2017, https://machinelearningmastery.com/gentle-introduction-autocorrelation-partial-autocorrelation/, Access Date: June 4, 2022.

[15] Ashish, Rajbhandari. Unit-root tests in Stata, 21 June 2016, https://blog.stata.com/2016/06/21/unit-root-tests-in-stata/, Access Date: June 1, 2022.

[16] Paul, S.P. Cowpertwait and Andrew V. Metcalfe. Introductory Time Series with R (Use R!), Publisher : Springer; 2009th edition (June 9, 2009), ISBN-13 : 978-0387886978.

[17] Mehmet, Caner and Lutz Kilian. Size Distortions of Tests of the Null Hypothesis of Stationarity: Evidence and Implications For Applied Work, 1999, https://edz.bib.uni-mannheim.de/www-edz/pdf/zei/b99-12.pdf, Access Date: June 4, 2022.

[18] DRob, J. Hyndman. Thoughts on the Ljung-Box test, January 2014, https://robjhyndman.com/hyndsight/ljung-box-test/, Access Date: June 4, 2022.

[19] Shweta, Introduction to Time Series Forecasting—Part 2 (ARIMA Models), Jul 30, 2021, https://towardsdatascience.com/introduction-to-time-series-forecasting-part-2-arima-models-9f47bf0f476b, Access Date: May 29, 2022.

[20] Susan, Li. A Quick Introduction on Granger Causality Testing for Time Series Analysis, Dec 23, 2020, https://towardsdatascience.com/a-quick-introduction-on-granger-causality-testing-for-time-series-analysis-7113dc9420d2, Access Date: June 3, 2022.

[21] Andrew, Lesniewski. Time Series Analysis Stationary ARMA models, Fall 2019, https://mfe.baruch.cuny.edu/wp-content/uploads/2014/12/TS_Lecture1_2019.pdf, Access Date: June 1, 2022.

[22] Birhan. Ambachew Taye, Alemayehu Amsalu Alene, Ashenaf Kalayu Negal and Bantie Getnet Yirsaw. Time series analysis of cow milk production at Andassa dairy farm, West Gojam Zone, Amhara Region, Ethiopia, 20 August 2020, https://d-nb.info/1220179523/34, Access Date: June 1, 2022.

[23] SaritMaitra. ForecastingusingGranger'sCausalityandVectorAuto-regressiveModel, Oct2019, https://towardsdatascience.com/granger-causality-and-vector-auto-regressive-model-for-time-series-forecasting-3226a64889a6, Access Date: June 1, 2022.

[24] Wang, Wei. Vertical specialization and increasing productive employment in Achieving Inclusive Growth in China Through Vertical Specialization, 2016, https://doi.org/10.1016/B978-0-08-100627-6.00004-7. Access Date: June 3, 2022.

[25] Gary, Keough. Evaluation of Time Series Model Forecasts for the Minnesot Awisconsin Milk Price, April 1991, https://www.nass.usda.gov/Education_and_Outreach/Reports,_Presentations_and_Conferences/Survey_Reports/Evaluation%20of%20Time%20Series%20Model%20Forecasts%20for%20Minnesota-Wisconsin%20Milk%20Price.pdf, Access Date: May 29, 2022.

[26] Helmut, Luetkepohl. 2011. Vector Autoregressive Models, Economics Working Papers ECO2011/30, European University Institute. https://ideas.repec.org/p/eui/euiwps/eco2011-30.html , Access Date: June 5, 2022.

[27] Linda, Kantor and Andrzej Blazejczyk. Food Availability (Per Capita) Data System, Last updated: Wednesday, July 21, 2021, https://www.ers.usda.gov/data-products/food-availability-per-capita-data-system/, Access Date: May 08, 2022.

[28] Angel Teran, Dairy Data, Last updated: Friday, May 06, 2022, https://www.ers.usda.gov/data-products/dairy-data/, Access Date: May 08, 2022.

[29] Ann, Kennedy-Perkins. Yoghurt cultures a rise in grocery prices, 11 February 2021, 1:45pm, https://www.stats.govt.nz/news/yoghurt-cultures-a-rise-in-grocery-prices, Access Date: June 6, 2022.

[30] Ken, Bailey. Dairy Risk-Management Education: Factors That Affect U.S. Farm-Gate Milk Prices, SEPTEMBER 29, 2017, https://extension.psu.edu/dairy-risk-management-education-factors-that-affect-u-s-farm-gate-milk-prices, Access Date: June 5, 2022.

[31] Ed Jesse and Bob Cropp, Basic Milk Pricing Concepts for Farmers, 2005, https://www.kydairy.org/uploads/2/4/0/0/24007917/session_3_resources_milk_pricing.pdf,Access Date: June 5, 2022.

[1] Handle Missing Values in Time Series For Beginners - https://www.kaggle.com/code/juejuewang/handle-missing-values-in-time-series-for-beginners/report

[2] Vegetation Index - https://www.star.nesdis.noaa.gov/smcd/emb/vci/VH/VH-Syst_10ap30.php

[3] STAR - Global Vegetation Health Products: Province-Averaged VH - https://www.star.nesdis.noaa.gov/smcd/emb/vci/VH/vh_adminMean.php?type=Province_Weekly_MeanPlot

[4] Forecasting using Granger's Causality and Vector Auto-regressive Model - https://towardsdatascience.com/granger-causality-and-vector-auto-regressive-model-for-time-series-forecasting-3226a64889a6

[5] USDA to Purchase Up to $3 Billion in Agricultural Commodities, Issue Solicitations for Interested Participants - https://www.ams.usda.gov/content/usda-purchase-3-billion-agricultural-commodities-issue-solicitations-interested

[6] Contribution of milk production to food and nutrition security - https://www.ilri.org/publications/contribution-milk-production-food-and-nutrition-security

[7] CPI: Average price data - https://www.bls.gov/cpi/factsheets/average-prices.htm

[8] One-Screen Data Search - https://data.bls.gov/PDQWeb/ap

[9] Average Price: Milk, Fresh, Whole, Fortified (Cost per Gallon/3.8 Liters) in U.S. City Average (APU0000709112). - https://fred.stlouisfed.org/series/APU0000709112

[10] CPI Average Data - https://www.bls.gov/cpi/factsheets/average-prices.htm

[11] Processing and Marketing Blunt the Impact of Volatile Farm Prices on Retail Dairy Prices - https://www.ers.usda.gov/amber-waves/2016/august/processing-and-marketing-blunt-the-impact-of-volatile-farm-prices-on-retail-dairy-prices/
[12] The 1996 grain price shock: how did it affect food inflation? - https://www.bls.gov/opub/mlr/1998/08/art1full.pdf
[13] The ups and downs of a farmer's milk price - https://www.wisfarmer.com/story/opinion/columnists/2020/06/24/ups-and-downs-farmers-milk-price/3247088001/
[14] United States: Milk pricing and its future - https://www.dairyglobal.net/industry-and-markets/market-trends/united-states-milk-pricing-and-its-future/
[15] Calculating Class 1 Price - https://www.ams.usda.gov/sites/default/files/media/ClassIworksheetfinal.pdf
[16] Federal Milk Marketing Order Price Formula - https://www.ers.usda.gov/webdocs/publications/42300/15266_aib761g_1_.pdf?v=8589.3
[17] what's the deal with cheese stocks? - https://www.voiceofmilk.com/news/444637/So-whats-the-deal-with-cheese-stocks.htm
[18] Cheese Pricing - https://ageconsearch.umn.edu/record/307889/files/aer462.pdf
[19] Calculating Class 1 Price - https://www.ams.usda.gov/sites/default/files/media/ClassIworksheetfinal.pdf
[20] Cheddar Cheese Pricing Rules the Dairy Industry. - Here are the Details and Dynamics. - https://www.dairybusiness.com/cheddar-cheese-pricing-rules-the-dairy-industry-here-are-the-details-and-dynamics/
[21] What's Gotten into the Price of Cheese? - https://www.nytimes.com/2020/06/22/business/cheese-cheddar-prices.html
[22] Average Price: Cheddar Cheese, Natural (Cost per Pound/453.6 Grams) in U.S. City Average - https://fred.stlouisfed.org/series/APU0000710212
[23] Average Price: Milk, Fresh, Whole, Fortified (Cost per Gallon/3.8 Liters) in U.S. City Average (APU0000709112) - https://fred.stlouisfed.org/series/APU0000709112
[24] Price Volatility in US Dairy Markets - https://www.fsa.usda.gov/Internet/FSA_File/8_andy_price_volat_diac_jun.pdfhttps://www.fsa.usda.gov/Internet/FSA_File/8_andy_price_volat_diac_jun.pdf
[25] A trademark of Hanumayamma Innovations and Technologies, inc., Agriculture & Data Analytics company - https://www.hanuinnotech.com/
[26] A Gentle Introduction to Autocorrelation and Partial Autocorrelation - https://machinelearningmastery.com/gentle-introduction-autocorrelation-partial-autocorrelation/
[27] eia - https://www.eia.gov/outlooks/steo/realprices/
[28] All real and nominal price series - https://www.eia.gov/outlooks/steo/realprices/real_prices.xlsx
[29] SIZE DISTORTIONS OF TESTS OF THE NULL HYPOTHESIS OF STATIONARITY: EVIDENCE AND IMPLICATIONS FOR APPLIED WORK - https://edz.bib.uni-mannheim.de/www-edz/pdf/zei/b99-12.pdf
[30] Thoughts on the Ljung-Box test - https://robjhyndman.com/hyndsight/ljung-box-test/
[31] Oil Seasons - https://en.macromicro.me/dynamic_chart/season_chart/854#:~:text=Crude%20oil%20is%20refined%20into,troughs%20in%20January%20and%20September.
[32] Dairy Product (DYMFG) - https://www.ers.usda.gov/webdocs/DataFiles/50472/dymfg.xls?v=739.7
[33] Dairy Data - https://www.ers.usda.gov/webdocs/DataFiles/50472/dymfg.xls?v=4653
[34] Food Availability Data System - https://www.ers.usda.gov/data-products/food-availability-per-capita-data-system/
[35] Time Series Analysis Stationary ARMA models - https://mfe.baruch.cuny.edu/wp-content/uploads/2014/12/TS_Lecture1_2019.pdf
[36] ADFuller Test - https://www.statsmodels.org/dev/generated/statsmodels.tsa.stattools.adfuller.html
[37] ADFuller - https://www.statsmodels.org/dev/generated/statsmodels.tsa.stattools.adfuller.html
[38] Stationarity and detrending (ADF/KPSS) - https://www.statsmodels.org/devel/examples/notebooks/generated/stationarity_detrending_adf_kpss.html
[39] KPSS - https://www.statsmodels.org/dev/_modules/statsmodels/tsa/stattools.html#kpss
[40] Kwiatkowski-Phillips-Schmidt-Shin test for stationarity- http://fmwww.bc.edu/repec/bocode/k/kpss.html
[41] Time series analysis of cow milk production at Andassa dairy farm, West Gojam Zone, Amhara Region, Ethiopia - https://d-nb.info/1220179523/34
[42] Dairy Product (DYMFG) - https://www.ers.usda.gov/webdocs/DataFiles/50472/dymfg.xls?v=739.7
[43] Dairy Data - https://www.ers.usda.gov/webdocs/DataFiles/50472/dymfg.xls?v=4653
[44] Food Availability Data System - https://www.ers.usda.gov/data-products/food-availability-per-capita-data-system/
[45] Unit-root tests in Stata - https://blog.stata.com/2016/06/21/unit-root-tests-in-stata/
[46] Random Walk Model - https://people.duke.edu/~rnau/411rand.htm
[47] Nominal Advanced Foreign Economies U.S. Dollar Index (TWEXAFEGSMTH) - https://fred.stlouisfed.org/series/TWEXAFEGSMTH
[48] Time Series - https://www.investopedia.com/articles/trading/07/stationary.asp
[49] eia - https://www.eia.gov/outlooks/steo/realprices/
[50] U.S. Recessionary periods - https://fredhelp.stlouisfed.org/fred/data/understanding-the-data/recession-bars/
[51] All real and nominal price series - https://www.eia.gov/outlooks/steo/realprices/real_prices.xlsx
[52] Vertical specialization and increasing productive employment in Achieving Inclusive Growth in China Through Vertical Specialization - https:https://www.sciencedirect.com/science/article/pii/B9780081006276000047
[53] A Quick Introduction On Granger Causality Testing For Time Series Analysis - https://towardsdatascience.com/a-quick-introduction-on-granger-causality-testing-for-time-series-analysis-7113dc9420d2
[54] Graph Node - https://observablehq.com/@mbostock/graph-o-matic
[55] Pandas Shift - https://pandas.pydata.org/docs/reference/api/pandas.DataFrame.shift.html

[56] Pandas Diff - https://pandas.pydata.org/docs/reference/api/pandas.DataFrame.diff.html
[57] 2018 Milk Production - https://www.fmmacentral.com/PDFdata/msb201903.pdf
[58] CPI Database - https://data.bls.gov/PDQWeb/ap
[59] CPI: Average price data - https://www.bls.gov/cpi/factsheets/average-prices.htm
[60] eia - https://www.eia.gov/outlooks/steo/realprices/
[61] All real and nominal price series - https://www.eia.gov/outlooks/steo/realprices/real_prices.xlsx
[62] Average Price: Cheddar Cheese, Natural (Cost per Pound/453.6 Grams) as U.S. City Averages (APU0000710212) - https://fred.stlouisfed.org/series/APU0000710212
[63] Average Price: Milk, Fresh, Whole, Fortified (Cost per Gallon/3.8 Liters) as U.S. City Averages (APU0000709112) - https://fred.stlouisfed.org/series/APU0000709112
[64] Average Price: Yogurt, Natural, Fruit Flavored (Cost per 8 Ounces/226.8 Grams) in U.S. City Average - https://fred.stlouisfed.org/series/APU0000710122
[65] USDA Dairy Product - https://www.ers.usda.gov/topics/animal-products/dairy/
[66] Dairy Data - https://www.ers.usda.gov/webdocs/DataFiles/50472/dymfg.xls?v=4653
[67] Dataset - https://www.ams.usda.gov/sites/default/files/media/DataFormModalShareStudy1978_2019.xlsx
[68] Dairy Data - https://www.ers.usda.gov/webdocs/DataFiles/50472/dymfg.xls?v=4653
[69] Food Availability Data System - https://www.ers.usda.gov/data-products/food-availability-per-capita-data-system/
[70] U.S. Dairy Consumption Beats Expectations in 2020 and Continues to Surge Upward Despite Disruption Caused by Pandemic - https://www.idfa.org/news/u-s-dairy-consumption-beats-expectations-in-2020-and-continues-to-surge-upward-despite-disruption-caused-by-pandemic
[71] Consumer Price Index for All Urban Consumers: Dairy and Related Products in U.S. City Average - https://fred.stlouisfed.org/series/CUUR0000SEFJ
[72] Dairy Data - https://www.ers.usda.gov/webdocs/DataFiles/50472/dymfg.xls?v=4653
[73] FAO Dairy Price Indices - https://www.fao.org/markets-and-trade/commodities/dairy/fao-dairy-price-index/en/
[74] FAO Dairy Prices - https://www.fao.org/fileadmin/user_upload/est_new2020/xlsFileToUpload/Indices/FAO_Dairy_Price_Indices.xlsx
[75] Yoghurt cultures a rise in grocery prices - https://www.stats.govt.nz/news/yoghurt-cultures-a-rise-in-grocery-prices
[76] eia - https://www.eia.gov/outlooks/steo/realprices/
[77] All real and nominal price series - https://www.eia.gov/outlooks/steo/realprices/real_prices.xlsx
[78] Crude oil prices spur gains for U.S. import and export price indexes, despite the appreciating dollar: 2016 annual summary - https://www.bls.gov/opub/mlr/2017/article/crude-oil-prices-spur-gains-for-u-s-import-and-export-price-indexes-despite-the-appreciating-dollar-2016-annual-summary.htm
[79] Dairy Product (DYMFG) - https://www.ers.usda.gov/webdocs/DataFiles/50472/dymfg.xls?v=739.7
[80] Dairy Data - https://www.ers.usda.gov/webdocs/DataFiles/50472/dymfg.xls?v=4653
[81] Food Availability Data System - https://www.ers.usda.gov/data-products/food-availability-per-capita-data-system/
[82] Prices Volatile - https://www.fsa.usda.gov/Internet/FSA_File/8_andy_price_volat_diac_jun.pdf
[83] Vector Autoregressive Models for Multivariate Time Series -https://faculty.washington.edu/ezivot/econ584/notes/varModels.pdf
[84] Dairy Risk-Management Education: Factors That Affect U.S. Farm-Gate Milk Prices - https://extension.psu.edu/dairy-risk-management-education-factors-that-affect-u-s-farm-gate-milk-prices
[85] Milk Pricing - https://www.kydairy.org/uploads/2/4/0/0/24007917/session_3_resources_milk_pricing.pdf
[86] NDM Usage - https://www.afpc.tamu.edu/research/publications/89/figure22.htm
[87] Chapter 4 - https://www.afpc.tamu.edu/research/publications/89/chap4.htm
[88] Farm Milk Priced - https://www.idfa.org/how-farm-milk-is-priced/
[89] AN ECONOMIC EVALUATION OF BASIC FORMULA PRICE (BFP) ALTERNATIVES - https://www.afpc.tamu.edu/research/publications/89/bfp1.htm
[90] Farm Milk Priced - https://www.idfa.org/how-farm-milk-is-priced/
[91] AN ECONOMIC EVALUATION OF BASIC FORMULA PRICE (BFP) ALTERNATIVES - https://www.afpc.tamu.edu/research/publications/89/bfp1.htm

CHAPTER 2

Data Engineering Techniques for Artificial Intelligence and Advanced Analytics

This chapter introduces the time series imputation strategies to fix missing data. Unlike datasets for training models using regression, time series models are very particular on time sequences and hence the imputation strategies need to be carved specifically to maintain statistical properties and should add value to model learning. As part of the chapter, we will start with mean/median imputation strategies, kNN, and Multivariate Imputation by Chained Equation (MICE) Strategy. Next, another important aspect of time series modeling is to overcome class or predictor bias. Synthetic Minority Oversampling Technique (SMOTE) & Adaptive Synthetic (ADASYN) Sampling are introduced in detail to fix some of the very useful agriculture datasets such as the one on Kansas Wheat Yield. Data enrichment is critical to preserve the time series signature. Finally, the chapter concludes with a food security wheat model for Kansas using heat yield with Adaptive synthetic (ADASYN) Sampling.

The importance of data and data collection & engineering techniques could be summarized with a single statement!

<table>
<tr>
<td></td>
<td>By the authority vested in me as President by the Constitution and the laws of the United States of America, it is hereby ordered—Executive Order on Ensuring a Data-Driven Response to COVID-19 and Future High-Consequence Public Health Threats [1].
JOSEPH R. BIDEN JR.

THE WHITE HOUSE,[92]
January 21, 2021.</td>
</tr>
</table>

Effective data collection and data engineering techniques have helped us advance in addressing the COVID-19 pandemic in an effective manner. The executive order emphasizes five aspects of the data:

- Effective collection & dissemination of data
- Enhancing Data Collection and Collaboration Capabilities
- Effectiveness, Interoperability, and Connectivity of data Systems
- Advancing Innovation in Public Health Data and Analytics
- Data Privacy & effective use.

Though the executive order is focused on health and public systems to deal with current and future threats, the principles of the data are still applicable across all data domains—Agriculture, Food Security, Cyber Security, and Public Health Systems.

Effective data is essential to develop machine learning and advanced analytics models and the demand for good-quality statistical data is necessary now than ever before. The argument is simple. Data plays an important role in ensuring food security. Timely and reliable statistics are key inputs to a broad development strategy not only for nations of the world but also very critical for commercial enterprises. On the same token, good-quality statistical data is important for small farmers and agricultural businesses. Improvements in the quality and quantity of data on all aspects of development are essential if we are to achieve the goal of a world without poverty[93] and reaching sustainability objectives. Good datasets are needed to set baselines, identify effective public and private actions, set goals and targets, monitor progress, and evaluate impacts. They are also an essential tool of good governance, providing means for people to assess what governments do and helping them to participate directly in the development process. For example, during Haiti Acute Food Insecurity National Situation in June–July 2013 and Southeastern Department in September 2013, it was recommended that data collection from agro-ecological zones is essential to prevent future food insecurity. The recommendation by the Integrated Food Security Phase Classification (IPC) is to establish a system for the regular collection of disaggregated information (rainfall, market prices, health and nutrition, water and sanitation, agricultural production, food preservation and processing) for future analysis purposes.[94] Similarly, to collect data related to prices in local markets, weekly wages per household, daily price, type, and frequency of survival strategies employed by households to enhance food security for Honduras.[95]

Food Security Data

Food Security is essential both for individuals and nations. Individuals are considered food secure if they meet 2100 calories food nutrition per day. Countries, on the other hand, are food secure if they've a risk-free food supply chain for ensuring sustainable food supplied.[96] Since Russia's invasion of Ukraine, governments from Turkey to Indonesia, Somalia to Lebanon, have scrambled to find new supplies and cope with rising prices.[97] But the stakes are especially high for Egypt, the most populous country in the Middle East, where the economy was in a precarious position before the war started [1]. Globally, climate change[98] is expected to threaten food production and certain aspects of food quality, as well as food prices and distribution systems. Bread, which Egyptians call A'ish-Arabic for life-is part of a social contract in Egypt with the government providing affordable food, gasoline and electricity. Egyptians eat more bread than most people in the world, around 330 pounds a year each on average, nearly triple the global figure. Bread-price increases would sting in a country where about 30% of people live on less than $2 a day, according to Egypt's official statistics. Unsubsidized baladi loaves now cost about 7 cents each, while subsidized loaves are less than 1 cent. Egyptian authorities secured 350,000 tons of wheat from France, Bulgaria and Russia, and say more shipments will come from India, even though the country's crops are suffering a heat wave.

Many crop yields are predicted to decline because of the combined effects of changes in rainfall, severe weather events, heat waves, and increasing competition from weeds and pests on crop plants. Livestock and fish production are also projected to decline. Prices are expected to rise in response to declining food production and associated trends such as increasingly expensive petroleum (used for agricultural inputs such as pesticides and fertilizers).

Like many important datasets, Food Security (please see Figure 1) data is a Time Series. The parameters that capture food security are bucketed into four groups: affordability, availability, quality & safety, and natural resources & resilience.[99] The dataset for food security is captured as part of FAOSTAT[100] and as part of the survey data of the National Health and Nutrition Examination Survey[101] (NHANES) (though it's not a direct agricultural dataset, it can function as a proxy for food security).

Please find below a graph on Trends in prevailing rates of food insecurity and very low food security in U.S. households, 1995–2020 (Source: Source: USDA, Economic Research Service, using data from the Current Population Survey Food Security Supplement[102]). The food security questionnaire provides several categories that describe the Household food security category (FSDHH): Household full food security:[103] no affirmative response in any of these items, Household marginal food security, Household low food security, and more on food security, and Household very low food security. I will cover it in the next chapter. Please refer Appendix A for Food Security Data Sources.

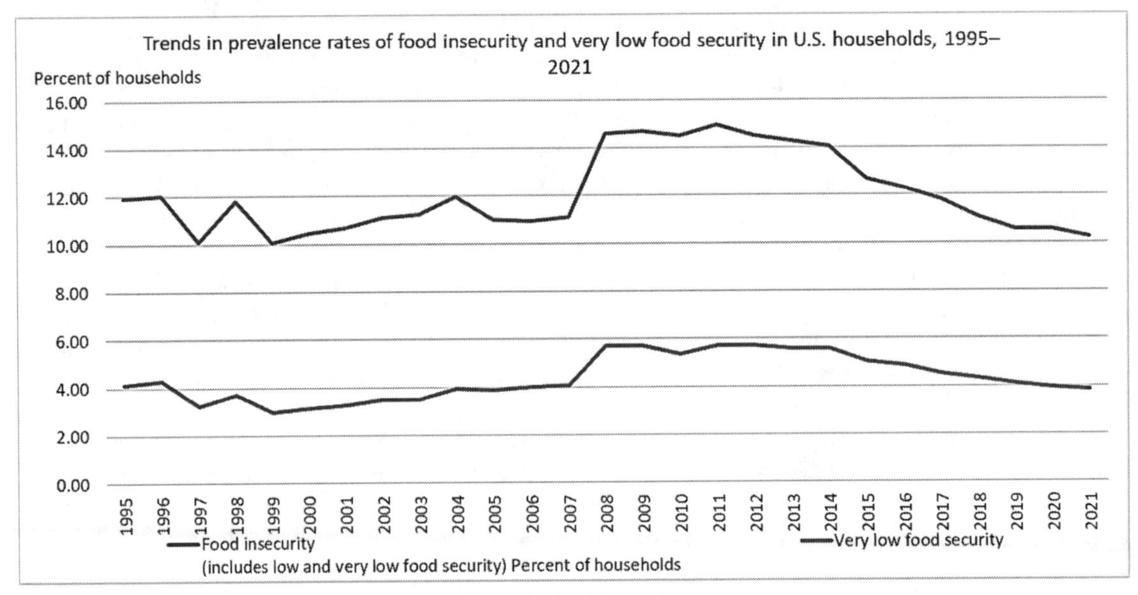

Figure 1: Food insecurity.

Agriculture—where the fight against hunger and climate change comes together

Agriculture and food systems are partly responsible for climate change, but they are also part of the solution. Appropriate actions in agriculture, forestry and fisheries can mitigate greenhouse gas emissions and promote climate adaptation—with efforts to reduce emissions and adjust practices to the new reality often enhancing and supporting one another. For millions of people, especially rural family farmers in developing countries, our actions can make the difference between poverty and prosperity, between hunger and food security and nutrition. Agriculture—where the fight against hunger and climate change come together—can unlock solutions[104] [2].

Data Encoding—Categorical to Numeric

Categorical values have signals, and the signals have a huge influence on Machine Learning models. To extract value & influence of these signals, for many models the categorical values are needed to transform into continuous signals.[105] Put it another way, conversion of categorical values to numerical is an integral process of analytics.

The following techniques are applied:

- One Hot Encoding
- Target Encoding
- Ordinal Encoding
- Hashing Encoding

I will cover one hot encoding in the code below. I will cover other encoding techniques in later chapters.

The software code for this model can be found on GitHub Link: WDI_Agriculture_and_Rural Development (Jupyter Notebook Code)

Consider the following Machine Learning Model that groups World countries based on the agriculture and rural development key indicators.[106] The featured indicators are:

- Arable land (% of land area):[107] Arable land includes land defined by the FAO as land under temporary crops (double-cropped areas are counted once), temporary meadows for mowing or for pasture, land under market or kitchen gardens, and land temporarily fallow. Land abandoned because of shifting cultivation is excluded.
- Arable land (hectares per person):[108] Arable land (hectares per person) includes land defined by the FAO as land under temporary crops (double-cropped areas are counted once), temporary meadows for mowing or for pasture, land under market or kitchen gardens, and land temporarily fallow. Land abandoned as a result of shifting cultivation is excluded.
- Cereal yield (kg per hectare):[109] Cereal yield, measured as kilograms per hectare of harvested land, includes wheat, rice, maize, barley, oats, rye, millet, sorghum, buckwheat, and mixed grains. Production data on cereals relates to crops harvested for dry grain only. Cereal crops harvested for hay or harvested green for food, feed, or silage and those used for grazing are excluded.
- Rural population (% of total population):[110] For 70 percent of the world's poor who live in rural areas, agriculture is the main source of income and employment. However, depletion and degradation of land and water pose serious challenges to producing enough food and other agricultural products to sustain livelihoods and meet the needs of urban populations.

Step 1: Load Agriculture and Rural Data

In the following code snippet, the libraries required to load agriculture and rural datasets are included.

```
import pandas as pd
import numpy as np
from sklearn.linear_model import LinearRegression
from sklearn.model_selection import train_test_split
import eli5
import matplotlib.pyplot as plt
from sklearn.preprocessing import MinMaxScaler
from sklearn.metrics import mean_squared_error, r2_score
```

Step 2: Load Agriculture and Rural Data from CSV files

Load Cereal Yield data for all countries.

```
dfCerealYield = pd.read_csv('API_AG.YLD.CREL.KG_DS2_en_csv_v2_4023913.csv', sep=',')
dfCerealYield=dfCerealYield.set_index('Country Code')
dfCerealYield.head()
```

Output:

	Country Name	Cereal yield (kg per hectare) , 2018
Country Code		
ABW	Aruba	NaN
AFE	Africa Eastern and Southern	1622.882922
AFG	Afghanistan	2164.900000
AFW	Africa Western and Central	1276.100388
AGO	Angola	753.300000

Load Agriculture Value add Percentage of GDP. Set Index on country code to later merge all required data frames.

```
dfAgriValueAddPerofGDP= pd.read_csv('API_NV.AGR.TOTL.ZS_DS2_en_csv_v2_4029227.csv', sep=',')
dfAgriValueAddPerofGDP=dfAgriValueAddPerofGDP.set_index('Country Code')
dfAgriValueAddPerofGDP.drop(['Country Name'], axis=1, inplace=True)
dfAgriValueAddPerofGDP.head()
```

Output:

	Agriculture, forestry, and fishing, value added (% of GDP), 2018
Country Code	
ABW	NaN
AFE	11.987675
AFG	22.042897
AFW	20.931406
AGO	8.607742

Similarly, load other key feature variables that include GDP Per Capita, Employment in agriculture, female (% of female employment), Fertilizer consumption (kilograms per hectare of arable land), Employment in agriculture, male (% of male employment) (modeled ILO estimate), Arable land (% of land area), Arable land (hectares per person), Rural population (% of total population), and Employment in agriculture (% of total employment).

Step 3: Load Country Income Group data

Load country meta data that consists of countries and their corresponding income group details.

```
dfIncomeGrp=
pd.read_csv('Metadata_Country_API_NV.AGR.TOTL.ZS_DS2_en_csv_v2_4029227.csv', sep=',')
dfIncomeGrp=dfIncomeGrp.set_index('Country Code')
dfIncomeGrp.head()
```

Output:

	Region	IncomeGroup	SpecialNotes	TableName	Unnamed: 5
Country Code					
ABW	Latin America & Caribbean	High income	NaN	Aruba	NaN
AFE	NaN	NaN	26 countries, stretching from the Red Sea in t...	Africa Eastern and Southern	NaN
AFG	South Asia	Low income	The reporting period for national accounts dat...	Afghanistan	NaN
AFW	NaN	NaN	22 countries, stretching from the westernmost ...	Africa Western and Central	NaN
AGO	Sub-Saharan Africa	Lower middle income	NaN	Angola	NaN

Drop Special Notes, Table Name, and Unnamed columns as they are not required for analysis.

Step 4: Concat Data Frames

Load country meta data

```
dataset = pd.concat([dfEmploymentInAgriculturePercentOftotalemployment,dfArableland
Percentageoflandarea,
dfArablelandhectaresperperson,dfCerealYield,dfAgriValueAddPerofGDP,dfGDPPerCapita,
dfEmpinAgriFemale,dfEmpinAgriMale,dfIncomeGrp,dfRuralPopPercentofTotPopulation,
dfFertilizerConsumption], axis=1)
print('Number of colums in Dataframe : ', len(dataset.columns))
print('Number of rows in Dataframe : ', len(dataset.index))
print(dataset)

dataset.drop(['SpecialNotes','TableName','Unnamed: 5'], axis=1, inplace=True)
dataset
```

Output:

	Employment in agriculture (% of total employment) (modeled ILO estimate) , 2018	Arable land (% of land area), 2018	Arable land (hectares per person) ,2018	Country Name	Cereal yield (kg per hectare) , 2018	Agriculture, forestry, and fishing, value added (% of GDP), 2018	GDP per capita, PPP (current international $) , 2018	Employment in agriculture, female (% of female employment) (modeled ILO estimate), 2018	Employment in agriculture, male (% of male employment) (modeled ILO estimate), 2018	Region	IncomeGroup
ABW	NaN	11.111111	0.018895	Aruba	NaN	NaN	NaN	NaN	NaN	Latin America & Caribbean	High income
AFE	59.574721	8.247616	0.184952	Africa Eastern and Southern	1622.882922	11.987675	3791.875407	62.537788	56.923578	NaN	NaN
AFG	43.130001	11.798854	0.207226	Afghanistan	2164.900000	22.042897	2082.635648	65.769997	37.259998	South Asia	Low income

Check for nulls and if found drop them.

```
dataset.isnull().sum() ## missing values
```

Output:

```
Employment in agriculture (% of total employment) (modeled ILO estimate) , 2018        31
Arable land (% of land area), 2018                                                     13
Arable land (hectares per person) ,2018                                                13
Country Name                                                                            0
Cereal yield (kg per hectare) , 2018                                                   38
Agriculture, forestry, and fishing, value added (% of GDP), 2018                       26
GDP per capita, PPP (current international $) , 2018                                   25
Employment in agriculture, female (% of female employment) (modeled ILO estimate), 2018  31
Employment in agriculture, male (% of male employment) (modeled ILO estimate), 2018    31
Region                                                                                 49
IncomeGroup                                                                            50
Rural population (% of total population), 2018                                          4
Fertilizer consumption (kilograms per hectare of arable land), 2018                    56
dtype: int64
```

There are rows present in the dataset. Drop them.

```
dataset=dataset.dropna()
```

The final dataset is ready for modeling.

Step 5: Visual Inspection of the data

To visualize the counts of the categorical column, we can plot the data frame. This provides the categorical values to counts.

```
import numpy as np

# plot a bar plot for each categorical feature count
categorical_features = [
  'Country Name','IncomeGroup','Region'
              ]
for col in categorical_features:
   counts = dataset[col].value_counts().sort_index()
   fig = plt.figure(figsize=(9, 6))
   ax = fig.gca()
   counts.plot.bar(ax = ax, color='steelblue')
   ax.set_title(col + ' counts')
   ax.set_xlabel(col)
   ax.set_ylabel("Frequency")
plt.show()
```

Output:

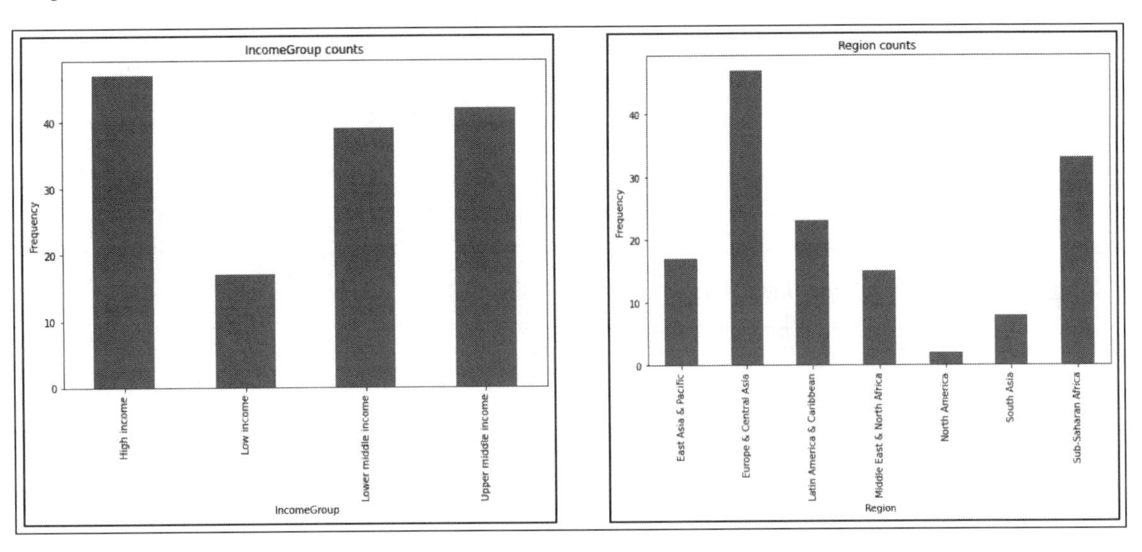

The income group has Low income, Lower middle income, Upper middle income, and High-income values. Similarly, the World Development Indicator regions include South Asia, Sub-Saharan Africa, Europe & Central Asia, Middle East & North Africa, Latin America & Caribbean, East Asia & Pacific, and North America.

Step 6: Agricultural Statistics

Before applying encoding, inferences are derived from agricultural datasets.

- **Employment in agriculture (% of total employment) (modeled ILO estimate), 2018**

For the year 2018, the World Bank Employment in agriculture (% of total employment) data has the low-income countries with the highest (59.03%) mean percentage followed by Lower Middle-Income countries (35.94%), Upper Middle-Income countries (17.50%), and finally High-Income countries (3.58%).

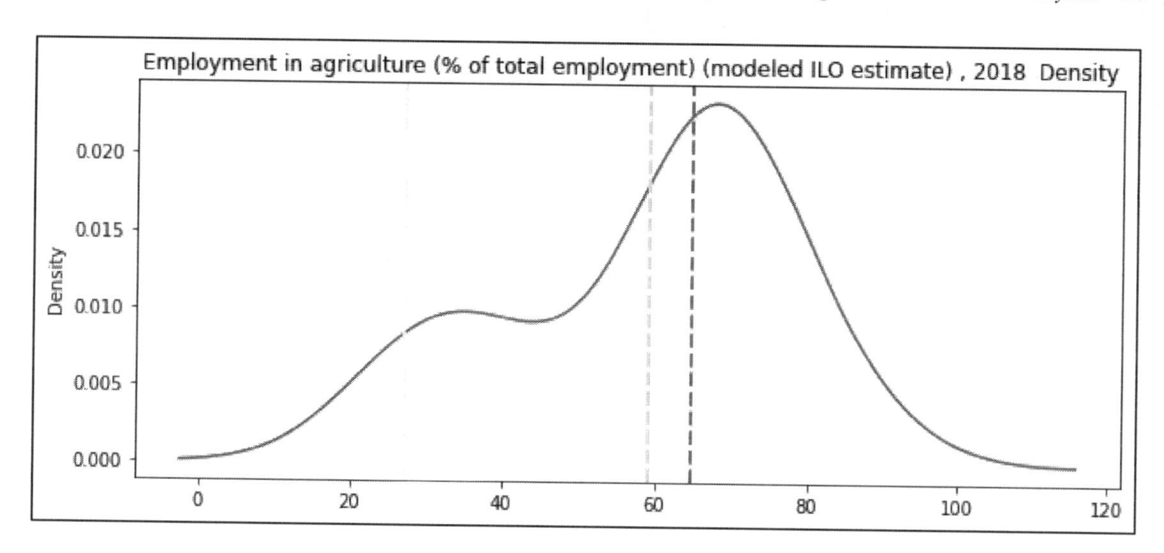

- **Employment in agriculture, male (% of male employment) (modeled ILO estimate), 2018**

For the year 2018, the World Bank Employment in agriculture, male (% of male employment) (modeled ILO estimate), 2018 data has the low-income countries with the highest (56.47%) mean percentage followed by Lower Middle-Income countries (52.69%), Upper Middle-Income countries (19.22%), and finally High-Income countries (4.64%).

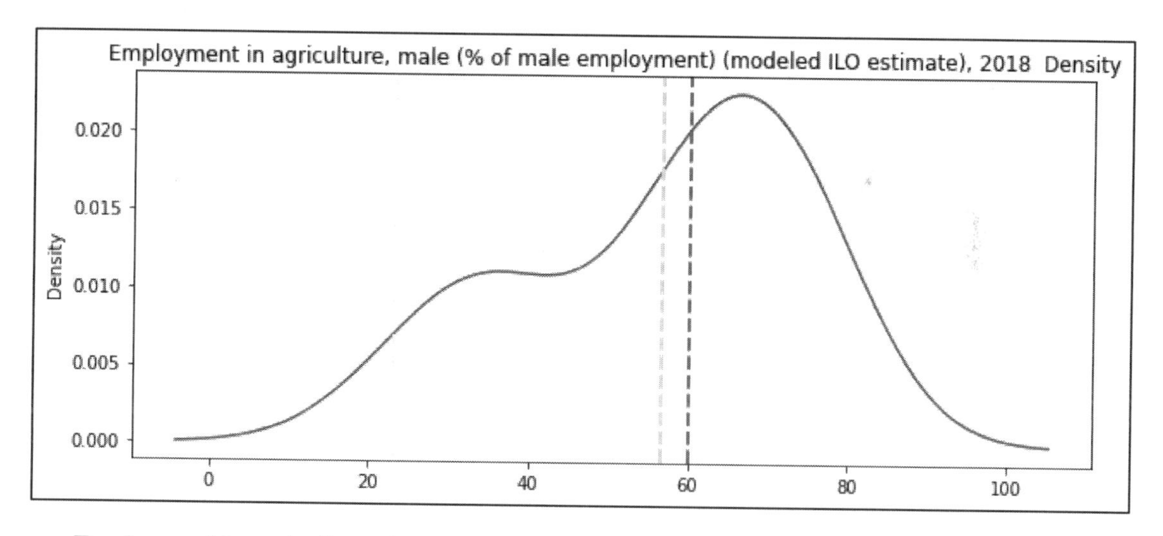

- **Employment in agriculture, female (% of female employment) (modeled ILO estimate), 2018**

For the year 2018, the World Bank Employment in agriculture, female (% of female employment) (modeled ILO estimate), data has the low-income countries with the highest (62.99%) mean percentage followed by Lower Middle-Income countries (36.10%), Upper Middle-Income countries (14.75%), and finally High-Income countries (2.18%). Service sector or outside farm employment opportunities are available in lower middle income countries as compared to low income countries.

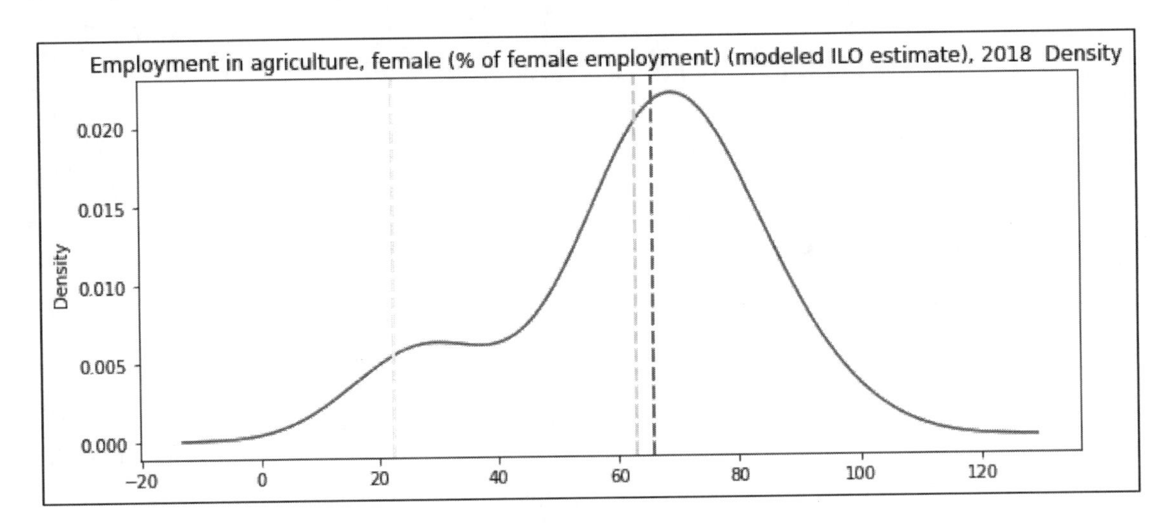

- **Rural population (% of total population), 2018**

For the year 2018, the World Bank reported Rural population (% of total population) data has the low-income countries with the highest (68.34%) mean percentage followed by Lower Middle-Income countries (52.69%), Upper Middle-Income countries (34.72%), and finally High-Income countries (20.47%). What it tells us is low-income countries have the highest rural population. New Census Data Shows Differences Between Urban and Rural Populations.

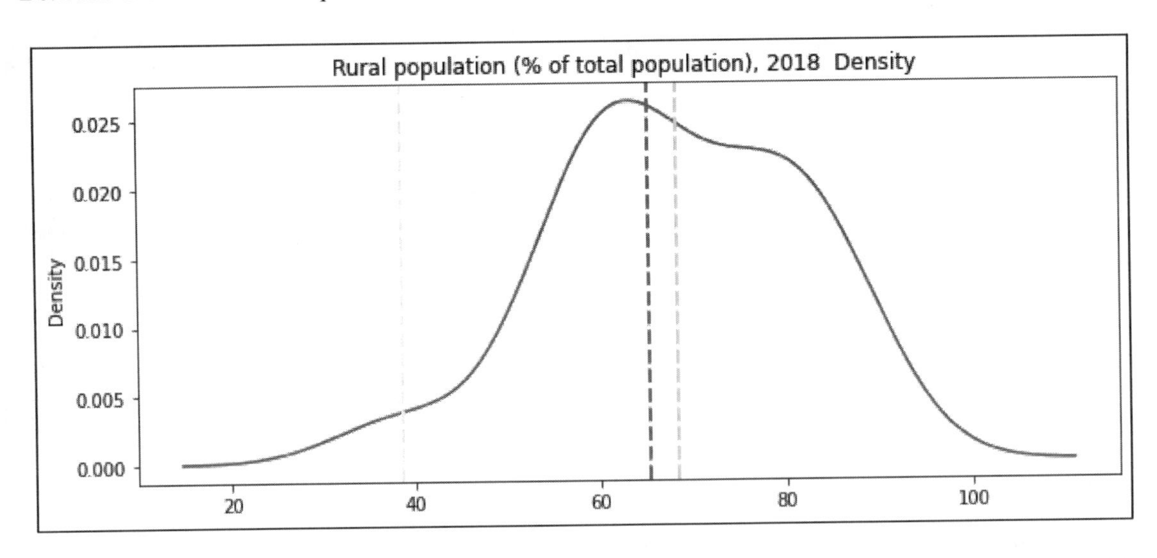

Likewise, recent U.S. population growth also has been uneven. Urban counties have grown at roughly the overall national rate of 13% since 2000. Suburban and small metropolitan areas have grown more briskly. Rural counties have lagged, and half of them have fewer residents now than they did in 2000.[111]

"Rural areas cover 97 percent of the nation's land area but contain 19.3 percent of the population (about 60 million people)".

2016, The U.S. Census Bureau Director[112]

- **Cereal yield (kg per hectare), 2018**

Cereal yield, measured in kilograms per hectare of harvested land, includes wheat, rice, maize, barley, oats, rye, millet, sorghum, buckwheat, and mixed grains. For the year 2018, Cereal yield (kg per hectare) world bank data reports the low-income countries with the lowest (1455.52 kg per hectare) mean value followed by Lower Middle-Income countries (2817.12 kg per hectare), Upper Middle-Income countries (3580.2 kg per hectare), and finally High-Income countries (5871.89 kg per hectare). What it says is that the low-income counties have the lowest cereal yield. Bringing in automation & agriculture smart climate technologies would improve the overall yield.

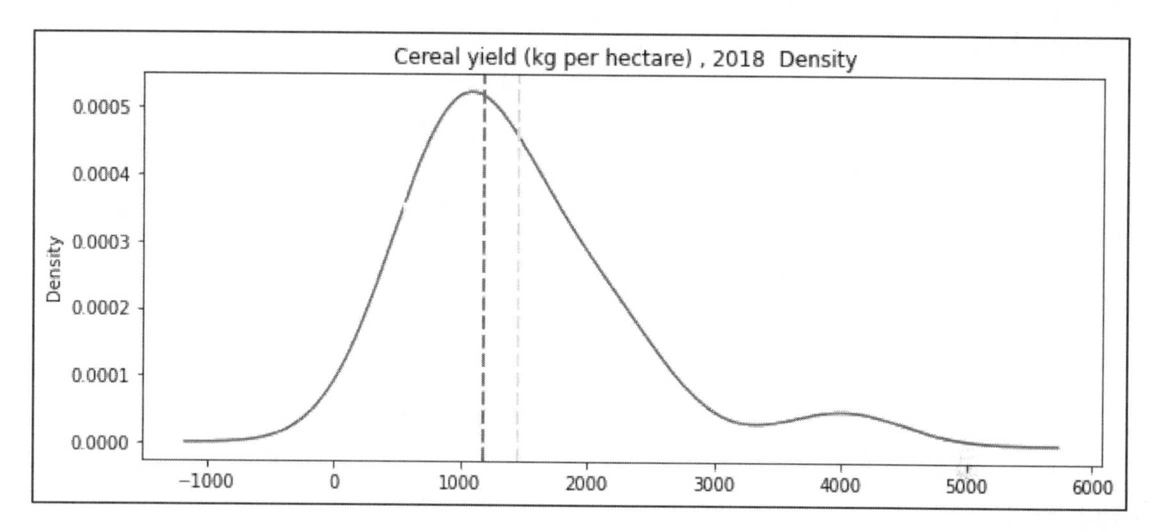

In terms of Cereal yield (kg per hectare), 2018: bell curve distribution, high-income countries have a lower tail (3σ) than that of low-income countries, suggesting that there are wider thresholds among the low-income countries' yields.

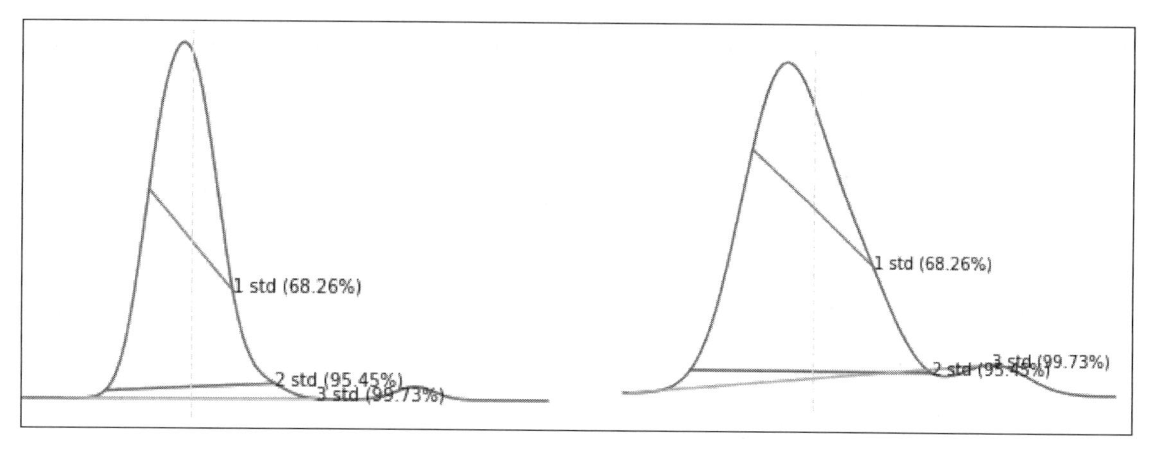

One-Hot Encoding

Dummy encoding and one-hot encoding are the same thing; the former term comes from statistics and the latter from electrical engineering (electronics) [3]. Let me explain the subtle difference. Because a regression model can only take numeric variables, statistics have long solved the problem by converting a categorical variable of n values into n-1 dummy variables. Why n-1? This is to avoid the issue of multicollinearity. *Multicollinearity* is a statistical concept where several independent variables in a model are correlated. Two variables are perfectly collinear if their correlation coefficient is +/–1.0. The problem is *multicollinearity among independent variables which will result in less reliable statistical inferences*.

One-hot encoding converts a categorical variable of n values into n dummy variable. All the variables created have values 1 and 0.

Step 7: One-Hot Encoding

To visualize the counts of the categorical column, we can plot the data frame. This provides the categorical values to counts.

```
import numpy as np

# plot a bar plot for each categorical feature count
categorical_features = [
  'Country Name','IncomeGroup','Region'
              ]
for col in categorical_features:
  counts = dataset[col].value_counts().sort_index()
  fig = plt.figure(figsize=(9, 6))
  ax = fig.gca()
  counts.plot.bar(ax = ax, color='steelblue')
  ax.set_title(col + ' counts')
  ax.set_xlabel(col)
  ax.set_ylabel("Frequency")
plt.show()
```

Data Enrichment

Survey data is a major component of agricultural datasets. For instance, consider the USDA Fertilizer Use and Price dataset.[113] This dataset summarizes fertilizer consumption in the United States by plant nutrient and major fertilizer products—as well as consumption of mixed fertilizers, secondary nutrients, and micronutrients—for 1960 through the latest year for which statistics are available. The share of planted crop acreage receiving fertilizer, and fertilizer applications per receiving acre (by nutrient), are presented for major producing States for corn, cotton, soybeans, and wheat (data on nutrient consumption by crop start in 1964). Fertilizer farm prices and indices of wholesale fertilizer prices are also available [4].

Another data set that would be useful and covered in this book is the USDA Costs and Returns database. Cost and return estimates[114] are reported for the United States and major production regions for corn, soybeans, wheat, cotton, grain sorghum, rice, peanuts, oats, barley, milk, hogs, and cow-calf. The series of commodity cost and return estimates for the U.S. and regions is divided into Recent and Historical estimates. Recent estimates date back to the most recent major revision in accounting methods, account format, and regional definitions for each commodity [5].

USDA Fertilizer Use

The fertilizer use dataset is a very important one consisting of fertilizer usage in states within the Unites States plus fertilizer prices.[115] Fertilizer price data is available through 2014. Fertilizer price indexes have been updated through 2018. To assess the impact of high inflation and current global scenarios, it would be prudent to compare with *historical fertilizer use data*. The combination of sanctions, shipping firms avoiding the Black Sea region and traders shunning Russian supplies has created a significant uncertainty for farmers regarding their ability to secure adequate fertilizer supplies. This is creating huge inflationary pressure on firm inputs. Natural gas, meanwhile, a key input for nitrogen fertilizers, is "critical and must be applied every growing season for most crops". Prices for natural gas around the globe have climbed, with Europe seeing record highs this year and U.S. prices reaching their highest since 2008. This has raised input costs to produce nitrogen fertilizers [6], and that gets passed along into higher food prices.

Later in the book, I have compiled machine learning models that predict the impact of fertilizer prices on farm costs and the impact of high inflation, due to the Ukraine war and global events, on the overall sustainability of small farms. The goal was to learn from crop-specific fertilizer application estimates that are derived from USDA surveys covering selected crops in selected years (1960–2018) to develop overall pricing models. Some of the years in the dataset are missing the data and hence would like to apply imputation strategies to overcome data NULLs.

A basic strategy, in general, to use incomplete datasets employed by many data scientists is to discard entire rows and/or columns containing missing values. A major issue with this approach is the discontinuity of the time series data. The missing data reduces statistical power,[116] which refers to the probability that the test will reject the null hypothesis when it is false. The missing data can cause bias in the estimation of parameters [7]. It can reduce the representativeness of the samples. Despite losing valuable information (please see figure), despite foregoing underlying data signals or a foreseeable prospective trend (for instance for years 2006–2009, 2011–2013, and 2017 fertilizer data is missing and hence the graph touches 0 values on Y-axis), this is a general operating practice, due to time taken by data transformations, followed by small and large data science teams. Lacking resources or even a theoretical framework, researchers, methodologists, and software developers resort to editing the data to lend an appearance of completeness. Unfortunately, ad hoc edits may do more harm than good, producing answers that are biased, inefficient (lacking in power), and unreliable[117] [7]. To maximize return on assets of information acquisition systems and to better serve customers, a novel strategy is to impute the missing values, i.e., to infer them from the known part of the data (please see Figure 2).

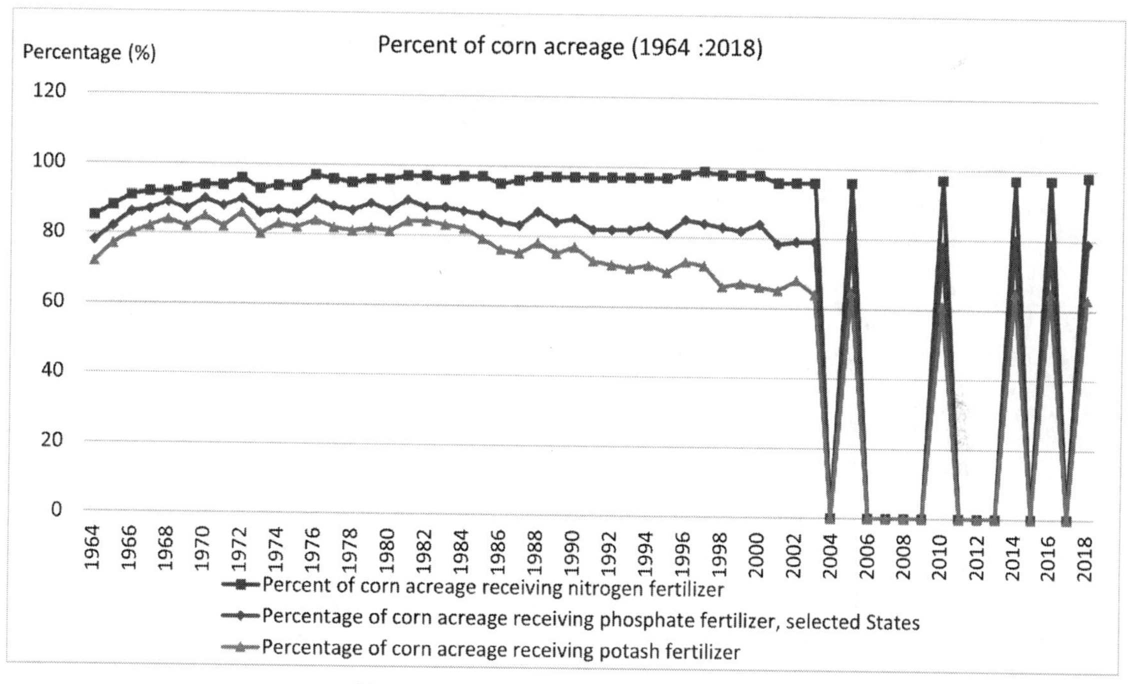

Figure 2: Percent of Corn Acreage (1964 : 2018).

The strategy for imputation could entail:

- Replace missing value with a placeholder value, mean, or other value[118] [8].
    - o Imputation using (mean/median) values.
    - o 3-Imputation Using (Most Frequent) or (Zero/Constant) Values.
    - o Imputation using k-NN.
    - o Imputation Using Multivariate Imputation by Chained Equation (MICE).
- Completely removing the rows and columns that have missing values.
- Inferring values based on statistical methods[119] (probabilistic principal component analysis (PPCA), Multivariate Imputation by Chained Equations (MICE)) [8].

- Imputation strategies for time series data have the following techniques[120] and rely on the assumption that adjacent observations are like one another. These methods do not work well when this assumption is not valid, especially in the presence of strong seasonality.
  - o Last observation carried forward (LOCF)—It is a common statistical approach for the analysis of longitudinal repeated measures data when some follow-up observations are missing.
  - o Next observation carried backward (NOCB).
  - o Linear interpolation [9]—It's the method of approximating a missing value by joining dots in increasing order along a straight line. In a nutshell, it calculates the unknown value in the same ascending order as the values that came before it. Because Linear Interpolation is the default method, we didn't have to specify it while utilizing it. It will almost always be utilized in a time-series dataset.[121] Interpolation is mostly used while working with time-series data because in time-series data we like to fill missing values with previous one to two values.
  - o Spline interpolation[122]—Time-series data follows a special trend or seasonality. Analyzing Time series data is a little different compared to the analysis of normal data frames. Whenever we have time-series data,[123] to deal with missing values we cannot use mean imputation techniques. Interpolation is a powerful method to fill missing values in time-series data [10]. For imputing future datasets interpolation is useful. One method that could be applied is Linear interpolation with variance could be to see future trends.

Mean/Median Strategy

This strategy works for missing column values and is highly effective for small numeric datasets. However, it does not factor in correlation between features and could lead to lower accuracies. Please consider fertilizer usage datasets, U.S. averages. Surging fertilizer costs are pushing food prices higher[124] and causing an impact on food security.

	"Corn is the most fertilizer-intensive crop and is likely to be the most impacted by rising input costs driven by the spike in fertilizer prices".

Simple Imputation Strategy

The simple imputation strategy entails replacing null values in the dataset with the mean value of a column. Replacement is based on the overall column mean values. It is not perfect but replaces NULLs with overall mean values of the respective columns.

 	Software code for this model: FertilizerUseAndPrice.ipynb (Jupyter Notebook Code)

Step 1: Load Fertilizer Usage Data

```
from sklearn.datasets import fetch_california_housing
from sklearn.linear_model import LinearRegression
from sklearn.model_selection import StratifiedKFold
from sklearn.metrics import mean_squared_error
from math import sqrt
import random
import numpy as np
random.seed(0)
#Fetching the dataset
import pandas as pd
fertilizerUseDF = pd.read_csv('USCropUsagePattern.csv')
fertilizerUseDF.tail(10)
```

Output:

	Year	Percent of corn acreage receiving nitrogen fertilizer, U.S. average	Percentage of corn acreage receiving phosphate fertilizer, U.S. average	Percentage of corn acreage receiving potash fertilizer, U.S. average	Percentage of cotton acreage receiving nitrogen fertilizer, U.S. average	Percentage of cotton acreage receiving phosphate fertilizer, U.S. average	Percentage of cotton acreage receiving potash fertilizer, U.S. average	Paid by farmers for fertilizer (2011=100)
45	2009	NaN	NaN	NaN	NaN	NaN	NaN	83.8
46	2010	97.0	78.0	61.0	90.0	62.0	52.0	76.8
47	2011	NaN	NaN	NaN	NaN	NaN	NaN	100.0
48	2012	NaN	NaN	NaN	NaN	NaN	NaN	101.4
49	2013	NaN	NaN	NaN	NaN	NaN	NaN	96.7
50	2014	97.0	80.0	65.0	NaN	NaN	NaN	94.5
51	2015	NaN	NaN	NaN	78.0	56.0	42.0	87.2
52	2016	97.0	79.0	65.0	NaN	NaN	NaN	71.8

To find total number of nulls in the data frame:

```
fertilizerUseDF.isna().sum()
```

Output:

```
Year                                                                    0
Percent of corn acreage receiving nitrogen fertilizer, U.S. average    10
Percentage of corn acreage receiving phosphate fertilizer, U.S. average 10
Percentage of corn acreage receiving potash fertilizer, U.S. average   10
Percentage of cotton acreage receiving nitrogen fertilizer,  U.S. average 12
Percentage of cotton acreage receiving phosphate fertilizer,  U.S. average 12
Percentage of cotton acreage receiving potash fertilizer,  U.S. average 12
Paid by farmers for fertilizer (2011=100)                              26
Received by farmers for all crops(2011=100)                            26
dtype: int64
```

Please note that percentage of Corn and Cotton U.S. averages have null values.

Step 2: List Null Values within the fertilizer usage Dataset.

To find time occurrences of nulls, please apply the following data frame technique:

```
fertilizerUseDF = fertilizerUseDF[fertilizerUseDF.isna().any(axis=1)]
fertilizerUseDF
```

Please note that you will find one or more null entries per row below:

	Year	Percent of corn acreage receiving nitrogen fertilizer, U.S. average	Percentage of corn acreage receiving phosphate fertilizer, U.S. average	Percentage of corn acreage receiving potash fertilizer, U.S. average	Percentage of cotton acreage receiving nitrogen fertilizer, U.S. average	Percentage of cotton acreage receiving phosphate fertilizer, U.S. average	Percentage of cotton acreage receiving potash fertilizer, U.S. average	Paid by farmers for fertilizer (2011=100)	Received by farmers for all crops(2011=100)
0	1964	85.0	78.0	72.0	77.0	58.0	43.0	NaN	NaN
1	1965	88.0	82.0	77.0	79.0	59.0	44.0	NaN	NaN
2	1966	91.0	86.0	80.0	76.0	59.0	45.0	NaN	NaN
3	1967	92.0	87.0	82.0	72.0	56.0	40.0	NaN	NaN
4	1968	92.0	89.0	84.0	73.0	56.0	43.0	NaN	NaN
5	1969	93.0	87.0	82.0	75.0	54.0	39.0	NaN	NaN
6	1970	94.0	90.0	85.0	72.0	48.0	36.0	NaN	NaN
7	1971	94.0	88.0	82.0	74.0	50.0	39.0	NaN	NaN
8	1972	96.0	90.0	86.0	77.0	55.0	41.0	NaN	NaN
9	1973	93.0	86.0	80.0	74.0	55.0	39.0	NaN	NaN
10	1974	94.0	87.0	83.0	79.0	58.0	46.0	NaN	NaN
11	1975	94.0	86.0	82.0	65.0	43.0	33.0	NaN	NaN
12	1976	97.0	90.0	84.0	75.0	53.0	37.0	NaN	NaN
13	1977	96.0	88.0	82.0	78.0	51.0	31.0	NaN	NaN
14	1978	95.0	87.0	81.0	69.0	45.0	31.0	NaN	NaN

Step 3: Visualize Data Density graph.

To visualize data density, draw density function.

```
def show_density(var_name,var_data):
  from matplotlib import pyplot as plt
  fig = plt.figure(figsize=(10,4))
  # Plot density
  var_data.plot.density()
  # Add titles and labels
  plt.title('Data Density ' + var_name, fontsize=12)
  plt.ylabel('Density', fontsize=16)
  plt.tick_params(axis = 'both', which = 'major', labelsize = 12)
  plt.tick_params(axis = 'both', which = 'minor', labelsize = 12)
  # Show the mean, median, and mode
  plt.axvline(x=var_data.mean(), color = 'black', linestyle='dashed', linewidth = 3)
  plt.axvline(x=var_data.median(), color = 'darkgray', linestyle='dashdot', linewidth = 3)
  plt.axvline(x=var_data.mode()[0], color = 'gray', linestyle='dotted', linewidth = 3)
  # Show the figure
  plt.show()
```

Output: Please find statistical values before imputation:

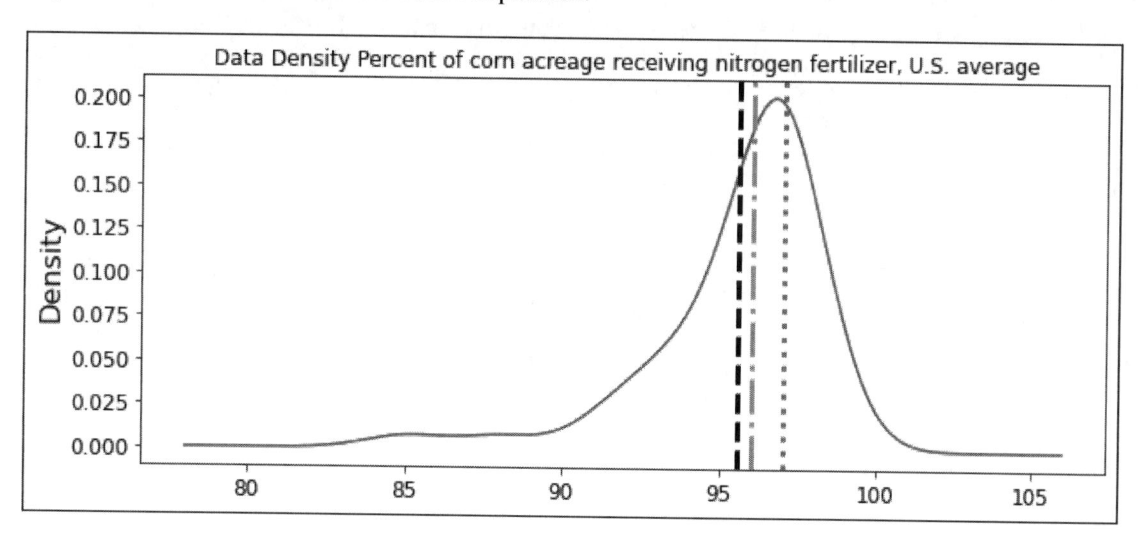

Data Distribution statistics for non-imputed column: Minimum: 85.00; Mean: 95.98 (dashed line); Median: 96.00 (dash dot line); Mode: 97.00 (dotted line); Maximum: 99.00; Range: 14.00; Variance: 7.16; Standard Deviation: 2.68.

Step 4: create train and target placeholders.

```
train = fertilizerUseDF[['Year',
   'Percent of corn acreage receiving nitrogen fertilizer, U.S. average',
   'Percentage of corn acreage receiving phosphate fertilizer, U.S. average',
   'Percentage of corn acreage receiving potash fertilizer, U.S. average',
   'Percentage of cotton acreage receiving nitrogen fertilizer, U.S. average',
   'Percentage of cotton acreage receiving phosphate fertilizer, U.S. average',
   'Percentage of cotton acreage receiving potash fertilizer, U.S. average']]
target = fertilizerUseDF[['Received by farmers for all crops(2011=100)']]
```

Output:

	Year	Percent of corn acreage receiving nitrogen fertilizer, U.S. average	Percentage of corn acreage receiving phosphate fertilizer, U.S. average	Percentage of corn acreage receiving potash fertilizer, U.S. average	Percentage of cotton acreage receiving nitrogen fertilizer, U.S. average	Percentage of cotton acreage receiving phosphate fertilizer, U.S. average	Percentage of cotton acreage receiving potash fertilizer, U.S. average
44	2008	NaN	NaN	NaN	NaN	NaN	NaN
45	2009	NaN	NaN	NaN	NaN	NaN	NaN
47	2011	NaN	NaN	NaN	NaN	NaN	NaN
48	2012	NaN	NaN	NaN	NaN	NaN	NaN
49	2013	NaN	NaN	NaN	NaN	NaN	NaN
50	2014	97.0	80.0	65.0	NaN	NaN	NaN
51	2015	NaN	NaN	NaN	78.0	56.0	42.0
52	2016	97.0	79.0	65.0	NaN	NaN	NaN
53	2017	NaN	NaN	NaN	78.0	59.0	45.0
54	2018	98.0	79.0	63.0	NaN	NaN	NaN

Step 5: Simple Imputer

In the following code, SCIKIT Simple Imputer to mean strategy to replace NaN with median values.

```
#Impute the values using scikit-learn SimpleImpute Class
from sklearn.impute import SimpleImputer
imp_mean = SimpleImputer( strategy='mean') #for median imputation replace 'mean' with 'median'
imp_mean.fit(train)
```

Output: total column size 328

```
[[1964.    85.    78.              72.    77.
   58.    43.    84.58666667]
 [1965.    88.    82.              77.    79.
   59.    44.    84.58666667]
328
```

Impute the trained dataset.

```
imputed_train_array = imp_mean.transform(train)
print(imputed_train_array)
column = imputed_train_array
print(column.size)
```

Convert array into Dataframe.[125]

```
imputed_traindf = pd.DataFrame(imputed_train_array)
```

Output:

	Year	Percent of corn acreage receiving nitrogen fertilizer, U.S. average	Percentage of corn acreage receiving phosphate fertilizer, U.S. average	Percentage of corn acreage receiving potash fertilizer, U.S. average	Percentage of cotton acreage receiving nitrogen fertilizer, U.S. average	Percentage of cotton acreage receiving phosphate fertilizer, U.S. average	Percentage of cotton acreage receiving potash fertilizer, U.S. average	Received by farmers for all crops(2011=100)
31	2008	94.806452	85.548387	78.290323	75.344828	52.241379	37.068966	95.9
32	2009	94.806452	85.548387	78.290323	75.344828	52.241379	37.068966	85.7
33	2011	94.806452	85.548387	78.290323	75.344828	52.241379	37.068966	100.0
34	2012	94.806452	85.548387	78.290323	75.344828	52.241379	37.068966	107.0
35	2013	94.806452	85.548387	78.290323	75.344828	52.241379	37.068966	105.7
36	2014	97.000000	80.000000	65.000000	75.344828	52.241379	37.068966	92.3
37	2015	94.806452	85.548387	78.290323	78.000000	56.000000	42.000000	86.5

Please note that percentage of acreage values are imputed with mean strategy.

```
imputed_traindf.isna().sum()
```

Output:

```
Year                                                                    0
Percent of corn acreage receiving nitrogen fertilizer, U.S. average     0
Percentage of corn acreage receiving phosphate fertilizer, U.S. average 0
Percentage of corn acreage receiving potash fertilizer, U.S. average    0
Percentage of cotton acreage receiving nitrogen fertilizer,  U.S. average   0
Percentage of cotton acreage receiving phosphate fertilizer,   U.S. average 0
Percentage of cotton acreage receiving potash fertilizer,   U.S. average    0
Received by farmers for all crops(2011=100)                             0
dtype: int64
```

Print imputed column distribution:

```
show_density('Percent of corn acreage receiving nitrogen fertilizer, U.S. average',
 imputed_trainMFdf['Percent of corn acreage receiving nitrogen fertilizer, U.S. average'])
```

Output:

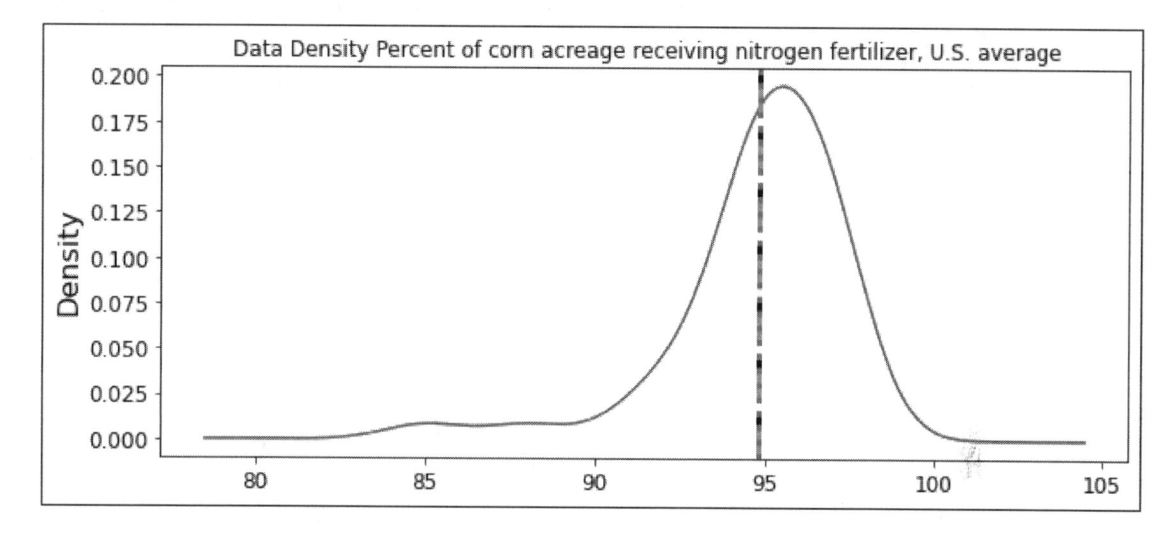

Data Distribution statistics for non-imputed column: Minimum: 85.00; **Mean: 94.81** (dashed line); Median: 94.81 (dash dot line); Mode: 94.81 (dotted line); Maximum: 98; Range: 13.00; Variance: 6.17; Standard Deviation: 2.48. With the mean imputation strategy, the statistical properties of the data changed. For Percentage of corn acreage receiving nitrogen fertilizer, U.S. average, the mean, median, mode and variance have changed. The data becomes less variant.

Most Frequent or (Zero/Constant) Strategy

Most frequent imputation strategy[126] fits with numerical/categorical (mostly recommended) values. That is, missing categorical column values can be filled with this strategy.[127]

	Software code for this model can be found on GitHub Link: FertilizerUseAndPrice.ipynb (Jupyter Notebook Code)

Step 1: create the most frequent imputer.

Create a simple imputer with the most frequent strategy.

```
#Impute the values using scikit-learn SimpleImpute Class
from sklearn.impute import SimpleImputer
imp_mean = SimpleImputer(missing_values=np.nan, strategy='most_frequent')
imp_mean.fit(train)
imputed_train_mf_array = imp_mean.transform(train)
print(imputed_train_mf_array)
column = imputed_train_mf_array
print(column.size)
```

Transform the imputed data series into a Data Frame.

```
imputed_train_mf_array = imp_mean.transform(train)
print(imputed_train_mf_array)
column = imputed_train_mf_array
print(column.size)
```

Step 2: Imputed Most Frequent Dataframe

Reconstruct the DataFrame.

```
imputed_trainMFdf = pd.DataFrame(imputed_train_mf_array,
        columns=['Year','Percent of corn acreage receiving nitrogen fertilizer, U.S. average',
            'Percentage of corn acreage receiving phosphate fertilizer, U.S. average',
            'Percentage of corn acreage receiving potash fertilizer, U.S. average',
            'Percentage of cotton acreage receiving nitrogen fertilizer, U.S. average',
            'Percentage of cotton acreage receiving phosphate fertilizer, U.S. average',
            'Percentage of cotton acreage receiving potash fertilizer, U.S. average','Received by
farmers for all crops(2011=100)'])
imputed_trainMFdf
```

Please find imputed column values.

	Year	Percent of corn acreage receiving nitrogen fertilizer, U.S. average	Percentage of corn acreage receiving phosphate fertilizer, U.S. average	Percentage of corn acreage receiving potash fertilizer, U.S. average	Percentage of cotton acreage receiving nitrogen fertilizer, U.S. average	Percentage of cotton acreage receiving phosphate fertilizer, U.S. average	Percentage of cotton acreage receiving potash fertilizer, U.S. average	Received by farmers for all crops(2011=100)
31	2008.0	97.0	87.0	82.0	76.0	48.0	30.0	95.9
32	2009.0	97.0	87.0	82.0	76.0	48.0	30.0	85.7
33	2011.0	97.0	87.0	82.0	76.0	48.0	30.0	100.0
34	2012.0	97.0	87.0	82.0	76.0	48.0	30.0	107.0
35	2013.0	97.0	87.0	82.0	76.0	48.0	30.0	105.7
36	2014.0	97.0	80.0	65.0	76.0	48.0	30.0	92.3
37	2015.0	97.0	87.0	82.0	78.0	56.0	42.0	86.5
38	2016.0	97.0	79.0	65.0	76.0	48.0	30.0	85.9
39	2017.0	97.0	87.0	82.0	78.0	59.0	45.0	86.2
40	2018.0	98.0	79.0	63.0	76.0	48.0	30.0	86.8

Step 3: Inspect statistical properties after most frequent strategy:

Data Distribution statistics for imputed column: Minimum: 85.00; **Mean: 95.34** (dashed line); Median: 96.00 (dash dot line); **Mode: 97.00 (dotted line); Maximum: 98; Range: 13.00; Variance: 8.23; Standard Deviation: 2.87.**

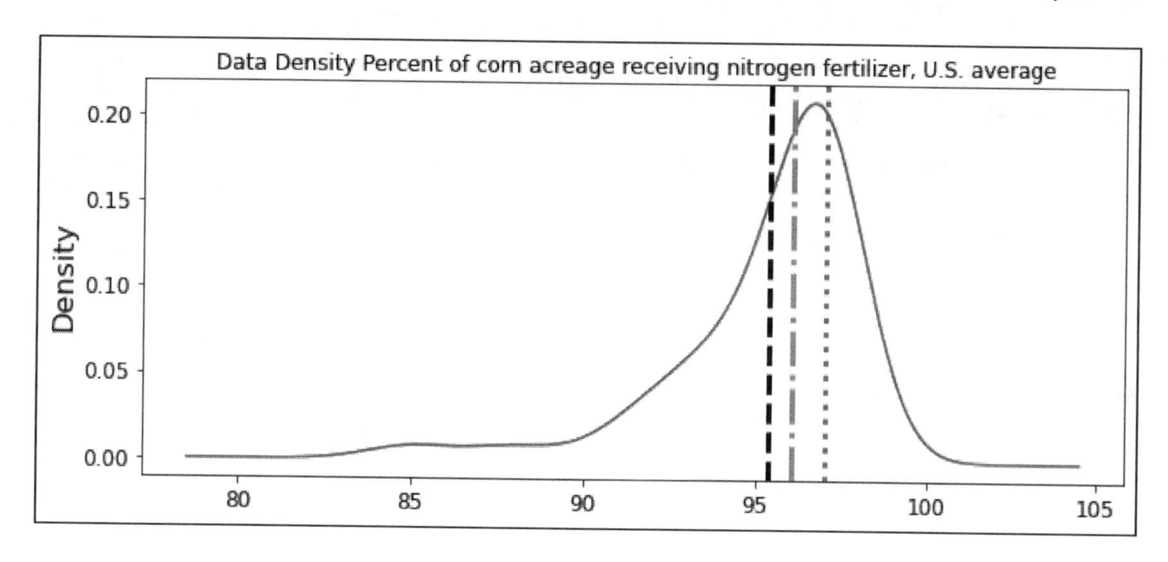

k-NN Strategy

Another imputation strategy is k nearest neighbors, an unsupervised machine learning technique. The algorithm uses 'feature similarity' from missing column values to closest approximations to predict the values of new data points.

The idea is Impute array with a passed in initial impute function (mean impute) and then use the resulting complete array to construct a KDTree.[128] Use this KDTree to compute with your nearest neighbors. After finding the `k` nearest neighbors, take their weighted average.

Lets see an example code using Impyute library which provides a simple and easy way to use KNN for imputation:

	Software code for this model can be found on GitHub Link: FertilizerUseAndPricekNN.ipynb (Jupyter Notebook Code)

Steps 1 & 2 are the same as in an imputation strategy—above steps created in mean imputation strategy.

Step 3

Apply kNN = 3. That is, average (by distance) of nearest neighbor 1 fills all null cells with the nearest cell value. Please increase system recursion time to 100 seconds; otherwise, time out could occur.

```
import sys
from impyute.imputation.cs import fast_knn
from math import sqrt
import random
import numpy as np

sys.setrecursionlimit(100000) #Increase the recursion limit of the OS

data = fertilizerUseDF.to_numpy()
# Weighted average (by distance) of nearest 3 neighbour
print('\naverage (by distance) of nearest 1 neighbour')
imputed_fertilizerUse_knn3_array =fast_knn(data, k=3)
print(imputed_fertilizerUse_knn3_array)
```

Output:

```
Weighted average of nearest 3 neighbour
[[1964.     85.      78.        72.      77.
   58.     43.      75.46666667  84.58666667 1964. ]
 [1965.     88.      82.        77.      79.
   59.     44.      75.46666667  84.58666667 1965. ]
 [1966.     91.      86.        80.      76.
   59.     45.      75.46666667  84.58666667 1966. ]
 [1967.     92.      87.        82.      72.
   56.     40.      75.46666667  84.58666667 1967. ]
 [1968.     92.      89.        84.      73.
   56.     43.      75.46666667  84.58666667 1968. ]
 [1969.     93.      87.        82.      75.
   54.     39.      75.46666667  84.58666667 1969. ]
 [1970.     94.      90.        85.      72.
   48.     36.      75.46666667  84.58666667 1970. ]
```

Step 4: convert kNN array back into Dataframe.

Convert kNN array back into data frame.

```
imputed_kNNtraindf = pd.DataFrame(imputed_fertilizerUse_knn3_array, columns = [
'Year',
    'Percent of corn acreage receiving nitrogen fertilizer, U.S. average',
    'Percentage of corn acreage receiving phosphate fertilizer, U.S. average',
    'Percentage of corn acreage receiving potash fertilizer, U.S. average',
    'Percentage of cotton acreage receiving nitrogen fertilizer, U.S. average',
    'Percentage of cotton acreage receiving phosphate fertilizer, U.S. average',
    'Percentage of cotton acreage receiving potash fertilizer, U.S. average',
    'Paid by farmers for fertilizer (2011=100)',
    'Received by farmers for all crops(2011=100)', 'fertlizierUsageYear'])
```

Output:

	Year	Percent of corn acreage receiving nitrogen fertilizer, U.S. average	Percentage of corn acreage receiving phosphate fertilizer, U.S. average	Percentage of corn acreage receiving potash fertilizer, U.S. average	Percentage of cotton acreage receiving nitrogen fertilizer, U.S. average	Percentage of cotton acreage receiving phosphate fertilizer, U.S. average	Percentage of cotton acreage receiving potash fertilizer, U.S. average	Paid by farmers for fertilizer (2011=100)	Received by farmers for all crops(2011=100)	fertlizierUsageYear
27	2004	95.693977	81.403108	69.552245	75.344828	52.241379	37.068966	42.700000	65.500000	2004
28	2005	96.000000	81.000000	65.000000	75.344828	52.241379	37.068966	49.900000	62.600000	2005
29	2006	95.742811	80.998106	69.336816	75.344828	52.241379	37.068966	53.500000	68.000000	2006
30	2007	94.806452	85.548387	78.290323	91.000000	67.000000	52.000000	65.600000	81.000000	2007
31	2008	94.806452	85.548387	78.290323	75.344828	52.241379	37.068966	119.200000	95.900000	2008
32	2009	96.532936	80.777626	67.829860	75.910184	53.041689	38.118913	83.800000	85.700000	2009
33	2011	96.019824	82.479268	70.938712	75.344828	52.241379	37.068966	100.000000	100.000000	2011
34	2012	94.806452	85.548387	78.290323	75.344828	52.241379	37.068966	101.400000	107.000000	2012
35	2013	96.166099	82.109279	70.052458	75.344828	52.241379	37.068966	96.700000	105.700000	2013
36	2014	97.000000	80.000000	65.000000	76.220314	53.480704	38.694869	94.500000	92.300000	2014
37	2015	95.613773	83.506338	73.398903	78.000000	56.000000	42.000000	87.200000	86.500000	2015
38	2016	97.000000	79.000000	65.000000	76.431178	55.006634	40.313907	71.800000	85.900000	2016
39	2017	96.553250	81.317680	69.044632	78.000000	59.000000	45.000000	66.200000	86.200000	2017
40	2018	98.000000	79.000000	63.000000	76.340271	54.775234	40.042367	66.700000	86.800000	2018

Lets plot the data distribution:

```
show_density('Percent of corn acreage receiving nitrogen fertilizer, U.S. average',
   imputed_kNNtraindf['Percent of corn acreage receiving nitrogen fertilizer, U.S. average'])
```

Output:

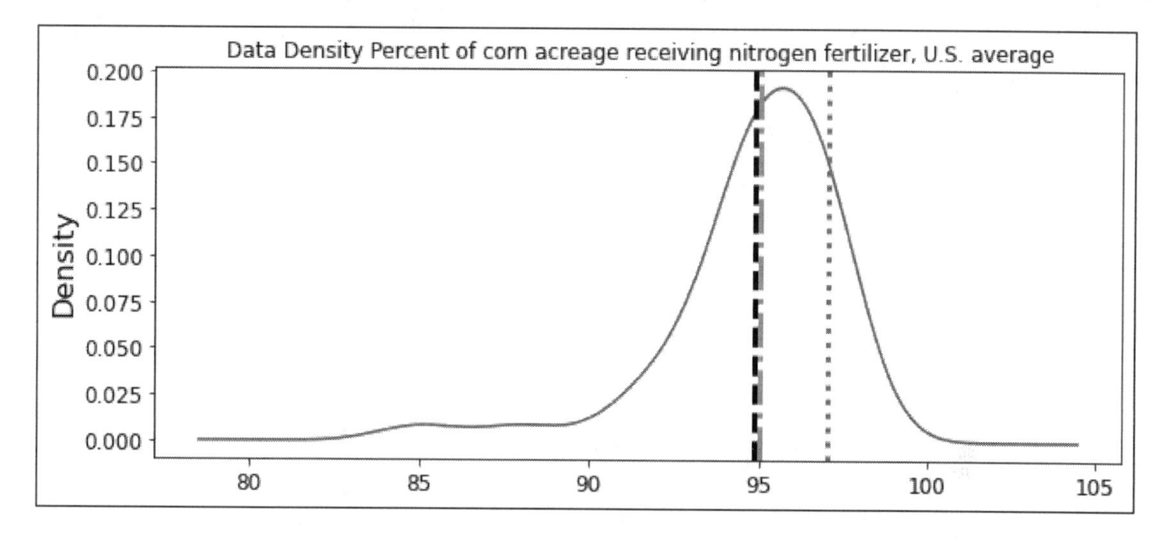

Data Distribution statistics for imputed column: Minimum: 85.00; Mean: 95.86 (dashed line); Median: 95.00 (dash dot line); Mode: 97.00 (dotted line); Maximum: 98; Range: 13.00; Variance: 6.29; Standard Deviation: 2.51. With kNN (k=3) imputation strategy, the statistical properties of the data changed. For Percentage of corn acreage receiving nitrogen fertilizer, U.S. average, the mean, median, mode and variance have changed. The data becomes less variant.

For K = 2 and higher, the weighted average values of the data are as below. As you can see the weighted average value computed for K is greater than 1.

```
# Weighted average of nearest 2 neighbour
print('\nWeighted average of nearest 2 neighbour')
imputed_fertilizerUse_knn2_array =fast_knn(data, k=2)
print(imputed_fertilizerUse_knn2_array)
```

Output:
K = 2

	Year	Percent of corn acreage receiving nitrogen fertilizer, U.S. average	Percentage of corn acreage receiving phosphate fertilizer, U.S. average	Percentage of corn acreage receiving potash fertilizer, U.S. average	Percentage of cotton acreage receiving nitrogen fertilizer, U.S. average	Percentage of cotton acreage receiving phosphate fertilizer, U.S. average	Percentage of cotton acreage receiving potash fertilizer, U.S. average	Paid by farmers for fertilizer (2011=100)	Received by farmers for all crops(2011=100)	fertilzierUsageYear
31	2008	94.806452	85.548387	78.290323	75.344828	52.241379	37.068966	119.2	95.9	2008
32	2009	96.216635	81.981452	69.746268	76.293048	53.583666	38.829947	83.8	85.7	2009
33	2011	94.806452	85.548387	78.290323	75.344828	52.241379	37.068966	100.0	100.0	2011
34	2012	94.806452	85.548387	78.290323	75.344828	52.241379	37.068966	101.4	107.0	2012
35	2013	94.806452	85.548387	78.290323	75.344828	52.241379	37.068966	96.7	105.7	2013
36	2014	97.000000	80.000000	65.000000	76.699201	54.158609	39.584230	94.5	92.3	2014
37	2015	96.167500	82.105735	70.043971	78.000000	56.000000	42.000000	87.2	86.5	2015
38	2016	97.000000	79.000000	65.000000	77.346378	57.336235	43.047623	71.8	85.9	2016
39	2017	97.510186	79.000000	63.979627	78.000000	59.000000	45.000000	66.2	86.2	2017
40	2018	98.000000	79.000000	63.000000	77.366249	57.386815	43.106977	66.7	86.8	2018

For K = 1 and higher, the weighted average values of the data are as below. As you can see the weighted average value computed for K is greater than 1.

```
# Weighted average of nearest 2 neighbour
print('\nWeighted average of nearest 2 neighbour')
imputed_fertilizerUse_knn2_array =fast_knn(data, k=2)
print(imputed_fertilizerUse_knn2_array)
```

Output: - K = 1

	Year	Percent of corn acreage receiving nitrogen fertilizer, U.S. average	Percentage of corn acreage receiving phosphate fertilizer, U.S. average	Percentage of corn acreage receiving potash fertilizer, U.S. average	Percentage of cotton acreage receiving nitrogen fertilizer, U.S. average	Percentage of cotton acreage receiving phosphate fertilizer, U.S. average	Percentage of cotton acreage receiving potash fertilizer, U.S. average	Paid by farmers for fertilizer (2011=100)	Received by farmers for all crops(2011=100)	fertilzierUsageYear
31	2008	94.806452	85.548387	78.290323	75.344828	52.241379	37.068966	119.2	95.9	2008
32	2009	96.216635	81.981452	69.746268	76.293048	53.583666	38.829947	83.8	85.7	2009
33	2011	94.806452	85.548387	78.290323	75.344828	52.241379	37.068966	100.0	100.0	2011
34	2012	94.806452	85.548387	78.290323	75.344828	52.241379	37.068966	101.4	107.0	2012
35	2013	94.806452	85.548387	78.290323	75.344828	52.241379	37.068966	96.7	105.7	2013
36	2014	97.000000	80.000000	65.000000	76.699201	54.158609	39.584230	94.5	92.3	2014
37	2015	96.167500	82.105735	70.043971	78.000000	56.000000	42.000000	87.2	86.5	2015
38	2016	97.000000	79.000000	65.000000	77.346378	57.336235	43.047623	71.8	85.9	2016
39	2017	97.510186	79.000000	63.979627	78.000000	59.000000	45.000000	66.2	86.2	2017
40	2018	98.000000	79.000000	63.000000	77.366249	57.386815	43.106977	66.7	86.8	2018

The K-NN imputation is more accurate than a Simple Imputer with mean/median, or frequent imputation strategies. Nonetheless, computation is as expensive as k-NN is in memory data wrangling and training. If the dataset has much higher outliers, the performance and accuracy of k-NN is affected immensely. Expense would be much higher as compared to datasets with minimum outliers.

Multivariate Imputation by Chained Equation (MICE) Strategy

Multivariate Imputation using Chained Equations" or "Multiple Imputation by Chained Equations" (MICE) accounts for the process that created missing data, *preserves the relations in the data*, and the uncertainty about relations, a stark contrast to what constant or mean/median simple imputer. Please find main steps used in MICE:[129] MICE perform various regressions to predict missing values from the dataset. The type of regression is based on the type of missing column. With a multiple imputation method, each variable with missing data is modeled conditionally using the other variables in the data before filling in the missing values. The MICE,[130] sometimes called "fully conditional specification" or "sequential regression multiple imputation" has emerged in the statistical literature as one principled method of addressing missing data. Creating multiple imputations, as opposed to single imputations, accounts for the statistical uncertainty in the imputations. In addition, the chained equations approach is very flexible and can handle variables of varying types (e.g., continuous, or binary) as well as complexities such as bounds, or survey skip patterns.

Software code for this model can be found on GitHub Link:
FertilizerUseAndPriceMICE.ipynb (Jupyter Notebook Code)

Step 1: Load Fertilizer Use

Load Fertilizer Use dataset.

```
from sklearn.datasets import fetch_california_housing
from sklearn.linear_model import LinearRegression
from sklearn.model_selection import StratifiedKFold
from sklearn.metrics import mean_squared_error
from math import sqrt
import random
import numpy as np
random.seed(0)
#Fetching the dataset
import pandas as pd
fertilizerUseDF = pd.read_csv('USCropUsagePattern.csv')
fertilizerUseDF.tail(10)
```

Output:

	Year	Percent of corn acreage receiving nitrogen fertilizer, U.S. average	Percentage of corn acreage receiving phosphate fertilizer, U.S. average	Percentage of corn acreage receiving potash fertilizer, U.S. average	Percentage of cotton acreage receiving nitrogen fertilizer, U.S. average	Percentage of cotton acreage receiving phosphate fertilizer, U.S. average	Percentage of cotton acreage receiving potash fertilizer, U.S. average	Paid by farmers for fertilizer (2011=100)	Received by farmers for all crops(2011=100)
45	2009	NaN	NaN	NaN	NaN	NaN	NaN	83.8	85.7
46	2010	97.0	78.0	61.0	90.0	62.0	52.0	76.8	87.0
47	2011	NaN	NaN	NaN	NaN	NaN	NaN	100.0	100.0
48	2012	NaN	NaN	NaN	NaN	NaN	NaN	101.4	107.0
49	2013	NaN	NaN	NaN	NaN	NaN	NaN	96.7	105.7
50	2014	97.0	80.0	65.0	NaN	NaN	NaN	94.5	92.3
51	2015	NaN	NaN	NaN	78.0	56.0	42.0	87.2	86.5
52	2016	97.0	79.0	65.0	NaN	NaN	NaN	71.8	85.9

Step 2: Inspect Data Distribution

By inspecting the data distribution, we can assess it.

```
show_density('Percent of corn acreage receiving nitrogen fertilizer, U.S. average',
    fertilizerUseDF['Percent of corn acreage receiving nitrogen fertilizer, U.S. average'])
```

Output:

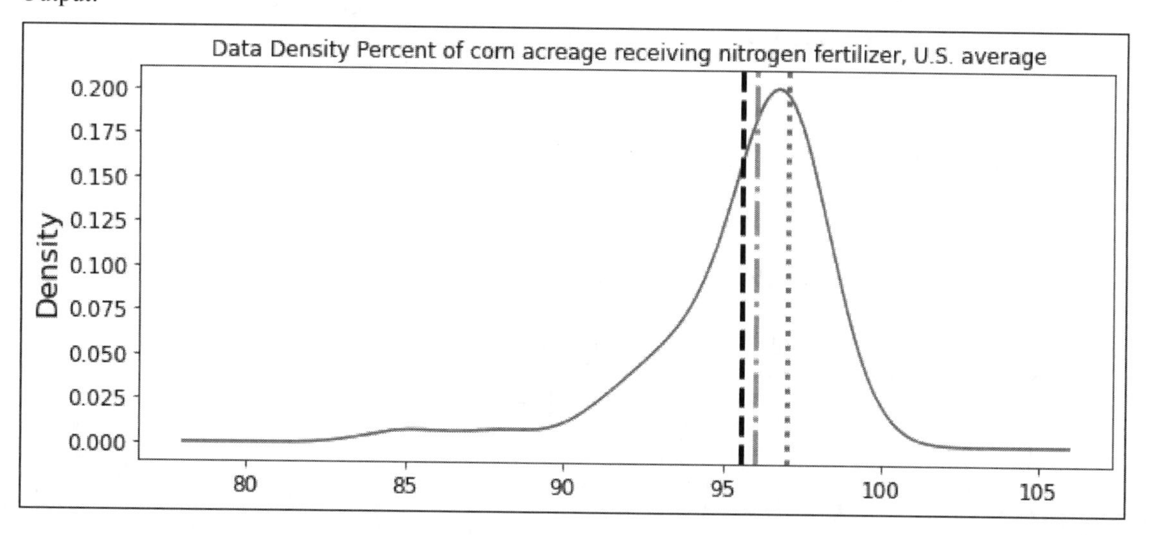

Data Distribution with the following details: Minimum: 85.00; Mean: 95.58 (dashed line); Median: 96.00 (dash dot line); Mode: 97.00 (dotted line); Maximum: 99; Range: 14.00; Variance: 7.16; Standard Deviation: 2.68.

Step 3: Apply the Multivariate Imputation by Chained Equation (MICE).

It is a sophisticated approach to use the IterativeImputer class, which models each feature with missing values as a function of other features and uses that estimate for imputation. It is done in an iterated manner and at each step, a feature column (*percent of corn acreage receiving nitrogen fertilizer, U.S. average*) is designated as output y and the other feature columns are treated as input X. A regressor is fit on (X,y) for a known y. Then, the regressor is used to predict the missing values of y. This is done for each feature in an iterative fashion, and then is repeated for max_iter imputation rounds.

```
import numpy as np
from sklearn.experimental import enable_iterative_imputer
from sklearn.impute import IterativeImputer
from sklearn.linear_model import LinearRegression
lr = LinearRegression()
imp = IterativeImputer(estimator=lr,missing_values=np.nan, max_iter=10, verbose=2, imputation_order='roman',random_state=0)
imputed_traindf=imp.fit_transform(fertilizerUseDF)
imputed_traindf
```

Output:

```
array([[1964.        ,   85.        ,   78.        ,   72.        ,
          77.        ,   58.        ,   43.        ,   68.05574159,
          54.58810853],
       [1965.        ,   88.        ,   82.        ,   77.        ,
          79.        ,   59.        ,   44.        ,   63.73660334,
          59.4608629 ],
       [1966.        ,   91.        ,   86.        ,   80.        ,
          76.        ,   59.        ,   45.        ,   46.12550384,
          53.75849075],
       [1967.        ,   92.        ,   87.        ,   82.        ,
          72.        ,   56.        ,   40.        ,   34.98686506,
          52.19339799],
       [1968.        ,   92.        ,   89.        ,   84.        ,
          73.        ,   56.        ,   43.        ,   59.87835246,
          63.84143961],
       [1969.        ,   93.        ,   87.        ,   82.        ,
          75.        ,   54.        ,   39.        ,   36.12258266,
          52.9209999 ],
       [1970.        ,   94.        ,   90.        ,   85.        ,
```

Step 4: Convert Imputed array to Dataframe.

Converted data to Fertilizer Data Frame.

```
fertilizerUseMICEDF = pd.DataFrame(imputed_traindf,
        columns=['Year',
    'Percent of corn acreage receiving nitrogen fertilizer, U.S. average',
    'Percentage of corn acreage receiving phosphate fertilizer, U.S. average',
    'Percentage of corn acreage receiving potash fertilizer, U.S. average',
    'Percentage of cotton acreage receiving nitrogen fertilizer, U.S. average',
    'Percentage of cotton acreage receiving phosphate fertilizer, U.S. average',
    'Percentage of cotton acreage receiving potash fertilizer, U.S. average',
    'Paid by farmers for fertilizer (2011=100)',
    'Received by farmers for all crops(2011=100)'])
fertilizerUseMICEDF.tail(10)
```

Output:

	Year	Percent of corn acreage receiving nitrogen fertilizer, U.S. average	Percentage of corn acreage receiving phosphate fertilizer, U.S. average	Percentage of corn acreage receiving potash fertilizer, U.S. average	Percentage of cotton acreage receiving nitrogen fertilizer, U.S. average	Percentage of cotton acreage receiving phosphate fertilizer, U.S. average	Percentage of cotton acreage receiving potash fertilizer, U.S. average	Paid by farmers for fertilizer (2011=100)	Received by farmers for all crops(2011=100)
45	2009.0	97.550868	84.050367	68.515710	78.752834	53.499826	42.865364	83.8	85.7
46	2010.0	97.000000	78.000000	61.000000	90.000000	62.000000	52.000000	76.8	87.0
47	2011.0	97.254403	84.207978	69.881065	79.083354	54.293825	41.573245	100.0	100.0
48	2012.0	97.755357	84.561277	71.448574	79.950980	55.078824	39.879130	101.4	107.0
49	2013.0	98.345909	84.686526	71.336259	80.371774	55.190134	39.677372	96.7	105.7
50	2014.0	97.000000	80.000000	65.000000	82.797566	51.532012	42.848006	94.5	92.3
51	2015.0	97.857940	84.161294	65.197156	78.000000	56.000000	42.000000	87.2	86.5
52	2016.0	97.000000	79.000000	65.000000	77.511360	48.530228	34.077436	71.8	85.9
53	2017.0	100.343459	85.460888	67.151719	78.000000	59.000000	45.000000	66.2	86.2
54	2018.0	98.000000	79.000000	63.000000	81.947974	55.189554	39.976580	66.7	86.8

Step 5: Inspect MICE Imputed Dataframe.

Let's inspect the statistical properties of the MICE data frame.

```
show_density('Percent of corn acreage receiving nitrogen fertilizer, U.S. average',
    fertilizerUseMICEDF['Percent of corn acreage receiving nitrogen fertilizer, U.S. average'])
```

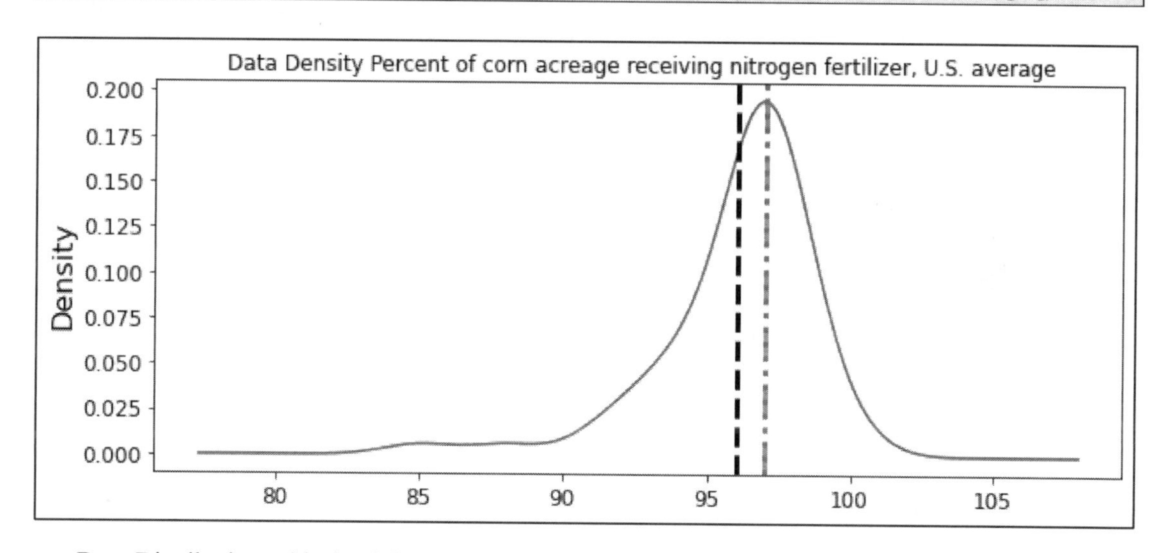

Data Distribution with the following details: Minimum: 85.00; Mean: 96.03 (dashed line); Median: 97.00 (dash dot line); Mode: 97.00 (dotted line); Maximum: 100.34; Range: 15.34; Variance: 7.20; Standard Deviation: 2.60.

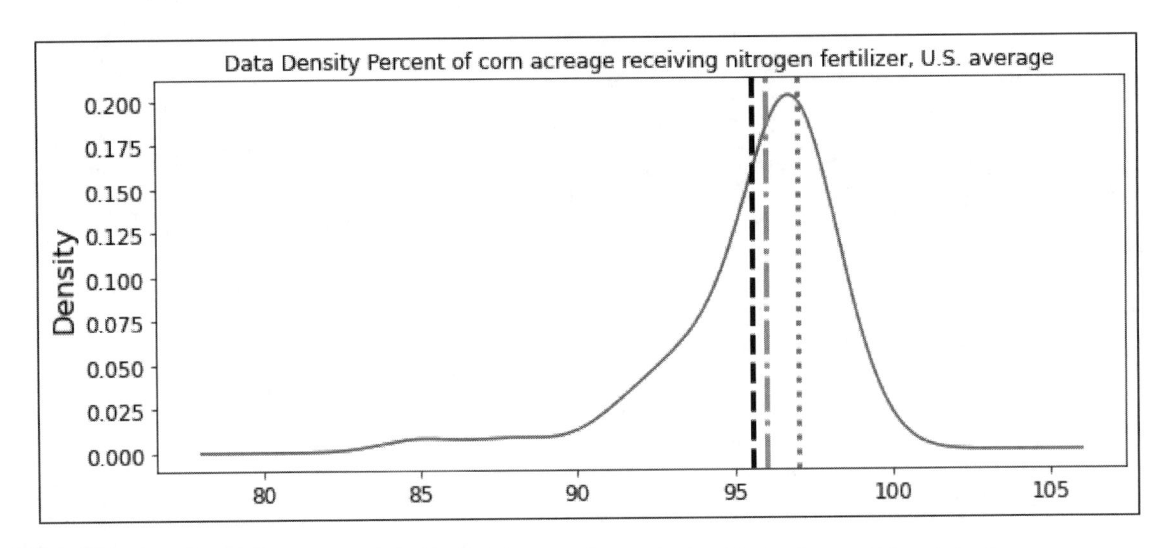

For reference, before imputation, the Data Distribution is as follows: Minimum: 85.00; Mean: 95.58 (dashed line); Median: 96.00 (dash dot line); Mode: 97.00 (dotted line); Maximum: 99; Range: 14.00; Variance: 7.16; Standard Deviation: 2.68.

	For perturbated datasets, application of appropriate imputation strategies, although an exogenous process, is as critical as identification of key influential feature variables to provide relevance of the model to the real-world aspects of the underlying data.

In summary,

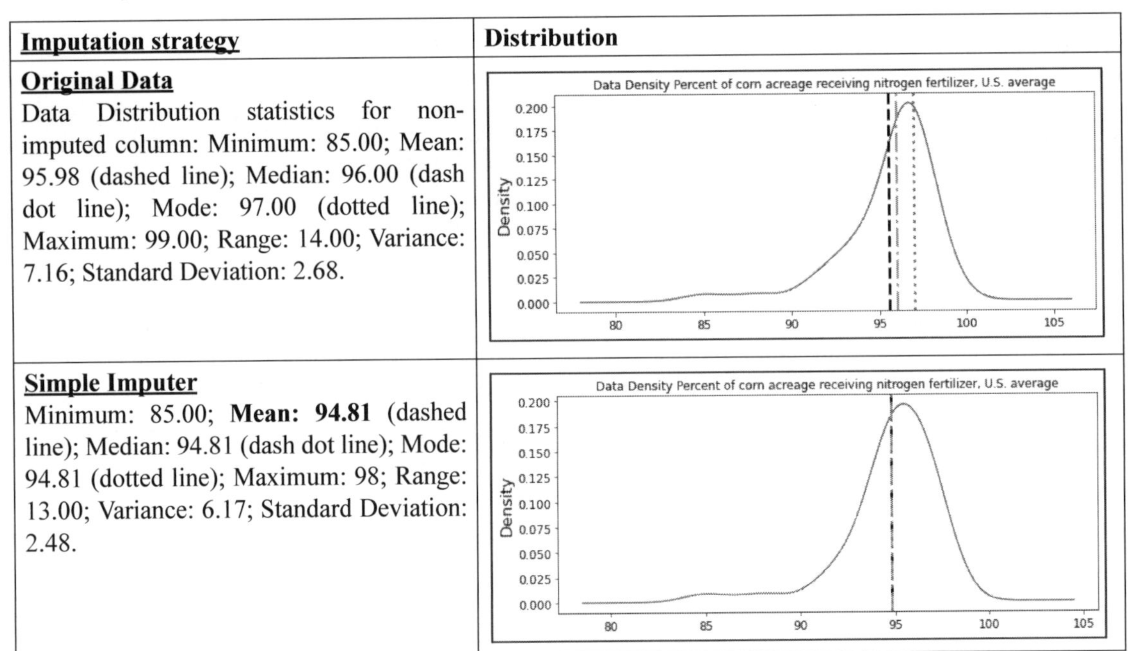

Imputation strategy	**Distribution**
Original Data Data Distribution statistics for non-imputed column: Minimum: 85.00; Mean: 95.98 (dashed line); Median: 96.00 (dash dot line); Mode: 97.00 (dotted line); Maximum: 99.00; Range: 14.00; Variance: 7.16; Standard Deviation: 2.68.	Data Density Percent of corn acreage receiving nitrogen fertilizer, U.S. average Density 0.200 0.175 0.150 0.125 0.100 0.075 0.050 0.025 0.000 80 85 90 95 100 105
Simple Imputer Minimum: 85.00; **Mean: 94.81** (dashed line); Median: 94.81 (dash dot line); Mode: 94.81 (dotted line); Maximum: 98; Range: 13.00; Variance: 6.17; Standard Deviation: 2.48.	Data Density Percent of corn acreage receiving nitrogen fertilizer, U.S. average Density 0.200 0.175 0.150 0.125 0.100 0.075 0.050 0.025 0.000 80 85 90 95 100 105

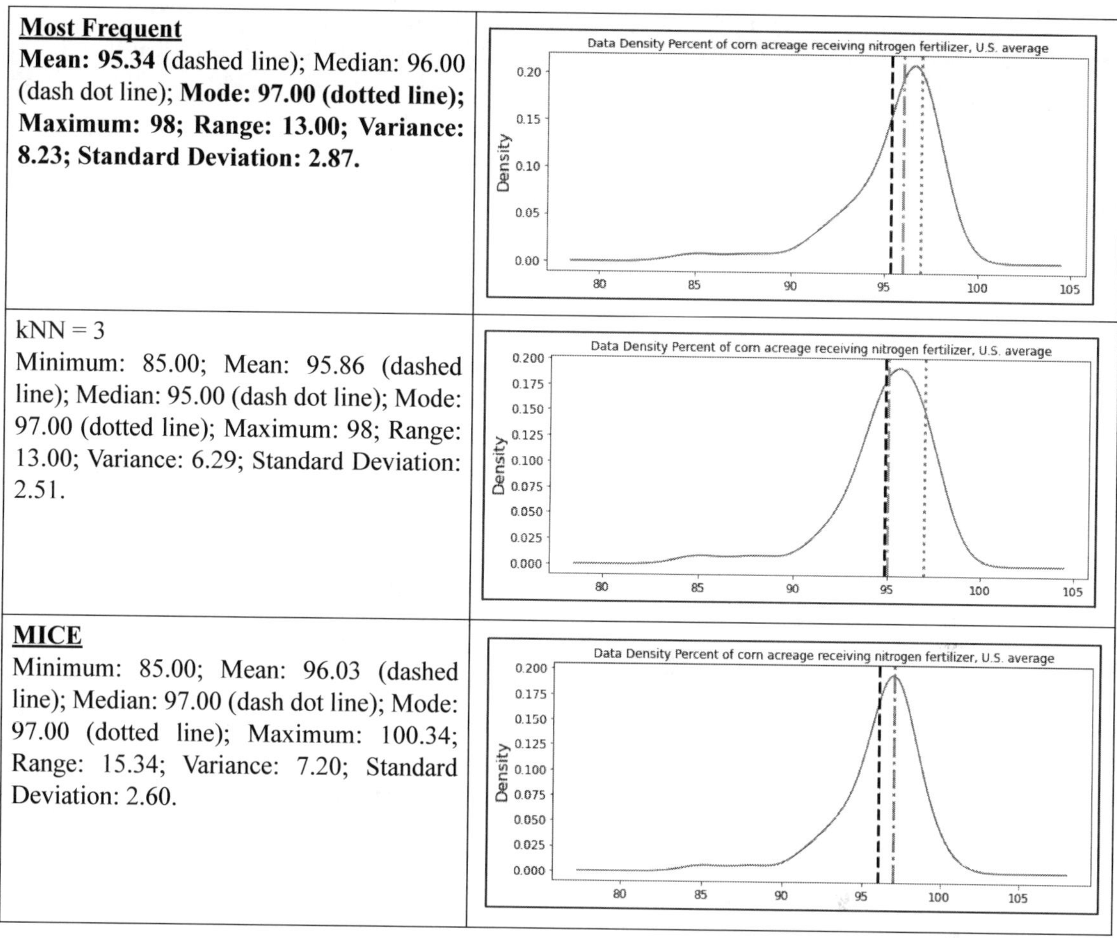

Most Frequent **Mean: 95.34** (dashed line); Median: 96.00 (dash dot line); **Mode: 97.00 (dotted line); Maximum: 98; Range: 13.00; Variance: 8.23; Standard Deviation: 2.87.**	Data Density Percent of corn acreage receiving nitrogen fertilizer, U.S. average
kNN = 3 Minimum: 85.00; Mean: 95.86 (dashed line); Median: 95.00 (dash dot line); Mode: 97.00 (dotted line); Maximum: 98; Range: 13.00; Variance: 6.29; Standard Deviation: 2.51.	Data Density Percent of corn acreage receiving nitrogen fertilizer, U.S. average
MICE Minimum: 85.00; Mean: 96.03 (dashed line); Median: 97.00 (dash dot line); Mode: 97.00 (dotted line); Maximum: 100.34; Range: 15.34; Variance: 7.20; Standard Deviation: 2.60.	Data Density Percent of corn acreage receiving nitrogen fertilizer, U.S. average

In summary, among all the techniques, most frequent preserves the original dataset data distribution and hence data attribute signatures.

Data Resampling

Data resampling is a necessary technique as part of exploratory data analysis (EDA). When it comes to time series analysis, resampling is a critical technique that allows you to flexibly define the *resolution* of the data you want. You can either increase the frequency like converting 5-minute data into 1-minute data (up-sample, increase in data points), or you can do the opposite (down-sample, decrease in data points) [11]. General Time Series data frequencies of the World Economic outlook include the following frequency list:[131]

Value	Description	Notes
A	Annual	To be used for data collected or disseminated every year.
B	Daily – business week	Similar to "daily", however there are no observations for Saturdays and Sundays (so, neither "missing values" nor "numeric values" should be provided for Saturday and Sunday). This treatment ("business") is one way to deal with such cases, but it is not the only option. Such a time series could alternatively be considered daily ("D"), thus, with missing values on the weekend.
D	Daily	To be used for data collected or disseminated every day.
H	Hourly	To be used for data collected or disseminated every hour.
M	Monthly	To be used for data collected or disseminated every month.
N	Minutely	While N denotes "minutely", usually, there may be no observations every minute (for several series the frequency is usually "irregular" within a day/days). And though observations may be sparse (not collected or disseminated every minute), missing values do not need to be given for the minutes when no observations exist: in any case the time stamp determines when an observation is observed.
Q	Quarterly	To be used for data collected or disseminated every quarter.
S	Half-yearly, semester	To be used for data collected or disseminated every semester.
W	Weekly	To be used for data collected or disseminated every week.

In general, there are 2 main reasons why resampling[132] is used.

1. **Resolutions vs. behavior**: To inspect how data behaves differently under different resolutions or frequency.
2. **Baseline resolution**: To join tables with different resolutions.

Datasets, generally, have different frequencies of collection. For instance, the Consumer Price Index (CPI) is computed monthly whereas Gross Domestic Product (GDP) is analyzed on a quarterly basis. To have all the features of a dataset in a common timeframe, resampling technique is applied.[133] Resampled time-series data is a convenient method for frequency conversion and resampling of a time series[134] [12]. Resampling generates a unique sampling distribution based on the actual data. We can apply various frequencies to resample our time series data. This is a very important technique in the field of analytics. When it comes to time series analysis, resampling is a critical technique that allows you to flexibly define the resolution of the data you want.

Resampling really opens a lot more possibilities for time series analysis. By inspecting the data behaviors under different resolutions, we can gain more insights than merely looking at the original data. Resampling could be performed daily, weekly, monthly, quarterly, and yearly with mean, minimum, maximum, and Sum strategies.

	Software code for this model: CommodityPricingReSample.ipynb (Jupyter Notebook Code)

In the following code, our goal is to load essential metal & oil commodities data and be able to shift the dataset from daily to weekly, monthly, and yearly data.

Step 1: Load Libraries

```
import logging
import pandas as pd
import numpy as np
from numpy import random
import nltk
from sklearn.model_selection import train_test_split
from sklearn.feature_extraction.text import CountVectorizer, TfidfVectorizer
from sklearn.metrics import accuracy_score, confusion_matrix
import matplotlib.pyplot as plt
from nltk.corpus import stopwords
import re
from bs4 import BeautifulSoup
import pandas as pd
import matplotlib.pyplot as plt
import matplotlib as mpl
import numpy as np
import scipy.stats as spstats
import seaborn as sns
%matplotlib inline
mpl.style.reload_library()
mpl.style.use('classic')
mpl.rcParams['figure.facecolor'] = (1, 1, 1, 0)
mpl.rcParams['figure.figsize'] = [6.0, 4.0]
mpl.rcParams['figure.dpi'] = 100
```

Output:
Load default libraries to process time series data.

Step 2: Load Gold Commodity Data from London Bullion Market Association
Load Gold dataset from LBMA.

```
#Gold price data
goldDF = pd.read_csv("lbma_gold_am_usd_1967-12-31_2022-03-31.csv",parse_dates=['Date'],
dayfirst=True)
goldDF = goldDF.set_index('Date')
goldDF.head()
goldDF.plot()
```

The data source for the gold data is London Bullion Market Association (LBMA) Precious Metal Prices.[135]
Output:

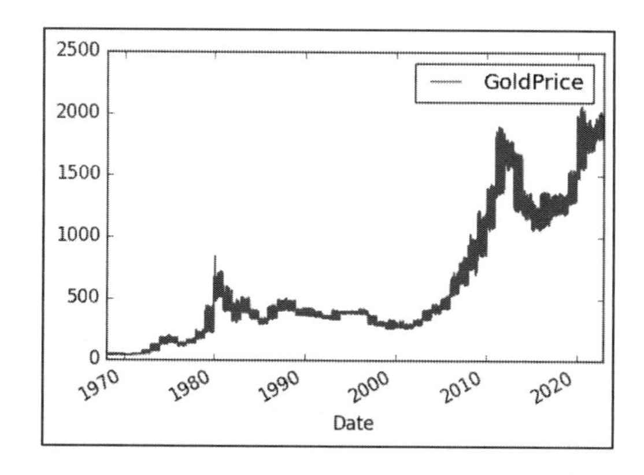

Step 3: Load Silver Commodity Data

```
#Silver price data
silverDF = pd.read_csv("lbma_silver_am_usd_1967-12-31_2022-03-31.csv",parse_dates=['Date'], dayfirst=True)
silverDF = silverDF.set_index('Date')
silverDF.head()
silverDF.plot()
```

The data source for the silver data is London Bullion Market Association (LBMA) Precious Metal Prices.[136]

Output:

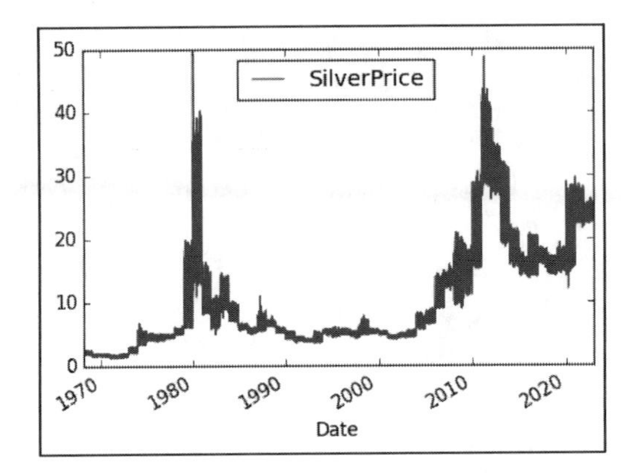

Step 4: Federal Funds Effective Rate Data

```
#Silver price data
usdFedFundsDF = pd.read_csv("FEDFUNDS_2022-03-31.csv",parse_dates=['Date'], dayfirst=True)
usdFedFundsDF = usdFedFundsDF.set_index('Date')
usdFedFundsDF.head()
usdFedFundsDF.plot()
```

The federal funds rate is the interest rate at which depository institutions trade federal funds (balances held at Federal Reserve Banks) with each other overnight. When a depository institution has surplus balances in its reserve account, it lends to other banks in need of larger balances. In simpler terms, a bank with excess cash, which is often referred to as liquidity, will lend to another bank that needs to quickly raise liquidity.

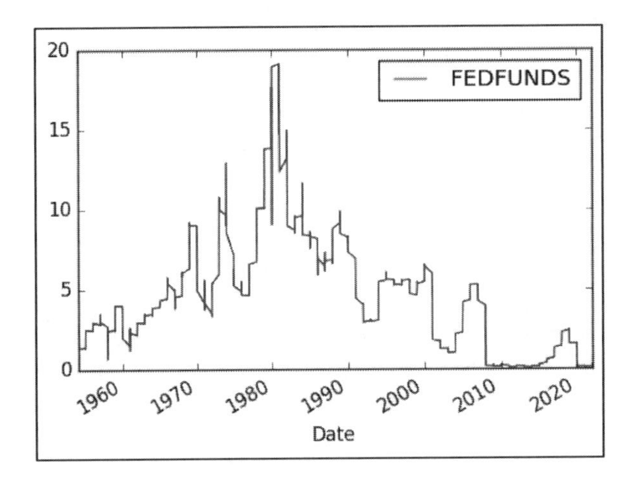

The Federal Open Market Committee (FOMC) meets eight times a year to determine the federal funds target rate. As previously stated, this rate influences the effective federal funds rate through open market operations or by buying and selling government bonds (government debt).

Step 5: Market Yield on U.S. Treasury Securities at 10-Year Constant Maturity Data

```
#Silver price data
interestRateDF = pd.read_csv("DGS10.csv",parse_dates=['DATE'], dayfirst=True)
interestRateDF = interestRateDF.rename(columns={"DATE":'Date'})
interestRateDF = interestRateDF.set_index('Date')
interestRateDF.head()
```

The Market Yield on U.S. Treasury Securities at 10-Year Constant Maturity[137] is the yield on U.S. Treasury securities.

Output:

	DGS10
Date	
2022-04-06	2.61
2022-04-07	2.66
2022-04-08	2.72
2022-04-11	2.79
2022-04-12	2.72
2022-04-13	2.70
2022-04-14	2.83
2022-04-15	.
2022-04-18	2.85
2022-04-19	2.93

DGS10 for the year 2022-04-15 has a non-numerical value. Data fixing is required!

```
interestRateDF['DGS10'] = interestRateDF['DGS10'].replace('.',np.nan)
```

Check for nulls:

```
interestRateDF.isnull().sum() ## missing values
```

Output: the data has 671 nulls. Fix nulls using the padding technique.

```
DGS10    671
dtype: int64
```

```
interestRateDF=interestRateDF.fillna(method='pad')
interestRateDF = interestRateDF.fillna(method = 'bfill')
interestRateDF.isnull().sum() ## missing values
```

Finally convert the column to Float.

```
interestRateDF['DGS10']=interestRateDF['DGS10'].str.replace(',', '.').astype('float')
interestRateDF.plot()
```

Output:

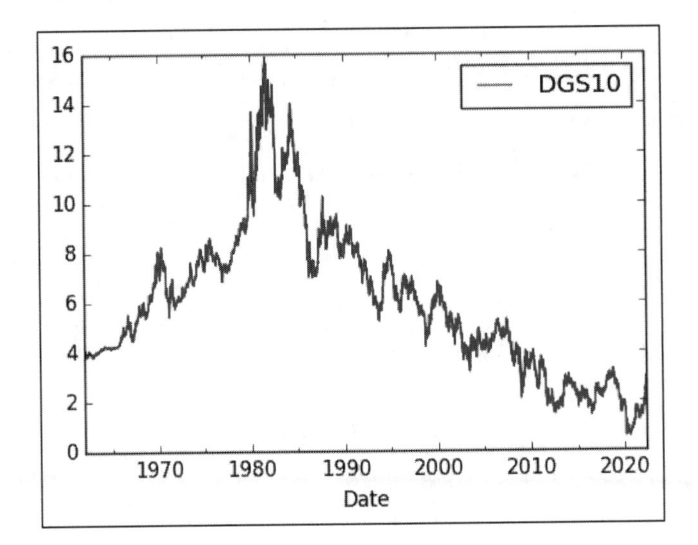

Step 6: S&P Index Data

```
#Silver price data
spindex500Index = pd.read_csv("SPIndex_2022-03-31.csv",parse_dates=['Date'], dayfirst=True)
spindex500Index = spindex500Index.set_index('Date')
spindex500Index.info()
dataset['CrudeOilPrices(WTI)'].plot()
```

The S&P 500[138] data is downloaded from yahoo.
Output:

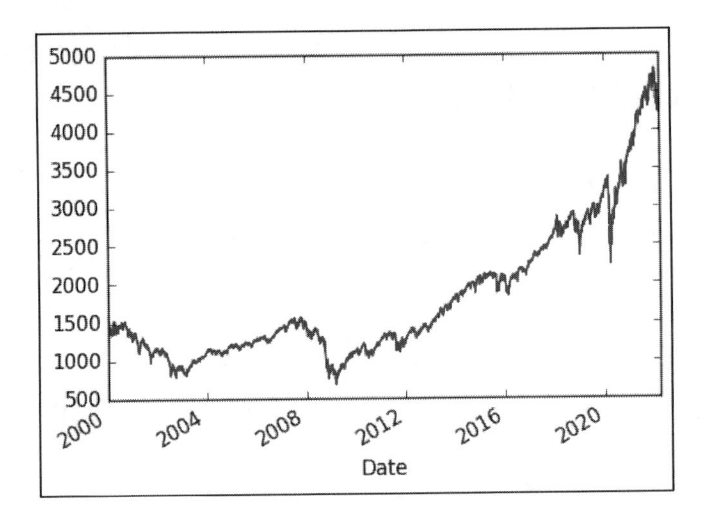

Step 7: Crude Oil Data

```
#Crude Oil data
oilPricesDF = pd.read_csv("DCOILWTICO_2022-03-31.csv",parse_dates=['Date'], dayfirst=True)
oilPricesDF = oilP
```

The S&P 500[139] data is downloaded from yahoo.

Output:

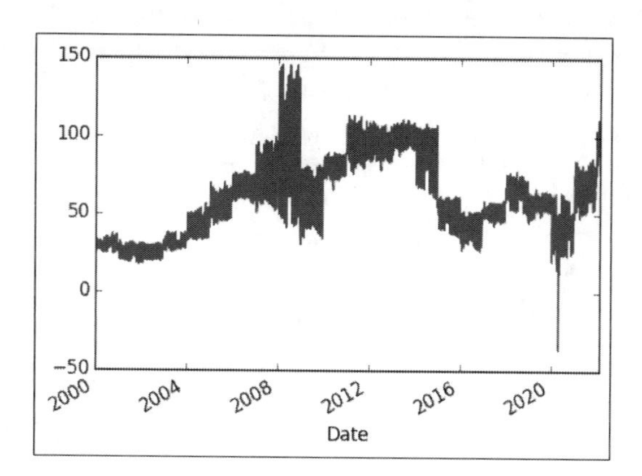

Step 8: Plot Data Frame features data

To visualize the time series of data frame features:

```
# Plot
fig, axes = plt.subplots(nrows=3, ncols=2, dpi=120, figsize=(10,6))
for i, ax in enumerate(axes.flatten()):
    data = dataset[dataset.columns[i]]
    ax.plot(data, color='red', linewidth=1)
    ax.set_title(dataset.columns[i])
    ax.xaxis.set_ticks_position('none')
    ax.yaxis.set_ticks_position('none')
    ax.spines["top"].set_alpha(0)
    ax.tick_params(labelsize=6)
plt.tight_layout();
```

Output:

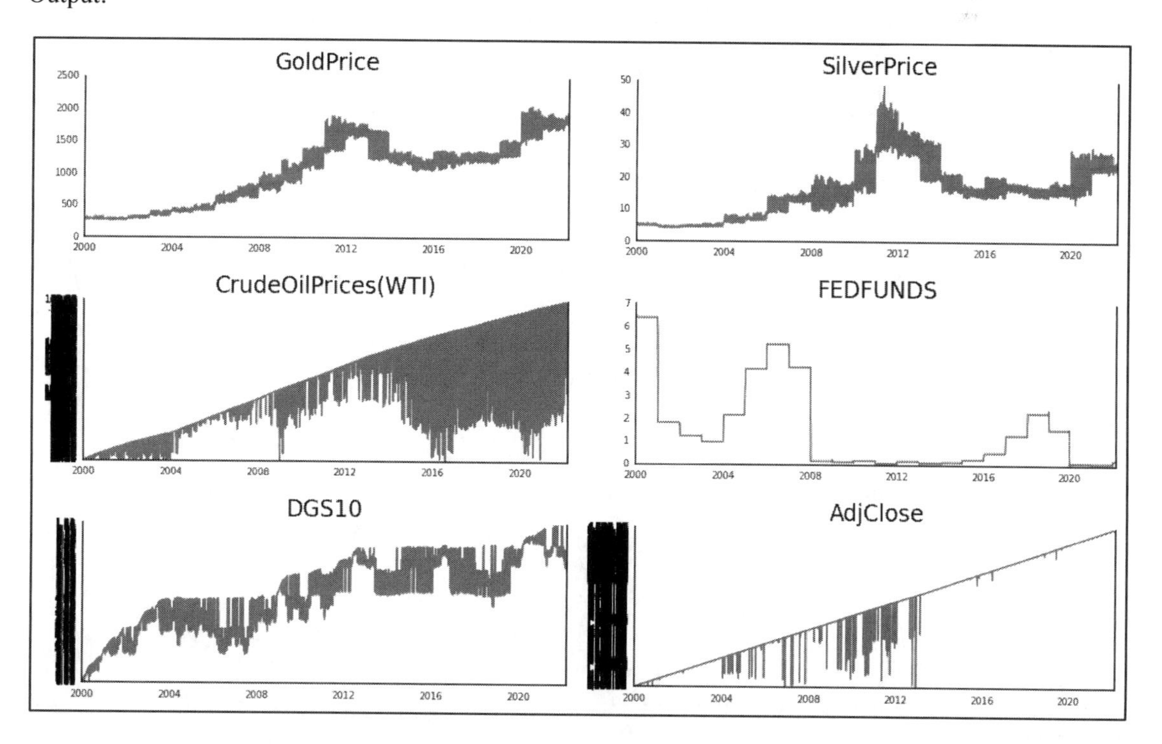

Step 9: Resample from daily to weekly data

Using Data Frame resampling technique, resample from daily to weekly data.

```
weekly_data = dataset.resample('W').mean()
```

The above code samples daily data to weekly. The plot of weekly data:

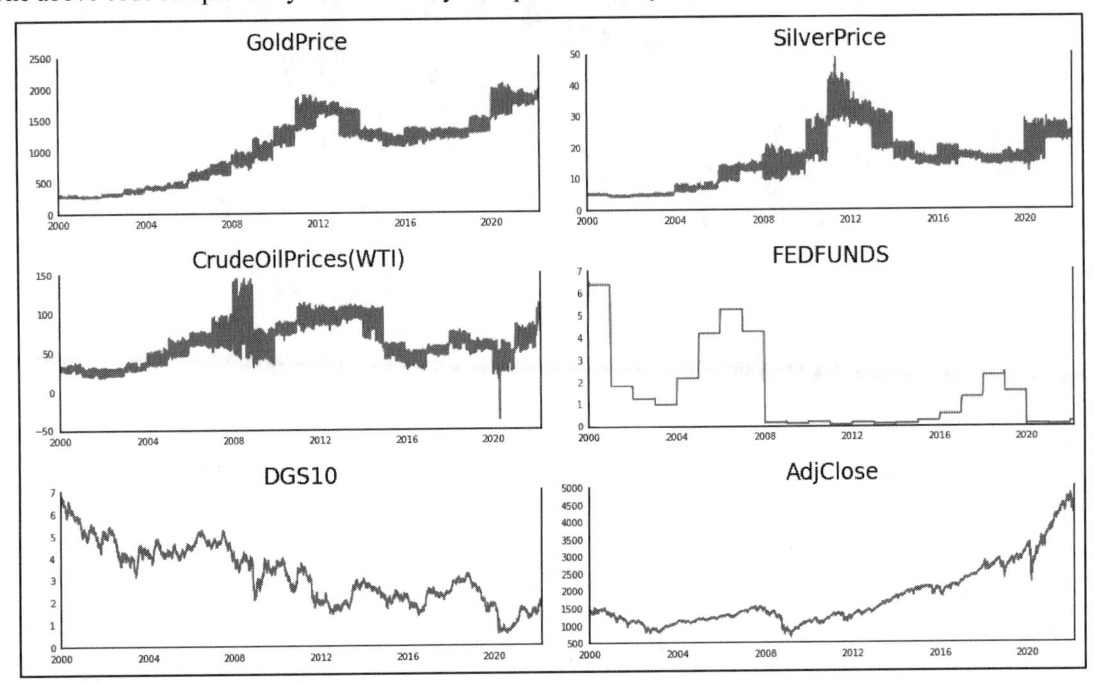

Step 10: Resample from daily to monthly data

Using Data Frame resampling technique, resample from daily to weekly data.

```
monthly_data = dataset.resample('M').mean()
```

Plot monthly data.

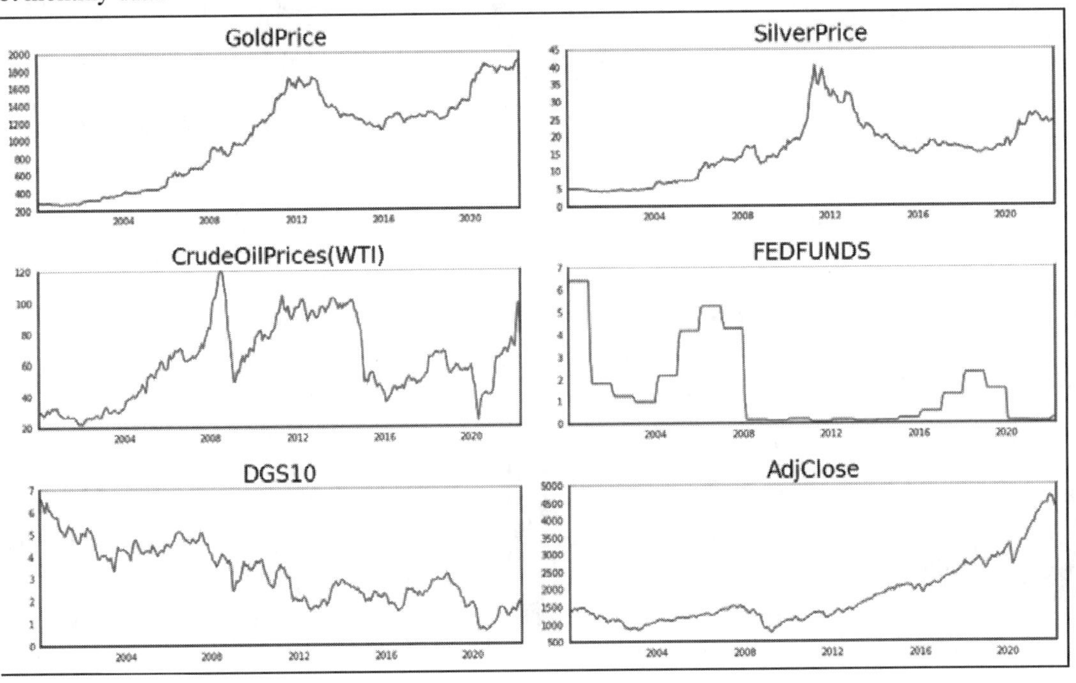

Step 11: Resample from daily to yearly data

Using Data Frame resampling technique, resample from daily to weekly data.

```
yearly_data = dataset.resample('Y').mean()
```

Plot yearly data

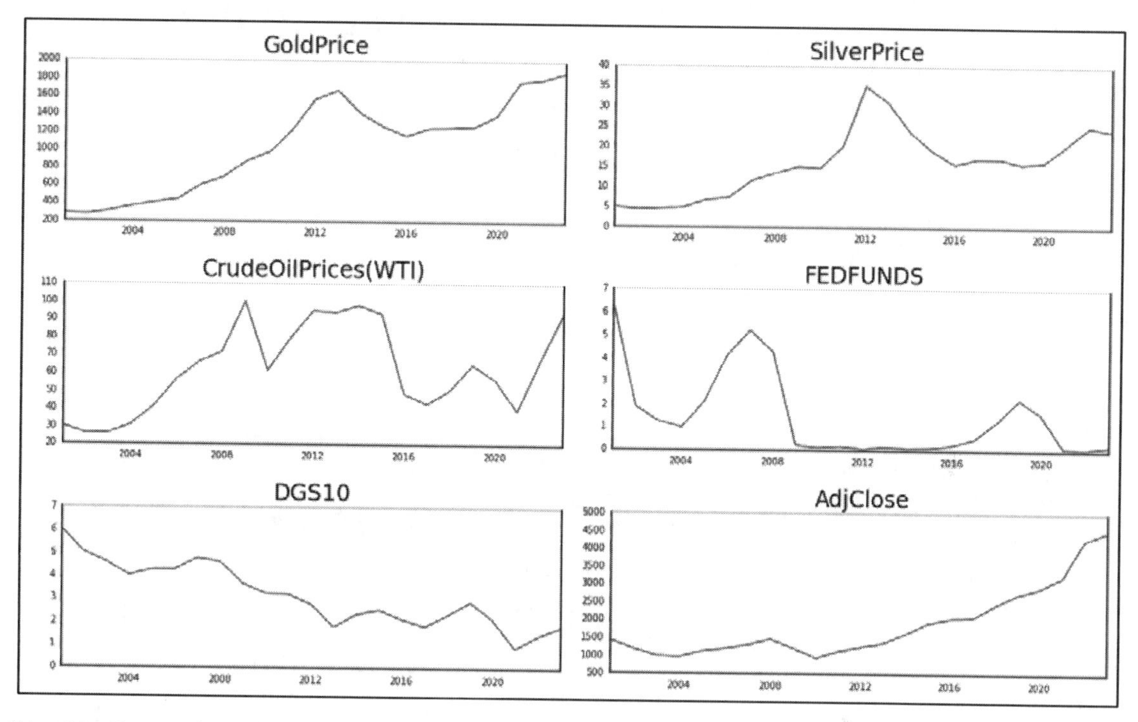

Step 12: Resample statistics

```
daily_data.describe()
```

Output: daily describe:

	GoldPrice	SilverPrice	CrudeOilPrices(WTI)	FEDFUNDS	DGS10	AdjClose
count	6437.000000	6437.000000	6437.000000	6437.00000	6437.000000	6437.000000
mean	1016.499759	15.684237	61.214174	1.51414	3.220492	1790.284097
std	510.296941	8.546880	25.724638	1.83803	1.356238	882.617880
min	256.700000	4.065000	-36.980000	0.05000	0.520000	676.530000
25%	438.250000	7.360000	40.570000	0.12000	2.110000	1175.430000
50%	1172.000000	15.770000	58.400000	0.65000	2.990000	1411.130000
75%	1346.850000	19.500000	80.300000	2.16000	4.290000	2139.760000
max	2061.500000	48.700000	145.310000	6.54000	6.790000	4796.560000

Except record count, all other statistic values (mean, std, min, 25%, 50%, 75%, and max) are the same. The resample only changes or aggregates the record on the time series.

```
weekly_data.describe()
```

Output: weekly resampled statistics data:

	GoldPrice	SilverPrice	CrudeOilPrices(WTI)	FEDFUNDS	DGS10	AdjClose
count	1158.000000	1158.000000	1158.000000	1158.000000	1158.000000	1158.000000
mean	1018.761824	15.710163	61.295153	1.506374	3.217212	1793.410376
std	509.664754	8.455284	25.131392	1.832378	1.355587	887.069367
min	258.500000	4.083000	3.324000	0.070000	0.553333	691.815714
25%	441.390000	7.298250	40.774000	0.120000	2.103500	1173.836000
50%	1174.737500	15.874000	58.429000	0.540000	2.970000	1412.742500
75%	1344.950000	19.597500	80.775429	2.160000	4.286357	2150.195500
max	1981.340000	44.664000	136.998000	6.444000	6.750000	4778.267143

```
monthly_data.describe()
```

Output: monthly resample statistics data:

	GoldPrice	SilverPrice	CrudeOilPrices(WTI)	FEDFUNDS	DGS10	AdjClose
count	267.000000	267.000000	267.000000	267.00000	267.000000	267.000000
mean	1021.443415	15.740049	61.372379	1.49987	3.210674	1801.556755
std	509.193224	8.349494	24.572299	1.82723	1.352850	897.606079
min	265.564583	4.204958	22.053077	0.07000	0.631154	750.981250
25%	437.798000	7.240064	41.936911	0.12000	2.093530	1182.380000
50%	1180.412000	15.976400	59.068750	0.54000	2.974400	1415.601739
75%	1368.612667	19.369484	79.616667	2.16000	4.280454	2151.444058
max	1903.300000	40.355417	119.851818	6.40000	6.644400	4670.178696

```
yearly_data.describe()
```

Output:

	GoldPrice	SilverPrice	CrudeOilPrices(WTI)	FEDFUNDS	DGS10	AdjClose
count	23.000000	23.000000	23.000000	23.000000	23.000000	23.000000
mean	1048.982844	16.004175	62.471045	1.457714	3.165953	1890.652746
std	532.310131	8.423965	24.796568	1.846958	1.341023	1021.761929
min	271.100347	4.371413	25.964792	0.071301	0.901478	946.017785
25%	524.132062	9.418051	42.235500	0.140412	2.138202	1201.232165
50%	1224.228966	15.738785	61.555986	0.533979	2.906944	1429.412694
75%	1401.740956	20.296552	86.420319	2.016412	4.280942	2270.071540
max	1874.004545	35.275034	100.008021	6.393367	6.031751	4513.870000

Synthetic Minority Oversampling Technique (SMOTE) & Adaptive Synthetic Sampling (ADASYN)

A dataset is imbalanced if the classification categories are not approximately equally represented. Often real-world data sets are predominately composed of "normal" examples with only a small percentage of "abnormal" or "interesting" examples. For example, in the example below Wheat[140] yield of Kansas (1982:2020) a drastic drop in yield is seen for the years 1989, 1995, and 2014 [13]. The drastic drop has a real-life event associated with it. For instance, the 1989 drop was due to drought conditions [14]. The principal grain affected by the drought is hard red winter wheat, which is produced mainly in Kansas, Colorado, Montana, Nebraska, Oklahoma, and South Dakota.[141] The hardest-hit state is Kansas, where the Agriculture Department projected a harvest of 202.4 million bushels, compared with 366.3 million in 1987 and 323 million in 1988. Kansas is the leading wheat producer; its output normally accounts for about a third of the winter wheat and 16 to 17 percent of all wheat (Figure 3).

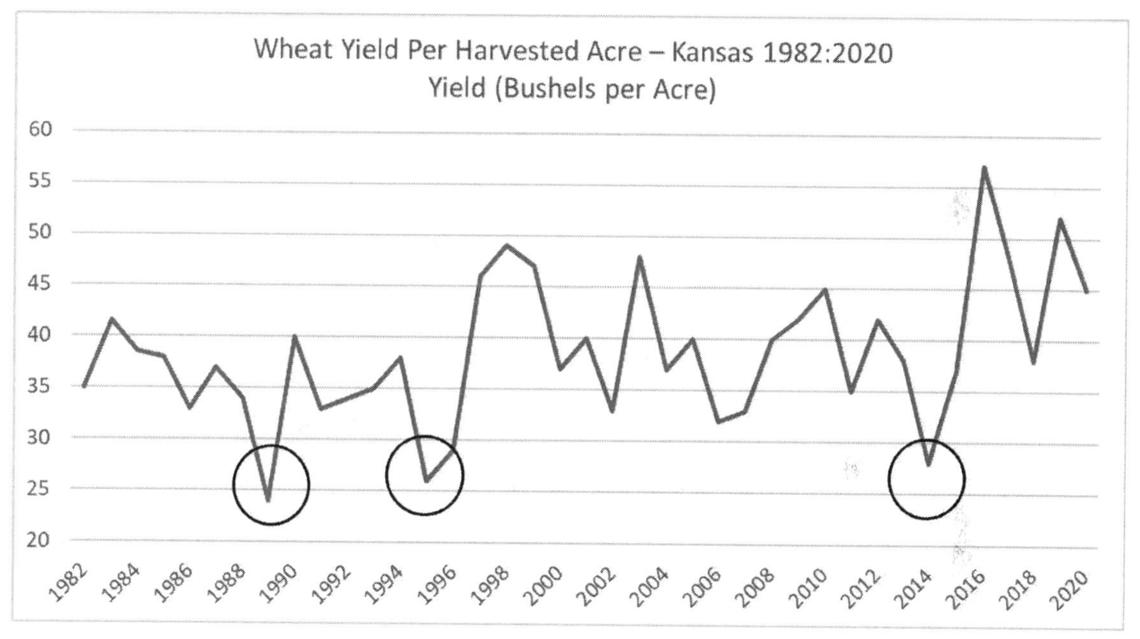

Figure 3: Wheat Yield (1982:2020).

The implication was a direct hit to consumers and food security. Winter wheat is planted in the fall, lies dormant in the winter, and matures in the spring. It is normally about 75 percent of the total national wheat crop. The hard red variety, the major wheat used in bread and the kind produced in Kansas and other parts of the Central and Southern Plains, normally represents more than two-thirds of all winter wheat [14]. In 1995–1996, drought[142] had caused severe losses to the farmers of Kansas and the impact was felt worldwide due to an increase in prices [15]. Droughts punctuated[143] by flood, and early rains, have all the contributed to the drop of wheat yield in 2014 [15]. The Kansas wheat harvest may be one of the worst on record, 2014, and the loss doesn't just hurt Kansas, according to a Kansas State University expert [15].

Kansas[144] grows winter wheat that is planted and sprouts in the fall, becomes dormant in the winter, grows again in the spring, and is harvested in early summer. All the wheat grown in Kansas in a single year would fit in a train stretching from western Kansas to the Atlantic Ocean. Kansas stores more wheat than any other state. On average, Kansas is the largest wheat producing state. Nearly one-fifth of all wheat grown in the United States is grown in Kansas. Therefore, it is called the "Wheat State" and "Breadbasket of the World". The net effect of wheat yield drop is on food availability across the nation and to the world as Kansas is the breadbasket. The purpose of the ML model that would be developed later is to tie the impact of weather

& satellite[145] imagery to predict future yield issues. The interesting parameters required to model the impact of drought and yield reduction are

- Vegetation Condition index (VCI)—VCI is based on the pre- and post-launch calibrated radiances converted to the no noise Normalized Difference Vegetation.[146]
- Temperature Condition index (TCI)—*TCI is a proxy for thermal conditions.*
- Vegetation Health index (VHI)—VHI is a proxy characterizing vegetation health or a combined estimation of *moisture and thermal conditions*.
- No noise (smoothed) Normalized Difference Vegetation Index (SMN)—The SMN is derived from the no noise NDVI, whose components were pre- and post-launch calibrated. SMN can be used to estimate the start and senescence of vegetation, start of the growing season and phenological phases.
- and No noise (smoothed) Brightness Temperature (SMT)—The SMT is the BT with complete removal of high frequency noise, SMT can be used for estimation of thermal conditions, cumulative degree days and others.

I will cover more on satellite data[147] in subsequent chapters.

Year	Yield (bushels)	FoodSecur	TCI_WK1	VHI_WK1	TCI_WK2	VHI_WK2	TCI_WK3	VHI_WK3	TCI_WK4	VHI_WK4
1982	35	Yes	53.35	62.36	60.09	63.94	64.8	63.68	68.1	62.12
1983	41.5	Yes	43.86	39.34	52.47	40.18	62.01	41.98	69.16	42.94
1984	38.5	Yes	85.71	51.45	81.82	52.01	74.01	51.71	65.43	51.54
1985	38	Yes	-1	-1	-1	-1	-1	-1	-1	-1
1986	33	Yes	26.96	36.43	25.29	36.65	21.44	36.39	20.97	37.17
1987	37	Yes	29.93	38.23	33.86	40.36	35.46	41.95	33.7	42.77
1988	34	Yes	56.15	45.59	60.72	47.37	61.23	47.97	59.83	48.54
1989	24	No	8.75	41.97	11.36	42.37	14.18		17.39	41.68
1990	40	Yes	8.7	33.94	9.69	34.49	12.03	35.08	17.42	37.11
1991	33	Yes	36.22	44.23	38.75	45.77	37.01	46.14	32.57	45.45

The purpose is to overlay SMN, SMT, VCI, TCI, and VHI for all years and be able to predict the next occurrence of food insecurity.

A problem with an imbalanced classification is that there are very few examples of the minority class, for instance 1989 food security value "No", for a model to effectively learn the decision boundary. One way to solve this problem is to oversample the examples in the minority class. In this case, we need to regenerate all relevant key attributes that are necessary. This can be achieved by simply duplicating examples from the minority class in the training dataset prior to fitting a model. This can balance the class distribution but does not provide any additional information to the model. One approach to addressing imbalanced datasets is to oversample the minority class and the Synthetic Minority Oversampling Technique or SMOTE for short.

Machine Learning Model: Kansas Wheat Yield SMOTE Model

In the following code, our goal is to use SMOTE technique to resample under-represented class labels to improve Food Security predictability.

Software code for this model:
Kansas_Wheat_Yield_NOAA_STAR.ipynb (Jupyter Notebook Code)

Data Sources

Kansas Wheat History[148]	Kansas Wheat Yield data (1982:2021) Important variables: Area Planted for all purposes Area Harvested for grains Yield per acre Production	USDA
NOAA STAR[149]—Global Vegetation Health Products: Province-Averaged VH	Mean data for USA Province = 17: Kansas, from 1982 to 2022, weekly for cropland[150] area only year, week, SMN, SMT, VCI, TCI, VHI	NOAA

Step 1: Load Libraries

Load required libraries to process the data.

```
import numpy as np
import pandas as pd
import functools
from datetime import date
import seaborn as sns; sns.set()
from statsmodels.tsa.arima_process import ArmaProcess
from scipy import stats
from scipy.stats import pearsonr
import matplotlib.pyplot as plt
```

Step 2: Load Kansas Wheat Yield production

Load Kansas Wheat Production from 1982–2020.

```
dfAverageKansasWheatProduction = pd.read_csv("AverageKansasWheatProduction.csv")
dfAverageKansasWheatProduction
```

Output:

index	Year	Yield (bushels)	FoodSecurity
0	1982	35.0	Yes
1	1983	41.5	Yes
2	1984	38.5	Yes
3	1985	38.0	Yes
4	1986	33.0	Yes
5	1987	37.0	Yes
6	1988	34.0	Yes
7	1989	24.0	No
8	1990	40.0	Yes
9	1991	33.0	Yes

Step 3: Show Data Distribution

Let's see data the distribution of Wheat yield.

```
def show_density(var_name,var_data):
    from matplotlib import pyplot as plt

    # Get statistics
    min_val = var_data.min()
    max_val = var_data.max()
    mean_val = var_data.mean()
    med_val = var_data.median()
    mod_val = var_data.mode()[0]
print('Minimum:{:.2f}\nMean:{:.2f}\nMedian:{:.2f}\nMode:{:.2f}\nMaximum:{:.2f}\n'.format(min_val,
                                                        mean_val,
                                                        med_val,
                                                        mod_val,
                                                        max_val))
    fig = plt.figure(figsize=(10,4))
    # Plot density
    var_data.plot.density()
    # Add titles and labels
    plt.title('Data Density ' + var_name, fontsize=12)
    plt.ylabel('Density', fontsize=16)
    plt.tick_params(axis = 'both', which = 'major', labelsize = 12)
    plt.tick_params(axis = 'both', which = 'minor', labelsize = 12)
    # Show the mean, median, and mode
    plt.axvline(x=var_data.mean(), color = 'black', linestyle='dashed', linewidth = 3)
    plt.axvline(x=var_data.median(), color = 'darkgray', linestyle='dashdot', linewidth = 3)
    plt.axvline(x=var_data.mode()[0], color = 'gray', linestyle='dotted', linewidth = 3)
    # Show the figure
    plt.show()
```

Output:

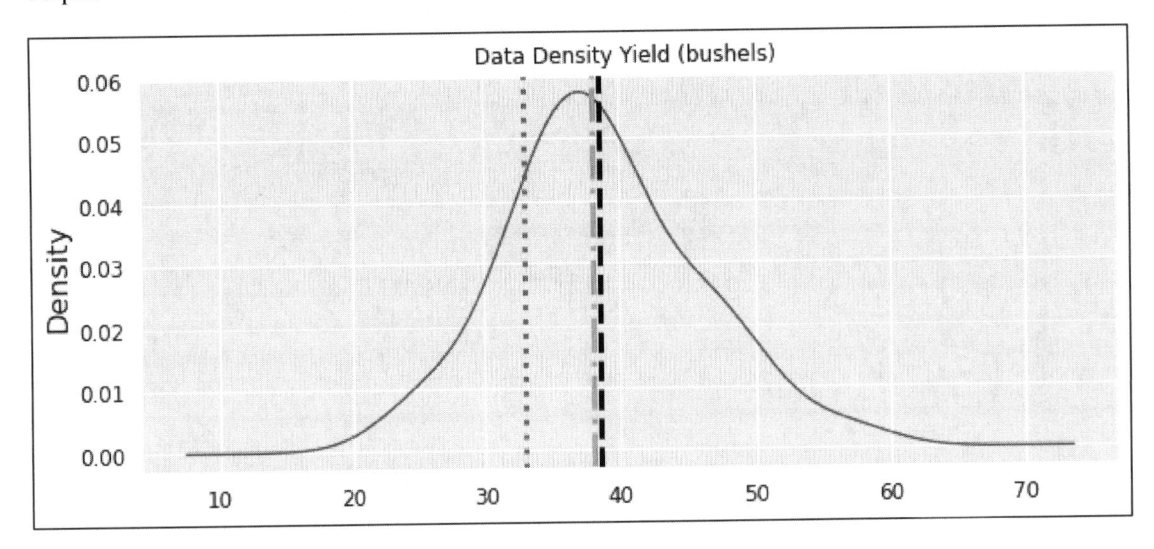

Data Distribution with the following details: Minimum: 24.00; Mean: 38.59 (dashed line); Median: 38.00 (dash dot line); Mode: 33.00 (dotted line); Maximum: 57.00; Range: 33.00; Variance: 49.68; Standard Deviation: 7.05. We will observe the distribution after SMOTE & ADASYN.
Plot Food Security label.

```
dfAverageKansasWheatProduction.boxplot(column='Yield (bushels)', by='FoodSecurity', figsize=(8,5))
```

Output:
Under representation of class label ("Food Security") with number of class labels: 4

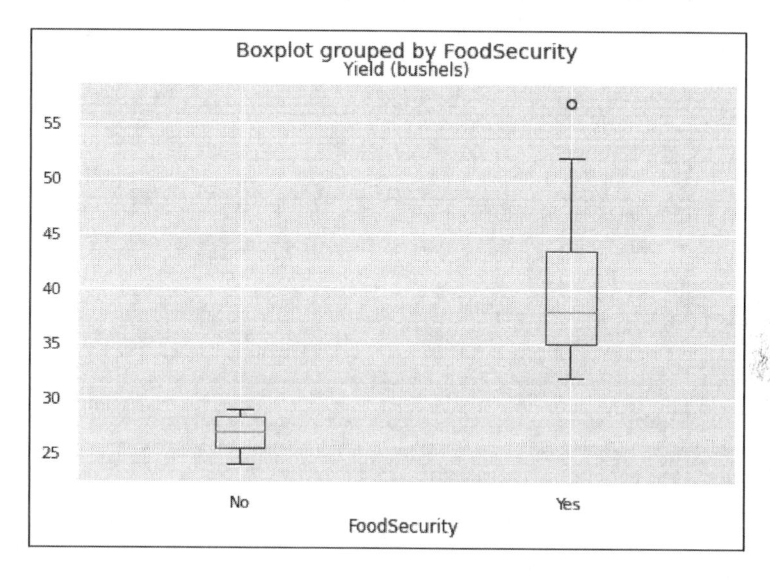

Class labels: "Yes" = 35; "No" = 4

```
pd.value_counts(dfKansasYieldwMODISVHITCI['FoodSecurity']).plot.bar()
plt.title('Food Security class histogram')
plt.xlabel('Class')
plt.ylabel('Frequency')
dfKansasYieldwMODISVHITCI['FoodSecurity'].value_counts()
```

Output:

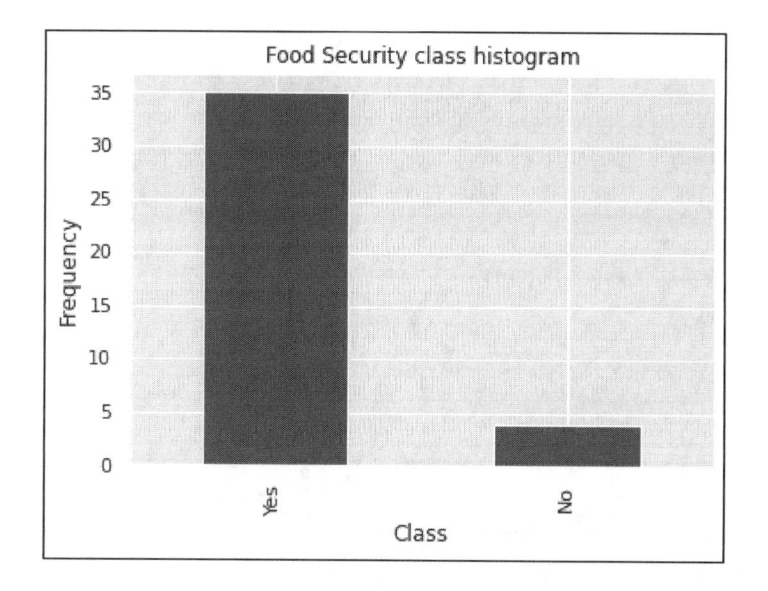

Step 4: Load NOAA STAR Kansas

Load Mean data for USA Province= 17: Kansas,[151] from 1982 to 2022, weekly for cropland area only.

```
dfMODISMeandataforUSAProvince17Kansas = pd.read_csv("MeandataforUSAProvince17Kansas.csv")
dfMODISMeandataforUSAProvince17Kansas
```

Output:

The data provides a complete view of Kansas NOAA STAR data from 1982:2020.

	year	week	SMN	SMT	VCI	TCI	VHI
0	1982	1	0.193	276.46	71.30	54.53	62.91
1	1982	2	0.184	276.20	67.89	61.29	64.59
2	1982	3	0.175	276.35	62.83	66.07	64.45
3	1982	4	0.166	276.81	56.64	69.35	63.00
4	1982	5	0.161	278.08	53.47	68.64	61.06
...	...	...	...	...	...	...	...
2127	2022	48	-1.000	-1.00	-1.00	-1.00	-1.00
2128	2022	49	-1.000	-1.00	-1.00	-1.00	-1.00
2129	2022	50	-1.000	-1.00	-1.00	-1.00	-1.00
2130	2022	51	-1.000	-1.00	-1.00	-1.00	-1.00
2131	2022	52	-1.000	-1.00	-1.00	-1.00	-1.00

2132 rows × 7 columns

Step 5: Transfer Week data into WK1 through WL_52

Convert WK data into WK 1 though WK 52.

```
week_dict = { 1 : "wk1",
        2: "wk2",
        3: "wk3",
        4: "wk4",
        5: "wk5",
        6: "wk6",
        7 : "wk7",
        8: "wk8",
        9: "wk9",
        10: "wk10",
        11: "wk11",
        12: "wk12",
        13 : "wk13",
        14: "wk14",
        15: "wk15",
        16: "wk16",
        17: "wk17",
        18: "wk18",
19 : "wk19",
        20: "wk20",
        21: "wk21",
        22: "wk22",
        23: "wk23",
        24: "wk24",
        25 : "wk25",
        26: "wk26",
        27: "wk27",
        28: "wk28",
        29: "wk29",
        30: "wk30",
        31 : "wk31",
        32: "wk32",
        33: "wk33",
        34: "wk34",
        35: "wk35",
        36: "wk36",
37 : "wk37",
        38: "wk38",
        39: "wk39",
        40: "wk40",
        41: "wk41",
        42: "wk42",
        43 : "wk43",
        44: "wk44",
        45: "wk45",
        46: "wk46",
        47: "wk47",
        48: "wk48",
        49 : "wk49",
        50: "wk50",
        51: "wk51",
        52: "wk52"

        }
```

```
dfMODISMeandataforUSAProvince17Kansas["week"] =
dfMODISMeandataforUSAProvince17Kansas["week"].apply(lambda x : week_dict[x])
```

```
dfMODISMeandataforUSAProvince17Kansas
```

Output:

	Year	week	SMN	SMT	VCI	TCI	VHI
0	1982	wk1	0.193	276.46	71.30	54.53	62.91
1	1982	wk2	0.184	276.20	67.89	61.29	64.59
2	1982	wk3	0.175	276.35	62.83	66.07	64.45
3	1982	wk4	0.166	276.81	56.64	69.35	63.00
4	1982	wk5	0.161	278.08	53.47	68.64	61.06
...	...	...	...	...	...	...	...
2127	2022	wk48	-1.000	-1.00	-1.00	-1.00	-1.00
2128	2022	wk49	-1.000	-1.00	-1.00	-1.00	-1.00
2129	2022	wk50	-1.000	-1.00	-1.00	-1.00	-1.00
2130	2022	wk51	-1.000	-1.00	-1.00	-1.00	-1.00
2131	2022	wk52	-1.000	-1.00	-1.00	-1.00	-1.00

2132 rows × 7 columns

Step 6: Extract Vegetation Health Index (VHI) Data from NOAA STAR for Kansas Data

Extract Vegetation Health Index (VHI) data. Copy VHI Data. Drop other columns – SMN, SMT, and VCI. Finally, PIVOT table to transfer into Week 1 – Week 52.

```
dfKansasMODISVHI_Weekly=dfMODISMeandataforUSAProvince17Kansas.copy()
```

```
dfKansasMODISVHI_Weekly = dfKansasMODISVHI_Weekly.drop([' SMN','SMT','VCI'], axis=1)
dfKansasMODISVHI_Weekly
```

```
dfKansasMODISVHI_Weekly = dfKansasMODISVHI_Weekly.pivot_table(' VHI', ['Year'], 'week')
dfKansasMODISVHI_Weekly
```

Output:

week	wk1	wk10	wk11	wk12	wk13	wk14	wk15	wk16	wk17	wk18	...	wk48	wk49	wk5	wk50	wk51	wk52	wk6	wk7	wk8	wk9
Year																					
1982	62.91	55.58	53.74	56.20	56.90	57.54	57.93	60.17	62.37	64.46	...	43.10	40.10	61.06	40.19	39.64	37.84	59.19	57.36	57.61	57.23
1983	38.58	54.95	59.07	62.39	63.65	65.46	66.47	67.93	67.77	67.63	...	56.62	55.86	43.13	54.46	52.27	50.16	44.86	46.30	48.16	51.07
1984	51.67	54.60	54.08	55.95	57.34	59.30	62.92	65.96	68.18	70.28	...	45.02	43.43	52.92	-1.00	-1.00	-1.00	55.11	55.14	55.69	56.48
1985	-1.00	49.04	50.14	51.31	51.79	52.81	53.95	55.22	56.56	58.08	...	45.66	42.08	-1.00	40.48	38.94	37.93	-1.00	-1.00	-1.00	49.66
1986	36.69	44.10	45.71	48.30	50.44	50.56	50.72	51.80	51.76	50.02	...	43.67	41.55	39.53	38.87	36.69	36.37	41.15	42.11	42.52	42.70
1987	37.45	52.61	51.64	51.05	51.43	52.25	52.06	50.68	50.00	50.31	...	46.54	43.57	43.95	42.31	43.31	44.22	46.66	48.97	49.64	50.66
1988	45.73	40.70	40.90	43.34	45.69	48.29	51.53	54.60	58.18	56.37	...	40.66	40.30	48.08	40.91	41.18	42.04	45.68	44.48	43.01	42.07
1989	42.26	34.38	28.48	23.96	21.21	19.53	17.80	17.77	18.92	19.23	...	43.86	41.61	42.55	39.11	36.86	35.49	43.51	44.09	43.44	40.06
1990	34.42	53.03	44.47	45.53	47.11	48.78	50.49	53.44	57.17	58.90	...	39.77	39.50	40.28	39.26	40.22	42.37	40.84	41.11	41.09	40.46
1991	44.23	35.20	34.75	34.22	34.82	36.55	37.33	40.40	43.81	44.29	...	28.48	28.79	45.24	28.23	27.96	28.00	44.35	41.69	39.40	37.69
1992	28.71	41.90	41.69	44.25	46.46	48.77	49.68	52.64	54.27	54.33	...	48.38	46.80	28.11	45.25	43.34	42.22	28.81	30.16	34.01	39.06
1993	42.93	49.68	48.69	49.78	51.99	54.36	57.77	61.48	64.51	67.58	...	45.29	42.51	43.08	41.35	40.84	39.74	42.99	44.41	46.30	48.78

The same transformation needs to be performed on the VHI and TCI columns.

Step 7: Merge Yield and NOAA STAR - Global Vegetation Health Products: Province-Averaged VH

Merge data frames: Wheat Production, TCI, VHI into one final data frame.

```
df_merged = dfAverageKansasWheatProduction.merge(dfKansasMODISTCI_Weekly,on= "Year", how= "inner")
df_merged
```

```
df_merged_final=df_merged.merge(dfKansasMODISVHI_Weekly,on= "Year", how= "inner")
df_merged_final
```

```
dfKansasYieldwMODISVHITCI=df_merged_final.copy()
```

```
dfKansasYieldwMODISVHITCI
```

Output:

	Year	Yield (bushels)	FoodSecurity	TCI_WK1	TCI_WK10	TCI_WK11	TCI_WK12	TCI_WK13	TCI_WK14	TCI_WK15	...	VHI_WK48	VHI_WK49	VHI_WK5
0	1982	35.0	Yes	54.53	57.70	50.44	50.46	47.22	44.95	51.52	...	43.10	40.10	61.06
1	1983	41.5	Yes	43.62	89.26	94.80	96.82	96.30	95.50	93.71	...	56.62	55.86	43.13
2	1984	38.5	Yes	85.48	64.22	65.82	70.52	75.74	78.89	84.74	...	45.02	43.43	52.92
3	1985	38.0	Yes	-1.00	23.58	22.08	20.78	19.37	18.83	19.17	...	45.66	42.08	-1.00
4	1986	33.0	Yes	27.12	32.48	30.71	33.40	33.99	31.86	36.26	...	43.67	41.55	39.53
5	1987	37.0	Yes	30.53	44.60	40.13	36.88	37.42	38.41	39.55	...	46.54	43.57	43.95
6	1988	34.0	Yes	57.39	29.94	27.09	28.40	29.52	32.36	40.90	...	40.66	40.30	48.08
7	1989	24.0	No	9.14	51.14	40.04	31.26	26.74	21.14	16.62	...	43.86	41.61	42.55
8	1990	40.0	Yes	8.95	37.06	46.04	43.20	45.62	49.84	56.14	...	39.77	39.50	40.28
9	1991	33.0	Yes	36.43	13.24	10.62	8.42	9.89	14.57	20.12	...	28.48	28.79	45.24
10	1992	34.0	Yes	26.09	31.76	24.91	26.51	31.30	36.61	42.41	...	48.38	46.80	28.11
11	1993	35.0	Yes	74.80	86.71	79.70	74.86	72.36	72.09	75.50	...	45.29	42.51	43.08
12	1994	38.0	Yes	27.24	43.08	45.77	49.24	52.50	58.50	64.96	...	-1.00	-1.00	41.43
13	1995	26.0	No	-1.00	27.42	30.60	33.90	37.19	48.45	69.89	...	27.56	29.64	49.65
14	1996	29.0	No	33.97	37.42	40.17	36.89	29.16	24.15	26.39	...	57.27	54.22	39.19

Step 8: Spilt Data into X & Y Columns

Split Wheat Yield data into Class Label and Feature variables.

```
X = np.array(dfKansasYieldwMODISVHITCI.iloc[:, dfKansasYieldwMODISVHITCI.columns != 'FoodSecurity'])
y = np.array(dfKansasYieldwMODISVHITCI.iloc[:, dfKansasYieldwMODISVHITCI.columns == 'FoodSecurity'])
print('Shape of X: {}'.format(X.shape))
print('Shape of y: {}'.format(y.shape))
```

Output:

```
Shape of X: (39, 106)
Shape of y: (39, 1)
```

Step 9: Apply SMOTE Technique to over sample the Class Labels "Food Security"
Use the SMOTE technique to train and the test data split.

```
from imblearn.over_sampling import SMOTE
from sklearn.model_selection import train_test_split
X_train, X_test, y_train, y_test = train_test_split(X, y, test_size=0.3, random_state=0)
print("Number transactions X_train dataset: ", X_train.shape)
print("Number transactions y_train dataset: ", y_train.shape)
print("Number transactions X_test dataset: ", X_test.shape)
print("Number transactions y_test dataset: ", y_test.shape)
```

Output:

```
Number transactions X_train dataset: (27, 106)
Number transactions y_train dataset: (27, 1)
Number transactions X_test dataset: (12, 106)
Number transactions y_test dataset: (12, 1)
```

Step 10: Apply SMOTE Technique to over sample the Class Labels "Food Security"
Use the SMOTE technique to train and test data split

```
from imblearn.over_sampling import SMOTE
print("Before OverSampling, counts of label '1': {}".format(sum(y_train==1)))
print("Before OverSampling, counts of label '0': {} \n".format(sum(y_train==0)))
#sm = SMOTE(random_state=2)
sm=SMOTE(k_neighbors=2)
X_train_res, y_train_res = sm.fit_resample(X_train, y_train.ravel()) #sm.fit_sample(X_train, y_train.
ravel())
print('After OverSampling, the shape of train_X: {}'.format(X_train_res.shape))
print('After OverSampling, the shape of train_y: {} \n'.format(y_train_res.shape))
print("After OverSampling, counts of label '1': {}".format(sum(y_train_res==1)))
print("After OverSampling, counts of label '0': {}".format(sum(y_train_res==0)))
```

Output:

Please note that to overcome "Expected n_neighbors <= n_samples, but n_samples = 4, n_neighbors = 6" errors, add Neighbors to SMOTE constructor. The nearest neighbors property is used to construct synthetic samples if the object, an estimator that inherits from KNeighborsMixin will be used to find the k_neighbors. For a smaller dataset, lower the number!

```
Before OverSampling, counts of label '1': [0]
Before OverSampling, counts of label '0': [0]
After OverSampling, the shape of train_X: (46, 106)
After OverSampling, the shape of train_y: (46,)
After OverSampling, counts of label '1': 0
After OverSampling, counts of label '0': 0
```

Step 11: Reconstruct Data Frame

```
import pandas as pd
import numpy as np
import pandas as pd
df1 = pd.DataFrame(y_train_res, columns = ['FoodSecurity'])
print(df)
print(type(df1))
```

Construct Feature variables:

```
mport pandas as pd
import numpy as np
import pandas as pd
df2 = pd.DataFrame(X_train_res, columns = ['Year',
 'Yield (bushels)',
 'TCI_WK1',
 'TCI_WK10',
 'TCI_WK11',
 'TCI_WK12',
 'TCI_WK13',
 'TCI_WK14',
 'TCI_WK15',
 'TCI_WK16',
 'TCI_WK17',
 'TCI_WK18',
 'TCI_WK19',
 'TCI_WK2',
 'TCI_WK20',
 'TCI_WK21',
 'TCI_WK22',
 'TCI_WK23',
 'TCI_WK24',
 'TCI_WK25',
 'TCI_WK26',
 'TCI_WK27',
 'TCI_WK28',
 'TCI_WK29',
 'TCI_WK3',
 'TCI_WK30',
 'TCI_WK31',
 'TCI_WK32',
 'TCI_WK33',
 'TCI_WK34',
 'TCI_WK35',
 'TCI_WK36',
 'TCI_WK37',
 'TCI_WK38',
 'TCI_WK39',
 'TCI_WK4',
 'TCI_WK40',
 'TCI_WK41',
 'TCI_WK42',
 'TCI_WK43',
 'TCI_WK44',
 'TCI_WK45',
 'TCI_WK46',
 'TCI_WK47',
 'TCI_WK48',
 'TCI_WK49',
 'TCI_WK5',
 'TCI_WK50',
 'TCI_WK51',
 'TCI_WK52',
 'TCI_WK6',
 'TCI_WK7',
 'TCI_WK8',
 'TCI_WK9',
 'VHI_WK1',
 'VHI_WK10',
 'VHI_WK11',
 'VHI_WK12',
 'VHI_WK13',
 'VHI_WK14',
 'VHI_WK15',
 'VHI_WK16',
 'VHI_WK17',
 'VHI_WK18',
 'VHI_WK19',
 'VHI_WK2',
 'VHI_WK20',
 'VHI_WK21',
 'VHI_WK22',
 'VHI_WK23',
 'VHI_WK24',
 'VHI_WK25',
 'VHI_WK26',
 'VHI_WK27',
 'VHI_WK28',
 'VHI_WK29',
 'VHI_WK3',
 'VHI_WK30',
 'VHI_WK31',
 'VHI_WK32',
 'VHI_WK33',
 'VHI_WK34',
 'VHI_WK35',
 'VHI_WK36',
 'VHI_WK37',
 'VHI_WK38',
 'VHI_WK39',
 'VHI_WK4',
 'VHI_WK40',
 'VHI_WK41',
 'VHI_WK42',
 'VHI_WK43',
 'VHI_WK44',
 'VHI_WK45',
 'VHI_WK46',
 'VHI_WK47',
 'VHI_WK48',
 'VHI_WK49',
 'VHI_WK5',
 'VHI_WK50',
 'VHI_WK51',
 'VHI_WK52',
 'VHI_WK6',
 'VHI_WK7',
 'VHI_WK8',
 'VHI_WK9'])
print(df2)
print(type(df2))
dfSMOTEKansasYieldwMODISVHITCI=df2.merge(df1,left_index=True, right_index=True)
dfSMOTEKansasYieldwMODISVHITCI
```

Output:

	Year	Yield (bushels)	TCI_WK1	TCI_WK10	TCI_WK11	TCI_WK12	TCI_WK13	TCI_WK14	TCI_WK15	TCI_WK16	...	VHI_WK49	VHI_WK5	VHI_WK50
0	1984.000000	38.500000	85.480000	64.220000	65.820000	70.520000	75.740000	78.890000	84.740000	88.100000	...	43.430000	52.920000	-1.000000
1	2020.000000	45.000000	17.300000	45.680000	43.290000	36.180000	32.660000	29.890000	29.160000	30.160000	...	34.870000	45.900000	35.270000
2	2002.000000	33.000000	17.940000	48.530000	51.640000	44.710000	32.260000	27.420000	30.290000	32.530000	...	46.810000	54.570000	46.550000
3	2018.000000	38.000000	13.090000	30.390000	31.240000	30.950000	32.030000	31.860000	32.610000	30.710000	...	42.630000	40.010000	40.990000
4	1998.000000	49.000000	35.000000	49.820000	46.990000	45.260000	42.020000	39.470000	41.210000	44.460000	...	44.370000	43.110000	43.740000
5	2017.000000	48.000000	25.170000	20.570000	23.960000	25.130000	26.220000	28.170000	33.430000	35.540000	...	34.500000	43.920000	34.810000
6	1990.000000	40.000000	8.950000	37.060000	46.040000	43.200000	45.620000	49.840000	56.140000	60.520000	...	39.500000	40.280000	39.260000
7	1995.000000	26.000000	-1.000000	27.420000	30.600000	33.900000	37.190000	48.450000	69.890000	83.280000	...	29.640000	49.650000	32.900000
8	1987.000000	37.000000	30.530000	44.600000	40.130000	36.880000	37.420000	38.410000	39.550000	33.810000	...	43.570000	43.950000	42.310000
9	1999.000000	47.000000	21.450000	42.240000	45.470000	46.300000	49.420000	54.710000	62.430000	70.100000	...	41.570000	47.070000	43.000000
10	1996.000000	29.000000	33.970000	37.420000	40.170000	36.890000	29.160000	24.150000	26.390000	29.690000	...	54.220000	39.190000	53.450000
11	2014.000000	28.000000	19.390000	36.400000	33.810000	28.630000	26.270000	24.740000	24.980000	25.020000	...	43.470000	44.900000	42.820000

Step 12: Inspect Data Frame

```
pd.value_counts(dfSMOTEKansasYieldwMODISVHITCI['FoodSecurity']).plot.bar()
plt.title('Food Security class histogram')
plt.xlabel('Class')
plt.ylabel('Frequency')
dfSMOTEKansasYieldwMODISVHITCI['FoodSecurity'].value_counts()
```

Output:

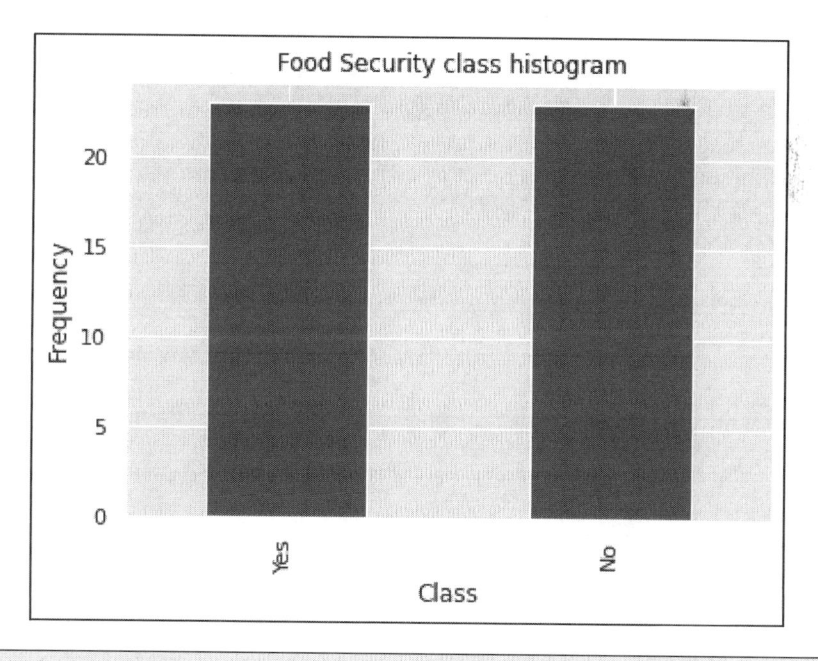


```
show_density('Yield (bushels)',
 dfSMOTEKansasYieldwMODISVHITCI['Yield (bushels)'])
```

Output:

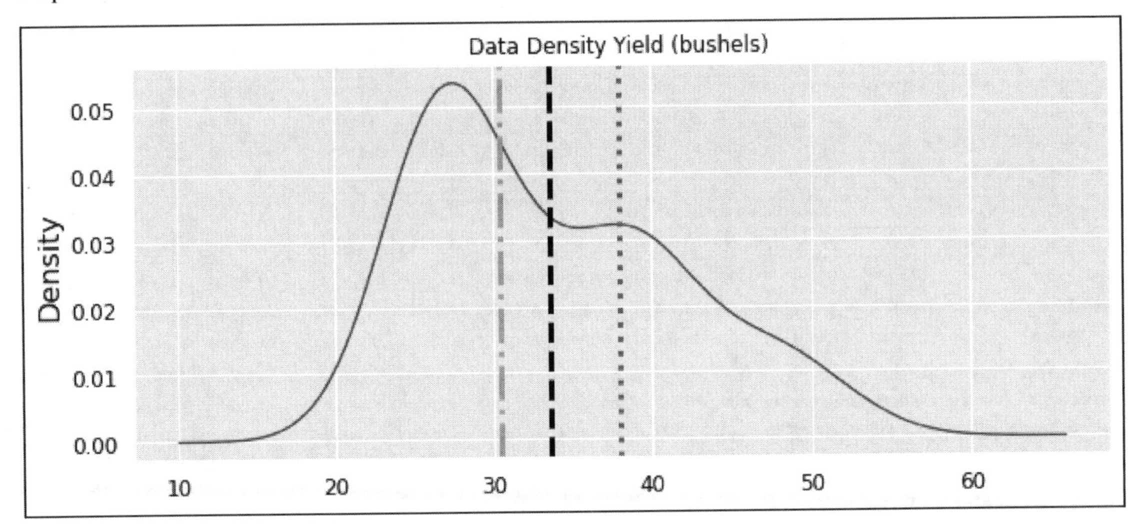

Data Distribution after SMOTE with the following details: Minimum: 24.00; Mean: 33.63 (dashed line); Median: 30.50 (dash dot line); Mode: 38.00 (dotted line); Maximum: 52.00; Range: 28.00; Variance: 61.88; Standard Deviation: 7.87. Many statistical properties have changed after SMOTE.

```
dfSMOTEKansasYieldwMODISVHITCI.boxplot(column='Yield (bushels)', by='FoodSecurity', figsize=(8,5))
```

Output:

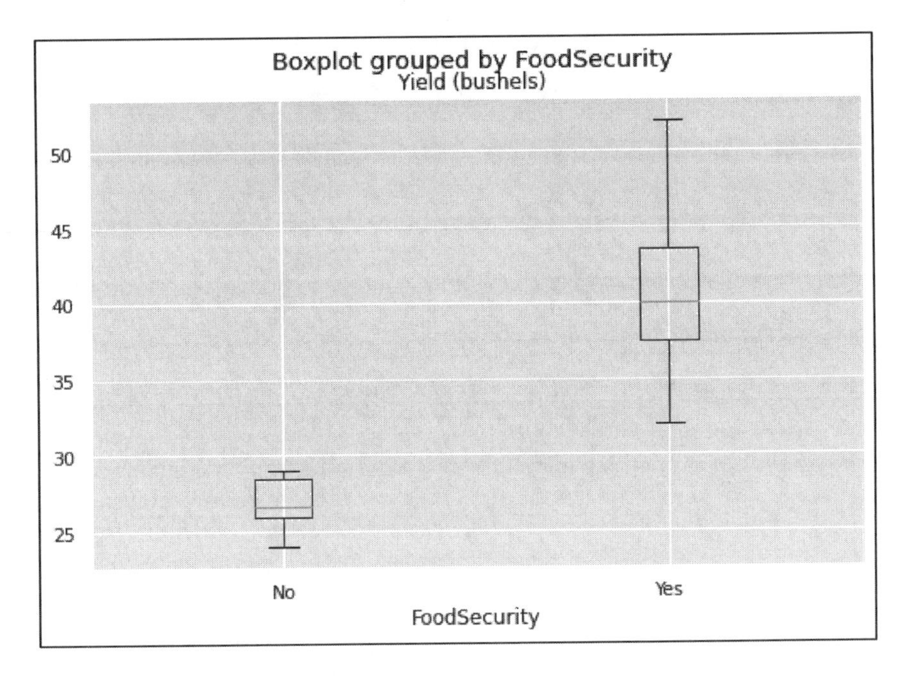

Machine Learning Model: Kansas Wheat Yield with Adaptive Synthetic (ADASYN) Sampling

The essential idea of Adaptive synthetic (ADASYN) is to use a weighted distribution for different minority class examples according to their learning difficulty level, where more synthetic data is generated for minority class examples that are harder to learn compared to those minority examples that are easier to learn; we can see this when sampling the following wheat dataset. As a result, the ADASYN approach improves learning with respect to the data distributions in two ways: (1) reducing the bias introduced by the class imbalance, and (2) adaptively shifting the classification decision boundary towards the difficult examples [17]. This method is like SMOTE, but it generates different sample numbers depending on an estimate of the local distribution of the class to be oversampled. In the following code, our goal is to use ADASYN technique to resample under-represented class labels to improve predictability of Food Security [18, 19].

The data for ADASYN is very similar to that of SMOTE—so we will reuse steps till 9.

Step 10: Apply ADASYN Technique to over sample the Class Labels "Food Security"

Use SMOTE technique to train and test data split.

```
from imblearn.over_sampling import SMOTE

print("Before OverSampling, counts of label '1': {}".format(sum(y_train==1)))
print("Before OverSampling, counts of label '0': {} \n".format(sum(y_train==0)))

#sm = SMOTE(random_state=2)
sm = ADASYN(n_neighbors=2)
#sm=SMOTE(k_neighbors=2)
X_train_res, y_train_res = sm.fit_resample(X_train, y_train.ravel()) #sm.fit_sample(X_train, y_train.
ravel())

print('After OverSampling, the shape of train_X: {}'.format(X_train_res.shape))
print('After OverSampling, the shape of train_y: {} \n'.format(y_train_res.shape))

print("After OverSampling, counts of label '1': {}".format(sum(y_train_res==1)))
print("After OverSampling, counts of label '0': {}".format(sum(y_train_res==0)))
```

Output:

Please note that to overcome "Expected n_neighbors <= n_samples, but n_samples = 4, n_neighbors = 6" errors, add Neighbors to SMOTE constructor. The nearest neighbor's property is to construct synthetic samples if object, an estimator that inherits from KNeighborsMixin that will be used to find the k_neighbors. For smaller datasets, lower the number!

```
Before OverSampling, counts of label '1': [0]
Before OverSampling, counts of label '0': [0]

After OverSampling, the shape of train_X: (47, 106)
After OverSampling, the shape of train_y: (47,)

After OverSampling, counts of label '1': 0
After OverSampling, counts of label '0': 0
```

ADASYN generates one more extra underrepresented class label.

Step 11: Reconstruct Data Frame

```
import pandas as pd
import numpy as np
import pandas as pd

df1 = pd.DataFrame(y_train_res, columns = ['FoodSecurity'])

print(df)
print(type(df1))
```

Construct Feature variables:

```
mport pandas as pd
import numpy as np
import pandas as pd
df2 = pd.DataFrame(X_train_res, columns =
[‘Year’,
 ‘Yield (bushels)’,
 ‘TCI_WK1’,
 ‘TCI_WK10’,
 ‘TCI_WK11’,
 ‘TCI_WK12’,
 ‘TCI_WK13’,
 ‘TCI_WK14’,
 ‘TCI_WK15’,
 ‘TCI_WK16’,
 ‘TCI_WK17’,
 ‘TCI_WK18’,
 ‘TCI_WK19’,
 ‘TCI_WK2’,
 ‘TCI_WK20’,
 ‘TCI_WK21’,
 ‘TCI_WK22’,
 ‘TCI_WK23’,
 ‘TCI_WK24’,
 ‘TCI_WK25’,
 ‘TCI_WK26’,
 ‘TCI_WK27’,
 ‘TCI_WK28’,
 ‘TCI_WK29’,
 ‘TCI_WK3’,
 ‘TCI_WK30’,
 ‘TCI_WK31’,
 ‘TCI_WK32’,
 ‘TCI_WK33’,
‘TCI_WK34’,
 ‘TCI_WK35’,
 ‘TCI_WK36’,
 ‘TCI_WK37’,
 ‘TCI_WK38’,
 ‘TCI_WK39’,
 ‘TCI_WK4’,
 ‘TCI_WK40’,
 ‘TCI_WK41’,
‘TCI_WK42’,
‘TCI_WK43’,
‘TCI_WK44’,
‘TCI_WK45’,
‘TCI_WK46’,
‘TCI_WK47’,
‘TCI_WK48’,
‘TCI_WK49’,
‘TCI_WK5’,
‘TCI_WK50’,
‘TCI_WK51’,
‘TCI_WK52’,
‘TCI_WK6’,
‘TCI_WK7’,
‘TCI_WK8’,
‘TCI_WK9’,
‘VHI_WK1’,
‘VHI_WK10’,
‘VHI_WK11’,
‘VHI_WK12’,
‘VHI_WK13’,
‘VHI_WK14’,
‘VHI_WK15’,
‘VHI_WK16’,
‘VHI_WK17’,
‘VHI_WK18’,
‘VHI_WK19’,
‘VHI_WK2’,
‘VHI_WK20’,
‘VHI_WK21’,
‘VHI_WK22’,
‘VHI_WK23’,
‘VHI_WK24’,
‘VHI_WK25’,
‘VHI_WK26’,
‘VHI_WK27’,
‘VHI_WK28’,
‘VHI_WK29’,
‘VHI_WK3’,
‘VHI_WK30’,
‘VHI_WK31’,
‘VHI_WK32’,
‘VHI_WK33’,
‘VHI_WK34’,
‘VHI_WK35’,
‘VHI_WK36’,
‘VHI_WK37’,
‘VHI_WK38’,
‘VHI_WK39’,
‘VHI_WK4’,
‘VHI_WK40’,
‘VHI_WK41’,
‘VHI_WK42’,
‘VHI_WK43’,
‘VHI_WK44’,
‘VHI_WK45’,
‘VHI_WK46’,
‘VHI_WK47’,
‘VHI_WK48’,
‘VHI_WK49’,
‘VHI_WK5’,
‘VHI_WK50’,
‘VHI_WK51’,
‘VHI_WK52’,
‘VHI_WK6’,
‘VHI_WK7’,
‘VHI_WK8’,
‘VHI_WK9’])
```

```
print(df2)
print(type(df2))
```

```
dfSMOTEKansasYieldwMODISVHITCI=df2.merge(df1,left_index=True, right_index=True)
dfSMOTEKansasYieldwMODISVHITCI
```

Output:

	Year	Yield (bushels)	TCI_WK1	TCI_WK10	TCI_WK11	TCI_WK12	TCI_WK13	TCI_WK14	TCI_WK15	TCI_WK16	...	VHI_WK49	VHI_WK5	VHI_WK50	VHI_WK51	VHI_WK52
0	1984.000000	38.500000	85.480000	64.220000	65.820000	70.520000	75.740000	78.890000	84.740000	88.100000	...	43.430000	52.920000	-1.000000	-1.000000	-1.000000
1	2020.000000	45.000000	17.300000	45.680000	43.290000	36.180000	32.660000	29.890000	29.160000	30.160000	...	34.870000	45.900000	35.270000	36.000000	36.920000
2	2002.000000	33.000000	17.940000	48.530000	51.640000	44.710000	32.260000	27.420000	30.290000	32.530000	...	46.810000	54.570000	46.550000	45.890000	46.740000
3	2018.000000	38.000000	13.090000	30.390000	31.240000	30.950000	32.030000	31.860000	32.610000	30.710000	...	42.630000	40.010000	40.990000	40.350000	40.860000
4	1998.000000	49.000000	35.000000	49.820000	46.990000	45.260000	42.020000	39.470000	41.210000	44.460000	...	44.370000	43.110000	43.740000	44.630000	45.420000
5	2017.000000	48.000000	25.170000	20.570000	23.960000	25.130000	26.220000	28.170000	33.430000	35.540000	...	34.500000	43.920000	34.810000	35.590000	36.340000
6	1990.000000	40.000000	8.950000	37.060000	46.040000	43.200000	45.620000	49.840000	56.140000	60.520000	...	39.500000	40.280000	39.260000	40.220000	42.370000
7	1995.000000	26.000000	-1.000000	27.420000	30.600000	33.900000	37.190000	48.450000	69.890000	83.280000	...	29.640000	49.650000	32.900000	33.830000	33.090000
8	1987.000000	37.000000	30.530000	44.600000	40.130000	36.880000	37.420000	38.410000	39.550000	33.810000	...	43.570000	43.950000	42.310000	43.310000	44.220000
9	1999.000000	47.000000	21.450000	42.240000	45.470000	46.300000	49.420000	54.710000	62.430000	70.100000	...	41.570000	47.070000	43.000000	42.520000	41.620000
10	1996.000000	29.000000	33.970000	37.420000	40.170000	36.890000	29.160000	24.150000	26.390000	29.690000	...	54.220000	39.190000	53.450000	52.900000	52.750000
11	2014.000000	28.000000	19.390000	36.400000	33.810000	28.630000	26.270000	24.740000	24.980000	25.020000	...	43.470000	44.900000	42.820000	41.640000	41.270000
12	1989.000000	24.000000	9.140000	51.140000	40.040000	31.260000	26.740000	21.140000	16.620000	15.490000	...	41.610000	42.550000	39.110000	36.860000	35.490000
13	2013.000000	38.000000	25.110000	56.570000	54.200000	49.510000	46.730000	48.040000	54.690000	58.090000	...	42.610000	43.520000	40.750000	40.330000	40.000000

Step 12: Inspect Data Frame

```
pd.value_counts(dfSMOTEKansasYieldwMODISVHITCI['FoodSecurity']).plot.bar()
plt.title('Food Security class histogram')
plt.xlabel('Class')
plt.ylabel('Frequency')
dfSMOTEKansasYieldwMODISVHITCI['FoodSecurity'].value_counts()
```

Output:

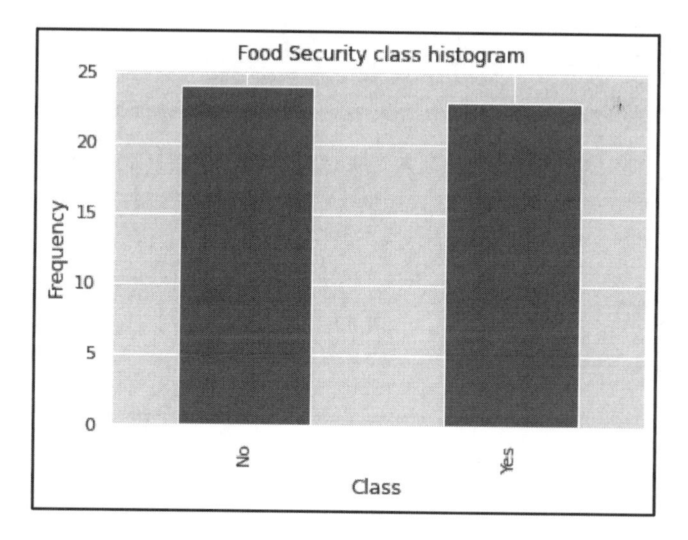


```
show_density('Yield (bushels)',
 dfSMOTEKansasYieldwMODISVHITCI['Yield (bushels)'])
```

Output:

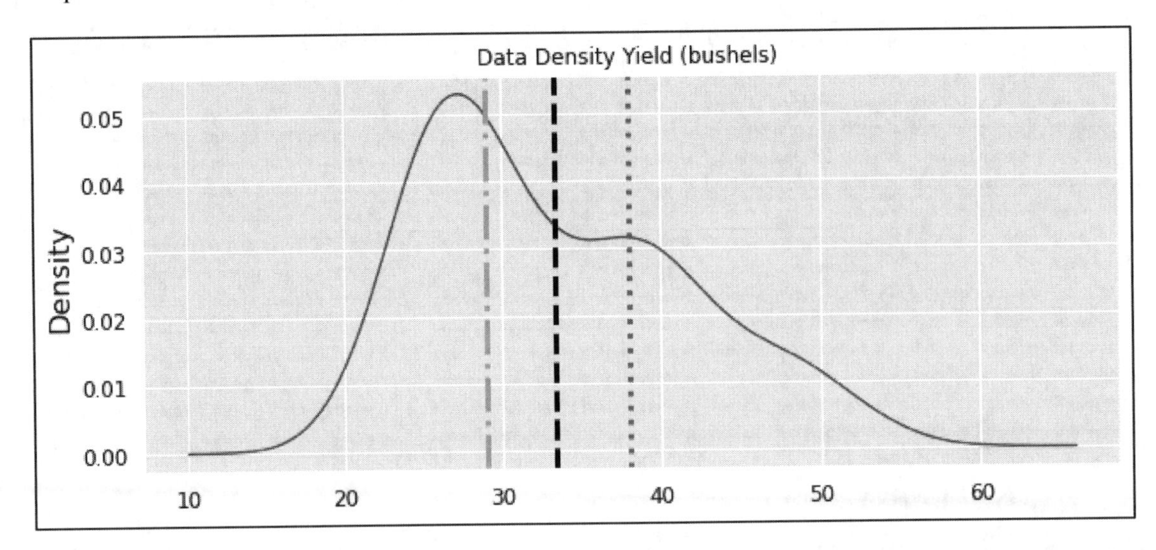

Data Distribution after ADASYN with the following details: Minimum: 24.00; Mean: 33.40 (dashed line); Median: 29.00 (dash dot line); Mode: 38.00 (dotted line); Maximum: 52.00; Range: 28.00; Variance: 63.00; Standard Deviation: 7.94. Most of the statistical properties have changed after ADASYN.

```
dfSMOTEKansasYieldwMODISVHITCI.boxplot(column='Yield (bushels)', by='FoodSecurity', figsize=(8,5))
```

Output:

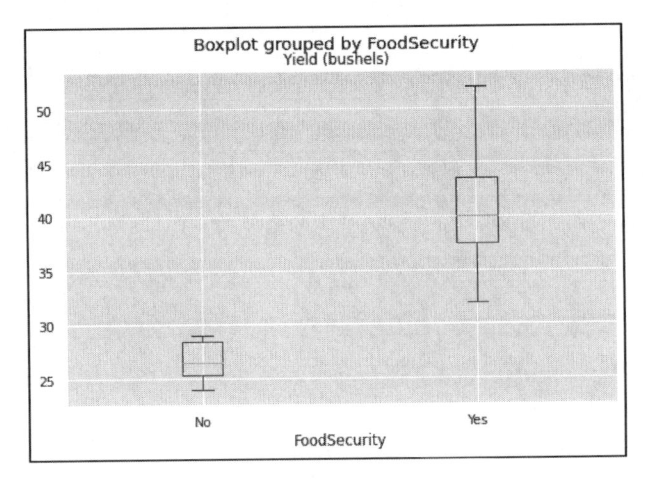

The key difference between ADASYN and SMOTE is that ADASYN uses a density distribution, as a criterion to automatically decide the number of synthetic samples that must be generated for each minority sample by adaptively changing the weights of the different minority samples to compensate for the skewness [20].

Original Data Data Distribution with the following details: Minimum: 24.00; Mean: 38.59 (dashed line); Median: 38.00 (dash dot line); Mode: 33.00 (dotted line); Maximum: 57.00; Range: 33.00; Variance: 49.68; Standard Deviation: 7.05.	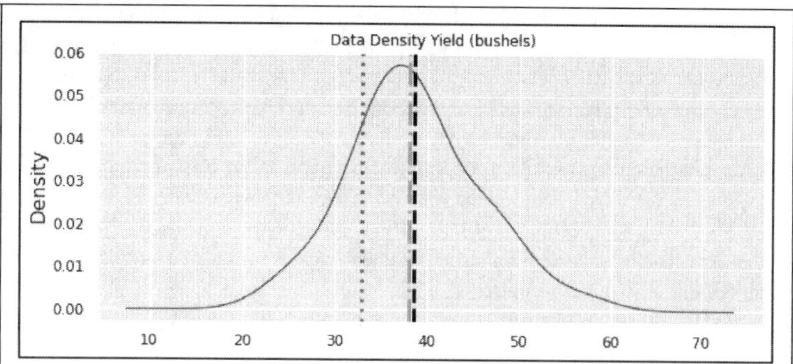
SMOTE Minimum: 24.00; Mean: 33.63 (dashed line); Median: 30.50 (dash dot line); Mode: 38.00 (dotted line); Maximum: 52.00; Range: 28.00; Variance: 61.88; Standard Deviation: 7.87. Many of statistical properties have changed after SMOTE.	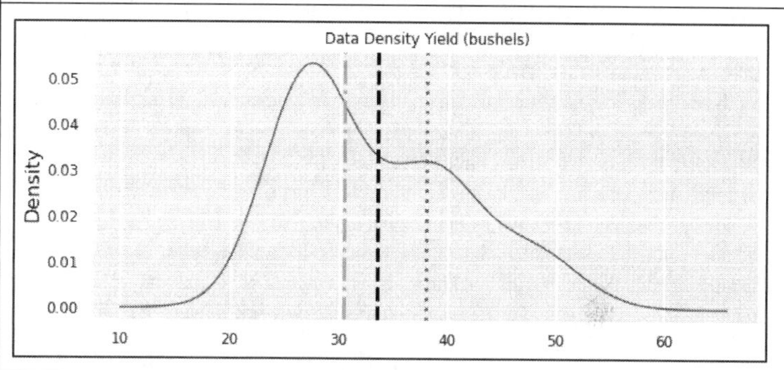
ADASYN Minimum: 24.00; Mean: 33.40 (dashed line); Median: 29.00 (dash dot line); Mode: 38.00 (dotted line); Maximum: 52.00; Range: 28.00; Variance: 63.00; Standard Deviation: 7.94	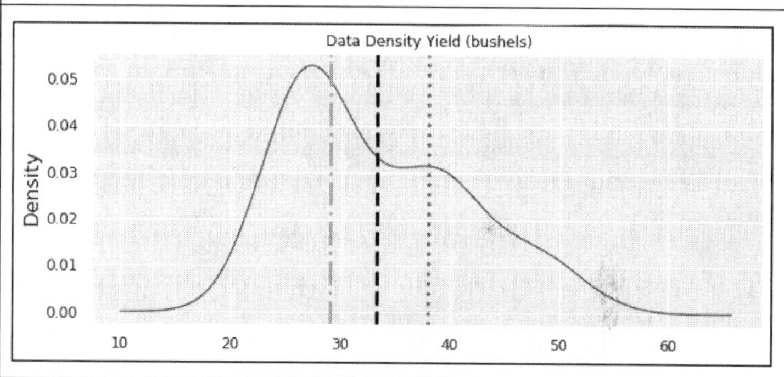

In the above model, increasing the food security label to 47 reduced the bias introduced by the class imbalance and adaptively shifted the classification decision boundary toward the difficult samples.

> After reading this chapter, you should be able to explain time series data enrichment strategies and imputation strategies. You should also be able to explain Multivariate Imputation by Chained Equation (MICE) Strategy and Synthetic Minority Oversampling Technique (SMOTE) & Adaptive Synthetic (ADASYN) techniques. Finally you should be able to apply the Synthetic Minority Oversampling Technique (SMOTE) & Adaptive Synthetic (ADASYN) techniques on food security Kansas wheat yield data.

References

[1] THE WHITE HOUSE, Executive Order on Ensuring a Data-Driven Response to COVID-19 and Future High-Consequence Public Health Threats, January 21, 2021, https://www.whitehouse.gov/briefing-room/presidential-actions/2021/01/21/executive-order-ensuring-a-data-driven-response-to-covid-19-and-future-high-consequence-public-health-threats/, Access Date: March 18, 2022.

[2] FAO, FAO's Work on Climate Change – Conference 2017, 2017, https://www.fao.org/3/i8037e/i8037e.pdf, Access Date: September 18, 2020.

[3] Vuppalapati, C., Vuppalapati, R., Kedari, S., Ilapakurti, A., Vuppalapati, J.S. and Kedari, S. 2018. Fuzzy logic infused intelligent scent dispenser for creating memorable customer experience of long-tail connected venues. International Conference on Machine Learning and Cybernetics (ICMLC), pp. 149–154, doi: 10.1109/ICMLC.2018.8527046.

[4] Chandrasekar, Vuppalapati. 2021. Machine Learning and Artificial Intelligence for Agricultural Economics: Prognostic Data Analytics to Serve Small Scale Farmers Worldwide, Publisher : Springer; 1st ed. 2021 edition (October 5, 2021), ISBN-13 : 978-3030774844.
[5] Ann, E. 2022. Stapleton, Data Science for Food and Agricultural Systems (DSFAS), https://nifa.usda.gov/grants/programs/data-science-food-agricultural-systems-dsfas, Access Date: February 27, 2022.
[6] T.W. Schultz Distortions of Agricultural Incentives. 1978. Relationship between milk production and price variations in the EC, July 1981, http://aei.pitt.edu/36837/1/A67.pdf, Access Date: November 22, 2019.
[7] Harshad, Hegde, Neel Shimpi, Aloksagar Panny, Ingrid Glurich, Pamela Christie and Amit Acharya. 2019. MICE vs. PPCA: Missing data imputation in healthcare, Informatics in Medicine Unlocked, Volume 17, 100275, ISSN 2352-9148, https://doi.org/10.1016/j.imu.2019.100275.
[8] Stef van Buuren and Karin Groothuis-Oudshoorn, mice: Multivariate Imputation by Chained Equations in R, December 2011, https://www.jstatsoft.org/article/view/v045i03/v45i03.pdf, Access Date: January 08, 2022.
[9] Melissa, J. Azur, Elizabeth A. Stuart. Constantine Frangakis, and Philip J. Leaf, Multiple imputation by chained equations: what is it and how does it work? 2011 Mar, https://www.ncbi.nlm.nih.gov/pmc/articles/PMC3074241/, Access Date: May 2020.
[10] Raghav Agrawal, 2022 Interpolation – Power of Interpolation in Python to fill Missing Values, Published on June 1, 2021 and Last Modified On July 22nd, https://www.analyticsvidhya.com/blog/2021/06/power-of-interpolation-in-python-to-fill-missing-values/, Access Date: March 01, 2022.
[11] Jeremy Chow, Using the Pandas "Resample" Function, Sep 10, 2019, https://towardsdatascience.com/using-the-pandas-resample-function-a231144194c4, Access Date: April 30, 2022.
[12] James, Ho, Time Series Data Analysis - Resample, Jan 13, 2021, https://towardsdatascience.com/time-series-data-analysis-resample-1ff2224edec9, Access Date: April 30, 2022.
[13] Zack Adkins, BREAD WHEAT, 2002, https://www.fao.org/3/y4011e/y4011e00.htm#Contents, Access Date: May 26, 2022.
[14] William Robbins, Lingering Drought Stunts Wheat Crop, May 12, 1989, https://www.nytimes.com/1989/05/12/us/lingering-drought-stunts-wheat-crop.html, Access Date: May 26, 2022.
[15] CNN, Wheat Belt drought taking heavy toll, April 26, 1996, http://www.cnn.com/US/9604/26/kansas.wheat.woes/index.html, Access Date: May 26, 2022.
[16] Kansas State University. Drought, poor wheat harvest in Kansas has effects on national economy, says climatologist. ScienceDaily. ScienceDaily, 10 July 2014, Access Date: May 26, 2022.
[17] Haibo, He, Yang Bai, E.A. Garcia and Shutao Li. ADASYN: Adaptive synthetic sampling approach for imbalanced learning," 2008 IEEE International Joint Conference on Neural Networks (IEEE World Congress on Computational Intelligence), 2008, pp. 1322–1328, doi: 10.1109/IJCNN.2008.4633969.
[18] Dr. Mukhisa Kituyi, Secretary-General of UNCTAD, Trade wars are huge threats to food security, 22 January 2020, https://unctad.org/news/trade-wars-are-huge-threats-food-security, Access Date: March 01, 2022.
[19] Hussein Mousa and Mark Ford. Food and Agricultural Import Regulations and Standards Country Report, January 03, 2022, https://apps.fas.usda.gov/newgainapi/api/Report/DownloadReportByFileName?fileName=Food%20and%20Agricultural%20Import%20Regulations%20and%20Standards%20Country%20Report_Riyadh_Saudi%20Arabia_12-31-2021.pdf, Access Date: March 10, 2022.
[20] Jiawei Han, Micheline Kamber and Jian Pei. Data Mining: Concepts and Techniques, Publisher: Morgan Kaufmann; 3 edition (June 15, 2011), ISBN-10: 9780123814791.

[92] Executive Order on Ensuring a Data-Driven Response to COVID-19 and Future High-Consequence Public Health Threats - https://www.whitehouse.gov/briefing-room/presidential-actions/2021/01/21/executive-order-ensuring-a-data-driven-response-to-covid-19-and-future-high-consequence-public-health-threats/

[93] The World Bank - https://data.worldbank.org/about

[94] Haiti: Acute Food Insecurity National Situation in June - July 2013 and Projection for Southeastern Department in September 2013 - https://www.ipcinfo.org/ipc-country-analysis/details-map/en/c/459529/?iso3=HTI

[95] Honduras: Acute Food Insecurity Situation in Southern Honduras December 2012 - January 2013 - https://www.ipcinfo.org/ipc-country-analysis/details-map/en/c/459515/?iso3=HND

[96] Hunger and Food Security - https://www.fao.org/hunger/en/

[97] Egypt's Bread Crisis Awakens Old Fears of Political Unrest - https://www.wsj.com/articles/egypts-bread-crisis-wheat-supplies-russia-ukrain-warawakens-old-fears-of-political-unrest-11653318765
[98] Food Security - https://www.cdc.gov/climateandhealth/effects/food_security.htm
[99] Agriculture and Food Security - https://www.usaid.gov/what-we-do/agriculture-and-food-security
[100] FAOSTAT Food Security - https://www.fao.org/faostat/en/#data/FS
[101] NHANES - https://www.cdc.gov/nchs/nhanes/index.htm
[102] Food Insecurity prevalence in US - https://www.ers.usda.gov/media/xtddtqat/trends.xlsx
[103] 2015-2016 Data Documentation, Codebook, and Frequencies Food Security (FSQ_I) - https://wwwn.cdc.gov/Nchs/Nhanes/2015-2016/FSQ_I.htm
[104] FAO'S WORK ON CLIMATE CHANGE - https://www.fao.org/3/i8037e/i8037e.pdf
[105] A Data Scientist's Toolkit to Encode Categorical Variables to Numeric - https://towardsdatascience.com/a-data-scientists-toolkit-to-encode-categorical-variables-to-numeric-d17ad9fae03f
[106] Agriculture & Rural Development - https://data.worldbank.org/topic/agriculture-and-rural-development
[107] Arable land - https://data.worldbank.org/indicator/AG.LND.ARBL.ZS
[108] Arable land (hectares per person) - https://data.worldbank.org/indicator/AG.LND.ARBL.HA.PC
[109] Cereal yield (kg per hectare) - https://data.worldbank.org/indicator/AG.YLD.CREL.KG
[110] Rural population (% of total population) - https://data.worldbank.org/topic/agriculture-and-rural-development
[111] Demographic and economic trends in urban, suburban and rural communities - https://www.pewresearch.org/social-trends/2018/05/22/demographic-and-economic-trends-in-urban-suburban-and-rural-communities/
[112] New Census Data Show Differences Between Urban and Rural Populations - https://www.census.gov/newsroom/press-releases/2016/cb16-210.html#:~:text=%E2%80%9CRural%20areas%20cover%2097%20percent,Census%20Bureau%20Director%20John%20H.
[113] Fertilizer Use and Price - https://www.ers.usda.gov/data-products/fertilizer-use-and-price/
[114] Commodity Costs and Returns - https://www.ers.usda.gov/data-products/commodity-costs-and-returns/
[115] All fertilizer use and price tables in a single workbook - https://www.ers.usda.gov/webdocs/DataFiles/50341/fertilizeruse.xls?v=8084.1
[116] Feature Engineering-Handling Missing Numeric Data with Python - https://medium.datadriveninvestor.com/feature-engineering-handling-missing-numeric-data-with-python-7be1e871f85e
[117] Missing Data: Our View of the State of the Art - https://pubmed.ncbi.nlm.nih.gov/12090408/
[118] Imputation of missing values - https://scikit-learn.org/stable/modules/impute.html
[119] MICE vs PPCA: Missing data imputation in healthcare - https://www.sciencedirect.com/science/article/pii/S2352914819302783
[120] Handle Missing Values in Time Series For Beginners - https://www.kaggle.com/code/juejuewang/handle-missing-values-in-time-series-for-beginners/report
[121] A Complete Guide to Dealing with Missing values in Python - https://www.analyticsvidhya.com/blog/2021/10/a-complete-guide-to-dealing-with-missing-values-in-python/
[122] Explain Forward and Backward Filling - https://datascience.stackexchange.com/questions/57776/explain-forward-filling-and-backward-filling-data-filling
[123] Interpolation – Power of Interpolation in Python to fill Missing Values - https://www.analyticsvidhya.com/blog/2021/06/power-of-interpolation-in-python-to-fill-missing-values/
[124] Surging Fertilizer Costs Are Pushing Food Prices Higher - https://www.barrons.com/articles/fertilizer-costs-food-prices-51650524400
[125] SimpleImputer - https://scikit-learn.org/stable/modules/generated/sklearn.impute.SimpleImputer.html
[126] sklearn SimpleImputer too slow for categorical data represented as string values - https://datascience.stackexchange.com/questions/66034/sklearn-simpleimputer-too-slow-for-categorical-data-represented-as-string-values
[127] 6 Different Ways to Compensate for Missing Values In a Dataset (Data Imputation with examples) - https://towardsdatascience.com/6-different-ways-to-compensate-for-missing-values-data-imputation-with-examples-6022d9ca0779
[128] Source code for impyute.imputation.cs.fast_knn - https://impyute.readthedocs.io/en/master/_modules/impyute/imputation/cs/fast_knn.html
[129] MICE - https://www.jstatsoft.org/article/view/v045i03/v45i03.pdf
[130] Multiple imputation by chained equations: what is it and how does it work? - https://www.ncbi.nlm.nih.gov/pmc/articles/PMC3074241/
[131] World Economic Outlook Database - https://www.imf.org/en/Publications/WEO/weo-database/2021/October/download-entire-database
[132] Time Series Data Analysis - Resample - https://towardsdatascience.com/time-series-data-analysis-resample-1ff2224edec9
[133] Using the Pandas "Resample" Function - https://towardsdatascience.com/using-the-pandas-resample-function-a231144194c4
[134] Time Series Resample - https://pandas.pydata.org/docs/reference/api/pandas.DataFrame.resample.html
[135] LBMA Precious Metal Prices - https://www.lbma.org.uk/prices-and-data/precious-metal-prices#/
[136] Silver Prices - https://www.lbma.org.uk/prices-and-data/precious-metal-prices#/
[137] Market Yield on U.S. Treasury Securities at 10-Year Constant Maturity - https://fred.stlouisfed.org/series/DGS10
[138] S&P Global Data - https://finance.yahoo.com/quote/%5EGSPC/history?period1=-628819200&period2=1650412800&interval=1d&filter=history&frequency=1d&includeAdjustedClose=true
[139] S&P Global Data - https://finance.yahoo.com/quote/%5EGSPC/history?period1=-628819200&period2=1650412800&interval=1d&filter=history&frequency=1d&includeAdjustedClose=true
[140] BREAD WHEAT - https://www.fao.org/3/y4011e/y4011e00.htm#Contents

[141] Lingering Drought Stunts Wheat Crop - https://www.nytimes.com/1989/05/12/us/lingering-drought-stunts-wheat-crop.html
[142] Wheat Belt drought taking heavy toll - http://www.cnn.com/US/9604/26/kansas.wheat.woes/index.html
[143] Drought, poor wheat harvest in Kansas has effects on national economy, says climatologist - https://www.sciencedaily.com/releases/2014/07/140710094340.htm
[144] Kansas Wheat - https://nationalfestivalofbreads.com/nutrition-education/wheat-facts
[145] Satellite Imagery Resources and Usage for the Farm Service Agency, April 2014, https://www.fsa.usda.gov/Internet/FSA_File/satellite_imageryresources.pdf
[146] STAR - Global Vegetation Health Products : Background and Explanation - https://www.star.nesdis.noaa.gov/smcd/emb/vci/VH/VH-Syst_10ap30.php
[147] Satellite Imagery Resources and Usage for the Farm Service Agency - https://www.fsa.usda.gov/Internet/FSA_File/satellite_imageryresources.pdf
[148] Kansas Wheat History - nass.usda.gov/Statistics_by_State/Kansas/Publications/Cooperative_Projects/KS-wheat-history21.pdf
[149] NOAA STAR - https://www.star.nesdis.noaa.gov/smcd/emb/vci/VH/vh_adminMean.php?type=Province_Weekly_MeanPlot
[150] Weekly Crop Land - https://www.star.nesdis.noaa.gov/smcd/emb/vci/VH/get_TS_admin.php?provinceID=17&country=USA&yearlyTag=Weekly&type=Mean&TagCropland=land&year1=1982&year2=2022
[151] Kansas Data - https://www.star.nesdis.noaa.gov/smcd/emb/vci/VH/get_TS_admin.php?provinceID=17&country=USA&yearlyTag=Weekly&type=Mean&TagCropland=crop&year1=1982&year2=2022

Section II: Food Security & Machine Learning

CHAPTER 3

Food Security

The chapter introduces food security, key drivers of food insecurity, and food security indicators & drivers. Additionally, it introduces the IPC Integrated Food Security and Nutrition Conceptual Framework, the USAID Office of Food for Peace Food Security Country Framework, the global food security index (GFSI) framework and the United Nations Suite of Food Security Indicators. Next, it introduces the Food Security Bell Curve Machine Learning Model to enable application on the Prevalence of moderate or severe food insecurity in the Population. Finally, the chapter concludes with Who In the World is Food Insecure with two food security machine learning models that model prevalence of moderate or severe food insecurity and Prevalence of severe food insecurity in the population.

Food is essential for our survival and a timely intake of food is a must to maintain a healthy lifestyle and to have a functioning body and mind. Food Security is an essential aspect of the governance of nations, states, cities, communities, and families. A household is not "food secure" unless it "feels" food secure.[152] Food security exists when all people, always, have physical and economic access to sufficient, safe and nutritious food that meets their dietary needs and food preferences for an active and healthy life.[153,154] The most frequently cited official definition, however, of food security comes from the 1996 United Nation World Food Summit:[155] "Food security[156] means always having both physical and economic access to sufficient food to meet dietary needs for a productive and healthy life". This definition covers four major dimensions of food security: availability, access, utilization, and stability. Prior to the 1996 UN Summit, academicians and policy makers referred to food hunger as lack of food security. Absence of an operational definition has complicated policy effectiveness for any government to undertake investments that citizens see as effective. In most Asian countries, the operational definition of food security has taken the form of domestic price stability relative to world prices. This divergence between domestic prices and world prices, at least on a day-to-day basis, then requires state control over trade flows in commodities, especially stable commodities such as rice and wheat. That is, the state must intervene in commodity marketing. To avoid food insecurity, finances should keep flowing[157] and governmental policies must serve as an effective defense to overcome price spikes.

Therefore, the concept of food security is broader than basic food production, ensuring availability of an adequate food supply. It *encompasses* numerous factors related to major dimensions of food security such as food production, distribution, market access, agricultural financial services, agricultural import tariffs, storage, spatial availability of food, nutritional adequacy, economic and social accessibility, preferences, non-food aspects such as clean water, sanitation and other healthcare services, which improve food utilization by individuals, as well as constancy of adequate food availability even during sudden shocks such as pandemics, wars or any seasonal, cyclical or other unexpected events. The last aspect of food security is paramount important and requires strategic planning. In this chapter, I would like to cover some aspects of it.

In a nutshell, food security is determined by four components [1]:

- Availability: "Does food exist near me?"
- Access: "Can I get to food easily?"
- Utilization: "Will this food contribute to my health and well-being?"
- Stability: "Will food be available tomorrow, next week, next month?"

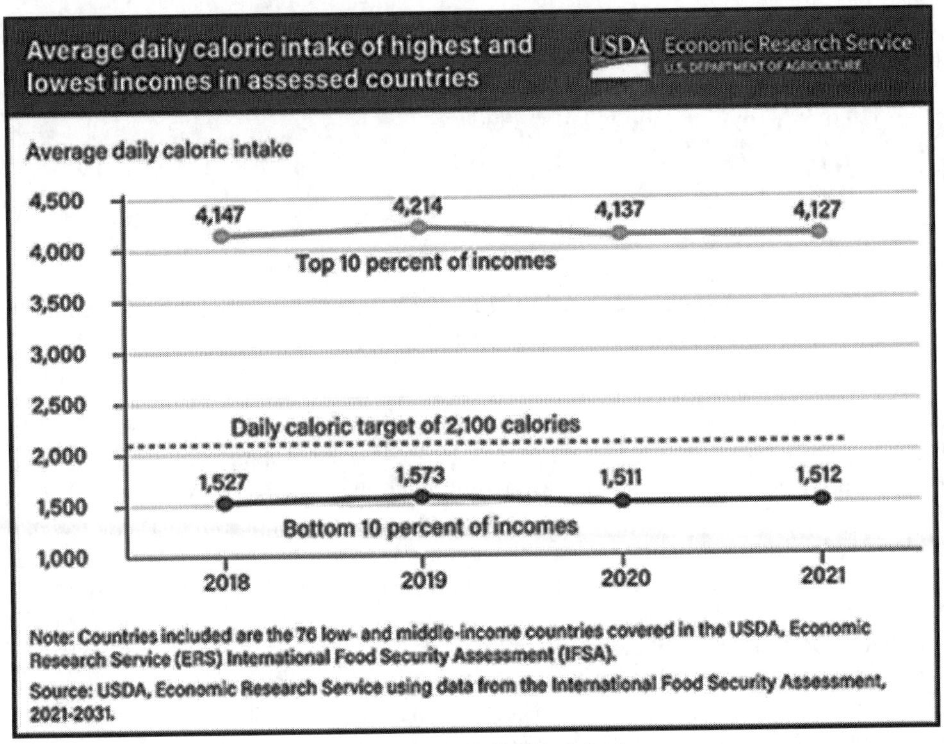

Figure 1: Average Daily Caloric Intake.

When any one of these components is stressed or unmet, it is considered food insecurity. Food insecurity persists across the world. For instance, the United States Department of Agriculture (USDA) Economic Research Service (ERS) monitors the food security of U.S. households through an annual, nationally representative survey. While most U.S. households are food secure, a minority of U.S. households experience food insecurity at times during the year, meaning that their access to adequate food for active, healthy living is limited by lack of money and other resources.[158] Some experience very low food security, a more severe range of food insecurity where food intake of one or more members is reduced and normal eating patterns are disrupted [2]. In 2020, 89.5 percent of U.S. households were food secure throughout the year. The *remaining 10.5 percent of households* were food insecure at least some time during the year, including 3.9 percent (5.1 million households) that had very low food security.[159] Prevalence[160] of food insecurity (please see Figure 1 and Figure 2) is not uniform across the country due to both the characteristics of populations and state-level policies and economic conditions [2,3,4].

Globally, food insecurity—defined as lacking access to at least 2,100 calories per day—has intensified during the Coronavirus (COVID-19) pandemic [5]. The world's poorest populations experienced a reduction in consistent access to food and increased food insecurity, in part because of pandemic-related income shocks. In the 76 low- and middle-income countries covered in USDA, Economic Research Service's (ERS) International Food Security Assessment (IFSA), average daily caloric consumption for 10 percent of these countries' populations with the lowest income falling by 3.9 percent, or 62 calories per day, in 2020 relative to 2019 [6]. Access to food for the wealthiest 10 percent of the 76 countries' populations was also affected, with daily caloric intake falling by 1.8 percent, or 77 calories, year to year. In part because of the persistent effects of COVID-19 on income levels, the number of food-insecure people in 2021 is estimated at 1.2 billion, an increase of 32 percent, or 291 million people, from the 2020 estimate. This suggests that in 2021 [6], nearly one-third of the population of the countries included in the IFSA study will lack access to sufficient food to meet the daily nutritional target of 2,100 calories set by the Food and Agricultural Organization of the United Nations (please see Figure).[161]

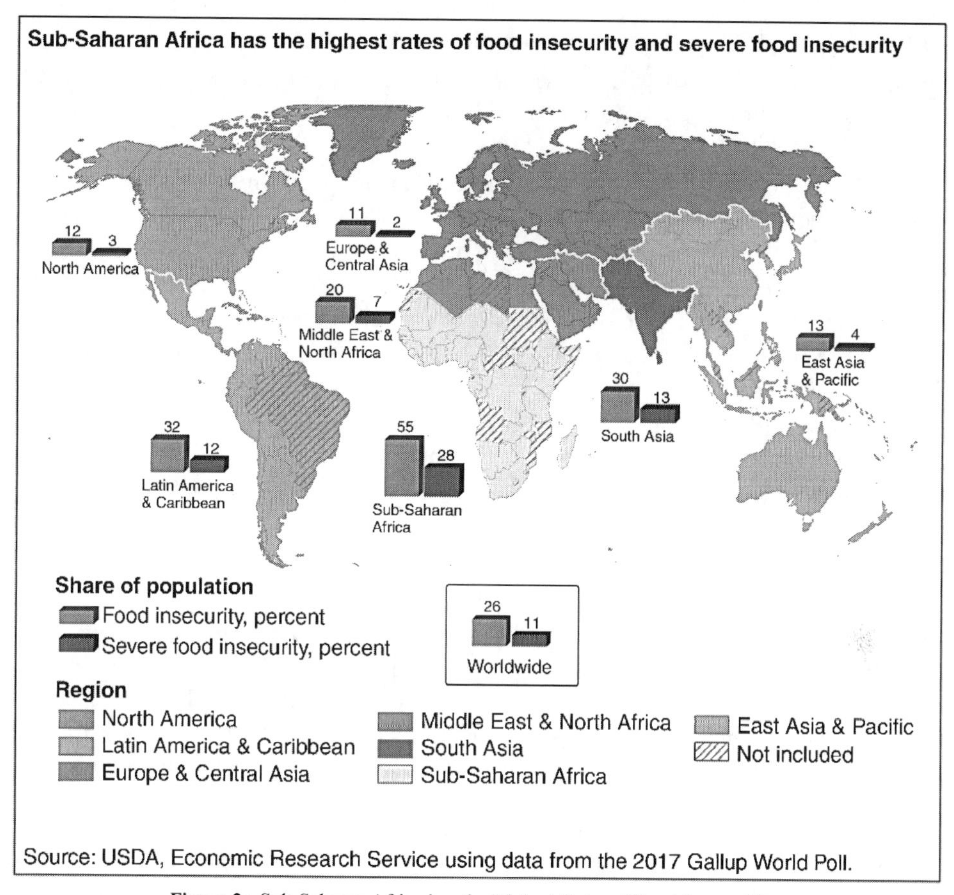

Figure 2: Sub-Saharan Africa has the Highest Rates of Food Insecurity.

Historically, in 2017, Sub-Saharan Africa had the highest prevalence of food insecurity (55 percent) and severe food insecurity (28 percent), followed by Latin America and the Caribbean (32 percent food insecure and 12 percent severely food insecure), and South Asia (30 percent food insecure and 13 percent severely food insecure). Food insecurity and severe food insecurity were lowest in North America, Eastern Europe and Central Asia [7].

In the discussion below, we will walk through the Machine Learning Model that analyzes food insecurity across the globe and provide a mathematical model for 2017 & 2019 food insecurity. It is necessary to compute food insecurity to provide better policy frameworks for public agencies & governing bodies to take care of people across the globe. The necessity is more human and economic as food insecurity inhibits the full potential of individuals and societies.

Food insecurity results in hunger in the short run, but persistent food insecurity causes malnutrition issues, such as the prevalence of underweight, stunting, wastage, and micronutrient deficiencies, and even death in acute cases.[162] Unhealthy food choices compelled by food insecurity, carbohydrate rich diets for example, can lead to obesity and non-communicable diseases like diabetes and hypertension. Generally, children and women are mostly affected by chronic food insecurity, resulting in lower educational outcomes for children that will limit their economic opportunities in adulthood and health complications for women. Family planning would also be severely affected by food insecurity [8]. Food insecurity drains national income and can lead to cyclic financial and social imbalance issues, with gross underperformance of countries for long periods of time. Hence, ensuring food security is crucial for reducing hunger and poverty in the world in line with the *Sustainable Development Goals* (SDGs), thereby improving the general wellbeing of the population and productivity of the labor force, and reducing persistent health and malnutrition issues.

Definitions (drawn from USAID Policy Determination 19, 1992) [8]

Food security. "Food security exists when all people at all times have both physical and economic access to sufficient food to meet their dietary needs for a productive and healthy life."

Food availability refers to food presence, e.g., when "sufficient quantities of appropriate food types from domestic production, commercial imports, commercial aid programs, or food stocks are consistently available or within individual reach." Hence, food availability is largely a function of macroeconomic factors.

Food access refers to availability of household resources for food production or purchase, i.e., "Individuals have sufficient assets or incomes to produce, purchase, or barter to maintain appropriate food levels for adequate dietary consumption/nutrition." Hence, food access is largely related to household income and food production resources.

Food utilization refers to the nutritional benefits derived from food consumption, i.e., "Food is properly used; food processing and storage techniques used are appropriate; existence and application of adequate nutritional knowledge, childcare techniques and adequate health and sanitation services." Hence food utilization is largely related to nutrition, health, and sanitation.

Domino Effect

Food security is a global need, and the world is actively working to ensure food security enhancement [9]. Recent political and economic global conflicts have changed the face of food security worldwide. Due to unprecedented disruptions caused by the Russia-Ukraine war, three negative influences on food security have been: the people of Ukraine and Russia are experiencing supply disruptions; countries relying heavily on Ukrainian and Russian exports (wheat, sunflower seeds/oil, and others) are suffering from scarcity; larger populations are already bearing the shock of higher food prices.[163] Record high natural gas prices have forced fertilizer companies to curtail production of ammonia and urea in many parts of the world,[164] e.g., Europe has cut down its production to 45% of its full capacity, and the shortage of these essential agricultural ingredients, will have unprecedented knock-on effects on global food supplies [10]. Agriculture and Food Security are national security issues and must be given utmost priority.

It is essential to study food security from global interconnect, supplier risk, and sustainability points of view for mainly protecting countries that are heavily dependent on food imports and are on the receiving side of unprecedented commodity disruptions. The shortage of essential food commodities due to the Russia-Ukraine war has triggered a *domino effect* that underscores how interconnected global commodity markets are raising the prices of other co-moving commodities produced elsewhere, including those of not ordinarily considered as substitutes.[165] We also learned that there simply isn't sufficient available capacity in the rest of the world to cover the Ukrainian shortfall [11]. Historically, we have seen such global disruptions due to weather, climate change, and political issues. For instance, the number of climate-related disasters, including extreme heat, droughts, floods and storms, has doubled since the early 1990s, with an average of 213[166] of these events occurring every year from 1990–2016. These disasters harm agricultural productivity, which means food availability suffers. With this comes a domino effect, causing food prices to skyrocket and income losses that prevent people from being able to afford food [11].

"When women have an income, [they] dedicate 90% to health, education, to food security, to the children, to the family, or to the community. So, when women have an income, everybody wins." [11]

- Michelle Bachelet, United Nations Commissioner on Human Rights

Food Security is National Security!

Food Insecurity wreaks national havoc and disrupts public order, creating unexpected challenges to maintain law and order. For instance, consider the causal relationship between Arab Spring and high food prices of essential staples.[167] While this research does not imply that all Arab Springs were triggered by food insecurity, it assumes that rising food prices increased the pre-existing social unrest, sparking protests in Egypt, Syria, and Morocco, and probably also in other Middle East and North Africa (MENA) countries affected by the riots [11]. Food scarcity also played a large role in the Arab Spring causing a series of anti-government protests in the Middle East in the early 2010s when international food prices shot up, unemployment rose, and frustration with corrupt political systems peaked. According to experts one of the driving forces behind the Arab Spring unrest was the high cost of food. Shrinking farmlands, bad weather, and poor water distribution contributed to higher prices and, as a result, anti-government sentiments [12]. Surging food prices have caused people to revolt in bygone centuries too. Bread shortages were flashpoints for unrest throughout history, as in the French Revolution in the late 18th century.

Food Insecurity could emerge from any threat angle. Consider the case of Sri Lanka. The World Bank has upgraded Sri Lanka to upper middle-income country. Being a strong economy, nonetheless, Sri Lanka had grappled with a depreciating currency, inflation and a crippling foreign debt burden, The country's debt burden has also been growing—from 39% of Gross National Income (GNI) in 2010 to 69% in 2019, according to the World Bank. Another reason was the slump in foreign tourism due to the COVID-19 pandemic. As a result, essential supplies of food items have become scarce [13].

To create a sustainable food security system, the principles of cybersecurity and application of related threat modeling need is required. Yes, Food Security and Cyber Security have one commonality: threats can emerge at any point from known and unknown sources. Food Security is sensitive to national sources both internal and external. For instance, sudden input costs to agricultural inputs such as oil, fertilizers, human labor could imbalance national and world food security at large. Risks that emerge due to policy changes related to import and export of biofuels, or credit to the agriculture segment could threaten the entire food security of a nation. Bilateral relationships between countries and economic events within large food producing nations, wars and conflicts could introduce risks to food security. Finally, weather, climate change, droughts, water & resource imbalances could result in food security risks that could propagate to the entire world food system, resulting in worldwide spikes in prices.

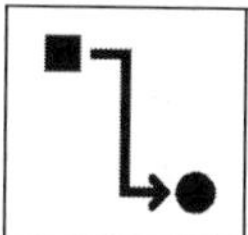

The world is realizing it's not easy to take a major commodity like sunflower oil and switch to an alternative." [14]

The onus on the policy makers is to consider sustainable food security as a central pillar to a country's national security. It would be too late to apply policy changes when a country or subnational is undergoing food insecurity. The simple reason is the policy effectiveness of food security takes a considerable amount of time to have a sustained impact. Long term vision is essential to create an environment for sustained food security.

From a pure cybersecurity point of view, food security is national security.[168] Agricultural and food productions systems ensure grocery shelves with items that close farm to fork. Any disruption to the agricultural production system jeopardizes the entire supply chain and increases the risk to Americans. Given the importance of Food Security, "Food Security is National Security Act of 2021" has been passed.[169]

Food security issues of the country need to be addressed with a holistic policy approach with near to long-term strategies to ensure food security to withstand possible future shocks, such as COVID-19 and the Russia-Ukraine conflict. While improving productivity levels in the agriculture sector, policy focus should also be on providing necessary farm inputs such as timely water supply, facilitating credit to farmers, developing off-farm opportunities at the rural level, improving marketing and distribution channels, through new technological tools such as digital platforms and home delivery systems targeting needy groups, which will support food accessibility even during extreme events. Further, data is a strategic asset to address food security and a sustainable food future for all. Data should be employed to survey the pulse of farm producers and consumers. Technology deployment at farm level should consider sensor-based data collection that will improve efficiency and robustness [8]. Food production databases and public sector distribution channels should also be strengthened with institutional support from local government and national authorities to prevent speculations, consumer exploitation, and malpractices by rouge agents during shock events. Moreover, improving warehousing and storage facilities such as cold storages, enhancing packaging and transportation services are important in the medium to long-term for reducing post-harvest losses, and food protection. Finally, developing strategic cooperative initiatives with renowned research facilities, international food agencies, innovative food technology startups, smart climate-based sensor development & analytics companies,[170] and global financial institutions to facilitate a smooth flow of technology and knowledge base enhancing food security for all is a must!

World Bank Classification: Country Income Groups

The World Bank classify world economies under four income groups—low, lower-middle, upper-middle, and high-income countries.[171] The classifications are updated each year on July 1 and are based on GNI per capita in current USD (using the Atlas method exchange rates) of the previous year (i.e., 2020 in this case) [15]. The purpose of the Atlas conversion factor is to reduce the impact of exchange rate fluctuations in the cross-country comparison of national incomes.[172]

Income classifications (historical list is available from World bank (OGHIST.XLS)[173] are set each year on July 1 for all World Bank member economies (please see Figure 3), and all other economies with populations of more than 30,000. These official analytical classifications are fixed during the World Bank's fiscal year (ending on June 30); thus, economies remain in the categories in which they are classified irrespective of any revisions to their per capita income data. The historical classifications shown are published on July 1 of each fiscal year. Classifications are essential to assess food insecurity as country classifications change dynamically. For instance, take the case of Sri Lanka's classification:[174] changed from Lower Middle Income (2017) to Upper Middle Income in 2018 and changed again back to Lower Middle income in 2019 [16], as the macroeconomic outlook changed. As per the World Bank, Sri Lanka is a lower-middle-income country with a GDP per capita of USD 3,852 (2019) and a total population of 21.8 million.[175]

The same goes with Algeria (the transition from an Upper Income (2018) to a Lower Income (2019) classification)—one primary reason is the economic crisis caused by the pandemic following a slowdown in GDP growth for five consecutive years (2015–2019), driven by a shrinking hydrocarbon sector, a labyrinthine public-led model of growth, and a private sector struggling to become the new economic growth engine. The hydrocarbon industry, which accounted for 20% of GDP, 41% of fiscal revenues, and 94% of export earnings in 2019, is experiencing a structural decline.[176] The World Bank Classifies Mauritius as a High-Income Country, upgraded from an Upper Income (UM) classification. This is based on its Gross National Income (GNI) per capita (current US $), calculated using the Atlas method.[177] A Country Income group classification is endogenous to the prevalence of a food insecurity machine learning model as it changes and though it is exogenous to the model it explains the increase in prevalence of food insecurity in moderate or severe terms.

World Bank Analytical Classifications
(presented in World Development Indicators)
GNI per capita in US$ (Atlas methodology)

Bank's fiscal year:	FY89	FY90	FY91	FY92	FY93	FY94	FY95	FY96	FY97	FY98	FY99	FY00	FY01
Data for calendar year:	1987	1988	1989	1990	1991	1992	1993	1994	1995	1996	1997	1998	1999
Low income (L)	<= 480	<= 545	<= 580	<= 610	<= 635	<= 675	<= 695	<= 725	<= 765	<= 785	<= 785	<= 760	<= 755
Lower middle income (LM)	481-1,940	546-2,200	581-2,335	611-2,465	636-2,555	676-2,695	696-2,785	726-2,895	766-3,035	786-3,115	786-3,125	761-3,030	756-2,995
Upper middle income (UM)	1,941-6,000	2,201-6,000	2,336-6,000	2,466-7,620	2,556-7,910	2,696-8,355	2,786-8,625	2,896-8,955	3,036-9,385	3,116-9,645	3,126- 9,655	3,031-9,360	2,996-9,26
High income (H)	> 6,000	> 6,000	> 6,000	> 7,620	> 7,910	> 8,355	> 8,625	> 8,955	> 9,385	> 9,645	> 9,655	> 9,360	> 9,265
BEL Belgium	H	H	H	H	H	H	H	H	H	H	H	H	H
BLZ Belize	LM	LM	LM	LM	LM	LM	LM	LM	LM	LM	LM	LM	LM
BEN Benin	L	L	L	L	L	L	L	L	L	L	L	L	L
BMU Bermuda	H	H	H	H	H	H	H	H	H	H	H	H	H
BTN Bhutan	L	L	L	L	L	L	L	L	L	L	L	L	L
BOL Bolivia	LM	LM	LM	LM	LM	LM	LM	LM	LM	LM	LM	LM	LM
BIH Bosnia and Herzegovina	..	..	..	..	..	LM	L	L	L	L	L	LM	LM
BWA Botswana	LM	LM	LM	LM	UM	UM	LM	LM	LM	LM	UM	UM	UM
BRA Brazil	UM	LM	UM	UM	UM	UM	UM	UM	UM	UM	UM	UM	UM
VGB British Virgin Islands	..	..	..	..	..	..	..	..	..	..	..	..	..
BRN Brunei Darussalam	H	..	..	H	H	H	H	H	H	H	H	H	H
BGR Bulgaria	..	..	LM	LM	LM	LM	LM	LM	LM	LM	LM	LM	LM
BFA Burkina Faso	L	L	L	L	L	L	L	L	L	L	L	L	L

Figure 3: Income Classification.

Country Income group classification is endogenous to the prevalence of a food insecurity machine learning model as the transition in classification, though exogenous to the model, explains the increase in prevalence of food insecurity in terms of moderate or severe!

For the current (2022) fiscal year,[178] low-income economies (please see Figure 4) are defined as those with a GNI per capita, calculated using the World Bank Atlas method, of $1,045 or less in 2020 (the progression of low income thresholds to higher values provides an insight into upbeat economies with decreasing food insecurity); lower middle-income economies are those with a GNI per capita between $1,046 and $4,095; upper middle-income economies are those with a GNI per capita between $4,096 and $12,695; high-income economies (please see Figure 5) are those with a GNI per capita of $12,696 or more. In each country, factors such as economic growth, inflation, exchange rates, and population growth influence GNI per capita. Revisions to national accounting methods and data can also have an influence in specific cases. The updated data on GNI per capita data for 2020 can be accessed [15].

Low-income economies ($1,045 OR LESS)

Twenty-seven countries have a per capita income of $1,045 or less. These countries are Afghanistan, Guinea-Bissau, Somalia, Burkina Faso, Korea, Dem. People's Rep, South Sudan, Burundi, Liberia, Sudan, Central African Republic, Madagascar, Syrian Arab Republic, Chad, Malawi, Togo, Congo Dem. Rep, Mali, Uganda, Eritrea, Mozambique, Yemen, Rep., Ethiopia, Niger, Gambia, Rwanda, Guinea and Sierra Leone.

Lower middle income economies ($1,046 TO $4,095)

Fifty-Five countries have a per capita income of $1,046 to $4,095. The country list includes[179] Algeria, Angola, Bangladesh, Benin, Bhutan, Bolivia, Cabo Verde, Cambodia, Cameroon, Comoros, Congo, Rep, Côte d'Ivoire, Djibouti, Egypt, Arab Rep., El Salvador, Eswatini, Ghana, Honduras, India, Kenya, Kiribati, Kyrgyz Republic, Lao PDR, Lesotho, Mauritania. Micronesia, Fed. Sts., Moldova Mongolia, Morocco, Myanmar, Nepal, Nicaragua, Nigeria, Pakistan, Papua New Guinea, Philippines, São Tomé and Principe, Senegal, Solomon Islands, Sri Lanka, Tanzania. Timor-Leste, Tunisia, Ukraine, Uzbekistan, Vanuatu, Vietnam, West Bank and Gaza, Zambia, and Zimbabwe.

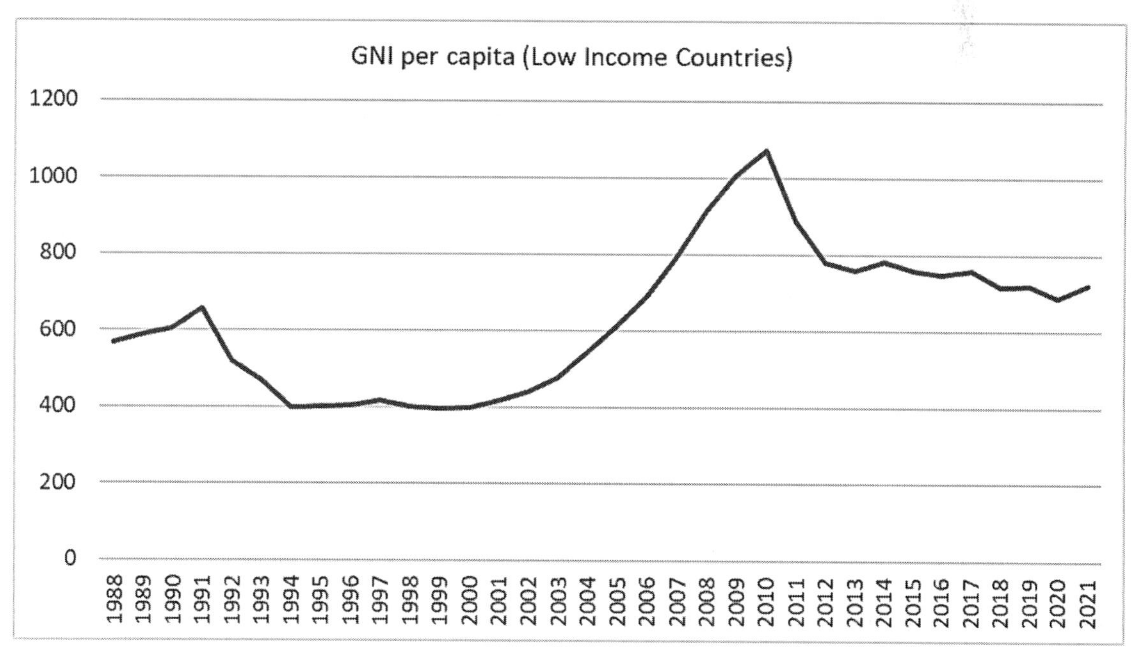

Figure 4: Low Income.

Upper-middle income economies ($4,096 TO $12,695)

Fifty-Five countries have a per capita income of $4,095 to $12, 695. The upper middle income countries include:[180] Albania, Gabon, Namibia, American Samoa, Georgia, North Macedonia, Argentina, Grenada, Panama, Armenia, Guatemala, Paraguay, Azerbaijan, Guyana, Peru, Belarus, Iraq, Romania, Bosnia and Herzegovina, Jamaica, Russian Federation, Botswana, Jordan, Serbia, Brazil, Kazakhstan, South Africa, Bulgaria, Kosovo, St. Lucia, China, Lebanon, St. Vincent and the Grenadines, Colombia, Libya, Surinam, Costa Rica ,Malaysia, Thailand, Cuba, Maldives, Tonga, Dominica, Marshall Islands, Turkey, Dominican Republic, Mauritius, Turkmenistan, Equatorial Guinea, Mexico, Tuvalu, Ecuador, Moldova, Fiji, Montenegro.

High income countries ($12,696 OR MORE)

Eighty countries (please see Figure 5) form this group.[181] The country list includes[182] Andorra, Greece, Poland, Antigua and Barbuda, Greenland, Portugal, Aruba, Guam, Puerto Rico, Australia, Hong Kong SAR, China, Qatar, Austria, Hungary, San Marino Bahamas The Iceland, Saudi Arabia, Bahrain, Ireland, Seychelles, Barbados, Isle of Man, Singapore, Belgium, Israel, Sint Maarten (Dutch part), Bermuda, Italy, Slovak Republic, British Virgin Islands, Japan, Slovenia, Brunei Darussalam, Korea, Rep., Spain, Canada, Kuwait, St. Kitts and Nevis, Cayman Islands, Latvia, St. Martin (French part), Channel Islands, Liechtenstein, Sweden, Chile, Lithuania, Switzerland, Croatia, Luxembourg, Taiwan, China, Curaçao, Macao SAR, China, Trinidad and Tobago, Cyprus, Malta, Turks and Caicos Islands, Czech Republic, Monaco, United Arab Emirates, Denmark, Nauru, United Kingdom, Estonia, Netherlands ,United States, Faroe Islands, New Caledonia, Uruguay, Finland, New Zealand, Virgin Islands (U.S.), France, Northern Mariana Island, French Polynesia, Norway, Germany, Oman, Gibraltar, and Palau.

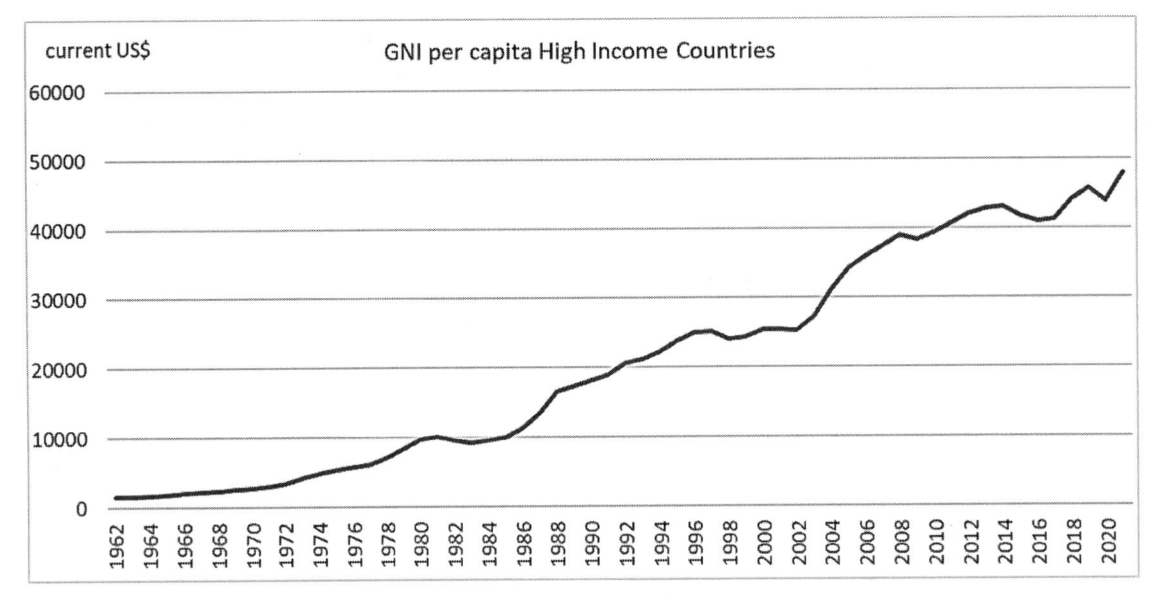

Figure 5: High Income (H) GNI.

Food Security Frameworks

As discussed above, for the proper functioning of the world, food security is a must. Importantly, sustainable food security empowers women, families, and future generations. To achieve complete food security and sustainability, it must be considered with the same rigor and discipline that is applied to science & engineering. In the following section, food security classical frameworks are studied and finally proposed as an innovative framework.

"The war in Ukraine is supercharging a three-dimensional crisis—food, energy and finance—with devastating impacts on the world's most vulnerable people, countries and economies" [17]

Antonio Guterres[183]
United Nations Secretary-General

The IPC Integrated Food Security and Nutrition Conceptual Framework

The IPC Integrated Food Security and Nutrition Conceptual Framework[184] expands on the well-known IPC Analytical Framework for Food Security and the UNICEF Analytical Framework for Malnutrition to contribute to a better understanding of the linkages between food security and nutrition [18] (please see Figure 6). Because classifications are performed separately for food insecurity and malnutrition, albeit considering their linkages ((Livelihood change, Food Consumption) → (Food Availability, Food Access, Household Food Utilization) → and (Food events and ongoing conditions)) this Conceptual Framework should not be used to guide IPC analysis, but rather to inform further analysis of linkages between the different conditions [18].

The IPC Food Security Framework focuses on variables that include Food Production, Food Reserves, Food Imports, Food Markets, and Transportation. The IPC Integrated Food Security and Nutrition Conceptual Framework considers the following [18]:

- The causal factors of food insecurity and malnutrition are common.
- Suboptimal care and feeding practices, together with low food availability, access, utilization, and stability, directly impact the food consumption of households and individuals.

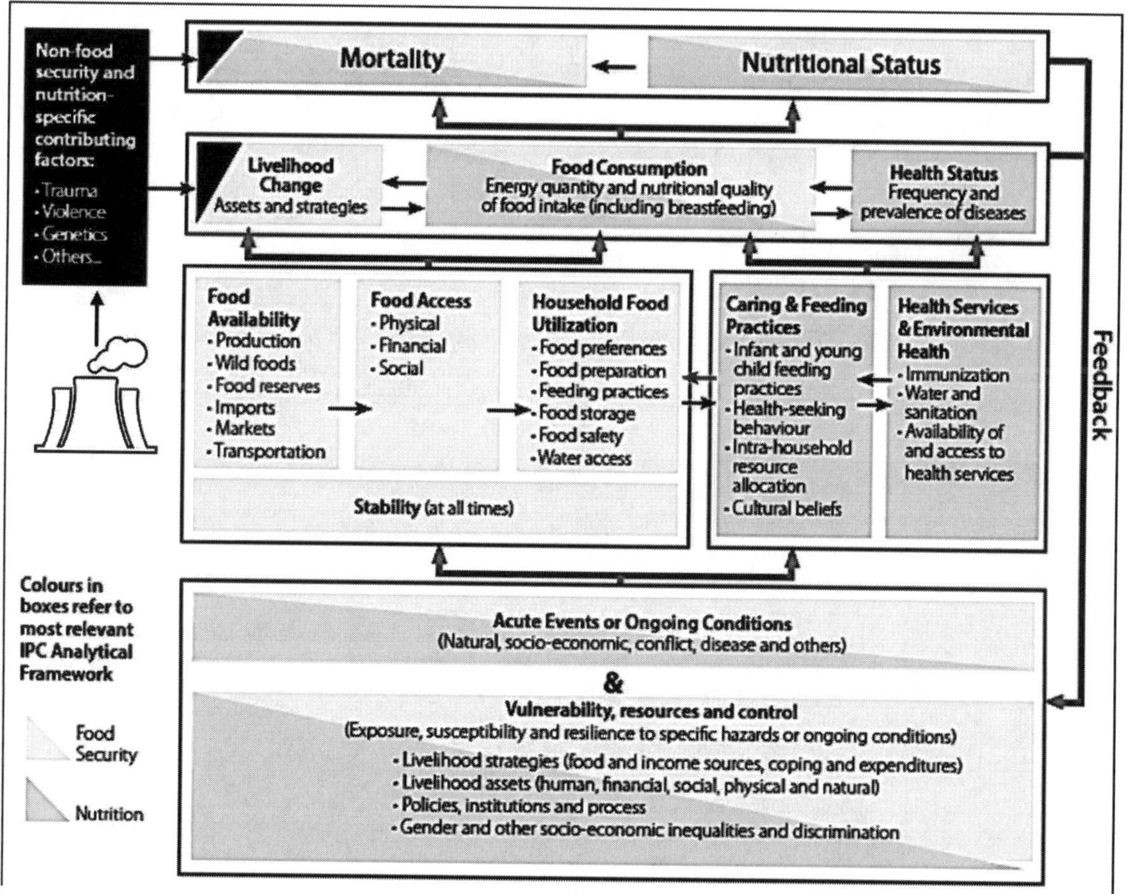

Figure 6: The IPC Integrated Food Security and Nutrition Conceptual Framework [18].

- There is a *reciprocal and complex relationship* between food consumption and health status. It is expected that people who live in households that have an inadequate quantity or quality of food for consumption are more likely to become ill. Furthermore, they are more likely to eat less, while their disease can impact the ability of households to access and utilize food, either because of the weakened immune system or because of a weakened ability to engage in productive activities.
- Food insecurity and malnutrition outcomes will contribute to overall vulnerability or may be a shock on their own, following their cyclical nature.

The USAID Office of Food for Peace Food Security Country Framework

Local country context, exposure to shocks & stresses, adaptive capacity strategies, and risk reduction strategies are the major aspects of the USAID Productive Safety Net Program (PNSP) Food security framework (please see Figure 7) analysis [19].

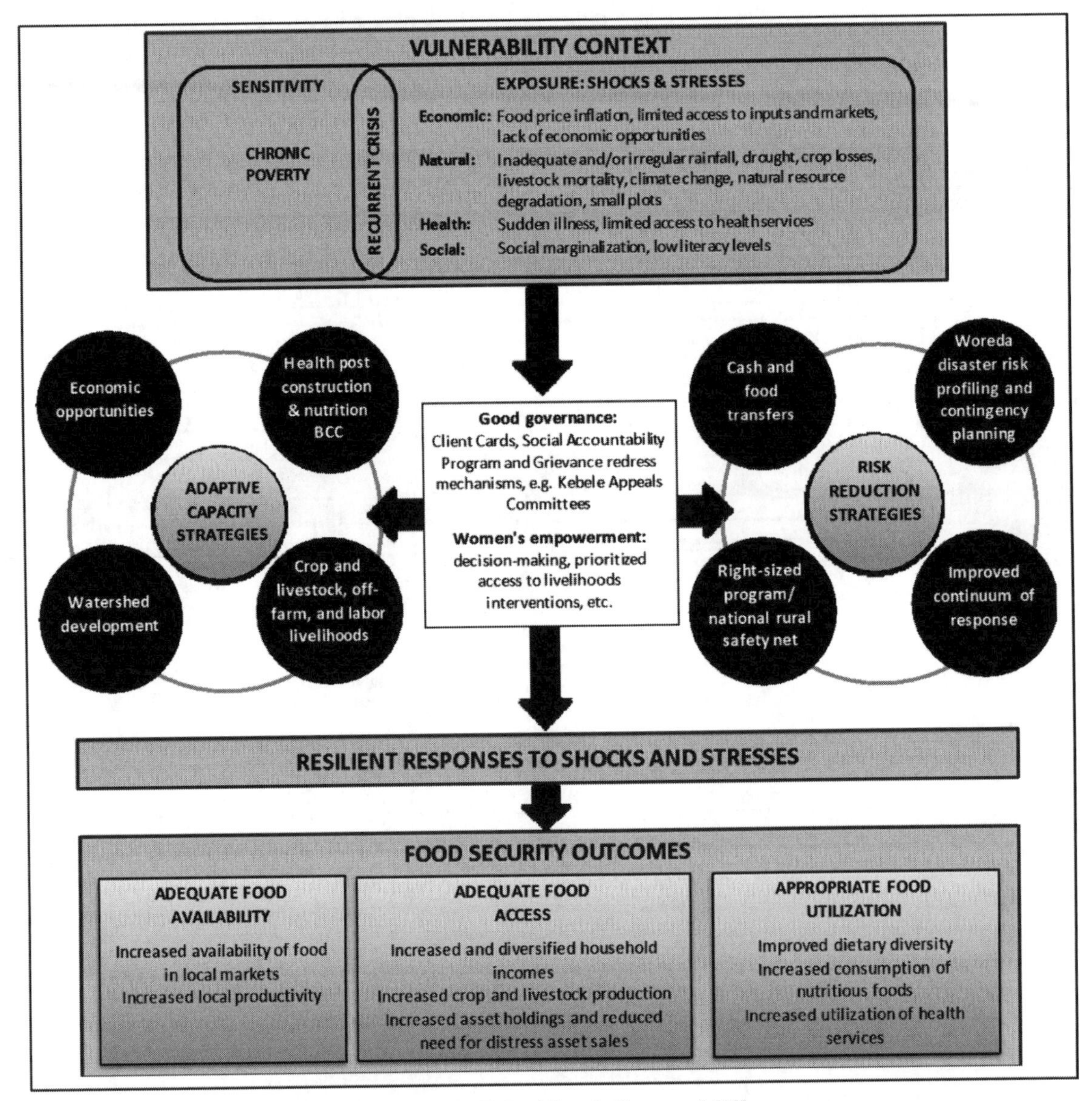

Figure 7: USAID Food Security Framework [19].

Although Food Security is a global need, country contexts play an important role in ensuring food security and are dependent on the country's agriculture, rural, economic, health, and social aspects. A country's population, imports, food dependencies, agriculture to weather pattern calendar, and climate change are the major variables. Adaptive Capacity Strategies look at economic opportunities, crops and livestock, off farm, and labor livelihood, watershed development through public works, and interventions designed to increase access to health services. Risk reduction strategies aim at cash and food transfer, the right-sized program for national/rural safety net programs, disaster risk profiling, and an improved continuum of response. The outcome of the above strategies is ensuring food security [19].

FAO—Food Insecurity and Vulnerability Information and Mapping Systems (FIVIMS)

The standard food security conceptual framework draws on the idea of *hierarchy of needs*. The assumption is "food first" where food security is a primary need that supersedes other human needs. However, evidence of people's behavior challenges the assumption. It is increasingly recognized that protecting food intake, especially in the short-term, is only one objective that people pursue.[185] To have sustained food security, it is important to realize that the long-term view & stability of key indicators at the national/subnational and international levels should be pursued (please see Figure 8). The key indicators to be mapped at national (*) and subnational level, by sector [20], include:[186] Nutritional status indicators, Food intake indicators, Basic health indicators, Reproductive health indicators, Economy—public finances, External trade, Local Markets, Income & entitlements, Agricultural Production Systems, and Environment, natural resources, and population.

Food Security should be considered as an important long term strategic view and planned continuously for changing demands of people, ongoing events, and future emerging events.

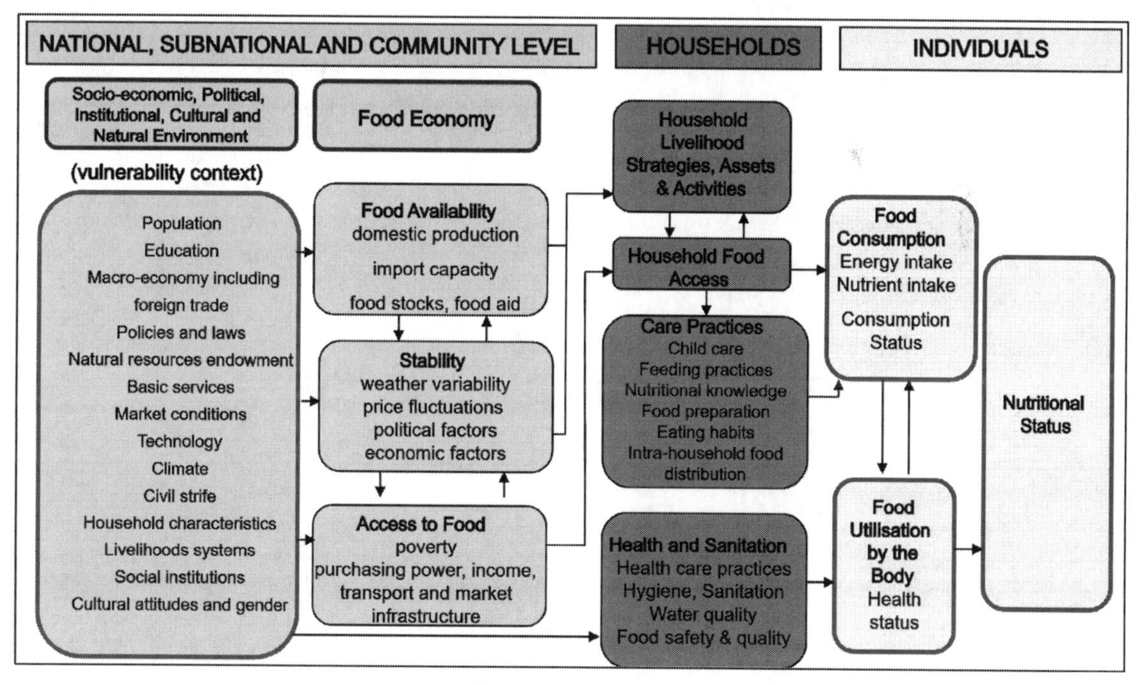

Figure 8: FIVIMS [20].

Food Security Monitoring System (FSMS)

The Food Security Monitoring System (FSMS), generally, covers 35 key indicators (please see Table 1), classified under the four dimensions of food security: availability (5 indicators), access (10 indicators), utilization (13 indicators) and stability (7 indicators). Indicators are classified according to the four dimensions of food security: availability, access, utilization, and stability [21].

Table 1: FSMS.

Number	Type of Indicator	Source
	Availability	
1	Average dietary energy supply adequacy	FAO
2	Average value of food production	FAO
3	Share of dietary energy supply derived from cereals, roots and tubers	FAO
4	Average vegetarian protein supply	FAO
5	Average supply of protein of animal origin	FAO
	Accessibility	
6	Gross domestic product per capita (in purchasing power equivalent)	WB
7	Domestic food price level index	WB/ILO/FAO
8	Percent of paved roads over total roads	WB
9	Road density	World Road Statistics
10	Rail lines density	WB
11	Prevalence of undernourishment (PoU)	FAO
12	Number of people undernourished	FAO
13	Share of food expenditure of the poor	FAO
14	Depth of the food deficit	FAO
15	Prevalence of food inadequacy	FAO
	Utilization	
16	Access to improved water sources	WHO/UNICEF
17	Access to improved sanitation facilities	WHO/UNICEF
18	Percentage of children under 5 years of age affected by wasting	WHO/UNICEF
19	Percentage of children under 5 years of age who are stunted	WHO/UNICEF
20	Percentage of children under 5 years of age who are underweight	WHO/UNICEF
21	Percentage of adults who are underweight	WHO
22	Prevalence of anaemia among pregnant women	WHO/WB
23	Prevalence of anaemia among children under 5 years of age	WHO/WB
24	Prevalence of vitamin A deficiency	WHO
25	Prevalence of iodine deficiency	WHO
26	Prevalence of undernourishment (PoU)	FAO
27	Mortality rate	WHO/WB/UNICEF
28	Food Consumption	
	Stability	
29	Cereal import dependency ratio	FAO
30	Percent of arable land equipped for irrigation	FAO
31	Value of food imports over total merchandise exports	FAO
32	Political stability and absence of violence/terrorism	WB/WWGI
33	Domestic food price volatility	WB/ILO/FAO
34	Per capita food production variability	FAO
35	Per capita food supply variability	FAO

Table 2: GSFI Table.

Number	Type of Indicator
	Affordability
1	Change in average food costs
2	Proportion of population under global poverty line
3	Inequality-adjusted income index
4	Agricultural import tariffs
5	Food safety net programmes 5.1) Presence of food safety net programmes 5.2) Funding for food safety net programmes 5.3) Coverage of food safety net programmes 5.4) Operation of food safety net program
6	Market access and agricultural financial services 6.1) Access to finance and financial products for farmers 6.2) Access to diversified financial products 6.3) Access to market data and mobile banking
	Availability
1	Sufficiency of supply 1.1) Food supply adequacy 1.2) Dependency on chronic food aid
2	Agricultural research and development 2.1) Public expenditure on agricultural research and development 2.2) Access to agricultural technology, education, and resources
3	Agricultural infrastructure 3.1) Crop storage facilities 3.2) Road infrastructure 3.3) Air, port and rail infrastructure 3.4) Irrigation infrastructure
4	Volatility of agricultural production
5	Political and Social barriers to access 5.1) Armed conflict 5.2) Political stability risk 5.3) Corruption 5.4) Gender inequality
6	Food Loss
7	Food security and access policy commitments 7.1) Food security strategy 7.2) Food security agency
	Quality and Safety
1	Dietary diversity
2	Nutritional standards 2.1) National dietary guidelines 2.2) National nutrition plan or strategy 2.3) Nutrition labeling 2.4) Nutrition monitoring and surveillance
3	Micronutrient availability 3.1) Dietary availability of vitamin A 3.2) Dietary availability of iron 3.3) Dietary availability of zinc
4	Protein quality
5	Food safety 5.1) Food safety mechanisms 5.2) Access to drinking water 5.3) Ability to store food safely

	Natural Resources & Resiliance
1	Exposure 1.1) Temperature rise 1.2) Drought 1.3) Flooding 1.4) Sea level rise
2	Water 2.1) Agricultural water risk – quantity 2.2) Agricultural water risk – quality
3	Land 3.1) Land degradation 3.2) Grassland 3.3) Forest change
4	Oceans, rivers, and lakes 4.1) Eutrophication 4.2) Marine biodiversity
5	Sensitivity 5.1) Food import dependency 5.2) Dependence on natural capital
6	Political commitment to adaptation 6.1) Early-warning measures/climate-smart Agriculture 6.2) Commitment to managing exposure 6.3) National agricultural adaptation policy 6.4) Disaster risk management
7	Demographic stress 7.1) Projected population growth 7.2) Urban absorption capacity

The Global Food Security Index (GFSI)

The GFSI[187] has 58 indicators (please see Table 2) that holistically look at Food Security across the nations. At a conceptual level, the GFSI looks at 4 major aspects: Affordability, Availability, Quality & Safety, and Natural Resources & Resilience. Each section has its own sub items. Affordability has six major driving indicators. Availability has six, Quality & Safety has five, and Natural Resources & Resilience has seven indicators. The aim is to attain ease of affordability, high sustained availability, accessibility across all levels, and preservation of natural resources & resilience for sustained food security.

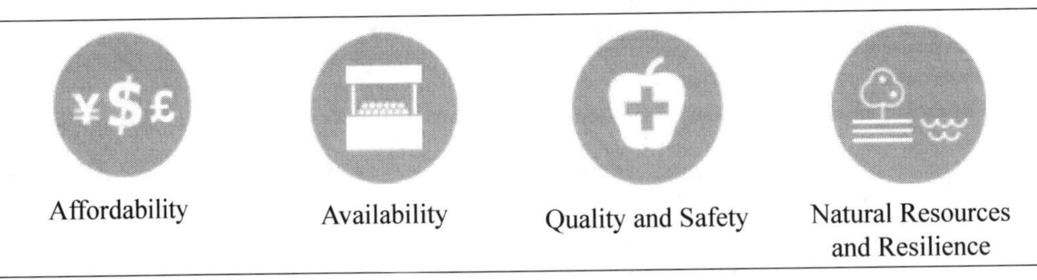

Macroeconomic measures for the International Food Security Assessment (please see Figure 9) of countries, 2021–2031[188] considers Annual Population (taken in millions) growth rate (percentage), Annual GDP growth rate (percentage), Per capita GDP (U.S $, 2015), Annual Per capita GDP growth rate (percentage), Annual CPI growth rate (percentage), Annual RER growth rate (percentage), and Real domestic price Annual growth rate (percentage) of major grains.

Country	Population (million)			Population: Annual growth rate (percentage)		GDP, million 2015 USD			GDP: Annual growth rate (percentage)		Per capita GDP (2015 USD)		
	2018-2020	2021	2031	2015-2020	2021-2031	2018-2020	2021	2031	2015-2020	2021-2031	2018-2020	2021	2031
Total IFSA Countries	3,814	3,938	4,566	1.7	1.5	8,341,232	8,486,164	13,572,641	2.8	4.8	2,187	2,155	2,973
ASIA	2,414	2,471	2,731	1.2	1.0	5,328,449	5,488,042	9,286,748	3.6	5.4	2,207	2,221	3,400
Commonwealth of Independent States	72	73.0	77.3	0.8	0.6	250,517	255,647	354,404	2.5	3.3	3,484	3,502	4,584
Armenia	3	3.0	2.9	-0.2	-0.5	12,419	12,918	19,318	3.3	4.1	4,099	4,290	6,739
Azerbaijan	10	10.3	10.8	0.9	0.5	52,105	51,174	63,641	-1.0	2.2	5,145	4,978	5,880
Georgia	5	4.9	4.9	0.0	-0.1	17,191	17,484	24,736	2.4	3.5	3,489	3,545	5,050
Kyrgyzstan	6	6.0	6.5	1.0	0.7	7,654	7,676	10,544	2.3	3.2	1,296	1,275	1,630
Moldova	3	3.3	2.9	-1.1	-1.2	8,916	9,062	12,696	2.5	3.4	2,622	2,723	4,309
Tajikistan	9	9.0	10.2	1.6	1.3	9,923	9,923	12,714	4.3	2.5	1,135	1,102	1,244
Turkmenistan	5	5.6	6.1	1.1	0.8	44,721	47,577	70,580	4.9	4.0	8,176	8,517	11,635
Uzbekistan	30	30.8	33.0	0.9	0.7	97,588	99,832	140,175	3.3	3.5	3,221	3,238	4,244
Central and Southern Asia	1,790	1,834	2,040	1.3	1.1	3,323,526	3,412,580	5,945,202	3.8	5.7	1,857	1,861	2,914.4
Afghanistan	36	38	47	2.4	2.2	21,944	23,551	35,310	2.6	4.1	613	628	758
Bangladesh	161	164	178	1.0	0.8	255,803	257,065	446,858	6.2	5.7	1,588	1,565	2,506
India	1,311	1,340	1,473	1.2	0.9	2,611,040	2,693,426	4,784,872	3.8	5.9	1,991	2,009	3,248
Nepal	30	31	33	1.1	0.6	26,180	28,620	54,985	4.8	6.7	872	935	1,685
Pakistan	229	238	285	2.1	1.8	318,511	320,957	491,860	3.1	4.4	1,393	1,347	1,725
Sri Lanka	23	23	24	0.7	0.5	90,047	88,961	131,317	1.7	4.0	3,961	3,862	5,422
Other Asia	58	60	66	1.4	1.1	66,536	63,614	81,820	-5.0	2.5	1,146	1,068	1,233
Korea, Democratic People's Republic of	26	26	27	0.5	0.3	29,986	31,698	36,769	0.3	1.5	1,168	1,224	1,377
Mongolia	3	3	3	1.1	0.7	13,718	14,634	21,902	3.0	4.1	4,374	4,577	6,403
Yemen	29	30	36	2.3	1.7	22,831	17,281	23,148	-14.4	3.0	780	567	639
South East Asia	494	504	548	1.0	0.8	1,687,870	1,756,202	2,905,323	3.8	5.2	3,414	3,484	5,304
Cambodia	17	17	19	1.5	1.1	23,429	25,750	46,640	6.2	6.1	1,404	1,501	2,427
Indonesia	265	269	287	0.8	0.6	1,022,108	1,051,848	1,712,723	3.4	5.0	3,858	3,909	5,975
Laos	7	8	9	1.5	1.3	18,211	19,818	31,618	5.5	4.8	2,481	2,624	3,690

Figure 9: IFSA.

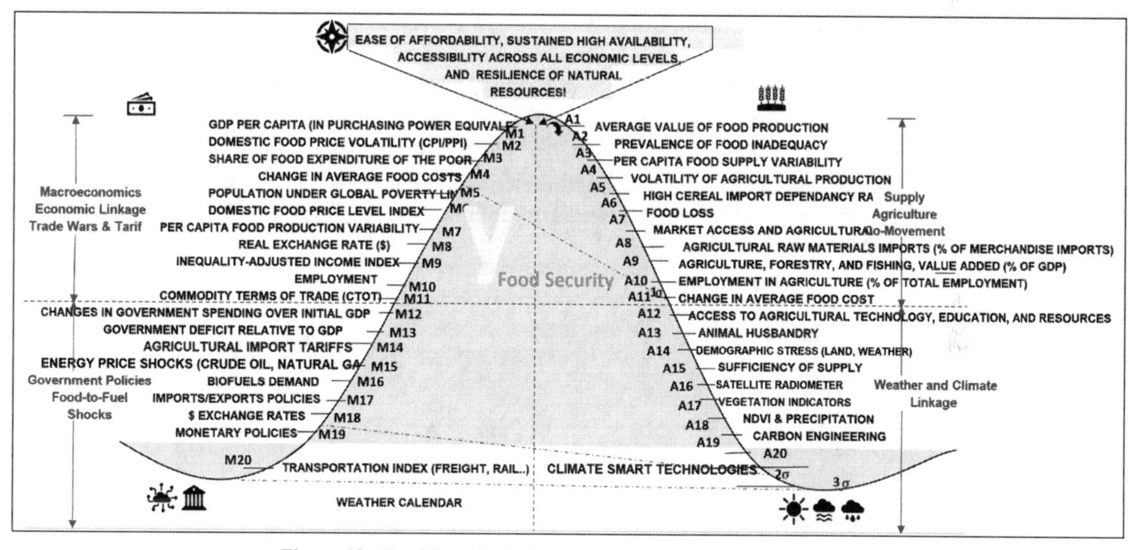

Figure 10: Food Security Bell Curve - Model for Machine Learning.

The Food Security Bell Curve—Machine Learning (FS-BCML) Framework

The prevalence of food insecurity can be modeled using macroeconomic indicators, linkages of market forces such as weather, energy price shocks, climate change, and changes in government policies (please see Figure 10) that are causing food insecurity.

FAO calculated prevalence of severe or moderate food insecurity in country populations (%) based on the Voice of Hunger[189] (VoH) methodology [22]. The VoH methodology measures the severity of food insecurity (lack of economic access to adequate food), based on conditions and behaviors reported in response to an 8-question survey module, the food insecurity experience scale survey module (FIES-SM; Figure 1). Response datasets are assessed and combined to create a measure using statistical methods based on the Item Response Theory, in which the severity of food insecurity experienced by an individual or household is modelled as a *latent trait*—a characteristic not directly observable. Item response theory is a *psychometric framework* used to simultaneously estimate the ability of the examine and the difficulty of the test.[190] This framework enables the comparison of test items by their mean difficulties and the comparison of examinees by their respective mean abilities [23].

The FAO FIES-SM (please see Figure 11) focuses on the applications of the item response theory in machine learning specifically for evaluating latent food insecurity characteristics. Observable conditions assumed to be caused by the latent trait and elicited in response to the FIES-SM questions provide the information for constructing the measure. Based on the measured food insecurity severity of a nationally representative sample of individual adults or households, a nationally prevalent food insecurity rate beyond a specified threshold is calculated. Thresholds and, thus, food insecurity prevalence rates are made comparable across countries using information on the severity of each item from a statistical model of each country's scale.[191] The scale has a list of questions (please see Figure 12).

FAO expects that national prevalence rates of food insecurity for monitoring progress toward the Sustainable Development Goal Target 2.1 will eventually be based on data from national surveys conducted by national statistical agencies in each country. FAO collects the food security prevalence data through Gallup and the Gallup World Poll (GWP) conducted annually since 2006 in about 150 countries, interviewing nationally representative samples of 1,000 adults (ages 15 and older) in each country on a range of topics including family economics, employment, human development, and well-being. The item responses theory measurement model used for FIES data assessment and scale construction is the single-parameter logistic measurement model commonly known as the Rasch Model (please see Figure 12).

The *Rasch Model is a Psychometric Assessment* of FIES Data [22]. The Item Response Theory (also known as Modern Test Theory) is a mathematical approach for quantifying latent traits based on the fundamental assumption that a subject's response to an item is a function of the difference between 1) his/her abilities and 2) the characteristics of the item. Within this class, the Rasch model[192] specifically defines difficulty/facility as the sole parameter of interest when evaluating items. This approach was developed in the 1950s by the Danish mathematician Georg Rasch as a tool for achievement testing among school children.[193] In addition to its continued use in educational assessment, the Rasch model (please see Figure 13) is widely applied by the social sciences, which depend heavily on patient-reported outcomes, and more recently, has been adopted by clinical and public health research fields as a tool for investigating diverse health outcomes, including rehabilitation and community violence [24].

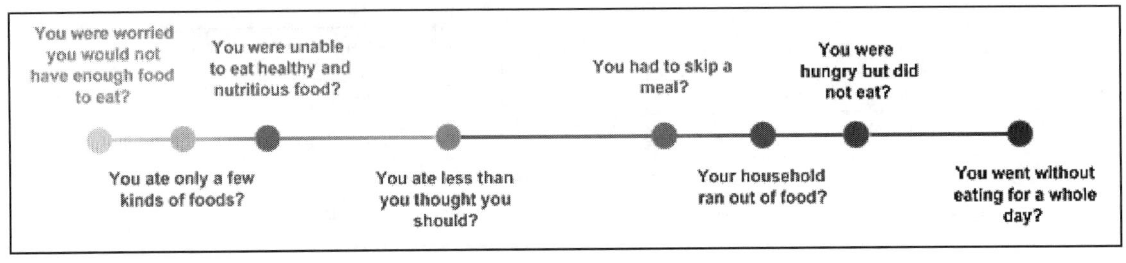

Figure 11: FIES Scale.

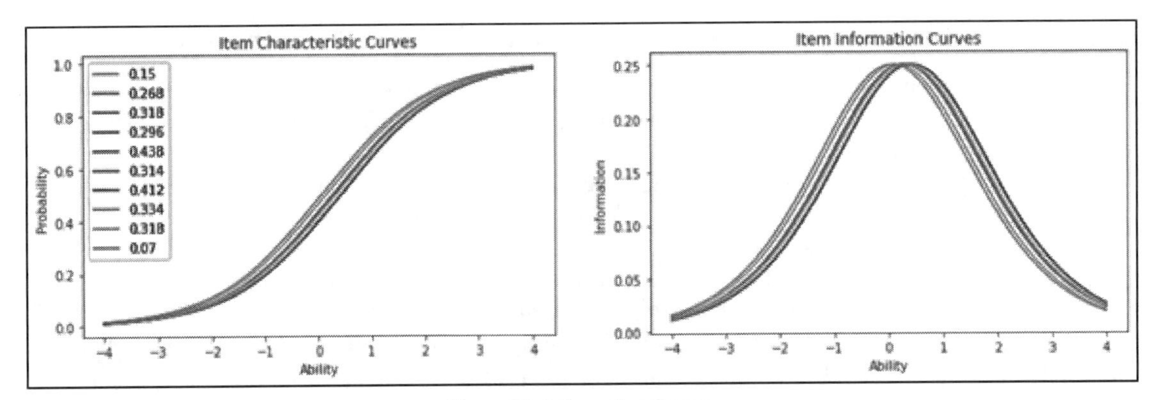

Figure 12: Information Curves.

Assessments & Sharable Content Object (SCO)

As a Software Publisher, I have used Psychometric kernel to assess responses from students for ACT standardized tests. Psychometric kernel was used to assess their expertise through testing and text-based assessments.

The FAO FIES-SM model constructs data that is comparable across 150 World countries and the model can be compared based on macroeconomic parameters. Country-level policies and programs have the potential to make large impacts on reducing inequality and food insecurity, beyond increasing individual-level income.[194] The FAO equating procedure maintains cross-country comparability by creating two standard food insecurity thresholds: moderate and severe food insecurity [25].

Individuals whose raw score equals zero are classified as food secure. Individuals with a raw score (please see Figure 13) of at least one and less than the country specific FIES Global Standard Scale (GSS), threshold for moderate food insecurity (typically a raw score of either 3 or 4) are classified as experiencing mild food insecurity. Those who report a raw score of at least the FIES GSS threshold for moderate food insecurity but less than the FIES GSS threshold for severe food insecurity (typically a raw score of 7) are deemed moderately food insecure. Those above the country specific FIES GSS threshold for severe food insecurity are classified as food secure [25].

The *multivariate panel regressions* with fixed yearly and country effects, thereby identify the relationship of the inter-year variation within countries that deviates from global changes common to all countries in the sample years (please see Equation 1). We estimate the relationship between food insecurity inequality (Y_{jt}) and prevalence (FS_{jt}) for country j in year t, exploiting the panel structure of the data by controlling for country (γ_j) and fixed yearly effects (δ_t):

$$Y_{jt} = \beta_0 + \beta_1 FS_{jt} + \beta_2 FS^2_{jt} + \gamma_j + \delta_t + \varepsilon_{jt} \qquad \text{(EQ. 1)}$$

The fixed country effects control the time-invariant features of countries (e.g., geographic, or historical factors) that explain persistent differences among countries, while the fixed yearly effects control the time varying *factors common to all countries* (e.g., global macroeconomic cycles) [25]. The focus of the following section is to analyze the impact of global macroeconomic parameters on overall food insecurity.

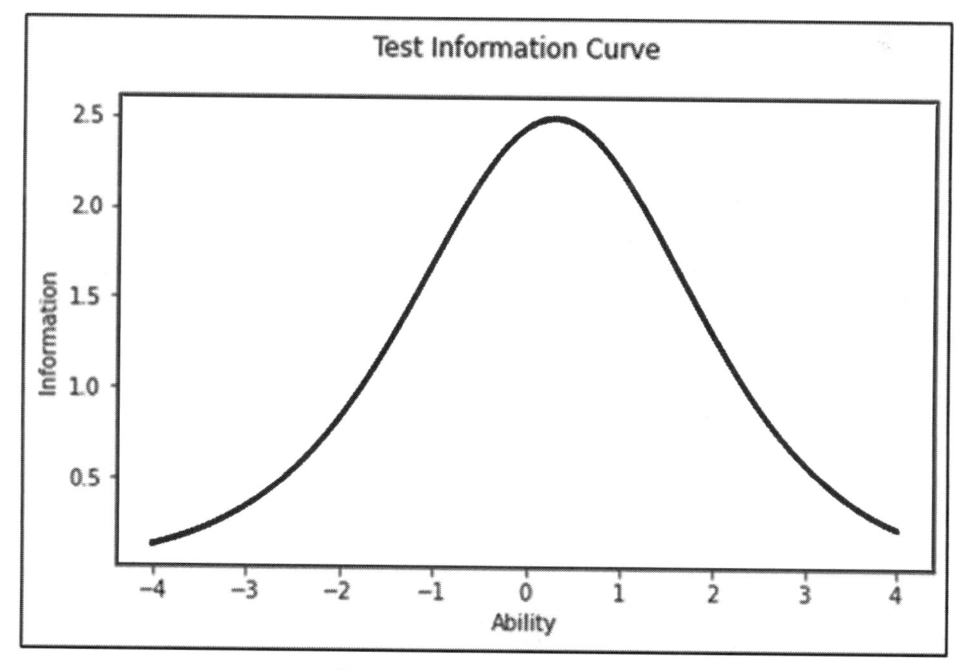

Figure 13: Test Information Curve.

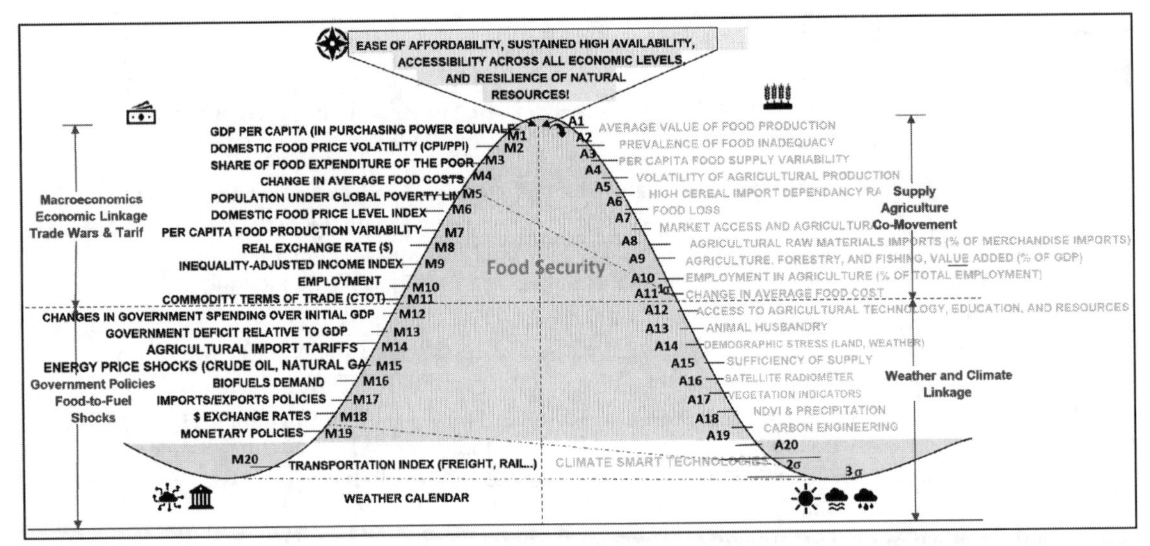

Figure 14: Food Security Bell Curve Model.

The Food Insecurity Prevalence model (please see Figure 14) comprises of four major key indicators: Macroeconomics Economic Linkage (demand side), Agricultural Co-movements (supply side), Governmental Policies, and Weather & Climate Linkages.

Identifying Top Risks for Global Food Insecurity

Previous research has shown that poverty has the strongest impact on adequate individual access to food. The Committee on World Food Security, a body set up in 1975 by the UN World Food Conference to oversee developments in food security, adopted in the early 1980s the recognition of food security as a tripartite concept, reflecting the criteria of availability, access, and stability. Similarly, the OECD suggests that food security has three dimensions: availability, access, and utilization, although this source indicates that there is a tendency to characterize it in terms of availability [26].

Early research emphasized the role played by economic growth in alleviating food insecurity. However, as discussed in the 2012 edition of FAO's State of Food Insecurity in the World, national economic growth is necessary but not sufficient for improving food security. Other factors, such as high food prices, income inequality, and the unequal distribution of food within countries and households, also affect food insecurity rates [27]. Food Insecurity is a complex and multi-faceted issue, and it requires an approach that cross connects several major factors due to multi-factor drivers! Identifying the world's food insecurity and the risk factors for food insecurity can help governments and aid organizations target their assistance to those most in need and develop more effective assistance programs.[195] We have developed Machine Learning Model that uses Gallup World Poll interviews, the World Bank Data, and the GFSI Food Security data to construct machine learning model that identifies the dominant key indicators of Food Insecurity [27].

The key indicators of food security, as depicted in the bell curve, are derived from GWP, World Bank, and EI GFSI. The framework proposed consists of four major domains: macroeconomics, Agriculture Supply, Governmental Policies, and Climate Change.

Macroeconomics: Economic Linkage & Trade

The bell curve lists the key macroeconomic indicators that analyze, correlate, and predict the emergence of food insecurity in a population in moderate or severe terms. The underlying theme is to see the rollup influence of these indicators on key indicators, status quo, and food nutrition gap. The *Status Quo* gap measures the difference between projected food supplies (calculated as domestic production plus commercial imports minus non-food uses) and a base period (2000–2020) per capita consumption. The *Nutrition gap* measures the difference between projected food supplies and the minimum amount of food needed to support per capita nutritional standards.[196] The Status Quo indicator provides a safety net criterion at the country level, whilst the Nutrition gap indicator gives a comparison of relative well-being at the population level [28]. Both are heavily influenced by the proposed bell curve indicators. The macroeconomic aggregate view considers both the demand and supply sides of the market [26]. The following key macroeconomic indicators play an important role in explaining the prevalence of food insecurity:

- Change in average food costs
- Food supply adequacy
- Dependency on chronic food aid
- Public expenditure on agricultural research and development
- Access to agricultural technology, education, and resources
- Crop storage facilities
- Volatility of agricultural production
- Food loss
- Agricultural raw materials imports (% of merchandise imports)
- Agriculture, forestry, and fishing, value added (% of GDP)
- Current account balance (% of GDP)
- Employment in agriculture (% of total employment) (modeled ILO estimate)
- Employment to population ratio, 15+, total (%) (modeled ILO estimate)
- Foreign direct investment, net inflows (% of GDP)
- GDP per capita, PPP (constant 2017 international $)
- Imports of goods and services (% of GDP)
- Official exchange rate (LCU per US$, period average)
- Population ages 0–14 (% of total population)
- Rural population (% of total population)
- Total reserves (includes gold, current US$)

In our model, we have collected around 113 countries' key macroeconomic indicators, along with key Global Food Security Index (GFSI) parameters and performed multivariate regression analysis to establish those that drive food insecurity. Two class prevalence labels are regressed across three groups of countries with different income groups.

	2019 World's Food Insecure Model				
	Low Income	**Lower Middle Income**	**Upper Middle Income**	**High Income**	**World**
Prevalence of moderate or severe food insecurity in the population	68.6%	NA	NA	95.44%	82.20%
Prevalence of severe food insecurity in the population	NA	NA	NA	89.73%	61.14% 20.57%

Models reveal the distinctive nature of food insecurity. The key drivers of food insecurity vary depending on a country's macroeconomic parameters, agricultural dependencies, weather, and supply related parameters.

The prevalence of moderate or severe food insecurity has increased from 2014 levels to 2019 averages (percentage).[197] Since 2014, the global prevalence of undernourishment (chronic food insecurity) has remained virtually unchanged at slightly below 9 per cent.[198] However, the total number of people going

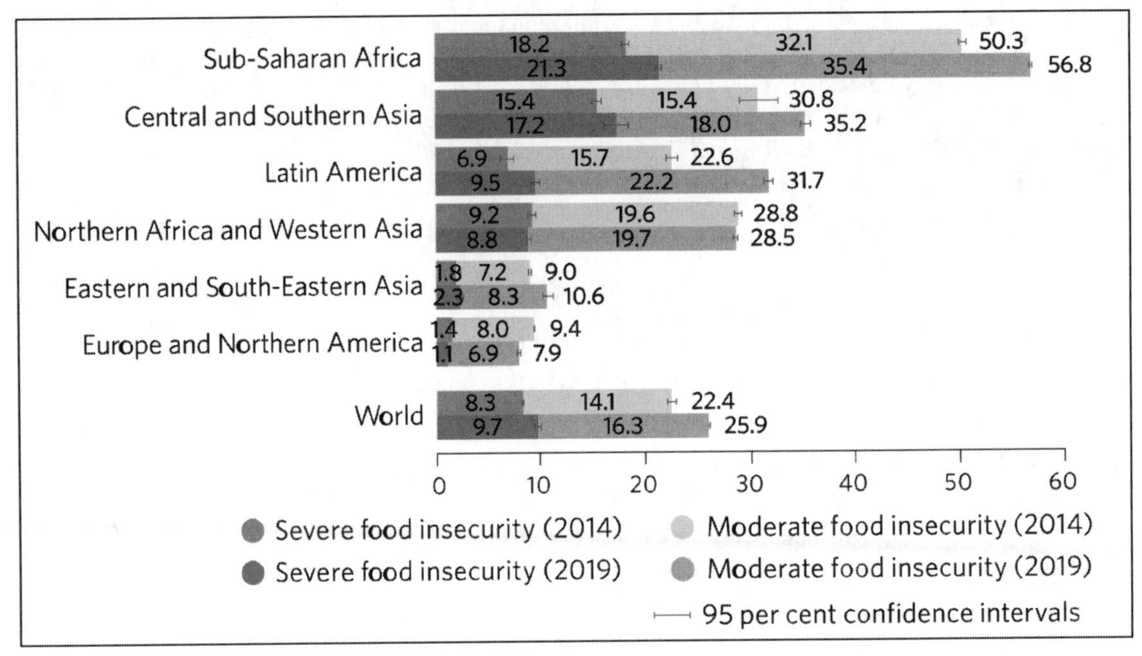

Figure 15: Food insecurity.

hungry has slowly increased for several successive years (please see Figure 15). Almost 690 million people were undernourished in 2019, up by nearly 60 million compared to 2014. Our Machine Learning Model identifies the key indicators. Along with conflict, climate shocks and the locust crisis, COVID-19 poses an additional threat to food systems, indirectly reducing purchasing power and the capacity to produce and distribute food, which affects the most vulnerable populations. In 2020, up to 132 million more people may suffer from undernourishment because of COVID-19 [29].

There is a nexus between market forces and commodity prices and to understand the relationship, the commodity prices are analyzed either fundamentally or technically. Fundamental analysis studies supply and demand relationships that define the price of a commodity at any given time. In this mode of analysis, the commodity price equilibrium, change in equilibrium price, price stability, and price levels are studied. In the figure below, both buyers and sellers are willing to exchange the quantity Q at the price P. At this point, supply and demand are in balance. Price determination depends equally on demand and supply (Figure 16).

It is truly a balance of the market components. To understand why the balance must occur, examine what happens when there is no balance, such as when the market price is below P in Figure 16.

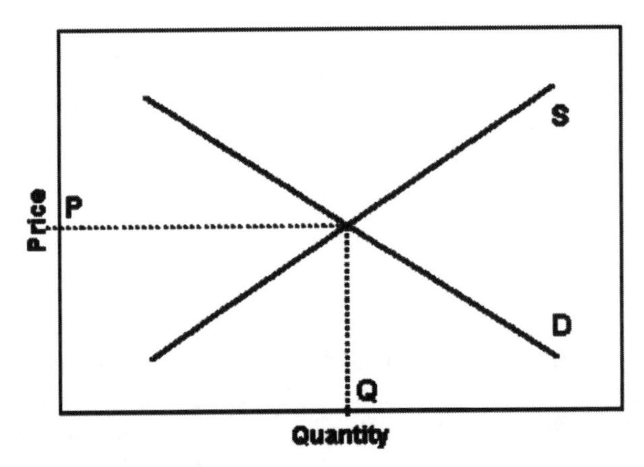

Figure 16: Price Equilibrium.

At any price below P, the quantity demanded is greater than the quantity supplied. In such a situation, consumers would clamor for a product that producers would not be willing to supply; a shortage would exist. In this event, consumers would choose to pay a higher price to get the desired product, while producers would be encouraged by a higher price to increase market supply. The result is a rise in the equilibrium price, P, at which supply, and demand are in balance. Similarly, if a price above P was chosen arbitrarily, the market would be in surplus with too much supply relative to demand. If that were to happen, producers would be willing to lower the selling price, and consumers would be induced by lower prices to buy more. Balance is restored only when the price falls.

Change in Equilibrium

Unusually good weather increases output. When a bumper crop develops, supply shifts towards the right from S_1 to S_2 as shown in Figure 17 and a higher product quantity is available at the new equilibrium price P_2. With no immediate change in consumer preferences for crops, there is a movement along the demand curve to a new equilibrium. Consumers will buy more but only at a lower price. By how much the price must fall to induce consumers to purchase more depends upon the price elasticity of demand (please see Figure 17).

In Figure 18, price falls from P1 to P2 if a bumper crop is produced. If the demand curve in this example was more vertical (more inelastic), the price-quantity adjustments needed to bring about a new supply-demand equilibrium would be different.

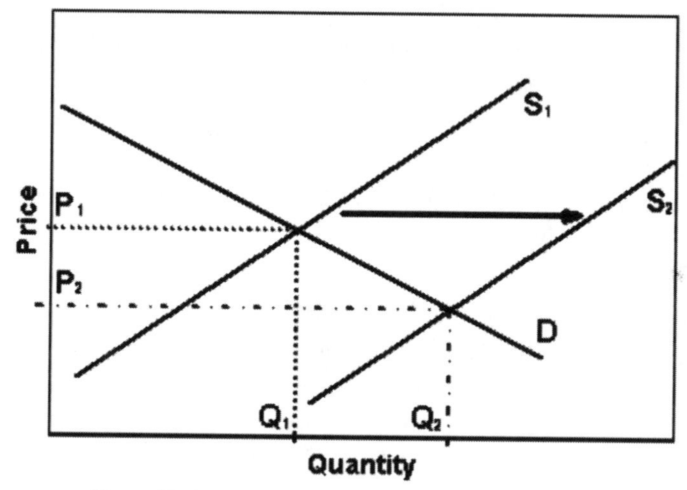

Figure 17: Movement Along Demand Curve (bumper crop).

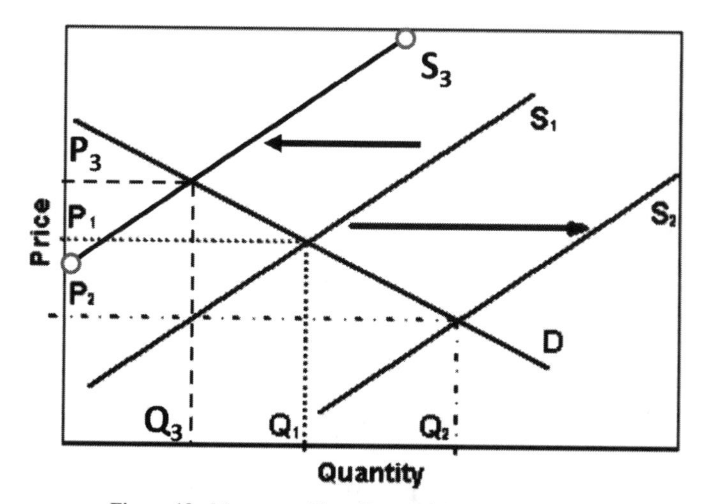

Figure 18: Movement Along Demand Curve (war events).

Conflicts, wars, extreme weather events, and lower rains will have a reverse effect on the above demand curve movement. When extreme weather events, wars/conflicts, or bad monsoon occur, supply shifts inward and upward, shown as in Figure 18 and a lower product quantity is available over the full range of prices. With scarce resources to buy crops, there is a movement along the demand curve to a new equilibrium. Consumers will pay more to buy a lower quantity a fact we can witness in price movements of essential commodities during the ongoing Ukraine—Russia war. Russia's invasion of Ukraine has disrupted supplies from one of the world's top grain-exporting regions, pushing up prices for wheat and corn. Bad weather afflicting other big crop-producing countries, including South America,[199] is also fueling the supply crunch. Meanwhile, demand remains robust for food, livestock feed and fuel made from grains, according to industry experts [30]. Sometimes shifting supply curves to the left requires unprecedented intervention. For instance, the White House is asking Congress to approve $500 million in food production assistance for U.S. crop farmers as part of a $33 billion aid package to address the effects of Russia's invasion of Ukraine.[200] The administration says the request "provides greater access to credit and lowers risk for farmers growing these food commodities, while lowering costs for American consumers." [31] The factsheet[201] released by the Biden administration has laid out necessary support for protecting agriculture, food security, and, of course, support for national security. The policy support included:

- Support for small- and medium- sized agrobusinesses during the fall harvest and for natural gas purchases by the Ukrainian state energy company to address critical food security, energy, and other emerging needs in Ukraine.

<table>
<tr>
<td></td>
<td>Price Equilibrium needs unprecedented actions: White House Calls on Congress to Provide Additional Support for Ukraine

The $3 billion in additional humanitarian assistance will provide critical resources to address food security needs around the globe, provide wheat and other commodities to people in need, build countries' resilience to global food supply and price shocks, and provide lifesaving aid to people displaced by or otherwise impacted by Putin's War in Ukraine [32].

An additional $500 million in domestic food production assistance will support the production of U.S. food crops that are experiencing a global shortage due to the war in Ukraine, for example, wheat and soybeans. Through higher loan rates and crop insurance incentives the request provides greater access to credit and lowers risk for farmers growing these food commodities, while lowering costs for American consumers.</td>
</tr>
</table>

The movement along the demand curve can be observed in Wheat futures' price movements, an essential food grain worldwide. Chicago wheat futures fell below $10.7 per bushel at the end of April 2022, the lowest in three weeks, as stronger supply projections eased shortage concerns.[202] The Biden administration proposed a $500 million bill for the farming sector, in an effort for wheat producers to increase the number of crop fields in response to surging grain prices. Support of stronger supply also came from India entering the export market as higher wheat prices enabled famers to divert new crop to private traders instead of state stockpilers. Egypt, the world's largest importer, is expected to purchase 1 million tonnes of wheat from India, including 240 thousand tonnes in April. Wheat prices surged to 14-year highs in March due to the war in Ukraine. Ukrainian production is expected to significantly stall, with 20% of winter plantations being wasted due to direct consequences of the war, while a declining economy should cut Russia's output [30].

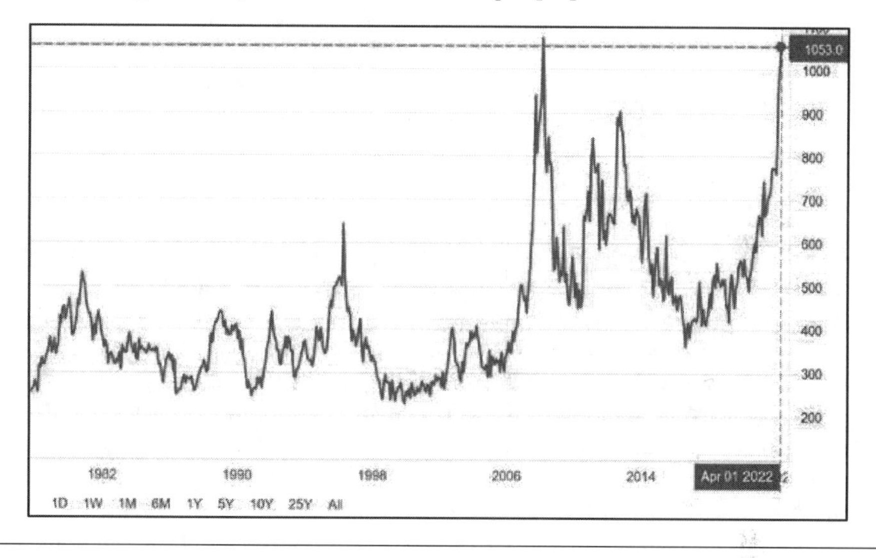

Demand shifts, like those of supply shifts shown below (please see Figure 19), are possible either due to changes in customer preferences, economic contraction, inflation, and other major economic/political events. Rapid inflation across the world could cause demand destruction in certain industries, one of the factors that could shift the demand curve inward or to the left ($D_1 \rightarrow D_2$) [30], and since agribusinesses serve a wide array of customers and industries, including food, feed, fuel, and industrial companies, a recession could significantly impact financial reports [30].

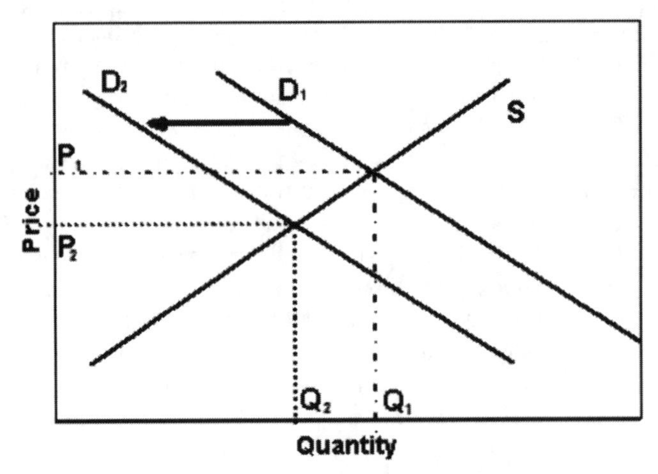

Figure 19: Movement Along Supply Curve.

The Paradox of Giffen Goods [33]

Giffen goods are rare forms of inferior goods that have no ready substitutes or alternatives, such as bread, rice, and potatoes. The only difference between Giffen goods and traditional inferior goods is that demand for the former increases even when their prices rise, regardless of a consumer's income. Giffen's paradox refers to the possibility that standard competitive demand, with nominal wealth held constant, can be upward sloping, violating the law of demand. For instance, take the case of wheat demand in Afghanistan during the great recession 2008. The increase in consumption of wheat products in urban areas following the wheat flour price increase of 2007/08 appears to contrast with the 'Law of Demand', which states that *quantity demanded decreases as price increases*. Though not conclusive, these results are broadly consistent with the paradox of Giffen goods—for which quantity demanded increases rather than decreasing as price increases. This paradox is driven by the fact that a Giffen good is an inferior good rather than a normal good. Holding prices constant, as income increases, the demand for a normal good increases, but the demand for an inferior good decreases.[203]

The findings from the National Risk and Vulnerability Assessment suggest that wheat products are both inferior goods and Giffen goods. The increase in wheat prices induced households to buy other goods that were relatively cheaper after the price increase; however, since household purchasing power declined due to the price increase of a major household necessity, households were induced to buy more of the inferior wheat products. In urban areas, the purchasing power effect outweighed the substitution effect and overall, households purchased more wheat products.

Technical Analysis

Technical analysis uses specialized methods (econometric, regressive, ensemble, and advanced machine learning) for predicting prices by analyzing past price patterns and levels. There is a pattern in the historical prices. A holistic view of the current pricing influence of commodities can be visualized using the Primary Commodity Price System[204] (PCPS). The PCPS captures commodity prices since 1992 and is a great source of historical information. While this has been described as driving a car using only the rear-view mirror, its wide acceptance by traders makes it a credible technique. Traders predict when price trends will change and how high or low prices will move by charting prices (usually futures) and looking for repeating patterns.[205] This is where machine learning and AI will help. The demand relationship is based on fundamental factors: equilibrium price. Price is dependent on the interaction between the demand and supply components of a market. Demand and supply represent the willingness of consumers and producers to engage in buying and selling. An exchange of a product takes place when buyers and sellers can agree upon a price.

The Producer Price Index (PPI) measures the average change in prices U.S. producers receive for the sale of their products (please see Figure 20). Higher producer prices mean consumers will pay more when they buy, whereas lower producer prices likely mean consumers will pay less at the retail level. March 3, 2016, marked the 125th anniversary of the PPI[206]—*one of the oldest economic time series compiled by the federal government.* The index, known as the Wholesale Price Index (WPI) until 1978, was established as part of a U.S. Senate resolution on March 3, 1891, the last day of the last session of the 51st U.S. Congress.2 This Congress was famously known as the "Billion-Dollar Congress", because of its expensive initiatives, such as expanding the Navy and creating pensions for families of military members who served in the Civil War. Please see PPI for all commodities 1926–1943 [34].

The sustainability of small farmers depends on market access and having historical prices play an important role.

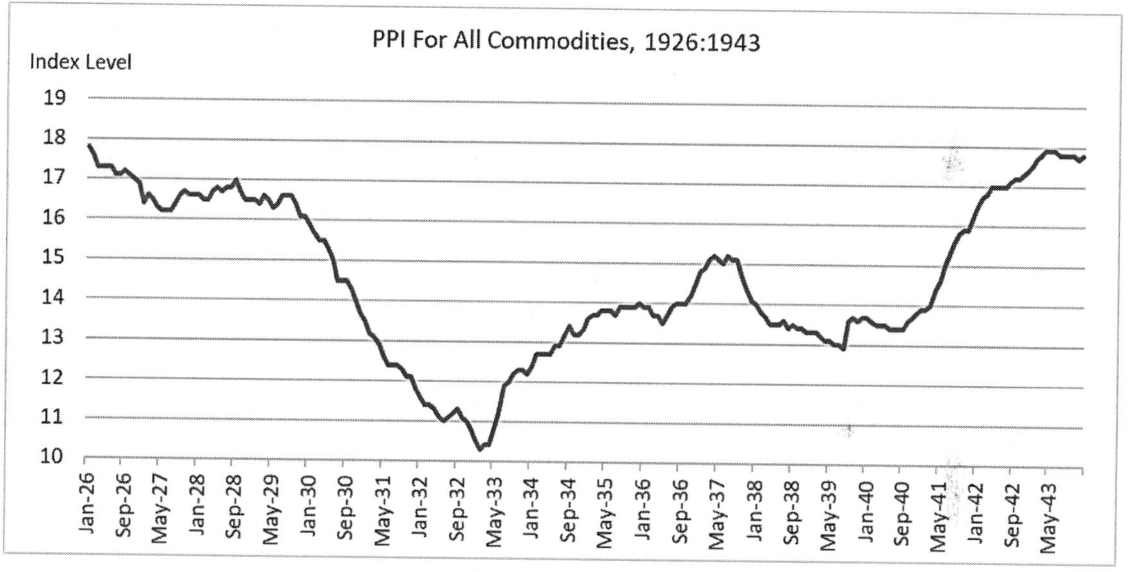

Figure 20: PPI for all Commodities.

On the other hand, the Consumer Price Index (CPI) is a measure of the average change over time in the prices paid by urban consumers for a market basket of consumer goods and services.[207] The consumer price index is a barometer measuring inflation. If the CPI declines, that means there's deflation, or a steady decrease in the prices of goods and services (Equation 2).

$$CPI_t = (C_t/C_o) * 100 \quad \text{(EQ. 2)}$$

Where CPI_t = Consumer price index in current period
C_t = Cost of market basket in current period
C_o = Cost of market basket in base period

Relationship between Consumer Price Index (CPI) and Producer Price Index (PPI) for Food.

The PPIs—measure changes in farm and wholesale prices—are typically far more volatile than the downstream CPIs.[208] Price volatility decreases as products move from the farm to the wholesale sector to the retail sector. Because of multiple processing stages in the U.S. food system, the *CPI typically lags movements in the PPI. The PPI is thus a useful tool for understanding what may soon happen to the CPI* [35].

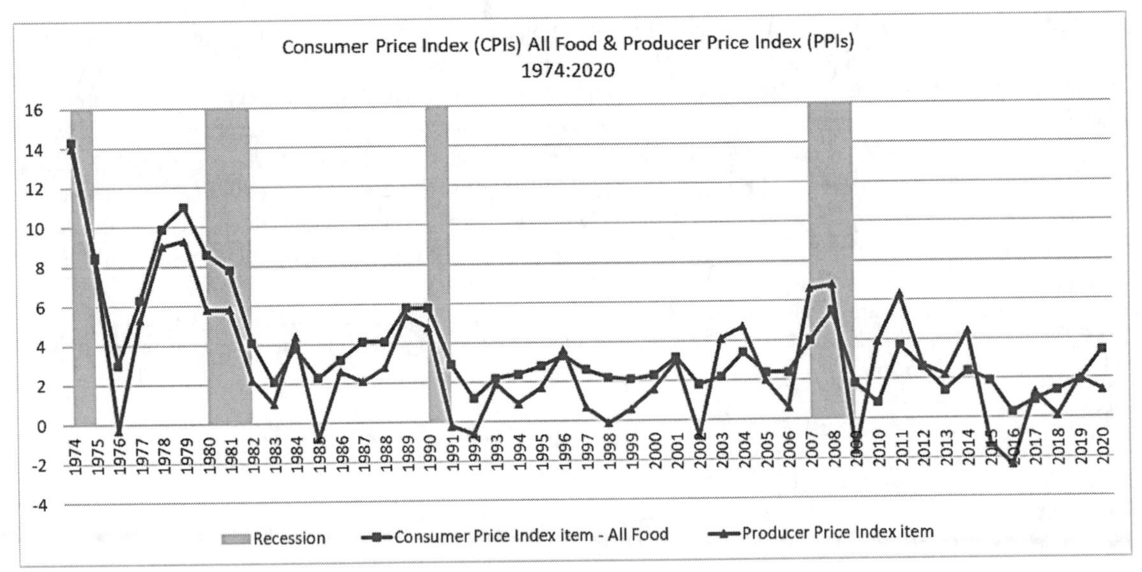

Figure 21: CPI amd PPI.

As you can see (Figure 21) the producer price index increase sets an upward trend for the CPI a fact we can witness in 1978, 1984,1989, 2008, and 2011.

The historical trends of PPI and CPI provide a trove of economic and comoving agricultural signals for developing machine learning models. A historical time series analysis with major predictors provides valuable insights. For example, In 1984, both PPI and CPI were below 5% and a variety of factors reinforced each other to hold inflation substantially in check as was the case in 1983 [36]:

- Good harvests for many agricultural crops, both in th United States and abroad (Figures 20 and 21);
- Continued weakness in world commodity markets for energy and many basic industrial materials;
- The unusually high value of the U.S. dollar in international currency markets, which encouraged a surge of imports that averted production and labor bottlenecks by tapping off much of the upswing in domestic demand;
- Weak export demand for most goods made in the U.S., also caused largely by the strength of the dollar;
- An excellent year for domestic capital investment projects designed to expand capacity with demand;
- Solid U.S. productivity improvements and general wage restraint, both of which held down rises in unit labor costs;
- American monetary policies which gave high priority to maintaining a low rate of inflation; and
- The slowing of domestic economic expansion in the latter half of the year.

As a result, inflation in 1984 at both the retail and the producer levels rose at a rate of less than 5 percent for the third consecutive year. This moderate performance coincided with the second year of strong economic recovery from a recession that ended in late 1982 [36].

Another technical analysis is commodity terms of trade. Global commodity prices greatly influence a country's terms of trade,[209] employment, income and ultimately inflation [37]. Commodity prices soared in 2021, with prices of several commodities reaching all-time highs.[210] The broad-based surge, led by energy and metals, partly reflected the strong rebound in demand from the global recession in 2020 [38]. It was amplified by weather-related supply disruptions for both fossil and renewable fuels.

Essentially, both market economic demand and weather-related changes have surged commodity processes. Commodity-exporting emerging markets and developing economies (EMDEs) generally need to take steps to better manage future commodity price shocks and to reduce their vulnerability to such shocks. This is one of the key factors in ensuring food security.

Commodities are critical sources of export and fiscal revenues for almost two-thirds of EMDEs and three-quarters of low-income countries (LICs). More than half of the world's poor reside in commodity exporting EMDEs. Dependence on commodities is particularly high for oil exporters. Countries whose exports are heavily concentrated on one or a few commodities tend to experience high volatility in their terms of trade and output growth. The full dataset[211] can be downloaded from the FAO site.

Commodity Super Cycles

Commodity prices tend to go through extended periods of boom and bust, known as super cycles [37]. In general, commodity price movements are important because they help determine the country's terms of trade, exchange rate, employment, income, and inflation.

Current account balance (% of GDP)

Current account balance is the sum of net exports of goods and services, net primary income, and net secondary income.[212] For predicting prevalence of moderate or severe food insecurity in the population (%), regression coefficient of current account balance (% of GDP) is negative (–0.70), indicating that there is a great deal of pressure on prevalence of moderate or severe food insecurity in the world. This may be explained by the need for a large level of food imports other consumables to satisfy the teeming and increasing population of the country.[213] What it says is, if all held constant, deficit current account balance of a country could heavily influence food security [39]! Countries with excessive current account deficits should, where appropriate, seek to reduce budget deficits over the medium term and make competitiveness-raising reforms, in areas including education and innovation policies. In economies with excessive current account surpluses remaining in the fiscal space,[214] policies should support recovery and medium-term growth through greater public investments [40].

Employment to population Ratio, 15+, total (%) (Modeled ILO Estimate)

The employment to population ratio is the proportion of a country's population that is employed. Employment is defined as persons of working age who, during a short reference period, are engaged in any activity to produce goods or provide services for compensation or profit, whether at work during the reference period (i.e., who worked in a job for at least one hour) or not at work due to a temporary job absence, or due to working-time arrangements. Ages 15 and older are generally considered the working-age population.[215] Unemployment affects food security.[216] Higher levels of unemployment at country level correlates with higher food insecurity [41]. USDA predicts food insecurity to increase[217] in 2021 at a higher rate in lower income countries due to income-related effects caused by employment, compounded by the COVID-19 fallout [42].

Foreign direct investment, net inflows (% of GDP)

Foreign direct investment[218] is the net inflows of investments to acquire a lasting management interest (10 percent or more of voting stock) in an enterprise operating in an economy other than that of the investor. It is the sum of equity capital, reinvestment of earnings, other long-term capital, and short-term capital as shown in the balance of payments. This series shows net inflows (new investment inflows less disinvestment) in the reporting economy from foreign investors and is divided by GDP. Foreign Direct Investment has both positive and negative regression coefficients while predicting prevalence of food insecurity. For moderate food insecurity prevalence, Foreign direct investment coefficient is negative [43]. Foreign Direct Investment has shown improvement[219] in food and financial security [44]. Hungary[220] is the country with the highest net foreign direct investment (% of GDP) and Netherland[221] & Ireland[222] have the most competitive FDI.

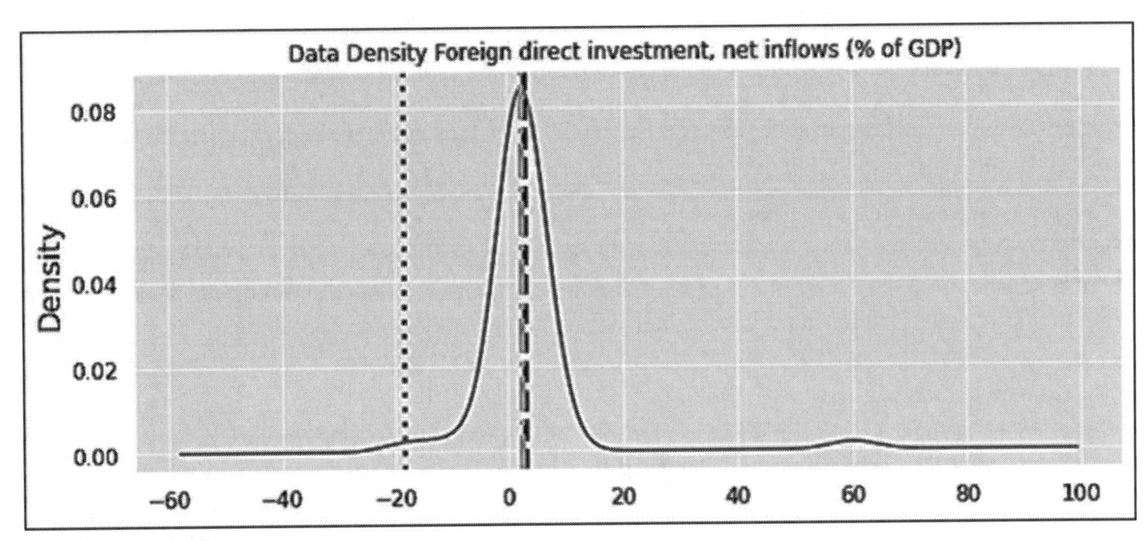

Figure 22: FDI Data Density.

The Data Distribution statistics (Figure 22): Minimum: –18.60; Mean: 2.81(dashed line); Median: 2.35 (dash dot line); Mode: –18.60 (dotted line); Maximum: 60.06 (based on Who Is the World's Food Insecure? Prevalence of moderate or severe food insecurity in the Population ML Model below).

To summarize, the distribution of data is skewed to the right—statistically, if the mode is less than the median, which is less than the mean (based on Who In the World is Food Insecure? Prevalence of moderate or severe food insecurity in the Population ML Model below).

GDP per capita, PPP (Constant 2017 International $)

GDP per capita based on purchasing power parity (PPP).[223] PPP GDP is gross domestic product converted to international dollars using purchasing power parity rates. An international dollar has the same purchasing power over GDP as the U.S. dollar has in the United States. GDP at purchaser prices is the sum of gross value added by all resident producers in the country plus any product taxes minus subsidies not included in the value of the products. It is calculated without making deductions for depreciation of fabricated assets or for depletion and degradation of natural resources. Data is constant in 2017 international dollars. As it is well known, food security is economy dependent, mostly, and higher purchasing power would help improve food security. Conversely, the highest levels of food insecurity and malnutrition are seen in the regions of the world that have a challenged real GDP per capita.[224]

Imports of goods and services (% of GDP)

Imports of goods and services represent the value of all goods and other market services received from the rest of the world.[225] They include the value of merchandise, freight, insurance, transport, travel, royalties, license fees, and other services, such as communication, construction, financial, information, business, personal, and government services. They exclude compensation of employees and investment income (formerly called factor services) and transfer payments. Imports of goods and services with governmental policies in place show that global trade improves the food and nutrition security of countries in Africa, Asia, and Latin America (Figure 23). Trade also promotes a healthier and more balanced diet,[226] as countries have access to an increased variety of food. The effect of trade in enhancing nutrition security, with an adequate supply of macro and micronutrients, is universal across nutrients and countries [45]. Trade protection, import of goods and services, enhance the agricultural and rural sectors. On the same token, the side effects of import protection,[227] while it increases income for producers, increases in the price of food can create a barrier for low-income consumers to access food. Also, many farmers themselves are net purchasers of food. This can be exacerbated when import protections serve to keep out competition, which can help to ensure efficiency and guard against corruption in markets. A balanced approach would help [46]!

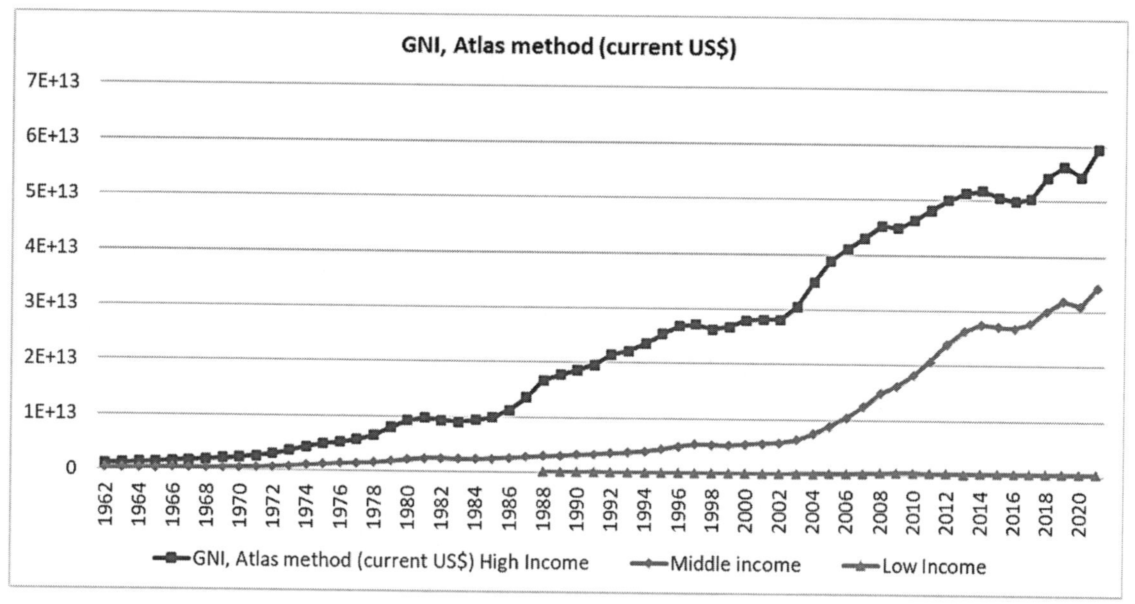

Figure 23: Income groups.

Official exchange rate (LCU per US$, Period Average)

Official exchange rate refers to the exchange rate determined by national authorities or to the rate determined in the legally sanctioned exchange market. It is calculated as an annual average based on monthly averages (local currency units relative to the U.S. dollar).[228] The International Monetary Fund (IMF) provides exchange rates on[229] and we use Representative Rates for Selected Currencies as exchange rate:

- Representative Rates for Selected Currencies - These rates, normally quoted as currency units per U.S. dollar, are reported daily to the Fund by the issuing central bank. Rates are reported for members whose currencies are used as a financial Fund.
- Special Drawing Rights (SDRs)[230] per Currency unit—(e.g., $ 1.00 = 0.67734 SDR). These rates are the official rates used by the Fund to conduct operations with member countries. The rates are derived from the currency's representative exchange rate, as reported by the central bank, normally against the U.S. dollar at spot market rates and rounded to six significant digits.
- Currency units per SDR—This rate, which is not used in Fund transactions, is the reciprocal of the SDR per currency unit rate, rounded to six significant digits (e.g., $ 1.47638 = 1 SDR).

Why monitor exchange rates? For the simple reason that exchange rates impact food prices which in turn impacts food security. In other words, as the FAO put it, exchange rate fluctuations can impact food prices and, consequently, access to food. If the local currency depreciates relative to the USD, food imports get more expensive. This leads to a rise in domestic food prices for imported items and, through substitution of these items with less expensive ones and consumption shifts to increase domestic food prices in general. At the same time, food exports become more competitive. Ensuing higher export demand can also drive-up domestic food prices. Moreover, if a currency rapidly loses value, holding on to food commodity stocks can be perceived as a more reliable form of saving than keeping local currency and leads to lower food availability in the market. The World Food Program (WFP) maintains an exchange rate dashboard[231] to monitor the changes. Looking at exchange rate movements together with food price inflation for Egypt, it is evident that the steep depreciation of the Egyptian pound in November 2016 (USD/EGP[232] jump can be seen in Figure 24) coincided with accelerated food price inflation (increase inflation)[233]. Here is the list of selected country exchange rates with respect to the USD (Foreign Exchange Rates—H.10 Weekly[234]).

Figure 24: USD/EGP Rates.

Population ages 0–14 (% of total population)

The population between the ages 0 to 14 is expressed as a percentage of the total population. Population is based on the de facto definition of population. This percentage drastically varies across country income groups.[235] For the year 2020, low-income countries lead with 42%, followed by lower middle-income countries with 30%, upper middle income (XT) countries with 20%, and high income (XD) countries with 16% (Figure 25). This key driver plays an important role in predicting the prevalence of food insecurity for the World [47].

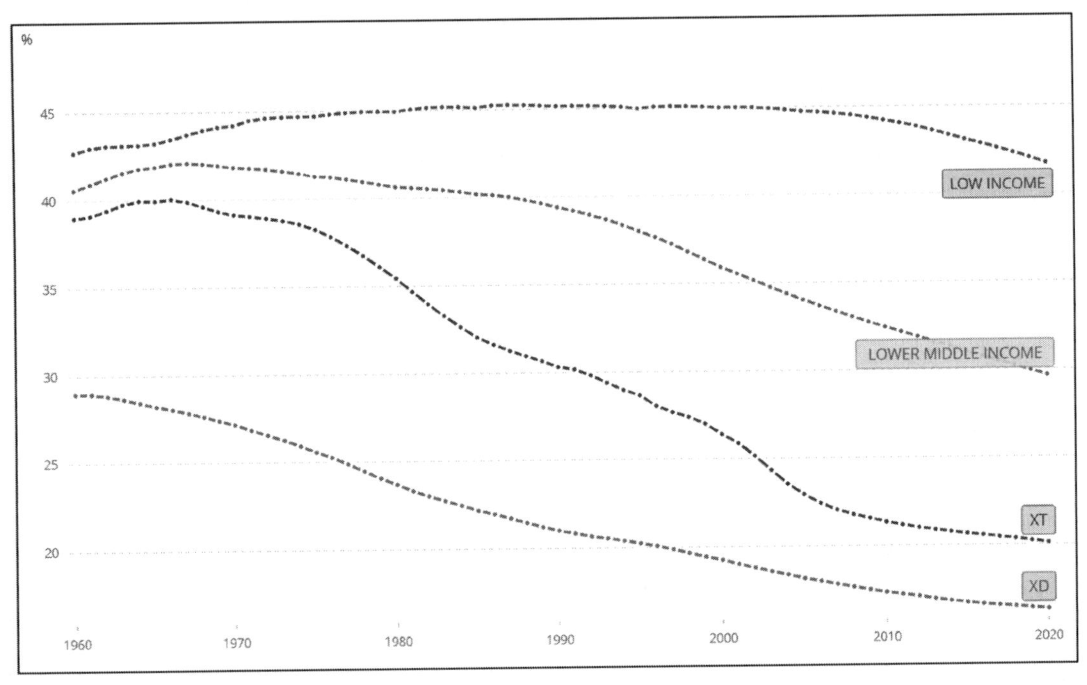

Figure 25: Population percentages.

An observation that can be made is, the percentage of children living in food-insecure households varies by urbanicity associated with country income groups.[236] Households in high income countries such as U.S., where children in large central metropolitan areas (13.2%) are more likely than those in large fringe metropolitan (7.4%) and medium and small metropolitan (10.5%) areas to live in households that are food insecure, for the year 2019–2020 [48]. In low-income countries the same problem is multi-fold. Urbanization[237] in low and middle-income nations presents both opportunities and immense challenges. As urban centers grow rapidly, inadequate housing and the lack of basic infrastructure and services affect a large and growing proportion of their population. All this contributes to food insecurity [49].

Rural population (% of Total Population)

Rural population refers to people living in rural areas as defined by national statistical offices. It is calculated as the difference between the total and urban population.[238] People living in urban and rural areas share similar concerns about food prices and availability—including the impact of the recent bird flu outbreaks. However, differences remain in food insecurity and diet satisfaction, according to the monthly Consumer Food Insights Report.[239] The rural population is decreasing at an alarming rate [50]. Countries with rural population greater than a mean 0f 32.04% include Austria, Azerbaijan, Ecuador, Guatemala, Guinea, Honduras, India, Indonesia, Ireland, Kenya, Myanmar, Nigeria, Paraguay, Philippines, Poland, Portugal, Romania, Rwanda, Serbia, South Africa, Thailand, Uzbekistan, Zambia. Labor is becoming scarce on developing Asian farms. India's agricultural workforce fell from 259 million in 2004–2005 to 228 million in 2011–2012. One potential reason is the disappearance of small farms due to lack of employment opportunities outside farms.[240] The same agricultural issues are prevailing in high income countries such as the U.S. As per Jayson Lusk,[241] the head and distinguished Professor of Agricultural Economics at Purdue "Rural Americans struggle more often than urban Americans to buy the food they want. As one might expect, current economic conditions, May 2022, appear to have further disadvantaged this group. For example, they face not only significantly higher food prices today than four months ago, but higher gas prices are likely affecting purchasing decisions among those who must drive many miles to a grocery store" [51].

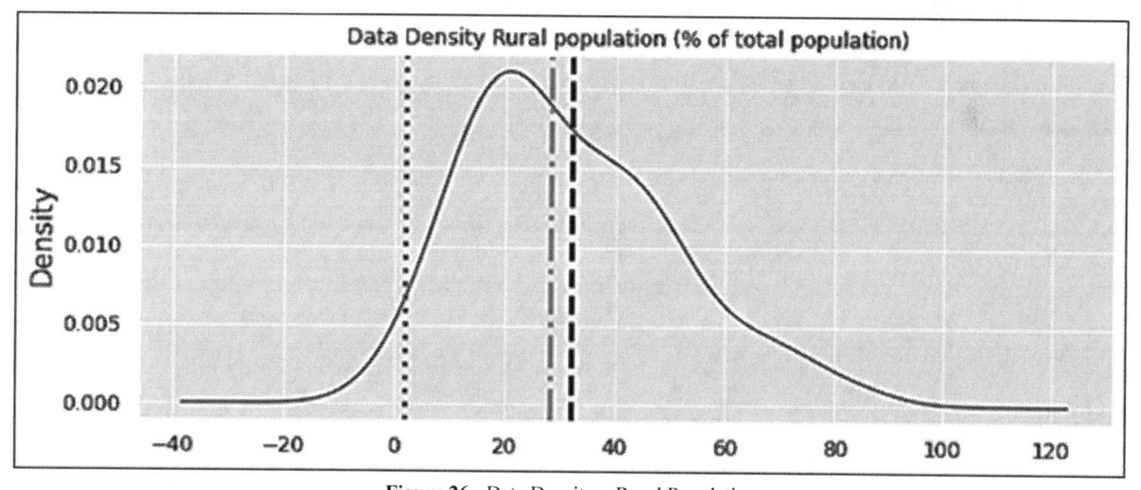

Figure 26: Data Density—Rural Population

The Data Distribution (Figure 26) statistics: Minimum:1.96; Mean:32.04(dashed line); Median:28.36 (dash dot line); Mode:1.96 (dotted line); Maximum:82.69 (based on Who In the World is Food Insecure? Prevalence of moderate or severe food insecurity in the Population ML Model below).

"Rural Americans struggle more often than urban Americans to buy the food they want. As one might expect, current economic conditions appear to have further disadvantaged this group. For example, they face not only significantly higher food prices today than four months ago, but higher gas prices are likely affecting purchasing decisions among those who must drive many miles to a grocery store".

Jayson Lusk,
the head and Distinguished Professor of Agricultural Economics at Purdue

Prevalence of moderate food insecurity is an inhibitor of day-to-day functional life whereas prevalence of severe food insecurity will have a long-term adverse impact and, in almost all cases, shunts a fully potential life.

Total reserves (Includes Gold, Current US$)

Total reserves[242] comprise holdings of monetary gold, special drawing rights (SDRs), reserves of IMF members held by the IMF, and holdings of foreign exchange under the control of monetary authorities. The gold component of these reserves is valued at year-end (December 31) London prices. Data are in current U.S. dollars.

Agriculture Supply & Co-Movement Drivers

Two genera of supply side key drivers are modeled to predict the prevalence of moderate and severe food insecurity. One is from the agricultural side and the other one is from price shocks of agriculture commodities (Figure 27).

Here is the list of key drivers with a brief description:

- Agricultural raw materials imports (% of merchandise imports)
- Agriculture, forestry, and fishing, value added (% of GDP)
- Employment in agriculture (% of total employment) (modeled ILO estimate)
- Cereal import dependency ratio
- Change in average food cost
- Food supply adequacy

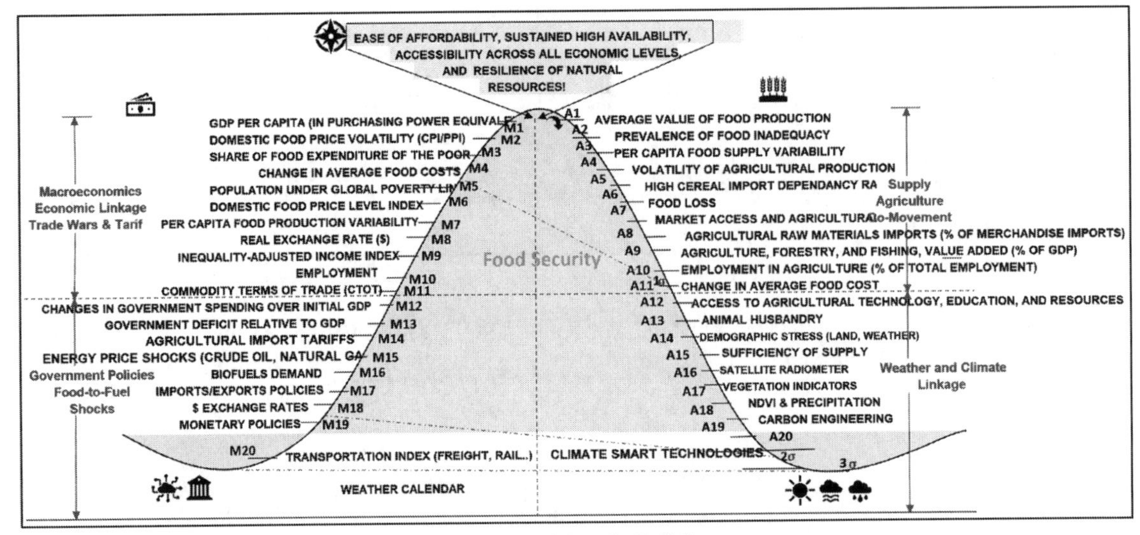

Figure 27: Food Security Bell Curve.

- Dependency on chronic food aid
- Public expenditure on agricultural research and development
- Access to agricultural technology, education, and resources
- Crop storage facilities
- Volatility of agricultural production
- Food loss
- Road Infrastructure
- Proportion of population under global poverty line

Agricultural raw materials imports (% of merchandise imports)

Agricultural raw materials comprise of the SITC section[243] 2 (crude materials except fuels) excluding divisions 22, 27 (crude fertilizers and minerals excluding coal, petroleum, and precious stones), and 28 (metalliferous ores and scrap). Agricultural raw materials[244] play an important role in food insecurity as the price trend for them is less predictable because weather-related shocks will continue to create annual price volatility. Price Rises will have the greatest impact on Import-Dependent countries and particularly grains and oilseeds crucial in developing country diets. Dependence on imports rises in many developing countries and high prices and rising import dependence lead to widening food gaps [52].

Agriculture, forestry, and fishing, value added (% of GDP)

Agriculture, forestry, and fishing correspond to ISIC divisions 1–3 and include forestry, hunting, and fishing, as well as cultivation of crops and livestock production. The value added is the net output of a sector after adding up all outputs and subtracting intermediate inputs. It is calculated without making deductions for depreciation of fabricated assets or depletion and degradation of natural resources. The origin of value added is determined by the International Standard Industrial Classification (ISIC), revision 4. Note: For VAB countries, gross value added at factor cost is used as the denominator.[245]

Employment in agriculture (% of total employment) (modeled ILO estimate)

Employment is defined as persons of working age who are engaged in any activity to produce goods or provide services in exchange for pay or profit, whether at work during the reference period or temporarily absent from a job, or due to working-time arrangements.[246] The agriculture sector consists of activities in agriculture, hunting, forestry, and fishing, in accordance with division 1 (ISIC 2) or categories A-B (ISIC 3) or category A (ISIC 4). Countries with employment in agriculture greater than the mean (16.99%) include Azerbaijan, Ecuador, Guatemala, Guinea, Honduras, India, Indonesia, Kenya, Myanmar, Nigeria, Paraguay, Peru, Philippines, Romania, Rwanda, Thailand, Uzbekistan, and Zambia. The regression coefficient of

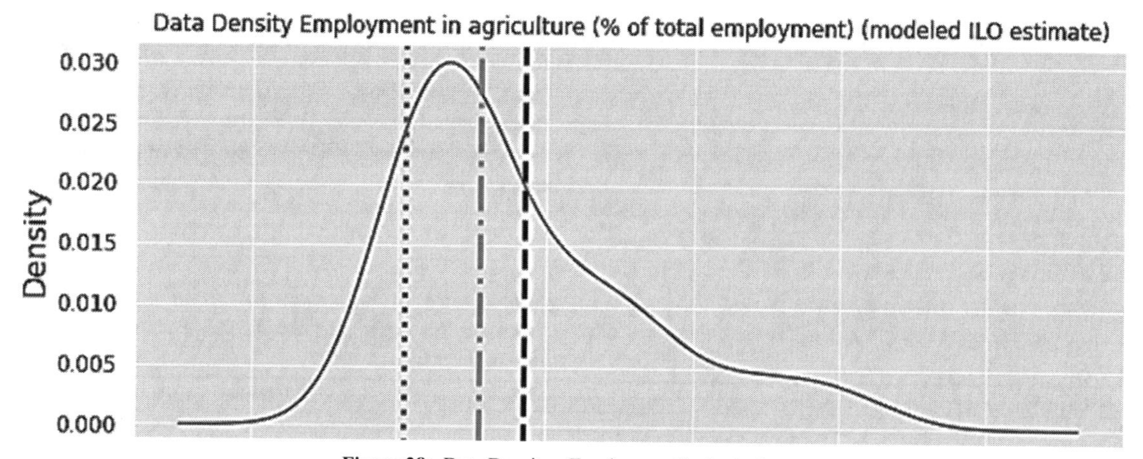

Figure 28: Data Density - Employment in Agriculture.

Employment in agriculture as a % of total employment is positive (0.30*) with respect to prevalent, moderate & severe food insecurity. However, the results differ with respect to country income groups. High-Income, Low-Income unemployment rates have a negative relationship with food security in the sample of developing countries, where high levels of unemployment (Figure 28) exacerbate the adverse effects of income inequality on food security. This is insignificant for advanced economies [53].

The Data Distribution (Figure 28) statistics: Minimum:0.92; Mean:16.99(dashed line); Median:11.06 (dash dot line); Mode:0.92 (dotted line); Maximum:62.29 (based on Who In the World is Food Insecure? Prevalence of moderate or severe food insecurity in the Population ML Model below). To summarize, the distribution of data is skewed to the right—statistically, if the mode is less than the median, which is less than the mean (based on Who In the World is Food Insecure? Prevalence of moderate or severe food insecurity in the Population ML Model below).

Cereal import dependency ratio

The cereal imports dependency ratio tells how much of the available domestic food supply of cereals has been imported and how much comes from the country's domestic production. It is computed as (cereal imports—cereal exports)/(cereal production + cereal imports - cereal exports) * 100 Given this formula the indicator assumes only values <= 100. Negative values indicate that the country is a net exporter of cereals.

Change in average food costs

A measure of the change in average food costs, as captured through the Food CPI which tracks changes in the price of the average food basket of goods since 2010. FAO Food Price Index (FFPI)[247] is the source. The FAO Food Price Index (FFPI) is a measure of the monthly change in international prices of a basket of food commodities. It consists of the average of five commodity group price indices weighted by the average export share of each group over the years, 2014–2016. FFPI comprises of the FAO Cereal Price Index, FAO Vegetable Oil Price Index, AO Dairy Price Index, FAO Meat Price Index, and FAO Sugar Price Index (Figure 29).

A measure of the change in average food costs, as captured through the Food CPI which tracks changes in the price of the average food basket of goods since 2010. FAO is the source of the data. The average value for year 2021 is 42.9 with minimum (Egypt (–0.1), Quatar (–0.1), Panama (–0.7), Ireland (–1.4)) and maximum 4027 (Venezuela). The International Monetary Fund (IMF) dataset also provides Food and Consumer Price Indexes.[248] Food Inflation influences the cost of living and health of a nation.

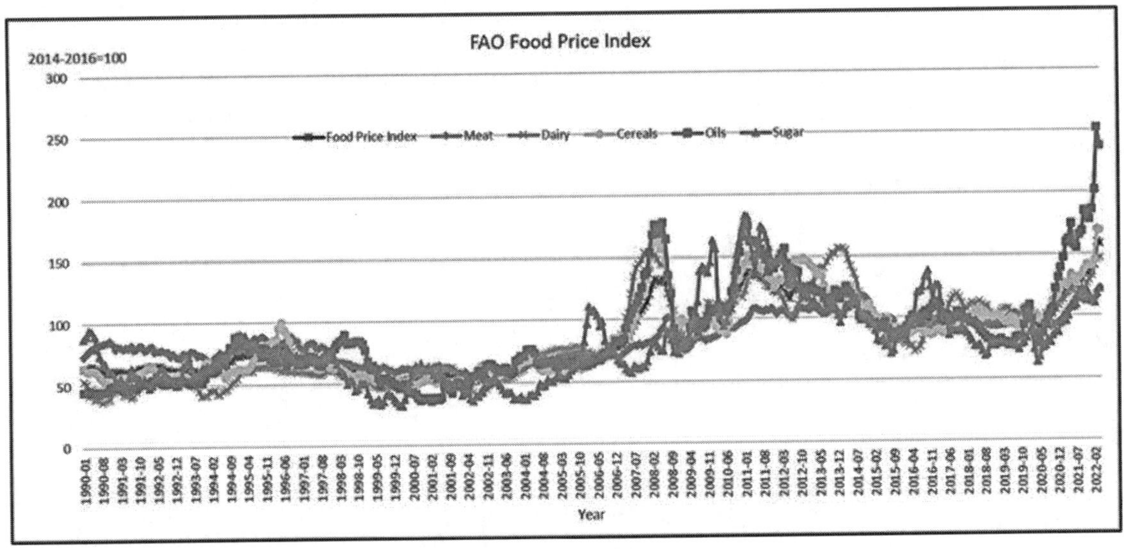

Figure 29: FFPI.

In Venezuela, during the 2016 hyperinflation,[249] the cost of a dozen eggs was $150 when inflation in Venezuela was predicted to hit 720%. As a result, many families and citizens sacrificed healthy food to maintain their food-income ratio (Figure 29). A sharp drop in global oil prices—on which Venezuela depends for most of its foreign currency— is a big part of the problem [54]. On a positive side, Venezuela's annual inflation rate hit of 686.4% in 2021, demonstrates a deceleration of consumer price growth versus the previous year when inflation was 2,959.8%.[250] Despite reduction in hyperinflation [55], prices remain high and continue to hit the earnings of Venezuelan families, limiting their ability to buy goods like food and medicine. A minimum monthly salary is equivalent to $1.50 [54]. Food access and availability are the major market access issues faced by the poor and underprivileged due to hyperinflation and can be witnessed in Venezuela where many food markets are empty Caracases.[251]

The inflation rate for consumer prices in Panama moved rose to somewhere between, 1.6% and 16.3% in the past 60 years. Consumer prices dropped by 0.2% in December 2020 on a month-on-month basis, following a flat reading in November. Falling prices for transport, furniture, recreation, and culture drove the overall reading [56]. For 2020, an inflation rate of –1.6% was calculated and it is –0.36% in 2019. During the observation period from 1960 to 2020, the average inflation rate was 2.7% per year. Overall, the price increase was 367.72 %. An item that cost 100 Balboa in 1960 was charged 467.72 Balboa in the beginning of 2021. In only a few countries negative inflation rates have been achieved (Figures 30, 31). This means that the general price level has been declining, and consumer prices have gone downwards. This case is called deflation. While deflation seems like a good thing,[252] it can signal an impending economic issue such as stricter monetary policy, recession, delayed wages, and other macroeconomic issues. When people feel prices are headed downward, they delay purchases in the hope that they can buy things for less later. But lower spending leads to less income for producers, which can lead to unemployment and higher interest rates [57]. Due to relatively stagnant worker wages as well as a hesitation from banks to distribute loans so easily to the ordinary citizen, inflation has remained considerably low in Quatar[253] (–2.72%).

	Sustained hyperinflation is a breeding ground for chronic diseases and food insecurity [58].	

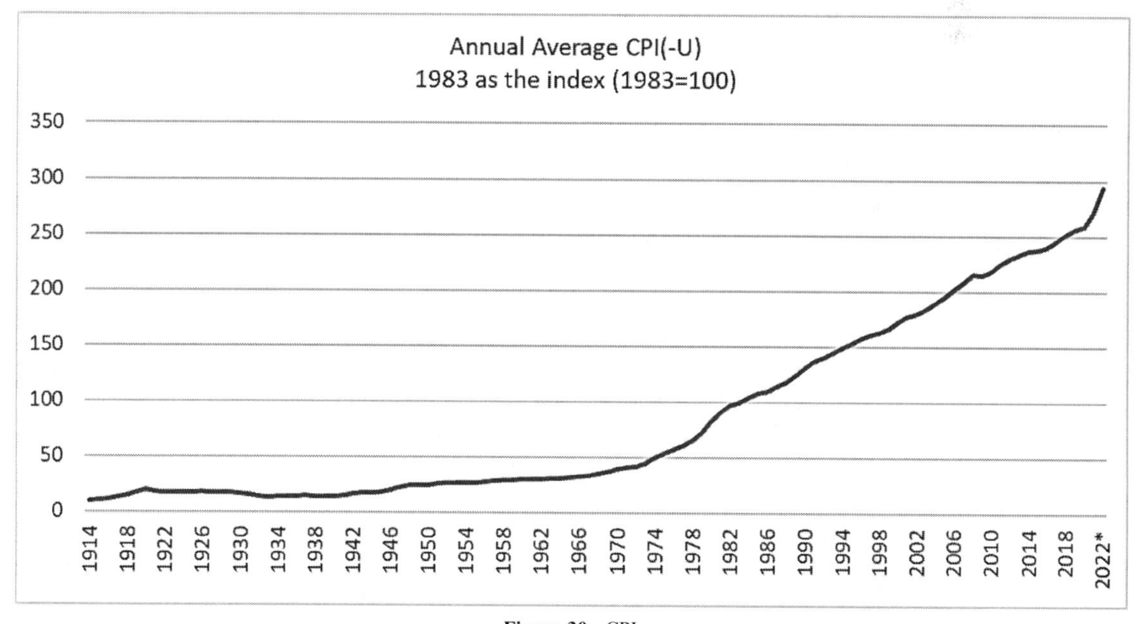

Figure 30: CPI.

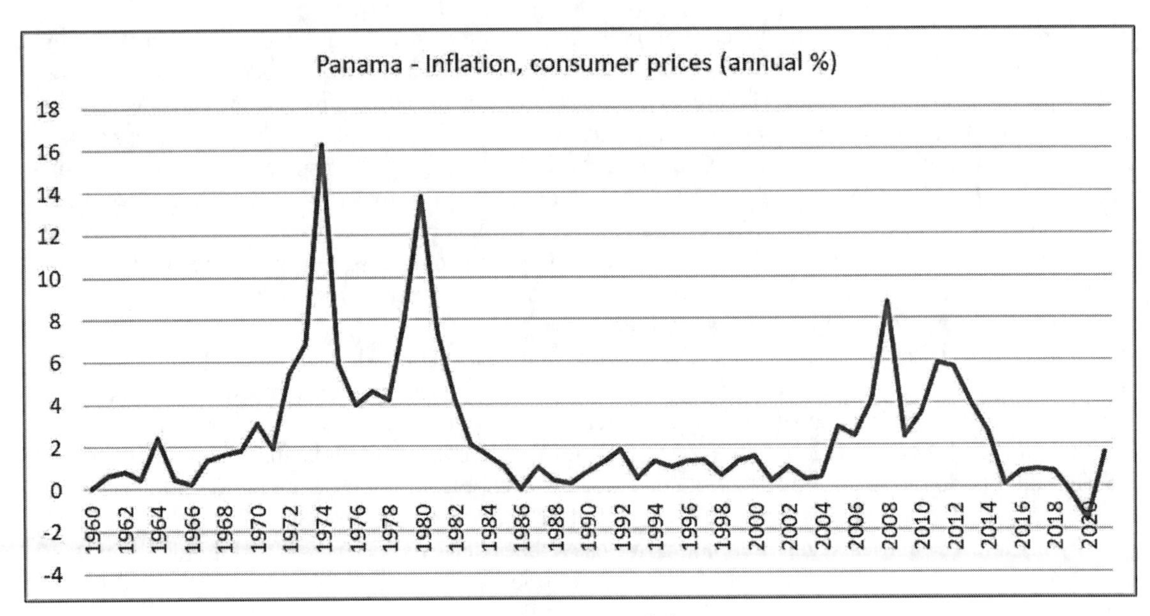

Figure 31: Panama Inflation.

Food supply adequacy

A score of 0–100 is calculated as the weighted average of the scores of the underlying sub-indicators for this indicator. Source: GFSI.

Dependency on chronic food aid

A measure of whether a country is a recipient of chronic food aid is made by assessing changes in average emergency food aid per capita received over the past 5 years. Source: OECD

Conventionally, food aid includes grants and concessional loans that conform to official development assistance (ODA). Food aid[254] is categorized and reported in terms of its uses and modes of supply. In terms of the use of food aid, three categories are distinguished:

1. Programmed food aid is supplied as a resource transfer providing balance of payments (BoP) or budgetary support.
2. Project food aid is usually provided to support specific poverty alleviation and disaster prevention activities, targeted on specific beneficiary groups or areas.
3. Relief food aid is targeted on, and freely distributed to, victims of natural or man-made disasters.

This indicator is measured in million USD constant prices, using 2018 as the base year [59]. To find the list of countries that need help, please visit FAO:[255] Countries requiring external assistance for food.

Road infrastructure

An assessment of the quality of road infrastructure is done on a 0–4 scale, where 4 is the best. Source: EIU Risk Briefing. Countries with road infrastructure that is greater than mean (2.15) include Austria, Belgium, Canada, Denmark, Finland, France, Germany, Hungary, Ireland, Italy, Netherlands, New Zealand, Norway, Panama, Portugal, Spain, Sweden, Switzerland, and Uruguay. Transportation issues, especially time and costs, make it difficult for many Americans to access healthy food.[256] Improving roads and transportation would enable more access to food and enable diversified options [60]. Investing in reliable road and railway connections is one of the best ways to curb hunger as better roads can cut down on food wastage that takes place while transporting it from the farms to markets, says a new study. Commissioned by the Copenhagen Consensus Centre, the study estimates that investing US $239 billion over the next 15 years in building good roads and electricity supplies to improve cold storage would yield benefits of US $3.1 trillion by safeguarding food, reports Reuter. This could curb hunger by 2030.[257]

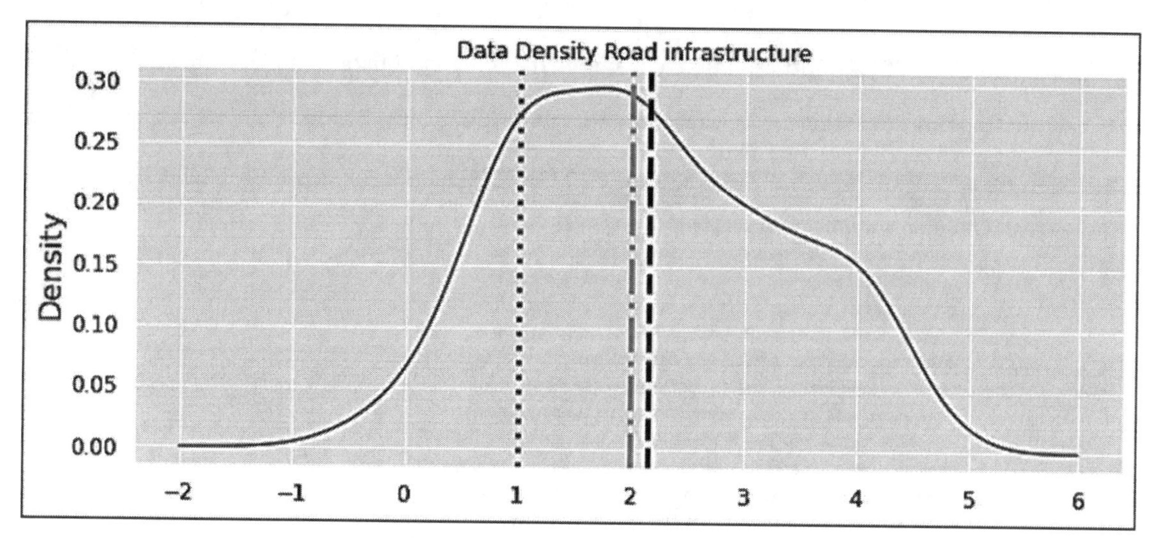

Figure 32: Data Density Road Infrastructure.

The Data Distribution (Figure 32) statistics: Minimum:0.00; Mean:2.15(dashed line); Median:2.00(dash dot line); Mode:1.00(dotted line); Maximum:4.00. To summarize, the distribution of data is skewed to the right—statistically, if the mode is less than the median, which is less than the mean 2.15 (based on Who In the World is Food Insecure? Prevalence of moderate or severe food insecurity in the Population ML Model below).

Financing infrastructure, including roads, storage and localized energy grids, will help provide food security for the millions of people living in hunger worldwide [61].

Governmental Policies and Tariffs

Tariffs are taxes paid on imported goods. In earlier eras tariffs were an important source of government revenue, but in recent decades (2000 and later) the U.S. government has used tariffs primarily to shield certain industries from foreign competition. Over the years, Congress has delegated substantial authority to impose tariffs on the executive branch,[258] which means presidents have considerable discretion to increase tariffs on specific products or imports from specific countries. President Trump used this power to increase tariffs on solar panels, washing machines, steel, and aluminum, as well as on a broad range of products from China. Overall, in 2019, the U.S. government brought in $79 billion in tariffs, twice the value from two years earlier and a sharp break from recent trends [62].

The U.S. is considering reducing tariffs in China to ease inflation. "I think some reductions may be warranted", Ms. Yellen, Treasury secretary, said, May 2022, of the tariffs, adding it could help to bring down prices. Tariffs were imposed on certain Chinese imports during the Trump administration. While some of the tariffs are important to protect U.S. national security, the cost of certain duties on China ended up being paid by Americans. When the tariffs were enacted, before the pandemic, 2019, annual inflation was trending near 2%. Now, U.S. has registered an inflation rate of 8.6% in May 2022, from the same month a year ago, marking its fastest pace since December 1981 [63]. Tariffs are common, and the World Integrated Trade Solution (WITS) web site provides a comprehensive view.[259]

Tariff Rate, Applied, Simple Mean, All Products

Consider the 2020 tariff rates for countries with higher values (Figure 33). The simple mean[260] applied tariff is the unweighted average of effectively applied rates for all products subject to tariffs calculated for all traded goods. Data is classified using the Harmonized System of trade at the six- or eight-digit level. Tariff line

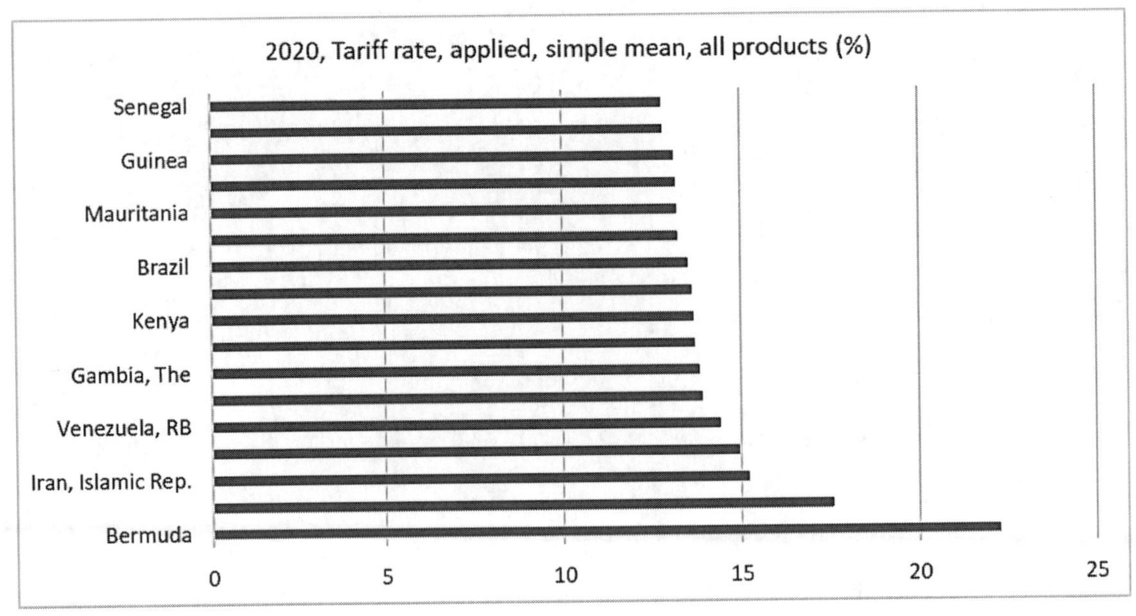

Figure 33: Top 15 countries with the highest Tariffs.

datasets were matched to Standard International Trade Classification (SITC) revision with 3 codes to define commodity groups.

Countries with the lowest tariffs are listed in the figure below. Governmental Policies, fiscal & monetary, will have an impact on macroeconomic parameters (Figure 34).

Although some policy impacts show immediately, many take time to percolate and show up on macroeconomic & agricultural key drivers [M1-M20 & A1-A20]. The relationship between trade reforms and food security status can be conceptualized at a general level, depicted in Figure 35, as a two-stage relationship where a set of causal factors impact on a series of intermediate indicators, which in turn determine

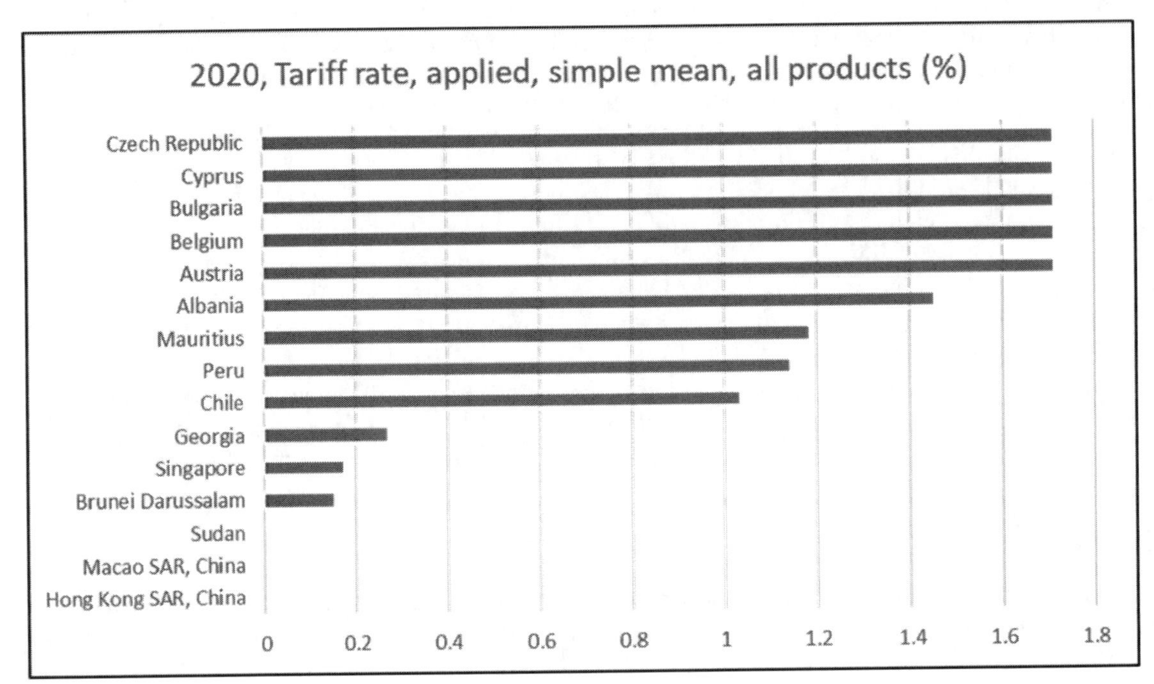

Figure 34: Tariff rate.

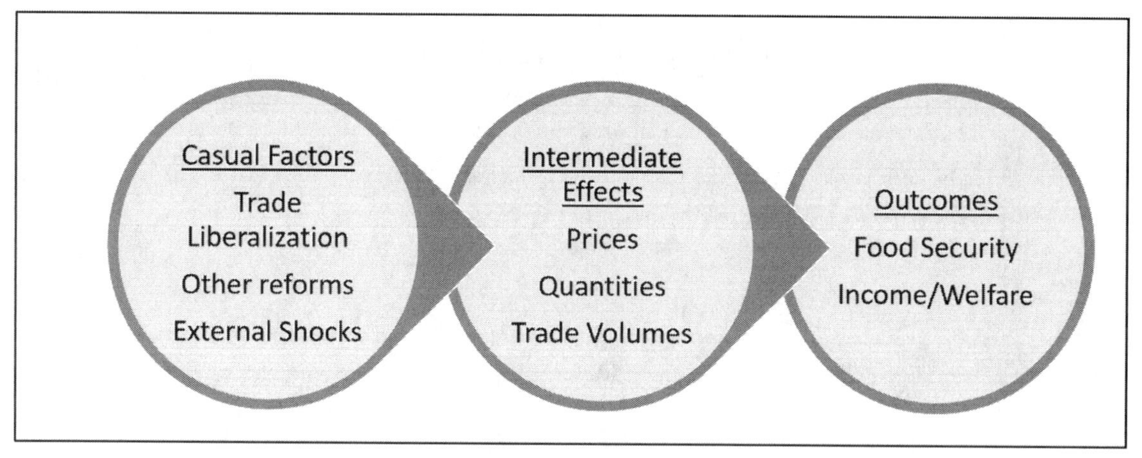

Figure 35: Reforms and Food Security.

the outcome in terms of changes in food security status.[261] It is recognized that an identical policy change in two different contexts, whether within a country at two discrete points in time, or across a set of countries, can result in very different outcomes, because of modifying factors [64].

Sometimes, extreme measures bypass the intermediate effects and directly result in a food crisis. Grain woes due to the Ukraine war, pose a policy dilemma for the U.S. and other nations. The looming global food crisis related to the Ukraine war divides the Biden administration. The issue is a plan calling for Belarus to help move grains to the market,[262] with a six-month waiver of sanctions on its potash fertilizer industry offered as an incentive [65]. At the center of a debate between administration officials is a proposal to offer a six-month waiver of sanctions on Belarus's potash fertilizer industry in the hope that it might compel Belarusian President Alexander Lukashenko to allow a rail corridor for Ukraine's grain. As the saying goes extreme situations call for extreme measures and the Ukraine war has put the entire world on a high food insecurity alert. International food markets are likely to face serious shortages this year as a result of the conflict, which has trapped much of the grain destined for export inside Ukraine and allegedly led to Russian looting. Russia and Ukraine together supply almost one-third of the world's wheat, a quarter of its barley and nearly three-quarters of its sunflower oil, according to the International Food Policy Research Institute. The Ukrainian government expects less than half of this year's harvest to make it out of the country [66].

Market Access and Agricultural Financial Services

Market access provides greater sustainability to small farmers. It provides a great way to convert farm to cash. Agricultural financial services in terms of loans provided to small farmers provide easier liquidity. A great way to look at it is credit to the agricultural macroeconomic parameter.

Credit to agriculture, another important aspect of facilitating cash to farmers, provides loans from the private/commercial banking sector to producers in agriculture, forestry, and fishing, including household producers, cooperatives, and agro-businesses.[263] The following figure provides the top 10 country credits to agriculture in 2020 (Figure 36).

Purchases by the federal government, in addition to agricultural credit, provide cash to farmers during economic downturns or shocks. For instance, take the example of Families First Coronavirus Response Act - USDA on April 19th, 2020, announced that it is exercising authority under the same to purchase and distribute up to $3 billion worth agricultural products to those in need. USDA will partner with regional and local distributors, whose workforce has been significantly impacted by the closure of many restaurants, hotels, and other food service entities, to purchase $3 billion in fresh produce, dairy, and meat products. USDA's Agricultural Marketing Service (AMS) will procure an estimated $100 million per month in fresh fruits and vegetables, $100 million per month in a variety of dairy products and $100 million per month in meat products to provide a pre-approved box of fresh produce, dairy, and meat products to food banks and other non-profits serving Americans in need.[264]

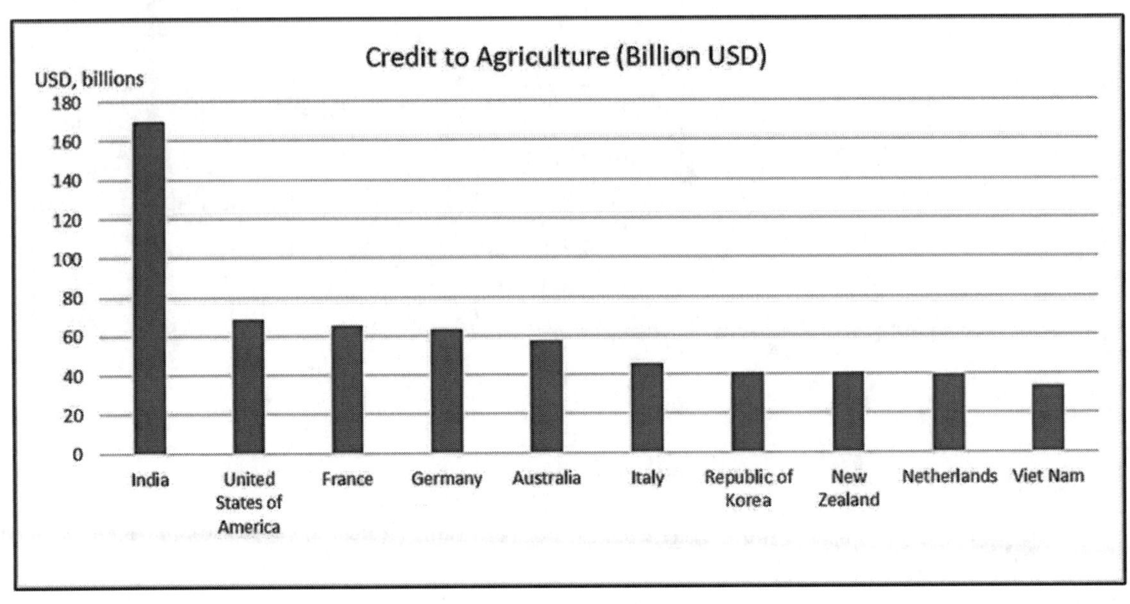

Figure 36: Credit to Agriculture 2020.

Public expenditure on agricultural research and development

A measure of government spending on agricultural R&D, as captured through the Agricultural Orientation Index, a proxy indicator assessing public investment in agriculture (Equation 1).
Ratio:

$$\frac{\text{Agriculture share of government expenditure (\%)}}{\text{Agriculture value added share of GDP (\%)}} \qquad \text{EQ. 1}$$

For the year 2019, Switzerland leads the countries [67] that have invested in agricultural R&D.[265] Among other countries whose investment is above the mean (0.428886) include Austria, Azerbaijan, Belarus, Bulgaria, Canada, Finland, Germany, Honduras, India, Ireland, Mexico, Norway, Panama, Romania, South Africa, Switzerland, Thailand and Zambia. Swiss farmers are *central to the Agricultural Innovation System*. They are represented by their professional organizations. The Swiss farmers' association represents

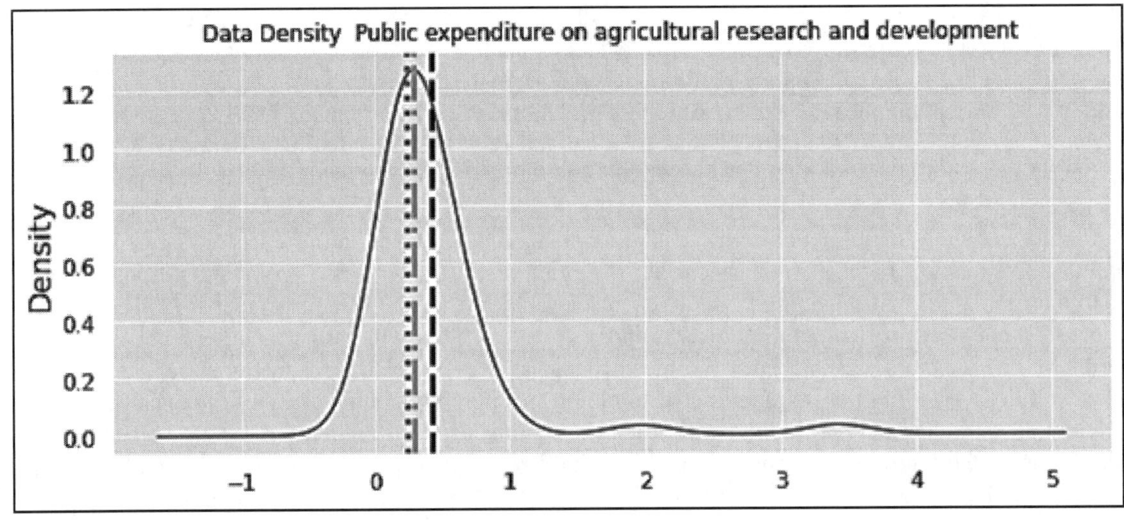

Figure 37: Data Density - Public expenditure.

57,000 farmer families. Bio Suisse, the federation of organic farmers in Switzerland, represents 32 Swiss organic farmers' associations.[266] Kenya was the lowest in public expenditure on agricultural research and development, 2019. Historically, the development of the agricultural sector has long been one of the main objectives of the Kenyan Government. "Kenya Vision 2030", the long-term economic development strategy adopted in 2008, promotes the transformation of subsistence agriculture into commercial agriculture as a key driver of income growth and increased food security. However, since 2006, price incentives for producers in Kenya have generally decreased across the major value chains (maize, tea, and coffee) a reason for the lowest as reported in 2019.[267]

The distribution (Figure 37) statistics: Mean:0.43 (dashed line); Median:0.29 (dash dot line); and Mode:0.24 (dotted line). To summarize, the data distribution is skewed to the right—statistically, if the mode is less than the median, which is less than the mean 36 (based on Who In the World is Food Insecure? and prevalence of moderate or severe food insecurity in the Population ML Model below).

Access to agricultural technology, education, and resources

A measure of access to agricultural technology, education and resources is the total factor productivity (TFP) of agriculture, which assesses the productivity of agricultural inputs (land, labor, investment) as captured by annual growth in agricultural output minus annual growth in agricultural inputs. Source: USDA and calculated Annual growth in agricultural output (%) minus annual growth in agricultural inputs (%).

Crop storage facilities

An assessment of whether there is evidence that the government has made investments through national funds or multilateral/donor funding to improve crop storage within the past five years. Source: Qualitative scoring by EIU analysts.

Volatility of agricultural production

Price fluctuations are a common feature of agricultural product markets functioning well. However, when these become large and unexpected—volatile—they can have a negative impact on the food security of consumers, farmers, and entire countries. Countries with volatility in agricultural production, as per the 2019 World Bank dataset, Austria, Azerbaijan, Belarus, Brazil, Bulgaria, Canada, Colombia, Ecuador, El Salvador, France, Hungary, Jordan, Malaysia, Norway, Paraguay, Romania, Serbia, South Africa, Spain, Switzerland, Ukraine, Uruguay and Zambia have greater volatilities than the mean of 0.12. Since 2007, world markets have seen a series of dramatic swings in commodity prices.[268] Food prices reached their highest levels in

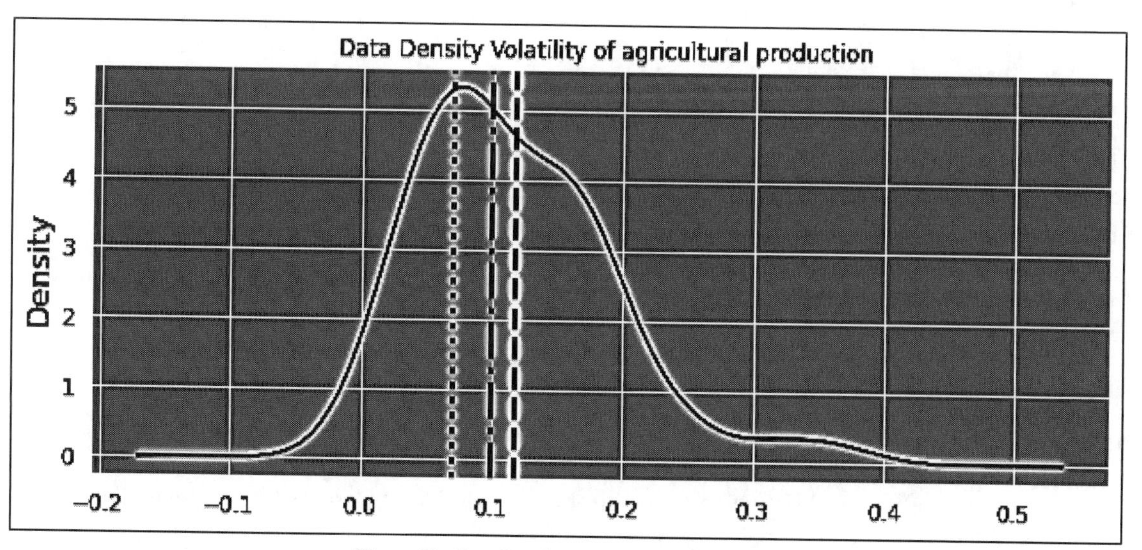

Figure 38: Data Density - Agriculture Production.

30 years during the summer of 2008, collapsing the following winter, before rapidly rising again in the months that followed. Food prices today remain high and are expected to remain volatile. A measure of the fluctuations in agricultural production, as captured by the standard deviation in the growth rates of cereal and vegetable production over the most recent 5-year period for which datasets are available. Source: FAO & Standard deviation of production growth rates. Food Price volatility is high in South Africa.[269]

The Data Distribution (Figure 38) statistics: Minimum:0.01; Mean:0.12(dashed line); Median:0.10 (dash dot line); Mode:0.07 (dotted line); Maximum:0.36 (based on Who In the World is Food Insecure? and prevalence of moderate or severe food insecurity in the Population ML Model below).

To summarize, the data distribution is skewed to the right—statistically, i the mode is less than the median, which is less than the mean.

Food loss

A measure of post-harvest and pre-consumer food loss as a ratio of the domestic supply (production, net imports, and stock changes) of crops, livestock, and fish commodities (in tonnes). Source: FAO and Total waste as a percentage of total domestic supply. FAO states 14% of the world's food is lost between harvest and retail.[270] Countries with food losses, as per 2019 FAO datasets, greater than the mean are Brazil, Bulgaria, Chile, Costa Rica, Dominican Republic, Greece, Guatemala, Guinea India, Jordan, Kenya, Mexico, Myanmar, New Zealand, Nigeria, Paraguay, Peru, Rwanda, South Africa, Ukraine, and Uruguay.

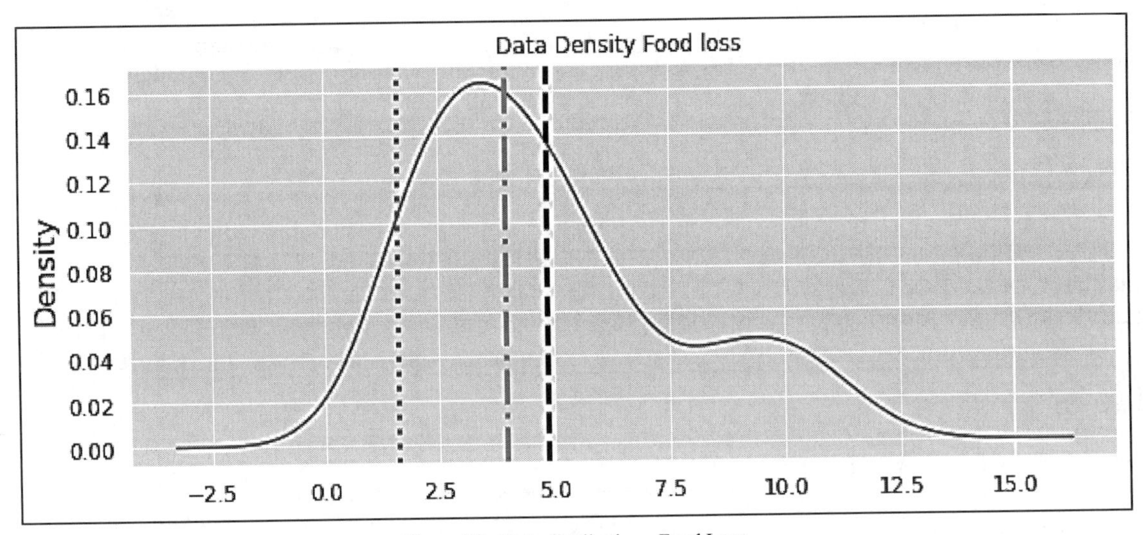

Figure 39: Data distribution - Food Loss.

The Data Distribution (please see Figure 39) statistics: Minimum:1.61; Mean:4.85 (dashed line); Median:3.97 (dash dot line); Mode:1.61 (dotted line); Maximum:11.37 (based on Who In the World is Food Insecure? And prevalence of moderate or severe food insecurity in the Population ML Model below).

Consumption and Income can be better articulated in terms of Engel's Law. According to Engel's Law, as income rises, the share of income spent on food decreases. Therefore, the higher a country's GDP per capita, the cheaper food becomes in relative terms [68]. Conversely, the poorer a family is, the greater is the portion of total outgoing income for food [69].

Engel's Law
The poorer a family is, the greater is the proportion of the total outgoing income used for food. ... The proportion of the outgoing income used for food, other things being equal, is the best measure of the material standard of living of a population.[271]

The implications of Engel's law are truly profound [69] as it clearly correlates the importance of agriculture to the marginalized communities. For poor countries, having a vibrant agricultural sector will be relatively more important, because agriculture will be a large proportion of the economy. The food budget share predicts well-being implies that economic growth is a solution to the calorie- or nutrient-deficit problem.

Software code for this model:
WHOisFoodInsecure_World_2019_Model.ipynb (Jupyter Notebook Code)

Machine Learning Model: Who In the World is Food Insecure? Prevalence of Moderate or Severe Food Insecurity in the Population

The purpose of the model is to predict prevalence of moderate or severe food insecurity in the population using 2019 data. The Data for the model (Figure 40) includes Global Food Security Indexes[272] and the FAO Suite of Food Security Indicators.[273] The model assesses the prevalence of severe food insecurity (characterized by individuals going without food for a whole day at a time during the year) and [22] prevalence of moderate-severe food insecurity. The model analyzes macroeconomic, agricultural, climatic, and key weather drivers across 113 countries for modeling prevalence of moderate or severe food insecurity in the population (%) and successfully predicts who is food insecure with an accuracy of 90.3%. The relationship between food insecurity and the key drivers of food insecurity—macroeconomic parameters, agricultural, and weather—vary by region, demonstrating that definitions of food insecurity depend on the regional context, and encompass more than monetary poverty alone [70].

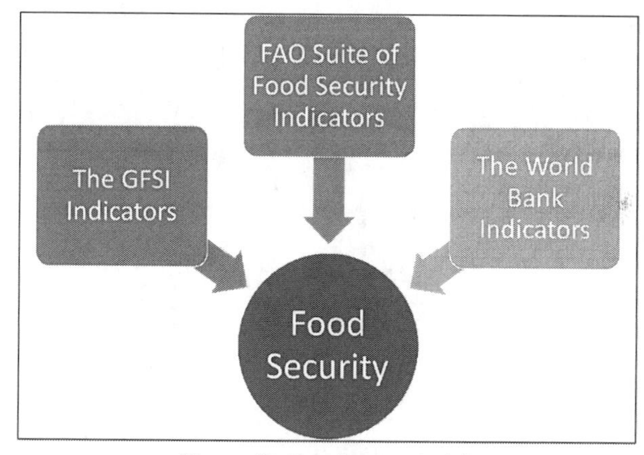

Figure 40: Data Sources model.

Step 1: Load FAO Food Security Indicator Data

The data source for food security indicators is from the GFSI 2021 Model.[274] The Global Food Security Index (GFSI) considers the issues of food affordability, availability, quality and safety, and natural resources and resilience across a set of 113 countries. The index is a dynamic quantitative and qualitative benchmarking model constructed from 58 unique indicators that measure the drivers of food security across both developing and developed countries.[275]

```
# Importing data
dfGFSIIndexFSIParams = pd.read_csv('GFSI_index_fsiParams_2019.csv',encoding = 'unicode_escape', engine ='python')
dfGFSIIndexFSIParams.head()
```

Output:

	Country Name	Change in average food costs	Food supply adequacy	Dependency on chronic food aid	Public expenditure on agricultural research and development	Access to agricultural technology, education and resources	Crop storage facilities	Road infrastructure	Irrigation infrastructure	Volatility of agricultural production	Food loss	Agricultural import tariffs
0	Algeria	3.4	149.0	1	0.469	-0.0306	1	1	3.30	0.066	8.36	23.6
1	Angola	15.9	116.0	1	0.050	-0.0352	1	1	0.15	0.389	11.63	18.9
2	Argentina	32.2	136.0	0	0.107	-0.1387	1	2	2.12	0.087	2.76	10.3
3	Australia	0.4	137.0	1	0.264	-0.0003	1	3	0.68	0.217	3.75	1.2
4	Austria	1.5	144.0	1	0.578	0.0140	1	4	3.76	0.178	3.97	12.0

Step 2: Load world bank indicators

```
dfWorldBankFSIPatams = pd.read_csv('worldbankData_fsiParams_AllCountries2019.csv')
dfWorldBankFSIPatams['Series Name'].unique()
```

Output: the following parameters are loaded as part of the Data Frame.

```
Prevalence of moderate or severe food insecurity in the population (%)',
    'Gini index',
    'Poverty headcount ratio at $1.90 a day (2011 PPP) (% of population)',
    'Poverty headcount ratio at $3.20 a day (2011 PPP) (% of population)',
    'Poverty headcount ratio at $5.50 a day (2011 PPP) (% of population)',
    'GDP per capita, PPP (constant 2017 international $)',
    'Rural population (% of total population)',
    'Population ages 0-14 (% of total population)',
    'GDP growth (annual %)',
    'Agriculture, forestry, and fishing, value added (% of GDP)',
    'Foreign direct investment, net inflows (% of GDP)',
    'Current account balance (% of GDP)',
    'Imports of goods and services (% of GDP)',
    'Employment in agriculture (% of total employment) (modeled ILO estimate)',
    'Total reserves (includes gold, current US$)',
    'Agricultural raw materials imports (% of merchandise imports)',
    'Official exchange rate (LCU per US$, period average)',
    'Employment to population ratio, 15+, total (%) (modeled ILO estimate)',
    'Prevalence of severe food insecurity in the population (%)
```

Step 3: Rename World Bank Column

The following code maps both FAO Security Indicators data and the GFSI index data. Please pivot the column, transposing from row to column values.

```
dfWorldBankFSIPatams.rename(columns = {'2019 [YR2019]':'Value'},inplace=True)
dfWorldBankFSIPatams_1 = dfWorldBankFSIPatams.copy()
dfWorldBankFSIPatams_1.drop(columns = {'Country Code','Series Code'},inplace=True)
    dfWorldBankFSIPatamsPivoted = dfWorldBankFSIPatams_1.pivot(index= 'Country Name',columns =
'Series Name',values ='Value').reset_index().rename_axis(None,axis=1)

dfWorldBankFSIPatamsPivoted
```

Output:

	Country Name	Agricultural raw materials imports (% of merchandise imports)	Agriculture, forestry, and fishing, value added (% of GDP)	Current account balance (% of GDP)	Employment in agriculture (% of total employment) (modeled ILO estimate)	Employment to population ratio, 15+, total (%) (modeled ILO estimate)	Foreign direct investment, net inflows (% of GDP)	GDP growth (annual %)	GDP per capita, PPP (constant 2017 international $)	Gini index	Imports of goods and services (% of GDP)	Official exchange rate (LCU per US$, period average)
0	Afghanistan	1.769460	25.773971	-20.170458	42.500000	42.118999	0.124496	3.911603	2065.036235	NaN	NaN	77.737949
1	Africa Eastern and Southern	1.082040	12.502325	NaN	59.253237	66.464624	1.458145	2.077898	3710.793575	NaN	27.256559	NaN
2	Africa Western and Central	1.214575	21.257161	NaN	41.537122	56.669362	2.150648	3.190336	4159.302886	NaN	25.936760	NaN
3	Albania	0.875030	18.391244	-7.914525	36.419998	53.391998	7.798722	2.113420	13656.592750	30.8	44.979562	109.850833
4	Algeria	NaN	12.336212	-9.870496	9.600000	37.411999	0.804111	1.000000	11521.984210	NaN	29.128253	119.353558
...	...	...	...	...	...	...	...	...	...	...	...	...
261	West Bank and Gaza	0.543971	7.054017	-10.388304	6.050000	33.051998	0.769720	1.362687	6245.448697	NaN	53.472437	NaN
262	World	1.366560	4.010493	NaN	26.691608	57.277506	1.696432	2.600878	16897.316020	NaN	27.758059	NaN
263	Yemen, Rep.	0.576346	5.000962	NaN	27.549999	32.283001	NaN	NaN	NaN	NaN	NaN	490.972500
264	Zambia	0.711466	2.860775	0.603691	49.639999	64.806000	2.350919	1.441306	3470.435511	NaN	34.155012	12.890000
265	Zimbabwe	0.375227	10.143657	4.773170	66.190002	80.391998	1.293799	-6.144236	3630.033985	50.3	32.030875	NaN

The World Bank and GFSI Indicator consist of both GFSI values indexed on the country name.

Step 4: View the Entire Data Frame Columns

To view World Bank and GFSI Columns:

```
dfWorldBankFSIPatamsPivotedFinal.columns
```

Output:

```
Index (['Country Name',
   'Agricultural raw materials imports (% of merchandise imports)',
   'Agriculture, forestry, and fishing, value added (% of GDP)',
   'Current account balance (% of GDP)',
   'Employment in agriculture (% of total employment) (modeled ILO estimate)',
   'Employment to population ratio, 15+, total (%) (modeled ILO estimate)',
   'Foreign direct investment, net inflows (% of GDP)',
   'GDP growth (annual %)',
   'GDP per capita, PPP (constant 2017 international $)', 'Gini index',
   'Imports of goods and services (% of GDP)',
   'Official exchange rate (LCU per US$, period average)',
   'Population ages 0–14 (% of total population)',
   'Poverty headcount ratio at $1.90 a day (2011 PPP) (% of population)',
   'Poverty headcount ratio at $3.20 a day (2011 PPP) (% of population)',
   'Poverty headcount ratio at $5.50 a day (2011 PPP) (% of population)',
   'Prevalence of moderate or severe food insecurity in the population (%)',
   'Prevalence of severe food insecurity in the population (%)',
   'Rural population (% of total population)',
   'Total reserves (includes gold, current US$)'],
   dtype='object')
```

Step 5: Check for nulls

The World Bank Dataset is completely parse and missing data elements in the given data must cover several countries with several field levels.

```
dfWorldBankFSIPatamsPivotedFinal.isna().sum()
```

Output:

Agricultural raw materials imports (% of merchandise imports)	75
Agriculture, forestry, and fishing, value added (% of GDP)	30
Current account balance (% of GDP)	94
Employment in agriculture (% of total employment) (modeled ILO estimate)	31
Employment to population ratio, 15+, total (%) (modeled ILO estimate)	31
Foreign direct investment, net inflows (% of GDP)	29
GDP growth (annual %)	15
GDP per capita, PPP (constant 2017 international $)	26
Gini index	208
Imports of goods and services (% of GDP)	40
Official exchange rate (LCU per US$, period average)	91
Population ages 0–14 (% of total population)	25
Poverty headcount ratio at $1.90 a day (2011 PPP) (% of population)	201
Poverty headcount ratio at $3.20 a day (2011 PPP) (% of population)	201
Poverty headcount ratio at $5.50 a day (2011 PPP) (% of population)	201
Prevalence of moderate or severe food insecurity in the population (%)	129
Prevalence of severe food insecurity in the population (%)	129
Rural population (% of total population)	4
Total reserves (includes gold, current US$)	96

As you can see many of the critical macroeconomic parameters are missing data.

Step 6: Imputation Strategies—using dataset mean

There are a couple of ways to overcome nulls with mean values of columns or replacing them with mean value of the metric. In this model experiment, replace with the column mean value.

```
dfWorldBankFSIPatamsPivotedFinal.fillna(dfWorldBankFSIPatamsPivotedFinal.mean(), inplace=True)
dfWorldBankFSIPatamsPivotedFinal.isna().sum()
```

Output:

Country Name	0
Agricultural raw materials imports (% of merchandise imports)	0
Agriculture, forestry, and fishing, value added (% of GDP)	0
Current account balance (% of GDP)	0
Employment in agriculture (% of total employment) (modeled ILO estimate)	0
Employment to population ratio, 15+, total (%) (modeled ILO estimate)	0
Foreign direct investment, net inflows (% of GDP)	0
GDP growth (annual %)	0
GDP per capita, PPP (constant 2017 international $)	0

Gini index	0
Imports of goods and services (% of GDP)	0
Official exchange rate (LCU per US$, period average)	0
Population ages 0–14 (% of total population)	0
Poverty headcount ratio at $1.90 a day (2011 PPP) (% of population)	0
Poverty headcount ratio at $3.20 a day (2011 PPP) (% of population)	0
Poverty headcount ratio at $5.50 a day (2011 PPP) (% of population)	0
Prevalence of moderate or severe food insecurity in the population (%)	0
Prevalence of severe food insecurity in the population (%)	0
Rural population (% of total population)	0
Total reserves (includes gold, current US$)	0
dtype: int64	

Step 7: Merge World Bank & GFSI

Merge both World Bank and GFSI datasets:

```
finaldfGFSIAndWB= pd.merge(dfGFSIIndexFSIParams, dfWorldBankFSIPatamsPivotedFinal)
finaldfGFSIAndWB.head()
```

Output:

	Country Name	Change in average food costs	Food supply adequacy	Dependency on chronic food aid	Public expenditure on agricultural research and development	Access to agricultural technology, education and resources	Crop storage facilities	Road infrastructure	Irrigation infrastructure	Volatility of agricultural production	...	Imports of goods and services (% of GDP)	Official exchange rate (LCU per US$, period average)
0	Algeria	3.4	149.0	1	0.469	-0.0306	1	1	3.30	0.066	...	29.128	119.354000
1	Angola	15.9	116.0	1	0.050	-0.0352	1	1	0.15	0.389	...	24.945	364.826000
2	Argentina	32.2	136.0	0	0.107	-0.1387	1	2	2.12	0.087	...	14.519	48.148000
3	Australia	0.4	137.0	1	0.264	-0.0003	1	3	0.68	0.217	...	21.675	1.439000
4	Austria	1.5	144.0	1	0.578	0.0140	1	4	3.76	0.178	...	52.036	957.962126

5 rows × 31 columns

Step 8: Load Employment by Age Data

Merge both World bank and GFSI datasets.

```
dfEmploymentByAge15plusRural=pd.read_csv('FAOSTAT_EmploymentByAge15plusRural_2019.csv')
dfEmploymentByAge15plusRural.head()
```

Output:

	Domain Code	Domain	Area Code (FAO)	Area	Indicator Code	Indicator	Sex Code	Sex	Year Code	Year	Source Code	Source	Unit	Value	Flag	Flag Description
0	OER	Employment Indicators: Rural	1	Armenia	21087	Employment by age, total (15+), rural areas	13	Total	2019	2019	3023	Labour force survey	1000 persons	448.602	X	International reliable sources
1	OER	Employment Indicators: Rural	11	Austria	21087	Employment by age, total (15+), rural areas	13	Total	2019	2019	3023	Labour force survey	1000 persons	1718.611	X	International reliable sources
2	OER	Employment Indicators: Rural	52	Azerbaijan	21087	Employment by age, total (15+), rural areas	13	Total	2019	2019	3023	Labour force survey	1000 persons	2454.455	X	International reliable sources
3	OER	Employment Indicators: Rural	57	Belarus	21087	Employment by age, total (15+), rural areas	13	Total	2019	2019	3023	Labour force survey	1000 persons	991.650	X	International reliable sources
4	OER	Employment Indicators: Rural	255	Belgium	21087	Employment by age, total (15+), rural areas	13	Total	2019	2019	3023	Labour force survey	1000 persons	675.691	X	International reliable sources

```
dfEmploymentByAge15plusRural_1.rename(columns={'Area':'Country Name','Value':'Employment by age,total(15+),rural areas(1000persons)'},inplace=True)
dfEmploymentByAge15plusRural_1.head()
```

Output:

	Country Name	Employment by age,total(15+),rural areas(1000persons)
0	Armenia	448.602
1	Austria	1718.611
2	Azerbaijan	2454.455
3	Belarus	991.650
4	Belgium	675.691

Step 9: Merge GFSI, World Bank, and Rural Employment Data Frames

Employment data needed to combine with World Bank and GFSI dataframes.

```
finaldfGFSIAndWBRuralEmp= pd.merge(finaldfGFSIAndWB, dfEmploymentByAge15plusRural_1)
print(finaldfGFSIAndWB.shape, finaldfGFSIAndWBRuralEmp.shape)
```

Output:

(103, 31) (53, 32)

Plot the countries count by each income group.

```
dfCountryIncomeLevel = pd.read_csv('worldcountries_incomeLevel_2019.csv',encoding = 'unicode_escape', engine ='python')
dfCountryIncomeLevel.head()
dfCountryIncomeLevel['IncomeGroup'] = dfCountryIncomeLevel['IncomeGroup'].replace(['..'],[np.nan])
dfCountryIncomeLevel['IncomeGroup'].value_counts ().plot(kind='barh', figsize=(8, 6))
plt.xlabel("Country Income Groups", labelpad=14)
plt.title("Count of the Countries", y=1.02);
```

As it turns out, 2019 World Bank country distribution: has the following range of incomes: High Income (68), Lower Middle Income (55), Low Income (54), and Upper Middle Income (40).

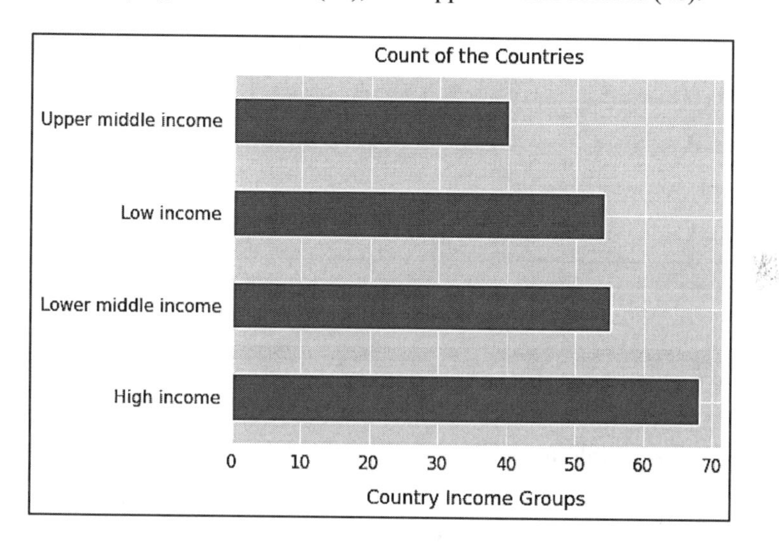

Convert categorical labels into numerical.

```
#Categorical values for Income group
dfCountryIncomeLevel_1['IncomeGroup'] = dfCountryIncomeLevel_1['IncomeGroup'].replace(['Low income','Lower middle income', 'Upper middle income', 'High income'],[0,1,2,3])
```

Step 10: Merge Country Income levels with World Bank and GFSI data

Load country's income group and combine with World Bank and GFSI Data.

```
finaldfGFSIAndWBRuralEmpCountryIncome= pd.merge(finaldfGFSIAndWBRuralEmp, dfCountryIncomeLevel_1)
finaldfGFSIAndWBRuralEmpCountryIncome.shape
```

Check for nulls to make sure all data is there to model.

```
FinaldfGFSIAndWBRuralEmpCountryIncome
```

Output:

	Country Name	Change in average food costs	Food supply adequacy	Dependency on chronic food aid	Public expenditure on agricultural research and development	Access to agricultural technology, education and resources	Crop storage facilities	Road infrastructure	Irrigation infrastructure	Volatility of agricultural production	...	Population ages 0-14 (% of total population)	Poverty headcount ratio at $1.90 a day (2011 PPP) (% of population)
0	Austria	1.5	144.0	1	0.578	0.0140	1	4	3.76	0.178	...	14.362	0.60
1	Azerbaijan	2.0	129.0	0	0.442	0.0213	1	2	30.26	0.122	...	23.442	4.66
2	Belarus	4.0	132.0	1	0.658	0.0501	1	2	0.36	0.147	...	17.036	0.00
3	Belgium	2.2	149.0	1	0.243	-0.0391	1	3	1.81	0.111	...	17.058	0.10
4	Brazil	0.9	133.0	0	0.237	-0.0240	1	1	3.20	0.200	...	21.009	4.90
5	Bulgaria	2.2	112.0	1	0.613	0.0135	1	2	2.70	0.145	...	14.685	0.90
6	Canada	0.8	138.0	1	0.557	0.1040	0	4	2.10	0.178	...	15.846	4.66
7	Chile	3.1	124.0	1	0.377	-0.0269	0	2	7.04	0.069	...	19.494	4.66
8	Colombia	0.8	124.0	1	0.314	0.0207	1	1	2.20	0.186	...	22.621	4.90
9	Costa Rica	1.6	121.0	0	0.235	0.0379	1	1	8.97	0.055	...	21.079	1.00
10	Denmark	0.1	132.0	1	0.231	-0.0169	1	3	11.55	0.075	...	16.401	0.30
11	Dominican Republic	3.6	118.0	0	0.301	0.0395	1	2	12.64	0.041	...	27.709	0.60
12	Ecuador	-1.6	113.0	0	0.124	-0.1241	1	2	30.59	0.155	...	27.707	3.60
13	El Salvador	0.5	115.0	0	0.206	0.0434	1	1	3.75	0.143	...	26.858	1.30

Step 11: Key Macroeconomic Distribution of World countries

Lets analyze the data for countries worldwide.

```
# Create a function that we can re-use
def show_distribution(column_name,var_data):
    from matplotlib import pyplot as plt

    # Get statistics
    min_val = var_data.min()
    max_val = var_data.max()
    mean_val = var_data.mean()
    med_val = var_data.median()
    mod_val = var_data.mode()[0]

    print('Minimum:{:.2f}\nMean:{:.2f}\nMedian:{:.2f}\nMode:{:.2f}\nMaximum:{:.2f}\n'.format(min_
val,
                                        mean_val,
                                        med_val,
                                        mod_val,
                                        max_val))

    # Create a figure for 2 subplots (2 rows, 1 column)
    fig, ax = plt.subplots(2, 1, figsize = (10,4))

    # Plot the histogram
    ax[0].hist(var_data)
    ax[0].set_ylabel('Frequency', fontsize=12)

    # Add lines for the mean, median, and mode
    ax[0].axvline(x=min_val, color = 'gray', linestyle='dashed', linewidth = 2)
    ax[0].axvline(x=mean_val, color = 'black', linestyle='dotted', linewidth = 2)
    ax[0].axvline(x=med_val, color = 'darkgray', linestyle='dashdot', linewidth = 2)
    ax[0].axvline(x=mod_val, color = 'yellow', linestyle='dashed', linewidth = 2)
    ax[0].axvline(x=max_val, color = 'gray', linestyle='dashed', linewidth = 2)
```

```
    # Plot the boxplot
    ax[1].boxplot(var_data, vert=False)
    ax[1].set_xlabel('Value', fontsize=12)

    # Add a title to the Figure
    fig.suptitle('Data Distribution - ' + column_name, fontsize=12)
    # Show the figure
    fig.show()
col = finaldfGFSIAndWBRuralEmpCountryIncomeCopy['Prevalence of moderate or severe food insecurity in the population (%)']
# Call the function
show_distribution('Prevalence of moderate or severe food insecurity in the population (%)', col)
```

```
for col in numeric_features:
    print(col)
    # Call the function
    show_distribution(col, finaldfGFSIAndWBRuralEmpCountryIncomeCopy[col])
```

Output:

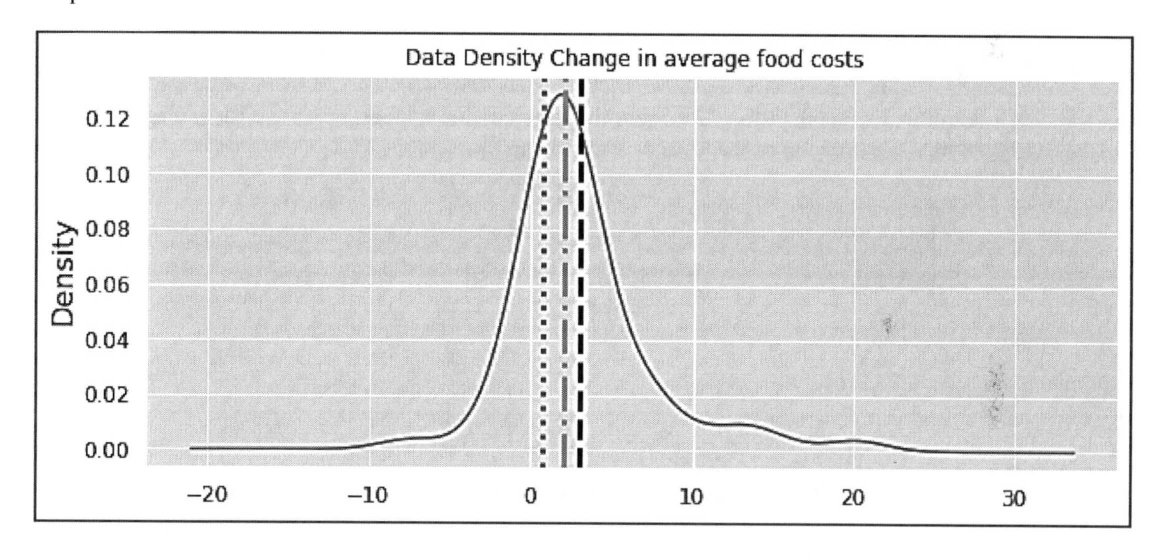

Change in average food costs: minimum: –7.30, mean: 3.06, median: 2.00, and mode: 0.80, and maximum: 20.10. Rwanda[276] has the lowest change in average food Inflation averaged at 5.45 percent from 2010 until 2022, reaching an all-time high of 26.60 percent in February, 2020 and a record low of –14.20 percent in October, 2018.[277] Highest inflation in Uzbekistan was reported in 2019 with change in average cost of 20.1%.

Rwanda	Uzbekistan

The lowest value of *Food Supply Adequacy* was reported for Zambia (minimum:94.00) in 2019 and with the highest value (maximum:153.00) for Ireland. Zambia's malnutrition rates remain among the highest in the world. 48 percent of the population are unable to meet their minimum calories' requirements and more than one-third of children under the age of five years are stunted. Limited knowledge of nutrition, poor feeding practices and limited and unhealthy diets are the main impairing contributing factors.[278]

In 2020, dietary energy supply adequacy for Ireland[279] was 158 %. Between 2006 and 2020, dietary energy supply adequacy of Ireland grew substantially from 141 to 158 % rising at an increasing annual

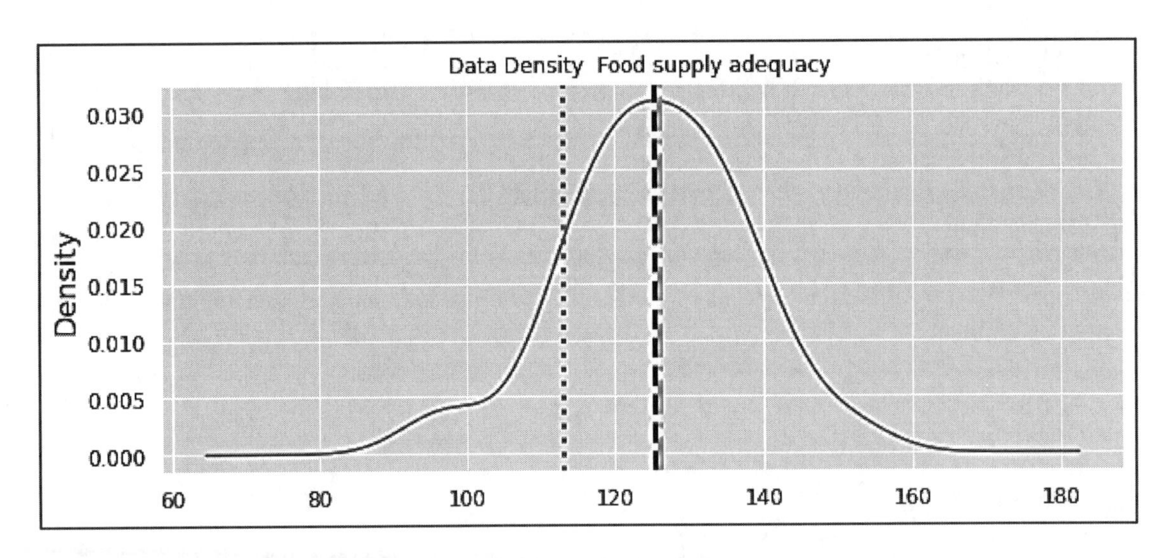

rate that reached a maximum of 1.40% in 2013 and then decreased to 0.64% in 2020. Russia's invasion of Ukraine, known as the "breadbasket of Europe", has far-reaching implications for food. But, as a country that produces and exports much more than the food required to feed it population, do we need to worry about Ireland's food security? Why did the Minister for Agriculture ask this week, "do all Irish farmers grow crops this year"? What does this say about Ireland's current food system and its food policies? Yet, we have to see the impact of the Ukraine war on one of the highest food adequacy countries.[280]

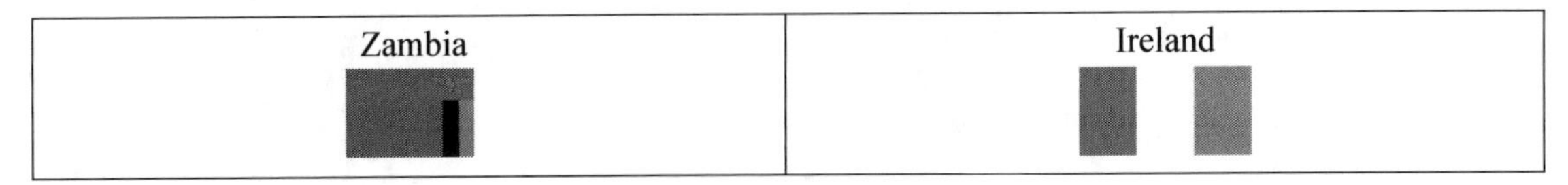

Step 12: Model Development

First, prepare key indicators to build the model.

```
model_features_prev_moderate_World= ['Change in average food costs', 'Food supply adequacy',
    'Dependency on chronic food aid',
    'Public expenditure on agricultural research and development',
    'Access to agricultural technology, education and resources',
    'Crop storage facilities',
    #'Road infrastructure',
#' Irrigation infrastructure',
    'Volatility of agricultural production',
    'Food loss',
    #'Agricultural import tariffs',
    'Agricultural raw materials imports (% of merchandise imports)',
    'Agriculture, forestry, and fishing, value added (% of GDP)',
    'Current account balance (% of GDP)',
    'Employment in agriculture (% of total employment) (modeled ILO estimate)',
    'Employment to population ratio, 15+, total (%) (modeled ILO estimate)',
    'Foreign direct investment, net inflows (% of GDP)',
    #'GDP growth (annual %)',
    'GDP per capita, PPP (constant 2017 international $)',
    'Imports of goods and services (% of GDP)',
    'Official exchange rate (LCU per US$, period average)',
    'Population ages 0–14 (% of total population)',
    'Rural population (% of total population)',
    'Total reserves (includes gold, current US$)']
```

Second, prepare independent and dependent variables.

```
X = finaldfGFSIAndWBRuralEmpCountryIncomeCopy[
   model_features_prev_moderate_World ]
y=finaldfGFSIAndWBRuralEmpCountryIncomeCopy['Prevalence of moderate or severe food insecurity
in the population (%)']
```

Third, model the Prevalence model.

```
X_train,X_test, y_train , y_test = train_test_split(X ,y, test_size = 0.15,random_state = 0)
print(X_train.shape)
print(X_test.shape)
model = LinearRegression()
model.fit(X_train,y_train)
# predicting the test
y_pred = model.predict(X_test)
mse = mean_squared_error(y_test, y_test)
print("MSE:", mse)
rmse = np.sqrt(mse)
print("RMSE:", rmse)
r2 = r2_score(y_test, y_pred)
print("R2:", r2)
```

Output:

```
(45, 20)
(8, 20)
MSE: 0.0
RMSE: 0.0
R2: 0.9012408596457369
```

The R^2 value is 90.12% with 0 Root Mean Square Error (RMSE) and Mean Square Error (MSE). Let's plot the model.

```
plt.scatter(y_test, y_pred)
plt.xlabel('Actual Value')
plt.ylabel('Predicted value')
plt.title('Prevelance of Food Insecurity')
z = np.polyfit(y_test, y_pred, 1)
p = np.poly1d(z)
plt.plot(y_test,p(y_test), color='magenta')
plt.show()
```

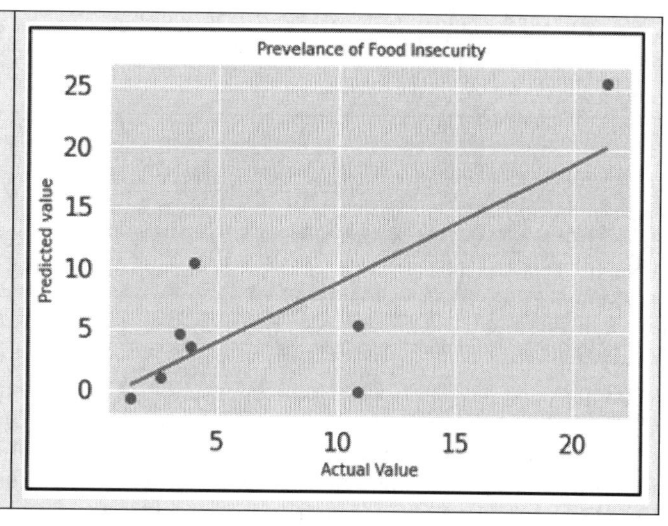

Step 13: Prepare Model Equation (y=mx+c).

Construct the model equation.

```
mx=""
for ifeature in range(len(X.columns)):
    if model.coef_[ifeature] <0:
        # format & beautify the equation
        mx += " - " + "{:.2f}".format(abs(model.coef_[ifeature])) + " * " + X.columns[ifeature]
    else:
        if ifeature == 0:
            mx += "{:.2f}".format(model.coef_[ifeature]) + " * " + X.columns[ifeature]
        else:
            mx += " + " + "{:.2f}".format(model.coef_[ifeature]) + " * " + X.columns[ifeature]

print(mx)
```

Output:

```
1.06 * Change in average food costs –0.13 * Food supply adequacy + 1.62 * Dependency on chronic food aid –4.24 * Public expenditure on agricultural research and development –30.30 * Access to agricultural technology, education and resources –2.86 * Crop storage facilities + 18.92 * Volatility of agricultural production + 0.62 * Food loss + 4.26 * Agricultural raw materials imports (% of merchandise imports) –0.72 * Agriculture, forestry, and fishing, value added (% of GDP) –0.70 * Current account balance (% of GDP) + 0.36 * Employment in agriculture (% of total employment) (modeled ILO estimate) + 0.15 * Employment to population ratio, 15+, total (%) (modeled ILO estimate) –0.07 * Foreign direct investment, net inflows (% of GDP) –0.00 * GDP per capita, PPP (constant 2017 international $) + 0.05 * Imports of goods and services (% of GDP) –0.00 * Official exchange rate (LCU per US$, period average) + 1.12 * Population ages 0–14 (% of total population) –0.07 * Rural population (% of total population) + 0.00 * Total reserves (includes gold, current US$)
```

```
# y=mx+c
if(model.intercept_ <0):
    print("The formula for the " + y.name + " linear regression line (y=mx+c) is = " + " - {:.2f}".format(abs(model.intercept_)) + " + " + mx )
else:
    print("The formula for the " + y.name + " linear regression line (y=mx+c) is = " + "{:.2f}".format(model.intercept_) + " + " + mx )
```

Output:

```
The formula for the Prevalence of moderate or severe food insecurity in the population (%) linear regression line (y=mx+c) is = –2.46 + 1.06 * Change in average food costs –0.13 * Food supply adequacy + 1.62 * Dependency on chronic food aid –4.24 * Public expenditure on agricultural research and development –30.30 * Access to agricultural technology, education and resources –2.86 * Crop storage facilities + 18.92 * Volatility of agricultural production + 0.62 * Food loss + 4.26 * Agricultural raw materials imports (% of merchandise imports) –0.72 * Agriculture, forestry, and fishing, value added (% of GDP) –0.70 * Current account balance (% of GDP) + 0.36 * Employment in agriculture (% of total employment) (modeled ILO estimate) + 0.15 * Employment to population ratio, 15+, total (%) (modeled ILO estimate) –0.07 * Foreign direct investment, net inflows (% of GDP) –0.00 * GDP per capita, PPP (constant 2017 international $) + 0.05 * Imports of goods and services (% of GDP) –0.00 * Official exchange rate (LCU per US$, period average) + 1.12 * Population ages 0–14 (% of total population) –0.07 * Rural population (% of total population) + 0.00 * Total reserves (includes gold, current US$)
```

Step 14: Explainability of the Model

Using Explainability, let's see the important variables.

```
eli5.show_weights(model,feature_names = model_features_prev_moderate_World)
```

y top features

Weight?	Feature
+18.920	Volatility of agricultural production
+4.265	Agricultural raw materials imports (% of merchandise imports)
+1.619	Dependency on chronic food aid
+1.121	Population ages 0-14 (% of total population)
+1.056	Change in average food costs
+0.619	Food loss
+0.360	Employment in agriculture (% of total employment) (modeled ILO estimate)
+0.151	Employment to population ratio, 15+, total (%) (modeled ILO estimate)
+0.048	Imports of goods and services (% of GDP)
+0.000	Total reserves (includes gold, current US$)
-0.000	GDP per capita, PPP (constant 2017 international $)
-0.002	Official exchange rate (LCU per US$, period average)
-0.066	Rural population (% of total population)
-0.067	Foreign direct investment, net inflows (% of GDP)
-0.135	Food supply adequacy
-0.699	Current account balance (% of GDP)
-0.716	Agriculture, forestry, and fishing, value added (% of GDP)
-2.461	<BIAS>
-2.862	Crop storage facilities
-4.236	Public expenditure on agricultural research and development
-30.296	Access to agricultural technology, education and resources

As it can be seen from the above model Explainability, prevalence of moderate or severe food insecurity in the population (%) is positively correlated with:

- Volatility of agricultural production
  - o All else held constant, a unit increase in volatility of agricultural production will increase prevalence of moderate or severe food insecurity by 18.92%.
- Agricultural raw materials imports (% of merchandise imports)
  - o Countries with high import dependencies are highly vulnerable to food insecurity.
- Dependency on chronic food aid
  - o Higher Dependency on chronic food aid would result in higher prevalence of moderate or severe food insecurity.
- Population ages 0–14 (% of total population)
  - o Women and children are highly susceptible to moderate or severe food insecurity.

To reduce prevalence of moderate or severe food insecurity, it is prudent to invest in the following:

- Access to agricultural technology, education, and resources
  - o Investing in agricultural technologies would prepare the World to counter prevalence of moderate or severe food insecurity. The relationship is simple and straightforward: enabling farm level productivity would improve agricultural productivity and thus provide access to more food reducing volatility of agricultural production.
- Public expenditure on agricultural research and development
  - o R& D investment in agriculture would prepare the world for climate change and threats to future food production.

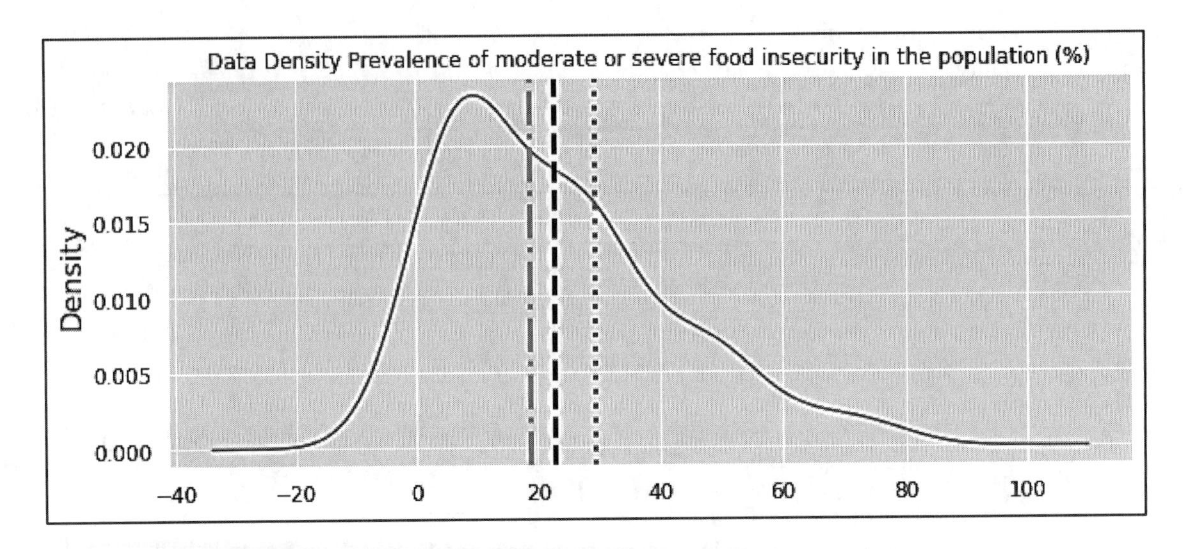

- Crop Storage Facilities:
  - o A huge benefactor to improve agricultural

Data Distribution statistics: Minimum:2.00; Mean:22.64(dashed line); Median:18.70(dash dot line); Mode:29.30(dotted line); Maximum:74.10.

Countries with prevalence of moderate or severe food insecurity (%) that is greater than a mean of 22.64% include Belarus, Brazil, Colombia, Dominican Republic, Ecuador, El Salvador, Guatemala, Guinea, Honduras, India, Jordan, Kenya, Mexico, Nigeria, Panama, Paraguay, Peru, Philippines, Rwanda, South Africa, Thailand, Ukraine, Uruguay, and Zambia. Kindly note that not all countries could be analyzed due to the non-availability of data. As it can be seen Several sub-Saharan African and south Asian countries with low to middle incomes should consider increasing investment in crop storage facilities, transport infrastructure, agricultural research and extension and food safety net programs in the future.

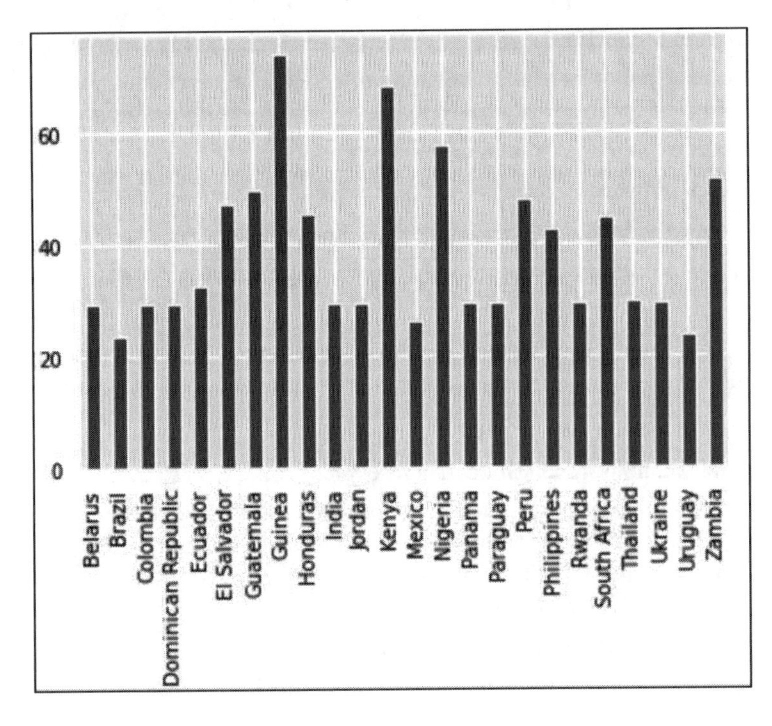

Software code for this model: WHOisFoodInsecure_World_2019_Model.ipynb (Jupyter Notebook Code)

Machine Learning Model: Who In the World is Food Insecure? Prevalence of Moderate or Severe Food Insecurity in the Population—Ordinary Least Squares (OLS) Model

Ordinary Least Squares regression (OLS) is a common technique for estimating coefficients of linear regression equations which describe the relationship between one or more independent quantitative variables and a dependent variable (simple or multiple linear regression).

We have developed an OLS model that analyzes macroeconomic, agricultural, climatic, and key weather drivers across 113 countries for the prevalence of moderate or severe food insecurity (%) and have successfully modelled one that can predict who is food insecure with an accuracy[281] of 80.4%. The relationship between food insecurity and the key drivers of food insecurity—macroeconomic parameters, agricultural, and weather—varies by region, demonstrating that definitions of food insecurity depend on regional context, and encompass more than monetary poverty alone [70].

Step 1: Build OLS Model

Using Explainability, let's see the important variables.

```
from statsmodels.formula.api import ols
#'BPXSY1','RIDAGEYR','BMXWT','BMXHT','RIAGENDR','RIDRETH3','DMDMARTL','
DMDEDUC2'
# 'RIDAGEYR', 'SDMVSTRA', 'INDFMPIR', 'INDFMIN2','INDHHIN2','FSDHH', 'FSDAD',
'FSDCH','ageCategory'
dataols = finaldfGFSIAndWBRuralEmpCountryIncomeCopy.copy()
dataols.rename({'Change in average food costs': 'Change_in_average_food_costs',
 'Food supply adequacy': 'Food_supply_adequacy',
     'Employment in agriculture (% of total employment) (modeled ILO estimate)': 'Employment_in_
agriculture',
    'Dependency on chronic food aid':'Dependency_on_chronic_foodaid',
    'Volatility of agricultural production':'Volatility_of_agricultural_production',
    'Employment to population ratio, 15+, total (%) (modeled ILO estimate)':'Employment_to_
population_ratio',
    'Foreign direct investment, net inflows (% of GDP)':'fdi_netflows',
    'Food loss': 'Food_loss',
    'GDP per capita, PPP (constant 2017 international $)': 'gdp_per_capita',
    'Access to agricultural technology, education and resources':'Access_to_agricultural_technology_
education_and_resources',
    'Public expenditure on agricultural research and development': 'Public_expenditure_on_agricultural_
research_and_development',
    'Crop storage facilities':'Crop_storage_facilities',
    'Imports of goods and services (% of GDP)':'importofgoods',
    'Official exchange rate (LCU per US$, period average)':'exchangerate',
    'Population ages 0-14 (% of total population)':'Population_ages',
    'Total reserves (includes gold, current US$)':'total_reserves',
    'Rural population (% of total population)':'rural_population',
    'Agricultural raw materials imports (% of merchandise imports)': 'Agricultural_raw_materials_
imports_per_of_merchandise_imports',
    'Agriculture, forestry, and fishing, value added (% of GDP)': 'Agriculture_forestry_andfishing_
    valueadded_per_of_GDP',
```

```
'Current account balance (% of GDP)':'Current_account_balance_per_of_GDP',
    'Prevalence of moderate or severe food insecurity in the population (%)': 'Prevalence_of_
moderate_or_severe_food_insecurity_in_the_population_percentage'}, axis=1, inplace=True)

formula='Prevalence_of_moderate_or_severe_food_insecurity_in_the_population_percentage ~ Access_
to_agricultural_technology_education_and_resources+total_reserves+rural_population + Population_
ages + exchangerate + importofgoods + gdp_per_capita + fdi_netflows + Employment_to_population_
ratio + Employment_in_agriculture + Agriculture_forestry_andfishing_valueadded_per_of_GDP +
Current_account_balance_per_of_GDP + Change_in_average_food_costs + Food_supply_adequacy +
Dependency_on_chronic_foodaid + Public_expenditure_on_agricultural_research_and_development +
Public_expenditure_on_agricultural_research_and_development + Crop_storage_facilities + Volatility_
of_agricultural_production + Food_loss + Current_account_balance_per_of_GDP'
model=ols(formula,dataols).fit()
model.summary()
```

Output:
OLS Regression Results:

OLS Regression Results			
Dep. Variable:	Prevalence_of_moderate_or_severe_food_insecurity_in_the_population_percentage	**R-squared:**	0.804
Model:	OLS	**Adj. R-squared:**	0.691
Method:	Least Squares	**F-statistic:**	7.131
Date:	Thu, 19 May 2022	**Prob (F-statistic):**	5.58e–07
Time:	04:21:44	**Log-Likelihood:**	–184.49
No. Observations:	53	**AIC:**	409.0
Df Residuals:	33	**BIC:**	448.4
Df Model:	19		
Covariance Type:	nonrobust		

	coef	std err	t	P>|t|	[0.025	0.975]
Intercept	–5.2845	35.738	–0.148	0.883	–77.993	67.424
Access_to_agricultural_technology_education_and_resources	–26.3028	39.939	–0.659	0.515	–107.559	54.954
total_reserves	9.875e–12	1.83e–11	0.541	0.592	–2.73e–11	4.7e–11
rural_population	–0.0459	0.189	–0.243	0.809	–0.430	0.338
Population_ages	1.1606	0.319	3.634	0.001	0.511	1.810
exchangerate	–0.0016	0.001	–2.310	0.027	–0.003	–0.000
importofgoods	0.0578	0.097	0.596	0.555	–0.140	0.255
gdp_per_capita	–0.0003	0.000	–1.522	0.137	–0.001	9.15e–05
fdi_netflows	–0.1007	0.174	–0.578	0.567	–0.456	0.254
Employment_to_population_ratio	0.2386	0.252	0.945	0.351	–0.275	0.752
Employment_in_agriculture	0.1872	0.328	0.571	0.572	–0.480	0.855
Agriculture_forestry_andfishing_valueadded_per_of_GDP	–0.4613	0.597	–0.772	0.445	–1.676	0.754
Current_account_balance_per_of_GDP	–0.3941	0.478	–0.824	0.416	–1.367	0.579
Change_in_average_food_costs	0.9317	0.604	1.543	0.132	–0.297	2.161
Food_supply_adequacy	–0.0785	0.234	–0.335	0.739	–0.555	0.398
Dependency_on_chronic_foodaid	3.2193	3.814	0.844	0.405	–4.540	10.979
Public_expenditure_on_agricultural_research_and_development	–3.7297	6.019	–0.620	0.540	–15.976	8.517
Crop_storage_facilities	–0.3848	4.609	–0.083	0.934	–9.763	8.993
Volatility_of_agricultural_production	3.0412	24.002	0.127	0.900	–45.792	51.874
Food_loss	0.2551	0.716	0.357	0.724	–1.201	1.711

Omnibus:	5.076	**Durbin-Watson:**	1.899
Prob(Omnibus):	0.079	**Jarque-Bera (JB):**	4.219
Skew:	0.479	**Prob(JB):**	0.121
Kurtosis:	3.997	**Cond. No.**	4.80e + 12

Machine Learning Model: Who In the World is Food Insecure? Prevalence of Severe Food Insecurity in the Population (%)

The purpose of the model is to predict prevalence of severe food insecurity (%) using 2019 data. The Data model for ML is from Global Food Security Indexes,[282] World Bank, and FAO Suite of Food Security Indicators. The model leverages key datasets mentioned in code walking through the above Machine Learning Model—Who In the World is Food Insecure? Prevalence of moderate or severe food insecurity in the Population; additionally, finds Road infrastructure, Volatility of agricultural production, and Agricultural import tariffs as key influencers in addition to the findings of the moderate prevalence model. The model analyzes macroeconomic, agricultural, climate, and key weather drivers across 113 countries for modeling prevalence of severe food insecurity (%) and have successfully modelled one that can predict who is food insecure with an accuracy of 32.13%. The relationship between food insecurity and the key drivers of food insecurity—macroeconomic parameters, agricultural, and weather—vary by region, demonstrating that definitions of food insecurity depend on regional contexts, and encompass more than monetary poverty alone [70].

Step 1: Populate Key Drivers of Severe food insecurity parameters
Please note that the same drivers are regressed as we did for moderate or severe food insecurity.

```
model_features_prev_severe_World= ['Change in average food costs', ' Food supply adequacy',
    'Dependency on chronic food aid',
    'Public expenditure on agricultural research and development',
    'Access to agricultural technology, education and resources',
    'Crop storage facilities',
    'Road infrastructure',
  #' Irrigation infrastructure',
    'Volatility of agricultural production',
    'Food loss',
    'Agricultural import tariffs',
    'Agricultural raw materials imports (% of merchandise imports)',
    'Agriculture, forestry, and fishing, value added (% of GDP)',
    'Current account balance (% of GDP)',
    'Employment in agriculture (% of total employment) (modeled ILO estimate)',
    'Employment to population ratio, 15+, total (%) (modeled ILO estimate)',
    'Foreign direct investment, net inflows (% of GDP)',
    #'GDP growth (annual %)',
    'GDP per capita, PPP (constant 2017 international $)',
    'Imports of goods and services (% of GDP)',
    'Official exchange rate (LCU per US$, period average)',
    'Population ages 0–14 (% of total population)',
    'Rural population (% of total population)',
    'Total reserves (includes gold, current US$)',

    ]
```

Step 2: Split Data
Perform regression. Please note that the dependent variable (Y) is "Prevalence of severe food insecurity in the population (%)".

```
X = finaldfGFSIAndWBRuralEmpCountryIncomeCopy[
  model_features_prev_severe_World

      ]
y=finaldfGFSIAndWBRuralEmpCountryIncomeCopy['Prevalence of severe food insecurity in the population (%)']
```

Step 3: Prepare the Model

```
X_train,X_test, y_train , y_test = train_test_split(X ,y, test_size = 0.15,random_state = 0)
print(X_train.shape)
print(X_test.shape)
model = LinearRegression()
model.fit(X_train,y_train)
# predicting the test
y_pred = model.predict(X_test)
mse = mean_squared_error(y_test, y_test)
print("MSE:", mse)
rmse = np.sqrt(mse)
print("RMSE:", rmse)
r2 = r2_score(y_test, y_pred)
print("R2:", r2)
```

Output:

```
(45, 22)
(8, 22)
MSE: 0.0
RMSE: 0.0
R2: 0.32123541264510447
```

As can be seen, the model explains 32.12% of the regression.

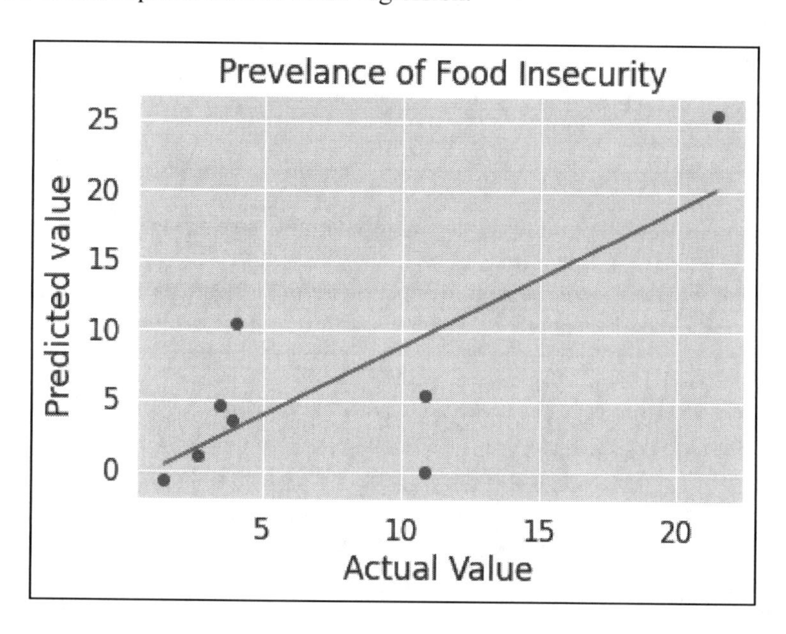

Step 4: Derive Model Equation (Y=mx+c)

The Model equation helps to see the relationship between dependent and independent variables.

```
mx=""
for ifeature in range(len(X.columns)):
    if model.coef_[ifeature] <0:
        # format & beautify the equation
        mx += " - " + "{:.2f}".format(abs(model.coef_[ifeature])) + " * " + X.columns[ifeature]
    else:
        if ifeature == 0:
            mx += "{:.2f}".format(model.coef_[ifeature]) + " * " + X.columns[ifeature]
        else:
            mx += " + " + "{:.2f}".format(model.coef_[ifeature]) + " * " + X.columns[ifeature]
print(mx)
```

Output:

```
0.25 * Change in average food costs – 0.01 * Food supply adequacy + 1.86 * Dependency on chronic food aid – 1.02 * Public expenditure on agricultural research and development – 10.83 * Access to agricultural technology, education and resources – 1.96 * Crop storage facilities – 0.58 * Road infrastructure + 17.02 * Volatility of agricultural production + 0.79 * Food loss – 0.10 * Agricultural import tariffs + 2.26 * Agricultural raw materials imports (% of merchandise imports) – 0.27 * Agriculture, forestry, and fishing, value added (% of GDP) – 0.00 * Current account balance (% of GDP) + 0.38 * Employment in agriculture (% of total employment) (modeled ILO estimate) – 0.15 * Employment to population ratio, 15+, total (%) (modeled ILO estimate) + 0.08 * Foreign direct investment, net inflows (% of GDP) + 0.00 * GDP per capita, PPP (constant 2017 international $) + 0.01 * Imports of goods and services (% of GDP) – 0.00 * Official exchange rate (LCU per US$, period average) + 0.81 * Population ages 0–14 (% of total population) – 0.08 * Rural population (% of total population) + 0.00 * Total reserves (includes gold, current US$)
```

```
# y=mx+c
if(model.intercept_ <0):
    print("The formula for the " + y.name + " linear regression line (y=mx+c) is = " + " - {:.2f}".format(abs(model.intercept_)) + " + " + mx )
else:
    print("The formula for the " + y.name + " linear regression line (y=mx+c) is = " + "{:.2f}".format(model.intercept_) + " + " + mx )
```

Output:

```
The formula for the Prevalence of severe food insecurity in the population (%) linear regression line (y=mx+c) is = – 14.59 + 0.25 * Change in average food costs – 0.01 * Food supply adequacy + 1.86 * Dependency on chronic food aid – 1.02 * Public expenditure on agricultural research and development – 10.83 * Access to agricultural technology, education and resources – 1.96 * Crop storage facilities – 0.58 * Road infrastructure + 17.02 * Volatility of agricultural production + 0.79 * Food loss – 0.10 * Agricultural import tariffs + 2.26 * Agricultural raw materials imports (% of merchandise imports) – 0.27 * Agriculture, forestry, and fishing, value added (% of GDP) – 0.00 * Current account balance (% of GDP) + 0.38 * Employment in agriculture (% of total employment) (modeled ILO estimate) – 0.15 * Employment to population ratio, 15+, total (%) (modeled ILO estimate) + 0.08 * Foreign direct investment, net inflows (% of GDP) + 0.00 * GDP per capita, PPP (constant 2017 international $) + 0.01 * Imports of goods and services (% of GDP) – 0.00 * Official exchange rate (LCU per US$, period average) + 0.81 * Population ages 0–14 (% of total population) – 0.08 * Rural population (% of total population) + 0.00 * Total reserves (includes gold, current US$)
```

Step 5: Model Explainability

```
eli5.show_weights(model,feature_names = model_features_prev_severe_World, top=100)
```

Output:

y top features

Weight?	Feature
+17.018	Volatility of agricultural production
+2.258	Agricultural raw materials imports (% of merchandise imports)
+1.860	Dependency on chronic food aid
+0.813	Population ages 0-14 (% of total population)
+0.795	Food loss
+0.376	Employment in agriculture (% of total employment) (modeled ILO estimate)
+0.245	Change in average food costs
+0.084	Foreign direct investment, net inflows (% of GDP)
+0.014	Imports of goods and services (% of GDP)
+0.000	GDP per capita, PPP (constant 2017 international $)
+0.000	Total reserves (includes gold, current US$)
-0.000	Official exchange rate (LCU per US$, period average)
-0.004	Current account balance (% of GDP)
-0.014	Food supply adequacy
-0.082	Rural population (% of total population)
-0.102	Agricultural import tariffs
-0.154	Employment to population ratio, 15+, total (%) (modeled ILO estimate)
-0.272	Agriculture, forestry, and fishing, value added (% of GDP)
-0.575	Road infrastructure
-1.025	Public expenditure on agricultural research and development
-1.960	Crop storage facilities
-10.835	Access to agricultural technology, education and resources
-14.589	<BIAS>

As it can be seen from the above model Explainability and prevalence of moderate or severe food insecurity in the population (%) are positively correlated with:

- Volatility of agricultural production
  - o All else held constant, a unit increase in volatility of agricultural production will increase 17.018% of prevalence of moderate or severe food insecurity.
- Agricultural raw materials imports (% of merchandise imports)
  - o Countries with high import dependencies are highly vulnerable to food insecurity.
- Dependency on chronic food aid
  - o Higher Dependency on chronic food aid would have higher prevalence of moderate or severe food insecurity in the population.
- Population ages 0–14 (% of total population)
  - o Women and children are highly susceptible to prevalence of moderate or severe food insecurity in the population.

To reduce the prevalence of moderate or severe food insecurity, it is prudent to invest in the following:

- Road Infrastructure[283]
  - o High negative correlation with the Percentage of Population with Severe Food Insecurity. What it tells is the inverse relationship, the decrease in road infrastructure leads to increase in Food insecurity. In essence, better rural transport is key to food security and zero hunger.[284] Financing infrastructure, including roads, storage and localized energy grids, will help provide food security for the millions of people living in hunger worldwide [61].

- Access to agricultural technology, education, and resources
    - o Investing in agricultural technologies would prepare the World to counter prevalence of moderate or severe food insecurity. The relationship is simple and straightforward: enabling farm level productivity would improve agricultural productivity and thus provide access to more food and reduce volatility of agricultural production.
- Public expenditure on agricultural research and development
    - o R& D investment in agriculture would prepare the world for climate change and threats to future food production.
- Crop Storage Facilities:
    - o A huge benefactor to improve agriculturally

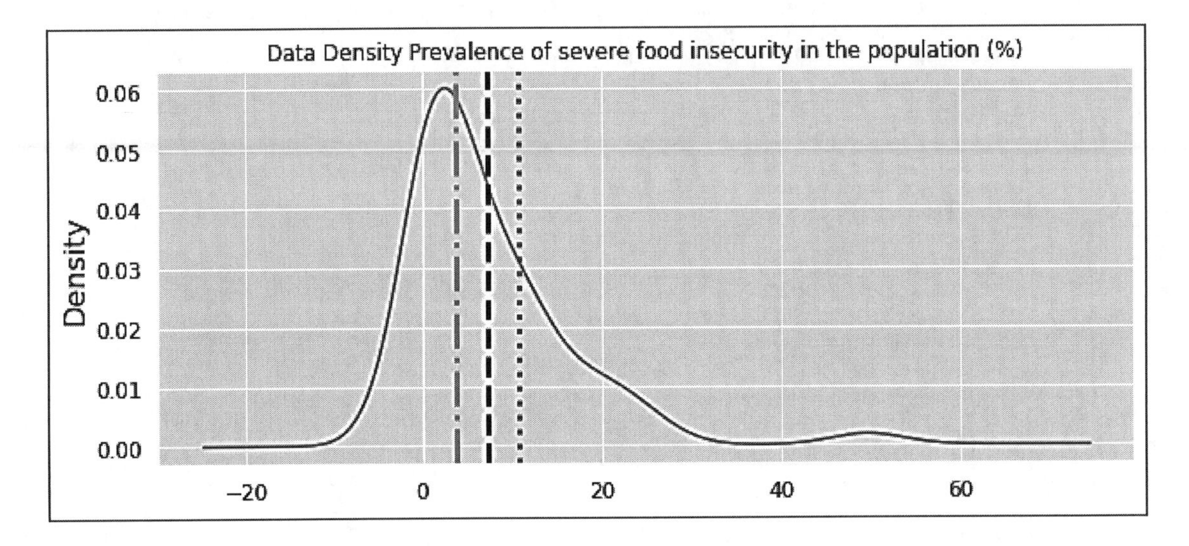

Data Distribution statistics: Minimum:0.00; Mean:7.42(dashed line); Median:3.90(dash dot line); Mode:10.83 (dotted line); Maximum:49.70.

Countries with prevalence of severe food insecurity in the population (%) that is greater than a mean of 22.64% include Belarus, Colombia, Dominican Republic, Ecuador, El Salvador, Guatemala, Guinea, Honduras, India, Jordan, Kenya, Malaysia, Nigeria, Panama, Paraguay, Peru, Rwanda, South Africa, Thailand and Zambia. Kindly note that not all countries could be analyzed due to the non-availability of data. As it can be seen Several sub-Saharan African and south Asian countries with low to middle incomes should consider increasing investment in crop storage facilities, transport infrastructure, agricultural research and extension and food safety net programs in the future.

Although rich in natural resources, Guinea[285] faces major socio-economic challenges. The poverty rate is alarming, and 21.8 percent of households are food insecure. Malnutrition remains high: 6.1 percent of children under 5 are affected by acute global malnutrition, 24.4 percent are stunted, and 12 percent are underweight. Rural populations are particularly vulnerable to food insecurity. Among those affected by severe food insecurity, 71.1 percent practice subsistence farming. Farmers with small holdings comprise the majority of the country's poor, showing a correlation between poverty and food insecurity. Besides, they have poor access to seeds and fertilizers, production and processing equipment, storage facilities, basic infrastructure, and affordable financial services. Although women play a crucial role in agriculture, particularly in food

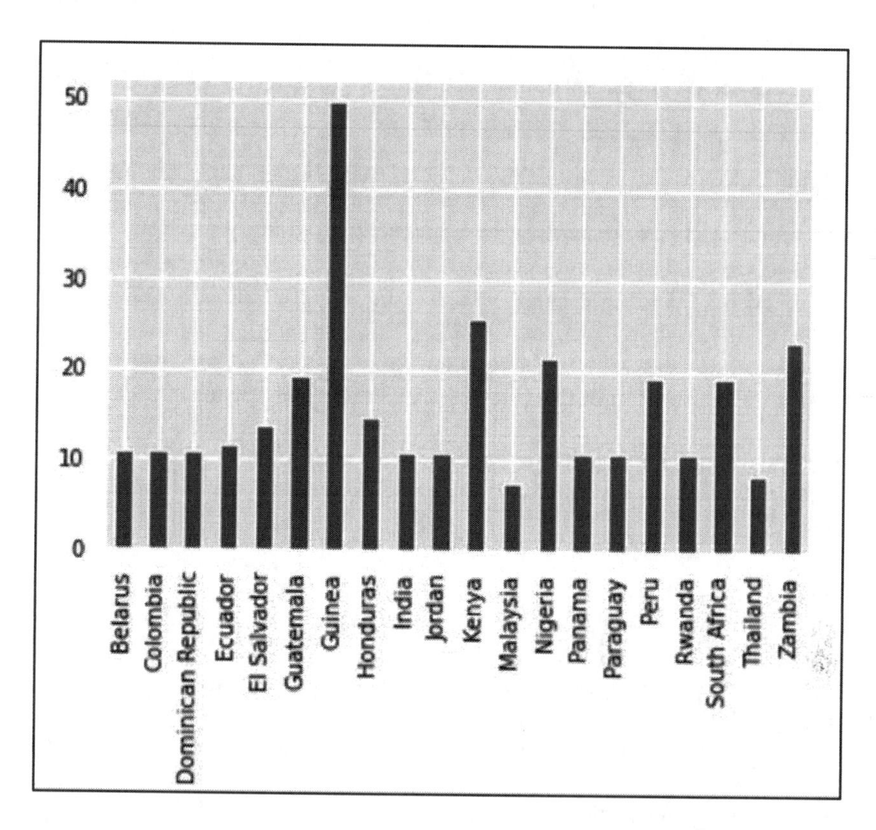

production, they have difficulties in accessing land and productive resources, education, formal employment, and income-generating activities. Their work is often unpaid for and undervalued. Women make up 60 percent of the people suffering from chronic hunger, and most rural people living in poverty.

Kenya: Even though the country has recently acquired lower-middle-income status, the increased wealth has not benefited all Kenyans equally.[286] Over one third of the population still lives under the international poverty line and social, economic and gender disparities remain. Access to adequate quantities of nutritious food remains a challenge for many, especially in arid and semi-arid regions which make up 80 percent of the country's land area. Families headed by women are more likely to be food insecure than those headed by men [71].

Machine Learning Model: Who In the World is Food Insecure? Prevalence of Severe Food Insecurity in the Population—Ordinary Least Squares (OLS) Model

We have developed an OLS model that analyzes macroeconomic, agricultural, climate, and key weather drivers across 113 countries for prevalence of moderate or severe food insecurity in the population (%) and have successfully model that can predict who's food insecure with the accuracy[287] of 80.4%. The relationship between food insecurity and the key drivers of food insecurity—macroeconomic parameters, agricultural, and weather—varies by region, demonstrating that definitions of food insecurity depend on regional context, and encompass more than monetary poverty alone [70].

Step 1: Build OLS Model

Using Explainability, let's see the important variables.

```
from statsmodels.formula.api import ols
#'BPXSY1','RIDAGEYR','BMXWT','BMXHT','RIAGENDR','RIDRETH3','DMDMARTL',
'DMDEDUC2'
# 'RIDAGEYR', 'SDMVSTRA', 'INDFMPIR', 'INDFMIN2', 'INDHHIN2', 'FSDHH', 'FSDAD',
'FSDCH','ageCategory'
dataols = finaldfGFSIAndWBRuralEmpCountryIncomeCopy.copy()
dataols.rename({'Change in average food costs': 'Change_in_average_food_costs',
   'Food supply adequacy': 'Food_supply_adequacy',
      'Employment in agriculture (% of total employment) (modeled ILO estimate)': 'Employment_in_
agriculture',
      'Dependency on chronic food aid':'Dependency_on_chronic_foodaid',
      'Volatility of agricultural production':'Volatility_of_agricultural_production',
      'Employment to population ratio, 15+, total (%) (modeled ILO estimate)':'Employment_to_
population_ratio',
      'Foreign direct investment, net inflows (% of GDP)':'fdi_netflows',
      'Food loss': 'Food_loss',
      'Agricultural import tariffs':'agri_import_tariff',
      'Road infrastructure':'Road_infrastructure',
      'GDP per capita, PPP (constant 2017 international $)': 'gdp_per_capita',
      'Access to agricultural technology, education and resources':'Access_to_agricultural_technology_
education_and_resources',
      'Public expenditure on agricultural research and development': 'Public_expenditure_on_agricultural_
research_and_development',
      'Crop storage facilities':'Crop_storage_facilities',
      'Imports of goods and services (% of GDP)':'importofgoods',
      'Official exchange rate (LCU per US$, period average)':'exchangerate',
      'Population ages 0–14 (% of total population)':'Population_ages',
      'Total reserves (includes gold, current US$)':'total_reserves',
      'Rural population (% of total population)':'rural_population',
      'Agricultural raw materials imports (% of merchandise imports)': 'Agricultural_raw_materials_
imports_per_of_merchandise_imports',
   'Agriculture, forestry, and fishing, value added (% of GDP)': 'Agriculture_forestry_andfishing_
valueadded_per_of_GDP',
   'Current account balance (% of GDP)':'Current_account_balance_per_of_GDP',
      'Prevalence of severe food insecurity in the population (%)': 'Prevalence_of_severe_food_insecurity_
in_the_population_percentage'}, axis=1, inplace=True)

formula='Prevalence_of_severe_food_insecurity_in_the_population_percentage ~ agri_import_
tariff+Road_infrastructure+Access_to_agricultural_technology_education_and_resources+total_
reserves+rural_population + Population_ages + exchangerate + importofgoods + gdp_per_
capita + fdi_netflows + Employment_to_population_ratio + Employment_in_agriculture +
Agriculture_forestry_andfishing_valueadded_per_of_GDP + Current_account_balance_per_of_GDP
+ Change_in_average_food_costs + Food_supply_adequacy + Dependency_on_chronic_foodaid +
Public_expenditure_on_agricultural_research_and_development + Public_expenditure_on_agricultural_
research_and_development + Crop_storage_facilities + Volatility_of_agricultural_production + Food_loss
+ Current_account_balance_per_of_GDP'
model=ols(formula,dataols).fit()
model.summary()
```

Output:
OLS Regression Results:

OLS Regression Results			
Dep. Variable:	Prevalence_of_severe_food_insecurity_in_the_population_percentage	**R-squared:**	0.686
Model:	OLS	**Adj. R-squared:**	0.473
Method:	Least Squares	**F-statistic:**	3.220
Date:	Thu, 19 May 2022	**Prob (F-statistic):**	0.00158
Time:	19:51:00	**Log-Likelihood:**	–160.25
No. Observation:	53	**AIC:**	364.5
Df Residuals:	31	**BIC:**	407.8
Df Model:	21		
Covariance Type:	nonrobust		

	coef	**std err**	**t**	**P>|t|**	**[0.025**	**0.975]**
Intercept	–9.3670	24.864	–0.377	0.709	–60.077	41.343
agri_import_tariff	–0.0562	0.180	–0.312	0.757	–0.423	0.311
Road_infrastructure	–0.2471	1.967	–0.126	0.901	–4.258	3.764
Access_to_agricultural_technology_education_and_resources	–0.3120	26.287	–0.012	0.991	–53.926	53.302
total_reserves	–7.108e-13	1.31e–11	–0.054	0.957	–2.75e–11	2.61e–11
rural_population	–0.1327	0.124	–1.072	0.292	–0.385	0.120
Population_ages	0.6744	0.209	3.229	0.003	0.248	1.100
exchangerate	–2.506e–05	0.000	–0.053	0.958	–0.001	0.001
importofgoods	0.0367	0.068	0.536	0.596	–0.103	0.176
gdp_per_capita	7.984e–05	0.000	0.560	0.580	–0.000	0.000
fdi_netflows	0.0238	0.131	0.182	0.857	–0.243	0.291
Employment_to_population_ratio	–0.1040	0.172	–0.604	0.551	–0.455	0.247
Employment_in_agriculture	0.3457	0.222	1.555	0.130	–0.108	0.799
Agriculture_forestry_andfishing_valueadded_per_of_GDP	–0.1224	0.420	–0.292	0.772	–0.978	0.733
Current_account_balance_per_of_GDP	0.0864	0.333	0.259	0.797	–0.594	0.766
Change_in_average_food_costs	0.1351	0.414	0.327	0.746	–0.709	0.979
Food_supply_adequacy	–0.0139	0.163	–0.085	0.933	–0.347	0.319
Dependency_on_chronic_foodaid	1.5752	2.529	0.623	0.538	–3.583	6.733

Public_expenditure_on_agricultur-al_research_and_development	-0.0289	4.141	–0.007	0.994	–8.475	8.417
Crop_storage_facilities	0.5802	3.128	0.185	0.854	–5.798	6.959
Volatility_of_agricultural_produc-tion	12.5966	15.759	0.799	0.430	–19.545	44.738
Food_loss	0.5033	0.484	1.040	0.307	–0.484	1.491

Omnibus:	15.036	**Durbin-Watson:**	1.986
Prob(Omnibus):	0.001	**Jarque-Bera (JB):**	23.637
Skew:	0.890	**Prob (JB):**	7.37e–06
Kurtosis:	5.745	**Cond. No.**	4.93e+12

Guidance to Policy Makers

Food Security is the most important aspect of society. Ensuring food security not only maintains the health status of citizens but also makes the country contribute to the World economy. In a way, food security is national security! Disruption to food security can emerge from many factors:

- Rising biofuel production adds to the demand for corn and rapeseeds oil spilling over to other foods.
- Strong food demand from emerging economies.
- Energy shocks, conflicts, or wars.
- **Drought conditions**[288] in major wheat-producing countries (e.g., Australia and Ukraine), **higher input costs** (animal feed, energy, and fertilizer), and restrictive trade policies in major net exporters of key food staples such as rice have also contributed.
- **Financial factors**: the depreciating US$ increases purchasing power of commodity users outside the dollar domain [72].

The predictive prevalence model developed to find out who is insecure due to severe food insecurity (%) correlates agriculture and change in food costs as major factors. As indicated by the model in low and middle-income countries, about 2.5 billion people's ability to make a living depends on climate-sensitive activities, like agriculture, pastoralism, and fisheries. And existing food production, supply adequacy, the need to invest in smart climate technologies, agricultural R&D, and macroeconomic parameters all are necessary to tackle the existing food insecurity. The policy makers and governments could invest in areas that prepare the entire world for a sustainable future and that would be the most important contribution of the model. Given population demands, food production needs to increase by at least 60% to adequately feed a growing global population by 2050. The prevalence models presented in this chapter are the first advanced analytics infused machine learning models that enable policy makers and governments to step in the right direction to tackle food insecurity. It is a new start!

After reading this chapter, you should be able to identify key drivers of food insecurity, and food security indicators & drivers as well. Additionally, you should be able to understand the IPC Integrated Food Security and Nutrition Conceptual Framework, the USAID Office of Food for Peace and Food Security Country Framework, the global food security index (GFSI) framework and the United Nations Suite of Food Security Indicators. Also, you should be able to apply the Food Security Bell Curve Machine Learning Model to analyze Prevalence of moderate or severe food insecurity in the Population. Finally, the chapter concludes with Who In the World is Food Insecure with two food security machine learning models that model prevalence of moderate or severe food insecurity and Prevalence of severe food insecurity in the population.

References

[1] Leanna Parekh, the basics of food security (and how it's tied to everything), Updated Mar 23, 2021, https://www.worldvision.ca/stories/food/the-basics-of-food-security, Access Date: April 08, 2022.

[2] Anikka Martin, Food Security and Nutrition Assistance, Last updated: Monday, November 08, 2021, https://www.ers.usda.gov/data-products/ag-and-food-statistics-charting-the-essentials/food-security-and-nutrition-assistance/, Access Date: December 12,2021.

[3] Alisha, Coleman-Jensen, Matthew P. Rabbitt, Laura Hales and Christian A. Gregory. The prevalence of food insecurity in 2020 is unchanged from 2019, https://www.ers.usda.gov/data-products/chart-gallery/gallery/chart-detail/?chartId=58378, Access Date: December 20, 2021.

[4] Alisha, Coleman-Jensen, Matthew P. Rabbitt, Laura Hales and Christian A. Gregory. Prevalence of food insecurity is not uniform across the country, Last updated: Monday, November 08, 2021, https://www.ers.usda.gov/data-products/chart-gallery/gallery/chart-detail/?chartId=58392, Access Date: December 21, 2021.

[5] Felix, G. Baquedano. World's most vulnerable populations consumed fewer calories a day during pandemic, Last updated: Friday, October 15, 2021, https://www.ers.usda.gov/data-products/chart-gallery/gallery/chart-detail/?chartId=102321, Access Date: May 16, 2022.

[6] Felix, G. Baquedano, Yacob Abrehe Zereyesus, Constanza Valdes and Kayode Ajewole. International Food Security Assessment 2021–31, July 2021, https://www.ers.usda.gov/publications/pub-details/?pubid=101732, Access Date: May 16, 2022.

[7] Michael, D. Smith and Birgit Meade. Who Are the World's Food Insecure? Identifying the Risk Factors of Food Insecurity Around the World, June 03, 2019, https://www.ers.usda.gov/amber-waves/2019/june/who-are-the-world-s-food-insecure-identifying-the-risk-factors-of-food-insecurity-around-the-world/, Access Date: May 15, 2022.

[8] CENTRAL BANK OF SRI LANKA, ANNUAL REPORT, 2020. https://www.cbsl.gov.lk/sites/default/files/cbslweb_documents/publications/annual_report/2020/en/13_Box_01.pdf, Access Date: April 11, 2022.

[9] Issues and Challenges of Inclusive Development: Essays in Honor of Prof. R. Radhakrishna by R. Maria Saleth, S. Galab and E. Revathi, Publisher : Springer; 1st ed. 2020 edition (June 19, 2020), ISBN-13 : 978-9811522284.

[10] Julia Horowitz, War has brought the world to the brink of a food crisis, 8:44 AM ET, Mon March 14, 2022, https://www.cnn.com/2022/03/12/business/food-crisis-ukraine-russia/index.html, Access Date: April 04, 2022.

[11] Giulia Soffiantini. Food insecurity and political instability during the Arab Spring, Global Food Security, Volume 26, 2020, 100400, ISSN 2211-9124, https://doi.org/10.1016/j.gfs.2020.100400.

[12] Jason Lalljee, Russia is disrupting food prices worldwide and it could cause civil unrest — it's happened before, Mar 9, 2022, 10:36 AM, https://www.businessinsider.com/ukraine-russia-bread-food-prices-civil-unrest-arab-spring-egypt-2022-3, Access Date: April 10, 2022.

[13] Shruti, Menon and Ranga Sirilal, Why is there a food emergency in Sri Lanka? 20 September 2021, https://www.bbc.com/news/world-asia-pacific-58485674, Access Date: April 10, 2022.

[14] Saabira Chaudhuri. Ukraine War Sparks Global Scramble for Cooking Oils, Apr. 6, 2022 6:02 am ET, https://www.wsj.com/articles/ukraine-war-sparks-global-scramble-for-cooking-oils-11649239342?page=1, Access Date: April 08, 2022.

[15] NADA HAMADEHCATHERINE VAN ROMPAEYERIC METREAU, New World Bank country classifications by income level: 2021–2022. JULY 01, 2021,https://blogs.worldbank.org/opendata/new-world-bank-country-classifications-income-level-2021-2022, May 14, 2022.

[16] The Associated Press, A political reckoning in Sri Lanka as economic crisis grows, April 28, 20222:46 AM ET, https://www.npr.org/2022/04/28/1095186509/a-political-reckoning-in-sri-lanka-as-economic-crisis-grows, Access Date: May 15, 2022.

[17] Megan Durisin, World Hunger 'Exploding' After 25% Spike Before Ukraine War, May 4, 2022, 1:04 AM PDT Updated on May 4, 2022, 7:43 AM PDT, https://www.bloomberg.com/news/articles/2022-05-04/world-hunger-spiked-before-ukraine-war-and-will-likely-get-worse, Access Date: May 20, 2022.

[18] IPC, The IPC Integrated Food Security and Nutrition Conceptual Framework, https://www.ipcinfo.org/ipc-manual-interactive/overview/17-the-ipc-conceptual-framework/en/, Access Date: April 07, 2022.

[19] Anderson, Stephen and Elisabeth Farmer. USAID Office of Food for Peace Food Security Country Framework for Ethiopia FY 2016 – FY 2020. Washington, D.C.: Food Economy Group, 2015.

[20] Dr. Biplab K. Nandi, REPORT OF THE REGIONAL EXPERT CONSULTATION OF THE ASIA-PACIFIC NETWORK FOR FOOD AND NUTRITION ON CONCRETIZING ACTIONS ON ESTABLISHMENT OF FOOD INSECURITY AND VULNERABILITY INFORMATION AND MAPPING SYSTEMS (FIVIMS), 12 November 1999, https://www.fao.org/3/x6623e/x6623e00.htm#TopOfPage, Access Date: April 07, 2022.

[21] Food Security Information and Knowledge Sharing System 2017 - Project Funded by the European Union, Food Security Information and Knowledge Sharing System - Sudan Federal Food Security Technical Secretariat, 2017, http://fsis.sd/SD/EN/FoodSecurity/Monitoring/, Access Date: April 24, 2024.

[22] Mark Nord, Carlo Cafiero and Sara Viviani, Methods for estimating comparable prevalence rates of food insecurity experienced by adults in 147 countries and areas, 2016, https://iopscience.iop.org/article/10.1088/1742-6596/772/1/012060/pdf, Access Date: April 25, 2022.

[23] BENEDITH MULONGO, analyzing music genre classification using item response theory, 2020, https://www.diva-portal.org/smash/get/diva2:1472081/FULLTEXT01.pdf, Access Date: April 28, 2022.
[24] OECD, The Rasch Model, 2009, https://www.oecd-ilibrary.org/docserver/9789264056251-6-en.pdf?expires=1650947236&id=id&accname=guest&checksum=6AC9EBAC550111E8456A7AFDEFA606FE, Access Date: April 25, 2022.
[25] Dennis Wesselbaum, Michael D. Smith, Christopher B. Barrett, Anaka Aiyar, A Food Insecurity Kuznets Curve? August 20 2021, http://barrett.dyson.cornell.edu/files/papers/FIKC%20Aug%202021.pdf, Access Date: April 26, 2022.
[26] FAO, Trade Reforms and Food Security, 2003, https://www.fao.org/3/y4671e/y4671e.pdf, Access Date: June 10, 2022.
[27] United Nations Statistics Division Development Data and Outreach, End hunger, achieve food security and improved nutrition and promote sustainable agriculture, 2020, https://unstats.un.org/sdgs/report/2020/goal-02/, Access Date: May 14, 2022.
[28] USDA, Food Security Assessment. USDA Economic Research Service. Situation and Outlook series GFA-11 Washington DC, 1999, https://www.ers.usda.gov/publications/pub-details/?pubid=37089, Access Date: June 11, 2022.
[29] United Nations Statistics Division Development Data and Outreach, End hunger, achieve food security and improved nutrition and promote sustainable agriculture, 2020, https://unstats.un.org/sdgs/report/2020/goal-02/, Access Date: May 14, 2022.
[30] Government of Alberta, Agricultural Marketing Guide, 2022, https://www.alberta.ca/agricultural-marketing-guide.aspx, Access Date: April 29, 2022.
[31] An official publication of Michigan Farm Bureau, Biden requests $500M for emergency US crop production incentives to offset Ukraine impact, April 28, 2022, https://www.michiganfarmnews.com/biden-requests-500m-for-emergency-u-s-crop-production-incentives-to-offset-ukraine-impact, Access Date: May 02, 2022.
[32] The White House, FACT SHEET: White House Calls on Congress to Provide Additional Support for Ukraine, APRIL 28, 2022, https://www.whitehouse.gov/briefing-room/statements-releases/2022/04/28/fact-sheet-white-house-calls-on-congress-to-provide-additional-support-for-ukraine/, Access Date: May 02, 2022.
[33] Anna D'Souza. Rising Food Prices and Declining Food Security: Evidence From Afghanistan, September 01, 2011, https://www.ers.usda.gov/amber-waves/2011/september/afghanistan-food-security/, Access Date: May 27, 2022.
[34] Lana Conforti. The first 50 years of the Producer Price Index: setting inflation expectations for today. Monthly Labor Review, U.S. Bureau of Labor Statistics, June 2016, https://doi.org/10.21916/mlr.2016.25.
[35] Matthew MacLachlan, Summary Findings Food Price Outlook 2022, Aptil 29 2022, https://www.ers.usda.gov/data-products/food-price-outlook/summary-findings/, Access Date: April 29, 2022.
[36] CRAIG HOWELL AND WILLIAM THOMAS, Inflation remained low during 1984, Monthly Labor Review 1985, https://www.bls.gov/opub/mlr/1985/04/art1full.pdf, Access Date: April 30, 2022.
[37] Bahattin Büyükşahin, Kun Mo and Konrad Zmitrowicz, Commodity Price Supercycles: What Are They and What Lies Ahead?, 2016, https://www.bankofcanada.ca/wp-content/uploads/2016/11/boc-review-autumn16-buyuksahin.pdf, Access Date: April 24, 2022.
[38] GARIMA VASISHTHA. Commodity price cycles in three charts, JANUARY 18, 2022, https://blogs.worldbank.org/developmenttalk/commodity-price-cycles-three-charts, Access Date: April 24, 2022.
[39] Temitayo, A. Adeyemo, Sikiru Ajijola, Samuel Kehinde Odetola and Victor O. Okoruwa. Impact of Agricultural Value Added on Current Account Balances in Nigeria,Journal of Economics and Sustainable Development www.iiste.org ISSN 2222-1700 (Paper) ISSN 2222-2855, https://www.iiste.org/Journals/index.php/JEDS/article/view/19999, Acces Date: May 16, 2022.
[40] Martin Kaufman and Daniel Leigh, How the Pandemic Widened Global Current Account Balances, AUGUST 2, 2021, https://blogs.imf.org/2021/08/02/how-the-pandemic-widened-global-current-account-balances/, Acces Date: May 16, 2022.
[41] Raifman, J., Bor, J. and Venkataramani, A. 2020. Unemployment insurance and food insecurity among people who lost employment in the wake of COVID-19. medRxiv : the preprint server for health sciences, 2020.07.28.20163618. https://doi.org/10.1101/2020.07.28.20163618.
[42] Felix, G. Baquedano, Yacob Abrehe Zereyesus, Constanza Valdes and Kayode Ajewole, Food Insecurity to Increase in 2021 at a Higher Rate in Lower Income Countries, September 07, 2021, https://www.ers.usda.gov/amber-waves/2021/september/food-insecurity-to-increase-in-2021-at-a-higher-rate-in-lower-income-countries/, Acces Date: May 16, 2022.
[43] Mehdi Ben Slimane, Marilyne Huchet, Habib Zitouna. Direct and indirect effects of FDI on food security: a sectoral approach. Workshop MAD Macroeconomics of Agriculture and Develoment - What challenges food security? Institut National de Recherche Agronomique (INRA). UMR Structures et Marchés Agricoles, Ressources et Territoires (1302). Nov 2013, Rennes, France. 27 p. ffhal-01189920.
[44] Fleming, K. 2019. The Effect of Foreign Direct Investment on Food Security: A Case Study from Rural Mozambique. Agricultural Economics and Agribusiness Undergraduate Honors Theses Retrieved from https://scholarworks.uark.edu/aeabuht/11.
[45] Ge Jiaqi, Polhill J. Gareth, Macdiarmid Jennie I., Fitton Nuala, Smith Pete et al. 2021. Food and nutrition security under global trade: a relation-driven agent-based global trade modelR. Soc. Open Sci. 8201587201587.

[46] QUNO. The Relationship between Key Food Security Measures and Trade Rules, NOVEMBER 2015, https://quno.org/resource/2015/11/relationship-between-key-food-security-measures-and-trade-rules, Acces Date: May 16, 2022.
[47] Choudhury, S. and Headey, D. 2017. What drives diversification of national food supplies? A cross-country analysis. Global Food Security 15: 85–93. https://doi.org/10.1016/j.gfs.2017.05.005.
[48] Heidi Ullmann, Ph.D., Julie D. Weeks, Ph.D. and Jennifer H. Madans, Ph.D. Children Living in Households That Experienced Food Insecurity: United States, 2019–2020, February 2022, https://www.cdc.gov/nchs/products/databriefs/db432.htm, Acces Date: May 16, 2022.
[49] Tacoli C. 2017. Food (In)Security in Rapidly Urbanising, Low-Income Contexts. International Journal of Environmental Research and Public Health 14(12): 1554. https://doi.org/10.3390/ijerph14121554, Acces Date: May 16, 2022.
[50] FAO, Poverty alleviation and food security in Asia: lessons and challenges, 1999, https://www.fao.org/publications/card/en/c/179e6317-eac5-539b-a470-4ad809e3ce8b/, Access Date: May 16, 2022.
[51] Sam Polzin and Jayson Lusk. Urban-rural food satisfaction, food security gaps show in new report, May 11, 2022, https://www.purdue.edu/newsroom/releases/2022/Q2/urban-rural-food-satisfaction,-food-security-gaps-show-in-new-report.html, Access Date: May 16, 2022.
[52] Stacey Rosen and Shahla Shapouri, Rising Food Prices Intensify Food Insecurity in Developing Countries, February 01, 2008, https://www.ers.usda.gov/amber-waves/2008/february/rising-food-prices-intensify-food-insecurity-in-developing-countries/, Acces Date: May 15, 2022.
[53] Haini, H., Musa, S.F.P.D., Wei Loon, P. and Basir, K.H. 2022. Does unemployment affect the relationship between income inequality and food security? International Journal of Sociology and Social Policy, Vol. ahead-of-print No. ahead-of-print. https://doi.org/10.1108/IJSSP-12-2021-0303.
[54] It costs $150 to buy a dozen eggs in Venezuela right now, MERY MOGOLLON AND ALEXANDRA ZAVIS, MAY 31, 2016, AT 5:00 AM, https://www.chicagotribune.com/nation-world/la-fg-venezuela-inflation-0531-snap-htmlstory.html, Access Date: April 09, 2022.
[55] Mayela Armas, Venezuela's Inflation Hit 686.4% in 2021 - Central Bank, Jan. 8, 2022, at 10:27 a.m.,https://money.usnews.com/investing/news/articles/2022-01-08/venezuelas-inflation-hit-686-4-in-2021-central-bank#:~:text=CARACAS%20(Reuters)%20%2D%20Venezuela's%20annual,central%20bank%20said%20on%20Saturday., Access Date: April 09, 2022.
[56] Stephen Vogado, Panama: Consumer prices end year in deflationary territory, January 15, 2020, https://www.focus-economics.com/countries/panama/news/inflation/consumer-prices-end-year-in-deflationary-territory, Access Date April 09, 2022.
[57] Kate Ashford and John Schmidt, What Is Deflation?,Updated: Aug 25, 2021, 9:48am, https://www.forbes.com/advisor/investing/what-is-deflation/, Access Date: April 09, 2022.
[58] Nicolle Yapur, Food Markets in Caracas Empty Out as Inflation Hits the Poorest, July 2, 2021, 4:00 AM PDT, https://www.bloomberg.com/news/articles/2021-07-02/food-markets-in-caracas-empty-out-as-inflation-hits-the-poorest, Access Date: April 09, 2022.
[59] Felix, G. Baquedano, World's most vulnerable populations consumed fewer calories a day during pandemic, Last updated: Friday, October 15, 2021, https://www.ers.usda.gov/data-products/chart-gallery/gallery/chart-detail/?chartId=102321, Access Date: May 16, 2022.
[60] Olivia Arena, Clare Salerno, January 28, 2020, Four Ways to Address Food Insecurity through Transportation Improvements, https://www.urban.org/urban-wire/four-ways-address-food-insecurity-through-transportation-improvements, Access Date: May 19, 2022.
[61] Laura Turley, David Uzsoki, Why Financing Rural Infrastructure Is Crucial to Achieving Food Security, January 9, 2019, https://www.iisd.org/articles/rural-infrastructure-food-security, Access Date: May 19, 2022.
[62] Geoffrey Gertz. Did Trump's tariffs benefit American workers and national security? SEPTEMBER 10, 2020, https://www.brookings.edu/policy2020/votervital/did-trumps-tariffs-benefit-american-workers-and-national-security/,Access Date: June 10, 2022.
[63] Amara Omeokwe and Richard Rubin, U.S. Considering Reducing Tariffs on China to Ease Inflation, Yellen Says, Updated June 8, 2022 4:16 pm ET, https://www.wsj.com/articles/yellen-expects-progress-on-global-tax-deal-11654705060?page=1,Access Date: June 10, 2022.
[64] FAO. Trade Reforms and Food Security, 2003, https://www.fao.org/3/y4671e/y4671e.pdf, Access Date: June 10, 2022.
[65] Vivian, Salama, Costas Paris and William Mauldin. Looming Global Food Crisis Related to Ukraine War Divides Biden Administration, Updated June 7, 2022 7:33 pm ET, https://www.wsj.com/articles/looming-food-crisis-related-to-ukraine-war-divides-biden-administration-11654639588?page=1,Access Date: June 10, 2022.
[66] Ryan, Dezember, Russia-Ukraine War Threatens Wheat Supply, Jolts Prices, March 13, 2022 8:00 am ET., https://www.wsj.com/articles/russia-ukraine-war-threatens-wheat-supply-jolts-prices-11647115099?mod=article_inline,Access Date: June 10, 2022.
[67] Paul, W. Heisey and Keith O. Fugliem. Agricultural Research Investment and Policy Reform in High-Income Countries,May 2018, https://www.ers.usda.gov/webdocs/publications/89114/err-249.pdf?v=0, Access Date: May 16, 2022.

[68] Aline D'Angelo Campos. Equity in Global Food Systems Change: A Cross-Country Analysis of the Drivers of Food Insecurity, April 2021, https://dukespace.lib.duke.edu/dspace/bitstream/handle/10161/22756/Regular%20MP_Aline%20D%27Angelo%20Campos.pdf?sequence=1, Access Date: April 10, 2022.
[69] Rulon Pope, Engel's Law. 2012. https://byustudies.byu.edu/article/engels-law/, Access Date: April 10, 2022.
[70] Pereira, A., Handa, S. and Holmqvist, G. 2021. Estimating the prevalence of food insecurity of households with children under 15 years, across the globe. Global Food Security, 28, 100482. https://doi.org/10.1016/j.gfs.2020.100482.
[71] WFP, Kenya, United Nations Complex, Gigiri, https://www.wfp.org/countries/kenya, Access Date: May 19, 2022.
[72] IMF. Impact of High Food and Fuel Prices on Developing Countries - Frequently Asked Questions, Last Updated: July 28, 2017, https://www.imf.org/external/np/exr/faq/ffpfaqs.htm, Access Date: May 19, 2022.

[152] Issues and Challenges of Inclusive Development: Essays in Honor of Prof. R. Radhakrishna by R. Maria Saleth, S. Galab and E. Revathi, Publisher : Springer; 1st ed. 2020 edition (June 19, 2020), ISBN-13 : 978-9811522284

[153] Food Security - https://www.fao.org/fileadmin/templates/faoitaly/documents/pdf/pdf_Food_Security_Cocept_Note.pdf

[154] USAID OFFICE OF FOOD FOR PEACE - Food Security Country Framework for Ethiopia - https://pdf.usaid.gov/pdf_docs/PBAAE621.pdf

[155] World Food Summit, Rome, 1996 - https://www.fao.org/3/w3548e/w3548e00.htm

[156] Agriculture and Food Security - https://www.usaid.gov/what-we-do/agriculture-and-food-security

[157] To avoid food insecurity, keep finance flowing - https://blogs.worldbank.org/psd/avoid-food-insecurity-keep-finance-flowing

[158] Food Security and Nutrition Assistance - https://www.ers.usda.gov/data-products/ag-and-food-statistics-charting-the-essentials/food-security-and-nutrition-assistance/

[159] The prevalence of food insecurity in 2020 is unchanged from 2019 - https://www.ers.usda.gov/data-products/chart-gallery/gallery/chart-detail/?chartId=58378

[160] Prevalence of food insecurity is not uniform across the country - https://www.ers.usda.gov/data-products/chart-gallery/gallery/chart-detail/?chartId=58392

[161] World's most vulnerable populations consumed fewer calories a day during pandemic - https://www.ers.usda.gov/data-products/chart-gallery/gallery/chart-detail/?chartId=102321

[162] CENTRAL BANK OF SRI LANKA - https://www.cbsl.gov.lk/sites/default/files/cbslweb_documents/publications/annual_report/2020/en/13_Box_01.pdf

[163] Food Is Just as Vital as Oil to National Security - https://www.bloomberg.com/opinion/articles/2022-03-07/food-is-just-as-vital-as-oil-to-national-security

[164] War has brought the world to the brink of a food crisis - https://www.cnn.com/2022/03/12/business/food-crisis-ukraine-russia/index.html

[165] Ukraine War Sparks Global Scramble for Cooking Oils - https://www.wsj.com/articles/ukraine-war-sparks-global-scramble-for-cooking-oils-11649239342?page=1

[166] The basics of food security (and how it's tied to everything) - https://www.worldvision.ca/stories/food/the-basics-of-food-security

[167] Food insecurity and political instability during the Arab Spring, Global Food Security - https://doi.org/10.1016/j.gfs.2020.100400.

[168] Food security is national security - https://www.ernst.senate.gov/news/press-releases/ernst-food-security-is-national-security
[169] S-3089: https://www.congress.gov/bill/117th-congress/senate-bill/3089/text?r=7&s=1
[170] Hanumayamma Innovations and Technologies, Inc., https://www.hanuinnotech.com
[171] New World Bank country classifications by income level: 2021-2022. - https://blogs.worldbank.org/opendata/new-world-bank-country-classifications-income-level-2021-2022
[172] The World Bank Atlas method - detailed methodology - https://datahelpdesk.worldbank.org/knowledgebase/articles/378832-what-is-the-world-bank-atlas-method
[173] World Bank – Historical Classification - http://databank.worldbank.org/data/download/site-content/OGHIST.xls
[174] A political reckoning in Sri Lanka as economic crisis grows - https://www.npr.org/2022/04/28/1095186509/a-political-reckoning-in-sri-lanka-as-economic-crisis-grows
[175] The World Bank In Sri Lanka - https://www.worldbank.org/en/country/srilanka
[176] The World Bank in Algeria - https://www.worldbank.org/en/country/algeria/overview#1
[177] World Bank Classifies Mauritius as High-Income Country - https://www.bom.mu/media/media-releases/world-bank-classifies-mauritius-high-income-country
[178] World Bank Country and Lending Groups, 2022 - https://datahelpdesk.worldbank.org/knowledgebase/articles/906519-world-bank-country-and-lending-groups
[179] World Bank Lower middle-income countries - https://data.worldbank.org/country/XN
[180] Upper Middle-Income Countries - https://data.worldbank.org/country/XT
[181] World Bank Country Classification, 2022 - https://datahelpdesk.worldbank.org/knowledgebase/articles/906519-world-bank-country-and-lending-groups
[182] High Income Countries - https://data.worldbank.org/country/XD
[183] World Hunger 'Exploding' After 25% Spike Before Ukraine War - https://www.bloomberg.com/news/articles/2022-05-04/world-hunger-spiked-before-ukraine-war-and-will-likely-get-worse
[184] The IPC Integrated Food Security and Nutrition Conceptual Framework - https://www.ipcinfo.org/ipc-manual-interactive/overview/17-the-ipc-conceptual-framework/en/
[185] REPORT OF THE REGIONAL EXPERT CONSULTATION OF THE ASIA-PACIFIC NETWORK FOR FOOD AND NUTRITION ON CONCRETIZING ACTIONS ON ESTABLISHMENT OF FOOD INSECURITY AND VULNERABILITY INFORMATION AND MAPPING SYSTEMS (FIVIMS) - https://www.fao.org/3/x6623e/x6623e00.htm#TopOfPage
[186] Food Security Key Indicators - https://www.fao.org/3/x6623e/x6623e11.htm#P1336_66469
[187] The GFSI Methodology -
[188] International Food Security Assessment, 2021-31 - https://www.ers.usda.gov/publications/pub-details/?pubid=101732
[189] Methods for estimating comparable prevalence rates of food insecurity experienced by adults in 147 countries and areas - https://iopscience.iop.org/article/10.1088/1742-6596/772/1/012060/pdf
[190] Analyzing music genre classification using item response theory - https://www.diva-portal.org/smash/get/diva2:1472081/FULLTEXT01.pdf
[191] FIFE Scale- https://www.fao.org/in-action/voices-of-the-hungry/fies/en/
[192] Rasch Modeling - https://www.publichealth.columbia.edu/research/population-health-methods/rasch-modeling#:~:text=The%20Rasch%20model%20provides%20a,for%20quantifying%20unobservable%20human%20conditions.
[193] The Rasch Model - https://www.oecd-ilibrary.org/docserver/9789264056251-6-en.pdf?expires=1650947236&id=id&accname=guest&checksum=6AC9EBAC550111E8456A7AFDEFA606FE
[194] A Food Insecurity Kuznets Curve? - http://barrett.dyson.cornell.edu/files/papers/FIKC%20Aug%202021.pdf
[195] Who Are the World's Food Insecure? Identifying the Risk Factors of Food Insecurity Around the World - https://www.ers.usda.gov/amber-waves/2019/june/who-are-the-world-s-food-insecure-identifying-the-risk-factors-of-food-insecurity-around-the-world/
[196] Food Security Assessment. USDA Economic Research Service. Situation and Outlook series GFA-11 Washington DC - https://www.ers.usda.gov/publications/pub-details/?pubid=37089
[197] United Nations Statistics Division Development Data and Outreach, End hunger, achieve food security and improved nutrition and promote sustainable agriculture - https://unstats.un.org/sdgs/report/2020/goal-02/
[198] Recent increases in food insecurity are likely to worsen because of COVID-19 - https://unstats.un.org/sdgs/report/2020/goal-02/
[199] Grain Traders' Profits Rise as Ukraine War Tightens Global Food Supply - https://www.wsj.com/articles/grain-traders-profits-rise-as-ukraine-war-tightens-global-food-supply-11651073868
[200] Biden requests $500M for emergency US crop production incentives to offset Ukraine impact - https://www.michiganfarmnews.com/biden-requests-500m-for-emergency-u-s-crop-production-incentives-to-offset-ukraine-impact
[201] FACT SHEET: White House Calls on Congress to Provide Additional Support for Ukraine - https://www.whitehouse.gov/briefing-room/statements-releases/2022/04/28/fact-sheet-white-house-calls-on-congress-to-provide-additional-support-for-ukraine/
[202] Wheat Futures - https://tradingeconomics.com/commodity/wheat
[203] Rising Food Prices and Declining Food Security: Evidence From Afghanistan - https://www.ers.usda.gov/amber-waves/2011/september/afghanistan-food-security/
[204] Primary Commodity Price System - https://data.imf.org/?sk=471DDDF8-D8A7-499A-81BA-5B332C01F8B9
[205] How to use charting to analyse commodity markets - https://www.alberta.ca/how-to-use-charting-to-analyse-commodity-markets.aspx
[206] The first 50 years of the Producer Price Index: setting inflation expectations for today - https://www.bls.gov/opub/mlr/2016/article/the-first-50-years-of-the-producer-price-index.htm

[207] The CPI - https://www.bls.gov/cpi/

[208] Summary Findings Food Price Outlook 2022 - https://www.ers.usda.gov/data-products/food-price-outlook/summary-findings/

[209] Commodity Price Super cycles: What Are They and What Lies Ahead? - https://www.bankofcanada.ca/wp-content/uploads/2016/11/boc-review-autumn16-buyuksahin.pdf

[210] Commodity price cycles in three charts - https://blogs.worldbank.org/developmenttalk/commodity-price-cycles-three-charts

[211] FFPI Dataset - https://www.fao.org/fileadmin/templates/worldfood/Reports_and_docs/Food_price_indices_data_may629.csv

[212] Current Account Balance - https://data.worldbank.org/indicator/BN.CAB.XOKA.GD.ZS

[213] Impact of Agricultural Value Added on Current Account Balances in Nigeria - https://www.iiste.org/Journals/index.php/JEDS/article/download/19999/20527

[214] How the Pandemic Widened Global Current Account Balances - https://blogs.imf.org/2021/08/02/how-the-pandemic-widened-global-current-account-balances/

[215] Employment to population ratio, 15+, total (%) (modeled ILO estimate) - https://data.worldbank.org/indicator/SL.EMP.TOTL.SP.ZS

[216] Unemployment insurance and food insecurity among people who lost employment in the wake of COVID-19 - https://www.ncbi.nlm.nih.gov/pmc/articles/PMC7402065/

[217] Food Insecurity to Increase in 2021 at a Higher Rate in Lower Income Countries - https://www.ers.usda.gov/amber-waves/2021/september/food-insecurity-to-increase-in-2021-at-a-higher-rate-in-lower-income-countries/

[218] Foreign Direct Investment - https://data.worldbank.org/indicator/BX.KLT.DINV.WD.GD.ZS

[219] The Effect of Foreign Direct Investment on Food Security: A Case Study from Rural Mozambique - https://scholarworks.uark.edu/cgi/viewcontent.cgi?article=1011&context=aeabuht

[220] Hungary - https://www.oecd.org/investment/HUNGARY-trade-investment-statistical-country-note.pdf

[221] 2021 Investment Climate Statements: Netherlands - https://www.state.gov/reports/2021-investment-climate-statements/netherlands/

[222] 2021 Investment Climate Statements: Ireland - https://www.state.gov/reports/2021-investment-climate-statements/ireland/

[223] GDP per capita, PPP (constant 2017 international $) - https://data.worldbank.org/indicator/NY.GDP.PCAP.PP.KD

[224] SOFS 2021 - https://www.fao.org/3/ca5162en/ca5162en.pdf

[225] Import of Goods and Services - https://data.worldbank.org/indicator/NE.IMP.GNFS.ZS

[226] Food and nutrition security under global trade: a relation-driven agent-based global trade model - https://royalsocietypublishing.org/doi/10.1098/rsos.201587

[227] The Relationship between Key Food Security Measures and Trade Rules - https://quno.org/resource/2015/11/relationship-between-key-food-security-measures-and-trade-rules

[228] Official exchange rate (LCU per US$, period average) - https://data.worldbank.org/indicator/PA.NUS.FCRF

[229] Exchange Rate Archives by Month - https://www.imf.org/external/np/fin/data/param_rms_mth.aspx

[230] SDRs - https://www.imf.org/en/About/Factsheets/Sheets/2016/08/01/14/51/Special-Drawing-Right-SDR

[231] Exchange rate monitoring - https://docs.wfp.org/api/documents/WFP-0000107215/download/

[232] USD/EGP Exchange rate - https://fxtop.com/en/historical-exchange-rates.php?A=1&C1=USD&C2=EGP&MA=1&DD1=01&MM1=01&YYYY1=1970&B=1&P=&I=1&DD2=25&MM2=02&YYYY2=2022&btnOK=Go%21

[233] IMF Data - https://data.imf.org/regular.aspx?key=61015892

[234] Foreign Exchange Rates -- H.10 Weekly - https://www.federalreserve.gov/releases/h10/current/default.htm

[235] What drives diversification of national food supplies? A cross-country analysis - https://www.ncbi.nlm.nih.gov/pmc/articles/PMC5727671/

[236] Children Living in Households That Experienced Food Insecurity: United States - https://www.cdc.gov/nchs/products/databriefs/db432.htm

[237] Food (In)Security in Rapidly Urbanising, Low-Income Contexts - https://www.ncbi.nlm.nih.gov/pmc/articles/PMC5750972/

[238] Rural Population (% of total population) - https://data.worldbank.org/indicator/SP.RUR.TOTL.ZS

[239] Urban-rural food satisfaction, food security gaps show in new report - https://www.purdue.edu/newsroom/releases/2022/Q2/urban-rural-food-satisfaction,-food-security-gaps-show-in-new-report.html

[240] 12 Things to Know: Food Security in Asia and the Pacific - https://www.adb.org/news/features/12-things-know-food-security-asia-and-pacific

[241] Urban-rural food satisfaction, food security gaps show in new report - https://www.purdue.edu/newsroom/releases/2022/Q2/urban-rural-food-satisfaction,-food-security-gaps-show-in-new-report.html

[242] Total reserves (includes gold, current US$) - https://data.worldbank.org/indicator/FI.RES.TOTL.CD

[243] Agricultural raw materials - https://data.worldbank.org/indicator/TM.VAL.AGRI.ZS.UN

[244] Rising Food Prices Intensify Food Insecurity in Developing Countries - https://www.ers.usda.gov/amber-waves/2008/february/rising-food-prices-intensify-food-insecurity-in-developing-countries/

[245] Agriculture, forestry, and fishing, value added (% of GDP) - https://data.worldbank.org/indicator/NV.AGR.TOTL.ZS

[246] Employment in agriculture - https://data.worldbank.org/indicator/SL.AGR.EMPL.ZS

[247] FAO Food Price Index - https://www.fao.org/worldfoodsituation/foodpricesindex/en/

[248] World Food Price Index - https://data.imf.org/regular.aspx?key=61015892

[249] It costs $150 to buy a dozen eggs in Venezuela right now - https://www.chicagotribune.com/nation-world/la-fg-venezuela-inflation-0531-snap-htmlstory.html

[250] Venezuela's Inflation Hit 686.4% in 2021 - Central Bank- https://money.usnews.com/investing/news/articles/2022-01-08/venezuelas-inflation-hit-686-4-in-2021-central-bank#:~:text=CARACAS%20(Reuters)%20%2D%20Venezuela's%20annual

[251] Food Markets in Caracas Empty Out as Inflation Hits the Poorest - https://www.bloomberg.com/news/articles/2021-07-02/food-markets-in-caracas-empty-out-as-inflation-hits-the-poorest
[252] What Is Deflation? - https://www.forbes.com/advisor/investing/what-is-deflation/
[253] Countries with the lowest inflation rate 2020 - https://www.statista.com/statistics/268190/countries-with-the-lowest-inflation-rate/
[254] Food Aid - https://data.oecd.org/oda/food-aid.htm
[255] Countries requiring external assistance for food - https://www.fao.org/giews/country-analysis/external-assistance/en/
[256] Four Ways to Address Food Insecurity through Transportation Improvements - https://www.urban.org/urban-wire/four-ways-address-food-insecurity-through-transportation-improvements
[257] Better roads must for food security - https://www.downtoearth.org.in/news/better-roads-must-for-food-security-48413
[258] Did Trump's tariffs benefit American workers and national security? - https://www.brookings.edu/policy2020/votervital/did-trumps-tariffs-benefit-american-workers-and-national-security/
[259] WITS - https://wits.worldbank.org/tariff/trains/country-byhs6product.aspx?lang=en
[260] Tariff rate - https://data.worldbank.org/indicator/TM.TAX.MRCH.SM.AR.ZS
[261] Trade Reforms and Food Security - https://www.fao.org/3/y4671e/y4671e.pdf
[262] Looming Global Food Crisis Related to Ukraine War Divides Biden Administration - https://www.wsj.com/articles/looming-food-crisis-related-to-ukraine-war-divides-biden-administration-11654639588?page=1
[263] Credit to agriculture - https://www.fao.org/faostat/en/#data/IC/visualize
[264] USDA to Purchase Up to $3 Billion in Agricultural Commodities, Issue Solicitations for Interested Participants - https://www.ams.usda.gov/content/usda-purchase-3-billion-agricultural-commodities-issue-solicitations-interested
[265] Agricultural Research Investment and Policy Reform in High-Income Countries - https://www.ers.usda.gov/webdocs/publications/89114/err-249.pdf?v=0
[266] Swiss Agricultural Investments - https://orgprints.org/id/eprint/29201/1/IMPRESA_Country_report_Switzerland_final.pdf
[267] MAFAP – Kenya Policy Coherence - https://www.fao.org/in-action/mafap/country-analysis/country-dashboard/en/?iso3=KEN
[268] Price volatility in agricultural markets - https://www.fao.org/economic/est/issues/volatility/en/#.YoSIw6jMKQI
[269] Food Price Volatility in Africa - https://ebrary.ifpri.org/digital/api/collection/p15738coll2/id/127343/download
[270] FAO: 14% of the world's food is lost between harvest and retail - https://www.globalagriculture.org/whats-new/news/en/33821.html
[271] Engel's Law - https://byustudies.byu.edu/article/engels-law/
[272] GFSI - https://my.corteva.com/GFSI?file=dl_index
[273] Suite of Food Security Indicators - https://www.fao.org/faostat/en/#data/FS
[274] GFSI Model 2021 - https://my.corteva.com/GFSI?file=dl_index
[275] GFSI Model - https://my.corteva.com/GFSI?file=dl_index
[276] Rwanda Food Inflation - https://tradingeconomics.com/rwanda/food-inflation#:~:text=Food%20Inflation%20in%20Rwanda%20averaged,percent%20in%20October%20of%202018.
[277] Accelerating Rwanda's Food Systems Transformation - https://www.rockefellerfoundation.org/wp-content/uploads/2022/02/Accelerating-Rwandas-Food-Systems-Transformation.pdf
[278] Zambia - https://www.wfp.org/countries/zambia
[279] Ireland - https://knoema.com/atlas/Ireland/Dietary-energy-supply-adequacy
[280] Can Ireland feed itself? Yes. A nutritious diet? Not at the moment - https://www.irishtimes.com/life-and-style/food-and-drink/can-ireland-feed-itself-yes-a-nutritious-diet-not-at-the-moment-1.4824313
[281] Estimating the prevalence of food insecurity of households with children under 15 years, across the globe - https://www.ncbi.nlm.nih.gov/pmc/articles/PMC8318352/
[282] GFSI - https://my.corteva.com/GFSI?file=dl_index
[283] Why Financing Rural Infrastructure Is Crucial to Achieving Food Security - https://www.iisd.org/articles/rural-infrastructure-food-security
[284] Better Rural Transport is Key to Food Security and Zero Hunger - https://slocat.net/1901-2/
[285] Guinea - https://www.wfp.org/countries/guinea
[286] Kenya - https://www.wfp.org/countries/kenya
[287] Estimating the prevalence of food insecurity of households with children under 15 years, across the globe - https://www.ncbi.nlm.nih.gov/pmc/articles/PMC8318352/
[288] Impact of High Food and Fuel Prices on Developing Countries-Frequently Asked Questions - https://www.imf.org/external/np/exr/faq/ffpfaqs.htm

Chapter 4

Food Security Drivers and Key Signal Pattern Analysis

The chapter introduces Food Security Drivers & Key Signal Pattern Analysis. Additionally, it introduces the Food Security Bell curve model and applied macroeconomic and key agricultural drivers to analyze the prevalence of Undernourishment in Afghanistan and Sri Lanka. Finally, the chapter concludes with the application of additional bell curve food security machine learning models on commodity prices.

Food Security is national security. Ensuring food security maintains national governance and social order. Higher food prices and food inflation are societal issues, an undue disadvantage for the poor and marginalized communities. Lack of food security can create unrest. Increased food prices challenge governments and test monetary policies.[289] Food insecurity could induce the worst living standards, a crisis for generations. It is essential to study market, macroeconomics, weather, and agricultural signals to create preventive models that could predict food insecurity issues [1].

Global Inflation has a self-inflicting effect. Increased food inflation, supply disruptions, due to the War in Ukraine for instance, fuels national protectionism, a move that can be seen by major commodity exporters[290] and dominant cooking oil producers [2]. Such actions, food protectionism, in essence, sadly rejuvenate severe food insecurity. For instance, India has curbed wheat exports just weeks after Indonesia announced a ban on palm oil shipments. India banned the export of wheat on May 14, 2022, largely owing to a record high domestic food inflation. Lower yield due to intense heat waves piled on the country's agony [3]. India, the world's second-largest wheat producer, said that factors including lower wheat production and sharply spiking global prices because of the war caused it worries about its own "food security". The decision came,[291] from the ministry of agriculture, as global agricultural markets were under severe stress due to Russia's invasion of Ukraine (please see Figure 1). The move sent wheat prices near a record high, threatening to further increase the cost of everything from bread to cakes and noodles [2]. The sharp spike in prices put undue pressure on import dependent countries such as Egypt and Afghanistan. The surprising thing is that India isn't even a prominent exporter on the world stage. The fact that it could have such a major impact[292] underscores the bleak prospect for global wheat supplies. War has crippled Ukraine's exports, and now droughts, floods and heat waves threaten crops for most major producers. To some extent, Indian wheat exports did rise. Countries like Egypt and Turkey, besides others in Asia, tapped India following the onset of the war. This, however, pushed prices to record highs at home. In the past few weeks, wheat prices have soared by 15%–20% in India, forcing government intervention. Now, with the ban, prices are rising globally, too [4].

Food protectionism will cause revolts in some cases in the host country itself. For instance, Indonesian farmers' revolted after prices dropped due to an export ban announced in the middle of April, 2022. Indonesia,[293] the world's biggest shipper of edible oils, lifted a ban on palm oil exports in a move that will bring relief to the global market after the war in Ukraine choked off critical supplies. Soybean oil, a *substitute* of palm oil, fell as much as 1.6% after Indonesia's announcement [5]. Global prices have increased drastically (Please see figure—data source UN FAO Food Price Index[294] (Figure 2)).

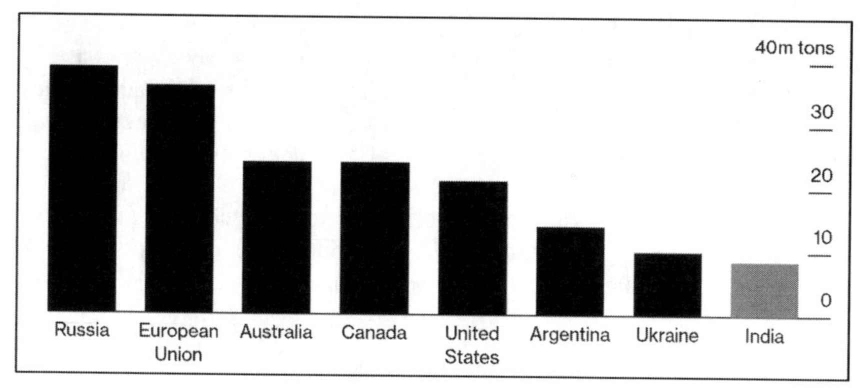

Figure 1: Wheat production.

Commodity markets are integral to the global economy. Understanding what drives developments in these markets is critical to the design of policy frameworks that facilitate the economic objectives of sustainable growth, inflation stability, poverty reduction, food security, and the mitigation of climate change. The purpose of this chapter is to analyze prevalence of undernourishment in countries worldwide, especially those with a low-income compared to high-income countries, for awareness of key undernourishment drivers. The ML models show that commodity markets are heterogeneous in terms of their drivers, price behavior, and macroeconomic impact on emerging markets and developing economies. The heterogeneity varies as widely as relationship between economic growth and commodity demands, depending on their stage of economic development [6]. Additionally, how these linkages influence food security is a critical aspect for creating world economies that can take care poor and marginalized communities more aptly through booms and busts. We will leverage the ML model developed in this chapter through the Food Security macroeconomic framework model.

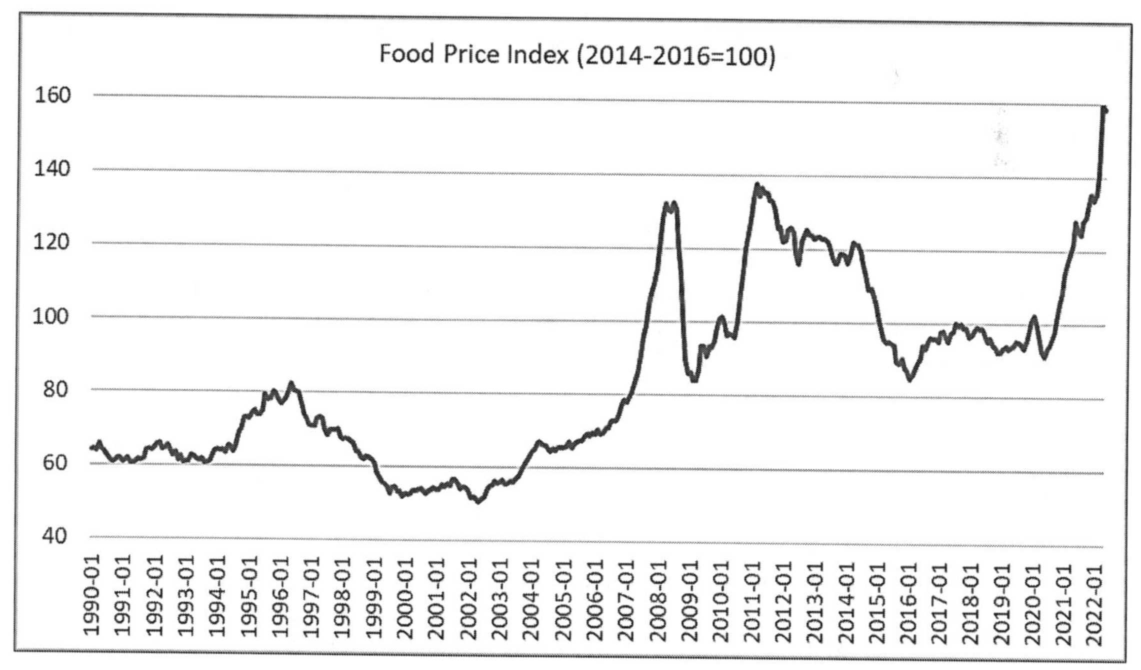

Figure 2: FAO Food Price Indices.

Analyzing food security gaps at a national level is a very intricate and difficult process as it encompasses historical data and, more important, market dynamic that happens on a day-to-day basis. For instance, changes in market values, demand economics, supply constraints, weather, global trade policies change, and many other factors. Nevertheless, for machine learning model development purposes, we have pursued two lines of angles in assessing the food security or prevalence of food under nourishment: status quo and food nutrition gap.

The *Status Quo* gap measures the difference between projected food supplies (calculated as domestic production plus commercial imports minus non-food uses) and a base period (2000–2020) per capita consumption. *Commercial imports* are where the macroeconomic factors play a decisive role (please see Table 1). Factors such as:

Table 1: Import & Food Security.

Primary Drivers	**Secondary Drivers**
• GDP GROWTH (ANNUAL %) [M2] • DOMESTIC FOOD PRICE LEVEL INDEX [M6] • REAL EXCHANGE RATE ($) [M8] • INFLATION, GDP DEFLATOR (ANNUAL %) [M14] • TOTAL DEBT SERVICE (% OF GNI) [M16] • FOREIGN DIRECT INVESTMENT, NET INFLOWS (% OF GDP) [M17] • TOTAL RESERVES (INCLUDES GOLD, CURRENT US$) [M18] • CURRENT ACCOUNT BALANCE (% OF GDP) [M19] • AGRICULTURAL IMPORT TARIFFS [A2] • VOLATILITY OF AGRICULTURAL PRODUCTION [A4] • AGRICULTURAL RAW MATERIALS IMPORTS (% OF MERCHANDISE IMPORTS) [A8] • EMPLOYMENT IN AGRICULTURE (% OF TOTAL EMPLOYMENT) [A10] • DEMOGRAPHIC STRESSORS (LAND, WATER, and WEATHER) [A16]	• PER CAPITA FOOD PRODUCTION VARIABILITY [M7] • EMPLOYMENT [M10] • CHANGES IN GOVERNMENT SPENDING OVER INITIAL GDP [M12] • GOVERNMENT DEFICIT RELATIVE TO GDP [M13] • ENERGY PRICE SHOCKS (CRUDE OIL, NATURAL GAS, and DIESEL) [M15] • PER CAPITA FOOD SUPPLY VARIABILITY [A3] • HIGH CEREAL IMPORT DEPENDANCY RATIO [A5] • ARABLE LAND EQUIPPED FOR IRRIGATION [A7] • AGRICULTURE, FORESTRY, AND FISHING, VALUE ADDED (% OF GDP) [A9] • CROP DIVERSIFICATION & SUFFICIENCY OF SUPPLY [A15] • AID, SOCIAL SECURITY/WELFARE (VULNERABLE & RURAL) [A17]

The primary factors have an immediate influence on food security. A fiscal deficit and a deficit on the current account of the balance of payments render a country to food state emergency in a short span of time as account balance impacts, among many others, a country's credit rating that has a direct implication on borrowing capacity.[295] A large current account balance drains a country's reserves and impacts the strength of Local Currency Units (LCU) against international currencies and impact the commercial sector that impacts employment and re-triggers the account balance, a vicious cycle, or a debt trap. High Cereal Import dependency[296] poses an external threat to a country's democratic institutions and could lead to social unrest if not properly managed. External factors such as bad crops in high producing countries, conflicts in the world that disrupt supply chains and the world commercial order, global economic meltdowns, and other shocks would lead to risks for countries with a high cereal dependency. Food and energy inflation can trigger inflation in economies that import food and energy. In Latin America and the Caribbean (LAC), food and energy were the main contributors to inflation in 2021. They accounted for more than 90% of inflation in Costa Rica, 75% in Paraguay, 66% in Brazil, and almost 60% in Colombia. Empirical analysis confirms that inflation in these regions was highly correlated with demand side pressures stemming from expansionary policies, as well as supply chain pressures, energy prices, and currency depreciation. Inflation is currently projected to increase further in 2022, due mainly to rising commodity prices and global supply disruptions. Surinam and Haiti, for example, are expected to experience double-digit inflation. Food-importing countries will be more quickly affected by further increases in international food prices[297] as there is causation between primary and secondary drivers. In real-world dynamics, the causation could be either of two types of drivers.

The *Nutrition gap*, which is the difference between projected food supplies and the amount of food needed to support minimum per capita nutritional (please see Table 2 & Figure 3) standards.[298]

Table 2: Nutrition gap.

Primary	Secondary
SHARE OF FOOD EXPENDITURE OF THE POOR [M3] POPULATION UNDER GLOBAL POVERTY LINE [M5]	DOMESTIC FOOD PRICE LEVEL INDEX [M6] DAIRY & ANIMAL HUSBANDRY [A13] CHANGE IN AVERAGE FOOD COST [A11] VOLATILITY OF AGRICULTURAL PRODUCTION[A4] PER CAPITA FOOD SUPPLY VARIABILITY [A3] ENERGY PRICE SHOCKS (CRUDE OIL, NATURAL GAS, and DIESEL) [M15]

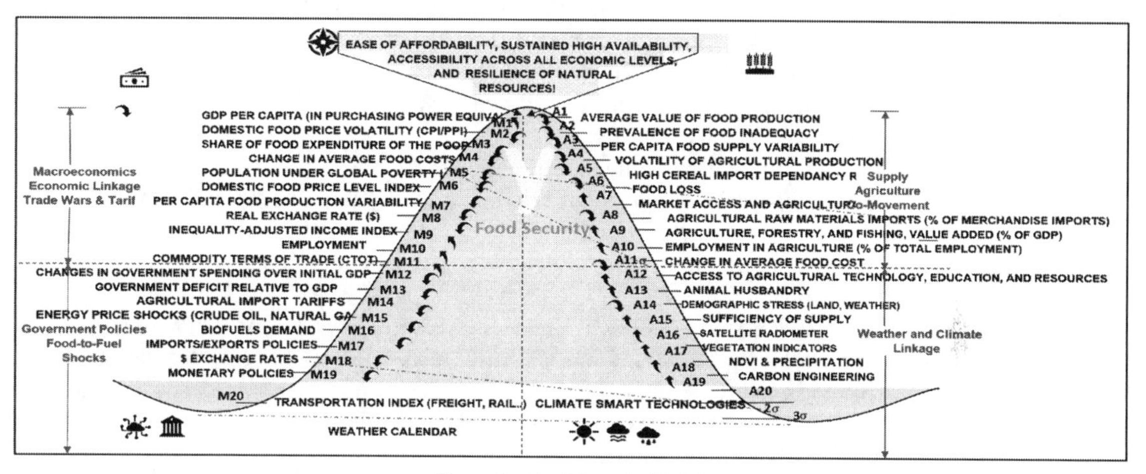

Figure 3: Food Security Model.

The Status Quo indicator provides a safety net criterion at the country level, whilst the Nutrition gap indicator gives a comparison of relative well-being at the population level [7].

Libyan consumers have been complaining about the sudden food price rises upon the outbreak of the Russia-Ukraine war.[299] Within days of the war starting, bakeries increased bread prices from four standard loaves per dinar to three loaves per dinar. Libya is heavily reliant on imported Ukrainian and Russian grain for its bread flour [8]. Prices of other basic foodstuffs such as flour, cooking oil, sugar, and tomato paste also rose. The public was outraged. Social media commentary complained that the war in Ukraine had barely started when bakeries raised prices. They complained that new grain supplies had not arrived in Libya yet to instantly affect bread prices. They demanded that the government, which subsidizes bread flour, act. War is tipping a fragile world towards mass hunger.[300] Fixing that is everyone's business.

Food Security Drivers & Signal Analysis

To analyze the drivers of food security, we have developed machine learning models for two distinct sets of countries: Low Income & High Income. Particularly, Afghanistan, Sri Lanka, and Australia.

Afghanistan

Afghanistan[301] is facing one of the world's largest food crises: more than half of the population—22.8 million people—has been facing acute food insecurity through March 2022, as per the World Food Program, branch of United Nations & world's largest humanitarian organization focused on hunger and food security. Eighty percent of those suffering live in rural communities. This record-high level of acute hunger has been driven by decades of conflict, political upheaval, declining economy and liquidity crisis, the impact of the COVID-19 pandemic, and severe ongoing drought.

The agriculture sector is the backbone of the economy in Afghanistan, and 80 percent of the population depends on it to support and feed its families. Yet, farmers' ability to produce food and earn an income is threatened as they continue to face drought, disrupted markets, and limited access to agricultural inputs. Many rural people are resorting to selling their animals and *abandoning their land* and livelihoods to migrate to urban areas. Decades of complex and protracted conflicts, combined with a changing climate, gender inequalities, rapid urbanization, underemployment, and the economic fallout of the COVID-19 pandemic pose considerable challenges to efforts to achieve Sustainable Development Goals (SDGs), including SDG 2 on Zero Hunger and improved nutrition.

Over half of the country's population lives below the poverty line, and food insecurity is on the rise, largely due to conflict and insecurity cutting off whole communities from livelihood opportunities. 22.8 million people are identified as acutely food insecure, including hundreds of thousands who have been displaced by conflict since the beginning of the year.

Undernutrition is of particular concern in women, children, displaced people, returnees, households headed by women, people with disabilities and the poor. Despite progress in recent years, undernutrition rates are now increasing, and 2 million children are malnourished.

Every year, some 250,000 people on average are affected by a wide range of environmental disasters including floods, droughts, avalanches, landslides, and earthquakes. The impact of disasters and dependency on water from rains or snowmelts severely limits the productivity of the agricultural sector, which is a source of income for 44 percent of the population.

Finally, Afghanistan's economic outlook is stark.[302] Under any scenario, Afghanistan will have a smaller economy, significantly higher rates of poverty, and more limited economic opportunities for the 600,000 Afghans reaching working age every year. Human development outcomes are likely to deteriorate in the context of substantial disruptions to basic services and increased poverty. The Russian invasion of Ukraine, war, and associated sanctions may have significant exacerbating impacts via increased prices for imported food and fuel.

The food security data related Afghanistan's macroeconomic variables, UN Data, and Food Security Index (FSI) data available since 2002. The limited data has posed limitations with respect to training and splitting it. The chapter, nevertheless, highlights the importance of exogenous and movement variables, and their transitions. The chapter enables us to construct both Regression and Time series models. The purpose of regression models is to establish coefficients of regression to see the impact of macroeconomic, co-movement, and other significant variables on prevalence of under nourishment. On the other hand, time series models are constructed to see future predictions as they differ from regression models; they depend on the order of observations. In regression, the order of the observations does not affect the model [9]. Time series models use historic values of a series to predict future values. Some time series models try to use seasonal and cyclic patterns.

The following section describes some of the critical macroeconomic and food security drivers:

"The world is exploding with food insecurity." [10]

Arif Husain[303]
Chief economist at the World Food Programme

Rural Population (% of total population)

Rural population (% of total population) in Afghanistan was reported at 73.97 % in 2020, according to the World Bank collection (please see Figure 5) of development indicators. Three-quarters of the population lives in rural areas, but faces rapid urbanization, providing both opportunities and challenges to the provision of education and economic development.[304] Poverty is widespread throughout the country, which has a high population growth rate.[305] An estimated 21 per cent of the rural population lives in extreme poverty and 38 per cent of rural household' face food shortages [11]. Agricultural production is the main source of rural livelihoods. Multiple years of conflict have hampered development of the agriculture sector, which also suffers from natural disasters and insufficient investment. Landholdings are small, so agriculture is rarely the

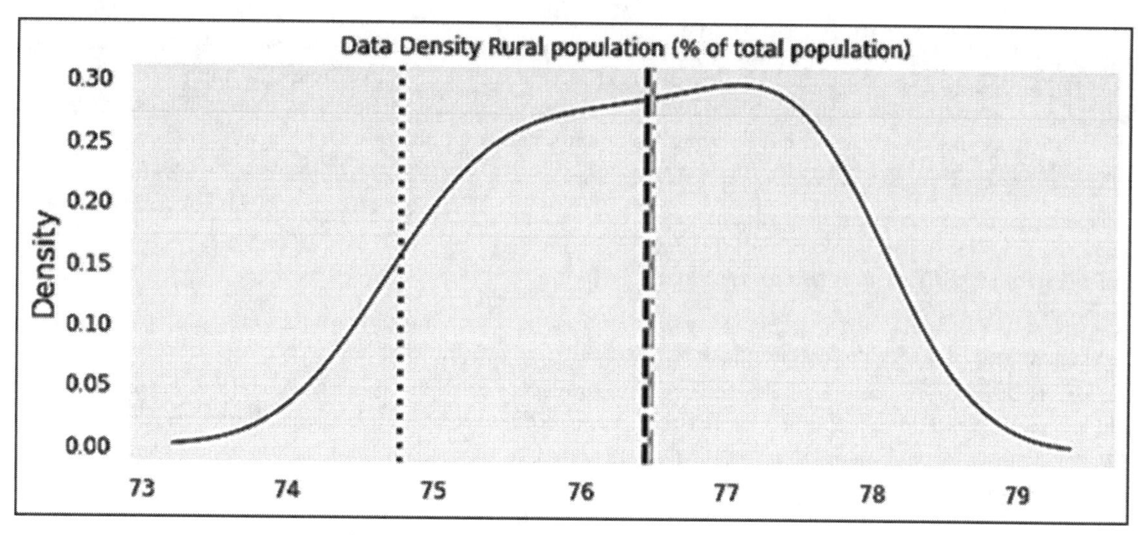

Figure 4: Data Density – Afghanistan Rural Population.

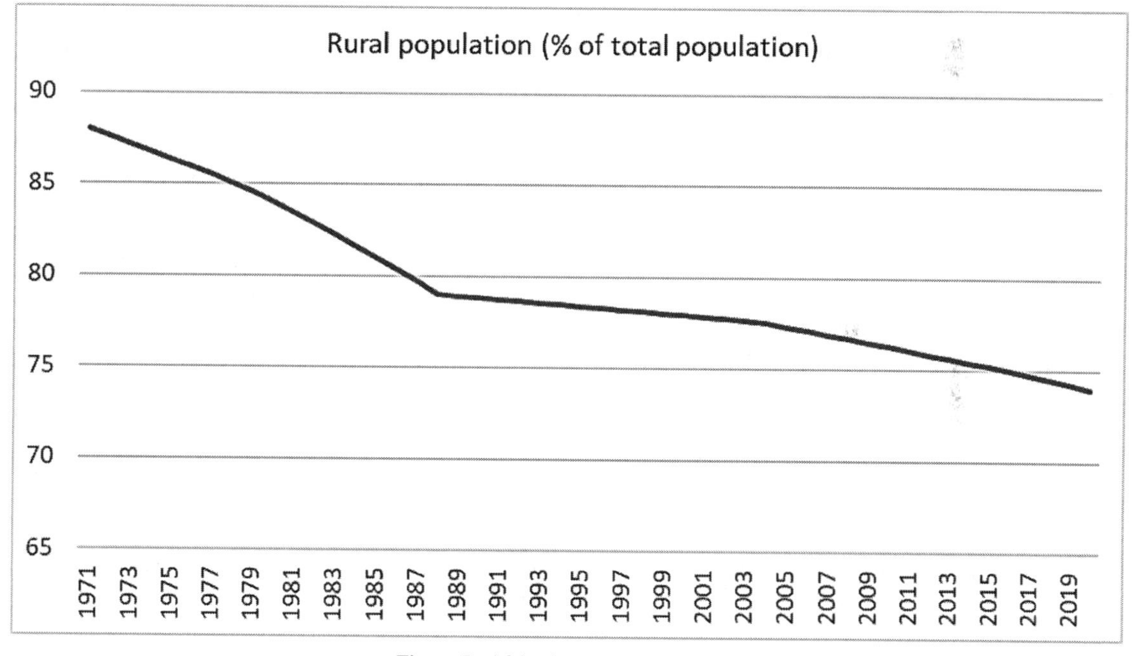

Figure 5: Afghanistan Rural Population.

family's main source of food or income. About two thirds of rural households' own livestock, and farmers also sell their labour. Loss of agriculture causes rural displacement [12].

The Data Distribution statistics (please see figure 4): Minimum:74.75; Mean:76.43 (dashed line); Median:76.47 (dash dot line); Mode: 74.75 (dotted line); Maximum:77.83. To summarize, the distribution of data is skewed to the right—statistically, if the mode is less than the median, which is less than the mean 76.43. The average rural population is 76.43%. A "skewed right" distribution is one in which the tail is on the right side. A "skewed left" distribution is one in which the tail is on the left side. Skewed data often occur due to lower or upper bounds on the data. That is, data that has a lower bound is often skewed right while data that has an upper bound is often skewed left. The above histogram is for a distribution that is skewed right[306] as higher percentages of rural population were frequent in the years [2002–2008].

	"We need to help Afghanistan avoid a hunger trap. Millions of Afghans are living on the edge of catastrophe—which will occur if their animals die, or fields go unplanted. Urgent investment in agriculture and livestock production is needed now, and it helps donors to save money down the road by putting the country back on track to food security." [12] QU Dongyu, FAO Director-General

GNI per capita, PPP (Current International $)

GNI per capita (formerly GNP per capita) is the gross national income, converted to U.S. dollars using the World Bank Atlas method, divided by the midyear population (please see Figure 6).

GNI is the sum of value added by all resident producers plus any product taxes (less subsidies) not included in the valuation of output plus net receipts of primary income (compensation of employees and property income) from abroad. GNI, calculated in national currency,[307] is usually converted to U.S. dollars at official exchange rates for comparisons across economies. GNI per capita, PPP (current international $) in Afghanistan was reported at 2100 USD, in 2022. It dropped to the lowest in 2008 to 1500 USD. The economic crisis in Afghanistan could be attributed to both political and economic factors.

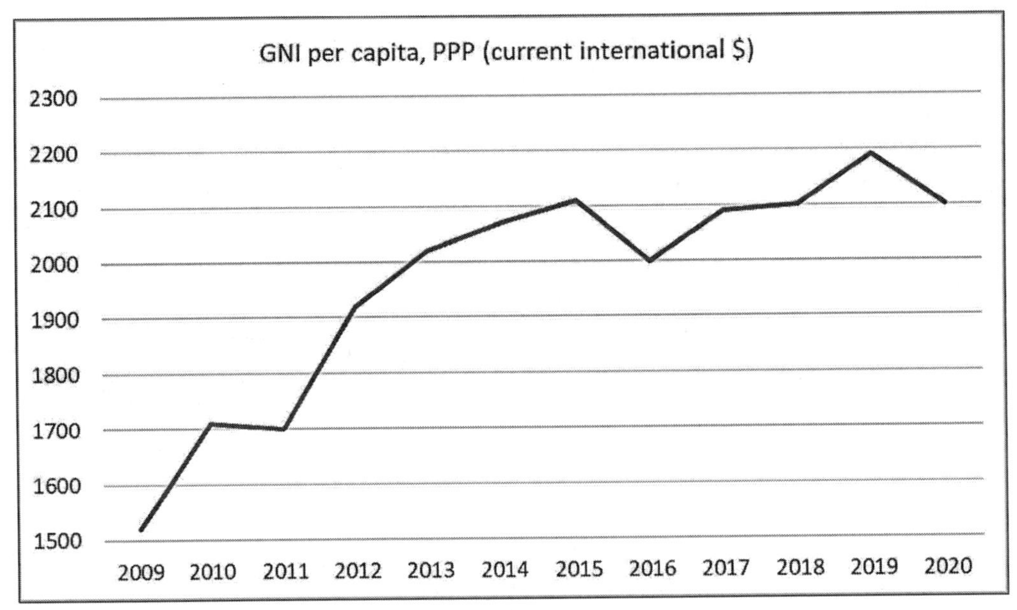

Figure 6: Afghanistan GNI

Some of the important drivers include [11]:

- cessation of aid (previously equal to 45 percent of GDP) driving a sharp fiscal contraction, leading to collapsing demand (total public spending is expected to have declined by around 60 percent)
- a loss of hard-currency aid inflows, which had previously financed a very large trade deficit (equal to around 30 percent of gross domestic product (GDP))
- loss of access to the overseas assets of the central bank (around US$9.2 billion)
- loss of central bank access to supplies of Afghani and USD cash notes, creating a *liquidity crisis* in the banking system, constraining firms', and households' access to working capital and savings held in commercial banks.
- rapid declines in investment confidence, given pervasive uncertainty and fear; and
- loss of human capital, as tens of thousands of highly skilled Afghans fled the country and new restrictions were imposed on women participation in the private and public sectors.

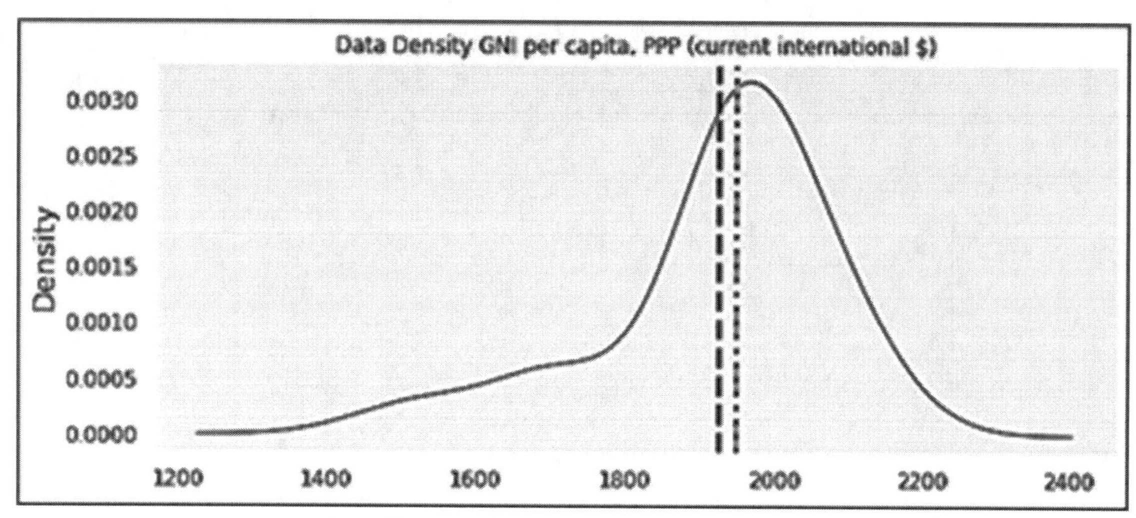

Figure 7: Afghanistan GNI Data Distribution.

Data Distribution statistics (please see Figure 7): Minimum:1520.00; Mean:1925.03 (dashed line); Median:1948.18 (dash dot line); Mode:1948.18 (dotted line); Maximum:2110.00. To summarize, the distribution of data is skewed to the right as GNI increased post 2008– statistically, if the mode is less than the median, which is less than the mean 1925.03.

Total Reserves

Afghanistan's economic outlook is stark [13]. Under any scenario, Afghanistan faces a smaller economy, significantly higher rates of poverty, and more limited economic opportunities for the 600,000 Afghans of working age every year. Human development outcomes are likely to deteriorate in the context of substantial disruptions to basic services and increased poverty. The Russian invasion of Ukraine, war, and associated sanctions may have significant exacerbating impacts via increased prices on imported food and fuel. To make the economy recover,[308] in 2008, the international community pledged more than $15 billion in aid to Afghanistan at a donors' conference in Paris, while Afghan President Hamid Karzai promised to fight corruption in the government.

Data Distribution statistics (please see Table 3): Minimum:3042,274,496.00 USD; Mean: 6,453,963,809.81 USD (dashed line); Median:6,653,774,877.25 USD (dash dot line); Mode:6,653,774,877.25 USD (dotted line); Maximum:8,097,280,956.00 USD. To summarize, the distribution of data is skewed to the left—statistically, if the mode is less than the median (in this case equal), which is less than the mean. A "skewed left" distribution is one in which the tail is on the left side. Skewed data often occurs due to lower or upper bounds on the data. That is data that has a lower bound is often skewed right while data that has an upper bound is often skewed left, a fact we can observe in the increase in total reserves after 2010. The above histogram is for a distribution that is skewed left[309] as higher total reserves frequent the years [2010–2016].

Per Capita Food Supply Variability (kcal/cap/day) & Per Capita food Production Variability (Constant 2004–2006 thousand int$ per capita)

Per capita food supply variability corresponds to the variability of the "food supply in kcal/capita/day".[310] In simple terms, the per capita food supply variability compares the variations of the food supply across countries and time (please see Table 4). Specifically, for Afghanistan, it provides a glimpse of food variability in the period 2002–2016. The variability is attributed to various factors such as a combination of instability and responses in production, trade, consumption, and storage [14]. As shown in the figure below, there is huge food supply variability before 2008 and it is dummied down after that. The possibilities could be increased food insecurity or prevalence of under nourishment due to unavailability! The indicator uses the data on dietary energy supply from the Food Balance Sheet (FBS) to measure annual fluctuations in the per

Table 3: Afghanistan Total Reserves.

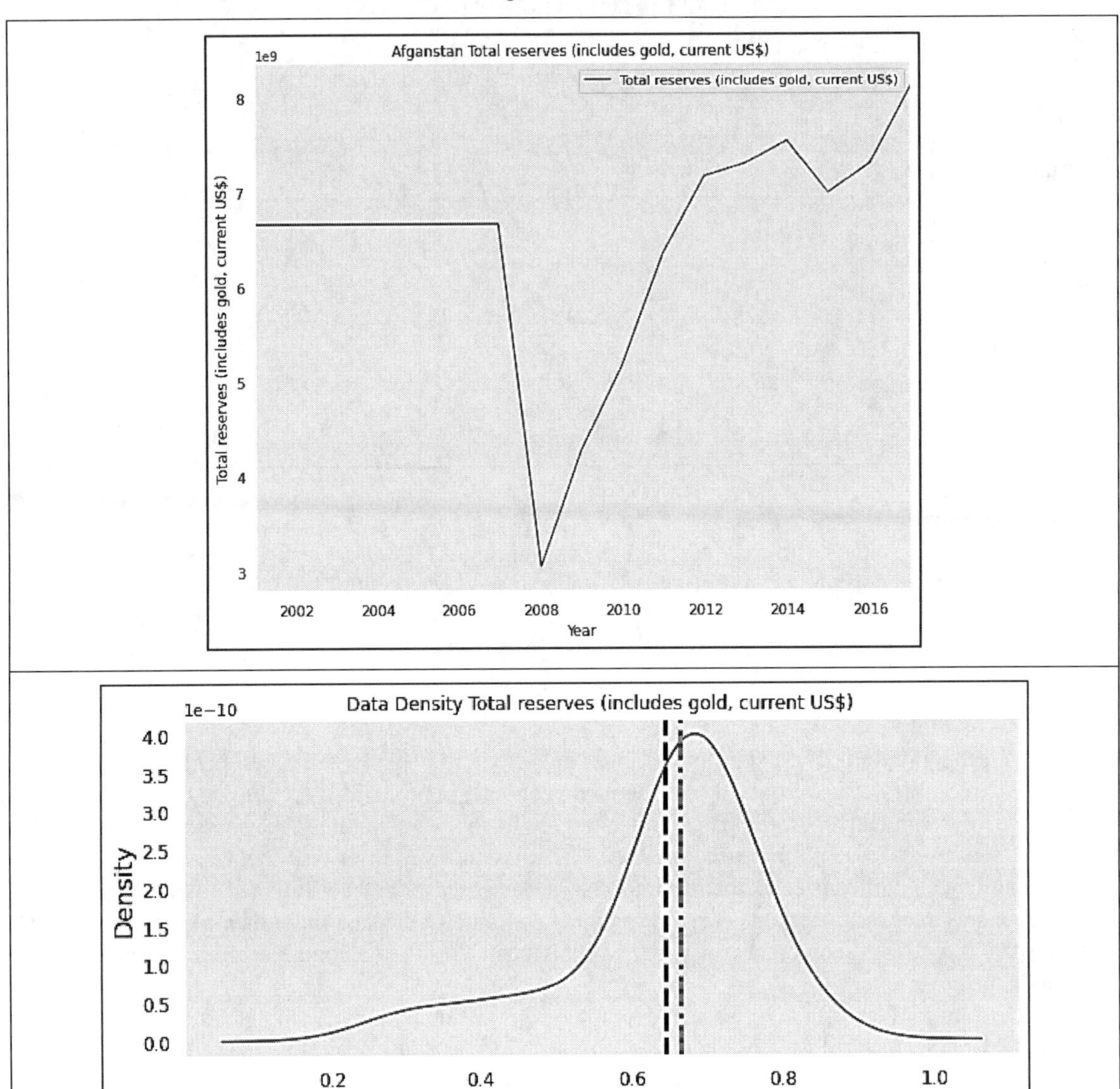

capita food supply (kcal), represented as the standard deviation over the previous five years per capita food supply. Food supply variability results from a combination of instability and responses in production, trade, consumption, and storage, in addition to changes in government policies such as trade restrictions, taxes and subsidies, stockholding, and public distribution.[311]

Afghanistan is one of the world's poorest, most food-insecure countries. With a long history of political instability and conflict, as well as weak infrastructure and mountainous terrain, Afghanistan is particularly vulnerable to economic and natural shocks.[312] Recent increases in the level and volatility of food prices pose a threat to countries like Afghanistan, where large populations often live in a state of food insecurity (defined as limited or uncertain availability of nutritionally adequate and safe foods) [15].

For skewed distributions, it is quite common to have one tail of the distribution considerably longer or drawn out relative to the other tail (please see Table 5). A "skewed right" distribution is one in which the tail is on the right side. A "skewed left" distribution is one in which the tail is on the left side. Skewed data often occurs due to lower or upper bounds on the data. That is, data that has a lower bound is often skewed

Table 4: Afghanistan Food Variability.

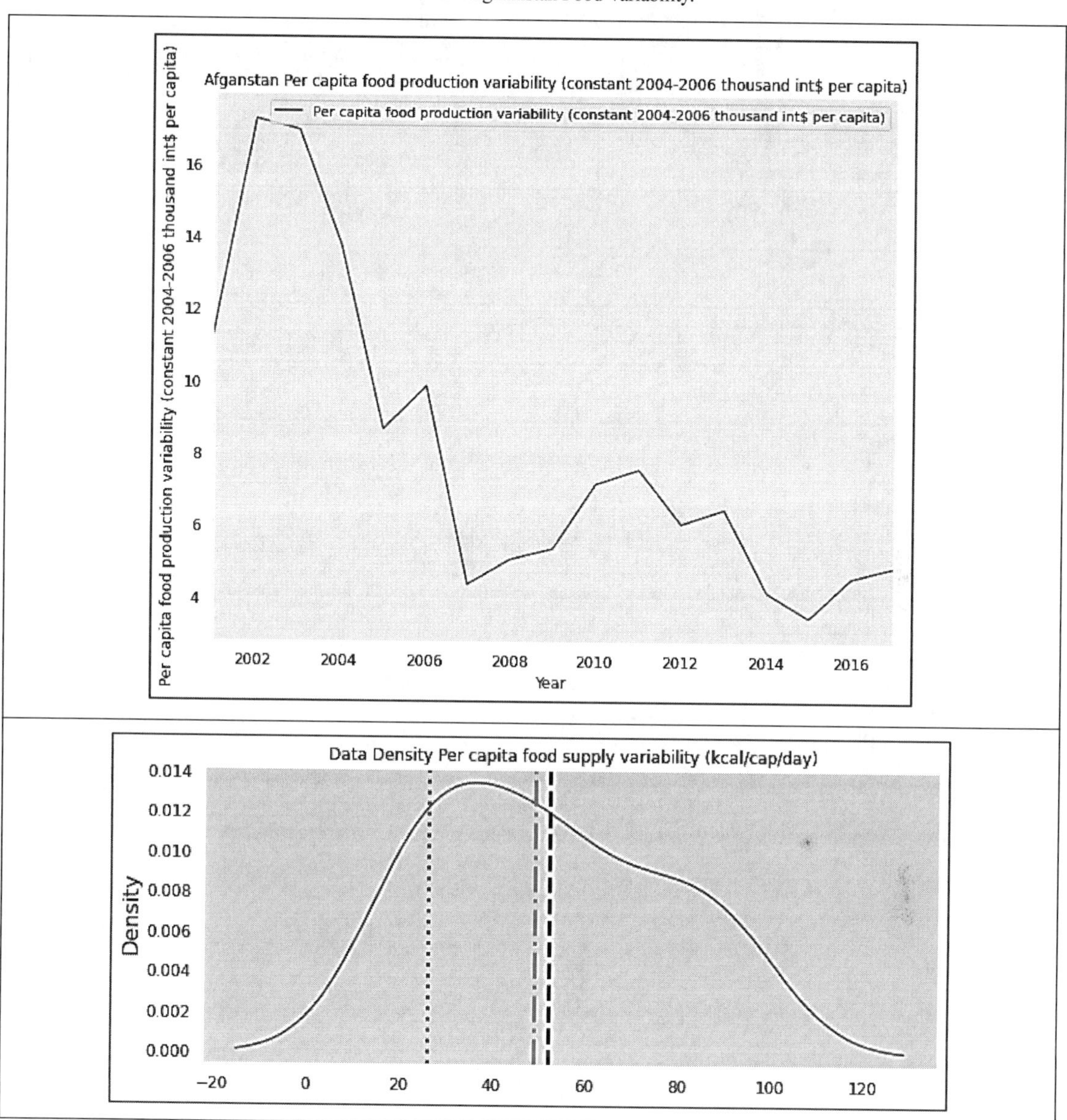

right while data that has an upper bound is often skewed left. The above histogram is for a distribution that is skewed right[313] as Per capita food supply variability (kcal/capita/day) was frequent in the lower values [2002–2012]. Data Distribution statistics: Minimum:21.00; Mean:52.12 (dashed line); Median:49.00 (dash dot line); Mode: 26.00 (dotted line); Maximum:93.00.

Afghanistan, as per the USDA ERS, had a nutrition gap—there was a difference between available food and food needed to support an intake of 2,100 calories per capita per day—of approximately 2 million tons in 2008; only North Korea had a larger estimated nutrition gap in that year. Using NRVA survey data, ERS researchers found that approximately 28 percent of the Afghan population did not meet the minimum daily energy requirements of 2,100 calories per day per person [15]. Wheat is the main staple. On average, Afghan households spend over 60 percent of their budgets on food, making them particularly vulnerable to declines in purchasing power brought on by increases in food prices.

Table 5: Afghanistan Food Supply Variability.

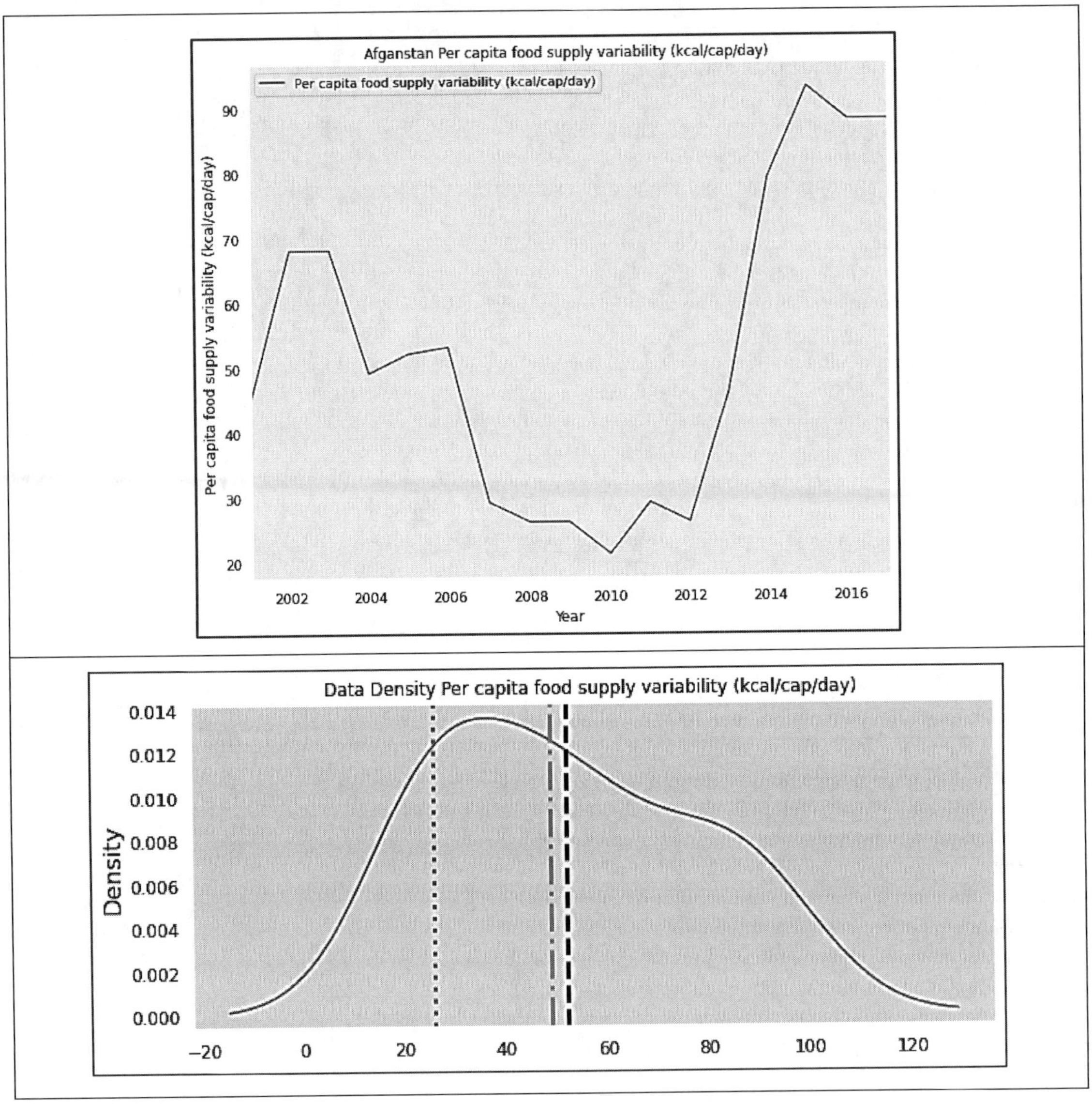

Cereal Import Dependency Ratio (percent) (3-year average)

Over 70 percent of cultivated crop area in Afghanistan is devoted to wheat. Due to large fluctuations in weather and sporadic conflicts, agricultural production is highly volatile, and the country is dependent on its trading partners to meet any shortfalls. According to ERS research, imports made up about 30 percent of annual consumption from 2000 to 2008 [15]. During 2007–2008, Afghanistan experienced several shocks that disrupted its food supply network, causing food prices to soar throughout the country. The 2008 wheat harvest of 1.5 million metric tons (1 metric ton=2,200 pounds) was the worst since 2000 due to drought and early snow melt. In February 2008, the Afghanistan Government eliminated import tariffs on wheat and wheat flour, but the downward effect on prices was small due to export bans in Pakistan (Afghanistan's biggest supplier), Iran, and Kazakhstan, as well as rising international food prices. Between fall 2007 and summer 2008, domestic wheat flour prices increased by over 100 percent, peaking around May–July 2008.

Cereal import requirements, mainly wheat and wheat flour,[314] in the July 2021to June 2022 marketing year are forecast at an above-average level of 3.4 million tonnes, over 20 percent more than in the previous year. However, even during the years with an above-average domestic wheat production, the country imported large quantities of wheat flour due to an inadequate domestic milling capacity. A signal we can see within the data is the distribution being skewed right {right tailed}. Imported flour is often blended with domestic flour to improve its protein content. In Pakistan and Kazakhstan, the main sources of wheat and wheat flour and export availabilities are estimated at average to slightly above-average levels in the current marketing year (please see Table 6).

Table 6: Afghanistan Cereal Import Dependency.

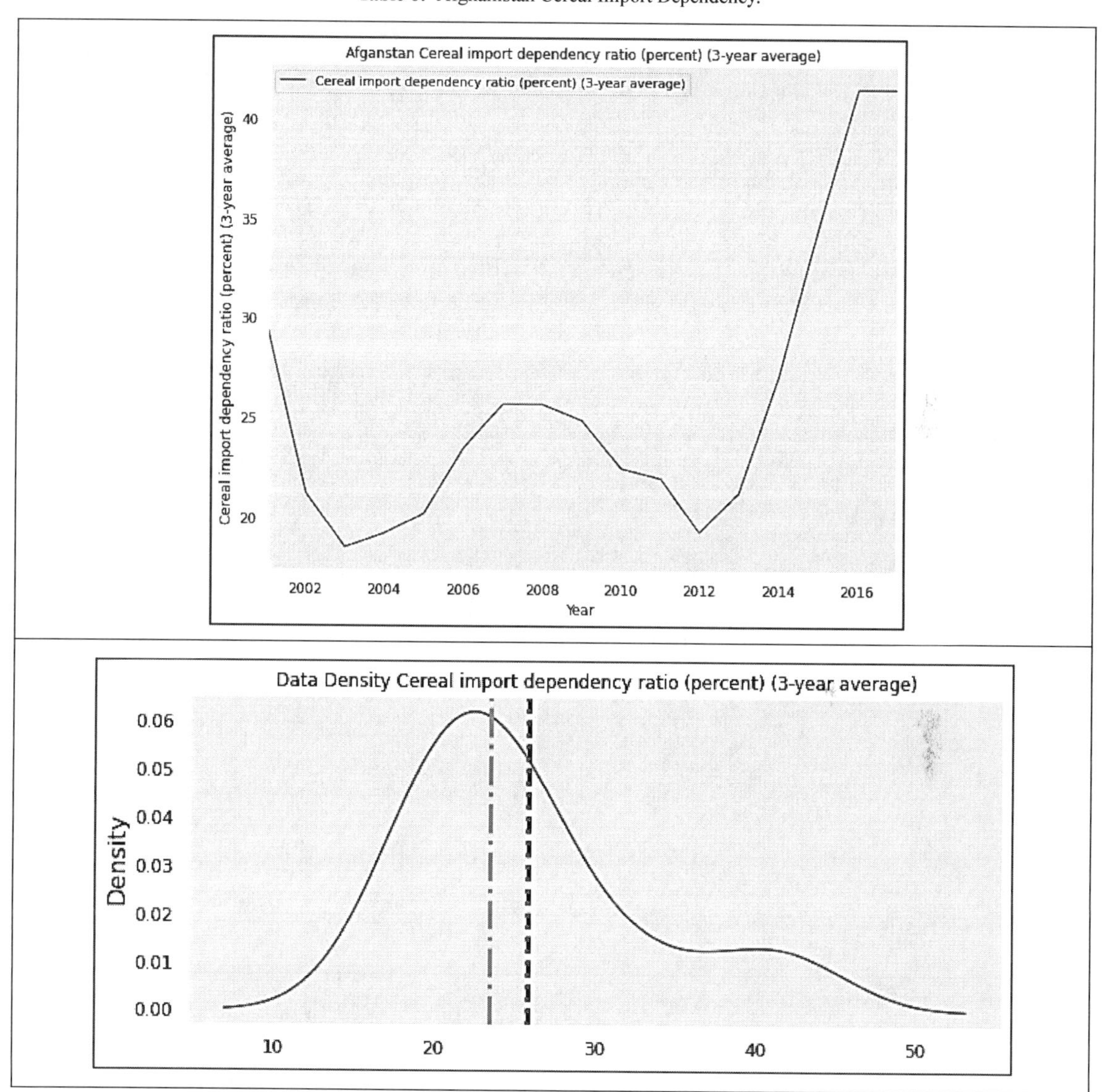

For skewed distributions, it is quite common to have one tail of the distribution considerably longer or drawn out relative to the other tail. A "skewed right" distribution is one in which the tail is on the right side. A "skewed left" distribution is one in which the tail is on the left side. Skewed data often occurs due to lower or upper bounds on the data. That is, data that has a lower bound is often skewed right while data that has an upper bound is often skewed left. The above histogram is for a distribution that is skewed right[315] as Cereal Import Dependency ratio was frequent in the lower values [2002–2012]. Data Distribution statistics: Minimum:18.50; Mean:25.79 (dashed line); Median:23.50 (dash dot line); Mode:25.70 (dotted line); Maximum:41.60.

	Price Shocks Led to Increases in Household Food Insecurity and Causing Households to Trade Off Quality for Quantity in Their Choice of Foods. [15].

Agricultural Irrigated Land (% of total agricultural land)

Agricultural irrigated land[316] (% of total agricultural land) in Afghanistan was 6.48 as of 2016. Its highest value over the past 15 years was 7.26 in 2003, while its lowest value was 4.62 in 2002. In 2019, total area equipped for irrigation in Afghanistan was 3,208 thousand hectares. Total area equipped for irrigation in Afghanistan[317] increased from 2,386 thousand hectares in 1970 to 3,208 thousand hectares in 2019 growing at an average annual rate of 0.61%. The land-cover maps, recently produced through the collaborative efforts of FAO, the United Nations Development Programme (UNDP) and the Afghan Geodesy and Cartography Office (Kabul), indicate that the extent of irrigated areas has not changed much in the last 35 years.[318] Worldwide, irrigated agriculture accounts for about four-fifths of global water withdrawals. The share of irrigated land ranges widely, from 4 percent of the total area cropped in Africa to 42 percent in South Asia. The leading countries are India and China with about 30 percent and 52 percent of all croplands irrigated, respectively. Without irrigation and drainage, much of the increases in agricultural output that have fed the world's growing population and stabilized food production would not have been possible. In the dry sub-humid countries, irrigation is critical for crop production. Due to highly variable rainfall, long dry seasons, and recurrent droughts, dry spells and floods, water management is a key determinant for agricultural production in these regions and is increasingly becoming more important with climate change. The World Bank estimates that rainfed agriculture is most significant in Sub-Saharan Africa where it accounts for about 96 percent of the cropland (please see Figure 8).

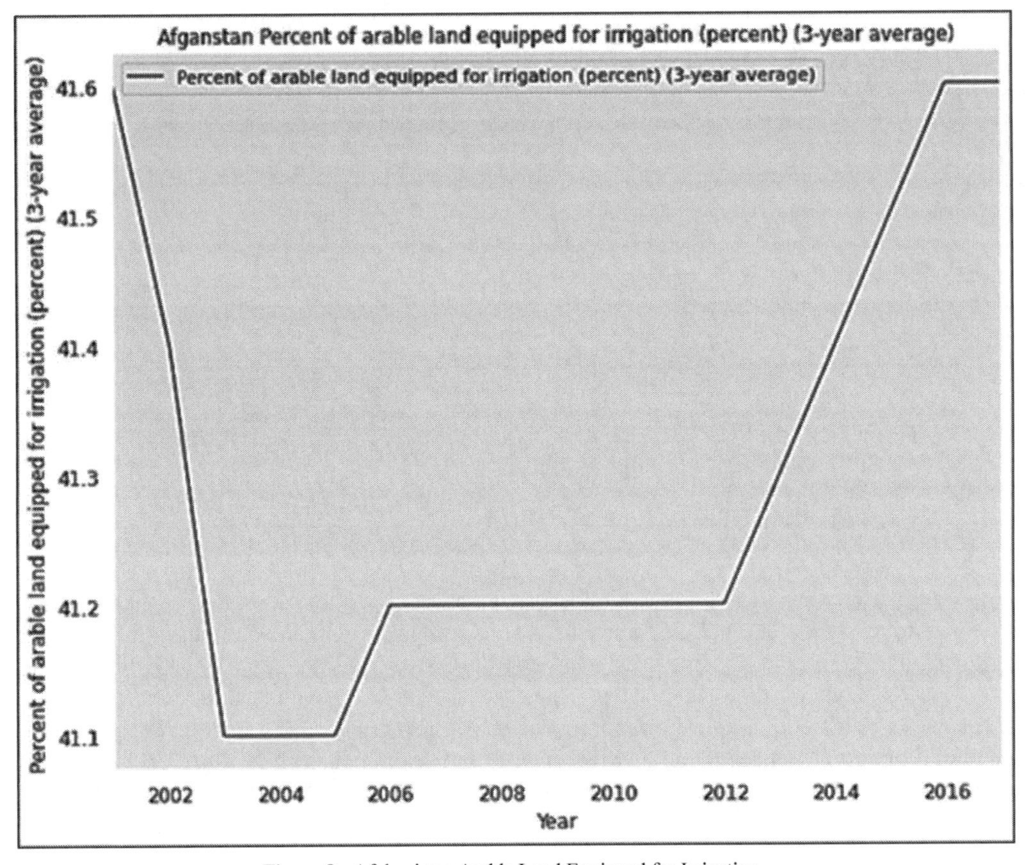

Figure 8: Afghanistan Arable Land Equipped for Irrigation.

Employment in Agriculture

Agriculture provides food security, a source of income, and stability to countless Afghan families. The Afghan economy is largely based on agriculture; major crops include wheat, rice, maize, barley, vegetables, fruit, and nuts. About 80 percent of the population lives in rural areas, where farming and agricultural labor are important sources of livelihood. About 67 percent of rural households and 15 percent of urban households have access to agricultural land—an important resource when food prices are high [15]. Despite being such a critical sector, employment in agriculture is trending downward. Employment in agriculture (% of total employment) (modeled ILO estimate) in Afghanistan was reported at 42.35 % in 2020 and it was above 65% in 2002. The reasons are numerous. Lack of investment, supply, and untimely markets are causing farmers to leave the agriculture fields (please see Table 7).

Table 7: Afghanistan Employment in Agriculture.

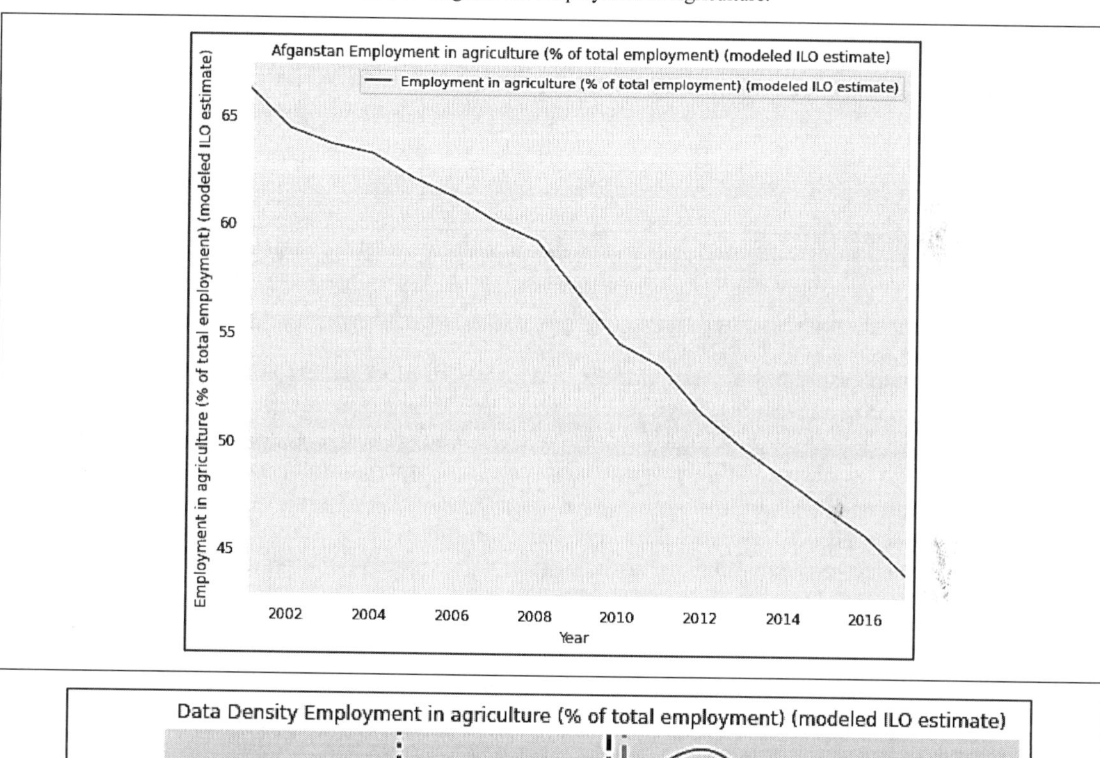

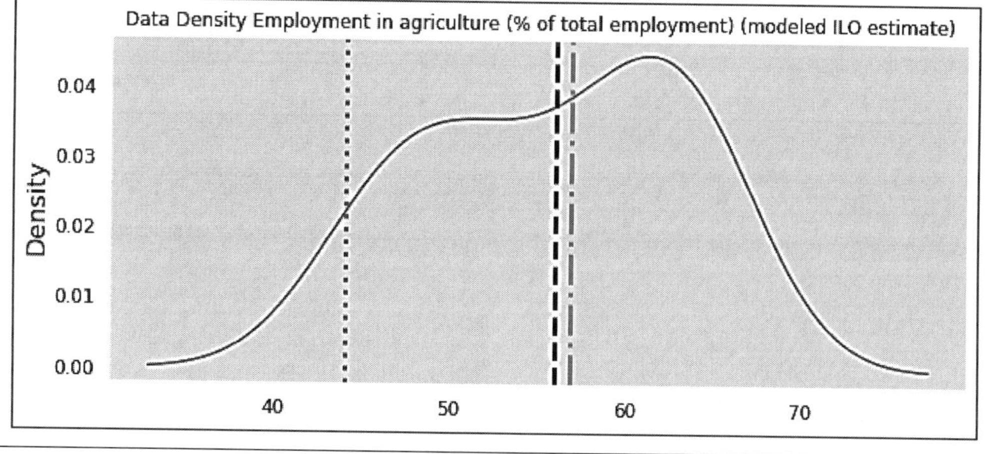

For skewed distributions, it is quite common to have one tail of the distribution considerably longer or drawn out relative to the other tail. A "skewed right" distribution is one in which the tail is on the right side. A "skewed left" distribution is one in which the tail is on the left side. Skewed data often occurs due to lower or upper bounds on the data. That is, data that has a lower bound are often skewed right while data that has an upper bound is often skewed left. The above histogram is for a distribution that is skewed right[319] as Employment in Agriculture (% of total employment) was frequent in the lower values [2002–2008]. Data Distribution statistics: Minimum:43.99; Mean:56.00 (dashed line); Median:56.89 (dash dot line); Mode: 43.99 (dotted line); Maximum:66.29.

"Afghan farmers need to be growing food again for their families and for Afghanistan and they need cash in their pockets. Agriculture can't wait and the people of Afghanistan can't wait. The situation is disastrous. All farmers spoken to have lost almost all their crops this year, many have been forced to sell their livestock, they have accumulated enormous debts and simply have no money. None of the farmers want to leave their land. But when you have no food, you have no grain from the previous harvest, there are no seeds in the fields and your livestock is gone, you have no choice." [12]

Richard Trenchard,
FAO Representative in Afghanistan

Prevalence of Undernourishment

Price shocks and household responses have a relationship—varied across rural and urban areas. Economic theory and empirical data suggest that urban or landless rural households are likely to be more *adversely affected* by food price increases than agricultural rural households that can produce their own food. When food prices increase, purchasing power—the amount of a good that a consumer can buy with a given sum of money—declines. Some households, however, are less dependent on food purchases than others. For example, if a household is a net seller of food (the total value of food produced is greater than the total value of food consumed), then its income will increase as food prices rise, mitigating the decline in welfare due to lower purchasing power.

In Afghanistan plotting Food Prices vs. Food consumption followed by analysis revealed, urban areas experienced a much greater decline in the real value of food consumption than rural areas. *For a 1-percent increase in wheat flour prices, the value of real monthly per capita food consumption in rural areas declined by 0.19 percent, while the decline in urban areas was 0.37 percent.* Other studies on the impacts of the high food price inflation in 2007/08 also found a disproportionate impact on urban areas in terms of poverty and total consumption [15].

Prevalence of undernourishment (please see Table 8) can better describe a farmer in Afghanistan [12]—"Drought, COVID-19 and armed conflict have brought us to this terrible situation. In our village people don't even have bread to eat. Our properties and livestock have been destroyed, and due to the lack of improved seeds, fertilizers, and water we couldn't harvest any wheat."

Prevalence of Undernourishment: Macroeconomic Parameter Causation

Causation, or causality, is the capacity of one variable to influence another.[320] The first variable may bring the second into existence or may cause the second variable to fluctuate [16]. The importance of market-based developments in the economic development of a country cannot be denied, and macroeconomic variables are important indicators that affect the overall market health of a country. The following framework provides the association of these variables (M1-M20 and A1-A20) with market economic factors. Techniques like Augmented Dickey-Fuller test, Regression Analysis and Granger Causality test have been applied to examine the causal relationship of selected macroeconomic variables affecting prevalence of food insecurity. Results of regression analysis indicate the presence of a strong positive relationship between applicable macroeconomic parameters and food security.

Table 8: Afghanistan Prevalence of undernourishment.

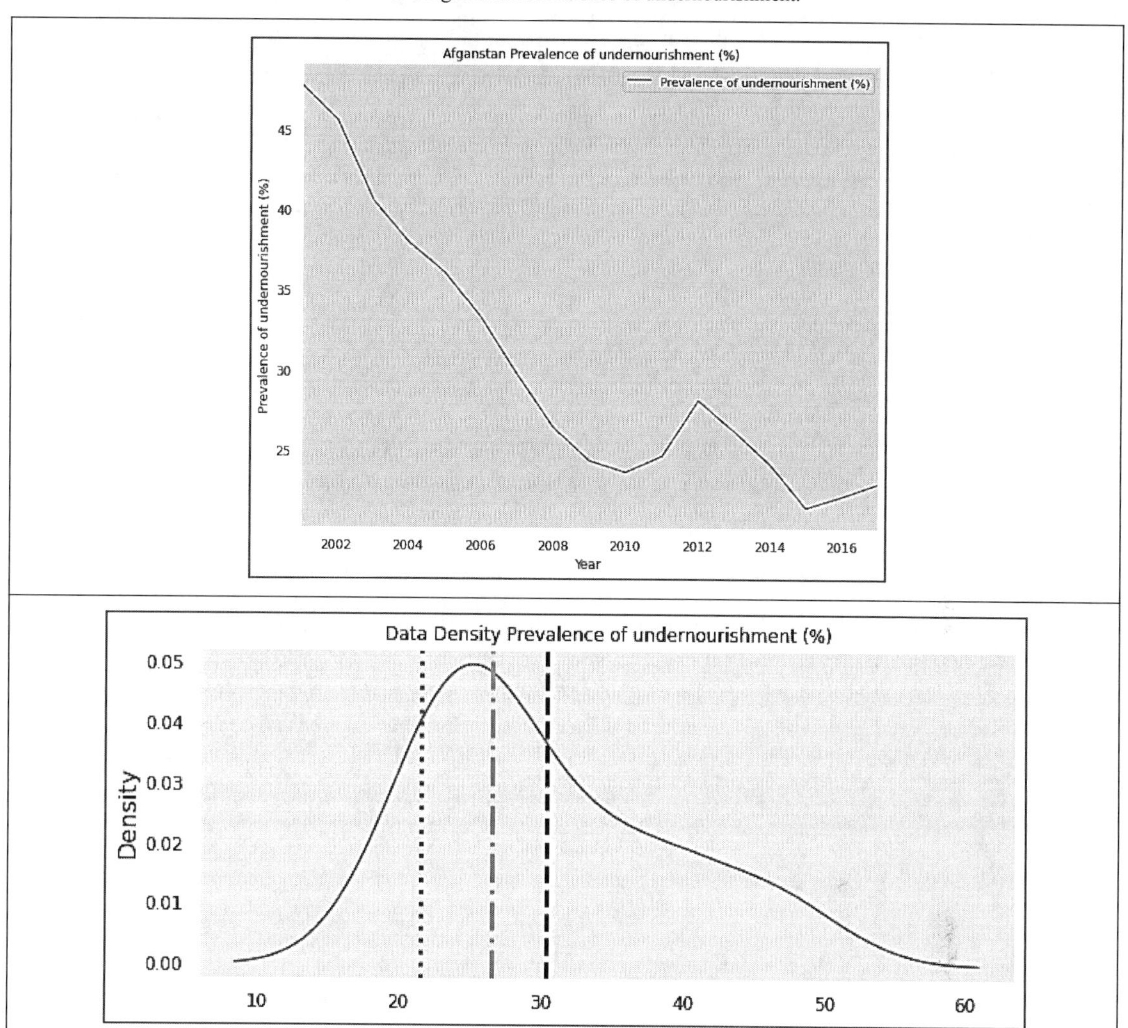

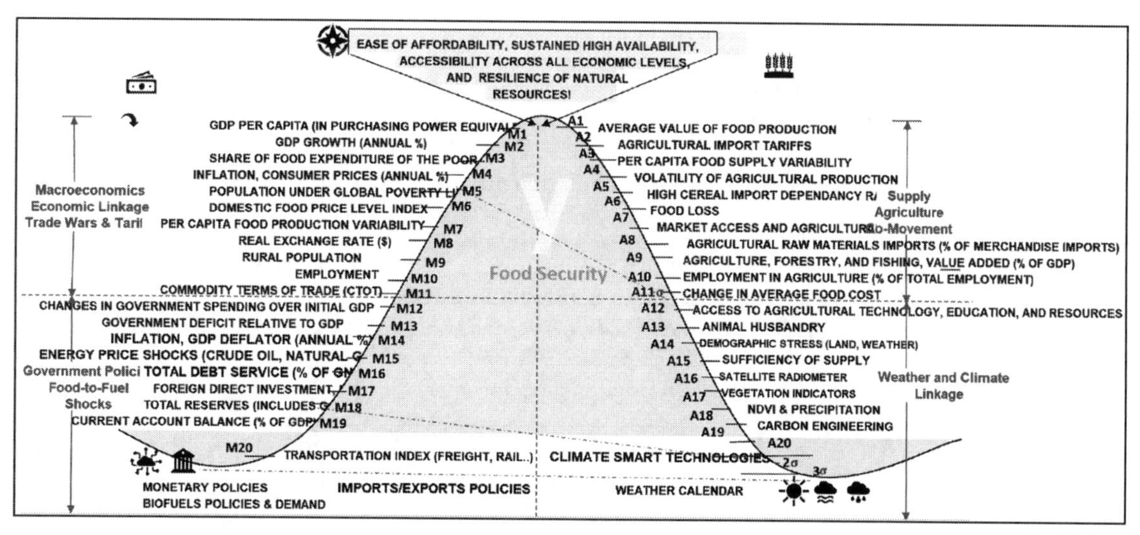

Figure 9: Food Security Model.

Macroeconomic economic drivers under review include (please see Figure 9 and Table 9):

Table 9: Macroeconomic drivers.

Macroeconomic drivers	LABEL	Framework Item
GNI per capita, PPP (current international $)	GNI	M1
GDP growth (annual %)	GDP	M2
Agriculture, forestry, and fishing, value added (% of GDP)	AGDP	A9
Total debt service (% of GNI)	DEBT	M16
Rural population (% of total population)	RPOP	M9
Foreign direct investment, net inflows (% of GDP)	FDI	M17
Agricultural raw materials exports (% of merchandise exports)	AEXP	A8
Employment in agriculture (% of total employment) (modeled ILO estimate)	AEMP	A10
Total reserves (includes gold, current US$)	RESV	M18
Inflation, consumer prices (annual %)	CPI	M4
Prevalence of undernourishment (% of population)	PUND	Tails of the bell curve!
Official exchange rate (LCU per US$, period average)	XLCU	M8
Inflation, GDP deflator (annual %)[321]	GDP-D	M14

Results of the Granger Causality test demonstrate that a bi-directional relationship exists between prevalence of undernourishment and Afghanistan's macroeconomic parameters. Except for the official exchange rate all other variables have a bi-directional relationship. This tells us, most macroeconomic factors play an important role in the prevalence of undernourishment (please see Figure 10).

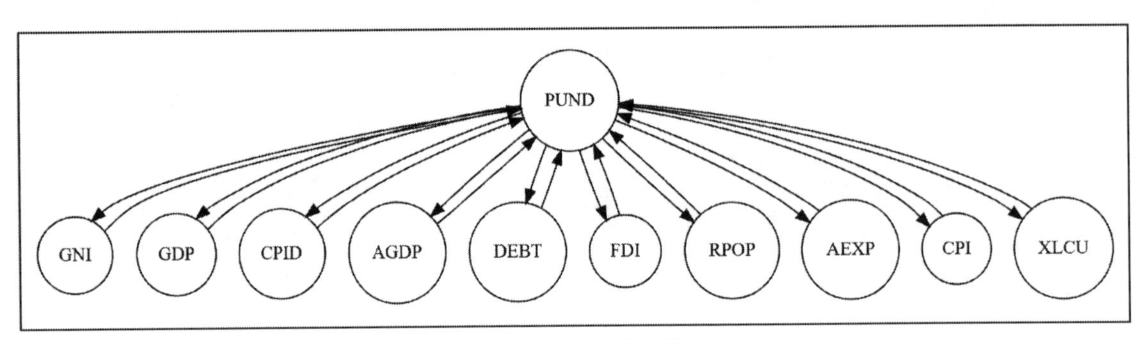

Figure 10: Granger Causality.

Here is the detailed directional causation of parameters. The highlighted cells indicate there is no influence of variable_x on the causation of variable_y or vice versa (please see Table 10).

Prevalence of Undernourishment: Food Security Indicators and Macroeconomic Parameter Causation

Results of Granger Causality test demonstrate that bi-directional relationship exists between prevalence of undernourishment and Sri Lanka's macroeconomic parameters (please see Figure 11). Except for official exchange rate (LCU per US$, period average) [M14], all other variables [M1, M2, M9, M14, M16, M17, M18, A8, A9, A10] have bi-directional relationship.

Table 10: Granger Causality.

	GNI_x	GDP_x	CPID_x	AGDP_x	DEBT_x	FDI_x	RPOP_x	AEXP_x	AEMP_x	RESV_x	CPI_x	XLCU_x	PUND_x
GNI_y	1	0	0	0.0127	0.0881	0	0	0	0.0424	0	0	0.0008	0
GDP_y	0	1	0	0.1846	0	0	0	0.2726	0	0	0	0.0279	0
CPID_y	0.046	0.1419	1	0	0.0479	0.2442	0	0.0068	0	0	0.3125	0.0001	0
AGDP_y	0	0	0.0009	1	0	0	0.0035	0.0101	0.0023	0	0.4879	0	0
DEBT_y	0.0081	0	0	0.0766	1	0.2114	0	0	0.0018	0	0	0.0751	0.0014
FDI_y	0	0.001	0	0.2821	0	1	0.3579	0	0	0	0	0.3002	0
RPOP_y	0.0004	0.0841	0	0	0	0.1151	1	0	0.0004	0.0519	0.3363	0	0
AEXP_y	0	0	0	0.0314	0	0	0	1	0	0	0.1054	0	0.0434
AEMP_y	0.0015	0	0	0.0015	0	0	0	0.2111	1	0	0	0.0871	0
RESV_y	0.0639	0	0	0	0.062	0.1395	0	0.0661	0	1	0	0.0113	0.019
CPI_y	0.4281	0.1134	0	0	0.0024	0.3045	0	0.0011	0	0	1	0.0527	0.0153
XLCU_y	0	0	0	0.0015	0	0	0	0.0003	0	0	0.0002	1	0
PUND_y	0	0.0054	0	0	0	0.0105	0.0015	0	0	0	0.0018	0	1

Macroeconomic economic drivers under review include (please see Figure 11 & Table 11):

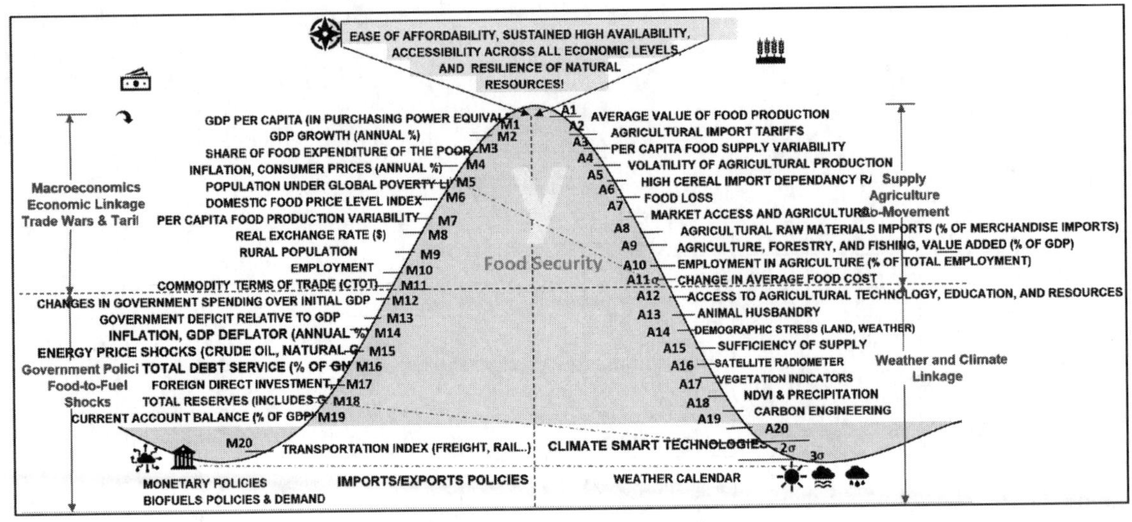

Figure 11: Food Security Model.

Table 11: Macroeconomic table.

Macroeconomic drivers	LABEL	Framework Item
Employment in agriculture (% of total employment) (modeled ILO estimate)	AEMP	A10
Foreign direct investment, net inflows (% of GDP)	FDI	M17
Rural population (% of total population)	RPOP	M9
Agriculture, forestry, and fishing, value added (% of GDP)	AGDP	A9
Official exchange rate (LCU per US$, period average)	XLCU	M8
Per capita food supply variability (kcal/cap/day)	CFSV	A3
Per capita food production variability (constant 2004-2006 thousand int$ per capita)	CFPV	M7
GNI per capita, PPP (current international $)	GNI	M1
Cereal import dependency ratio (percent) (3-year average)	CID	A5
Percent of arable land equipped for irrigation (percent) (3-year average)	ALEI	A7
Prevalence of undernourishment (% of population)	PUND	Tails of the bell curve!

Except for Official exchange rate (LCU per US$, period average), Per capita food production variability & Cereal import dependency ratio [M14], all other variables [M1, M7, M8, M9, M17, M18, A3, A5, A7, A9, and A10] have bi-directional relationship (please see Table 12 & Figure 12).

Table 12: Granger Causality test.

	AEMP_x	FDI_x	RPOP_x	AGDP_x	XLCU_x	CFPV_x	CFSV_x	CID_x	GNI_x	ALEI_x	RESV_x	PUND_x
AEMP_y	1	0	0	0	0.307	0	0	0	0	0	0	0
FDI_y	0	1	0	0	0	0	0	0	0	0	0	0
RPOP_y	0	0.0009	1	0	0	0.0032	0	0	0	0	0	0
AGDP_y	0	0	0	1	0	0	0	0	0	0	0	0
XLCU_y	0	0	0	0	1	0	0	0.0001	0	0	0	0
CFPV_y	0	0	0	0	0	1	0	0	0	0	0	0
CFSV_y	0	0	0	0.0031	0	0.2914	1	0.0228	0	0.0054	0	0.0233
CID_y	0	0	0	0	0	0	0	1	0	0	0	0
GNI_y	0	0	0	0	0.0009	0	0	0.0009	1	0	0	0
ALEI_y	0	0	0	0	0	0	0	0	0	1	0	0
RESV_y	0	0	0	0	0	0	0	0	0.1408	0	1	0
PUND_y	0	0	0	0	0	0	0	0	0	0.0075	0	1

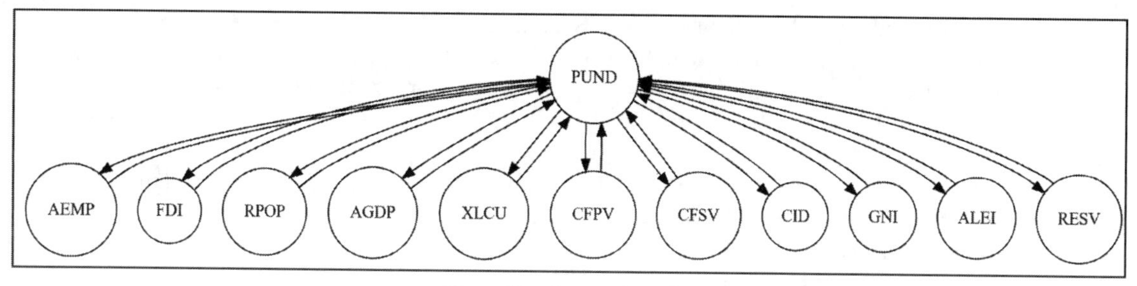

Figure 12: Granger Causality.

Software code for this model: Prevalence of severe food insecurity in the total population (percent) (3-year average).ipynb (Jupyter Notebook Code)

Afghanistan Macroeconomic Key Drivers & Linkage Model - Prevalence of Undernourishment

To model Afghan food security, following parameters are used:

Parameter	References	Source
GNI per capita, PPP (current international $)	NY.GNP.PCAP.PP.CD[322]	THE WORLD BANK IBRD • IDA
GDP growth (annual %)	NY.GDP.MKTP.KD. ZG[323]	
Inflation, GDP deflator (annual %)	NY.GDP.DEFL.KD. ZG[324]	
Agriculture, forestry, and fishing, value added (% of GDP)	NV.AGR.TOTL. ZS[325]	
Total debt service (% of GNI)	DT.TDS.DECT.GN.ZS[326]	
Foreign direct investment, net inflows (BoP, current US$)	BX.KLT.DINV.WD.GD.ZS[327]	
Foreign direct investment, net (BoP, current US$)	BN.KLT.DINV.CD[328]	
Rural population (% of total population)	SP.RUR.TOTL. ZS[329]	
Foreign direct investment, net inflows (% of GDP)	BX.KLT.DINV.WD.GD.ZS[330]	
Foreign direct investment, net outflows (% of GDP)	BM.KLT.DINV.WD.GD.ZS[331]	
Agricultural raw materials exports (% of merchandise exports)	SL.AGR.EMPL. ZS[332]	
Employment in agriculture (% of total employment) (modeled ILO estimate)	SL.AGR.EMPL. ZS[333]	
Total reserves (includes gold, current US$)	FI.RES.TOTL.CD[334]	
Inflation, consumer prices (annual %)	FP.CPI.TOTL. ZG[335]	
Prevalence of undernourishment (%)	Prevalence of undernourishment[336]	
Official exchange rate (LCU per US$, period average)	PA.NUS.FCRF[337]	
Per capita food production variability (constant 2004–2006 thousand int$ per capita)	Per capita food production variability[338]	UNdata A world of information
Per capita food supply variability (kcal/capita/day)	Per capita food supply variability (kcal/capita/day)[339]	
Cereal import dependency ratio (percent) (3-year average)	Cereal import dependency ratio[340]	
Percent of arable land equipped for irrigation (percent) (3-year average)	Percent of arable land equipped for irrigation[341]	
Value of food imports in total merchandise exports (percent) (3-year average)	Value of food imports in total merchandise exports[342]	

The Data for the model is from Global Food Security Index[343] and FAO Suite of Food Security Indicators.[344] The model assesses the prevalence of severe food insecurity (characterized by individuals not eating for a whole day at times during the year) and [17] prevalence of moderate-severe food insecurity. The High-Income countries list includes:[345] Andorra, Greece, Poland, Antigua and Barbuda, Greenland, Portugal, Aruba, Guam, Puerto Rico, Australia, Hong Kong SAR, China, Qatar, Austria, Hungary, San Marino Bahamas The Iceland, Saudi Arabia, Bahrain, Ireland, Seychelles, Barbados, Isle of Man, Singapore, Belgium, Israel, Sint Maarten (Dutch part), Bermuda, Italy, Slovak Republic, British Virgin Islands, Japan, Slovenia, Brunei Darussalam, Korea, Rep., Spain, Canada, Kuwait, St. Kitts and Nevis, Cayman Islands, Latvia, St. Martin (French part), Channel Islands, Liechtenstein, Sweden, Chile, Lithuania, Switzerland, Croatia, Luxembourg, Taiwan, China, Curaçao, Macao SAR, China, Trinidad and Tobago, Cyprus, Malta, Turks and Caicos Islands, Czech Republic, Monaco, United Arab Emirates, Denmark, Nauru, United Kingdom, Estonia, Netherlands,United States, Faroe Islands, New Caledonia, Uruguay, Finland, New Zealand, Virgin Islands (U.S.), France, Northern Mariana Island, French Polynesia, Norway, Germany, Oman, Gibraltar, and Palau.

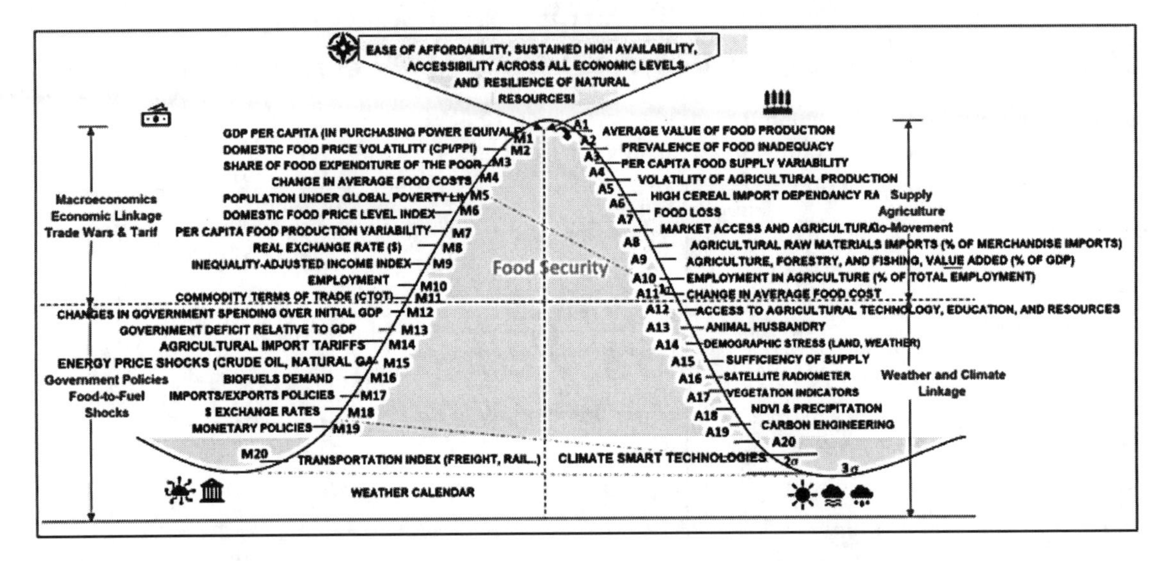

Step 1: Load Libraries

Load Python libraries to analyze the data.

```
import numpy as np
import pandas as pd
import functools
from datetime import date
import plotly.graph_objects as go
import seaborn as sns; sns.set()
from sklearn.impute import SimpleImputer
from scipy import stats
from scipy.stats import pearsonr
import matplotlib.pyplot as plt
from sklearn.preprocessing import MinMaxScaler
from sklearn.model_selection import train_test_split
from sklearn.linear_model import LinearRegression
from sklearn.metrics import r2_score
from sklearn import metrics
from sklearn.impute import SimpleImputer
```

```
from fbprophet import Prophet
from fbprophet.diagnostics import cross_validation, performance_metrics
from statsmodels.tsa.stattools import grangercausalitytests
from statsmodels.tsa.stattools import adfuller
import statsmodels.api as sm
from statsmodels.tsa.api import VAR
import matplotlib as mpl
mpl.rcParams['figure.figsize'] = (10,8)
mpl.rcParams['axes.grid'] = False
```

Step 2: Load Afghanistan specific FAO parameters

Load FAO specific Global | FAO, Statistics Division Afghanistan parameters.[346] The data can be downloaded from the Suite of Food Security Indicators.[347] The Suite of Food Security Indicators presents the core set of food security indicators. Following the recommendation of experts gathered in the Committee on World Food Security (CFS) Round Table on hunger measurement, hosted at FAO headquarters in September 2011.

```
Pdf = pd.read_csv("Afghanistan.csv")
Predf.head()
```

Output:

Prevalence of undernourishment (percent) (3-year average) is three year moving average data.

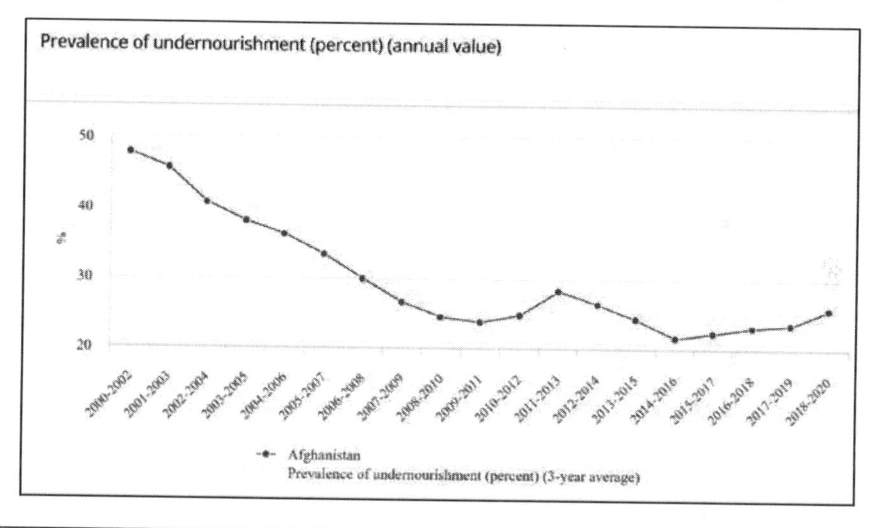

	Domain Code	Domain	Area Code (M49)	Area	Element Code	Element	Item Code (SDG)	Item	Year Code	Year	Unit	Value	Flag	Flag Description	Note
0	SDGB	SDG Indicators	4	Afghanistan	6121	Value	SN_ITK_DEFC	2.1.1 Prevalence of undernourishment (%)	2001	2001	%	47.8	_	NaN	Global | FAO, Statistics Division | Estimated ...
1	SDGB	SDG Indicators	4	Afghanistan	6121	Value	SN_ITK_DEFC	2.1.1 Prevalence of undernourishment (%)	2002	2002	%	45.6	_	NaN	Global | FAO, Statistics Division | Estimated ...
2	SDGB	SDG Indicators	4	Afghanistan	6121	Value	SN_ITK_DEFC	2.1.1 Prevalence of undernourishment (%)	2003	2003	%	40.6	_	NaN	Global | FAO, Statistics Division | Estimated ...
3	SDGB	SDG Indicators	4	Afghanistan	6121	Value	SN_ITK_DEFC	2.1.1 Prevalence of undernourishment (%)	2004	2004	%	38.0	_	NaN	Global | FAO, Statistics Division | Estimated ...

```
predf1 = Predf[["Year","Value"]]
predf1.rename(columns = {"Value": "Prevalence of undernourishment (%)"},inplace = True)
```

Step 3: Load Afghanistan world bank indicators

```
df = pd.read_csv("AfghanistanWorldIndicators.csv")
df
```

Output: the following parameters are loaded as part of the DataFrame.

- Poverty headcount ratio at national poverty lines (% of population),
- GNI per capita, PPP (current international $),
- Mortality rate, under-5 (per 1,000 live births),
- School enrollment, primary and secondary (gross), gender parity index (GPI),
- GDP growth (annual %)', 'Inflation, GDP deflator (annual %),
- Agriculture, forestry, and fishing, value added (% of GDP),
- Total debt service (% of GNI),
- Foreign direct investment, net inflows (BoP, current US$),
- Foreign direct investment, net (BoP, current US$),
- Rural population (% of total population),
- Foreign direct investment, net inflows (% of GDP),
- Foreign direct investment, net outflows (% of GDP),
- Agricultural raw materials exports (% of merchandise exports),
- Employment in agriculture (% of total employment) (modeled ILO estimate),
- Total reserves (includes gold, current US$), and
- Inflation, consumer prices (annual %).

Foreign direct investment, net inflows (BoP, current US$)	Foreign direct investment, net (BoP, current US$)	Rural population (% of total population)	Foreign direct investment, net inflows (% of GDP)	Foreign direct investment, net outflows (% of GDP)	Agricultural raw materials exports (% of merchandise exports)	Employment in agriculture (% of total employment) (modeled ILO estimate)	Total reserves (includes gold, current US$)	Inflation, consumer prices (annual %)	Year	Prevalence of undernourishment (%)
680000	..	77.831	..	..	..	66.29000092	..	..	2001	47.8
50000000	..	77.739	1.232991023	..	..	64.41999817	..	..	2002	45.6
57799999.9	..	77.647	1.280018761	0.022145653	..	63.70000076	..	..	2003	40.6
186900000	..	77.5	3.575816135	-0.01339257	..	63.27000046	..	..	2004	38.0
271000000	..	77.297	4.364535244	0.024157944	..	62.15000153	..	12.68626872	2005	36.1
238000000	..	77.093	3.414004444	..	..	61.27999878	..	6.78459655	2006	33.3
188690000	..	76.887	1.935703036	..	..	60.13999939	..	8.680570785	2007	29.8
46033740	-47951776.13	76.68	0.455360078	-0.018972977	5.033139549	59.29000092	3042274496	26.41866415	2008	26.5
56107246.5	-55860135.47	76.472	0.451888843	0.001990237	7.559673049	56.88999939	4265888740	-6.811161069	2009	24.4
190774432	-192022479.5	76.263	1.203117228	-0.0078708	10.76704557	54.58000183	5162440807	2.178537524	2010	23.7
5.21734e+07	-5.10363e+07	76.052	0.293025	0.00638658	12.879	53.56	6.34464e+09	11.8042	2011	24.7
56823660	-65684511.34	75.84	0.285441076	-0.044510525	0.236604474	51.40000153	7152304398	6.441212809	2012	28.2
48311346	-47774027.01	75.627	0.239801324	0.002667071	0.036992232	49.81999969	7288702711	7.385771784	2013	26.3
42975262.5	-42994416.21	75.413	0.209664813	-9.34E-05	0.047523022	48.40999985	7528549880	4.673996035	2014	24.2
169146608	-166983554.9	75.197	0.88400092	0.011304637	14.79906597	47.06999969	6976968154	-0.661709165	2015	21.5
9.35913e+07	-7.95853e+07	74.98	0.516606	0.0773107	16.3591	45.81	7.28191e+09	4.38389	2016	22.2
51533896.77	-40273161	74.75	0.274796599	0.060046146	17.06363382	43.99000168	8097280956	4.975951506	2017	23.0
119435105.7	-80631404.07	74.505	0.661572002	0.21494051	16.69459936	43.13000107	8206682106	0.626149149	2018	23.4
23404553.65	2916492.929	74.246	0.124495944	0.140009657	21.03505915	42.5	8497655795	2.302372515	2019	25.6

As can be seen, there are nulls within the dataset.

Step 4: Apply Imputation Strategy (Forward Fills) to remove NULLs

Apply imputers to fill NULLs. When it comes to macroeconomic drivers, it is always important to apply imputation values closer to the time horizon of a year when data or NULLs are missing. This would represent imputation data that is closer to reality. One could impute with most frequent values, but the only limitation would be their being out of sync with time.

```
imp_mean = SimpleImputer(missing_values=np.nan, strategy='mean')
imparray = imp_mean.fit_transform(df5)
final_df = pd.DataFrame(data = imparray, columns = df5.columns)
final_df
```

Output:

	GNI per capita, PPP (current international $)	GDP growth (annual %)	Inflation, GDP deflator (annual %)	Agriculture, forestry, and fishing, value added (% of GDP)	Total debt service (% of GNI)	Foreign direct investment, net inflows (BoP, current US$)	Foreign direct investment, net (BoP, current US$)	Rural population (% of total population)	Foreign direct investment, net inflows (% of GDP)	Foreign direct investment, net outflows (% of GDP)	Agricultural raw materials exports (% of merchandise exports)
0	1948.181818	6.664775	6.343872	27.206524	0.200889	6.800000e+05	-7.232338e+07	77.831	1.189047	0.031741	10.209288
1	1948.181818	6.664775	6.343872	38.627892	0.200889	5.000000e+07	-7.232338e+07	77.739	1.232991	0.031741	10.209288
2	1948.181818	8.832278	11.655238	37.418855	0.200889	5.780000e+07	-7.232338e+07	77.647	1.280019	0.022146	10.209288
3	1948.181818	1.414118	11.271432	29.721067	0.200889	1.869000e+08	-7.232338e+07	77.500	3.575816	-0.013393	10.209288
4	1948.181818	11.229715	10.912774	31.114855	0.200889	2.710000e+08	-7.232338e+07	77.297	4.364535	0.024158	10.209288
5	1948.181818	5.357403	7.199751	28.635969	0.200889	2.380000e+08	-7.232338e+07	77.093	3.414004	0.031741	10.209288
6	1948.181818	13.826320	22.527756	30.105011	0.200889	1.886900e+08	-7.232338e+07	76.887	1.935703	0.031741	10.209288
7	1948.181818	3.924984	2.096289	24.892270	0.200889	4.603374e+07	-4.795178e+07	76.680	0.455360	-0.018973	5.033140
8	1520.000000	21.390528	-2.163404	29.297501	0.086740	5.610725e+07	-5.586014e+07	76.472	0.451889	0.001990	7.559673
9	1710.000000	14.362441	3.814630	26.210069	0.065671	1.907744e+08	-1.920225e+08	76.263	1.203117	-0.007871	10.767046
10	1700.000000	0.426355	16.593347	23.743664	0.058063	5.217342e+07	-5.103628e+07	76.052	0.293025	0.006387	12.879009
11	1920.000000	12.752287	7.301756	24.390874	0.074421	5.682366e+07	-6.568451e+07	75.840	0.285441	-0.044511	0.236604
12	2020.000000	5.600745	4.822785	22.810663	0.124356	4.831135e+07	-4.777403e+07	75.627	0.239801	0.002667	0.036992
13	2070.000000	2.724543	0.566945	22.137041	0.205922	4.297526e+07	-4.299442e+07	75.413	0.209665	-0.000093	0.047523
14	2110.000000	1.451315	2.447563	20.634323	0.292926	1.691466e+08	-1.669836e+08	75.197	0.884001	0.011305	14.799066
15	2000.000000	2.260314	-2.197526	25.740314	0.328834	9.359132e+07	-7.958527e+07	74.980	0.516606	0.077311	16.359109
16	2090.000000	2.647003	2.403656	26.420199	0.334931	5.153390e+07	-4.027316e+07	74.750	0.274797	0.060046	17.063634
17	2100.000000	1.189228	2.071349	22.042897	0.346979	1.194351e+08	-8.063140e+07	74.505	0.661572	0.214941	16.694599

Step 5: Add Afghan Exchange Rate

One of the key drivers of inflation is the Afghani currency exchange rate.[348]

1	`ExRate = pd.read_csv("ExchangeRate.csv")` `ExRate`
2	`ExRate.drop(["Series Code","Series Code","Country Name","Country Code"],axis=1,inplace = True)`

3	`ExRate1 = ExRate.T`
	`ExRate1.reset_index(inplace = True)`
	`ExRate1.columns = ExRate1.iloc[0]`
	`ExRate1["Year"] = ExRate1["Series Name"].apply(lambda x :x.split("[")[0])`
	`ExRate1.drop("Series Name", axis = 1 , inplace = True)`
	`ExRate1 = ExRate1[1:]`
	`ExRate1`

4	`ExRate1["Year"] = ExRate1["Year"].astype(int)`
5	`ExRate1["Official exchange rate (LCU per US$, period average)"] = ExRate1["Official exchange rate (LCU per US$, period average)"].astype(float)`
6	`final_df1 = final_df.merge(ExRate1,on ="Year")`

The above code performs the following: 1. Loads exchange rate, 2. Replaces country name, 3. Transposes indexes and columns, 4. Converts Year to INT, 5. Converts "Official exchange rate (LCU per US$, period average)" column to floating, and 6. Merges exchange rate into the main DataFrame.

Output:

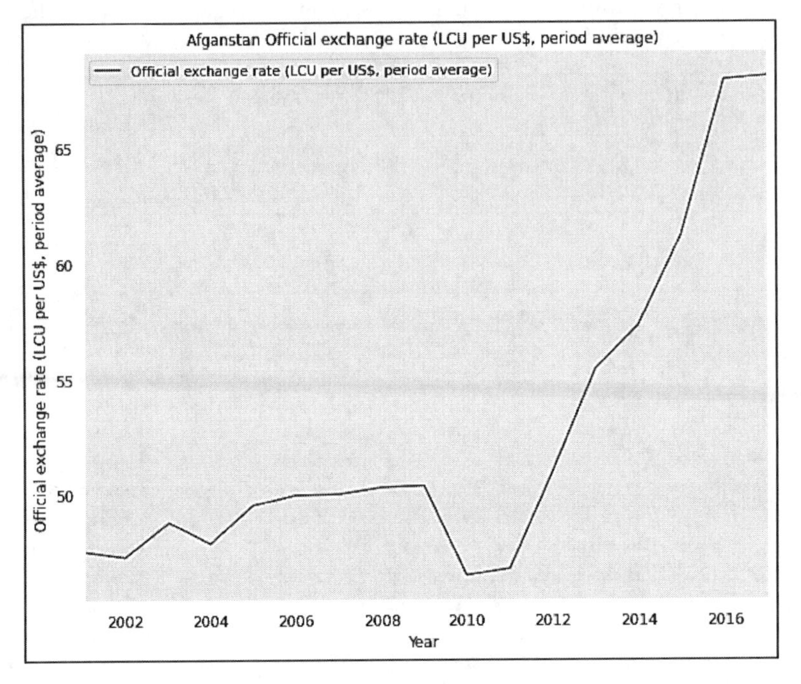

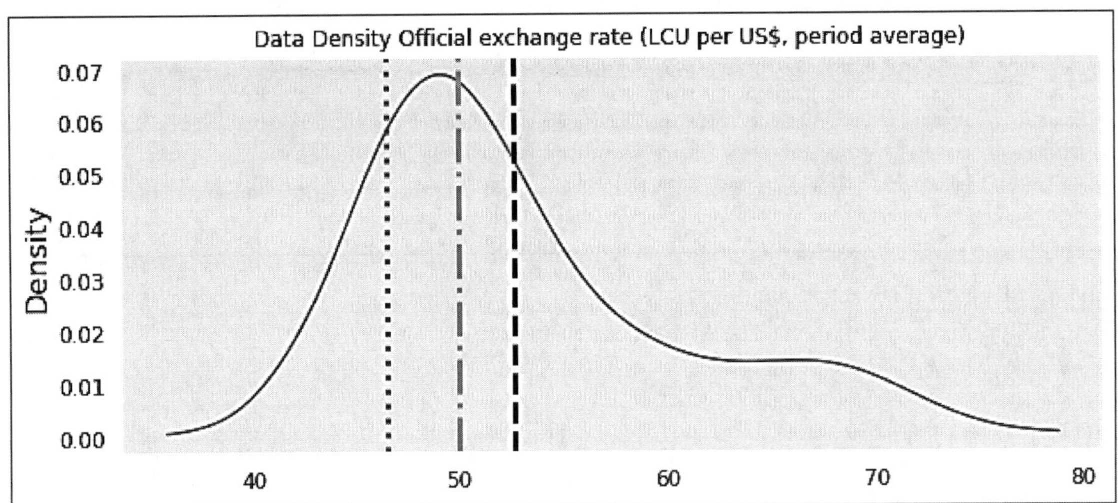

The value for Official exchange rate (LCU per US$, period average) in Afghanistan was 67.87 as of 2016. As the graph below shows, over the past 20 years this indicator reached a maximum value of 72.08 in 2018 and a minimum value of 47.26 in 2002. In 2020, exchange rate for Afghanistan[349] was 76.8 LCU per US dollars. Earlier the exchange rate of Afghanistan started to increase to reach a level of 76.8 LCU per US dollars in 2020, then, it went through a trough reaching a low of 33.8 LCU per US dollars in 1979.

	“If everyone starts to impose export restrictions or to close markets, that will worsen the crisis”. Cem Ozdemir German agriculture minister

Step 6: Plot Correlation Map

Develop correlation ship value map.

```
figure = plt.figure(figsize = (24,10))
sns.heatmap(final_df1.corr(),annot=True)
```

Output:

We can observe Prevalence of undernourishment (%) highly correlated with agriculture, forestry, and fishing, value added (% of GDP) (73%), rural population (% of total population) (83%), employment in agriculture (% of total employment) (modeled ILO estimate) (82.6%) and negatively correlated with Official exchange rate (LCU per US$, period average) (–53.6%).

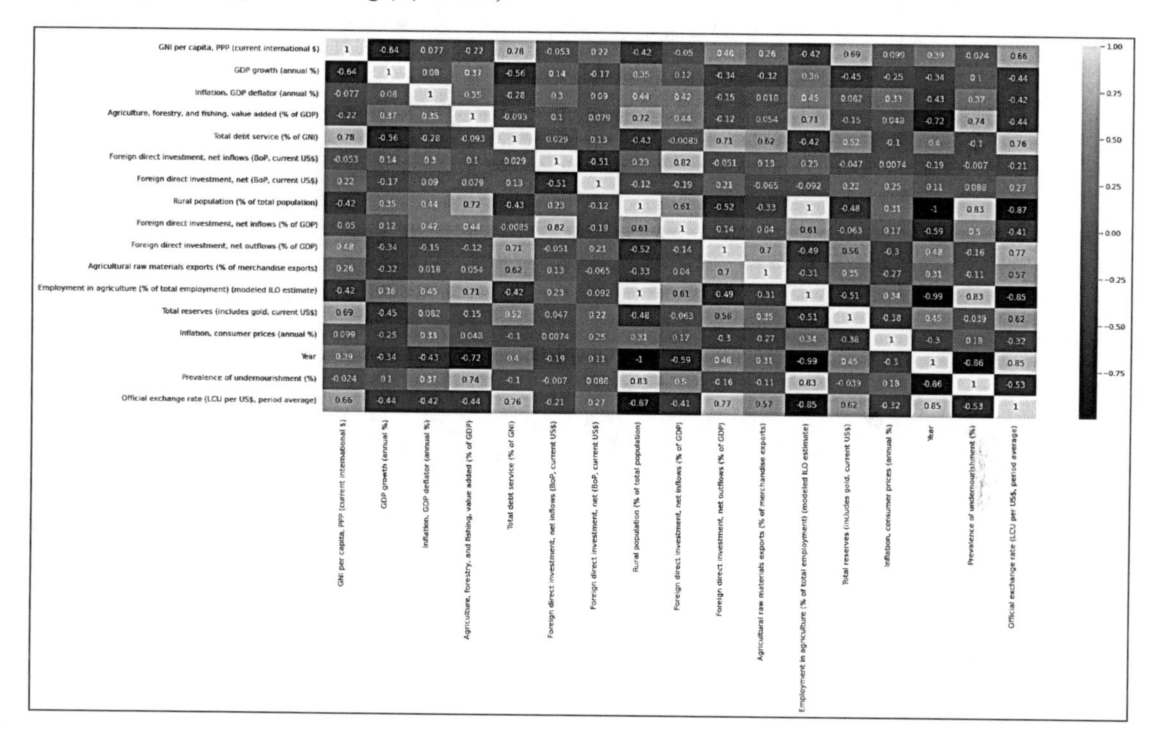

Step 7: Model - Prevalence of undernourishment (%).

Build the model - Prevalence of undernourishment (%).

```
1  X1 = final_df1[["Employment in agriculture (% of total employment) (modeled ILO estimate)",
        "Foreign direct investment, net inflows (% of GDP)",
        "Rural population (% of total population)",
        "Agriculture, forestry, and fishing, value added (% of GDP)",
        "Inflation, GDP deflator (annual %)",
          "Official exchange rate (LCU per US$, period average)"]]
2  y1 = final_df1['Prevalence of undernourishment (%)']
3  X_train1,X_test1, y_train1 , y_test1 = train_test_split(X1 ,y1, test_size = 0.2,random_state = 0)
4
```

Develop Independent (X) and dependent (y) variable series and create train/test split. Next step is to construct a Linear Regression and predicting the model.

```
regressor1 = LinearRegression()
regressor1.fit(X_train1,y_train1)
```

```
y_pred1 = regressor1.predict(X_test1)
```

```
import matplotlib.pyplot as plt
%matplotlib inline
plt.scatter(y_test1,y_pred1)
plt.xlabel('Actual')
plt.ylabel("Predicted")
plt.title('Actual vs Predicted')
# overlay the regression line
z = np.polyfit(y_test1, y_pred1, 1)
p = np.poly1d(z)
plt.plot(y_test1,p(y_test1), color='magenta')
plt.show()
```

Output:

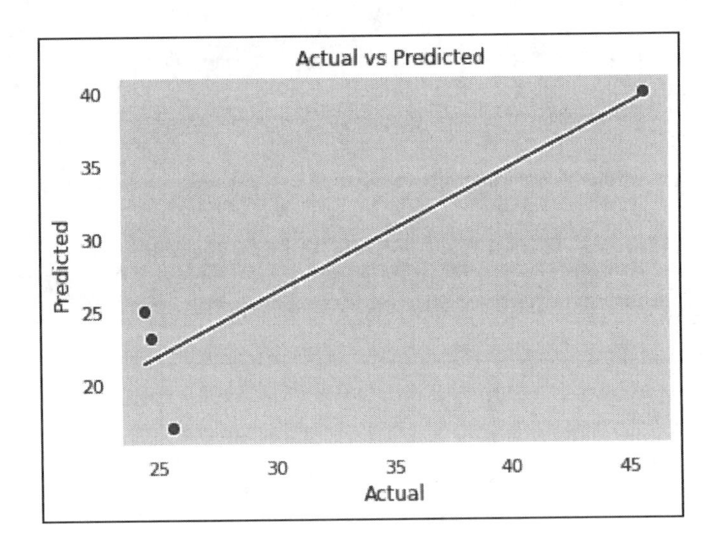

Step 8: Regression Score Function - R^2 (coefficient of determination)

The best possible score is 1.0 and it can be negative (because the model can be arbitrarily worse). In the general case when the true y is not constant, a constant model that always predicts the average y disregarding the input features would give a score of 0.0.

```
r2_score(y_test1,y_pred1)
```

Output:

R^2= 0.6672784210134468

Step 9: Develop Model Equation (y=mx+c) to predict Prevalence of undernourishment (%)
Develop model equation (y=mx+c).

```
mx=""
for ifeature in range(len(X1.columns)):
  if regressor1.coef_[ifeature] <0:
    # format & beautify the equation
    mx += " - " + "{:.2f}".format(abs(regressor1.coef_[ifeature])) + " * " + X1.columns[ifeature]
  else:
    if ifeature == 0:
      mx += "{:.2f}".format(regressor1.coef_[ifeature]) + " * " + X1.columns[ifeature]
    else:
      mx += " + " + "{:.2f}".format(regressor1.coef_[ifeature]) + " * " + X1.columns[ifeature]
print(mx)
```

```
# y=mx+c
if(regressor1.intercept_ <0):
   print("The formula for the " + y1.name + " linear regression line (y=mx+c) is = " + " - {:.2f}".
format(abs(regressor1.intercept_)) + " + " + mx )
else:
   print("The formula for the " + y1.name + " linear regression line (y=mx+c) is = " + "{:.2f}".
format(regressor1.intercept_) + " + " + mx )
```

Output:

```
The formula for the Prevalence of undernourishment (%) linear regression line (y=mx+c) is = – 2845.79
+ – 3.52 * Employment in agriculture (% of total employment) (modeled ILO estimate) – 1.07 * Foreign
direct investment, net inflows (% of GDP) + 39.81 * Rural population (% of total population) – 0.69 *
Agriculture, forestry, and fishing, value added (% of GDP) + 0.14 * Inflation, GDP deflator (annual %)
+ 0.95 * Official exchange rate (LCU per US$, period average)
```

Step 10: Model Explainability
To see model important parameters, run model Explainability.

```
import eli5
eli5.show_weights(regressor1,feature_names = model_features_prev_moderate_afghan_m1)
```

Output:
Please find a summary of the top features.

Weight?	Feature
+39.813	Rural population (% of total population)
+0.946	Official exchange rate (LCU per US$, period average)
+0.136	Inflation, GDP deflator (annual %)
-0.687	Agriculture, forestry, and fishing, value added (% of GDP)
-1.073	Foreign direct investment, net inflows (% of GDP)
-3.523	Employment in agriculture (% of total employment) (modeled ILO estimate)
-2845.793	<BIAS>

Summary—Macroeconomic Key Drivers - Forces Model

The reduction in Prevalence of undernourishment (%) in Afghanistan is due to the following:

Independent variable	Forces Model	Statistical significance	Description
Rural population (% of total population)	Drop in Rural population improves food security.	Positively correlated 39.81	The drop in rural population as percentage of population in Afghanistan is positively correlated with drop in Prevalence of undernourishment (%). All held constant, a unit decrease in rural population as percentage of total population would reduce prevalence of undernourishment (%) by 3.9% [12,15].
Official exchange rate (LCU per US$, period average)	Increase in Agriculture, forestry, and fishing would improve food security.	Positively correlated, 0.95	The drop-in official exchange rate (LCU per US$, period average) in Afghanistan is positively correlated with drop in Prevalence of undernourishment (%). All held constant, a unit decrease in official exchange rate (LCU per US$, period average) would reduce prevalence of undernourishment (%) by 9.5% [12,15].
Inflation, GDP deflator (annual %)	Drop in Rural population improves food security.	Positively correlated, 0.14	The drop-in Inflation, GDP deflator (annual %) in Afghanistan is positively correlated with drop in Prevalence of undernourishment (%). All held constant, a unit decrease in Inflation, GDP deflator (annual %) would drop prevalence of undernourishment (%) by 1.4% [12,15].
Agriculture, forestry, and fishing, value added (% of GDP)	Increase in Agriculture, forestry, and fishing would improve food security.	Negatively correlated, 0.69	Agriculture, forestry, and fishing, value added (% of GDP) negatively correlated with drop in Prevalence of undernourishment (%). A drop-in agriculture, forestry, and fishing, value added (% of GDP) in Afghanistan is negatively correlated with drop in Prevalence of undernourishment (%). All held constant, a unit decrease in Agriculture, forestry, and fishing, value added (% of GDP) would increase prevalence of undernourishment (%) by 69% [12,15].
Foreign direct investment, net inflows (% of GDP)	Drop in Foreign direct investment, net inflows (% of GDP) decreases food security.	Negatively correlated, 1.073	Foreign direct investment, net inflows (% of GDP) are negatively correlated with drop in Prevalence of undernourishment (%). A drop-in Foreign direct investment, net inflows (% of GDP) in Afghanistan is negatively correlated with drop in Prevalence of undernourishment (%). All held constant, a unit decrease in Foreign direct investment, net inflows (% of GDP) would increase prevalence of undernourishment (%) by 107.3% [12,15].
Employment in agriculture (% of total employment) (modeled ILO estimate)	An increase in Employment in agriculture (% of total employment) would improve food security.	Negatively correlated, 3.523	Employment in agriculture (% of total employment) (modeled ILO estimate) is negatively correlated with drop in Prevalence of undernourishment (%). A drop-in Employment in agriculture (% of total employment) (modeled ILO estimate) in Afghanistan is negatively correlated with drop in Prevalence of undernourishment (%). All held constant, a unit decrease in Employment in agriculture (% of total employment) (modeled ILO estimate) would increase prevalence of undernourishment (%) by 352.3% [12,15].

One item to observe is BIAS which is very high in the model – the model is unable to explain the variation in the dependent variable.

Afghanistan Macroeconomic Key Drivers, Food Security Parameters, & Linkage Model - Prevalence of Undernourishment

We will extend the above model of Afghanistan with added food security indicators.

Step 11: Add FAO STAT Food Security Indicators

```
foodSecdf = pd.read_csv("Afghanistan_foodSecurityIndicators.csv")
foodSecdf
```

Output:

Load Food Security Indicators

```
Cereal import dependency ratio (percent) (3-year average)',
   'Percent of arable land equipped for irrigation (percent) (3-year average)',
   'Value of food imports in total merchandise exports (percent) (3-year average)',
   'Per capita food production variability (constant 2004-2006 thousand int$ per capita)',
   'Per capita food supply variability (kcal/cap/day)'
```

Capture food security per capita parameters. Pivot the parameters to arrange them based on the year index.

```
foodSecdf1 = foodSecdf[foodSecdf["Item"] != 'Per capita food production variability (constant 2004-2006 thousand int$ per capita)']
foodSecdf2 = foodSecdf1[foodSecdf1["Item"] != 'Per capita food supply variability (kcal/cap/day)']
```

```
fooddf_final = foodSecdf2.pivot( index = "Year",columns = "Item" , values = "Value")
fooddf_final
```

Output:

Item	Cereal import dependency ratio (percent) (3-year average)	Percent of arable land equipped for irrigation (percent) (3-year average)	Value of food imports in total merchandise exports (percent) (3-year average)
Year			
2000-2002	34.2	41.7	266.0
2001-2003	29.5	41.6	308.0
2002-2004	21.2	41.4	225.0
2003-2005	18.5	41.1	211.0
2004-2006	19.2	41.1	202.0
2005-2007	20.2	41.1	200.0
2006-2008	23.5	41.2	226.0
2007-2009	25.7	41.2	238.0
2008-2010	25.7	41.2	285.0
2009-2011	24.9	41.2	346.0
2010-2012	22.5	41.2	406.0

As can be seen, cereal imports, arable land, and food imports are three-year specific drivers. For null values, we will apply forward fill imputation strategies.

```
food_part = fooddf_final1[["year","Cereal import dependency ratio (percent) (3-year average)","Percent of arable land equipped for irrigation (percent) (3-year average)",
  "Value of food imports in total merchandise exports (percent) (3-year average)" ]]
food_part.dtypes
```

```
food_part["Year"] = food_part["year"].astype(int)
```

```
foodSecdf3 = foodSecdf[foodSecdf["Item"] != 'Cereal import dependency ratio (percent) (3-year average)']
foodSecdf4 = foodSecdf3[foodSecdf3["Item"] != 'Percent of arable land equipped for irrigation (percent) (3-year average)']
foodSecdf5 = foodSecdf4[foodSecdf3["Item"] != 'Value of food imports in total merchandise exports (percent) (3-year average)']
```

Merge Macroeconomic and food security data frame.

```
final_df2 = final_df1.merge(foodSecdf6, on = "Year")
final_df3 = final_df2.merge(food_part , on = "Year")
```

```
final_df4 = final_df3.ffill()
final_df4
dfAfghanistan=final_df4.copy()
```

Output:

Foreign direct Investment, net inflows (% of GDP)	Foreign direct Investment, net outflows (% of GDP)	...	Inflation, consumer prices (annual %)	Year	Prevalence of undernourishment (%)	Official exchange rate (LCU per US$, period average)	Per capita food production variability (constant 2004-2006 thousand int$ per capita)	Per capita food supply variability (kcal/cap/day)	year	Cereal import dependency ratio (percent) (3-year average)	Percent of arable land equipped for irrigation (percent) (3-year average)
1.189047	0.031741	...	6.124620	2001.0	47.8	47.500015	11.2	45.0	2001	29.5	41.6
1.232991	0.031741	...	6.124620	2002.0	45.6	47.263000	17.3	68.0	2002	21.2	41.4
1.280019	0.022146	...	6.124620	2003.0	40.6	48.762754	17.0	68.0	2003	18.5	41.1
3.575816	-0.013393	...	6.124620	2004.0	38.0	47.845312	13.8	49.0	2004	19.2	41.1
4.364535	0.024158	...	12.686269	2005.0	36.1	49.494597	8.7	52.0	2005	20.2	41.1
3.414004	0.031741	...	6.784597	2006.0	33.3	49.925331	9.9	53.0	2006	23.5	41.2
1.935703	0.031741	...	8.680571	2007.0	29.8	49.962018	4.4	29.0	2007	25.7	41.2
0.455360	-0.018973	...	26.418664	2008.0	26.5	50.249615	5.1	26.0	2008	25.7	41.2
0.451889	0.001990	...	-6.811161	2009.0	24.4	50.325000	5.4	26.0	2009	24.9	41.2
1.203117	-0.007871	...	2.178538	2010.0	23.7	46.452461	7.2	21.0	2010	22.5	41.2
0.293025	0.006387	...	11.804186	2011.0	24.7	46.747008	7.6	29.0	2011	22.0	41.2
0.285441	-0.044511	...	6.441213	2012.0	28.2	50.921400	6.1	26.0	2012	19.3	41.2
0.239801	0.002667	...	7.385772	2013.0	26.3	55.377500	6.5	46.0	2013	21.3	41.3
0.209665	-0.000093	...	4.673996	2014.0	24.2	57.247500	4.2	79.0	2014	27.2	41.4
0.884001	0.011305	...	-0.661709	2015.0	21.5	61.143462	3.5	93.0	2015	34.6	41.5

Step 12: Prepare Afghanistan Model with Food Security Parameters

Prepare numerical features.

```
model_features_prev_moderate_afghan=["Employment in agriculture (% of total employment) (modeled ILO estimate)",
    "Foreign direct investment, net inflows (% of GDP)",
    "Rural population (% of total population)",
    "Agriculture, forestry, and fishing, value added (% of GDP)",
    "Official exchange rate (LCU per US$, period average)",
  'Per capita food production variability (constant 2004-2006 thousand int$ per capita)',
  'Per capita food supply variability (kcal/cap/day)',
  'Cereal import dependency ratio (percent) (3-year average)',
  'Percent of arable land equipped for irrigation (percent) (3-year average)',
               'Total reserves (includes gold, current US$)',
               'GNI per capita, PPP (current international $)']
```

Prepare independent and dependent variables.

```
X3 = dfAfghanistan[["Employment in agriculture (% of total employment) (modeled ILO estimate)",
    "Foreign direct investment, net inflows (% of GDP)",
    "Rural population (% of total population)",
    "Agriculture, forestry, and fishing, value added (% of GDP)",
    "Official exchange rate (LCU per US$, period average)",
  'Per capita food production variability (constant 2004-2006 thousand int$ per capita)',
  'Per capita food supply variability (kcal/cap/day)',
  'Cereal import dependency ratio (percent) (3-year average)',
      'GNI per capita, PPP (current international $)',
  'Percent of arable land equipped for irrigation (percent) (3-year average)',
      'Total reserves (includes gold, current US$)'
]]
y3 = dfAfghanistan['Prevalence of undernourishment (%)']
```

Train and split data frame.

```
X_train3,X_test3, y_train3 , y_test3 = train_test_split(X3 ,y3, test_size = 0.2,random_state = 0)
```

Step 13: Regress the Model

Develop Linear regression model.

```
regressor2 = LinearRegression()
regressor2.fit(X_train3,y_train3)
y_pred3 = regressor2.predict(X_test3)
```

Predict the model.

```
import matplotlib.pyplot as plt
%matplotlib inline
plt.scatter(y_test3,y_pred3)
plt.xlabel('Actual')
plt.ylabel("Predicted")
plt.title('Actual vs Predicted')
# overlay the regression line
z = np.polyfit(y_test3, y_pred3, 1)
p = np.poly1d(z)
plt.plot(y_test3,p(y_test3), color='magenta')
plt.show()
```

Output:

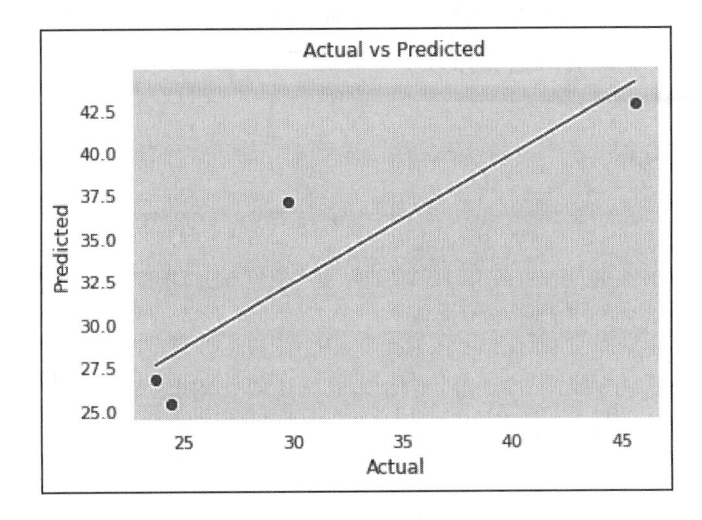

Step 14: Regression Score Function - R^2 (coefficient of determination)

The best possible score is 1.0 and it can be negative (because the model can be arbitrarily worse). In the general case when the true y is not constant, a constant model that always predicts the average y disregarding the input features would get a score of 0.0.

```
r2_score(y_test3,y_pred3)
```

Output:
R^2= 0.7664244425734887

Step 15: Develop Model Equation (y=mx+c) to predict Prevalence of undernourishment (%)

Develop model equation (y=mx+c).

```
mx=""
for ifeature in range(len(X3.columns)):
   if regressor2.coef_[ifeature] <0:
      # format & beautify the equation
      mx += " - " + "{:.2f}".format(abs(regressor2.coef_[ifeature])) + " * " + X3.columns[ifeature]
   else:
      if ifeature == 0:
         mx += "{:.2f}".format(regressor2.coef_[ifeature]) + " * " + X3.columns[ifeature]
      else:
         mx += " + " + "{:.2f}".format(regressor2.coef_[ifeature]) + " * " + X3.columns[ifeature]
print(mx)
```

```
# y=mx+c
if(regressor2.intercept_ <0):
    print("The formula for the " + y3.name + " linear regression line (y=mx+c) is = " + " - {:.2f}".
format(abs(regressor2.intercept_)) + " + " + mx )
else:
    print("The formula for the " + y3.name + " linear regression line (y=mx+c) is = " + "{:.2f}".
format(regressor2.intercept_) + " + " + mx )
```

Output:

```
The formula for the Prevalence of undernourishment (%) linear regression line (y=mx+c) is = – 2508.03
+ – 3.20 * Employment in agriculture (% of total employment) (modeled ILO estimate) – 0.93 * Foreign
direct investment, net inflows (% of GDP) + 34.38 * Rural population (% of total population) + 0.06 *
Agriculture, forestry, and fishing, value added (% of GDP) + 0.12 * Official exchange rate (LCU per US$,
period average) – 0.37 * Per capita food production variability (constant 2004–2006 thousand int$ per
capita) – 0.11 * Per capita food supply variability (kcal/cap/day) + 0.39 * Cereal import dependency ratio
(percent) (3-year average) + 0.01 * GNI per capita, PPP (current international $) + 1.25 * Percent of arable
land equipped for irrigation (percent) (3-year average) + 0.00 * Total reserves (includes gold, current US$)
```

Step 16: Model Explainability

To see important model parameters, run model Explainability.

```
import eli5
eli5.show_weights (regressor1,feature_names = model_features_prev_moderate_afghan_m1)
```

Output:

Please find a summary of the top features.

Weight?	Feature
+34.376	Rural population (% of total population)
+1.247	Total reserves (includes gold, current US$)
+0.395	Cereal import dependency ratio (percent) (3-year average)
+0.117	Official exchange rate (LCU per US$, period average)
+0.060	Agriculture, forestry, and fishing, value added (% of GDP)
+0.009	Percent of arable land equipped for irrigation (percent) (3-year average)
+0.000	GNI per capita, PPP (current international $)
-0.115	Per capita food supply variability (kcal/cap/day)
-0.375	Per capita food production variability (constant 2004-2006 thousand int$ per capita)
-0.930	Foreign direct investment, net inflows (% of GDP)
-3.197	Employment in agriculture (% of total employment) (modeled ILO estimate)
-2508.030	<BIAS>

Summary—Macroeconomic Key Drivers - Forces Model

The reduction in Prevalence of undernourishment (%) in Afghanistan is due to the following:

Independent variable	Forces or drivers impacting model	Statistical significance	Description
Rural population (% of total population)	Drop in Rural population improves food security. 	Positively correlated 34.376	The drop in rural population as percentage of population in Afghanistan is positively correlated with drop in Prevalence of undernourishment (%). All held constant, a unit decrease in rural population as percentage of total population would reduce prevalence of undernourishment (%) by 34.376% [12,15].
Total reserves (includes gold, current US$)	Increase in Total reserves (includes gold, current US$) positively correlated with drop in Prevalence of undernourishment (%). 	Positively correlated 1.247	The increase in Total reserves (includes gold, current US$) in Afghanistan is positively correlated with drop in Prevalence of undernourishment (%). All held constant, a unit increase in Total reserves (includes gold, current US$) would reduce prevalence of undernourishment (%) by 124.7 [12,15].
Cereal import dependency ratio (percent) (3-year average)	Increase in cereal import dependency ratio. 	Positively correlated 0.395	The increase in Cereal import dependency ratio (percent) (3-year average) in Afghanistan is positively correlated with drop in Prevalence of undernourishment (%). One reason could be lack of infrastructure and climate change that were detrimental to agricultural production. Thus, cereal dependency increase has alleviated prevalence of undernourishment (%) [12,15].
Official exchange rate (LCU per US$, period average)	Drop in official exchange rate (LCU per US$, period average) would improve food security. 	Positively correlated, 0.117	The drop-in official exchange rate (LCU per US$, period average) in Afghanistan is positively correlated with drop in Prevalence of undernourishment (%). All held constant, a unit decrease in official exchange rate (LCU per US$, period average) would drop prevalence of undernourishment (%) by 11.7% [12,15].
Inflation, GDP deflator (annual %)	Drop in Rural population improves food security. 	Positively correlated, 0.14	The drop-in Inflation, GDP deflator (annual %) in Afghanistan is positively correlated with drop in Prevalence of undernourishment (%). All held constant, a unit decrease in Inflation, GDP deflator (annual %) would drop prevalence of undernourishment (%) by 1.4% [12,15].

Agriculture, forestry, and fishing, value added (% of GDP)	Increase in Agriculture, forestry, and fishing would improve food security. 	Positively correlated, 0.060	Agriculture, forestry, and fishing, value added (% of GDP) positively correlated with drop in Prevalence of undernourishment (%). A drop-in Agriculture, forestry, and fishing, value added (% of GDP) in Afghanistan is positively correlated with drop in Prevalence of undernourishment (%). Due to high cereal import dependencies and climate & drought, drop in agricultural value as percentage of GDP improved decrease in prevalence of undernourishment [12,15].
Foreign direct investment, net inflows (% of GDP)	Drop in Foreign direct investment, net inflows (% of GDP) decreases food security. 	Negatively correlated, 0.93	Foreign direct investment, net inflows (% of GDP) negatively correlated with drop in Prevalence of undernourishment (%). A drop-in Foreign direct investment, net inflows (% of GDP) in Afghanistan is negatively correlated with drop in Prevalence of undernourishment (%). All held constant, a unit decrease in Foreign direct investment, net inflows (% of GDP) would increase prevalence of undernourishment (%) by 93% [12,15].
Employment in agriculture (% of total employment) (modeled ILO estimate)	Increase Employment in agriculture (% of total employment) would improve food security. 	Negatively correlated, –3.19	Employment in agriculture (% of total employment) (modeled ILO estimate) negatively correlated with drop in Prevalence of undernourishment (%). A drop-in Employment in agriculture (% of total employment) (modeled ILO estimate) in Afghanistan is negatively correlated with drop in Prevalence of undernourishment (%). All held constant, a unit decrease in Employment in agriculture (% of total employment) (modeled ILO estimate) would increase prevalence of undernourishment (%) by 319.7% [12,15].
Per capita food supply variability (kcal/cap/day)	Increase in per capita food supply variability (kcal/cap/day) is negatively correlated with drop in prevalence of undernourishment  (%). Desired (drop) in per capita food supply variability.	Negatively correlated, –0.375	Increase in per capita food supply variability (kcal/cap/day) is negatively correlated with *drop in* prevalence of undernourishment (%). An increase in per capita food supply variability (kcal/cap/day) is negatively correlated with drop in Prevalence of undernourishment (%). All held constant, a unit increase in per capita food supply variability (kcal/cap/day) would increase prevalence of undernourishment (%) by 37.5% [12,15].

Per capita food production variability (constant 2004–2006 thousand int$ per capita)	Decrease in per capita food production variability (constant 2004–2006 thousand int$ per capita) is positively correlated with drop in prevalence of undernourishment (%). Desired (drop) in per capita food production variability.	Negatively correlated, –0.375	Increase in per capita food production variability (constant 2004–2006 thousand int$ per capita) is negatively correlated with *drop in* prevalence of undernourishment (%). Conversely, a drop in per capita food production variability (constant 2004–2006 thousand int$ per capita) improves *drop in* prevalence of undernourishment (%) [12,15].

Summary to Policy Makers & Further Steps

It is essential that we study signal patterns of macroeconomic forces that contribute to the prevalence of food insecurity and take appropriate proactive actions to prevent global apocalyptic [1] food inflation issues. Understanding how households respond to price shocks can help national and local governments and aid agencies design interventions and respond to local needs during economic crises or natural disasters. Such efforts are particularly important in poor, conflict ridden countries where often there is limited data and analysis pertaining to food consumption and household coping mechanisms [15]. The United Nations Food and Agriculture Organization estimated that in 2010, nearly a billion people in the world were undernourished with calorie intake below the minimum dietary energy requirement. Recent wildfires and export bans (Russia), flooding (Pakistan), and political instability (Middle East) have added to international commodity price volatility and raised concerns about potential increases in food insecurity and global poverty. For households that spend most of their budgets on food, large increases in food prices erode purchasing power, disproportionately affecting poor households and threatening their nutrition and health. Potential policy interventions to mitigate the effects of food price shocks could include micronutrient supplementation programs based on food inflation indicators; agricultural employment generation programs; reduction in supply variability; improvement in downward trend of production variability; agricultural investments; incentives to encourage the adoption of yield-increasing agricultural practices; improvements in transportation and irrigation infrastructure; targeted food distribution programs; fertilizer supply; and wheat fortification programs; all these would enable Afghanistan to reduce future food insecurity. Food insecurity Implications for Afghanistan are beyond Afghans, and they will impact the entire world. Investment in infrastructure and improvement of agricultural practices are much needed! [12,14,15]

Higher Food Inflation & Monetary Policy Test

"Food Inflation in UK due to War in Ukraine and COVID-19 is the biggest test of the monetary policy framework in 25 years. There is no question about that."

The Bank of England governor

Sri Lanka

Sri Lanka is a lower-middle-income country (please see below table on income group changes - OGHIST. XLS)[350] with a GDP per capita of USD 3,852 (2019) and a total population of 21.8 million.[351] Sri Lanka's macroeconomic challenges are linked to years of high fiscal deficits, driven primarily by low revenue collection, and erosion of export competitiveness due to a restrictive trade regime and weak investment climate. As per the March 2022[352] IMF report[353] the fertilizers ban (reversed in November 2021) had also hurt tea and rubber exports, leading to "substantial" losses. When Sri Lanka's foreign currency shortages became a serious problem in early 2021, the government tried to limit the outflow by banning imports of chemical fertilizers, telling farmers to use locally sourced organic fertilizers instead. The switch to organic fertilizers resulted in widespread crop failure, exacerbating foreign currency shortages due to export declines. This led to widespread crop failure. Sri Lanka had to supplement its food stocks from abroad, which made its foreign currency shortage even worse.

Income Group	Years (From – To)
Low income (L)	1987–1996
Low income (L)	1997–2017
Upper Middle Income (UM)	2018
Lower middle income (LM)	2019

Growth slowed to an average of 3.1 percent between 2017 and 2019 from 6.2 percent between 2010 and 2016, as a peace dividend and a policy thrust toward reconstruction faded away and macroeconomic shocks adversely impacted growth. Consumer prices (CPI) have skyrocketed in 2021 (12.1%) and projected[354] to remain high for the year 2022 (17.2%). As a result, Sri Lanka has seen weeks of protests due to rising food prices and acute shortages of fuel.[355] Due to the shortage of foreign reserves for months (current account balance was projected[356] to decline –7.1% from 2021 –4.3%, and 2020 –1.3%), Sri Lanka has lacked the foreign currency to buy all that it needs from abroad. Sri Lanka's foreign currency reserves have virtually run dry, and it can no longer afford to pay for imports of staple foods and fuel [C4520220530]. *Shortages of food and fuel have caused prices to soar.* Inflation is now running at 30% [18] and projected to be 17.2% by the year 2022.

Economic downturn[357] and high food prices affect purchasing power of vulnerable households and that is what Sri Lankans are going through! The macroeconomic situation in the country has deteriorated since 2021 mostly reflecting dwindling foreign currency reserves after revenues from merchandise exports and from the tourist sector were affected by the COVID-19 pandemic and its containment measures. The reduction in economic activities has caused widespread loss of income and livelihoods, sharply reducing households' purchasing power. This is of particular concern as domestic prices of basic food items have been rising since mid-2021, seriously limiting households' access to food [19]. Public investments and future borrowings should prioritize key sectors and address immediate needs and induce sustainable and resilient growth through an economic transformation.

Fish make up about 50% of Sri Lankans' animal protein intake, a ratio three times the global average. The fisheries sector also supports close to one million fishers, workers, and their families.[358]

Rural Population (% of total population)

Rural population (% of total population) in Sri Lanka was reported at 81.29 % in 2020, according to the World Bank collection of development indicators, compiled from officially recognized sources. Sri Lanka's 26-year conflict—which uprooted hundreds of thousands of people, often more than once—ended in 2009.[359] The hardest hit areas were the northern and eastern provinces, where most people earn a living from agriculture. Most displaced families have returned to their places of origin, but the conflict wiped out their homes and assets. Many lost family members, and about 15 percent of the households are now headed by women.

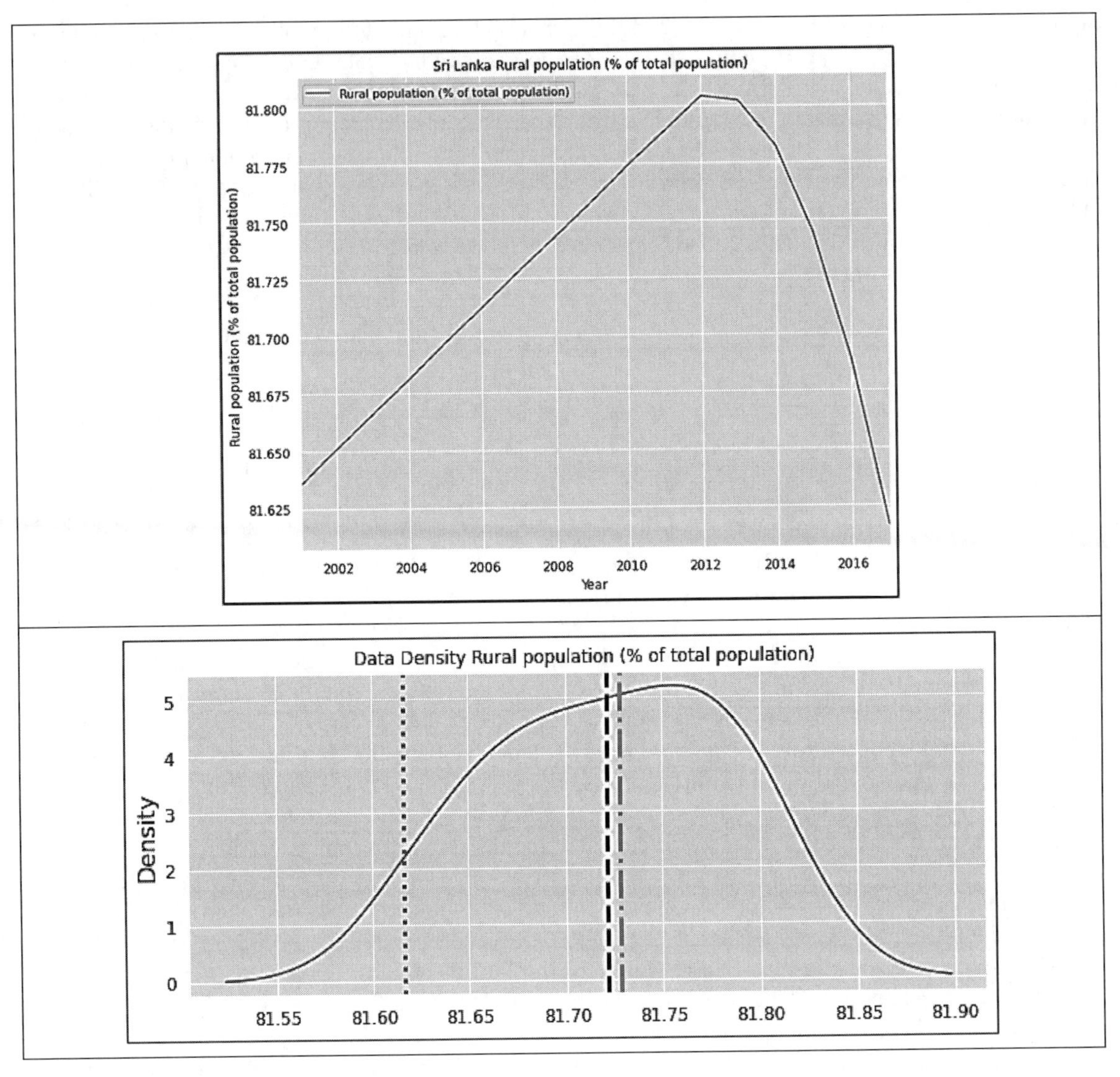

Data Distribution statistics: Minimum:81.62; Mean:81.72 (dashed line); Median:81.73 (dash dot line); Mode: 81.62(dotted line); Maximum:81.80. To summarize, the distribution of data is skewed to the right—statistically, if the mode is less than the median, which is less than the mean. The average rural population is 81.72%. A "skewed right" distribution is one in which the tail is on the right side. A "skewed left" distribution is one in which the tail is on the left side. Skewed data often occurs due to lower or upper bounds on the data. That is, data that has a lower bound is often skewed right while data that has an upper bound is often skewed left. The above histogram is for a distribution that is skewed left[360] as higher percentages of rural population were frequent in the years [2002–2012].

Employment in Agriculture

Agriculture provides food security, a source of income, and stability to countless Sri Lankan families. The agriculture sector[361] contributes about 7.4 percent to the national GDP, out of which the fisheries sector contributes around 1.3 percent, and the livestock sector accounts for 0.9 percent. Over 30 percent of Sri Lankans are employed in the agricultural sector Sri Lanka's primary food crop is rice. Rice is cultivated during two seasons. Tea is cultivated in the central highlands and is a major source of foreign exchange. Fruits, vegetables, and oilseed crops are also cultivated in the country. Despite being such a critical sector, employment in agriculture is trending downward. Employment in agriculture (% of total employment) (modeled ILO estimate) in Sri Lanka was reported at above 26.00 % in 2016 and it was above 45% in 2002. The reasons are numerous. Import dependency, Lack of investment, supply, and untimely market are causing farmers to leave the agriculture fields [20].

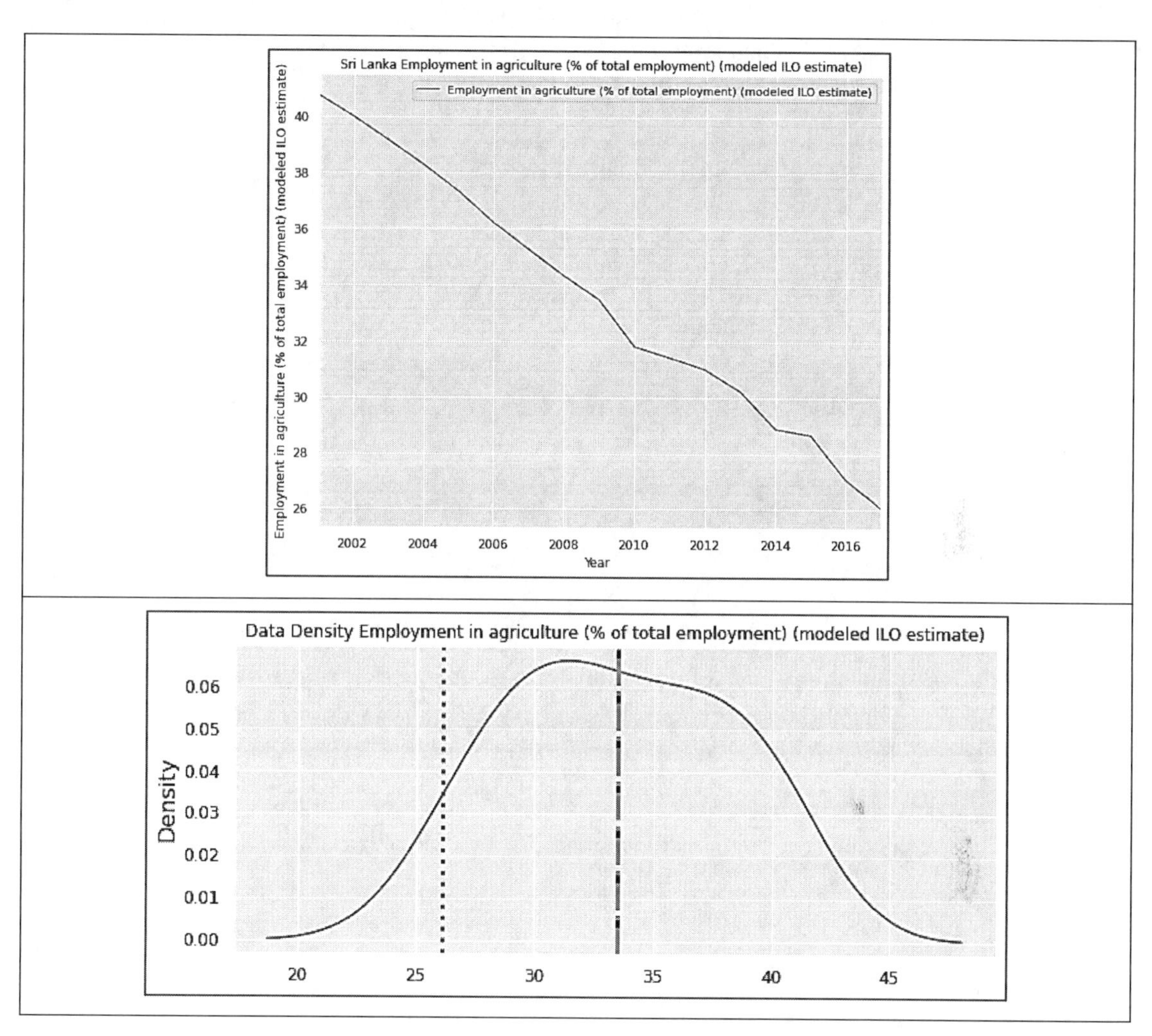

For skewed distributions, Skewed data often occur due to lower or upper bounds on the data. That is, data that has a lower bound is often skewed right while data that has an upper bound is often skewed left. The above histogram is for a distribution that is skewed left[362] as Employments in Agriculture (% of total employment) were frequent in the higher values [2002–2008]. Data Distribution statistics: Minimum:26.07; Mean:33.53 (dashed line); Median:33.49 (dash dot line); Mode: 26.07 (dotted line); Maximum:40.76.

Agriculture is the bed rock of many farming nations, and it is fundamentally assuring national security. Investment in agriculture in terms of farm productivity is essentially an investment into the countries' future national savings!

Author

Percent of Arable Land Equipped for Irrigation (percent) (3-year average)

Total arable land equipped for irrigation is defined as the area equipped to provide water (via irrigation) to the crops. It includes areas equipped for full and partially controlled irrigation, equipped lowland areas, pastures, and areas equipped for spate irrigation. This indicator provides a measure of the dependence of a country or region's agriculture on irrigation. *It shows the vulnerability of agriculture to water stress and climatic shocks (such as droughts), which has implications for national food security depending on production and trade patterns.*[363] Sri Lanka is almost self-sufficient in rice and imports large quantities only when local production is not sufficient to cover domestic consumption. Although Sri Lanka is a fertile tropical land with the potential for the cultivation and processing of a variety of crops, issues such as productivity and profitability hamper the growth of the sector [20].

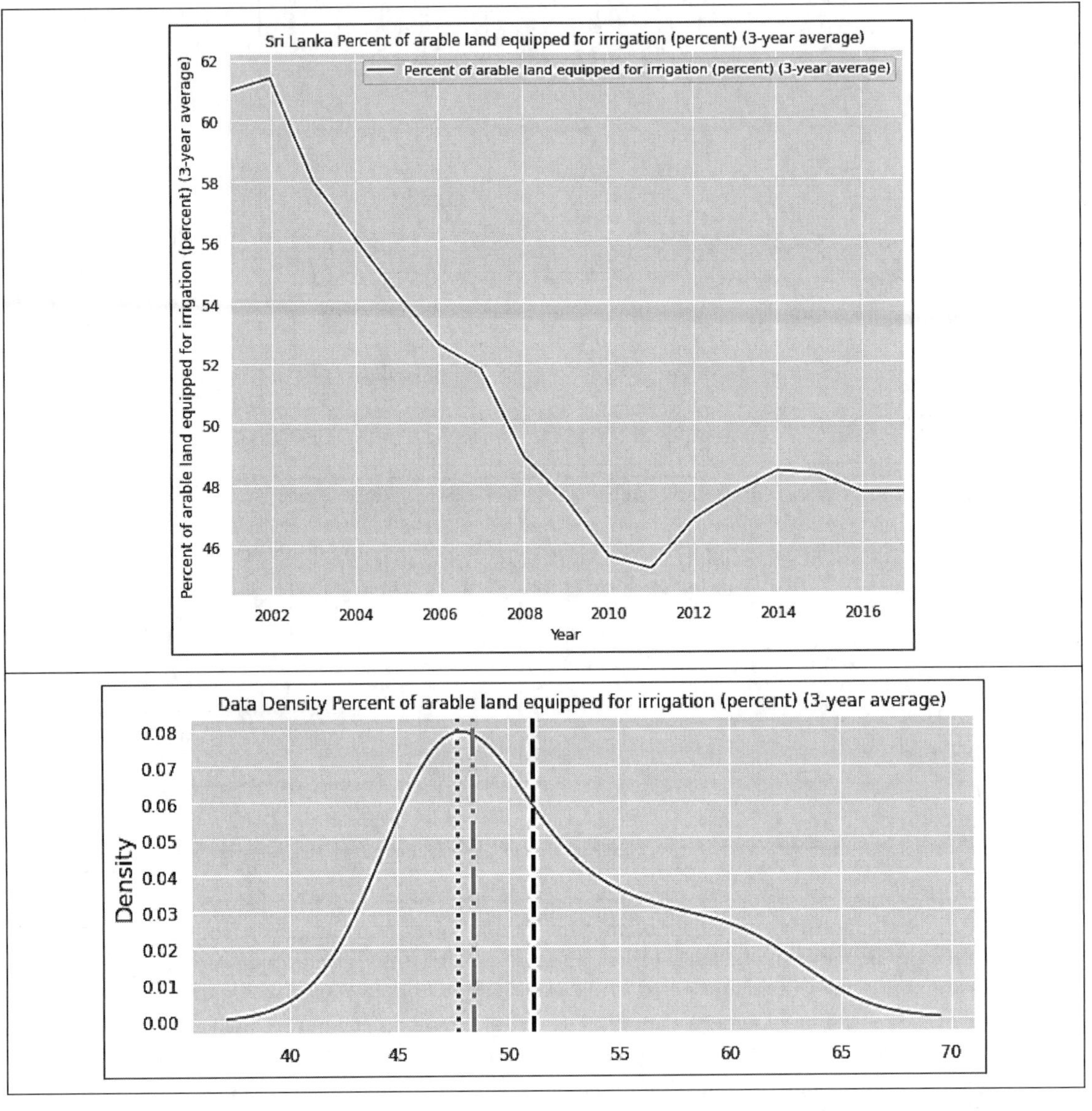

That is, data that has a lower bound is often skewed right while data that has an upper bound is often skewed left. The above histogram is for a distribution that is skewed right[364] as Percent of arable land equipped for irrigation was trending lower since 2003 onwards. Data Distribution statistics: Minimum:26.07; Mean:33.53 (dashed line); Median:33.49 (dash dot line); Mode:26.07 (dotted line); Maximum:40.76.

Agriculture, Forestry, and Fishing, Value Added (% of GDP)

The agriculture sector[365] contributes about 7.4 percent to the national GDP, 2021, out of which the fisheries sector contributes around 1.3 percent, and the livestock sector accounts for 0.9 percent [20]. In 2020, the share of agriculture in Sri Lanka's gross domestic product was 8.36 percent, industry contributed approximately 26.25 percent and the services sector contributed about 59.67 percent. The agriculture sector contributed above 20% in 2002 and in 2016 that share fell to just above 8%.

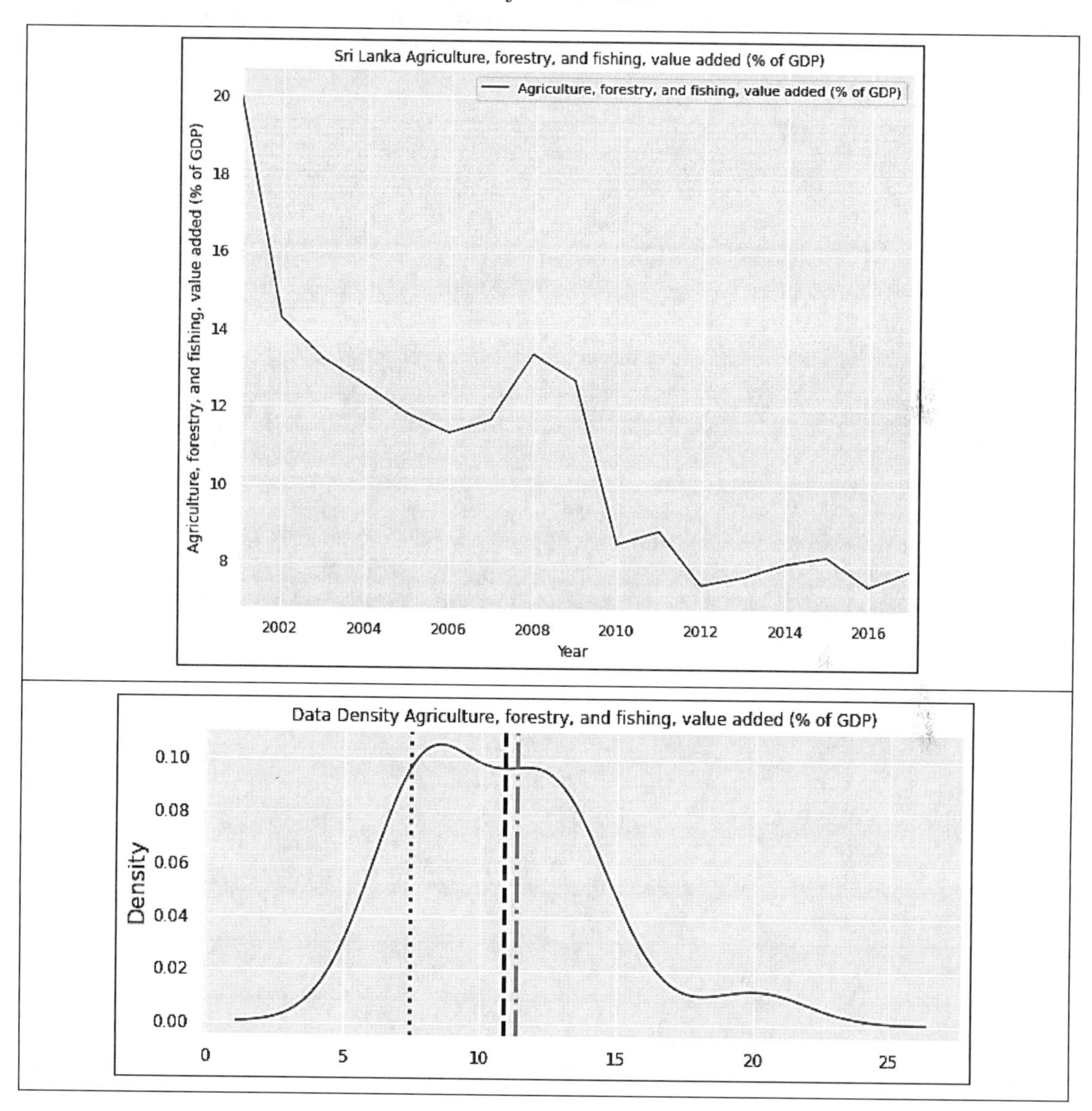

The above histogram is for a distribution that is skewed right[366] as Agriculture, forestry, and fishing, value added (% of GDP) were trending lower since 2003 onwards. Data Distribution statistics: Minimum:7.43; Mean:10.88 (dashed line); Median:11.34 (dash dot line); Mode:7.43 (dotted line); Maximum:20.07.

Official Exchange Rate (LCU per US$, period average)

In a market-based economy, household, producer, and government choices about resource allocation are influenced by relative prices, including the real exchange rate, real wages, real interest rates, and other prices in the economy. Relative prices also largely reflect these agents' choices. Thus, relative prices convey vital information about the interaction of economic agents in an economy and with the rest of the world.[367] Though not a direct PPP, it is a good candidate to purchase buying power of a currency. The value for Official exchange rate (LCU per US$, period average) in Sri Lanka was 162.46 as of 2018. As the graph below shows,[368] over the past 58 years this indicator reached a maximum value of 162.46 in 2018 and a minimum value of 4.76 in 1960.

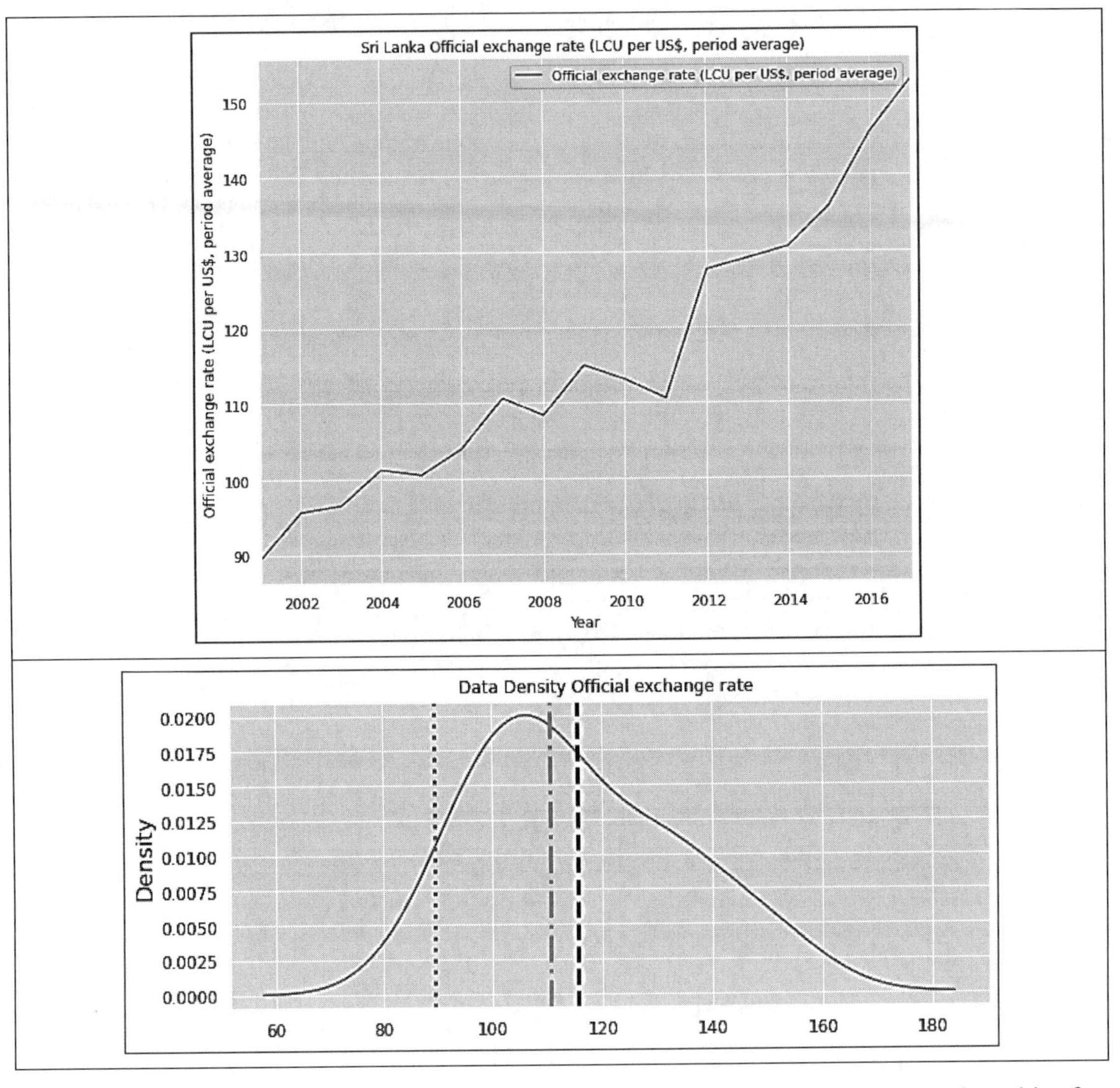

That is, data that has a lower bound is often skewed right while data that has an upper bound is often skewed left. The above histogram is for a distribution that is skewed right as the Official exchange rate (LCU per US$, period average) was trending higher 2011 onwards. Data Distribution statistics: Minimum:89.38; Mean:115.64 (dashed line); Median:110.62 (dash dot line); Mode: 89.38 (dotted line); Maximum:152.45.

Cereal Import Dependency Ratio

Sri Lanka is almost self-sufficient in rice and imports large quantities only when local production is not sufficient to cover the domestic needs [19]. On the other hand, imports of wheat (not produced in the country) and wheat flour account for the largest share of national cereal imports. The imports of food and beverages accounted for 9.7 percent of total imports in 2020 with total agriculture, food, and beverage imports reaching $1.6 billion. Since the start of the COVID-19 pandemic, the Government of Sri Lanka has placed temporary restrictions, albeit without a stated expiration, on many agricultural imports. Some government officials have publicly called for restrictions to be made permanent and to be expanded to cover all food products [20]. Sri Lanka is a net importer of dairy products and Cotton, Yarn, and Fabric products.

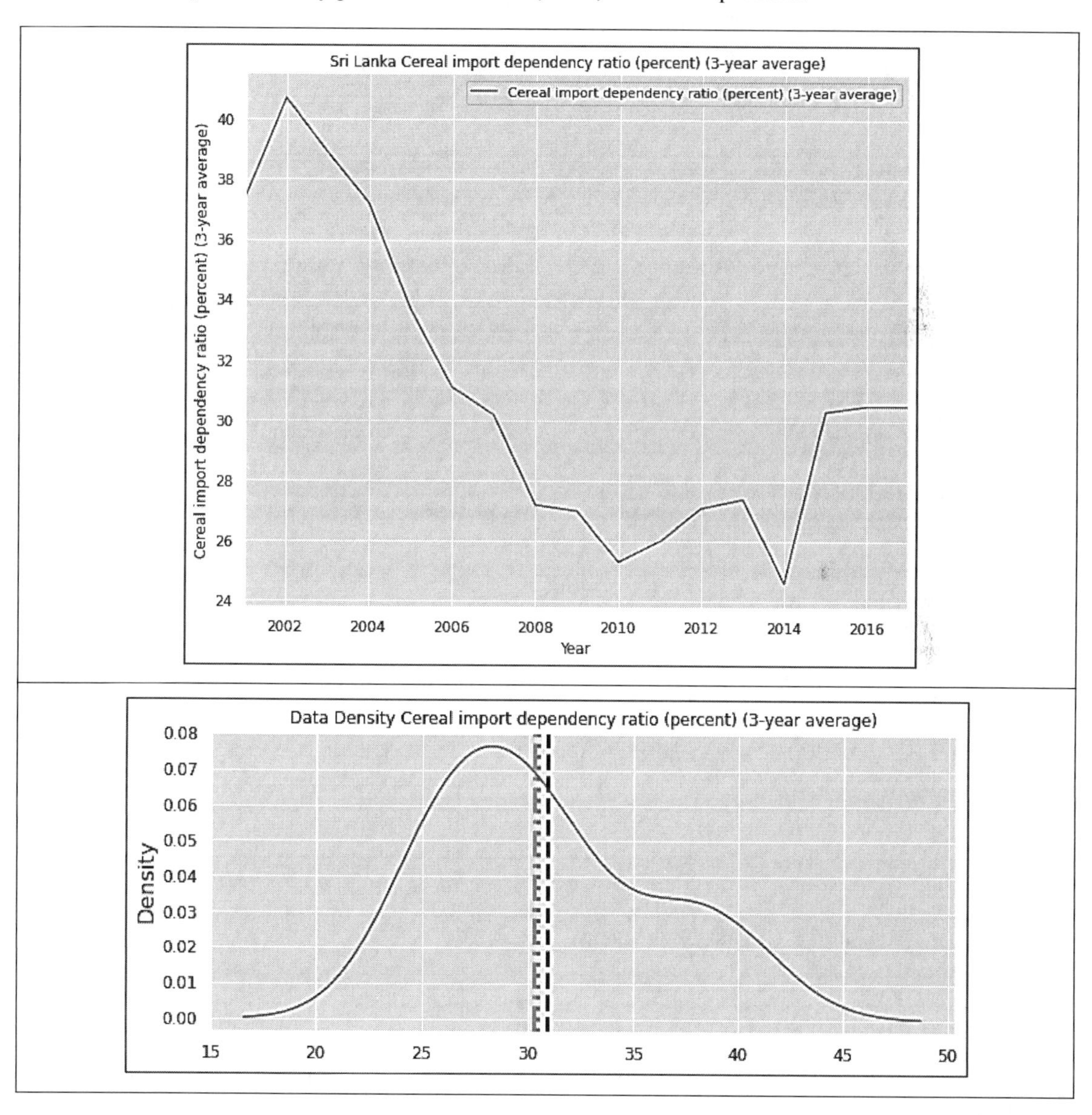

For skewed distributions, Skewed data often occurs due to lower or upper bounds on the data. That is, data that has a lower bound is often skewed right while data that has an upper bound is often skewed left. The above histogram is for a distribution that is skewed right[369] as import dependency ratios (percent) (three-averages) were frequent in the lower values [2010–2016]. Data Distribution statistics: Minimum:24.60; Mean:30.89 (dashed line); Median:30.30 (dash dot line); Mode: 30.50 (dotted line); Maximum:40.70.

Prevalence of Undernourishment: Macroeconomic Parameter Causation

Causation, or causality, is the capacity of one variable to influence another. The first variable may bring the second into existence or may cause the incidence of the second variable to fluctuate. The importance of market-based development in the economic development of a country cannot be denied, and macroeconomic variables are important indicators that affect the overall health of the country's market. The following framework provides the association of these variables (M1-M20 and A1-A20) with economic market factors. Techniques of Augmented Dickey-Fuller test, Regression Analysis and Granger Causality test have been applied to examine the causal relationship of selected macroeconomic variables with the prevalence of food insecurity. Results of regression analysis indicate the presence of a strong positive correlation between applicable macroeconomic parameters and food security.

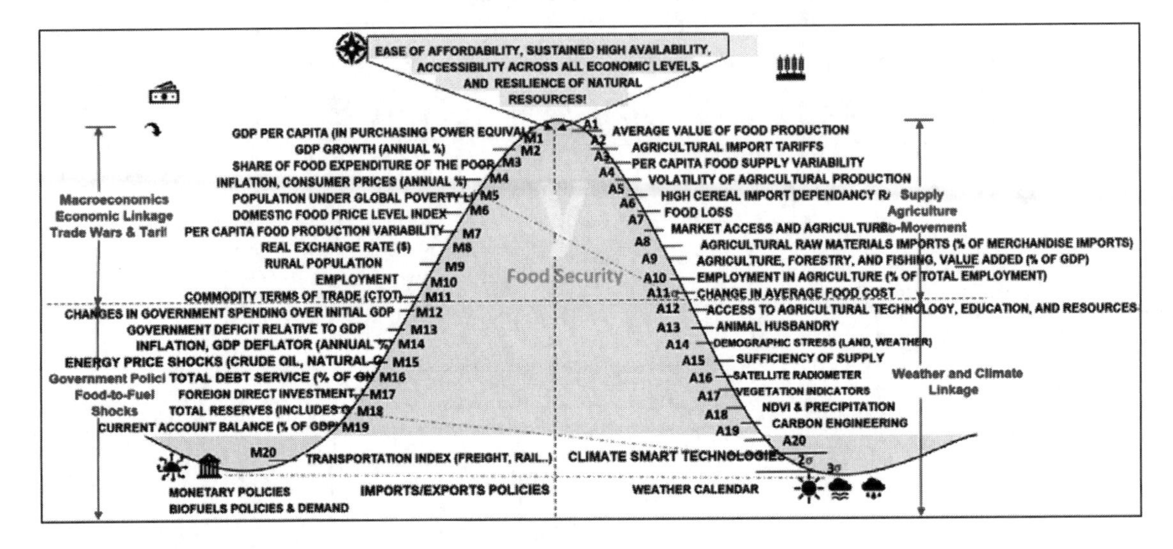

Macroeconomic economic drivers under review include:

Macroeconomic drivers	LABEL	Framework Item
GNI per capita, PPP (current international $)	GNI	M1
GDP growth (annual %)	GDP	M2
Agriculture, forestry, and fishing, value added (% of GDP)	AGDP	A9
Total debt service (% of GNI)	DEBT	M16
Rural population (% of total population)	RPOP	M9
Foreign direct investment, net inflows (% of GDP)	FDI	M17
Agricultural raw materials exports (% of merchandise exports)	AEXP	A8
Employment in agriculture (% of total employment) (modeled ILO estimate)	AEMP	A10
Total reserves (includes gold, current US$)	RESV	M18
Inflation, consumer prices (annual %)	CPI	M4
Prevalence of undernourishment (% of population)	PUND	Tails of the bell curve!
Official exchange rate (LCU per US$, period average)	XLCU	M8
Inflation, GDP deflator (annual %)[370]	GDP-D	M14

Results of the Granger Causality test demonstrate that a bi-directional relationship exists between the prevalence of undernourishment and Sri Lanka's macroeconomic parameters (please see Figure 13). Except for official exchange rate (LCU per US$, period average) [M14], all other variables [M1, M2, M9, M14, M16, M17, M18, A8, A9, A10] have a bi-directional relationship.

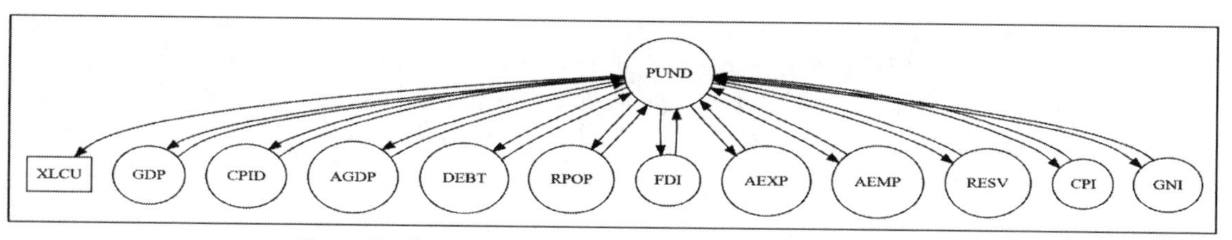

Figure 13: Granger Causality - Macroeconomic Parameter Causation.

Prevalence of Undernourishment: Food Security Indicators and Macroeconomic Parameter Causation

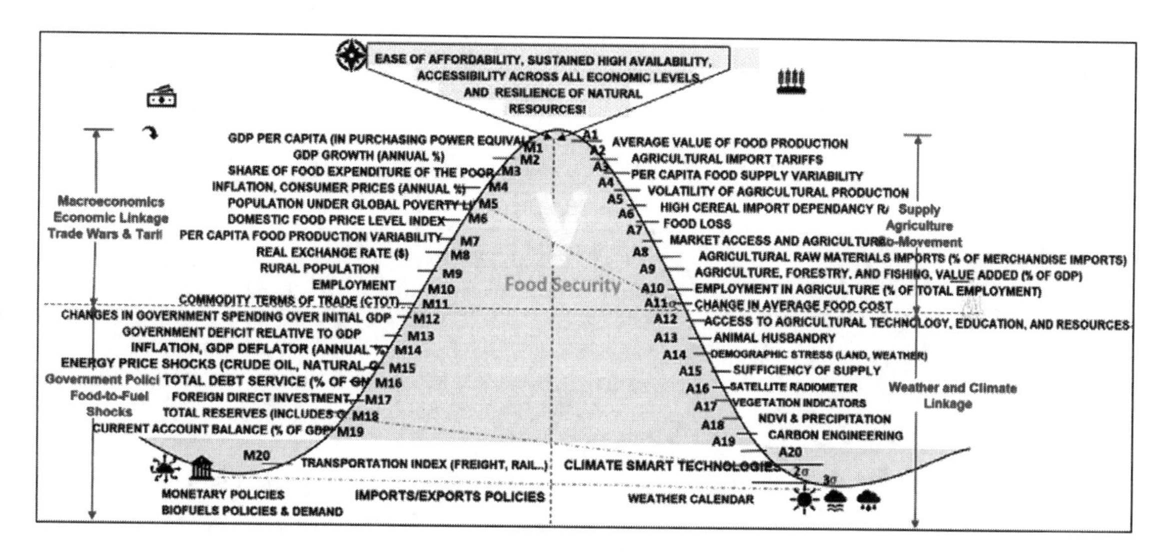

Macroeconomic economic drivers under review include:

Macroeconomic drivers	LABEL	Framework Item
Employment in agriculture (% of total employment) (modeled ILO estimate)	AEMP	A10
Foreign direct investment, net inflows (% of GDP)	FDI	M17
Rural population (% of total population)	RPOP	M9
Agriculture, forestry, and fishing, value added (% of GDP)	AGDP	A9
Official exchange rate (LCU per US$, period average)	XLCU	M8
Per capita food supply variability (kcal/cap/day)	CFSV	A3
Per capita food production variability (constant 2004-2006 thousand int$ per capita)	CFPV	M7
GNI per capita, PPP (current international $)	GNI	M1
Cereal import dependency ratio (percent) (3-year average)	CID	A5
Percent of arable land equipped for irrigation (percent) (3-year average)	ALEI	A7
Prevalence of undernourishment (% of population)	PUND	Tails of the bell curve!

Results of Granger Causality test demonstrate that a bi-directional relationship exists between prevalence of undernourishment and Sri Lanka's macroeconomic parameters (please see Figure 14). Except for Official exchange rate (LCU per US$, period average), Per capita food production variability & Cereal import dependency ratio [M14], all other variables [M1, M7, M8, M9, M17, M18, A3, A5, A7, A9, and A10] have a bi-directional relationship.

	AEMP_x	FDI_x	RPOP_x	AGDP_x	XLCU_x	CFPV_x	CFSV_x	CID_x	ALEI_x	PUND_x
AEMP_y	1	0	0	0	0	0	0	0	0	0
FDI_y	0	1	0	0	0	0	0	0	0	0
RPOP_y	0	0	1	0	0	0	0	0	0	0
AGDP_y	0	0	0	1	0.0094	0	0	0	0	0
XLCU_y	0	0	0	0	1	0	0.0001	0.0007	0	0
CFPV_y	0	0	0	0	0	1	0	**0.1849**	0	0
CFSV_y	0	0.0034	0	0.0282	0	0	1	0	0	0
CID_y	0	0.0225	0	0	0	0	0	1	0	0
ALEI_y	0	0	0	0	0	0	0	0	1	0
PUND_y	0	0.0593	0	0	**0.1597**	0	0.0072	0	0	1

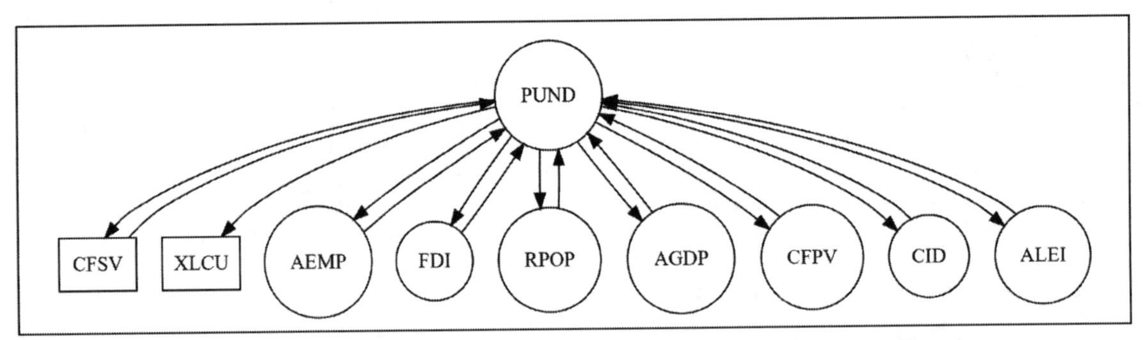

Figure 14: Granger Causality - Food Security Indicators and Macroeconomic Parameter Causation.

Inflation, GDP deflator (annual %)

The Consumer Price Index (CPI) and the gross domestic product (GDP) price index and implicit price deflator are measures of inflation in the U.S. economy. The CPI measures price changes in goods and services purchased out of pocket by urban consumers, whereas the GDP price index and implicit price deflator measure price changes in goods and services purchased by consumers, businesses, government, and foreigners, but not importers. Thus, which one to use in each scenario depends on one's purpose [21], Inflation or GDP deflator (annual %): The fixed basket used in CPI calculations is static and sometimes misses changes in prices of goods outside the basket of goods. Since GDP isn't based on a fixed basket of goods and services, the GDP deflator has an advantage over the CPI. The GDP deflator measures prices of all goods and services produced, whereas the CPI measures the prices of only the goods and services bought by consumers.

Sri Lanka Macroeconomic Key Drivers & Linkage Model - Prevalence of Undernourishment

To model Sri Lanka's food security, we will load World Bank macroeconomic parameters—similar to those of the Afghanistan use case except that we load Sri Lanka specific parameters:

Parameter	References	Source
GNI per capita, PPP (current international $)	NY.GNP.PCAP.PP.CD	https://data.worldbank.org/indicator THE WORLD BANK IBRD · IDA
GDP growth (annual %)	NY.GDP.MKTP.KD. ZG	
Inflation, GDP deflator (annual %)	NY.GDP.DEFL.KD. ZG	
Agriculture, forestry, and fishing, value added (% of GDP)	NV.AGR.TOTL. ZS	
Total debt service (% of GNI)	DT.TDS.DECT.GN.ZS	
Foreign direct investment, net inflows (BoP, current US$)	BX.KLT.DINV.WD.GD.ZS	
Foreign direct investment, net (BoP, current US$)	BN.KLT.DINV.CD	
Rural population (% of total population)	SP.RUR.TOTL. ZS	
Foreign direct investment, net inflows (% of GDP)	BX.KLT.DINV.WD.GD.ZS	
Foreign direct investment, net outflows (% of GDP)	BM.KLT.DINV.WD.GD.ZS	
Agricultural raw materials exports (% of merchandise exports)	SL.AGR.EMPL. ZS	
Employment in agriculture (% of total employment) (modeled ILO estimate)	SL.AGR.EMPL. ZS	
Total reserves (includes gold, current US$)	FI.RES.TOTL.CD	
Inflation, consumer prices (annual %)	FP.CPI.TOTL. ZG	
Prevalence of undernourishment (%)	Prevalence of undernourishment	
Official exchange rate (LCU per US$, period average)	PA.NUS.FCRF	
Per capita food production variability (constant 2004–2006 thousand int$ per capita)	Per capita food production variability	http://data.un.org/ UNdata A world of information
Per capita food supply variability (kcal/cap/day)	Per capita food supply variability (kcal/cap/day)	
Cereal import dependency ratio (percent) (3-year average)	Cereal import dependency ratio	
Percent of arable land equipped for irrigation (percent) (3-year average)	Percent of arable land equipped for irrigation	
Value of food imports in total merchandise exports (percent) (3-year average)	Value of food imports in total merchandise exports	

The data for the model is from Global Food Security Index[371] repository and FAO Suite of Food Security Indicators.[372] The model assesses the prevalence of severe food insecurity (characterized by individuals not eating for a whole day at times during the year) and [17] prevalence of moderate-severe food insecurity.

Step 1: Load Libraries

Load Python libraries to analyze the data.

```
import numpy as np
import pandas as pd
import functools
from datetime import date
import plotly.graph_objects as go
import seaborn as sns; sns.set()
from sklearn.impute import SimpleImputer
from scipy import stats
from scipy.stats import pearsonr
import matplotlib.pyplot as plt
from sklearn.preprocessing import MinMaxScaler
from sklearn.model_selection import train_test_split
from sklearn.linear_model import LinearRegression
from sklearn.metrics import r2_score
from sklearn import metrics
from sklearn.impute import SimpleImputer
from fbprophet import Prophet
from fbprophet.diagnostics import cross_validation, performance_metrics
from statsmodels.tsa.stattools import grangercausalitytests
from statsmodels.tsa.stattools import adfuller
import statsmodels.api as sm
from statsmodels.tsa.api import VAR
import matplotlib as mpl
mpl.rcParams['figure.figsize'] = (10,8)
mpl.rcParams['axes.grid'] = False
```

Step 2: Load Sri Lanka specific FAO parameters

Load FAO specific Global | FAO, Statistics Division Sri Lanka parameters.[373] The data can be downloaded from the Suite of Food Security Indicators.[374] The Suite of Food Security Indicators presents the core set of food security indicators. Following the recommendation of experts gathered in the Committee on World Food Security (CFS) Round Table on hunger measurement, hosted at FAO headquarters in September 2011.

```
Predf = pd.read_csv("SRILANKA_SDG_Indicators.csv")
Predf.head()
```

Output:
Prevalence of undernourishment (percent) (3-year average) is three year moving average data.

	Domain Code	Domain	Area Code (M49)	Area	Element Code	Element	Item Code (SDG)	Item	Year Code	Year	Unit	Value	Flag
0	SDGB	SDG Indicators	144	Sri Lanka	6121	Value	SN_ITK_DEFC	2.1.1 Prevalence of undernourishment (%)	2001	2001	%	16.9	_
1	SDGB	SDG Indicators	144	Sri Lanka	6121	Value	SN_ITK_DEFC	2.1.1 Prevalence of undernourishment (%)	2002	2002	%	16.3	_
2	SDGB	SDG Indicators	144	Sri Lanka	6121	Value	SN_ITK_DEFC	2.1.1 Prevalence of undernourishment (%)	2003	2003	%	16.1	_
3	SDGB	SDG Indicators	144	Sri Lanka	6121	Value	SN_ITK_DEFC	2.1.1 Prevalence of undernourishment (%)	2004	2004	%	15.6	_
4	SDGB	SDG Indicators	144	Sri Lanka	6121	Value	SN_ITK_DEFC	2.1.1 Prevalence of undernourishment (%)	2005	2005	%	14.7	_

```
predfl = Predf[["Year","Value"]]
predfl.rename(columns = {"Value": "Prevalence of undernourishment (%)"},inplace = True)
```

Step 3: Load Sri Lanka world bank indicators

```
df = pd.read_csv("Sri_Lanka_World_Indicators_mod.csv")
df
```

Output: the following parameters are loaded as part of the DataFrame.

- Poverty headcount ratio at national poverty lines (% of population),
- GNI per capita, PPP (current international $),
- Mortality rate, under-5 (per 1,000 live births),
- School enrollment, primary and secondary (gross), gender parity index (GPI),
- GDP growth (annual %), Inflation, GDP deflator (annual %),
- Agriculture, forestry, and fishing, value added (% of GDP),
- Total debt service (% of GNI),
- Foreign direct investment, net inflows (BoP, current US$),
- Foreign direct investment, net (BoP, current US$),
- Rural population (% of total population),
- Foreign direct investment, net inflows (% of GDP),
- Foreign direct investment, net outflows (% of GDP),
- Agricultural raw materials exports (% of merchandise exports),
- Employment in agriculture (% of total employment) (modeled ILO estimate),
- Total reserves (includes gold, current US$), and
- Inflation, consumer prices (annual %).

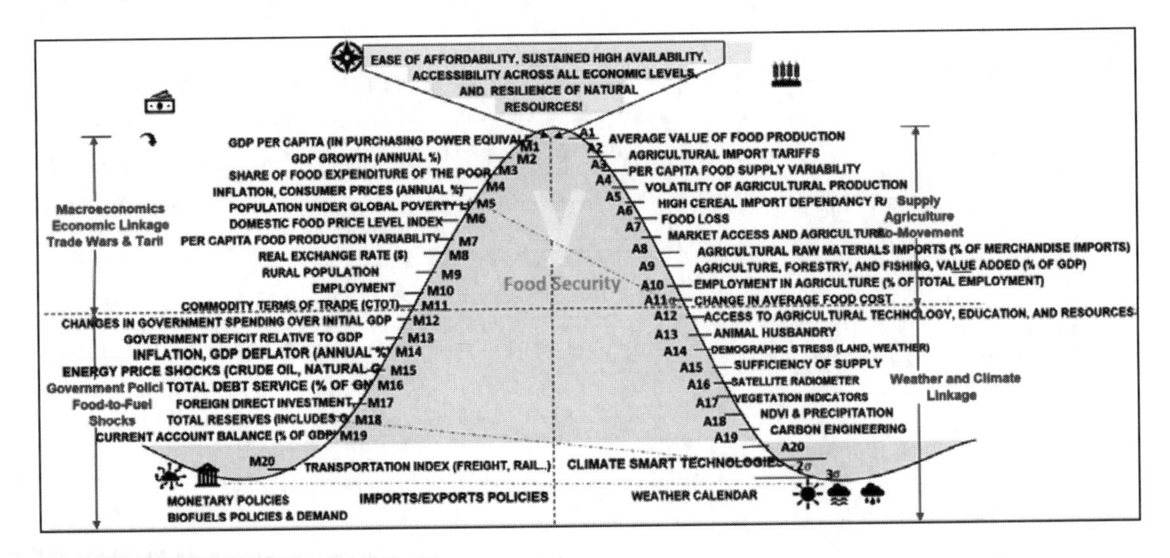

Fix NULLs with imputation strategy.

```
imp_mean = SimpleImputer(missing_values=np.nan, strategy='mean')
imparray = imp_mean.fit_transform(df5)
```

```
final_df = pd.DataFrame(data = imparray, columns = df5.columns)
final_df
```

Output:

	GNI per capita, PPP (current international $)	GDP growth (annual %)	Inflation, GDP deflator (annual %)	Agriculture, forestry, and fishing, value added (% of GDP)	Total debt service (% of GNI)	Foreign direct investment, net inflows (BoP, current US$)	Foreign direct investment, net (BoP, current US$)	Rural population (% of total population)	Foreign direct investment, net inflows (% of GDP)	Foreign direct investment, net outflows (% of GDP)	Agricultural raw materials exports (% of merchandise exports)
0	4270.0	-1.545408	13.664800	20.066605	4.809524	1.717900e+08	-1.717900e+08	81.635	1.090747	0.000006	1.602221
1	4480.0	3.964676	8.111570	14.279323	4.360681	1.965000e+08	-1.850500e+08	81.651	1.188278	0.069241	1.665949
2	4820.0	5.940269	8.748664	13.230515	3.265069	2.287200e+08	-2.014100e+08	81.666	1.211327	0.144637	2.042100
3	5170.0	5.445061	8.801492	12.543804	3.803695	2.328000e+08	-2.270100e+08	81.681	1.126677	0.028022	2.201608
4	5610.0	6.241748	10.418727	11.819477	1.830425	2.724000e+08	-2.340000e+08	81.697	1.116129	0.157340	2.050664
5	6170.0	7.668292	11.277029	11.336280	3.425921	4.797000e+08	-4.504000e+08	81.712	1.696263	0.103607	2.669374
6	6730.0	6.796826	14.028443	11.683164	2.725304	6.030000e+08	-5.480000e+08	81.728	1.863973	0.170014	2.289763
7	7140.0	5.950088	16.327016	13.379201	3.132171	7.522000e+08	-6.905000e+08	81.743	1.847530	0.151546	3.070384
8	7470.0	3.538912	5.879883	12.691971	3.422321	4.040000e+08	-3.840000e+08	81.758	0.960391	0.047544	3.233134
9	8110.0	8.015967	22.799261	8.496140	2.510045	4.775590e+08	-4.350590e+08	81.774	0.841873	0.074922	3.949703
10	8920.0	8.404733	3.831394	8.831664	2.034286	9.559200e+08	-8.959200e+08	81.789	1.464052	0.091894	4.023059
11	10130.0	9.144572	10.828432	7.449326	3.160798	9.411166e+08	-8.771909e+08	81.804	1.375210	0.093412	3.325215
12	10600.0	3.395733	6.236913	7.666517	2.947064	9.325513e+08	-8.674780e+08	81.802	1.254815	0.087561	2.622709

Next, load Sri Lanka's exchange rate.

Step 4: Load Currency Exchange Rate

Lets load Sri Lanka's currency exchange rate.

```
ExRate = pd.read_csv("SriLanka_Exchange_rates.csv")
ExRate
```

Step 5: Perform Grangers causality

To see the variables' time series' influence each other, let's perform grangers causality!

```
granger_cols=['GNI per capita, PPP (current international $)',
   'GDP growth (annual %)', 'Inflation, GDP deflator (annual %)',
   'Agriculture, forestry, and fishing, value added (% of GDP)',
   'Total debt service (% of GNI)',
   'Rural population (% of total population)',
   'Foreign direct investment, net inflows (% of GDP)',
   'Agricultural raw materials exports (% of merchandise exports)',
   'Employment in agriculture (% of total employment) (modeled ILO estimate)',
   'Total reserves (includes gold, current US$)',
   'Inflation, consumer prices (annual %)',
   'Prevalence of undernourishment (% of population)',
   'Official exchange rate (LCU per US$, period average)',
   'Prevalence of undernourishment (%)']
grangers_causality_matrix(dataset, variables = granger_cols)
```

The output of granger's causation confirms that except Local Currency Exchange rates, all other variables exhibit bi-directional causation, and this is sufficient to perform modeling.

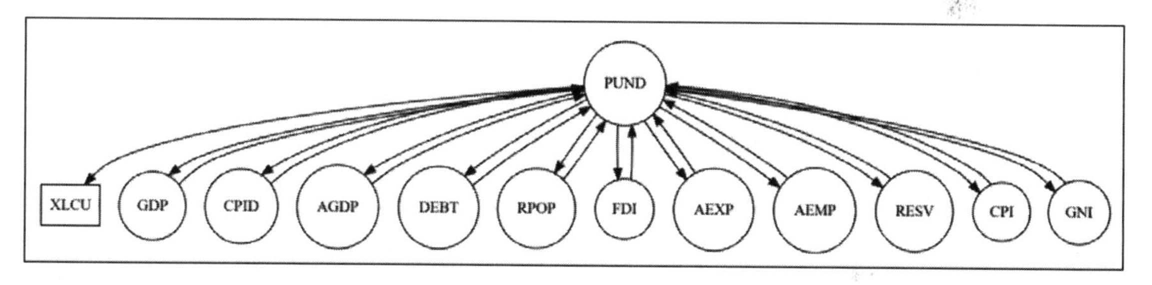

Step 6: Perform Linear Regression

Split the data frame into independent and dependent variables.

```
X1 = final_df1[["Employment in agriculture (% of total employment) (modeled ILO estimate)",
   "Foreign direct investment, net inflows (% of GDP)",
   "Rural population (% of total population)",
   "Agriculture, forestry, and fishing, value added (% of GDP)",
   "Inflation, GDP deflator (annual %)",
     "Official exchange rate (LCU per US$, period average)"]]
y1 = final_df1['Prevalence of undernourishment (%)']
X_train1,X_test1, y_train1 , y_test1 = train_test_split(X1 ,y1, test_size = 0.2,random_state = 0)
```

Split the data into training and test data.

```
regressor1 = LinearRegression()
regressor1.fit(X_train1,y_train1)
y_pred1 = regressor1.predict(X_test1)
```

Plot the prediction.

```
plt.scatter(y_test1, y_pred1)
plt.xlabel('Actual Labels')
plt.ylabel('Predicted Labels')
plt.title('Predictions')
# overlay the regression line
z = np.polyfit(y_test1, y_pred1, 1)
p = np.poly1d(z)
plt.plot(y_test1,p(y_test1), color='magenta')
plt.show()
```

Output:

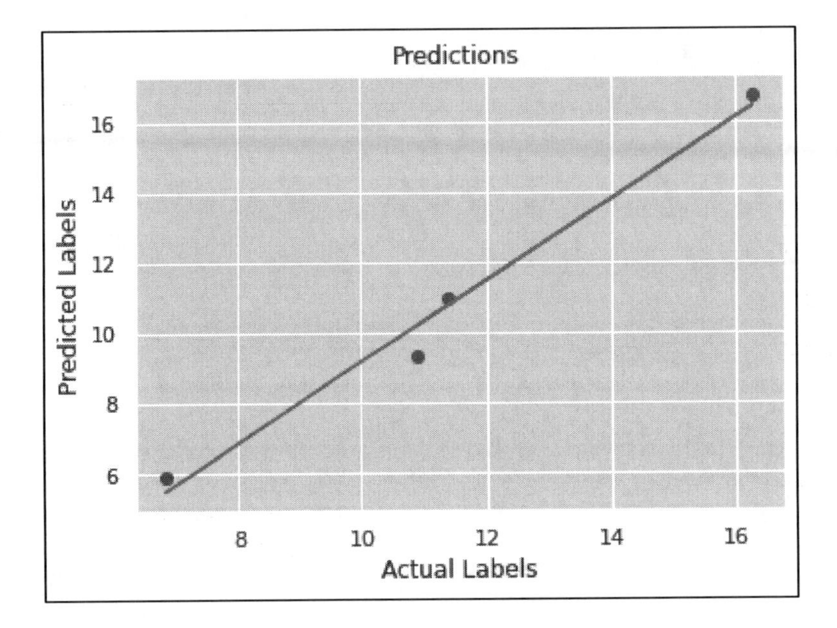


```
r2_score (y_test1, y_pred1)
```

Output:
0.9191 or 91.91%

Step 7: Generate Model Equation (y=mx+c)

Generate regression coefficients and develop y=mx+c

```
mx=""
for ifeature in range(len(X1.columns)):
  if regressor1.coef_[ifeature] <0:
     # format & beautify the equation
     mx += " - " + "{:.2f}".format(abs(regressor1.coef_[ifeature])) + " * " + X1.columns[ifeature]
  else:
     if ifeature == 0:
        mx += "{:.2f}".format(regressor1.coef_[ifeature]) + " * " + X1.columns[ifeature]
     else:
        mx += " + " + "{:.2f}".format(regressor1.coef_[ifeature]) + " * " + X1.columns[ifeature]
print(mx)
```

Output:

```
0.95 * Employment in agriculture (% of total employment) (modeled ILO estimate) – 0.08 * Foreign direct investment, net inflows (% of GDP) – 2.21 * Rural population (% of total population) – 0.18 * Agriculture, forestry, and fishing, value added (% of GDP) + 0.07 * Inflation, GDP deflator (annual %) + 0.02 * Official exchange rate (LCU per US$, period average)
```

```
# y=mx+c
if(regressor1.intercept_ <0):
 print("The formula for the " + y1.name + " linear regression line (y=mx+c) is = " + " - {:.2f}".format(abs(regressor1.intercept_)) + " + " + mx )
else:
 print("The formula for the " + y1.name + " linear regression line (y=mx+c) is = " + "{:.2f}".format(regressor1.intercept_) + " + " + mx )
```

Output:

```
The formula for the Prevalence of undernourishment (%) linear regression line (y=mx+c) is = 159.35 + 0.95 * Employment in agriculture (% of total employment) (modeled ILO estimate) – 0.08 * Foreign direct investment, net inflows (% of GDP) – 2.21 * Rural population (% of total population) – 0.18 * Agriculture, forestry, and fishing, value added (% of GDP) + 0.07 * Inflation, GDP deflator (annual %) + 0.02 * Official exchange rate (LCU per US$, period average)
```

Guidance to Policy Makers

Compounding the damage from the COVID-19 pandemic, the Russian invasion of Ukraine has magnified the slowdown in the global economy, which is entering what could become a protracted period of feeble growth and elevated inflation,[375] according to the World Bank's latest Global Economic Prospects report. This raises the risk of stagflation—stagflation happens when economic growth is sluggish while inflation is high—with potentially harmful consequences for middle and low income economies alike [22]. As shown in the Afghanistan and Sri Lanka models, the improvement of food security is dependent upon increased Gross Domestic Product (GDP) and decreased Consumer Prices and the prognosis for the low-income countries is increase of food insecurity.

"The war in Ukraine, lockdowns in China, supply-chain disruptions, and the risk of stagflation are hammering growth. For many countries, recession will be hard to avoid. Markets look forward, so it is urgent to encourage production and avoid trade restrictions. Changes in fiscal, monetary, climate and debt policies are needed to counter capital misallocation and inequality." [22]

David Malpass
World Bank President[376]

Policymakers, moreover, should refrain from distortionary policies such as price controls, subsidies, and export bans, which could worsen the recent increase in commodity prices. Against the challenging backdrop of higher inflation, weaker growth, tighter financial conditions, and limited fiscal policy space, governments will need to reprioritize spending toward targeted relief for vulnerable populations.

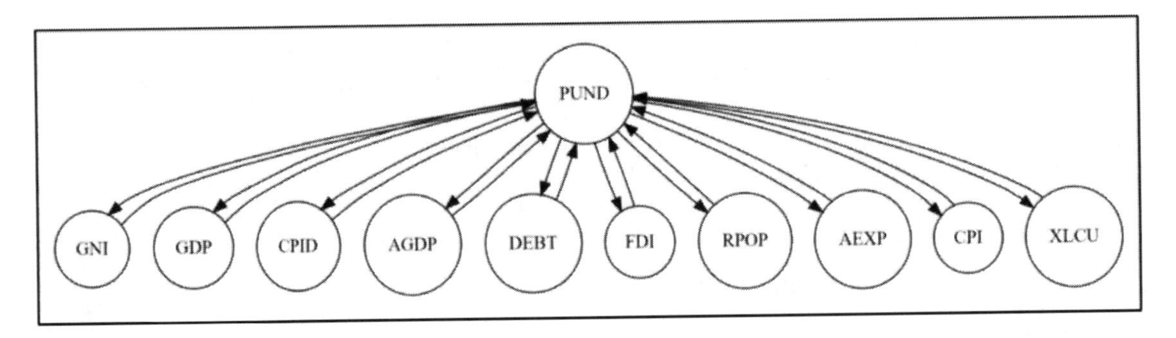

"Developing economies will have to balance the need to ensure fiscal sustainability with the need to mitigate the effects of today's overlapping crises on their poorest citizens. Communicating monetary policy decisions clearly, leveraging credible monetary policy frameworks, and protecting central bank independence can effectively anchor inflation expectations and reduce the amount of policy tightening required to achieve the desired effects on inflation and activity."
Ayhan Kose,
Director of the World Bank's Prospects Group.

The guidance to policy makers is simple and straight forward—encourage demand and enable free monetary& cash flows in the market. Policymakers, moreover, should refrain from distortionary policies such as price controls, subsidies, and export bans, which could worsen the recent increase in commodity prices. Against the challenging backdrop of higher inflation [22], weaker growth, tighter financial conditions, and limited fiscal policy space, governments will need to reprioritize spending toward targeted relief for vulnerable populations. Exclusive aid, social security and welfare programs have to be pursued by low & middle income countries to maintain economic opportunities at rural levels. Exclusive care has to be taken to improve demand at the rural level.

After reading this chapter, you should be able to answer queries on Food Security Drivers & Key Signal Pattern Analysis. Additionally, it introduces the Food Security Bell curve model and applied key macroeconomic and agricultural drivers to analyze prevalence of Undernourishment in Afghanistan and Sri Lanka. Finally, the chapter concludes with the application of additional bell curve food security machine learning models on commodity prices.

References

[1] Bitew, F., Sparks, C. and Nyarko, S. 2022. Machine learning algorithms for predicting undernutrition among under-five children in Ethiopia. Public Health Nutrition 25(2): 269–280. doi:10.1017/S1368980021004262.

[2] Jasmine Ng, Food Inflation Risks Flare on India's Wheat Export Curbs, May 20, 2022, 4:00 AM PDT, https://www.bloomberg.com/news/newsletters/2022-05-20/supply-chain-latest-india-wheat-inflation, Access Date: May 20, 2022.

[3] Manavi Kapur, India's wheat export ban is another reality check for its lofty soft power goals, Published May 16, 2022, https://qz.com/india/2165965/indias-wheat-export-ban-sends-global-prices-soaring/?utm_source=YPL, Access Date: May 20, 2022.

[4] Pratik Parija. Wheat Surges Amid Fears of Shortages as India Restricts Exports, May 14, 2022, 11:01 PM PDTUpdated onMay 16, 2022, 8:27 AM PDT, https://www.bloomberg.com/news/articles/2022-05-15/world-s-food-problems-piling-up-as-india-restricts-wheat-exports, Access Date: May 20, 2022.

[5] Eko Listiyorini. Indonesia Lifts Palm Oil Export Ban in Relief to Global Market, May 19, 2022, 3:07 AM PDT Updated on May 19, 2022, 3:45 AM PDT, https://www.bloomberg.com/news/articles/2022-05-19/indonesia-lifts-palm-oil-export-ban-in-relief-to-global-market, Access Date: May 20, 2022.

[6] Nandita Roy. Commodity Markets: Evolution, Challenges, and Policies, https://www.worldbank.org/en/research/publication/commodity-markets, Access Date: May 20, 2022.

[7] USDA. Food Security Assessment. USDA Economic Research Service. Situation and Outlook series GFA-11 Washington DC, 1999, https://www.ers.usda.gov/publications/pub-details/?pubid=37089, Access Date: June 11, 2022.

[8] Sami Zaptia. Rising Libyan food prices on the eve of Ramadan: caused by Ukraine war and other external factors or simple profiteering? March 17, 2022, https://www.libyaherald.com/2022/03/rising-libyan-food-prices-on-the-eve-of-ramadan-caused-by-ukraine-war-and-other-external-factors-or-simple-profiteering/, Access Date: June 14, 2022.

[9] Gary Keough. Evaluation of Time Series Model Forecasts for the Minnesot Awisconsin Milk Price, April 1991, https://www.nass.usda.gov/Education_and_Outreach/Reports,_Presentations_and_Conferences/Survey_Reports/Evaluation%20of%20Time%20Series%20Model%20Forecasts%20for%20Minnesota-Wisconsin%20Milk%20Price.pdf, Access Date: May 29, 2022.

[10] Megan Durisin. World Hunger 'Exploding' After 25% Spike Before Ukraine War, May 4, 2022, 1:04 AM PDTUpdated onMay 4, 2022, 7:43 AM PDT, https://www.bloomberg.com/news/articles/2022-05-04/world-hunger-spiked-before-ukraine-war-and-will-likely-get-worse, Access Date: May 20, 2022.

[11] The World Bank. Afghanistan Country Profile, Last Updated: Apr 13, 2022, https://www.worldbank.org/en/country/afghanistan/overview#1,Access Date: May 27, 2022.

[12] FAO. Afghanistan: To avert a catastrophe, agricultural assistance is urgently needed, 19/11/2021, https://www.fao.org/newsroom/detail/afghanistan--agricultural-assistance-farmers-drought/en, Access Date: May 28, 2022.

[13] Jerry E. Pacturan. Afghanistan, 2022, https://www.ifad.org/en/web/operations/w/country/afghanistan, Access Date: May 27, 2022.

[14] Krishna Prasad Vadrevu (Editor), Thuy Le Toan (Editor), Shibendu Shankar Ray (Editor), and Chris Justice (Editor), Remote Sensing of Agriculture and Land Cover/Land Use Changes in South and Southeast Asian Countries, Publisher : Springer; 1st ed. 2022 edition (March 29, 2022), ISBN-13 : 978-3030923648.

[15] Anna D'Souza, Rising Food Prices and Declining Food Security. Evidence from Afghanistan, September 01, 2011, https://www.ers.usda.gov/amber-waves/2011/september/afghanistan-food-security/, Access Date: May 27, 2022.

[16] Rafay, Abdul, Naz, Farah and Rubab, Saman. Causal Relationship between Macroeconomic Variables: Evidence from Developing Economy (December 23, 2013). Journal of Contemporary Issues in Business Research, Vol. 3, No. 2, 88-99, 2014, Available at SSRN: https://ssrn.com/abstract=2371263 or http://dx.doi.org/10.2139/ssrn.2371263.

[17] Mark Nord, Carlo Cafiero and Sara Viviani. Methods for estimating comparable prevalence rates of food insecurity experienced by adults in 147 countries and areas, 2016, https://iopscience.iop.org/article/10.1088/1742-6596/772/1/012060/pdf, Access Date: April 25, 2022.

[18] Ayeshea Perera. Sri Lanka: Why is the country in an economic crisis? May 20, 2022, https://www.bbc.com/news/world-61028138, Access Date: May 20, 2022.

[19] GIEWS - Global Information and Early Warning System, Country Briefs - Sri Lanka, Publish Date: 03-March-2022, https://www.fao.org/giews/countrybrief/country.jsp?lang=en&code=LKA, Access Date: May 20, 2022.

[20] U.S. Trade. Sri Lanka - Country Commercial Guide, Publish DateL 2021-09-28, https://www.trade.gov/country-commercial-guides/sri-lanka-agricultural-sector, Access Date: May 20, 2022.

[21] Jonathan D. Church. Comparing the Consumer Price Index with the gross domestic product price index and gross domestic product implicit price deflator. Monthly Labor Review, U.S. Bureau of Labor Statistics, March 2016, https://doi.org/10.21916/mlr.2016.13, Access Date: June 06, 2022.

[22] Nandita Roy and David Young. Stagflation Risk Rises Amid Sharp Slowdown in Growth, June 07, 2022, https://www.worldbank.org/en/news/press-release/2022/06/07/stagflation-risk-rises-amid-sharp-slowdown-in-growth-energy-markets, Access Date: June 10, 2022.

[289] "Apocalyptic" food prices will be disastrous for world's poor, says Bank governor-https://www.theguardian.com/business/2022/may/16/apocalyptic-food-prices-will-be-disastrous-for-worlds-poor-says-bank-governor

[290] Food Inflation Risks Flare on India's Wheat Export Curbs - https://www.bloomberg.com/news/newsletters/2022-05-20/supply-chain-latest-india-wheat-inflation

[291] "If Everyone Starts...": G7 Criticizes India's Move To Stop Wheat Exports - https://www.ndtv.com/india-news/g7-criticises-indias-decision-to-stop-wheat-exports-2975981

[292] Wheat Surges Amid Fears of Shortages as India Restricts Exports - https://www.bloomberg.com/news/articles/2022-05-15/world-s-food-problems-piling-up-as-india-restricts-wheat-exports

[293] Indonesia Lifts Palm Oil Export Ban in Relief to Global Market - https://www.bloomberg.com/news/articles/2022-05-19/indonesia-lifts-palm-oil-export-ban-in-relief-to-global-market

[294] UN Food Price Index - https://www.fao.org/worldfoodsituation/foodpricesindex/en/
[295] From Crisis to Inclusive Economic Growth in Sri Lanka - https://blogs.worldbank.org/endpovertyinsouthasia/crisis-inclusive-economic-growth-sri-lanka
[296] Russia-Ukraine crisis poses a serious threat to Egypt – the world's largest wheat importer - https://theconversation.com/russia-ukraine-crisis-poses-a-serious-threat-to-egypt-the-worlds-largest-wheat-importer-179242
[297] Inflation, a rising threat to the poor and vulnerable in Latin America and the Caribbean - https://blogs.worldbank.org/latinamerica/inflation-rising-threat-poor-and-vulnerable-latin-america-and-caribbean
[298] Food Security Assessment. USDA Economic Research Service. Situation and Outlook series GFA-11 Washington DC - https://www.ers.usda.gov/publications/pub-details/?pubid=37089
[299] Rising Libyan food prices on the eve of Ramadan: caused by Ukraine war and other external factors or simple? - profiteering - https://www.libyaherald.com/2022/03/rising-libyan-food-prices-on-the-eve-of-ramadan-caused-by-ukraine-war-and-other-external-factors-or-simple-profiteering/
[300] Food Catastrophe - https://www.economist.com/leaders/2022/05/19/the-coming-food-catastrophe
[301] Afghanistan - https://www.wfp.org/countries/afghanistan
[302] World Bank - https://www.worldbank.org/en/country/afghanistan/overview#1
[303] World Hunger 'Exploding' After 25% Spike Before Ukraine War - https://www.bloomberg.com/news/articles/2022-05-04/world-hunger-spiked-before-ukraine-war-and-will-likely-get-worse
[304] Afghanistan - https://www.unicef.org/rosa/media/5491/file/Afghanistan
[305] Afghanistan - https://www.ifad.org/en/web/operations/w/country/afghanistan
[306] Histogram Interpretation: Skewed (Non-Normal) Right - https://www.itl.nist.gov/div898/handbook/eda/section3/eda33e6.htm
[307] GNI Per Capita - https://www.macrotrends.net/countries/AFG/afghanistan/gni-per-capita
[308] Afghanistan - https://www.pbs.org/newshour/politics/asia-jan-june11-timeline-afghanistan
[309] Histogram Interpretation: Skewed (Non-Normal) Right - https://www.itl.nist.gov/div898/handbook/eda/section3/eda33e6.htm
[310] Food Security Information and Knowledge Sharing System - https://fsis.sd/SD/EN/FoodSecurity/Indicators/SSI/21/
[311] Per capita food supply variability - http://fsis.sd/SD/EN/FoodSecurity/Indicators/SSI/20/
[312] Rising Food Prices and Declining Food Security: Evidence From Afghanistan - https://www.ers.usda.gov/amber-waves/2011/september/afghanistan-food-security/
[313] Histogram Interpretation: Skewed (Non-Normal) Right - https://www.itl.nist.gov/div898/handbook/eda/section3/eda33e6.htm
[314] GIEWS - https://www.fao.org/giews/countrybrief/country.jsp?lang=en&code=AFG
[315] Histogram Interpretation: Skewed (Non-Normal) Right - https://www.itl.nist.gov/div898/handbook/eda/section3/eda33e6.htm
[316] Agricultural irrigated land - https://www.indexmundi.com/facts/afghanistan/indicator/AG.LND.IRIG.AG.ZS
[317] Afghanistan - Total area equipped for irrigation - https://knoema.com/atlas/Afghanistan/topics/Land-Use/Area/Total-area-equipped-for-irrigation
[318] AQUASTAT - https://www.fao.org/aquastat/fr/geospatial-information/global-maps-irrigated-areas/irrigation-by-country/country/AFG
[319] Histogram Interpretation: Skewed (Non-Normal) Right - https://www.itl.nist.gov/div898/handbook/eda/section3/eda33e6.htm
[320] Causal Relationship between Macroeconomic Variables: Evidence from Developing Economy - https://papers.ssrn.com/sol3/papers.cfm?abstract_id=2371263
[321] Comparing the Consumer Price Index with the gross domestic product price index and gross domestic product implicit price deflator - https://www.bls.gov/opub/mlr/2016/article/comparing-the-cpi-with-the-gdp-price-index-and-gdp-implicit-price-deflator.htm
[322] NY.GNP.PCAP.PP.CD - https://data.worldbank.org/indicator/NY.GNP.PCAP.PP.CD
[323] GDP Growth - https://data.worldbank.org/indicator/NY.GDP.MKTP.KD.ZG
[324] NY.GDP.DEFL.KD.ZG - https://data.worldbank.org/indicator/NY.GDP.DEFL.KD.ZG
[325] Agriculture, forestry, and fishing, value added (% of GDP) - https://data.worldbank.org/indicator/NV.AGR.TOTL.ZS
[326] DT.TDS.DECT.GN.ZS - https://data.worldbank.org/indicator/DT.TDS.DECT.GN.ZS
[327] Foreign Direct Investment - https://data.worldbank.org/indicator/BX.KLT.DINV.WD.GD.ZS
[328] BN.KLT.DINV.CD - https://data.worldbank.org/indicator/BN.KLT.DINV.CD
[329] Rural Population (% of total population) - https://data.worldbank.org/indicator/SP.RUR.TOTL.ZS
[330] BX.KLT.DINV.WD.GD.ZS - https://data.worldbank.org/indicator/BX.KLT.DINV.WD.GD.ZS
[331] BM.KLT.DINV.WD.GD.ZS - https://data.worldbank.org/indicator/BM.KLT.DINV.WD.GD.ZS
[332] SL.AGR.EMPL.ZS - https://data.worldbank.org/indicator/SL.AGR.EMPL.ZS
[333] SL.AGR.EMPL.ZS - https://data.worldbank.org/indicator/SL.AGR.EMPL.ZS
[334] Total reserves (includes gold, current US$) - https://data.worldbank.org/indicator/FI.RES.TOTL.CD
[335] FP.CPI.TOTL.ZG - https://data.worldbank.org/indicator/FP.CPI.TOTL.ZG
[336] Prevalence of undernourishment - https://www.fao.org/sustainable-development-goals/indicators/211/en/
[337] Official exchange rate (LCU per US$, period average) - https://data.worldbank.org/indicator/PA.NUS.FCRF
[338] Per capita food production variability http://data.un.org/Data.aspx?q=economy&d=FAO&f=itemCode%3A21030
[339] Per capita food supply variability - http://data.un.org/Data.aspx?q=Economy&d=FAO&f=itemCode%3A21031
[340] Cereal - http://data.un.org/Data.aspx?q=dependency+ratio&d=FAO&f=itemCode%3A21035
[341] Percent of arable land equipped for irrigation - http://data.un.org/Data.aspx?q=Ethiopia+including+Eritrea&d=FAO&f=itemCode%3A21034%3BcountryCode%3A178%2C238%2C5810

[342] Value of food imports in total merchandise exports - http://data.un.org/Data.aspx?q=Value+of+food+imports+over+total+merchandise+exports+(%25)+(3-year+average)&d=FAO&f=itemCode%3A21033%3BcountryCode%3A5815
[343] GFSI - https://my.corteva.com/GFSI?file=dl_index
[344] Suite of Food Security Indicators - https://www.fao.org/faostat/en/#data/FS
[345] High Income Countries - https://data.worldbank.org/country/XD
[346] GFSI Model - https://my.corteva.com/GFSI?file=dl_index
[347] Suite of Food Security Indicators - https://www.fao.org/faostat/en/#data/FS
[348] Official exchange rate (LCU per US$, period average) - https://data.worldbank.org/indicator/PA.NUS.FCRF?locations=AF
[349] Afghanistan - Official exchange rate - https://knoema.com/atlas/Afghanistan/topics/Economy/Financial-Sector-Exchange-rates/Exchange-rate
[350] World Bank - Historical Classification - http://databank.worldbank.org/data/download/site-content/OGHIST.xls
[351] The World Bank In Sri Lanka - https://www.worldbank.org/en/country/srilanka/overview#4
[352] The World Economic report - https://www.imf.org/en/Publications/WEO/Issues/2022/04/19/world-economic-outlook-april-2022
[353] World Economic Outlook - https://www.imf.org/-/media/Files/Publications/WEO/2022/April/English/text.ashx
[354] Table A7 Consumer Prices - https://www.imf.org/-/media/Files/Publications/WEO/2022/April/English/text.ashx
[355] Sri Lanka: Why is the country in an economic crisis? - https://www.bbc.com/news/world-61028138
[356] Table A12: Current Account Balances - https://www.imf.org/-/media/Files/Publications/WEO/2022/April/English/text.ashx
[357] GIEWS - Global Information and Early Warning System - Sri Lanka - https://www.fao.org/giews/countrybrief/country.jsp?lang=en&code=LKA
[358] Towards Improved Livelihoods and Higher Revenues From Sustainable Fisheries in Sri Lanka - https://www.worldbank.org/en/news/feature/2022/03/02/towards-improved-livelihoods-higher-revenues-from-sustainable-fisheries-srilanka
[359] Sri Lanka - https://www.fao.org/emergencies/countries/detail/en/c/161522/
[360] Histogram Interpretation: Skewed (Non-Normal) Right - https://www.itl.nist.gov/div898/handbook/eda/section3/eda33e6.htm
[361] Sri Lanka - https://www.trade.gov/country-commercial-guides/sri-lanka-agricultural-sector
[362] Histogram Interpretation: Skewed (Non-Normal) Right - https://www.itl.nist.gov/div898/handbook/eda/section3/eda33e6.htm
[363] Percent of arable land equipped for irrigation - http://fsis.sd/SD/EN/FoodSecurity/Indicators/SEI/16/
[364] Histogram Interpretation: Skewed (Non-Normal) Right - https://www.itl.nist.gov/div898/handbook/eda/section3/eda33e6.htm
[365] Sri Lanka - https://www.trade.gov/country-commercial-guides/sri-lanka-agricultural-sector
[366] Histogram Interpretation: Skewed (Non-Normal) Right - https://www.itl.nist.gov/div898/handbook/eda/section3/eda33e6.htm
[367] Official Exchange rate - https://data.worldbank.org/indicator/PA.NUS.FCRF
[368] Sri Lanka – Official Exchange Rate - https://www.indexmundi.com/facts/sri-lanka/official-exchange-rate#:~:text=The%20value%20for%20Official%20exchange,value%20of%204.76%20in%201960.
[369] Histogram Interpretation: Skewed (Non-Normal) Right - https://www.itl.nist.gov/div898/handbook/eda/section3/eda33e6.htm
[370] Comparing the Consumer Price Index with the gross domestic product price index and gross domestic product implicit price deflator - https://www.bls.gov/opub/mlr/2016/article/comparing-the-cpi-with-the-gdp-price-index-and-gdp-implicit-price-deflator.htm
[371] GFSI - https://my.corteva.com/GFSI?file=dl_index
[372] Suite of Food Security Indicators - https://www.fao.org/faostat/en/#data/FS
[373] GFSI Model - https://my.corteva.com/GFSI?file=dl_index
[374] Suite of Food Security Indicators - https://www.fao.org/faostat/en/#data/FS
[375] Stagflation Risk Rises Amid Sharp Slowdown in Growth - https://www.worldbank.org/en/news/press-release/2022/06/07/stagflation-risk-rises-amid-sharp-slowdown-in-growth-energy-markets
[376] Stagflation Risk Rises Amid Sharp Slowdown in Growth - https://www.worldbank.org/en/news/press-release/2022/06/07/stagflation-risk-rises-amid-sharp-slowdown-in-growth-energy-markets

Section III: Prevalence of Undernourishment and Severe Food Insecurity in the Population Models

CHAPTER 5

Commodity Terms of Trade and Food Security

This chapter introduces Commodity Terms of Trade & Food Security linkage. As part of the linkage, the chapter details out Mechanics of food inflation. Additionally, it introduces Food Price Inflation vs. General Inflation vs. Food Security. Next it introduces commodity costs & returns models and develops Machine Learning Models for Food Grain Producer Prices & Consumer Prices linkage. Finally, the chapter concludes with the Prophet—Food Grain Producer Prices & Consumer Prices & the role of food security as national security and the role of trade in food security.

The link between commodity prices and macroeconomic performance has been hotly debated in the literature, with some studies finding that commodity booms raise growth while others suggest a "resource curse" that undercuts sustainable growth [1].

Movements in commodity prices affect different countries differently depending on the composition of both their exports and imports; many developing countries export non-fuel primary commodities but import energy. For countries with Low Income and Lower Middle income, spikes in commodity prices mean an increase in prevalence of moderate or severe food insecurity. Booms in commodity prices do not therefore translate directly into terms of trade booms for all commodity exporters and busts for all commodity importers. To explore the country-specific dimension of global commodity price movements, it is useful to consider the commodity terms of trade: the ratio of commodity export prices to commodity imports prices, with each price weighted by the share of the relevant commodity in the country's GDP or total trade. Using commodity terms of trade allows us to define country specific commodity price cycles, and therefore complement the literature describing cycles in specific commodities.

Commodity price indexes are important as they encompass boom-and-bust cycles:

- First, commodity-price booms tend to be larger than commodity-price busts.
- Second, around 1/3 of all booms (busts) are followed by busts (booms) and the larger the boom, the larger the subsequent bust [1].
- Third, median annual growth is nearly 2 percentage points higher during commodity-price booms than during busts.
- Fourth, during both booms and busts, large real appreciations are associated with significantly lower growth.
- Fifth, the larger the pre-boom government deficit, the smaller the growth during the subsequent boom, possibly because larger initial deficits increase the potential (a factor we have modeled in the previous chapter taking into consideration the Current account balance (% of GDP)) for crowding out of private spending.

What drives commodity price booms and busts?

Analysis of the dynamic effects of commodity demand shocks, commodity supply shocks, and inventory or other commodity-specific demand shocks on real commodity prices establishes that commodity demand shocks strongly dominate commodity supply shocks in driving prices over a broad set of commodities and over a broad period while commodity demand shocks have gained importance over time, commodity supply shocks have become less relevant [2]. Nonetheless, with COVID-19, Ukraine war, and high inflation across the world, supply shocks come to have a considerable long-term impact and will have an impact on demand as well.

Commodity prices are major influences on food security. The linkage of food insecurity and commodity prices has been thoroughly studied and analyzed. Understanding the drivers of commodity price booms and busts is of first-order importance for the global economy. A significant portion of incomes and welfare of both commodity-consuming and commodity-producing nations hinges upon these prices [2]. They also vitally affect the distribution of incomes within nations as the ownership of natural resources varies widely. What is more, the long-run drivers of commodity prices also have serious implications for the formation and persistence of both growth-enhancing and growth-detracting institutions. The U.S. Department of Agriculture (USDA) study clearly corroborates the linkage. Household-level determinants of food insecurity, like education levels and household income, are well documented in earlier research. But the effects of national-level economic factors such as inflation and food prices on food security—and the extent to which changes in both household and macroeconomic factors account for year-to-year changes in the national prevalence of food insecurity now clearly known. The extent to which nationally aggregated economic measures such as the unemployment rate can proxy for household-level employment and labor force data in explaining year-to-year changes in the prevalence of food insecurity is also known.[377] In addition, as proposed in the earlier chapter, there are several macroeconomic, agricultural, weather, and policies related key drivers that influence food security. The proposed framework drivers [A1-A20 & M1-M20] look from both demand (economic), supply (agriculture), and linkage points of view and model the food insecurity causes (please see Figure 1). The framework is extensible to append changes in drivers, and to adapt to future needs.

The *unemployment rate, inflation, and relative prices* of food are three major national-level economic measures taken together accounted for 92 percent of the year-to-year variation in the national prevalence of food insecurity. Though this insight was derived from the U.S. national survey it is equally applicable to the Worldwide prevalence of food insecurity [3]. Increase in oil prices is characterized by low growth, high unemployment, and high inflation (also often referred to as periods of stagflation) [4]. It is no wonder that changes in oil prices have been viewed as an important source of economic fluctuations.[378] Increase in oil prices is a breeding ground to increase in food insecurity—the relationship is completely associative.

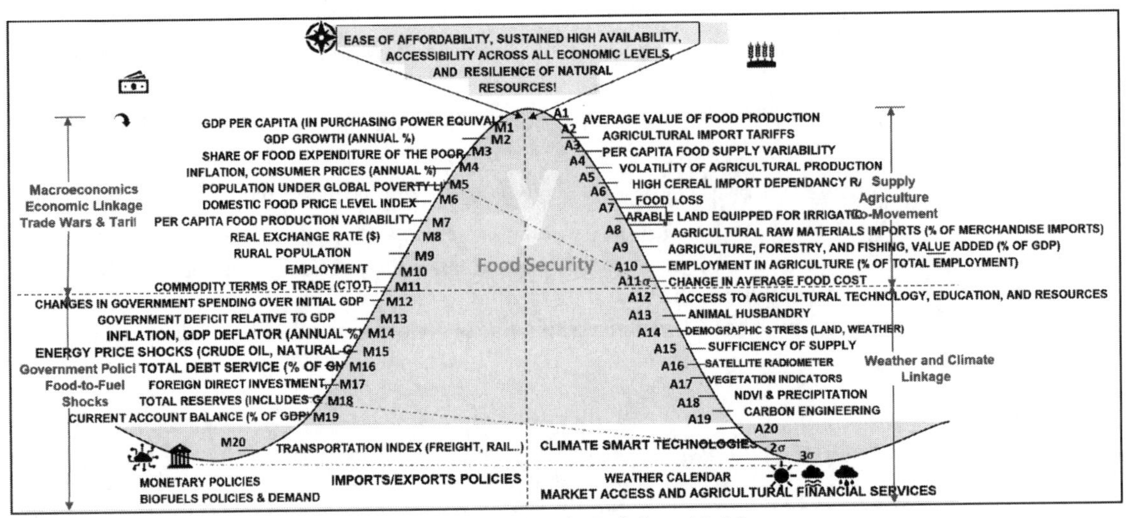

Figure 1: Food Security Bell Curve Model.

The summary of the linkages between the three national level indicators to food insecurity are as under:

- An increase of 1 percentage point in the *unemployment rate* (measured as the highest monthly unemployment rate in the past calendar year) was associated with an increase of 0.5 percentage point in the prevalence of food insecurity.
- An increase of 1 percentage point in *annual inflation*, as measured by the Consumer Price Index (CPI-U), was associated with a 0.5-percentage-point increase in the prevalence of food insecurity.
- An increase of 1 percent in the annual *relative price of food* (i.e., the ratio of food price to the price of all goods and services) was associated with a 0.6-percentage-point increase in the prevalence of food insecurity (please see Figure 2).

The shallower the supply of workers gets the more wages rise[379] and the harder it is to bring inflation back under control.[380] This is true especially if the economy is hot and labor participation is low [5].

As seen from the graph above, a precipitous drop in employment participation due to the COVID-19 pandemic was witnessed. The impact on the commodities could be analyzed later.

The study of commodity-price fluctuations plays an important role in predicting and modeling food security. Commodity demand shocks strongly dominate commodity supply shocks in driving prices over a broad set of commodities over a broad period. While commodity demand shocks have gained importance over time, commodity supply shocks have become less relevant [2]. For this the study of CTOT is crucial. To focus on the effects of commodity-price fluctuations on countries that both export and import primary commodities, we use a country-specific measure of the commodity terms of trade (CTOT) (please see Equation 1) [1].

$$CTOT_{jt} = \Pi_i (P_{it}/MUV_t)^{Xij} / \Pi_i (P_{it}/MUV_t)^{Mij} \quad \text{EQ. 1}$$

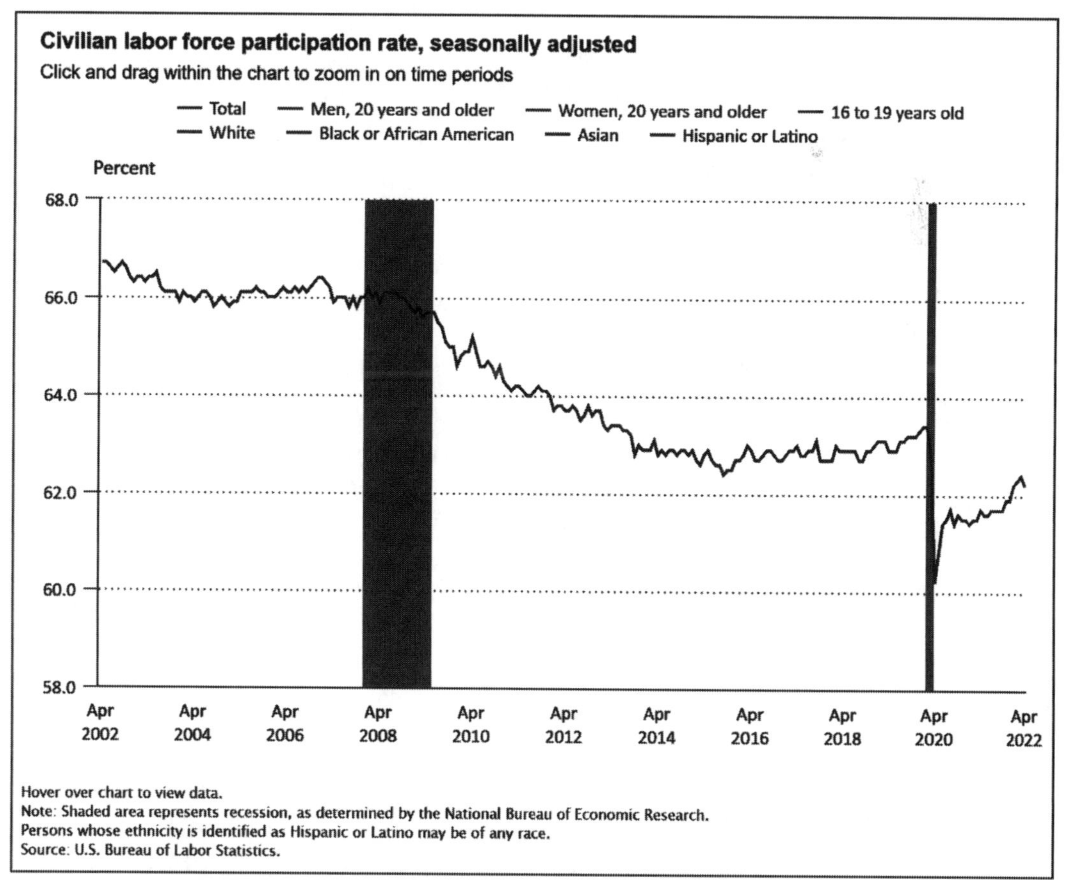

Figure 2: Civilian Labor Force Participation Rate.

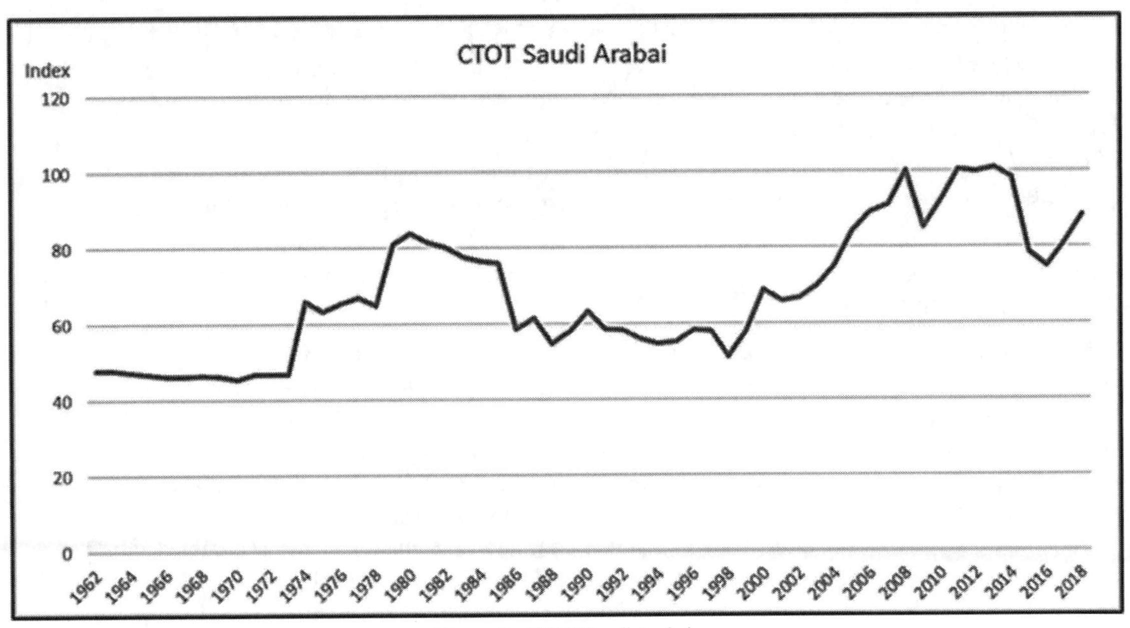

Figure 3: CTOT Saudi Arabai.

where,

- P_{it} is the individual commodity price
- MUV_t is a manufacturing unit value index used as a deflator
- X_{ij} is the share of exports of commodity i in country j's GDP
- M_{ij} is the share of imports of commodity i in country j's GDP

To insulate CTOT from the changes in export and import volumes and only to respond to commodity prices, the weights (that is, the export and import shares) are time-averaged and set to remain constant over time.

The CTOT index is based on the prices of 32 individual commodities: Shrimp; Beef; Lamb; Wheat; Rice; Corn (Maize); Bananas; Sugar; Coffee; Cocoa; Tea; Soybean Meal; Fish Meal; Hides; Soyabean; Natural Rubber; Hard log; Cotton; Wool; Iron Ore; Copper; Nickel; Aluminum; Lead; Zinc; Tin; Soy Oil; Sunflower Oil; Palm Oil; Coconut Oil; Gold; Crude Oil.

CTOT specifically distinguishes economies in two categories, broadly: fuel and non-fuel commodity exporters. The commodity terms of trade have moved in different ways in fuel and non-fuel commodity exporters. The most recent boom in energy prices gave a sizable boost to the commodity terms of trade of fuel exporters. Those of non-fuel commodity exporters on average also rose, but more modestly.

For instance, the CTOT graph of Saudi Arabia shows (please see Figure 3) booms during 2008 & recent ones due to increased prices of gas & the export of gas. We will analyze CTOT on FAO in the emergencies section.

Before modeling the CTOT on the prevalence of food insecurity (as explained in previous chapter), let's see the influence of inflation drivers on the overall fundamental aspects of commodity markets.

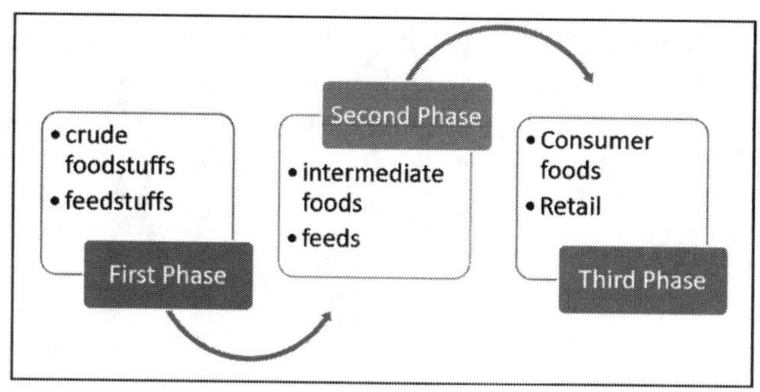

Figure 4: Food Inflation Transmission.

Mechanics of Food Inflation

Prices for agricultural commodities such as feed grains are inherently volatile because they are susceptible to both supply and demand shocks.[381] Commodity supply shocks, which can be interpreted as a disruption in the physical production of a particular commodity through natural disasters, cartel action, strikes, weather, or disease related shortcomings, for example, only affect world real GDP temporarily [2]. Commodity Demand shocks represent an unexpected expansion in global GDP, e.g., periods of rapid industrialization have long-run effects not only on global GD but also on the production of individual commodities. Related, but nevertheless demand shocks could emerge in the form of unexpected purchases by foreign buyers, for example, the unexpected purchase of large amounts of American wheat in the 1970s by the Soviet Union [6,7]. The long-term effect is that an increase in price due to a shift in commodity demand for all commodities triggers technological advances and investment in new production capacities [2] (e.g., new discoveries of mineral deposits, expansion of arable land) to augment future demands.

Any large change in agricultural prices can have a significant impact on the Producer Price Index (PPI) for crude foodstuffs and feedstuffs (both for human and animal husbandry), the first of the PPI's three stages-of-processing indexes for foods. The impact of the price shock can then pass from crude foodstuffs and feedstuffs on to intermediate foods and feeds (warehouses, cooperatives, and others), and then to finished consumer foods (individuals, families, and countries), along the PPI's stage-of-processing model. So higher input costs will have a global impact, especially if the country is a major commodity producer and prime exporter.

Nonetheless, as price shocks pass from one stage of processing to the next, the amplitude of the shocks tends to diminish somewhat at each stage of processing.

Because feed grains are inputs to so many food products, higher feed grain prices can cause inflation to spread throughout almost the entire food industry. Higher wheat costs can cause higher flour prices, which in turn can cause higher bread, pasta, and cereal prices. Higher corn and soyabean costs can cause higher prices for animal feeds, cooking oil, and margarine. By affecting animal feed prices, higher soyabean and corn prices can also have very significant consequences for the meat, poultry, egg, and dairy markets. A Net-net, increase in producer price index is a leading indicator of inflation. The Producer Price Index varies at the national level.

Food inflation is generally lower than general inflation (please see Figure 4). Nonetheless, we have seen historically food inflation exceeds general consumer inflation. The all-items Consumer Price Index (CPI) measures overall price changes across seven major household spending categories: housing, transportation, food, medical care, apparel, education and communication, and recreation.

Producer Price Index (PPI)—one of the oldest economic time series compiled by the federal government

The Producer Price Index (PPI) is the nation's primary measure of price changes in the domestic supply chain, allowing us to monitor how price increases or decreases are passed through from producers to consumers. This Principal Federal Economic Indicator marked its 125th anniversary on March 3, 2016[382]—one of the oldest economic time series compiled by the federal government. The index, known as the Wholesale Price Index (WPI) until 1978, was established as part of a U.S. Senate resolution on March 3, 1891, the last day of the last session of the 51st U.S. Congress.2 This Congress was famously known as the "Billion-Dollar Congress", because of its expensive initiatives, such as Naval expansion the and creating pensions for families of military members who served in the Civil War.

A poster used during World War I to indicate compliance with price controls. Exceptions to the typical wartime inflation trend occurred in the WPIs for fuels, metals, and chemicals as a result of the introduction of government price controls.
Source: U.S. National Archives and Records Administration.

Looking Back at Historical Prices

Since the beginning of the 20th century, there have been several periods of dramatic crop price increases in the United States, including those experienced during the two World Wars. Two periods of rising agricultural prices are of particular interest, the early 1970s and the mid-1990s. Both periods saw record-breaking prices of at least two of the three principal field crops—wheat, corn, and soybeans—and the price increases were sustained for two or more consecutive years. Each period was followed by declines in prices as the conditions that prompted the rapid increase in prices were reversed [8].

Before the great recession of 2008: wheat, corn, and soybean prices[383] began rising rapidly in 1971. Prices peaked and reached record highs in 1974 and then declined (please see Figure 5), settling at a higher level than during the 1960s. Prices for most crops again started to climb slowly in 1990 and escalated rapidly beginning in 1994, peaking in 1995 (corn and wheat) and 1996 (soybeans) before declining sharply. While the increases in this period were not as dramatic as those in the 1970s, corn and wheat prices reached record levels.

After the great recession in 2008: the prices of commodities have become very volatile and show swings especially due to energy shocks. The COIV-19 pandemic, and macroeconomic factors including tariffs and trade wars including changes in supply and demand conditions put pressure on agricultural crop prices (please see Table 1 & Figure 6):

While history provides some insights into current and future economic phenomena, the past does not necessarily predict the future, nor does it fully explain events occurring in the markets today. The current financial and economic structure in the agricultural sector is different from that in the past and policy options

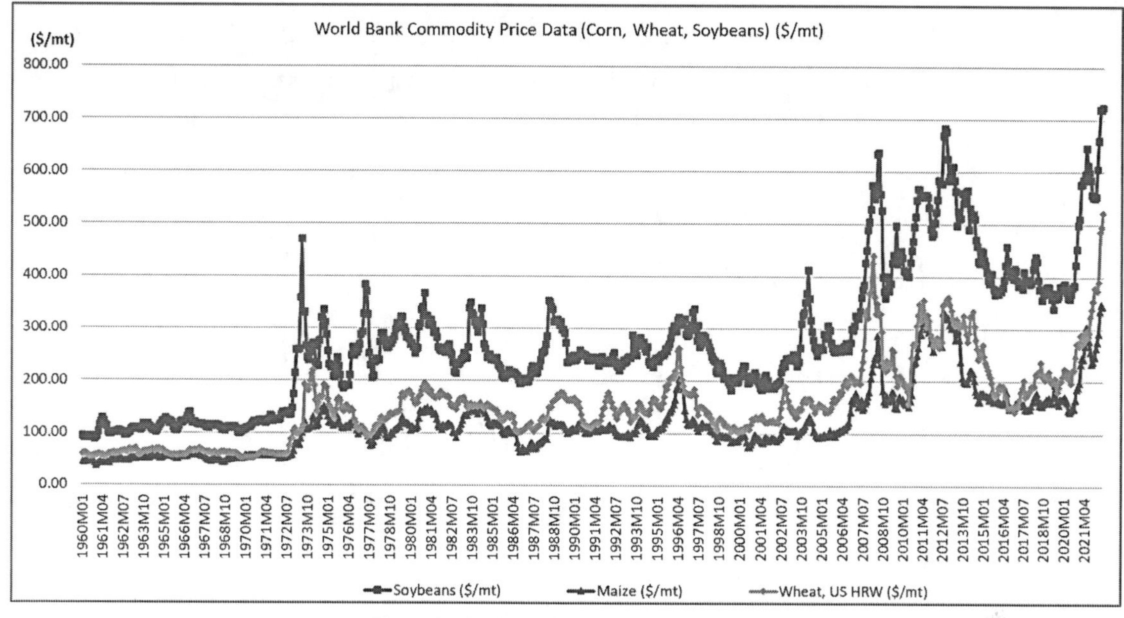

Figure 5: World Bank Commodity Price Data.

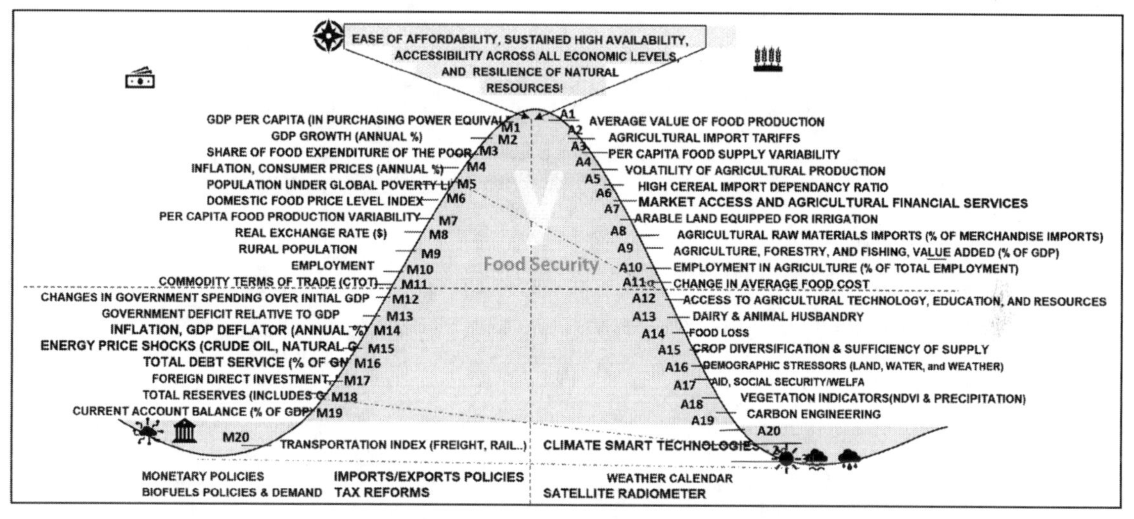

Figure 6: Food Security Model.

and actions have changed as well. Nonetheless, future global income growth and policy developments will have a substantial impact on the demand for agricultural commodities. Although movements in the value of the dollar will influence the demand for U.S. agricultural exports, it is expected that food demand growth will resume and stimulate gains in global agricultural trade as the world economy recovers.

Food demand in developing economies will likely accelerate since incomes in these countries are far from levels where food demand becomes saturated. Additionally, developing countries, which accounted for over 80 percent of the global population in 2007, will continue to experience large population gains along with increased urbanization and expansion of the middle class; in addition, populations in developing countries tend to be younger than those in developed countries, further supporting the potential for increased food demand and sustained growth in export demand [9].

Table 1: Demand Drivers.

Contributing factor [ML Drivers] (please see Figure 6)	1970s	1990s	2008–2020
Long run			
Demand			
Export demand growth	✓	✓	✓
Due to food demand growth [**M1, M2, M9A9, A10**]	NA	✓	✓
Due to population growth [**M9**]	NA	✓	✓
New use/innovation: biofuels	NA	✓	✓
Supply			
Slow production growth	✓	✓	✓
Declining R&D investment [**A12**]	NA	✓	✓
Land retirement [**A7**]	NA	✓	✓
Data & Insights	NA	NA	✓
Machine Learning & AI [**A20**]			✓
Climate-Smart Sensor Technologies [**A20**]	NA	NA	✓
Short run			
Demand			
Government food policies [**A17**]		✓	✓
Supply			
Government food policies (Tariffs & Trade Wars)	✓	✓	✓
Weather-induced crop losses/failure [**A16, A18, A19**]	✓	✓	✓
Pandemic & Shocks [**M15**]	NA	NA	✓
Regional Conflicts & Wars (SSP5)	✓	✓	✓
Macroeconomic Factors			
Economic growth [**M1 & M2**]			
Depreciation of U.S. dollar [**M8**]			
Rising oil prices [**M15**]	✓	✓	✓
Accumulation of petrodollars/foreign reserves [**M18**]	✓	NA	✓
Futures market/speculation [**A6**]	✓	✓	✓
Inflation [M4, **A11**]	✓	✓	✓
Financial crisis [**M15**]	NA	✓	✓

Food Prices Inflation vs. General Inflation vs. Food Security[384]

In some of the past 50 years, increases in food prices have outpaced general inflation; in other years, they have not. From 1986 through 1990, food price inflation was higher than economy-wide inflation (a fact we can witness seeing the month over month change in both food inflation and economy wide inflation indexes) in 4 out of 5 years. For example, in 1989, food prices rose 5.8 percent over 1988 price levels due to large increases in prices of meats, eggs, and fresh fruit and vegetables. Economy-wide inflation (please see Figure 7) was 4.8 percent that year [6,10].

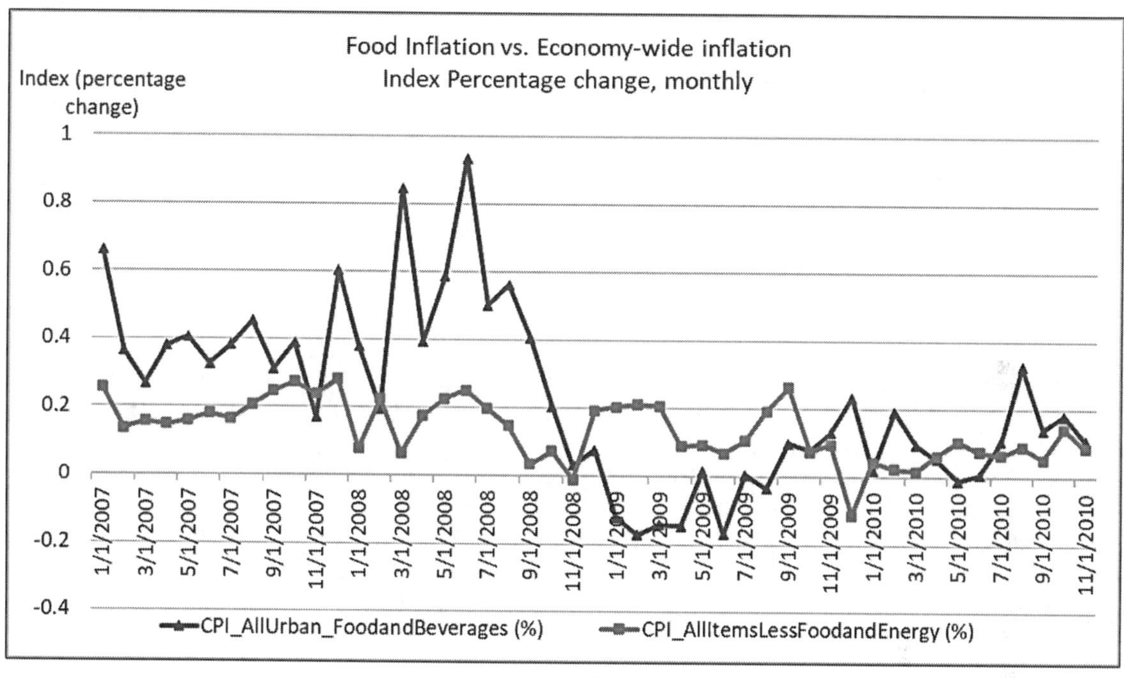

Figure 7: Food Inflation.

However, throughout much of the 1990s and early 2000s, food price inflation was lower than economy-wide inflation, including 5 consecutive years from 1991 to 1995. Rising prices for medical care and housing helped push up the all-items CPI during those years [10]. Annual price increases for these two categories in 1991–95 ranged from 4.5 to 8.7 (CPILFESL[385] index changed from 138.5–163) percent for medical care and from 2.5 to 4.0 percent for housing.

More recently, food price inflation has outpaced economy-wide inflation in 7 of the last 9 years (please see Figure 8). While economy-wide inflation rose 2.8 percent in 2007 and 3.8 percent in 2008 (CPILFESL index changed from 2009 to 2019), food prices rose 4.0 percent in 2007 and 5.5 percent in 2008 (CPIFABSL[386] index change) due to a rapid increase in farm-level prices for rice, grains, and oilseeds.

Campbell Soup Says Consumers Are Curbing Purchases Due to Higher Prices

Food makers' costs are rising rapidly as pandemic-driven supply-chain challenges are exacerbated by the war in Ukraine. Suppliers of meat and packaged foods have said rising costs for grain, fuel and other inputs are denting their bottom lines. Higher prices have helped improve profits for many food companies. Soup maker's sales volumes[387] were also hit by labor and material shortages!

"We've got to be very prudent about it, where we may be seeing elasticities begin creeping up, we obviously have to be a lot more careful there" [11].

Mark Clouse
Campbell Chief Executive

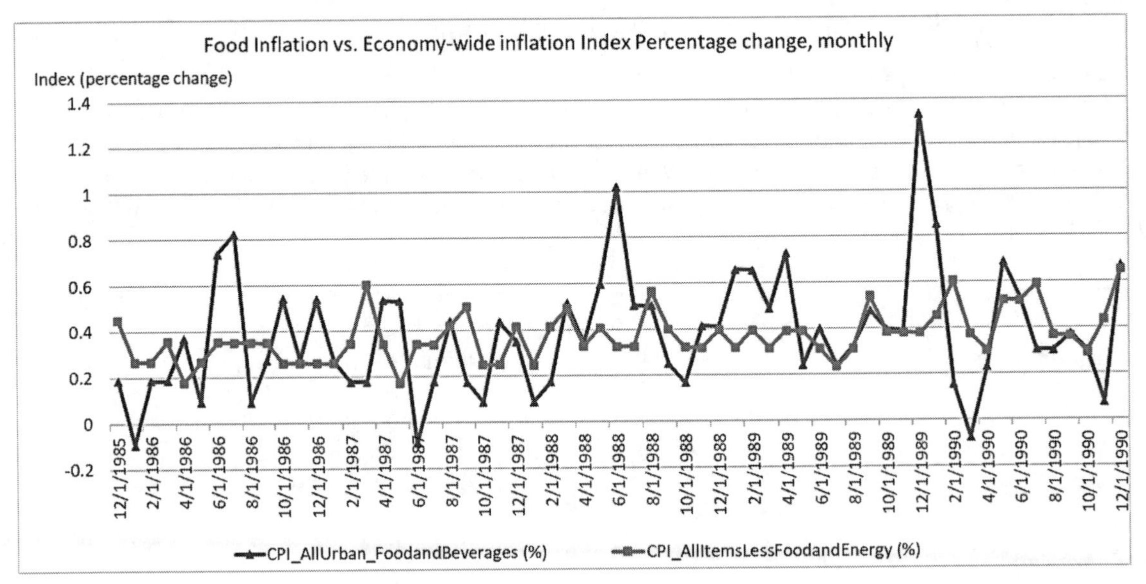

Figure 8: Food Inflation vs. Economy wide inflation.

The effects of food price inflation can be seen on the increase in food insecurity that has increased significantly in 2008 (to 14.6 percent), and remained essentially unchanged (that is, the difference was not statistically significant) at that level in 2009 and 2010. Comparatively, food insecurity was in cumulative decline mode from 2011 (14.9 percent) to 2014 (14.0 percent) (as can be seen food inflation was lower than economic-wide inflation) was statistically significant (please see Figure 9).

In a nutshell, there is a direct relationship between producer food prices and overall economy wide inflation. Using a simple econometric model, we found some evidence of a statistical relationship between feed grain prices and consumer food prices. The tests suggested that we would expect a 100-point, 1-month

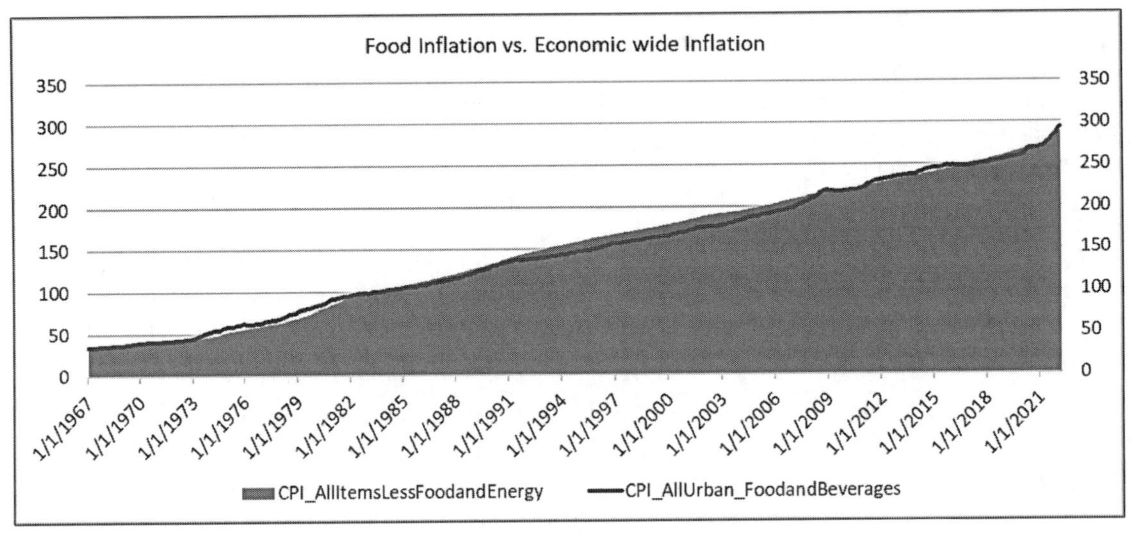

Figure 9: Food Inflation vs. Economic wide inflation.

increase in the feed grain index to be associated with a 4.1-point cumulative increase in the CPI-U for food and beverages over a 12-month period (please see EQ 2). The tests also suggested that, in the long run, a 100-point increase in the feed grain index should be associated with an 11.4-point increase in the CPI-U for food and beverages, and that an increase in the CPI-U for food and beverages tends to follow an increase in the PPI for feed grain. Hence, the tests were wholly supportive of the theory that feed grain shocks at the producer level can cause food inflation to rise relative to core inflation at the consumer level [6] (please see Equation 2).

$$\text{CPI for food and beverages} = 3.772 + 0.832\ \text{CPI core} + 0.114\ \text{feed grains} \qquad \text{EQ. 2}$$

Commodity Costs and Returns

Cost and return estimates are reported for the United States and major production regions for corn, soyabeans, wheat, cotton, grain sorghum, rice, peanuts, oats, barley, milk, hogs, and cow-calf (please see Appendix B). The series of commodity cost and return estimates for the U.S. and regions is divided into Recent and Historical estimates (please see Figure 10). Recent estimates date back to the most recent major revision in accounting methods, account format, and regional definitions for each commodity. Historical estimates date back to when the series began. Cost-of-Production Forecasts are also available for major U.S. field crops. Organic Costs and Returns for corn, milk, wheat, and soyabeans are also available.[388] Please find Corn cost-returns.[389]

Corn production costs and returns per planted acre, excluding Government payments

(dollars per planted acre, except where indicated)

U.S. total | Southern Seaboard | Prairie Gateway | Northern Great Plains | Northern Crescent

	Base survey of 2016							
	2021	**2020**	**2019**	**2018**	**2017**	**2016**	**2015**	**2014**
Gross value of production								
Primary product, grain	927.36	642.58	662.59	625.86	598.00	602.07	611.22	601.80
Secondary product, silage	2.34	2.13	2.35	2.25	2.03	1.85	1.38	1.38
Total, gross value of production	929.70	644.71	664.94	628.11	600.03	603.92	612.60	603.18
Operating costs								
Seed	91.42	91.84	93.48	95.96	97.07	98.36	101.62	101.04
Fertilizer [1]	116.57	116.93	115.86	108.97	113.46	126.53	137.33	149.23
Chemicals	31.48	32.63	34.01	34.02	34.77	35.65	27.95	29.20
Custom services [2]	23.49	22.94	22.74	22.48	22.05	22.69	19.04	18.24
Fuel, lube, and electricity	26.96	27.19	32.41	30.71	27.21	24.08	21.28	32.80
Repairs	37.96	35.56	35.13	33.91	32.75	32.20	26.18	26.17
Purchased irrigation water	0.28	0.28	0.29	0.27	0.26	0.26	0.12	0.12
Interest on operating capital	0.10	0.69	3.46	3.39	1.72	0.78	0.28	0.12
Total, operating costs	328.26	328.06	337.38	329.71	329.29	340.55	333.80	356.92
Allocated overhead								
Hired labor	5.64	5.33	5.25	4.85	4.58	4.49	3.28	3.16
Opportunity cost of unpaid labor	32.34	31.01	29.68	28.12	26.40	25.76	25.63	24.75
Capital recovery of machinery and equipment	134.64	126.06	127.40	122.67	120.30	117.96	102.63	100.15

Figure 10: Corn production costs.

Stagflation Risks Rise Amid Sharp Slowdowns in Growth

The war in Ukraine, lockdowns in China, supply-chain disruptions in June 2022, and the risk of stagflation are hammering growth. The Global Economic Prospects[390] report offers the first systematic assessment of how current global economic conditions compare with the stagflation of the 1970s—with a particular emphasis on how stagflation could affect emerging markets and developing economies. The recovery from the stagflation of the 1970s required steep increases in interest rates in major advanced economies, which played a prominent role in triggering a string of financial crises in emerging markets and developing economies.

The current juncture resembles the 1970s in three key aspects: persistent supply-side disturbances fueling inflation, preceded by a protracted period of a highly accommodative monetary policy in major advanced economies, prospects for weakening growth, and vulnerabilities that emerging markets and developing economies face with respect to the monetary policy tightening that will be needed to rein in inflation [12]. However, the ongoing episode also differs from the 1970s in multiple dimensions: the dollar is strong, a sharp contrast with its severe weakness in the 1970s; the percentage increases in commodity prices are smaller; and the balance sheets of major financial institutions are generally strong. More importantly, unlike the 1970s, central banks in advanced economies and many developing economies now have clear mandates for price stability, and, over the past three decades, they have established a credible track record of achieving their inflation targets.

World Economic Survey, 1979–1980[391] observes that the present world economic situation is characterized by a slow pace of economic advancement in most countries, which is expected to weaken further in the coming months, particularly in developed market economies. This is accompanied by high rates of price inflation (which are pervading all economies) and substantial changes in the pattern of current-account balances, occasioned principally by the doubling of the price of oil between the end of 1978 and the early months of 1980.

Interactive Visualization: U.S. Commodity Costs and Returns by Region and by Commodity

A view of commodity costs and returns (please see Figure 11) by region and by commodity can be visualized.[392]

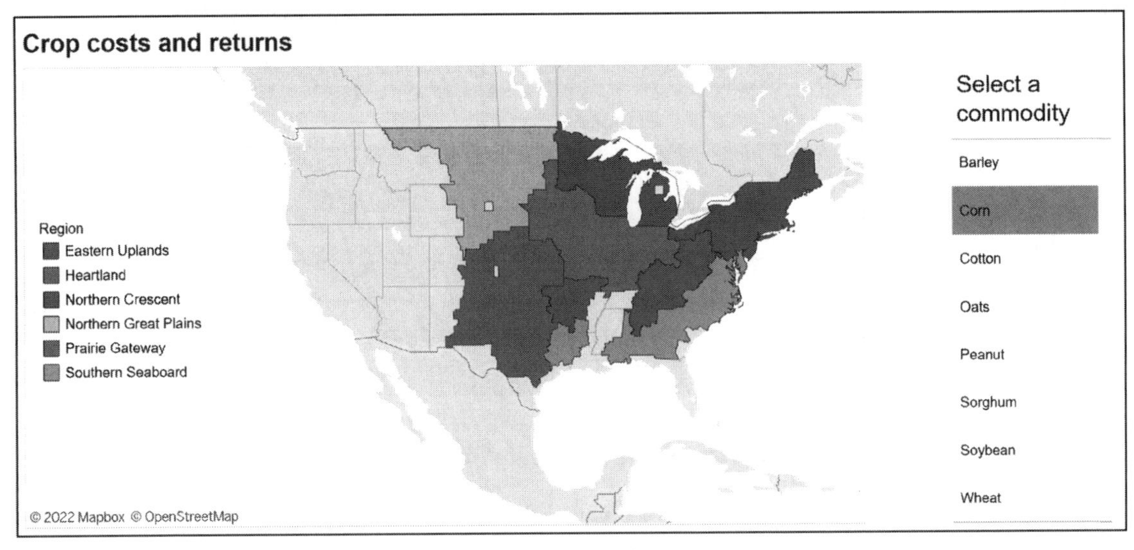

Figure 11: Crop costs and returns.

U.S. average: Corn (please see Figure 12):

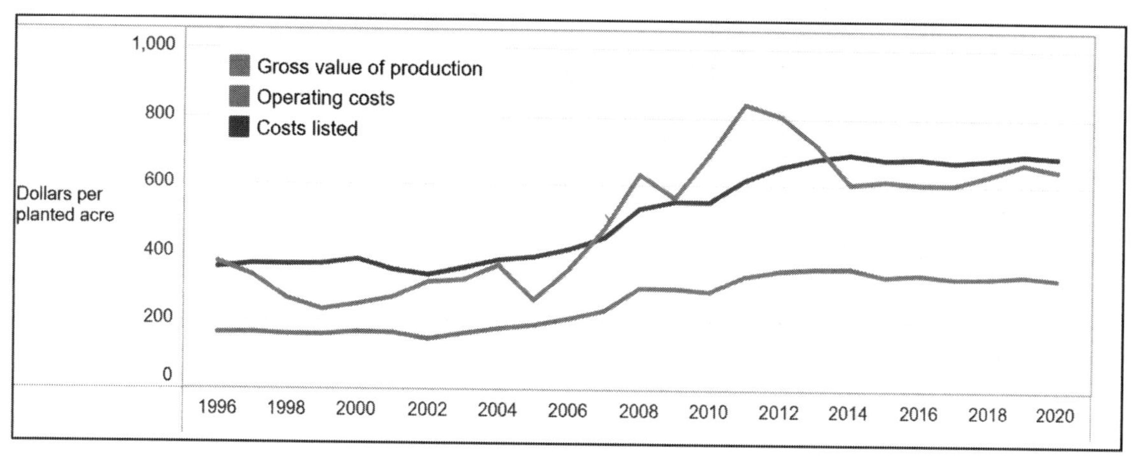

Figure 12: U.S. Corn average.

Machine Learning Model: Food Grain Producer Prices & Consumer Prices

To investigate the effects of feed grain prices on the price of consumer food, a Vector-autoregression model,[393] Linear Regressive Model, and Time Series model have been constructed. The first step in constructing the models was to take a simple average of the monthly index levels of the Producer Price Index (PPI)'s for corn, soyabeans, and wheat.[394] For modeling purposes, we named this the 'feed grain' index and tested for a relationship between it and the CPI for food and beverages. To control general inflation, I have included the CPI for all items less food and energy (the CPI core index) [13,14].

In summary, the model therefore included the three-time series (the feed grain index, CPI for food and beverages, and the CPI for All Items Less Food and Energy) and a constant. The dataset contains monthly index levels over the January 01, 1975–March 01, 2022, period.

To test the stationarity of the data, the model applied Augmented Dickey-Fuller tests with thirty lags. The test failed to reject the null hypothesis of nonstationarity for the index levels of all three series at the 95-percent confidence level. The test rejected the null hypothesis for the first difference of all three series.

To create the model with the most appropriate number of lags, plot_acf was applied. This method found that a model with 30 lags had the best fit. Therefore, it employed a vector autoregression of 30 lagged endogenous variable coefficients.

The results of the model supported the hypothesis that the feed grain index tends to have a positive relationship with the CPI for food and beverages (0.025676), controlling for core inflation. The sum of the coefficients of the 30 lags of the feed grain index on the CPI for food and beverages was + 0.025676. In addition, the cointegrating equation suggests a long-term equilibrating relationship of the form (please see Equation 3):

CPI for food and beverages = – 4.95 + 0.92 * CPI_AllItemsLessFoodandEnergy + 0.11 * GrainIndex EQ. 3

Furthermore, a pair-wise Granger Causality test found feed Grains to Granger-cause CPI: food and beverages at the 95-percent confidence level. (Pair-wise tests also found CPI: food and beverages to Granger-cause CPI: all items less food and energy and found that CPI: all items less food and energy to Granger-cause CPI: food and beverages. No other pair-wise tests suggested Granger Causality.)

The regression analysis, therefore, was wholly supportive of the hypothesis that an increase in feed grain prices tends to increase the price for retail food prices, over some time lag. The sum of the lagged coefficients of feed grains on the CPI for food and beverages was both positive and jointly significant. Moreover, the cointegrating equation established that in the long run, a 1-point increase in feed grains should be associated with an 11point increase in the CPI for food and beverages. Finally, the model found that feed grains do Granger-cause the CPI for food and beverages [6].

Datasets:

- Producer Price Index by Commodity: Farm Products: Corn (WPS012202)[395]
- All Wheat Data[396] (Wheat Data-All Years.xlsx[397])
- Wheat Grain Producer Prices[398]
- Prices Received: Soyabean Prices Received by Month, US[399]
- Feed Grains Database[400]

	VARFeedGrainsAndCPI.ipynb (Jupyter Notebook Code)

Step 1: Load Producer Price Index by Commodity: Farm Products: Corn (WPS012202)

Load Producer Price Index by Commodity: Farm Products: Corn (WPS012202) and DATE index that would be used later to combine datasets.

```
#Corn price data
# https://fred.stlouisfed.org/series/WPS012202
ppiCornDF = pd.read_csv("WPS012202.csv",parse_dates=['DATE'], dayfirst=True)
ppiCornDF = ppiCornDF.set_index('DATE')
ppiCornDF.head(50)
```

Output:

	WPS012202
DATE	
1975-01-01	131.7
1975-02-01	124.2
1975-03-01	118.1
1975-04-01	118.2
1975-05-01	114.6
1975-06-01	112.4
1975-07-01	112.8
1975-08-01	115.6
1975-09-01	114.5
1975-10-01	112.0
1975-11-01	105.8
1975-12-01	104.1
1976-01-01	.
1976-02-01	.
1976-03-01	.

Rename the columns from 'WPS012202':'PPI_Corn' and fix data issues.

```
ppiCornDF= ppiCornDF.rename(columns={'WPS012202':'PPI_Corn'})
ppiCornDF['PPI_Corn'] = ppiCornDF['PPI_Corn'].replace('.',np.nan)
ppiCornDF['PPI_Corn'] = ppiCornDF['PPI_Corn'].astype('float64')
```

Step 2: Load Producer Price Index by Commodity: Farm Products: Wheat (WPU0121)

Load Producer Price Index by Commodity: Farm Products: Wheat (WPU0121)[401] and DATE index that would be used later to combine datasets.

```
# Producer Price Index by Commodity: Farm Products: Wheat (WPU0121)
# https://fred.stlouisfed.org/series/WPU0121
ppiWheatDF = pd.read_csv("WPU0121.csv",parse_dates=['DATE'], dayfirst=True)
ppiWheatDF = ppiWheatDF.set_index('DATE')
ppiWheatDF.head()
```

Output:

DATE	WPU0121
1947-01-01	53.1
1947-02-01	56.5
1947-03-01	68.0
1947-04-01	66.0
1947-05-01	66.7

```
ppiWheatDF= ppiWheatDF.rename(columns={'WPU0121':'PPI_Wheat'})
ppiWheatDF['PPI_Wheat'] = ppiWheatDF['PPI_Wheat'].astype('float64')
ppiWheatDF.info()
```

Output:

```
<class 'pandas.core.frame.DataFrame'>
DatetimeIndex: 903 entries, 1947-01-01 to 2022-03-01
Data columns (total 1 columns):
 #   Column     Non-Null Count  Dtype
---  ------     --------------  -----
 0   PPI_Wheat  903 non-null    float64
dtypes: float64(1)
memory usage: 14.1 KB
```

Step 3: Producer Price Index by Commodity: Farm Products: Soyabeans (WPU01830131)

Load Producer Price Index by Commodity: Farm Products: Wheat (WPU0121)[402] and DATE index that would be used later to combine datasets.

```
# Producer Price Index by Commodity: Farm Products: Soybeans (WPU01830131)
# https://fred.stlouisfed.org/series/WPU01830131
ppiSoyBeansDF = pd.read_csv("WPU01830131.csv",parse_dates=['DATE'], dayfirst=True)
ppiSoyBeansDF = ppiSoyBeansDF.set_index('DATE')
ppiSoyBeansDF.head()
```

Output:

DATE	WPU01830131
1947-01-01	51.8
1947-02-01	53.5
1947-03-01	65.0
1947-04-01	62.7
1947-05-01	48.7

Rename the column: from "WPU01830131" to "'PPI_SoyBeans".

```
ppiSoyBeansDF= ppiSoyBeansDF.rename(columns={'WPU01830131':'PPI_SoyBeans'})
```

Soybeans Index data has data issues for the year:

```
ppiSoyBeansDF['PPI_SoyBeans'] = ppiSoyBeansDF['PPI_SoyBeans'].replace('.',np.nan)
ppiSoyBeansDF['PPI_SoyBeans'] = ppiSoyBeansDF['PPI_SoyBeans'].astype('float64')
```

Step 4: Consumer Price Index for All Urban Consumers: All Items Less Food and Energy in U.S. City Average (CPILFESL)

Load Consumer Price Index for All Urban Consumers: All Items Less Food and Energy in U.S. City Average (CPILFESL)[403] and DATE index that would be used later to combine datasets.

```
# Consumer Price Index for All Urban Consumers: All Items Less Food and Energy in U.S. City Average
(CPILFESL)
# https://fred.stlouisfed.org/series/CPILFESL

cpiU_AllItemsLessEFDF = pd.read_csv("CPILFESL.csv",parse_dates=['DATE'], dayfirst=True)
cpiU_AllItemsLessEFDF = cpiU_AllItemsLessEFDF.rename(columns={"DATE":'Date'})
cpiU_AllItemsLessEFDF = cpiU_AllItemsLessEFDF.set_index('Date')
cpiU_AllItemsLessEFDF.head()
```

Change column name and print data frame.

```
cpiU_AllItemsLessEFDF=
cpiU_AllItemsLessEFDF.rename(columns={'CPILFESL':'CPI_AllItemsLessFoodandEnergy'})
```

Output:

	CPI_AllItemsLessFoodandEnergy
Date	
2021-06-01	277.922
2021-07-01	278.794
2021-08-01	279.306
2021-09-01	280.017
2021-10-01	281.705
2021-11-01	283.179
2021-12-01	284.770
2022-01-01	286.431
2022-02-01	287.878
2022-03-01	288.811

Check for nulls.

```
cpiU_AllItemsLessEFDF.isnull().sum() ## missing values
```

Output:

```
CPI_AllItemsLessFoodandEnergy 0
dtype: int64
```

Step 5: Consumer Price Index for All Urban Consumers: Food and Beverages in U.S. City Average (CPIFABSL)

Consumer Price Index for All Urban Consumers: All Items Less Food and Energy in U.S. City Average (CPIFABSL)[404] and DATE index that would be used later to combine datasets.

```
# Consumer Price Index for All Urban Consumers: Food and Beverages in U.S. City Average (CPIFABSL)
# https://fred.stlouisfed.org/series/CPIFABSL
cpiU_FoodandBeveragesDF = pd.read_csv("CPIFABSL.csv",parse_dates=['DATE'], dayfirst=True)
cpiU_FoodandBeveragesDF = cpiU_FoodandBeveragesDF.set_index('DATE')
cpiU_FoodandBeveragesDF.info()
```

Rename the columns.

```
cpiU_FoodandBeveragesDF=
cpiU_FoodandBeveragesDF.rename(columns={'CPIFABSL':'CPI_AllUrban_FoodandBeverages'})
cpiU_FoodandBeveragesDF['CPI_AllUrban_FoodandBeverages'] = cpiU_FoodandBeveragesDF['CPI_AllUrban_FoodandBeverages'].astype('float64')
```

Step 6: Exploratory analysis

Let's load the data and do some analysis with visualization to have insights to the data. Exploratory data analysis is quite extensive in multivariate time series. I will cover some areas here to get insights to the data. However, it is advisable to conduct all statistical tests to ensure a clear understanding of the data distribution. Prepare the dataset that contains Corn, Wheat, Soy Beans, Food and Beverages CPI, and All items CPI with no food (Economy-wise inflation).

```
dataset =
pd.concat([ppiCornDF,ppiWheatDF,ppiSoyBeansDF,cpiU_AllItemsLessEFDF,cpiU_FoodandBeveragesDF], axis=1)
print('Number of colums in Dataframe : ', len(dataset.columns))
print('Number of rows in Dataframe : ', len(dataset.index))
print(dataset)
```

Output:

```
Number of colums in Dataframe :  5
Number of rows in Dataframe :  903
            PPI_Corn  PPI_Wheat  PPI_SoyBeans  CPI_AllItemsLessFoodandEnergy  \
1947-01-01       NaN     53.100        51.800                            NaN
1947-02-01       NaN     56.500        53.500                            NaN
1947-03-01       NaN     68.000        65.000                            NaN
1947-04-01       NaN     66.000        62.700                            NaN
1947-05-01       NaN     66.700        48.700                            NaN
...              ...        ...           ...                            ...
2021-11-01   235.222    254.482       203.615                        283.179
2021-12-01   242.750    262.467       213.190                        284.770
2022-01-01   246.579    253.978       235.206                        286.431
2022-02-01   251.138    257.358       262.004                        287.878
2022-03-01   287.660    319.999       279.991                        288.811

            CPI_AllUrban_FoodandBeverages
1947-01-01                            NaN
1947-02-01                            NaN
1947-03-01                            NaN
1947-04-01                            NaN
1947-05-01                            NaN
...                                   ...
2021-11-01                        284.676
2021-12-01                        286.018
2022-01-01                        288.436
2022-02-01                        291.325
2022-03-01                        294.119

[903 rows x 5 columns]
```

Let's fix the dates for all the series—January 01, 1975–March 01, 2022.

```
### dataset=dataset.loc['20100101':'20200824']
dataset=dataset.loc['19750101':'20220301']
dataset.head()
```

Output:

	PPI_Corn	PPI_Wheat	PPI_SoyBeans	CPI_AllItemsLessFoodandEnergy	CPI_AllUrban_FoodandBeverages
1975-01-01	131.7	111.4	111.9	52.3	58.8
1975-02-01	124.2	106.0	100.6	52.8	59.0
1975-03-01	118.1	98.4	92.7	53.0	58.9
1975-04-01	118.2	94.7	98.7	53.3	58.8
1975-05-01	114.6	88.4	87.2	53.5	59.0

Check for nulls:

```
dataset.isnull().sum() ## missing values
```

Output:

```
PPI_Corn                        120
PPI_Wheat                         0
PPI_SoyBeans                      0
CPI_AllItemsLessFoodandEnergy     0
CPI_AllUrban_FoodandBeverages     0
dtype: int64
```

Now we found missing values for PPI_Corn, and filled it with previous values. This should be fine as it would maintain the coherence of the data.

```
dataset=dataset.fillna(method='pad')
dataset = dataset.fillna(method = 'bfill')
```

Now, re-run any nulls which are still there.

```
dataset.isnull().sum() ## missing values
```

Output:

```
PPI_Corn                         0
PPI_Wheat                        0
PPI_SoyBeans                     0
CPI_AllItemsLessFoodandEnergy    0
CPI_AllUrban_FoodandBeverages    0
dtype: int64
```

Step 7: Generate Grain Index

Generate Grain index that is based on the corn, wheat, and soyabeans values.

```
dataset['GrainIndex'] = dataset[['PPI_Corn', 'PPI_Wheat','PPI_SoyBeans']].mean(axis=1)
```

Drop corn, wheat, and soyabeans columns.

```
dataset.drop(['PPI_Corn', 'PPI_Wheat','PPI_SoyBeans'], axis=1, inplace=True)
```

The NaN values in the data are filled with data from previous days. After doing some necessary pre-processing, the dataset was cleaned for further analysis.

```
# Plot
fig, axes = plt.subplots(nrows=3, ncols=1, dpi=120, figsize=(10,16))
for i, ax in enumerate(axes.flatten()):
    data = dataset[dataset.columns[i]]
    ax.plot(data, color='red', linewidth=1)
    ax.set_title(dataset.columns[i])
    ax.xaxis.set_ticks_position('none')
    ax.yaxis.set_ticks_position('none')
    ax.spines["top"].set_alpha(0)
    ax.tick_params(labelsize=6)
plt.tight_layout();
```

Output:

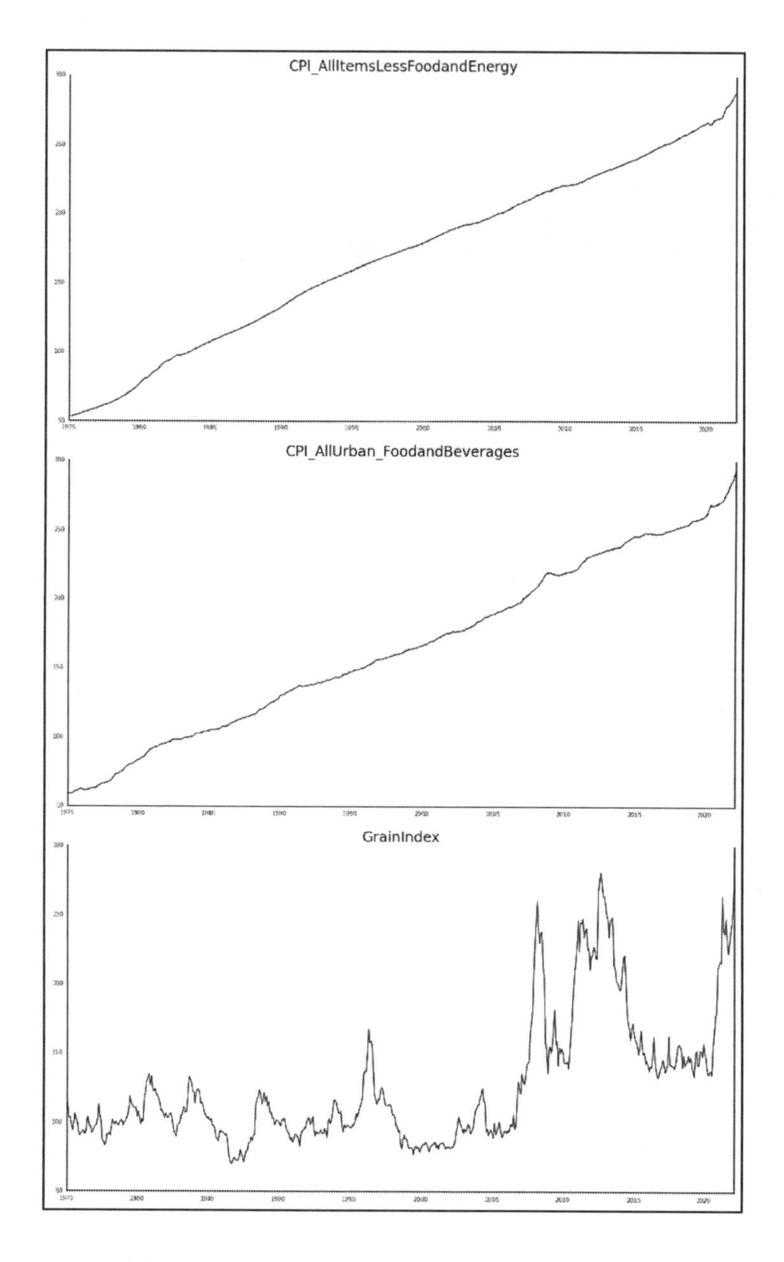

From the plot, we can visibly conclude that, Grain Index contains a unit root with a stochastic trend showing a systematic pattern that is unpredictable.

To extract maximum information from our data, it is important to have a normal or Gaussian distribution of the data. To check for that, we have done a normality test based on the Null and Alternate Hypothesis intuition.

Step 8: Normality Test

To extract maximum information from our data, it is important to have a normal or Gaussian distribution of the data. To check for that, we have done a normality test based on the Null and Alternate Hypothesis intuition. The function stats.normaltest tests the null hypothesis that a sample comes from a normal distribution. It is based on D'Agostino and Pearson's[405,406] test that combines skew and kurtosis to produce an omnibus test of normality.

```
from scipy import stats
CPI_AllItemsLessFoodandEnergy=dataset.CPI_AllItemsLessFoodandEnergy.values
stat,p = stats.normaltest(CPI_AllItemsLessFoodandEnergy)
print("CPI_AllItemsLessFoodandEnergy Statistics = %.3f, p=%.3f" % (stat,p))
alpha = 0.05
if p> alpha:
   print('Data looks Gaussian (fail to reject null hypothesis)')
else:
   print('Data looks non-Gaussian (reject null hypothesis)')
```

Output:

CPI_AllItemsLessFoodandEnergy Statistics = 158.986, p=0.000
Data looks non-Gaussian (reject null hypothesis)

```
from scipy import stats
CPI_AllUrban_FoodandBeverages=dataset.CPI_AllUrban_FoodandBeverages.values
stat,p = stats.normaltest(CPI_AllUrban_FoodandBeverages)
print("CPI_AllUrban_FoodandBeverages Statistics = %.3f, p=%.3f" % (stat,p))
alpha = 0.05
if p> alpha:
   print('Data looks Gaussian (fail to reject null hypothesis)')
else:
   print('Data looks non-Gaussian (reject null hypothesis)')
```

Output:

CPI_AllUrban_FoodandBeverages Statistics = 230.253, p=0.000
Data looks non-Gaussian (reject null hypothesis)

```
from scipy import stats
GrainIndex=dataset.GrainIndex.values
stat,p = stats.normaltest(GrainIndex)
print("GrainIndex Statistics = %.3f, p=%.3f" % (stat,p))
alpha = 0.05
if p> alpha:
   print('Data looks Gaussian (fail to reject null hypothesis)')
else:
   print('Data looks non-Gaussian (reject null hypothesis)')
```

Output:

GrainIndex Statistics = 134.953, p=0.000
Data looks non-Gaussian (reject null hypothesis)

All the variables are normally distributed.

Step 8: Kurtosis and Skewness

The all Economy-wise inflation (CPI_AllItemsLessFoodandEnergy) distribution gives us some intuition about the distribution of our data. The kurtosis of this dataset is –1.05. It means it has Moderate skewness. Generally, a kurtosis value between –1 and –0.5 or 0.5 and 1 is representative of moderate skewness.

```
import matplotlib.pyplot as plt
import scipy.stats as stats
from scipy.stats import kurtosis
dataset.CPI_AllItemsLessFoodandEnergy.plot(kind = 'density')
print('CPI_AllItemsLessFoodandEnergy: Kurtosis of normal distribution:
{}'.format(stats.kurtosis(dataset.CPI_AllItemsLessFoodandEnergy)))
print('CPI_AllItemsLessFoodandEnergy:Skewness of normal distribution:
{}'.format(stats.skew(dataset.CPI_AllItemsLessFoodandEnergy)))
```

Output:

CPI_AllItemsLessFoodandEnergy: Kurtosis of normal distribution: -1.0597134273479736
CPI_AllItemsLessFoodandEnergy:Skewness of normal distribution: -0.1529109836314409

Similarly:

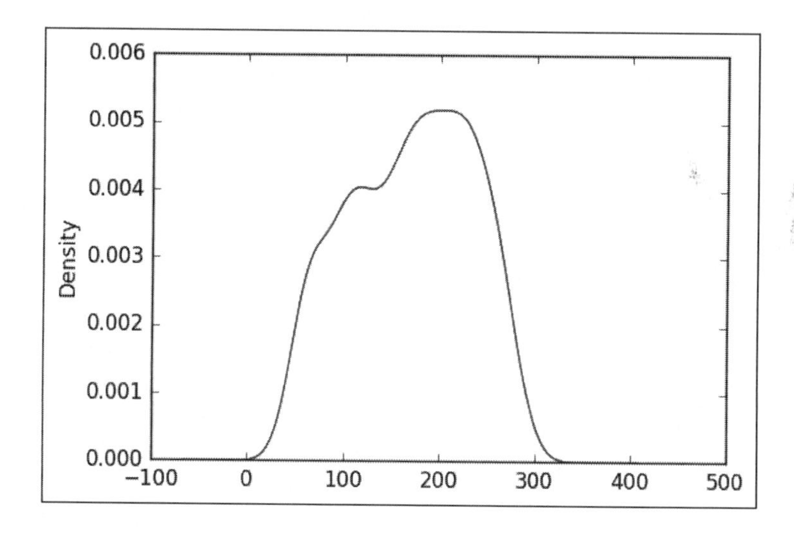

CPI_AllUrban_FoodandBeverages: Kurtosis of normal distribution: –1.1272598168828003 **CPI_AllUrban_ FoodandBeverages**:Skewness of normal distribution: 0.07392880488928787	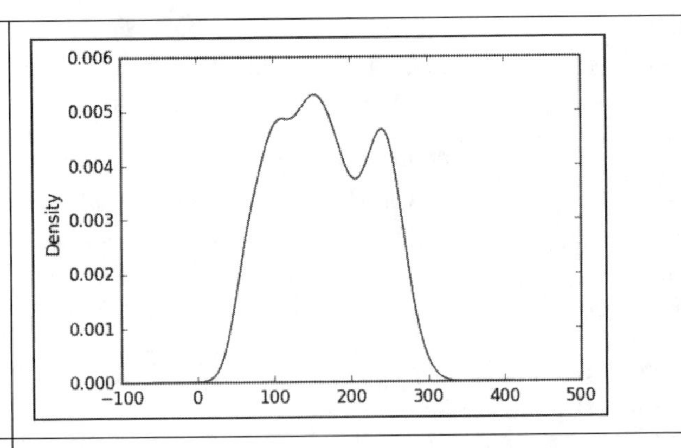
GrainIndex: Kurtosis of normal distribution: 1.3264403453808624 **GrainIndex**:Skewness of normal distribution: 1.4430826436404	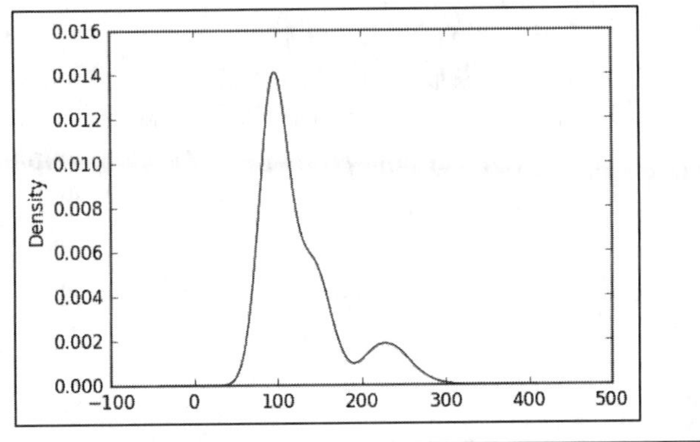

```
plt.figure(figsize=(14,6))
plt.subplot(1,2,1)
dataset['GrainIndex'].hist(bins=50)
plt.title('GrainIndex')
plt.subplot(1,2,2)
stats.probplot(dataset['GrainIndex'], plot=plt);
dataset.GrainIndex.describe().T
```

Output:

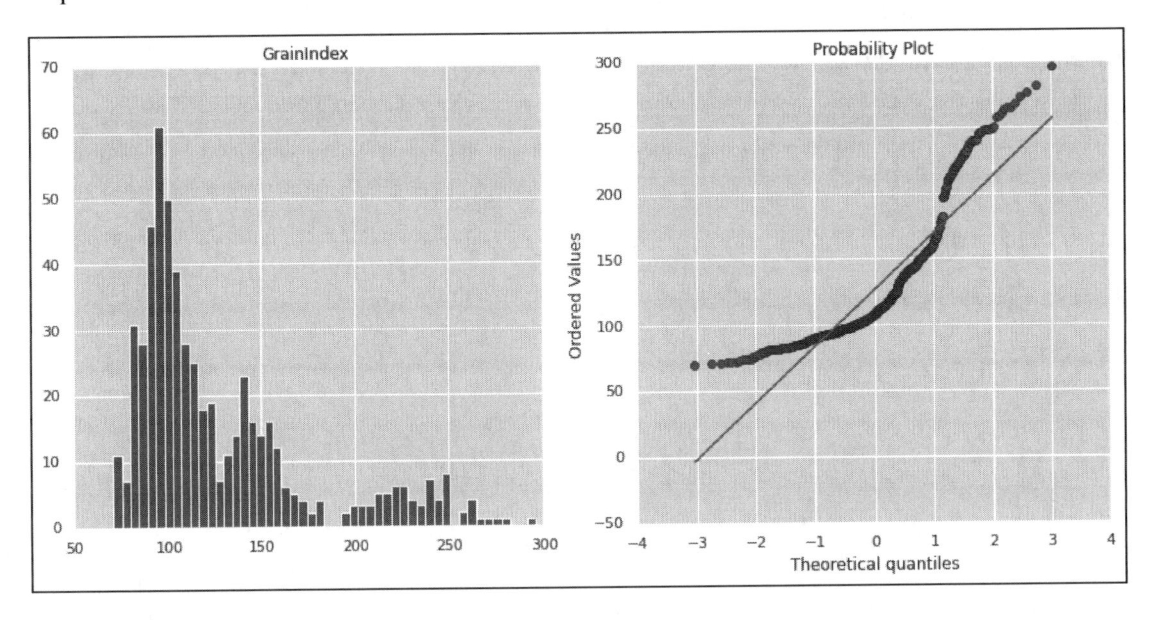

Normal probability plot also shows the data is far from normally distributed.

Step 9: Autocorrelation

Autocorrelation measures the degree of similarity between a time series and a lagged version of itself over successive time intervals. Its sometimes also referred to as "serial correlation" or "lagged correlation" since it measures the relationship between a variable's current values and its historical values. When the autocorrelation in a time series is high, it becomes easy to predict future values by simply referring to past values. Auto-correlation or serial correlation can be a significant problem in analyzing historical data if we do not know how to look for it. Informally, autocorrelation is the similarity between observations as a function of the time lag between them.

```
import pandas as pd
import matplotlib.pyplot as plt
import statsmodels.api as sm
MAX_LAGS=25
import pandas as pd
import matplotlib.pyplot as plt
import statsmodels.api as sm
MAX_LAGS=25
# plots the autocorrelation plots for each stock's price at 150 lags
for i in dataset:
  sm.graphics.tsa.plot_acf(dataset[i])
  plt.title('ACF for %s' % i)
plt.show()
```

Lags: An int or array of lag values, used on the horizontal axis. Uses np.arange(lags) when lags is an int. If not provided, lags=np.arange(len(corr)) autocorrelation of +1 which represents a perfect positive correlation which means, an increase seen in one time series leads to a proportionate increase in the other time series. We need to apply transformation and neutralize this to make the series stationary. It measures linear relationships; even if the autocorrelation is minuscule, there may still be a nonlinear relationship between a time series and a lagged version of itself.

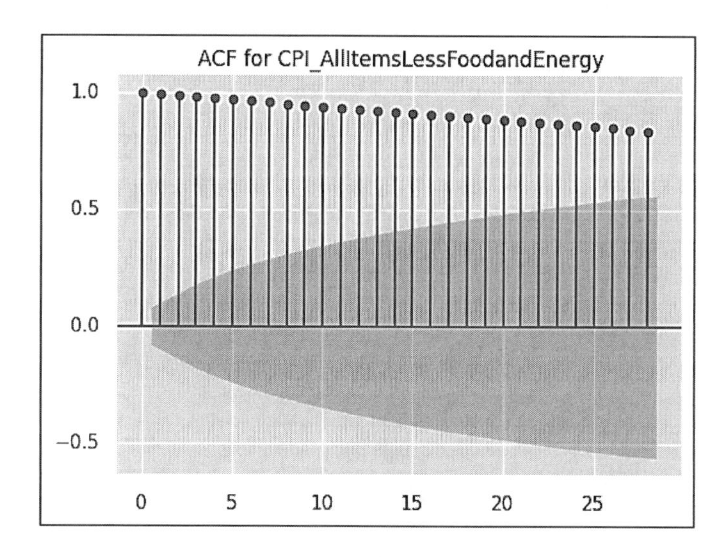

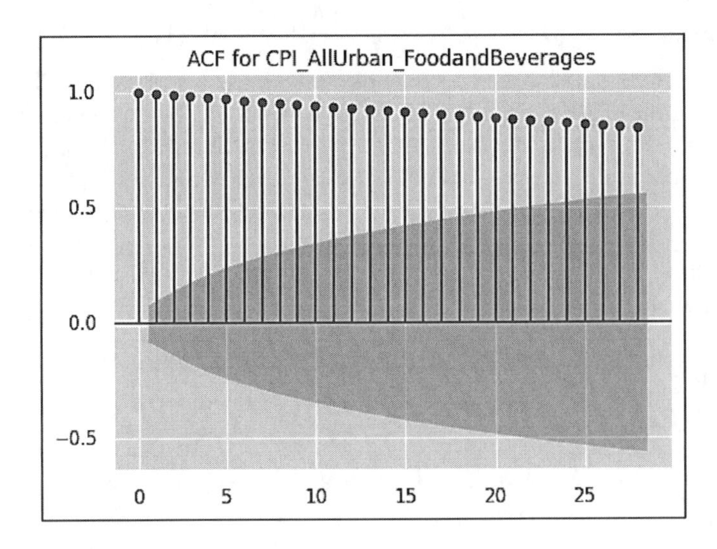

Step 10: Train and Test Data

The VAR model will be fitted on X_train and then used to forecast the next 30 observations. These forecasts will be compared against the actual present in test data.

```
n_obs=30
X_train, X_test = dataset[0:-n_obs], dataset[-n_obs:]
print(X_train.shape, X_test.shape)
```

Output:

(537, 3) (30, 3)

Step 11: Transformation

We apply first differencing on the training set to make all the series stationary. However, this is an iterative process where we after the first differencing, the series may still be non-stationary. We shall have to apply a second difference or log transformation to standardize the series in such cases.

```
transform_data = X_train.diff().dropna()
transform_data.head()
```

Output:

	CPI_AllItemsLessFoodandEnergy	CPI_AllUrban_FoodandBeverages	GrainIndex
1975-02-01	0.5	0.2	-8.066667
1975-03-01	0.2	-0.1	-7.200000
1975-04-01	0.3	-0.1	0.800000
1975-05-01	0.2	0.2	-7.133333
1975-06-01	0.3	0.6	-2.800000

```
transform_data.describe()
```

Output:

	CPI_AllItemsLessFoodandEnergy	CPI_AllUrban_FoodandBeverages	GrainIndex
count	536.000000	536.000000	536.000000
mean	0.396060	0.372743	0.044216
std	0.186536	0.367551	7.424779
min	-0.248000	-0.600000	-45.333333
25%	0.300000	0.100000	-3.200000
50%	0.400000	0.314500	0.133333
75%	0.500000	0.600000	3.033333
max	1.200000	1.997000	48.133333

Step 12: Augmented Dickey-Fuller Unit Root Test - Stationarity check

Stationarity is checked using the Augmented Dickey-Fuller (ADFuller) unit root test. The Augmented Dickey-Fuller (ADFuller) test can be used to test for a unit root in a univariate process in the presence of serial correlation. There are several unit-root tests and the Augmented Dickey-Fuller may be one of the more widely used tests. It uses an autoregressive model and optimizes an information criterion across different multiple lag values.

The null hypothesis of the test states that the time series can be represented by a unit root, that it is not stationary (has some time-dependent structure). The alternate hypothesis (rejecting the null hypothesis) states that the time series is stationary.

- Null hypothesis: non-Stationarity exists in the series.
- Alternative Hypothesis: Stationarity exists in the series

Once the Adfuller test is applied on the regression residue it is checked for any heterocedasticity, in other words, for stationarity.

Since your Adfuller p-value is lower than a certain specified alpha (i.e., 5% or 0.05), then you may reject the null hypothesis (Ho), because the probability of getting a p-value as low as that by mere luck (random chance) is very unlikely.

Once the Ho is rejected, the alternative hypothesis (Ha) can be accepted, which in this case would be: the residue series is stationary.

Here is the hypothesis relation for you:

Ho: the series is not stationary, it presents heterocedasticity. In another words, your residue depends on itself (i.e., yt depends on y_{t-1}, y_{t-1} depends on y_{t-2} ..., and so on)

Ha: the series is stationary (That is normally what we desire in regression analysis). Nothing more needs to be done.

```
import statsmodels.tsa.stattools as sm
def augmented_dickey_fuller_statistics(time_series):
   result = sm.adfuller(time_series.values, autolag='AIC')
   print('ADF Statistic: %f' % result[0])
   print('p-value: %f' % result[1])
   print('Critical Values:')
   for key, value in result[4].items():
      print('\t%s: %.3f' % (key, value))
print('Augmented Dickey-Fuller Test: CPI_AllItemsLessFoodandEnergy Time Series')
augmented_dickey_fuller_statistics(transform_data['CPI_AllItemsLessFoodandEnergy'])
print('Augmented Dickey-Fuller Test: CPI_AllUrban_FoodandBeverages Time Series')
augmented_dickey_fuller_statistics(transform_data['CPI_AllUrban_FoodandBeverages'])
print('Augmented Dickey-Fuller Test: GrainIndex Time Series')
augmented_dickey_fuller_statistics(transform_data['GrainIndex'])
```

Output:

```
Augmented Dickey-Fuller Test: CPI_AllItemsLessFoodandEnergy Time Series
ADF Statistic: -3.566296
p-value: 0.006439
Critical Values:
        1%: –3.443
        5%: –2.867
        10%: –2.570
Augmented Dickey-Fuller Test: CPI_AllUrban_FoodandBeverages Time Series
ADF Statistic: –7.010919
p-value: 0.000000
Critical Values:
        1%: –3.443
        5%: –2.867
        10%: –2.570
Augmented Dickey-Fuller Test: GrainIndex Time Series
ADF Statistic: –7.791811
p-value: 0.000000
Critical Values:
        1%: –3.443
        5%: –2.867
        10%: –2.570
```

For CPI_AllItemsLessFoodandEnergy, the ADF Statistic is –3.566296. The more negative this statistic is, the more likely we are to reject the null hypothesis (we have a stationary dataset). We can see that our statistic value of –3.566296 is less than the value of –3.443 at 1%.

This suggests that we can reject the null hypothesis with a significance level of less than 1% (i.e., a low probability that the result is a statistical fluke).

Rejecting the null hypothesis means that the process has no unit root, and in turn that the time series is stationary or does not have time-dependent structure.

Similarly, CPI_AllUrban_FoodandBeverages and GrainIndex Time Series all exhibit stationarity.

Plot the data:

```
fig, axes = plt.subplots(nrows=3, ncols=1, dpi=120, figsize=(10,6))
for i, ax in enumerate(axes.flatten()):
    d = transform_data[transform_data.columns[i]]
    ax.plot(d, color='red', linewidth=1)
    # Decorations
    ax.set_title(dataset.columns[i])
    ax.xaxis.set_ticks_position('none')
    ax.yaxis.set_ticks_position('none')
    ax.spines['top'].set_alpha(0)
    ax.tick_params(labelsize=6)
plt.tight_layout();
```

Output:

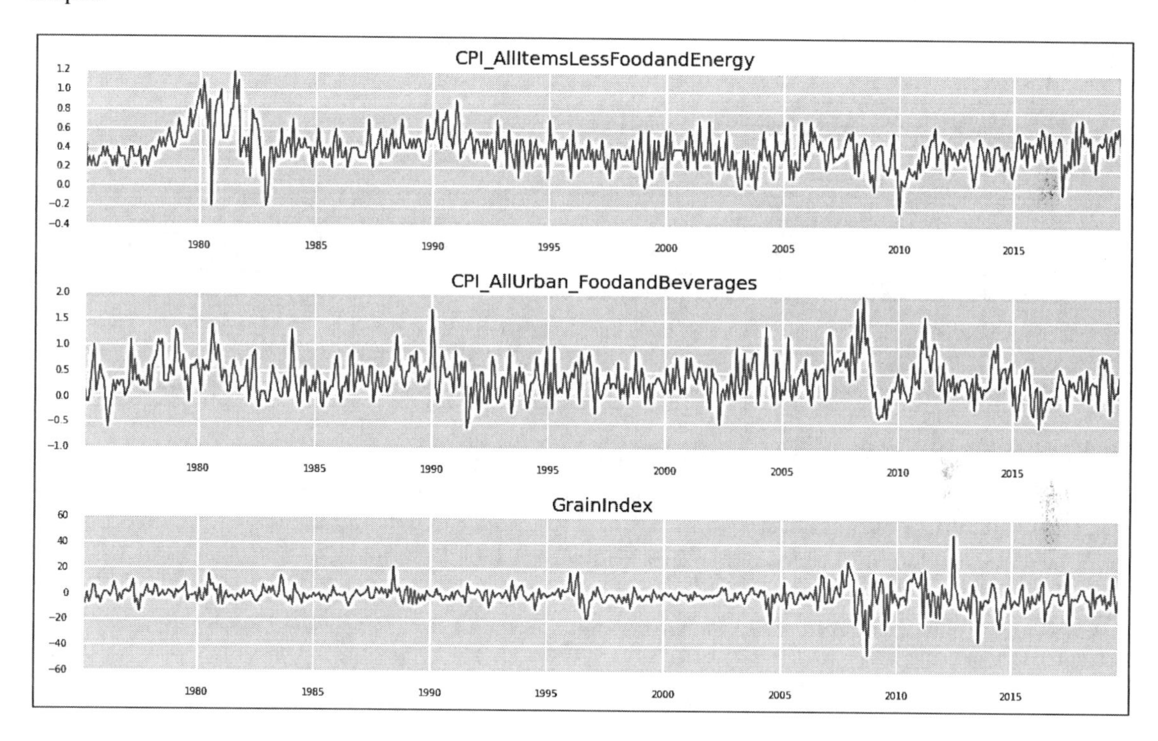

Step 13: Granger's Causality Test

The formal definition of Granger causality can be explained as whether past values of x aid in the prediction of y_t, conditional on having already accounted for the effects on y_t of past values of y (and perhaps of past values of other variables). If they do, the x is said to "Granger cause" y. So, the basis behind VAR is that each of the time series in the system influences each other.

Granger's causality Tests the null hypothesis that the coefficients of past values in the regression equation are zero. So, if the p-value obtained from the test is lesser than the significance level of 0.05, then, you can safely reject the null hypothesis. This has been performed on the original dataset.

```
import statsmodels.tsa.stattools as sm
maxlag=MAX_LAGS
test = 'ssr-chi2test'
def grangers_causality_matrix(X_train, variables, test = 'ssr_chi2test', verbose=False):
   dataset = pd.DataFrame(np.zeros((len(variables), len(variables))), columns=variables, index=variables)
     for c in dataset.columns:
       for r in dataset.index:
          test_result = sm.grangercausalitytests(X_train[[r,c]], maxlag=maxlag, verbose=False)
          p_values = [round(test_result[i+1][0][test][1],4) for i in range(maxlag)]
          if verbose: print(f'Y = {r}, X = {c}, P Values = {p_values}')
          min_p_value = np.min(p_values)
          dataset.loc[r,c] = min_p_value
          dataset.columns = [var + '_x' for var in variables]
     dataset.index = [var + '_y' for var in variables]
     return dataset
```

Invoke Grangers test.

```
grangers_causality_matrix(dataset, variables = dataset.columns)
```

Output:

	CPI_AllItemsLessFoodandEnergy_x	CPI_AllUrban_FoodandBeverages_x	GrainIndex_x
CPI_AllItemsLessFoodandEnergy_y	1.000	0.0000	0.0
CPI_AllUrban_FoodandBeverages_y	0.000	1.0000	0.0
GrainIndex_y	0.003	0.0001	1.0

The rows are the responses (y values), and the columns are the predictor series (x values).

If we take the value 0.0000 in (row 1, column 2), it refers to the p-value of the Granger's Causality test for CPI_AllUrban_FoodandBeverages_x causing CPI_AllItemsLessFoodandEnergy_y. The 0.0000 in (row 2, column 1) refers to the p-value of CPI_AllItemsLessFoodandEnergy_x causing CPI_AllUrban_FoodandBeverages_y and so on.

So, looking at the p-Values, we can assume that all the other variables (time series) in the system are interchangeably causing each other. This justifies the VAR modeling approach for this system of multi time-series for forecasting.

Step 14: VAR Model

VAR requires stationarity of the series which means the means to the series do not change over time (we can see this from the plot drawn next to the Augmented Dickey-Fuller Test).

```
import numpy as np
import pandas
import statsmodels.api as sm
from statsmodels.tsa.api import VAR
mod = VAR(transform_data)
res = mod.fit(maxlags=MAX_LAGS, ic='aic')
print(res.summary())
```

Output:

Correlation matrix of residuals			
	CPI_ AllItemsLessFoodandEnergy	CPI_AllUrban_ FoodandBeverages	GrainIndex
CPI_ AllItemsLessFoodandEnergy	1.000000	0.043808	0.025676
CPI_AllUrban_ FoodandBeverages	0.043808	1.000000	0.029310
GrainIndex	0.025676	0.029310	1.000000

The biggest correlations are 0.043808 (CPI_AllItemsLessFoodandEnergy & CPI_AllUrban_ FoodandBeverages); however, these are small enough to be ignored in this case.

Step 15: Residual Plots

The residual plot looks normal with a constant mean throughout apart from some large fluctuations during 1996,1997, and 2006 & later.

```
y_fitted = res.fittedvalues
y_fitted
y_fitted = res.fittedvalues
residuals = res.resid
plt.figure(figsize = (15,5))
plt.plot(residuals, label='resid')
plt.plot(y_fitted, label='VAR prediction')
plt.xlabel('Date')
plt.xticks(rotation=45)
plt.ylabel('Residuals')
plt.grid(True)
```

Output:

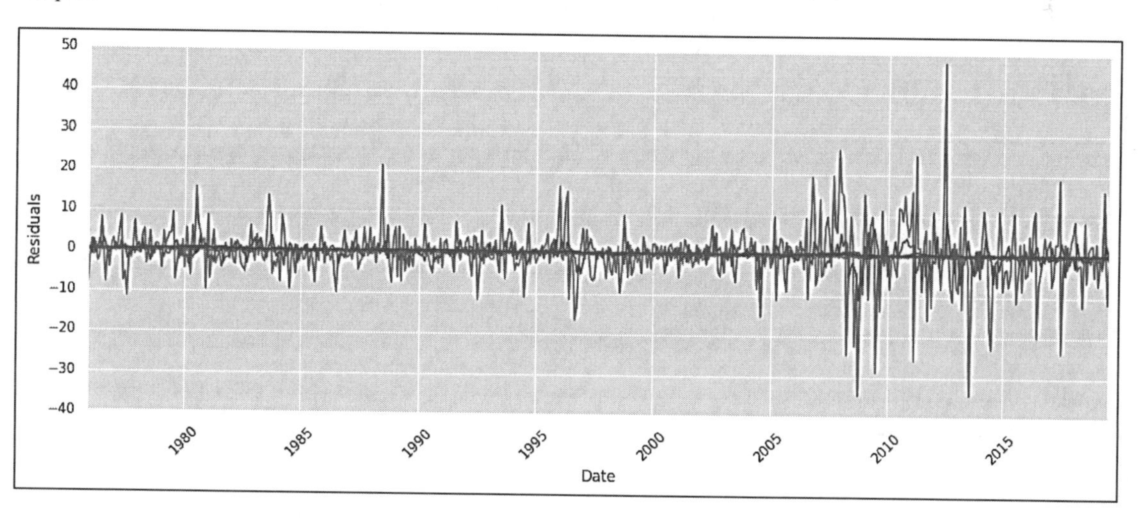

Step 16: Durbin-Watson Statistic

The Durbin-Watson Statistic is related to auto correlation. The Durbin-Watson statistic will always have a value between 0 and 4. A value of 2.0 means that there is no autocorrelation detected in the sample. Values from 0 to less than 2 indicate positive autocorrelation and values from 2 to 4 indicate negative autocorrelation. *A rule of thumb is that test statistic values in the range of 1.5 to 2.5 are relatively normal. Any value outside this range could be a cause for concern.*

```
from statsmodels.stats.stattools import durbin_watson
out = durbin_watson(res.resid)
for col,val in zip(transform_data.columns, out):
  print((col), ":", round(val,2))
```

Output:
CPI_AllItemsLessFoodandEnergy : 2.0
CPI_AllUrban_FoodandBeverages : 2.02
GrainIndex : 2.0

There is no autocorrelation (2.0) existing; so, we can proceed with the forecast.

Step 17: Prediction:

To forecast, the VAR model expects up to the lag order number of observations from the past data. This is because the terms in the VAR model are essentially the lags of the various time series in the dataset, so we need to provide as many of the previous values as indicated by the lag order used by the model.

```
# Get the lag order
lag_order = res.k_ar
print(lag_order)
# Input data for forecasting
input_data = transform_data.values[-lag_order:]
print(input_data)
# forecasting
pred = res.forecast(y=input_data, steps=n_obs)
pred = (pd.DataFrame(pred, index=X_test.index, columns=X_test.columns + '_pred'))
print(pred)
```

Output:

```
9
[[ 0.6     0.671   -3.86666667]
 [ 0.351   0.832    2.66666667]
 [ 0.497   0.457   -7.6 ]
 [ 0.576  -0.274   -1.36666667]
 [ 0.363   0.54    -4.8 ]
 [ 0.624   0.065    16.5 ]
 [ 0.546   0.162    2.3 ]
 [ 0.644   0.17    -10.23333333]
 [ 0.453   0.492   -0.06666667]]
```

Step 18: Invert the transformation:

The forecasts are generated but on the scale of the training data used by the model. So, to bring it back up to its original scale, we need to de-difference it. The way to convert the differencing is to add these differences consecutively to the base number. An easy way to do it is to first determine the cumulative sum at index and then add it to the base number.

This process can be reversed by adding the observation at the prior time step to the difference value.

inverted(ts) = differenced(ts) + observation(ts-1)

```
# inverting transformation
def invert_transformation(X_train, pred):
    forecast12 = pred.copy()
    columns = X_train.columns
    #col='CPI_AllUrban_FoodandBeverages'
    #forecast[str(col)+'_pred'] = dataset[col].iloc[-1] + forecast[str(col)+'_pred'].cumsum()
    for col in columns:
        print('+++++++++++++++')
        print(col)
        forecast12[str(col)+'_pred'] = X_train[col].iloc[-1] + forecast12[str(col)+'_pred'].cumsum()
        print('+++++++++++++++')
    return forecast12
output = invert_transformation(X_train, pred)
output
```

Output:

	CPI_AllItemsLessFoodandEnergy_pred	CPI_AllUrban_FoodandBeverages_pred	GrainIndex_pred
2019-10-01	265.091117	259.000022	142.103309
2019-11-01	265.523307	259.343795	141.223509
2019-12-01	265.991688	259.583131	141.538200
2020-01-01	266.441642	260.008886	141.810006
2020-02-01	266.903500	260.313153	140.342343
2020-03-01	267.354151	260.712997	139.653543
2020-04-01	267.818070	261.095508	140.848172
2020-05-01	268.294947	261.431181	141.046152
2020-06-01	268.743843	261.793309	141.142729
2020-07-01	269.189103	262.178885	140.849445
2020-08-01	269.628727	262.540369	141.193839
2020-09-01	270.061839	262.906176	140.896752
2020-10-01	270.501136	263.275799	141.129889
2020-11-01	270.937793	263.649203	141.261751
2020-12-01	271.374093	264.029003	141.222783
2021-01-01	271.808447	264.414461	141.174076

Step 19: Accuracy:

The following code generates accuracy of the predicted column.

```
#combining predicted and real data set
combine = pd.concat([output['CPI_AllUrban_FoodandBeverages_pred'], X_test['CPI_AllUrban_FoodandBeverages']], axis=1)
combine['accuracy'] = round(combine.apply(lambda row: row.CPI_AllUrban_FoodandBeverages_pred / row.CPI_AllUrban_FoodandBeverages * 100, axis = 1),2)
combine['accuracy'] = pd.Series(["{0:.2f}%".format(val) for val in combine['accuracy']],index = combine.index)
combine = combine.round(decimals=2)
```

Output:

	CPI_AllUrban_FoodandBeverages_pred	CPI_AllUrban_FoodandBeverages	accuracy
2019-10-01	259.00	259.02	99.99%
2019-11-01	259.34	259.48	99.95%
2019-12-01	259.58	259.85	99.90%
2020-01-01	260.01	260.67	99.75%
2020-02-01	260.31	261.49	99.55%
2020-03-01	260.71	262.32	99.39%
2020-04-01	261.10	265.88	98.20%
2020-05-01	261.43	267.69	97.66%
2020-06-01	261.79	268.95	97.34%
2020-07-01	262.18	267.98	97.84%
2020-08-01	262.54	268.29	97.86%
2020-09-01	262.91	268.43	97.94%
2020-10-01	263.28	268.90	97.91%
2020-11-01	263.65	269.00	98.01%
2020-12-01	264.03	269.82	97.85%
2021-01-01	264.41	270.35	97.81%
2021-02-01	264.80	270.70	97.82%

Step 19: Evaluation:

To evaluate the forecasts, a comprehensive set of metrics, such as MAPE, ME, MAE, MPE and RMSE can be computed. We have computed some of these below.

```
from sklearn.metrics import mean_absolute_error
from sklearn.metrics import mean_squared_error
import math
#Forecast bias
forecast_errors = [combine['CPI_AllUrban_FoodandBeverages'][i]- combine['CPI_AllUrban_FoodandBeverages_pred'][i] for i in range(len(combine['CPI_AllUrban_FoodandBeverages']))]
bias = sum(forecast_errors) * 1.0/len(combine['CPI_AllUrban_FoodandBeverages'])
print('Bias: %f' % bias)
print('Mean absolute error:', mean_absolute_error(combine['CPI_AllUrban_FoodandBeverages'].values, combine['CPI_AllUrban_FoodandBeverages_pred'].values))
print('Mean squared error:', mean_squared_error(combine['CPI_AllUrban_FoodandBeverages'].values, combine['CPI_AllUrban_FoodandBeverages_pred'].values))
print('Root mean squared error:', math.sqrt(mean_squared_error(combine['CPI_AllUrban_FoodandBeverages'].values, combine['CPI_AllUrban_FoodandBeverages_pred'].values)))
```

Output:

```
Bias: 8.181667
Mean absolute error: 8.181666666666674
Mean squared error: 107.9934966666668
Root mean squared error: 10.391991948931967
```

Mean absolute error tells us how big of an error we can expect from the forecast on average. Our error rates are quite low here, indicating we have the right fit for the model.

The VAR model is a popular tool for the purpose of predicting the joint dynamics of multiple time series based on linear functions of past observations. More analysis, e.g., impulse response (IRF) and forecast error variance decomposition (FEVD) can also be done along-with VAR for assessing the impacts of shocks from one asset on another.

Machine Learning Model: Prophet—Food Grain Producer Prices & Consumer Prices

 	VARFeedGrainsAndCPI.ipynb (Jupyter Notebook Code)

Prophet is a procedure for forecasting time series data based on an additive model where non-linear trends are fit with yearly, weekly, and daily seasonality, plus holiday effects.[407] It works best with time series' that have strong seasonal effects and several seasons of historical data. Prophet is robust to missing data and shifts in the trend, and typically handles outliers well.

Step 1: Setup Dataset

Setup the data frame for the Prophet model. While using the Prophet model, we don't have to go through the analysis that we did with VAR. The Prophet model performs most of the work internally.

```
dataset.info()
dfProphetDataset=dataset
```

Output:

```
<class 'pandas.core.frame.DataFrame'>
DatetimeIndex: 567 entries, 1975-01-01 to 2022-03-01
Freq: MS
Data columns (total 3 columns):
 # Column Non-Null Count Dtype
--- ------ -------------- -----
 0 CPI_AllItemsLessFoodandEnergy 567 non-null float64
 1 CPI_AllUrban_FoodandBeverages 567 non-null float64
 2 GrainIndex 567 non-null float64
dtypes: float64(3)
memory usage: 17.7 KB
```

Step 2: Setup Dataset

Model the dataset by including the label column "y":

```
from fbprophet import Prophet
model = Prophet()
train_df = dfProphetDataset.rename(columns={"CPI_AllUrban_FoodandBeverages":'y'})
train_df["ds"] = train_df.index
model.fit(train_df)
```

Output:

<fbprophet.forecaster.Prophet at 0x7f4f33081da0>

Step 3: Perform Forecast

Model the dataset by including the label column "y":

```
pd.plotting.register_matplotlib_converters()
# We want to forecast over the next 5 months
future = model.make_future_dataframe(6, freq='M', include_history=True)
forecast = model.predict(future)
model.plot(forecast)
```

Output:

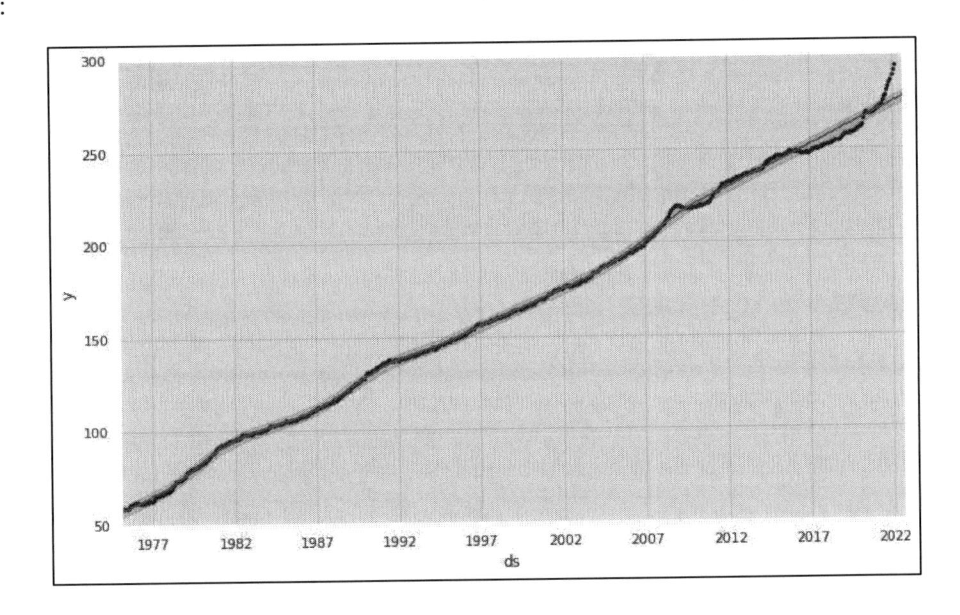

```
Forecast
```

Output:

	ds	trend	yhat_lower	yhat_upper	trend_lower	trend_upper	additive_terms	additive_terms_lower	additive_terms_upper	yearly	yearly_low
0	1975-01-01	56.949925	54.139497	59.794843	56.949925	56.949925	0.184640	0.184640	0.184640	0.184640	0.1846
1	1975-02-01	57.289006	54.282859	60.415419	57.289006	57.289006	0.247181	0.247181	0.247181	0.247181	0.247
2	1975-03-01	57.595273	55.380426	61.188796	57.595273	57.595273	0.668694	0.668694	0.668694	0.668694	0.6686
3	1975-04-01	57.934354	54.766355	60.857950	57.934354	57.934354	-0.136441	-0.136441	-0.136441	-0.136441	-0.1364
4	1975-05-01	58.262497	55.345396	61.178272	58.262497	58.262497	-0.059998	-0.059998	-0.059998	-0.059998	-0.0599
...	...	...	...	...	...	...	...	...	...	...	
568	2022-04-30	275.642802	272.462783	278.432146	275.642802	275.642802	-0.160829	-0.160829	-0.160829	-0.160829	-0.1608
569	2022-05-31	276.023887	272.894552	278.783255	276.023887	276.023887	-0.007561	-0.007561	-0.007561	-0.007561	-0.0075

```
combineProphetLower = pd.concat([forecast.yhat, forecast.yhat_lower], axis=1)
combineProphetLower['accuracy'] = round(combineProphetLower.apply(lambda row: row.yhat /row.yhat_lower * 100, axis = 1),2)
combineProphetLower['accuracy'] = pd.Series(["{0:.2f}%".format(val) for val in combineProphetLower['accuracy']],index = combineProphetLower.index)
combineProphetLower = combineProphetLower.round(decimals=2)
```

Output:

	yhat	yhat_lower	accuracy
0	57.13	54.14	105.53%
1	57.54	54.28	105.99%
2	58.26	55.38	105.21%
3	57.80	54.77	105.54%
4	58.20	55.35	105.16%
...	...	...	...
568	275.48	272.46	101.11%
569	276.02	272.89	101.14%
570	276.40	273.35	101.12%
571	276.63	273.45	101.17%
572	277.44	274.40	101.11%

573 rows × 3 columns

FAO in Emergencies—CTOT & Food Inflation

Food security is a global need and every country in the world is actively working to ensure enhancement of food security [1]. Recent global political and economic conflicts have changed the face of food security. Due to unprecedented disruptions caused by the Russia-Ukraine war, three negative influences of food security are: first, on the people of Ukraine and Russia who are experiencing supply disruptions; second, on countries relying heavily on their exports (wheat, sunflower seeds/oil, and others); and third, on broader populations that already are feeling the shock of higher food prices.[408] Record high natural gas prices have forced the fertilizer companies to curtail production of ammonia and urea in many parts of the world,[409] e.g., Europe to 45% of capacity, and with less of these two essential agricultural ingredients, the knock-on effects for global food supplies will be unprecedented [12]. Agriculture and Food Security are national security issues and must be given unprecedented importance.

It is essential to study food security from global interconnect, supplier risk, and sustainability points of view for mainly protecting the countries that are heavily dependent on food imports and would be on receiving end of unprecedented commodity disruptions. The shortage of essential food commodities due to the Russia-Ukraine war has triggered a *domino effect* that underscores how interconnected global commodity markets are pushing up the price of other co-movement commodities produced elsewhere, including those not ordinarily considered substitutes.[410] We have also learned that there simply isn't sufficient available capacity in the rest of the world to cover the Ukrainian shortfall [15]. Historically, we have seen such global disruptions due to weather, climate change, and political issues. For instance, the number of climate-related disasters, including extreme heat, droughts, floods, and storms, has doubled since the early 1990s, with an average of 213[411] of these events occurring every year from 1990–2016. These disasters harm agricultural productivity, which means food availability suffers. With this comes a domino effect, causing food prices to skyrocket and income losses that prevent people from being able to afford food [16].

"When women have an income, [they] dedicate 90% to health, education, food security, to the children, to the family, or to the community. So, when women have an income, everybody wins" [16].

- Michelle Bachelet, United Nations Commissioner on Human Rights

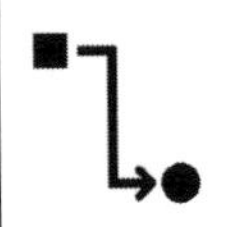

The world is realizing it's not easy to take a major commodity like sunflower oil and switch to an alternative" [15].

Trade and Food Security

There are two different approaches to problems regarding the relationship between trade, specifically imports of food, and food security (please see Figure 13). The first is to argue that it is unimportant that a country be able to grow the food it needs, all that is necessary is that it should be able to acquire the food it needs, i.e., to export goods to earn enough to pay for food imports. This has been defined above as self-reliance. Others argue that countries should be self-sufficient so that they meet their food needs fully from domestic production. This may imply supporting, if not protecting, farmers. Not all countries can expect to be self-sufficient in food. Some countries may not even be able to be self-reliant if they have very limited export opportunities and high food needs relative to local production (e.g., many small island economies). Thus, governments should not begin by choosing a strategy of self-sufficiency or self-reliance. Rather, they should start by establishing an efficient (undistorted) agriculture sector and identify the extent to which this meets food needs.

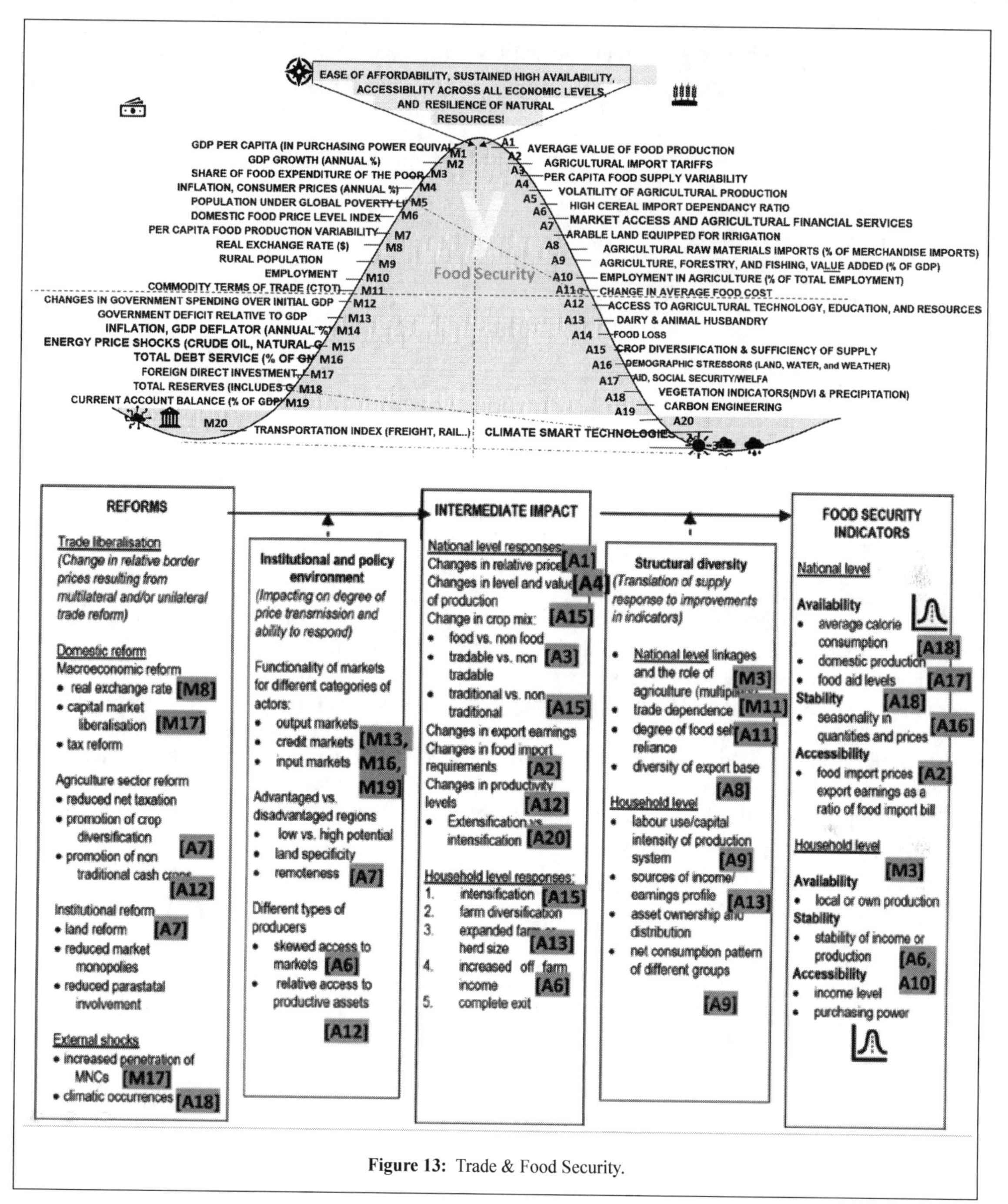

Figure 13: Trade & Food Security.

It was true that many developing countries started with a bias against agriculture,[412] prior to the 2000s, but most recent agriculture sector reforms have made agriculture a core export sector [17,18] a fact we can see from tariff rates (1993:2017). As can be seen in the figure below, the import tariffs have been gradually reduced (please see Figure 14).

Agriculture sector reforms are intended to increase productivity. In general, these will increase farm incomes or profit margins, and allow prices (especially of foods) to be reduced (at least in real terms). In this sense, agriculture sector reforms confer widespread benefits. Climate Smart sensors and Research & Development of new technologies could provide huge benefits, especially infusion of artificial intelligence in agriculture.

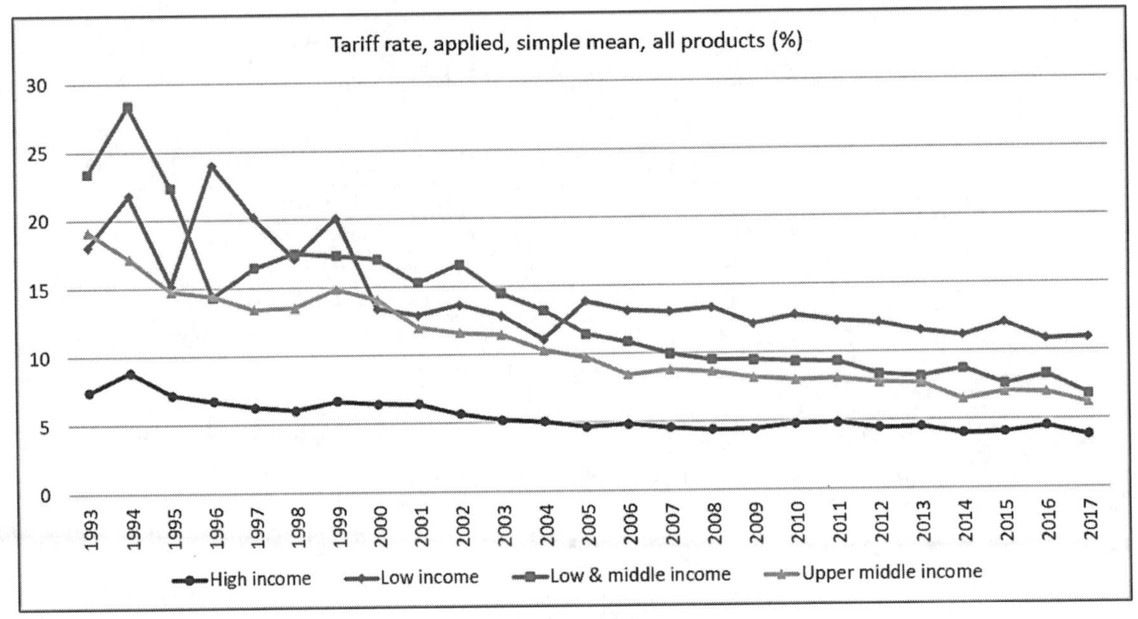

Figure 14: Tariff Rate.

Trade reform has mixed benefits. Import liberalization (easier access at lower prices) benefits those using imported inputs. This may include producers—farmers using imported fertilizer—or consumers (e.g., lower prices for food). However, it increases competition against those competing with imports, and this may include food producers. Commercial farmers, for example, may benefit from cheaper imported inputs but face increased competition from cheaper food. Measures that favor exporters are generally beneficial, but from the self-sufficiency perspective a problem arises if farmers substitute food for cash crops. However, if the cash crops earn the revenue to import food, it is consistent with self-reliance. Assuming that appropriate reforms have been implemented so that policies are not biased against agriculture, countries can find themselves in several situations. Four cases are presented here (please see Table 2):

Table 2: Food Export & Food Security.

	Food Exporter/Importer	**Food Security**
Some countries with efficient agricultural producers will be net food exporters. Food security will not be an issue, but they will be concerned with open access to foreign markets.	Food Exporter	Not an issue
Some countries will be naturally self-sufficient. At prevailing domestic prices, which should be equivalent to true world prices, domestic producers are capable of meeting local food needs at least in normal years. In good years they could export food, or stock food as insurance against a bad year.	self-sufficient; meeting local food needs at least in normal years; In good years they could export food, or stock food as insurance against a bad year	Not an issue
Some countries will not be self-sufficient but will have export earnings that allow them to meet food import needs, i.e., they are self-reliant.	self-reliant; earn through exports to import food	They may be exposed to risk if they are export dependent on primary commodities. Therefore, it is preferable that export earnings are from a diversified, especially manufacturing, portfolio.
Some countries will naturally be food insecure. Strictly speaking, it is only in respect of such countries that the issue of an active food security policy arises.	Low agricultural productivity can also lead to food-aid.	A costly option is to provide subsidies to farmers, and this may not be viable. Several countries are likely to remain dependent on food aid, or aid that can be used to finance food imports.

The relationship between trade[413] reforms and food security status can be conceptualized at a general level, depicted in Figure 15, as a two-stage relationship where a set of causal factors impact on a series of intermediate indicators, which in turn determine the outcome in terms of changes in food security status. It is recognized that an identical policy change in two different contexts, whether within a country at two discrete points in time, or across a set of countries, can result in quite different outcomes, because of modifying factors [18]. The trade influence on food security can be analyzed from the above framework [18].

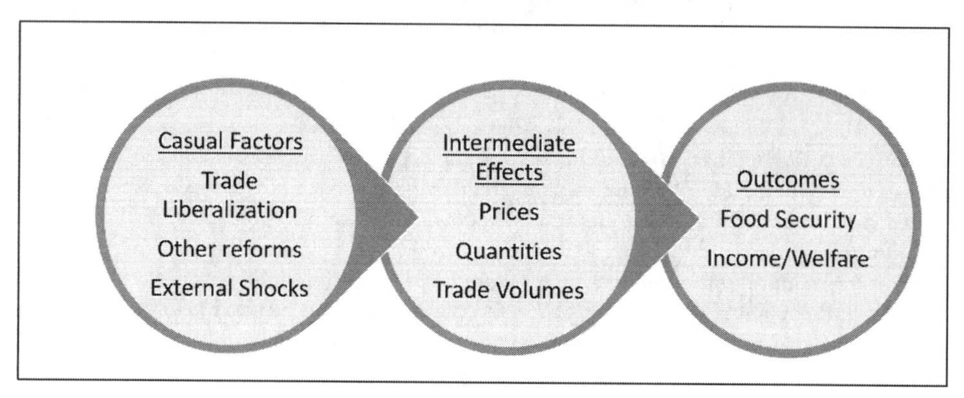

Figure 15: Trade and Food Security.

Food Security is National Security!

Food Insecurity creates national havoc and could disrupt public order, creating unprecedented challenges to maintain the law and order. For instance, consider a casual relationship between Arab Spring and high food prices of essential staple.[414] While this research does not imply that all Arab Springs were triggered by food insecurity, it assumes that rising food prices increased the pre-existing social unrest, sparking protests in Egypt, Syria, and Morocco, and probably also in other MENA countries affected by the riots [19]. Food scarcity also played a large role in the Arab Spring, a series of anti-government protests in the Middle East in the early 2010s after international food prices shot up, unemployment rose, and frustration with corrupt political systems peaked. One of the driving forces behind the Arab Spring, experts say, was the high cost of food. Shrinking farmlands, bad weather, and poor water distribution contributed to higher prices and, as a result, anti-government sentiment [20]. Surging food prices have caused people to revolt in past centuries too. Bread shortages were flashpoints for unrest throughout history, as in the French Revolution in the late 18th century.

Possible Effects of Three Orthogonal Shocks on Three Endogenous Variables

We establish that commodity demand shocks and inventory or other commodity-specific demand shocks strongly dominate commodity supply shocks in driving the fluctuation in real commodity prices over a broad set of commodities and over a broad period.

Additionally, we find that the contribution of commodity demand shocks to real prices varies across different commodities. However, commodity demand shocks exhibit common patterns with respect to timing across the markets for agricultural, metal, and soft commodities. Inventory or other commodity-specific demand shocks are the most important drivers in commodity price fluctuations for most of our agricultural and soft commodities. Commodity supply shocks play some role in explaining fluctuations for commodities, but mainly, their influence on real commodity prices is limited in impact and transitory in duration. There are significant differences in the persistence across the different types of shocks. While commodity demand and inventory or other commodity-specific demand shocks affect prices up to 10 years, supply shocks only have an effect for up to 5 years. Finally, commodity demand shocks have gained importance over time; commodity supply shocks have become less relevant. To this end, we assembled a new data set on the level of prices and production for 12 commodities, spanning the categories of agricultural, metal, and soft commodities from 1870 to 2013. In marked contrast to the literature on crude oil prices which generally uses monthly data over multiple years or a few decades, we use annual data over the past century and a half. This context makes it hard for us to rationalize a steep—that is, an inelastic—short-run supply curve which is one of the basic identifying assumptions of SVARs based on short-run restrictions (see, e.g., Kilian 2009), or to impose bounds on the short run price elasticity of supply, as used in models with sign restrictions (e.g., Kilian and Murphy 2014) [2].

Our results indicate that commodity demand shocks strongly dominate commodity supply shocks as drivers of commodity price booms and busts over a broad set of commodities and over a broad period.

The chapter introduces Commodity Terms of Trade & Food Security linkage. As part of the linkage, the chapter details the Mechanics of food inflation. Additionally, it introduces Food Price Inflation vs. General Inflation vs. Food Security. Next it introduces commodity costs & returns models and develops Machine Learning Models for Food Grain Producer Prices & Consumer Prices linkages. Finally, the chapter concludes with the Prophet—Food Grain Producer Prices & Consumer Prices & the role of food security as national security and role of trade in food security.

References

[1] Nikola Spatafora and Irina Tytell1, Commodity Terms of Trade: The History of Booms and Busts, September 2009, https://www.imf.org/external/pubs/ft/wp/2009/wp09205.pdf, Access Date: May 02, 2022.

[2] David, S. Jacks and Martin Stuermer. What Drives Commodity Price Booms and Busts? January 2017, https://www.dallasfed.org/research/papers/2016/wp1614.aspx, Access Date: May 06, 2022.

[3] Mark Nord, Alisha Coleman-Jensen and Christian Gregory. Prevalence of U.S. Food Insecurity Is Related to Changes in Unemployment, Inflation, and the Price of Food, June 2016, https://www.ers.usda.gov/webdocs/publications/45213/48166_err167_summary.pdf?v=0, Access Date: May 03, 2022.

[4] Federal Reserve Bank of San Francisco. What are the possible causes and consequences of higher oil prices on the overall economy? November 2007, https://www.frbsf.org/education/publications/doctor-econ/2007/november/oil-prices-impact-economy/, Access Date: May 12, 2022.

[5] Justin Lahart. The U.S. Economy Is Desperately Seeking Workers, May 6, 2022, 11:39 am ET, https://www.wsj.com/articles/desperately-seeking-workers-11651851566, Access Date: May 07, 2022.

[6] Jerry Light and Thomas Shevlin, the 1996 grain price shock: how did it affect food inflation? August 1998, https://www.bls.gov/opub/mlr/1998/08/art1full.pdf, Access Date: May 04, 2022.

[7] Jennifer Fadoul. How Satellite Maps Help Prevent Another 'Great Grain Robbery', Last Updated: Aug 23, 2021, https://www.nasa.gov/feature/how-satellite-maps-help-prevent-another-great-grain-robbery, Access Date: May 04, 2022.

[8] May Peters, Suchada Langley and Paul Westcott. Agricultural Commodity Price Spikes in the 1970s and 1990s: Valuable Lessons for Today, March 01, 2009, https://www.ers.usda.gov/amber-waves/2009/march/agricultural-commodity-price-spikes-in-the-1970s-and-1990s-valuable-lessons-for-today/, Access Date: June 11, 2022.

[9] May Peters, Suchada Langley and Paul Westcott. Agricultural Commodity Price Spikes in the 1970s and 1990s: Valuable Lessons for Today, March 01, 2009, https://www.ers.usda.gov/amber-waves/2009/march/agricultural-commodity-price-spikes-in-the-1970s-and-1990s-valuable-lessons-for-today/, Access Date: June 11, 2022.
[10] Annemarie Kuhns and Ryan Kuhns. Food Price Inflation Has Outpaced Economy-Wide Inflation in Recent Years, October 03, 2016, https://www.ers.usda.gov/amber-waves/2016/october/food-price-inflation-has-outpaced-economy-wide-inflation-in-recent-years/, Access Date: May 04, 2022.
[11] Annie Gasparro and Will Feuer. Campbell Soup Says Consumers Are Curbing Purchases Due to Higher Prices, Updated June 8, 2022, 3:05 pm ET, https://www.wsj.com/articles/campbell-sales-rise-amid-higher-prices-11654691474, Access Date: June 10, 2022.
[12] Nandita Roy and David Young. Stagflation Risk Rises Amid Sharp Slowdown in Growth, June 07, 2022, https://www.worldbank.org/en/news/press-release/2022/06/07/stagflation-risk-rises-amid-sharp-slowdown-in-growth-energy-markets, Access Date: June 10, 2022.
[13] Todd E. Clark. Do Producer Prices Lead Consumer Prices? 1995, https://www.kansascityfed.org/documents/1005/1995-Do%20Producer%20Prices%20Lead%20Consumer%20Prices%3F.pdf, Access Date: May 04, 2022.
[14] S. Brock Blomberg and Ethan S. Harris. The Commodity-Consumer Price Connection: Fact or Fable? October 1995 Volume 1, Number 3, https://www.newyorkfed.org/research/epr/95v01n3/9510blom.html#:~:text=In%20examining%20the%20empirical%20relationship,more%20on%20fable%20than%20fact., Access Date: May 04, 2022.
[15] Saabira Chaudhuri. Ukraine War Sparks Global Scramble for Cooking Oils, Apr. 6, 2022 6:02 am ET, https://www.wsj.com/articles/ukraine-war-sparks-global-scramble-for-cooking-oils-11649239342?page=1, Access Date: April 08, 2022.
[16] Leanna Parekh, the basics of food security (and how it's tied to everything), Updated Mar 23, 2021, https://www.worldvision.ca/stories/food/the-basics-of-food-security, Access Date: April 08, 2022.
[17] ALBERTO VALDÉS, why developing countries should stop discriminating against agriculture, June 11 2015, https://www.ifpri.org/blog/why-developing-countries-should-stop-discriminating-against-agriculture, Access Date: June 10, 2022.
[18] FAO. Trade Reforms and Food Security, 2003, https://www.fao.org/3/y4671e/y4671e.pdf, Access Date: June 10, 2022.
[19] Giulia Soffiantini. Food insecurity and political instability during the Arab Spring, Global Food Security, Volume 26, 2020, 100400, ISSN 2211-9124, https://doi.org/10.1016/j.gfs.2020.100400.
[20] Jason Lalljee. Russia is disrupting food prices worldwide and it could cause civil unrest — it's happened before, Mar 9, 2022, 10:36 AM, https://www.businessinsider.com/ukraine-russia-bread-food-prices-civil-unrest-arab-spring-egypt-2022-3, Access Date: April 10, 2022.

[377] Prevalence of U.S. Food Insecurity Is Related to Changes in Unemployment, Inflation, and the Price of Food - https://www.ers.usda.gov/webdocs/publications/45213/48166_err167_summary.pdf?v=0

[378] What are the possible causes and consequences of higher oil prices on the overall economy? - https://www.frbsf.org/education/publications/doctor-econ/2007/november/oil-prices-impact-economy/

[379] The U.S. Economy Is Desperately Seeking Workers - https://www.wsj.com/articles/desperately-seeking-workers-11651851566

[380] Civilian labor force participation rate - https://www.bls.gov/charts/employment-situation/civilian-labor-force-participation-rate.htm

[381] The 1996 grain price shock: how did it affect food inflation? - https://www.bls.gov/opub/mlr/1998/08/art1full.pdf

[382] The first 50 years of the Producer Price Index: setting inflation expectations for today - https://www.bls.gov/opub/mlr/2016/article/the-first-50-years-of-the-producer-price-index.htm

[383] World Commodity Prices – Pink Sheet Data - https://www.worldbank.org/en/research/commodity-markets

[384] Food Price Inflation Has Outpaced Economy-Wide Inflation in Recent Years - https://www.ers.usda.gov/amber-waves/2016/october/food-price-inflation-has-outpaced-economy-wide-inflation-in-recent-years/

[385] Consumer Price Index for All Urban Consumers: All Items Less Food and Energy in U.S. City Average (CPILFESL) - https://fred.stlouisfed.org/series/CPILFESL

[386] Consumer Price Index for All Urban Consumers: Food and Beverages in U.S. City Average (CPIFABSL)- https://fred.stlouisfed.org/series/CPIFABSL#

[387] Campbell Soup Says Consumers Are Curbing Purchases Due to Higher Prices - https://www.wsj.com/articles/campbell-sales-rise-amid-higher-prices-11654691474

[388] Commodity Cost Returns - https://www.ers.usda.gov/data-products/commodity-costs-and-returns/

[389] Corn-Cost-Returns - https://www.ers.usda.gov/webdocs/DataFiles/47913/CornCostReturn.xlsx?v=105.6

[390] The Global Economic Prospects, June 2022 - https://openknowledge.worldbank.org/bitstream/handle/10986/37224/9781464818431.pdf

[391] World Economic And Social Survey Archive: 1970-1979 - https://www.un.org/development/desa/dpad/publication/world-economic-and-social-survey-archive-1970-1979/

[392] Visualization - https://www.ers.usda.gov/data-products/commodity-costs-and-returns/interactive-visualization-u-s-commodity-costs-and-returns-by-region-and-by-commodity/

[393] Do Producer Prices Lead Consumer Prices? - https://www.kansascityfed.org/documents/1005/1995-Do%20Producer%20Prices%20Lead%20Consumer%20Prices%3F.pdf

[394] The Commodity-Consumer Price Connection: Fact or Fable? - https://www.newyorkfed.org/research/epr/95v01n3/9510blom.html#:~:text=In%20examining%20the%20empirical%20relationship,more%20on%20fable%20than%20fact

[395] Producer Price Index by Commodity: Farm Products: Corn (WPS012202) - https://fred.stlouisfed.org/series/WPS012202

[396] Wheat Data - https://www.ers.usda.gov/data-products/wheat-data/

[397] All Wheat - https://www.ers.usda.gov/webdocs/DataFiles/54282/Wheat%20Data-All%20Years.xlsx?v=6454

[398] Wheat Producer Prices - https://www.nass.usda.gov/Charts_and_Maps/Agricultural_Prices/pricewh.php

[399] Prices Received: Soybean Prices Received by Month, US - https://www.nass.usda.gov/Charts_and_Maps/Agricultural_Prices/pricesb.php

[400] Feed Grains Database - https://www.ers.usda.gov/data-products/feed-grains-database/

[401] Producer Price Index by Commodity: Farm Products: Wheat (WPU0121) - https://fred.stlouisfed.org/series/WPU0121

[402] Producer Price Index by Commodity: Farm Products: Wheat (WPU0121) - https://fred.stlouisfed.org/series/WPU0121

[403] Consumer Price Index for All Urban Consumers: All Items Less Food and Energy in U.S. City Average (CPILFESL) - https://fred.stlouisfed.org/series/CPILFESL

[404] Consumer Price Index for All Urban Consumers: Food and Beverages in U.S. City Average (CPIFABSL) - https://fred.stlouisfed.org/series/CPIFABSL

[405] D'Agostino, R.B. 1971. An omnibus test of normality for moderate and large sample size, Biometrika 58: 341–348

[406] D'Agostino, R. and Pearson, E.S. 1973. Tests for departure from normality. Biometrika 60: 613–622

[407] Prophet - https://facebook.github.io/prophet/

[408] Food Is Just as Vital as Oil to National Security - https://www.bloomberg.com/opinion/articles/2022-03-07/food-is-just-as-vital-as-oil-to-national-security

[409] War has brought the world to the brink of a food crisis - https://www.cnn.com/2022/03/12/business/food-crisis-ukraine-russia/index.html

[410] Ukraine War Sparks Global Scramble for Cooking Oils - https://www.wsj.com/articles/ukraine-war-sparks-global-scramble-for-cooking-oils-11649239342?page=1

[401] The basics of food security (and how it's tied to everything) - https://www.worldvision.ca/stories/food/the-basics-of-food-security

[412] Why developing countries should stop discriminating against agriculture - https://www.ifpri.org/blog/why-developing-countries-should-stop-discriminating-against-agriculture

[413] Trade and Food Security - https://www.fao.org/3/y4671e/y4671e.pdf

[414] Food insecurity and political instability during the Arab Spring, Global Food Security - https://doi.org/10.1016/j.gfs.2020.100400

CHAPTER 6

Climate Change and Agricultural Yield Analytics

The chapter introduces the impact of climate change on agricultural productivity. It introduces the Phenological Stages of wheat production and applies time series techniques to analyze the wheat yield. Additionally, NOAA Star Global Vegetation Health, Vegetation Condition index (VCI), Temperature Condition index (TCI), Vegetation Health index (VHI), No noise (smoothed) Normalized Difference Vegetation Index (SMN) are deployed. Next it introduces Coupled Model Intercomparison Project climate projections (CMIP) and Shared Socioeconomic Pathway (SSP) Projection Models. Finally, the chapter concludes with the development of Machine Learning Models for Kansas wheat and major Indian wheat producing states with the application of climate change impacts such as drought and heatwaves.

A household is not "food secure" unless it "feels" food secure.[415] Food security exists when all people, always, have physical and economic access to sufficient, safe, and nutritious food that meets their dietary needs and food preferences for an active and healthy life.[416] Food security is a global need and every country in the world is actively working to ensure enhancing its food security [1]. For instance, the United States Department of Agriculture (USDA) Economic Research Service (ERS) monitors the food security of U.S. households through an annual, nationally representative survey. While most U.S. households are food secure, a minority of U.S. households experience food insecurity at times during the year, meaning that their access to adequate food for active, healthy living is limited by lack of money and other resources.[417] Some experience very low food security, a more severe range of food insecurity where food intake of one or more members is reduced and normal eating patterns are disrupted [2]. In 2020, 89.5 percent of U.S. households were food secure throughout the year. The remaining 10.5 percent of households were food insecure at least some time during the year, including 3.9 percent (5.1 million households) that had very low food security.[418] Prevalence[419] of food insecurity (please see Figure 1 and Figure 2) is not uniform across the country due to both the characteristics of populations and State-level policies and economic conditions [2,3,4].

Extremely hot weather is a new normal!

The extreme heat is impacting hundreds of millions of people in one of the most densely populated parts of the world, threatening to damage whole ecosystems.[420]

There is a cascading impact. Extreme heat has multiple and cascading impacts not just on human health, but also on ecosystems, agriculture, water and energy supplies and key sectors of the economy.

Large fluctuations in energy prices, recently, have been a distinguishing characteristic of the World economy. The drivers of price fluctuations could be due to increased global conflicts, war in Ukraine, climatic events, droughts, and the COVID-19 pandemic. The U.S. economy has experienced price fluctuations since the 1970s. Turmoil in the Middle East, rising energy prices in the U.S. and evidence of global warming recently have reignited interest[421] in the link between energy prices and economic performance [1] and the impact of the exogenous nature of energy prices on agricultural farm inputs like fertilizers.

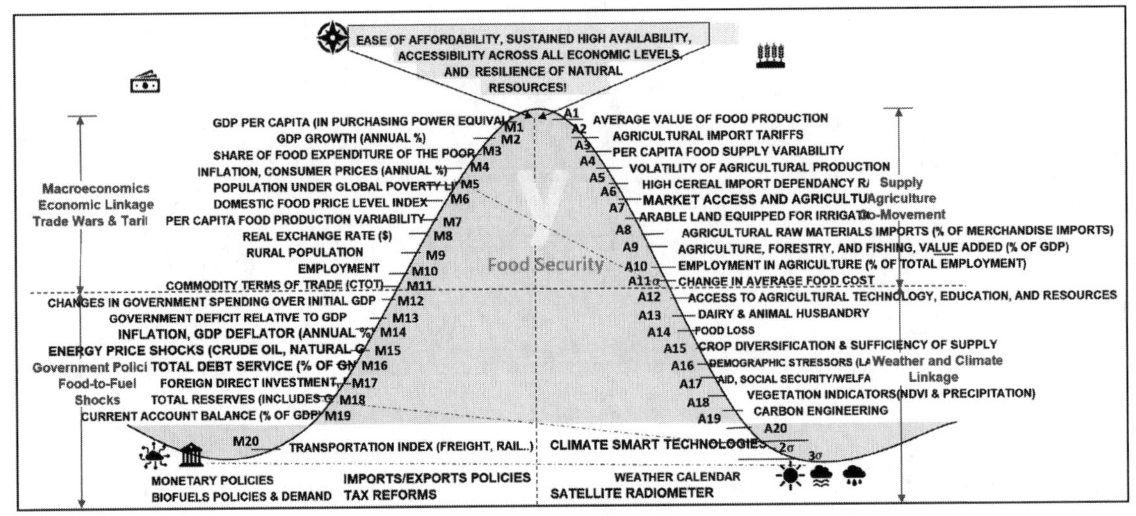

Figure 1: Food Security ML Model.

In this chapter, I would like to model events that are drivers of large fluctuations and their influence on food security. Specifically, I would like to focus on events that have a considerable impact on the global economy: extreme weather, climate change, and fertilizer price spikes due to wars. It is widely accepted that energy prices in general and crude oil & natural gas prices have been endogenous with respect to European, U.S. and the World macroeconomic conditions (please see Figure 1). *Endogeneity* here refers to the fact that not only do energy prices affect the U.S. economy, but that there is reverse causality from the U.S. and more generally global macroeconomic aggregates to the price of energy [1]. Clearly, both the supply and demand for energy depend on global macroeconomic aggregates such as real global economic activity and interest rates and could be affected by conflicts and wars.

Wheat is a widely adapted crop.[422] It grows in temperate, irrigated to dry and high-rain-fall areas and warm, humid to dry, cold environments. Undoubtedly, this wide adaptation has been possible due to the complex nature of the plant's genome, which provides great plasticity to the crop. Wheat[423] is a C3 plant and as such it thrives in cool environments. Much has been written about its physiology, growth, and development, which at present is reasonably well understood [2]. About 85% of the plant species on the planet are C3 plants, including rice, wheat, soybeans, and all trees. The C3 plant is a "normal" plant-one that doesn't have photosynthetic adaptations to reduce photorespiration. Photosynthesis is the process that plants use to turn light, carbon dioxide, and water into sugars that fuel plant growth, using the primary photosynthetic enzyme Rubisco. Lack of photosynthesis could lead to yield reduction.[424]

Most plant species on Earth use C3 photosynthesis, in which the first carbon compound produced contains three carbon atoms. In this process, carbon dioxide enters a plant through its stomata (microscopic pores on plant leaves), where amidst a series of complex reactions, the enzyme Rubisco fixes carbon into sugar through the Calvin-Benson cycle that plants and algae use to turn carbon dioxide from the air into sugar, the food autotrophs need to grow. By the way, every living thing on Earth depends on the *Calvin cycle.*[425] However, two key restrictions slow down photosynthesis [2].

1. Rubisco aims to fix carbon dioxide, but can also fix oxygen molecules, which creates a toxic two-carbon compound. Rubisco fixes oxygen about 20 percent of the time, initiating a process called photorespiration that recycles toxic compounds. Photorespiration costs the plant energy it could have used to photosynthesize.
2. When stomata are open to let carbon dioxide in, they also let water vapor out, leaving C3 plants at a disadvantage in drought, high-temperature environments, and climate change.

Table 1: Wheat Development & Time.

Development stage	Time *(days)*	
	Spring 1[a]	Winter 2[b]
Emergence	0	0
Floral initiation (double ridge)	20	35
Terminal spikelet	45	60
First node	60	80
Heading	90	120
Anthesis	100	130
Physiological maturity	140	170

a: Yecora, low sensitivity to vernalization and moderate sensitivity to photoperiod.
b: WW33G, high sensitivity to vernalization and moderate sensitivity to photoperiod. (Source: Adapted from Stapper and Fischer, 1990).

The need to improve photosynthesis (lack of sufficient light results in agricultural production)[426] in C3 crops to ensure greater food security under future climate scenarios is one of the research projects that are being conducted worldwide. C3 plants are limited by carbon dioxide and may benefit from increasing levels of atmospheric carbon dioxide resulting from the climate crisis. However, this benefit may be offset by a simultaneous increase in temperature that may cause stomatal stress.

Coming to wheat development, organ differentiation defines the various stages of wheat development. Physiologically, the following stages are usually distinguished: germination, emergence, tillering, floral initiation or double ridge, terminal spikelet, first node or beginning of stem elongation, boot, spike emergence, anthesis and maturity. These stages may be grouped into germination to emergence (E); growth stage 1 (GS1) from emergence to double ridge; growth stage 2 (GS2) from double ridge to anthesis; and growth stage 3 (GS3), which includes the grain filling period, from anthesis to maturity [3].

The timespan of each development phase essentially depends on genotype, temperature, day-length, and sowing date. Table 1 shows typical time-lapse values for the various stages in spring and winter type genotypes shown in May at 34°S. Various environmental stresses, particularly heat but also water and salinity, may shorten the wheat growth phases.

The phyllochron is defined as the interval between similar growth stages of two successive leaves in the same culm. It has been used extensively to understand and describe cereal development. The phyllochron is strongly dependent on temperature, but severe water deficits and strong nitrogen deficiency retard the leaf emergence rate in spring wheat. There is genetic variation (differences) in the phyllochron of genotypes of bread wheat and durum wheat [3].

Cereal development is normally expressed in degree-days (GDD), using 0° or 4°C as the base temperature for physiological processes in wheat, such that:

$$GDD = [(T_{max} + T_{min})/2] - T_b \qquad \text{EQ. 1}$$

where T_{max} and T_{min} are the maximum and minimum daily temperatures and T_b is the base temperature (Cao and Moss 1989a, 1989b). The GDD (please see Figure 1) varies with the growing stage and allows a rough estimation of when a given growth stage is going to occur at a particular site.

The world over, wheat[427] is produced in the following countries (please see Figure 2). The source of crop distribution map is, https://www.mapspam.info/data/

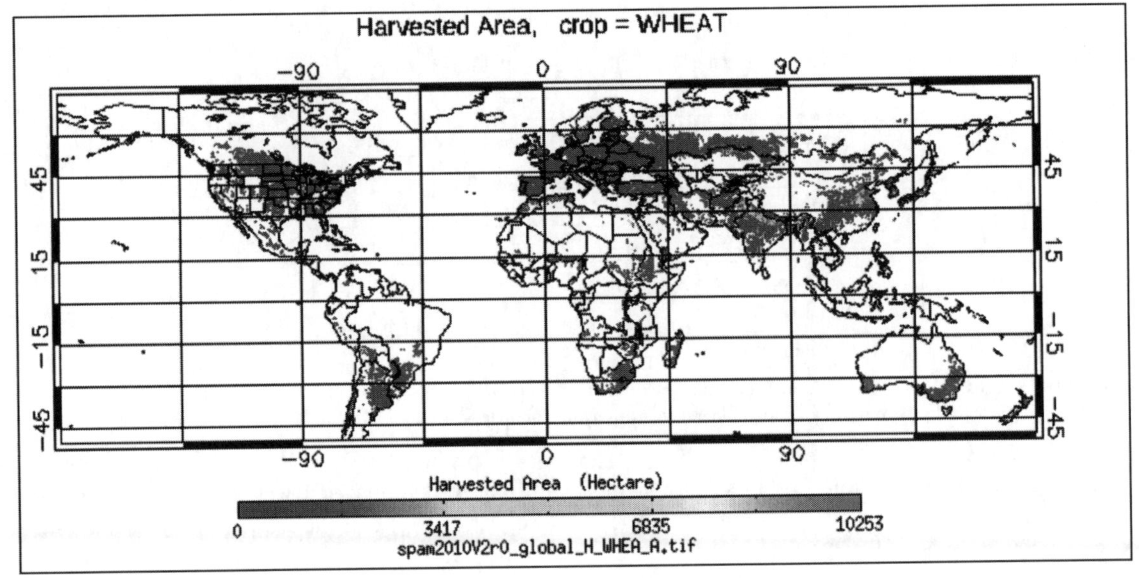

Figure 2: Worldwide Wheat Harvest Areas.

Wheat Phenological Stages

Wheat is a great short season grass.[428] With a short seasonal crop like wheat, one doesn't face the need to keep water for a full-fledged crop. Typically, here in North Dakota, *hard red spring wheat is planted in the spring*

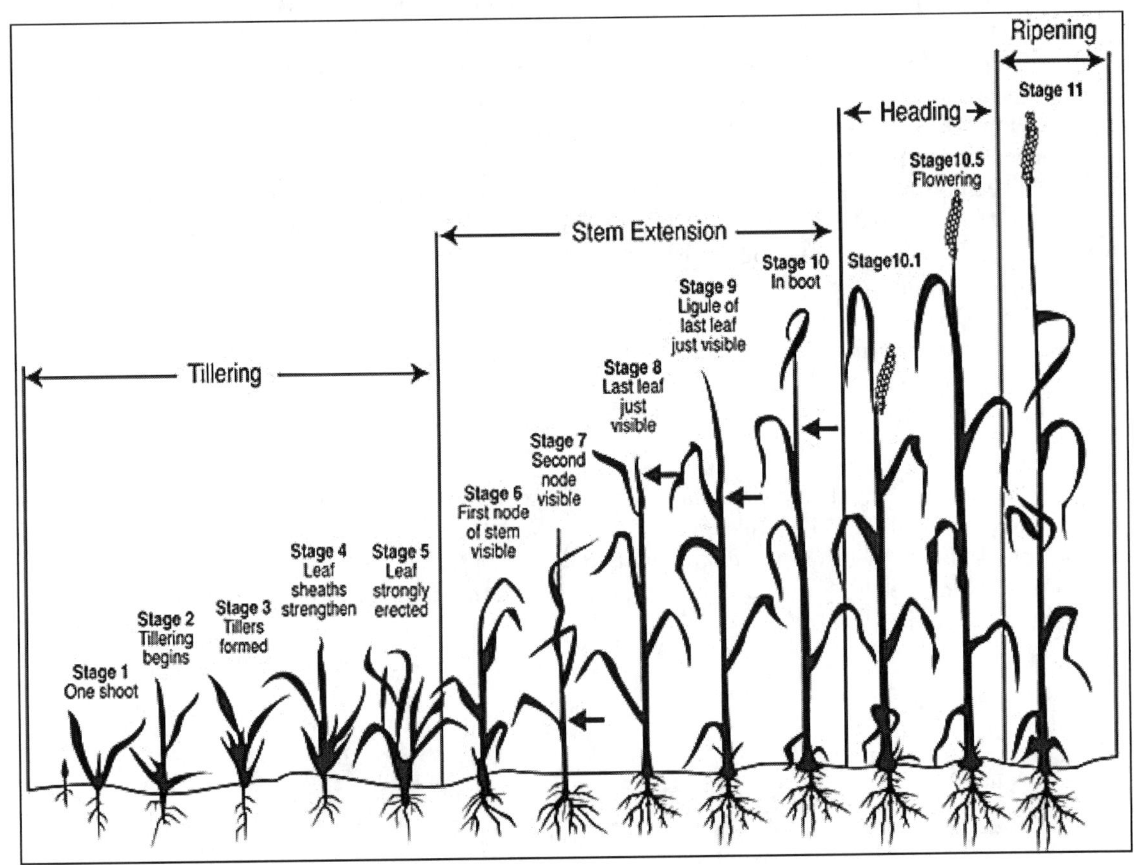

Figure 3: Wheat Phenological Stages.

from late April to the end of May and we harvest from early August to late September.[429] It's incredible that the wheat grows for about three months before it is ready to be harvested [3]. Wheat growth can be broadly divided into different stages (please see Figure 3): germination/emergence, tillering, stem elongation, boot, heading/flowering, and grain-fill/ripening. Several systems have been developed to identify wheat growth stages, the two most popular being the Feekes and the Zadoks scales. Being able to know and recognize what stage your wheat crop is in is vital to producing a good wheat crop. Wheat responds best to certain inputs at certain stages of development.

Germination/Seedling State

During the germination stage, an appropriate temperature and adequate moisture are needed for wheat seeds to germinate. Wheat seeds enjoy an optimum temperature *between 54° and 77°F*. Under favorable conditions, seedling emergence usually occurs within seven days. Until the first leaf becomes functional, the seedling depends on energy and nutrients stored in the seed (please see Figure 4).

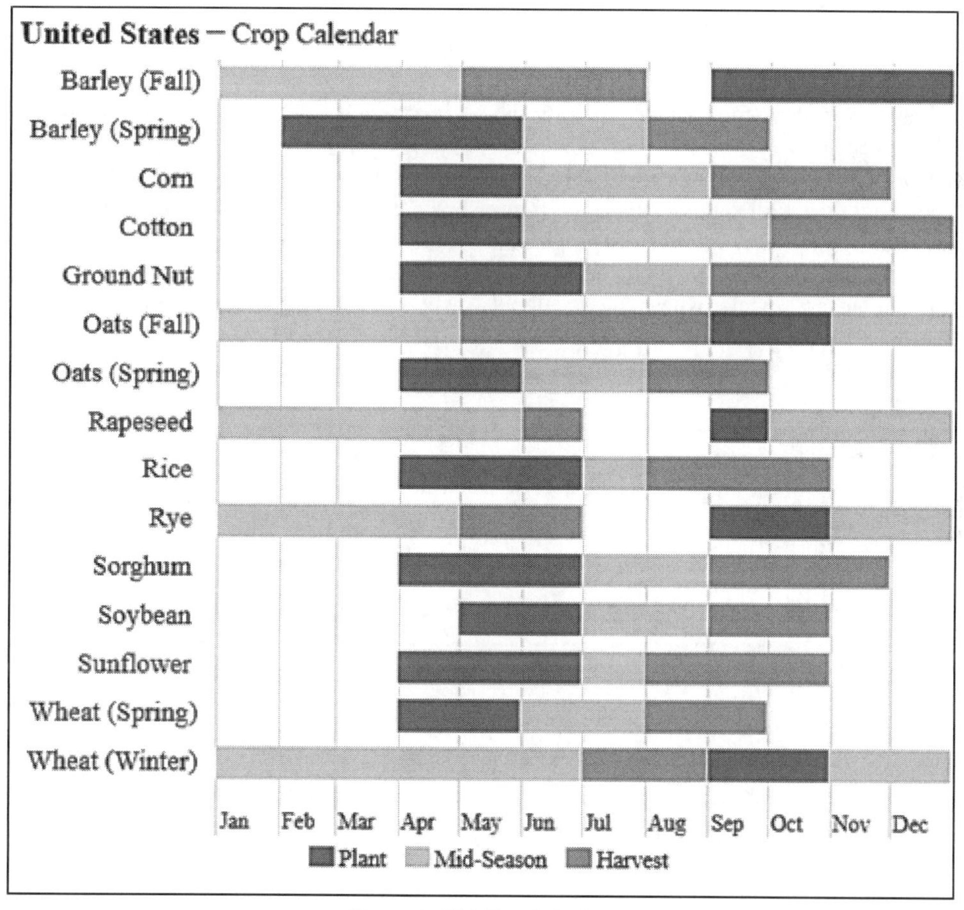

Figure 4: U.S. Crop Calendar.

Tillering & Stem Extension

The next stage is tillering. Tillering usually starts when a plant has 3 to 4 leaves. After the wheat plant finishes forming tillers, it begins elongation of its internodes or the stem extension stage of growth. Most short season wheat will typically produce 7–8 leaves on the main stem before stem elongation occurs. The boot stage begins when the head begins to form inside the flag leaf.

Heading & Flowering (Pollination)

The next is "heading" where the head fully emerges from the stem. After this takes place, the plant starts reproductive growth or flowering. Pollination is normally very quick, lasting only *about three to five days*. Wheat is self-pollinated and it is during this time that kernels per head are determined by the number of flowers that are pollinated. High *temperatures and drought stress during heading and flowering can reduce kernel numbers or yield.*

Ripening & Maturity

After pollination, the ripening stage begins. Ripening is divided into four levels of maturity: milk, soft dough, hard dough, and finally maturity. It is during this time that the wheat plant turns to a straw color and the kernel becomes very hard. The kernel becomes difficult to divide with a thumbnail, cannot be crushed between fingernails, and can no longer be dented by a thumbnail. Harvest can begin when the grain has reached a *suitable moisture level*. Many farmers can tell *maturity by chewing on a kernel to determine hardness and approximate moisture level*.

Harvesting

This is the final stage—the harvest stage. Overall, the optimal temperature for wheat development differs (please see Figure 5).

For wheat,[430] base and optimum temperatures aren't always 0°C (32°F) and 25°C (77°F) respectively. They start lower and rise with development. The figure shows that although plants can grow at 0°C (32°F) during the seedling stages, they make slow progress at the heading stage if temperatures are much below 10°C (50°F). Fortunately, varieties differ in their base and optimum temperatures by as much as 7°C (44.6°F) at any stage [4]. In general, winter wheat can develop at lower temperatures than spring wheat.

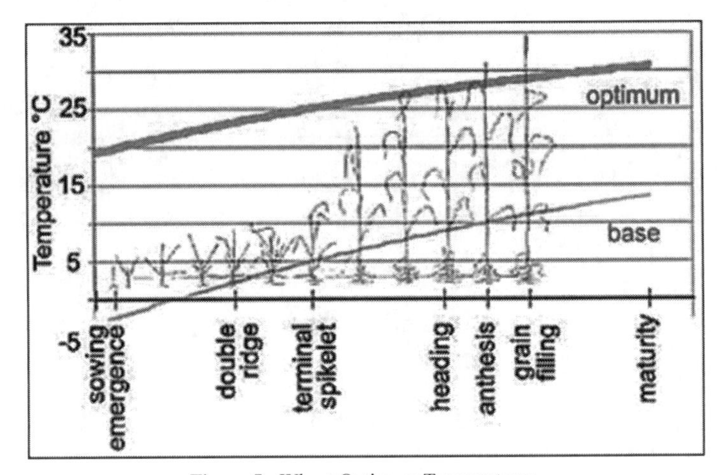

Figure 5: Wheat Optimum Temperatures.

Other Items

Water stress and heat stress affects wheat yield. Water stress is a common and wide occurrence in nature. It occurs whenever water absorption by the crop is lower than the evaporative demand of the atmosphere. Two major processes are involved [3]:

(1) water absorption by the crop, which is controlled by root characteristics and soil physical properties; and

(2) crop evapotranspiration, which depends on atmospheric properties, notably net radiation and vapor pressure deficit (vpd), and crop characteristics, such as crop ground cover and stomatal conductance. Notwithstanding that, wheat may experience water stress in any environment.

Heatwaves and high temperatures

Spain is in the grip of its first heatwave of the year, with temperatures in parts of the west and south expected to reach 44ºC (111.2ºF). Heatwaves—defined as at least three consecutive days of temperatures above the average recorded for July and August from 1971–2000—are becoming more frequent and are beginning earlier, according to Aemet, the Spanish meteorological office. "We are facing unusually high temperatures in June", said Rubén del Campo, an Aemet spokesperson. Climate change is making heatwaves more frequent and more intense.

High temperatures severely limit wheat yield. They accelerate plant development and specifically affect the floral organs, fruit formation and the functioning of the photosynthetic apparatus [3]. Although recognizing the fundamental linkage between water and heat stresses in plants, attention here will focus on one of them, heat stress, and assume that the wheat plants do not suffer water shortages. For breeding purposes, however, resistance to these two stresses usually must be combined.

Heat waves that occur during the planting season render a detrimental yield effect (at least 60% drop [5]) on the wheat production.[431] The reason is simple. During stages of germination & seedling stage to tillering & stem extension, it is desired to have a gradual temperature hike for a good crop [5]. However, if heat waves during these stages (please see Figure 6) disrupt gradual temperature increase and force sudden jumps the drop in crop yield is substantial. For example, take the case of heat waves that occurred during March 2022 in India. India and US produce a third of the wheat supply to the world. In addition, during the Ukraine war, Wheat from India was a good risk mitigator of the World wheat supply. However, heatwaves in the early time frame of the wheat season (please see Figure 6—agriculture calendar of India),[432,433,434] March, especially, major wheat growing areas of the country, have substantially increased the risk to the entire world. As a result of heat waves, the wheat crop was down 60 percent compared to normal harvests, with temperatures in March 2022 hitting 112ºF in Lucknow, Uttar Pradesh; 120ºF in Chandigarh, Punjab; and 109ºF in Bhopal, Madhya Pradesh. In March, when the ideal temperature should rise gradually, according to wheat farmers, we saw it jump suddenly from 32ºC to 40ºC [90ºF to 104ºF]; temperatures shot up abruptly which was like an electric shock. As per farmers, if such unreasonable weather patterns continue year after year, farmers will suffer badly.

Climate change is a risk to the entire humanity and the planet. When we try to compensate production loss of wheat due to Ukraine war[435] food security concerns apply across agricultural activities by smallholders and agribusinesses & the immediate food security dimension of this conflict is related to food access and not food availability [6]. Heat waves could inhibit this balance. As pointed out by a senior fellow at the Delhi-based Centre for Policy Research, "we don't know if India will be able to meet export demand because it is going to create issues in domestic supply, as wheat prices go up. India cannot replace Russia and Ukraine with its wheat exports, mainly because of the heat shock in March 2022." The future will reveal the judgment; as of writing of this book, the warning from the United Nations Food and Agriculture Organization the Russia-Ukraine conflict could leave an additional 8 million to 13 million people undernourished by [5,7], 2023 and the prognosis is dire.

"Heat waves have a "horrific" short and long term impact on people in India and further ahead [5]. Wheat prices will be driven up, and if you look at what is happening in Ukraine, with many countries relying on wheat from India to compensate shortfalls, the impact will be felt well beyond India."

Senior adviser to Climate Action Network International

The U.N.'s Intergovernmental Panel on Climate Change in a bombshell February report[436,437,438,439,440] said global warming meant hundreds of millions were already or would soon be at the risk of extreme heat exposure, depending on carbon emissions. By 2100, the report clearly mentioned half to three-quarters of humanity would be exposed "to periods of life-threatening climatic conditions arising from coupled impacts of extreme heat and humidity" [8].

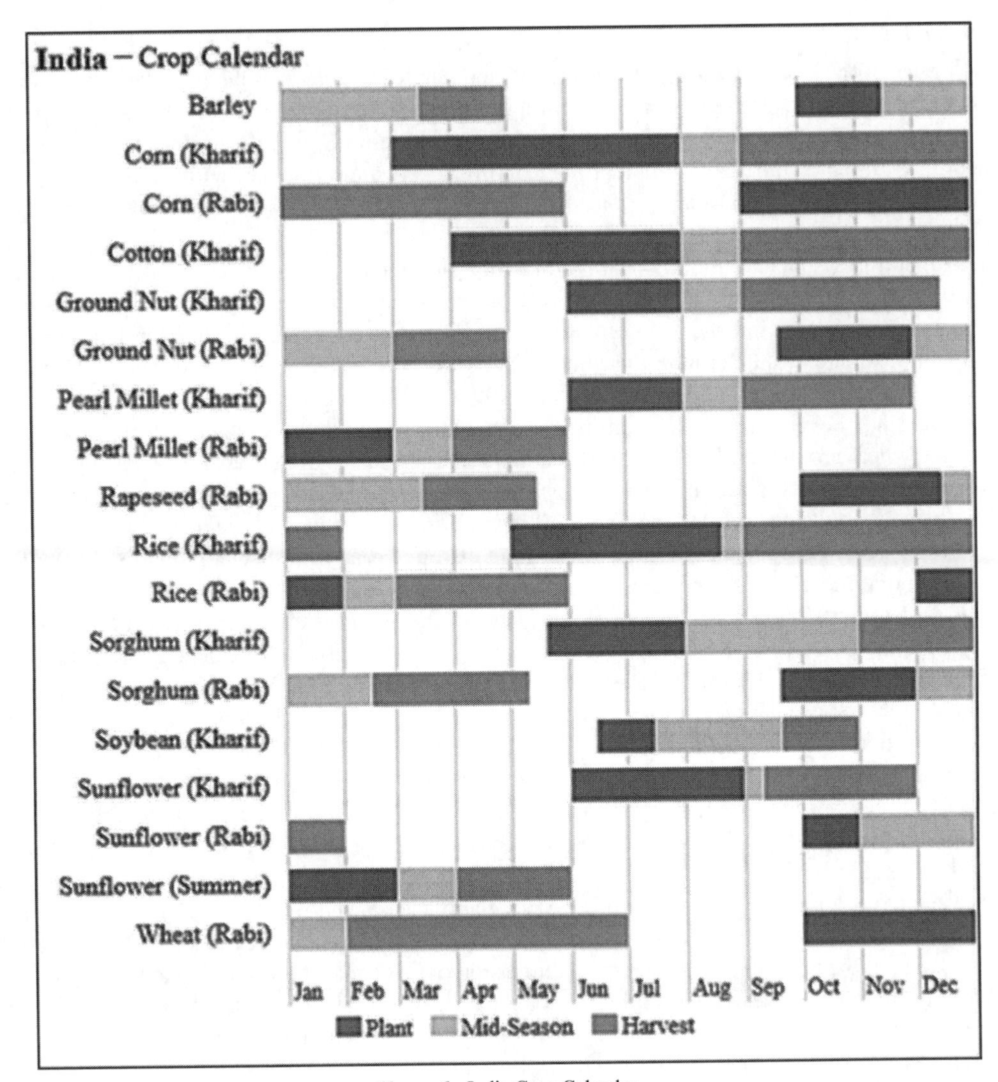

Figure 6: India Crop Calendar.

Wet-bulb temperatures

Wet-bulb temperatures, where air cannot be cooled by evaporating water perspiration alone is unable to cool a human body. A wet-bulb temperature of 95° F is umantainable for more than six hours, even for healthy adults resting in the shade.

"Climate Change is the number one challenge of our time."

Odile Renaud-Basso[441]

European Bank for Reconstruction and Development (EBRD) President

Cold and freezing

Most cultivated plants are sensitive to low temperatures, showing negative effects in yield at around 12°C (53.6°F). Light saturation at lower photosynthetic rates and photoinhibition are commonly observed phenomena at low temperatures [9] along with increased chlorophyll 'a' fluorescence. Prolonged exposure to light at low temperatures may produce severe and irreversible photoinhibition, followed by chlorophyll destruction from photo-oxidation and finally death of the tissue. The major effect of cold damage is the decrease in photosynthesis. Furthermore, the export of C from the leaf decreases and soluble carbohydrates accumulates. An advantage of sugar accumulation is that it protects the cells exposed to low temperatures. After exposure to low temperatures, it has been found that rubisco activity increases [9].

Frost can be particularly damaging between flag leaf emergence and ten days after anthesis. The damage appears as an erratic occurrence of aborted spikelet at the base, center or tip of the spikes. It also manifests as sterile florets in parts of the entire spike. This is due to an initial supercooling of plant tissues and an erratic spread of the freezing front through stems and ears later on. The damage during this period occurs at minimum screen temperatures below 0°C and tissue temperatures of around –4°C and is associated with radiative cooling on calm clear nights. No genetic resistance for this type of damage is usually found, and the only way to deal with the problem is through escape at the flowering stage, either by early or late flowering. Early flowering may be a better strategy in environments where terminal heat stress and drought are common stresses [9].

Climate Change & Wheat Yield

Wheat crops have wide adaptability. They can be grown not only in the tropical and sub-tropical zones, but also in temperate zones and cold tracts. Wheat can tolerate severe cold and snow and resume growth with the setting in of warm weather in spring. It can be cultivated at sea level and heights up to 3300 meters.[442]

The best wheat is produced in areas favored with cool, moist weather during the major portion of the growing period followed by dry, warm weather to enable the grain to ripen properly. The optimum temperature range for ideal germination of wheat seeds is 20–25°C (68–77° F) though they can germinate in the temperature range 3.5 to 35°C (38.3–95°F). Rains just after sowing hamper germination and encourage seedling blight. Areas with a warm and damp climate are not suited for wheat growing.

During the heading and flowering stages, excessively high or low temperatures and drought are harmful to wheat. Cloudy weather, with high humidity and low temperatures is conducive to rust attack. Wheat plants require about 14–15°C (57.2–59°) optimum average temperature at the time of ripening. The temperature conditions at the time of grain filling and development are very crucial for yield. Temperatures above 25°C (77°F) during this period tend to depress grain weight. When temperatures are high, too much energy is lost through the process of transpiration by the plants and the reduced residual energy results in poorer grain formation and lower yields. In some parts of the world, wheat is mainly a rabi (winter) season crop, particularly in India.

"The root cause for increasing heat waves in the Indo-Pakistan region is global warming due to human activities resulting in carbon emissions".

Roxy Mathew Koll,
A climate scientist at the Indian Institute of Tropical Meteorology and
a lead IPCC author

Climate change and extreme weather force farmers to compact their planting season and which would result in lower yields. This is particularly true when the weather is adverse during the cropping season. Wet and cool temperatures, during Spring 2022 season, in key parts of the Midwest have delayed farmers' planting plans, leaving them with a few days to get crops in the ground before they start to lose out on a bigger harvest.[443] The causation is food inflation as already high prices for agricultural commodities could rise even more, with supplies thinning as farmers worldwide grapple with tough weather. For corn the situation is particularly tenuous because corn planted after the planting season in the 2nd week of May runs an increased risk of yielding less. With global grain markets already tight due to poor weather in key growing areas and Russia's war in Ukraine, further disruptions to U.S. crops could push crop prices beyond current near-record highs [10].

Corn crops are very sensitive to timing. Corn crops usually produce less grain when planted in the middle to late May. The reason is simple. Planting when the ground is too wet or saturated can smother seeds that need oxygen to germinate. Wet soil also can rot emerging crops and form a crust too tough [11] for sprouts to break through.[444] When corn is planted after May 12, yields start to slip, but can stay high until around May 20. It all depends on the weather. If it's just rainy enough and overcast, there's not much you can do, and we need to obey mother nature. Wet soil in Corn Belt states, Illinois, Indiana, Minnesota, and North Dakota, have prevented farmers from getting their machinery into their fields due to the above-average precipitation over the early three months of Spring 2022. Months leading to Spring season play a critical role as to what to expect. Even if farmers get a late crop on the ground, corn may not fully mature by the time the fall's first frost hits, producing yields one-fifth or more below the norm, according to agricultural economists [11].

Weather is a driver of agricultural yields. The lack of moisture in the winter wheat and excessive moisture in the spring will affect yields and quality especially during the early spring planting season. Not only do farmers depend on the weather but also the first quarter revenues of farming companies are weather dependent. The inclement weather adds another challenge to a punishing period for farmers, seed and chemical suppliers, and tractor makers. Planting delays due to challenging weather conditions in the beginning of the 2022 spring season pushed back purchases by the farmers and since company's do not realize revenues until after the seeds make it to the farmer, weather cut into the company's first-quarter seed sales [10]. While some farmers may gamble on a late-planted crop that could yield a fraction of a typical harvest, many will turn to crop insurance that pays out when farmers are unable to plant seeds by a preset deadline.

It's important we have a good crop for reasonable food prices. We need to have a good crop especially with what's happening in Ukraine.

Voice of Farmer

Climate Impacts on Agriculture and Food Supply

Our lifestyle, as we know, would change due to climate change and it pushes economic opportunities to marginalized communities, especially farmers, to the brink of sustainability issues. Agriculture is an important sector of the U.S. and the World economy. The crops, livestock, and seafood produced in the United States contribute more than $300 billion to the economy each year. When food services and other agriculture related industries are included, the agriculture and food sectors contribute more than $750 billion to the gross domestic product.

Agriculture and fisheries are highly dependent on the climate. Increases in temperature and carbon dioxide (CO_2) can have a significant impact on agriculture (please see Figure 7). Changes in the ozone layer, greenhouse gases and climate change affect agricultural producers significantly as agriculture and fisheries depend on specific climate conditions. Temperature changes can cause habitat ranges and crop planting dates to shift;[445] droughts and floods due to climate change may hinder farming practices [24]. Climate changes may affect the production of maize (corn) and wheat as early as 2030, according to a new NASA study.[446] The net effect is severe for the World as with the interconnectedness of the global food system, impacts in even one region's breadbasket will is felt worldwide. Climate change is a global issue. Finally, paleoclimate data reveals that climate change is not just about temperature. As carbon dioxide has changed in the past, many

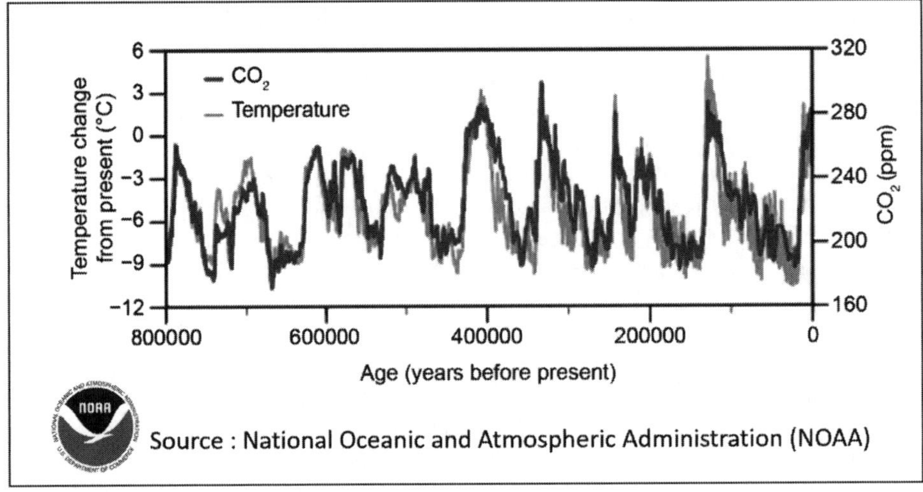

Figure 7: CO_2 vs. Temperature.

other aspects of climate have changed too. Temperature change (light blue) and carbon dioxide change (dark blue) are measured from the EPICA Dome C[447] ice core in Antarctica [25].

Rice yields in the Philippines, Thailand, and Viet Nam could decline by as much as 50% by the end of this century without efforts to combat climate change.[448]

How War in Ukraine Is Tearing Apart the Global Food System

Meanwhile, record input costs are hitting farmers' margins across the world. From crop nutrients to seeds, freight and diesel, the rising prices have set alarm bells ringing about the potential impact on agricultural production at a time when global food costs are already at all-time highs.

Farmers around the world may cut fertilizer usage—possibly by around 10%—to cope with higher prices, which will affect yields, though it's hard to quantify the effect. Still, it's a symptom of a much larger problem. "If global yields suddenly start dropping by 5%, 10% or whatever number it is, even small percentage drops get compounded across the whole world and thereby start to have fairly significant impacts" [26].

Wheat Futures

Both climate changes and the Ukraine war have a considerable impact on wheat prices a fact that can be seen in wheat futures. Supply-and-demand shocks specific to the wheat market were the dominant cause of price spikes between 1991 and 2011 in the three U.S. wheat futures markets (hard red winter, hard red spring, and soft red winter wheat). Focusing specifically on the February 2008 wheat price spike (please see Figure 8), the study finds that, depending on the market, wheat prices in that month would have been 40–62 percent lower in the absence of current crop year supply and demand shocks, such as unexpected weather events that lower wheat yields. Wheat prices would have been forecasting commodities providing immense help to farmers on better supply and demand side management.[449]

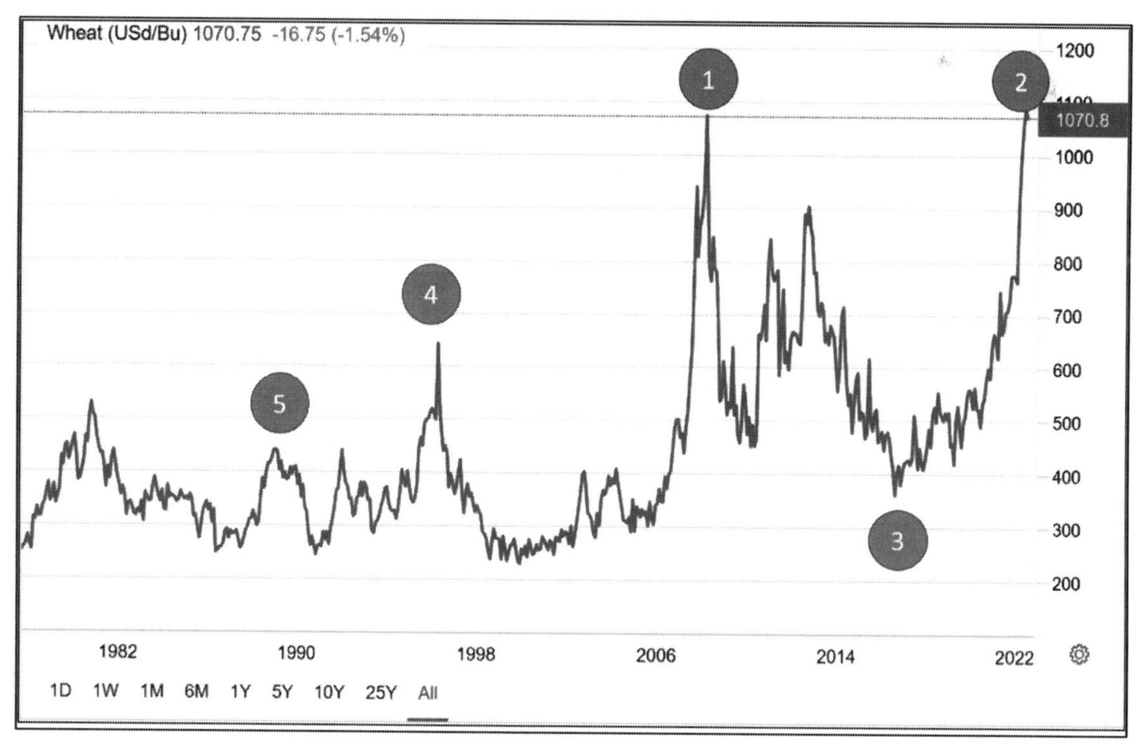

Figure 8: Wheat Futures.

(1) 2008 Spike in Wheat Prices	
	The rapid increase in crop prices between 2006 and mid-2008, while unprecedented in magnitude, was not unique. Two other periods of major rapid runups in prices occurred in 1971–74 and 1994–96. Each price surge resulted from a combination of factors, including depreciation of the U.S. dollar, strong worldwide demand for agricultural products, supply shocks, and policy responses by major trading countries.[450] The initial causes of the late-2006 price spikes included droughts in grain-producing nations and rising oil prices. Oil price increases also caused general escalations in the costs of fertilizers, food transportation, and industrial agriculture.
(2) June 2022 – Wheat supplies	
	June 2022—Chicago wheat futures eased to $10.5 per bushel mark, not far from the 7-week lows of $10.3 extending their volatile momentum as concerns of shortages eased, while investors continued to monitor diplomatic developments regarding possible Ukrainian grain shipments. Forecasts of improved weather in North America and Europe lifted output projections for the summer harvest, while favorable growing conditions in Australia lifted production forecasts to well above the 10-year average. Still, expectations for the resumption of shipments from Ukraine remained subdued as the West is unlikely to relax sanctions on Moscow, a necessary requirement by the Kremlin to open trade corridors in the Ukrainian Black Sea and Sea of Azov ports. 22 million tonnes of Ukrainian grain are estimated to be stuck in port silos since shipments were halted on February 24.[451]
(3) 2015 – Wheat prices fell	
	Cheap oil contributed to abundant global supplies of food in 2014 and prospects of bumper wheat, maize and rice crops in 2015—factors that are driving the sharp decline in international food prices. The agriculture and food sector continue to benefit from less expensive chemical fertilizers, fuel and transportation costs brought on by the previous year's oil price declines, with food prices holding steady despite recent oil price hikes.[452] Between August 2014 and May 2015, wheat prices plunged by 18%, rice prices dropped by 14% and maize prices declined by 6%. However, the arrival of El Nino, the appreciation of the U.S. dollar and the recent increase in oil prices could drive up food prices in the coming months. The demand for maize by the biofuel industry and developments in rice support policies among major producers are other factors that could impact food prices.
(4) 1996 Grain Price shocks	
	The dynamics of food inflation appear to have changed, such that the 1996 grain price shocks had a smaller impact than shocks in the past.[453]

NOAA Star Global Vegetation Health (VH)

The satellite-based global VH System is designed to monitor, diagnose, and predict long- and short-term land environmental conditions and climate-dependent socioeconomic activities. The System is based on earth satellite observations, biophysical theories of vegetation response to the environment, set of algorithms for satellite data processing, interpretation, product development, validation, calibration, and applications.

Satellite observations are principally represented by the Advanced Very High-Resolution Radiometers (AVHRR) flown onto NOAA polar-orbiting satellites. Data is global with a 4 km and 7-day resolution composite. In addition, the System uses data and products from GOES, METEOSAT, MTSAT and DMSP. The System contains the following vegetation health indices: Vegetation Condition index (VCI), Temperature Condition index (TCI), Vegetation Health index (VHI), Soil Saturation index (SSI), no noise Normalized Difference Vegetation Index (SMN), no noise Brightness Temperature (SMT), Fire risk index (FRI); it also contains Drought, Malaria, Vegetation health, Ecosystem and Land sensitivity (to ENSO) products.

El Niño and La Niña are the warm and cool phases of a recurring climate pattern across the tropical Pacific—the El Niño-Southern Oscillation, or "ENSO" for short. The pattern shifts back and forth irregularly every two to seven years, and each phase triggers predictable disruptions of temperature and precipitation. These changes disrupt the large scale air movements in the tropics, triggering a cascade of global side effects.[454]

Vegetation Condition Index (VCI)—Global, 4 km, 7-day Composite, Validated

VCI is based on the pre and post launch calibrated radiances converted to the no noise Normalized Difference Vegetation Index (NDVI=(NIR-VIS)/(NIR+VIS)). The VCI was expressed as a NDVI anomaly relative to the 25-year climatology estimates based on bio-physical and ecosystem laws (law-of-minimum, law-of-tolerance and carrying capacity). The VCI is a proxy for moisture conditions.[455]

Temperature Condition Index (TCI)—Global, 4 km, 7-day Composite, Validated

The TCI is based on 10.3 to 11.3 μm AVHRR radiance measurements converted to the brightness temperature (BT), which was improved through the complete removal of high frequency noise. The BT was expressed as an anomaly relative to 25-year climatology estimates based on bio-physical and ecosystem laws (law-of-minimum, law-of- tolerance and carrying capacity). The TCI is a proxy for thermal conditions.[456]

Vegetation Health Index (VHI)—Global, 4 km, 7-day Composite, Validated

The *Vegetation Health Index (VHI) illustrates the severity of droughts based on the vegetation health and the influence of temperature on plant conditions*. The VHI is a composite index and the elementary indicator used to compute the ASI. It combines both the Vegetation Condition Index (VCI) and the Temperature Condition Index (TCI). The TCI is calculated using a similar equation to the VCI but relates the current temperature to the long-term maximum and minimum, as it is assumed that higher temperatures tend to cause a deterioration in vegetation conditions. A decrease in the VHI would, for example, indicate relatively poor vegetation conditions and warmer temperatures, signifying stressed vegetation conditions, and over a longer period would be indicative of drought. The VHI images are computed for two main seasons and three modalities: dekadal, monthly and annual.

VHI=a * VCI + (1- a) * TCI, where "a" is a coefficient determining the contribution of the two indices. The VHI is a proxy characterizing vegetative health or a combined estimation of moisture and thermal conditions.

Normalized Difference Vegetation Index (NDVI)

The Normalized Difference Vegetation Index (NDVI) measures[457] the "greenness" of the ground cover and is used as a proxy to indicate the vegetative density and health. The NDVI values range from +1 to –1, with high positive values corresponding to dense and healthy vegetation, and low and/or negative values indicating poor vegetative conditions or a sparse vegetative cover. The NDVI anomaly indicates the variation of the current dekad to the long-term average, where a positive value (for example 20 percent) would signify enhanced vegetative conditions compared to the average, while a negative value (for instance –40 percent) would indicate comparatively poor conditions.

No noise (smoothed) Normalized Difference Vegetation Index (SMN)—Global, 4 km, 7-day Composite, Validated

The SMN is derived from the no noise NDVI with pre and post launch calibrated components. The SMN can be used to estimate the start and senescence of vegetation and the start of the growing season and phenological phases.

Historical Drought Frequency

The maps depict the frequency of severe droughts[458] in areas where, (i) 30 percent of the cropped land; or (ii) 50 percent of the cropped land has been affected (please see Figure 9). The historical frequency of severe droughts (as defined by ASI)[459] is based on the entire times series for the years 1984 to 2020.[460]

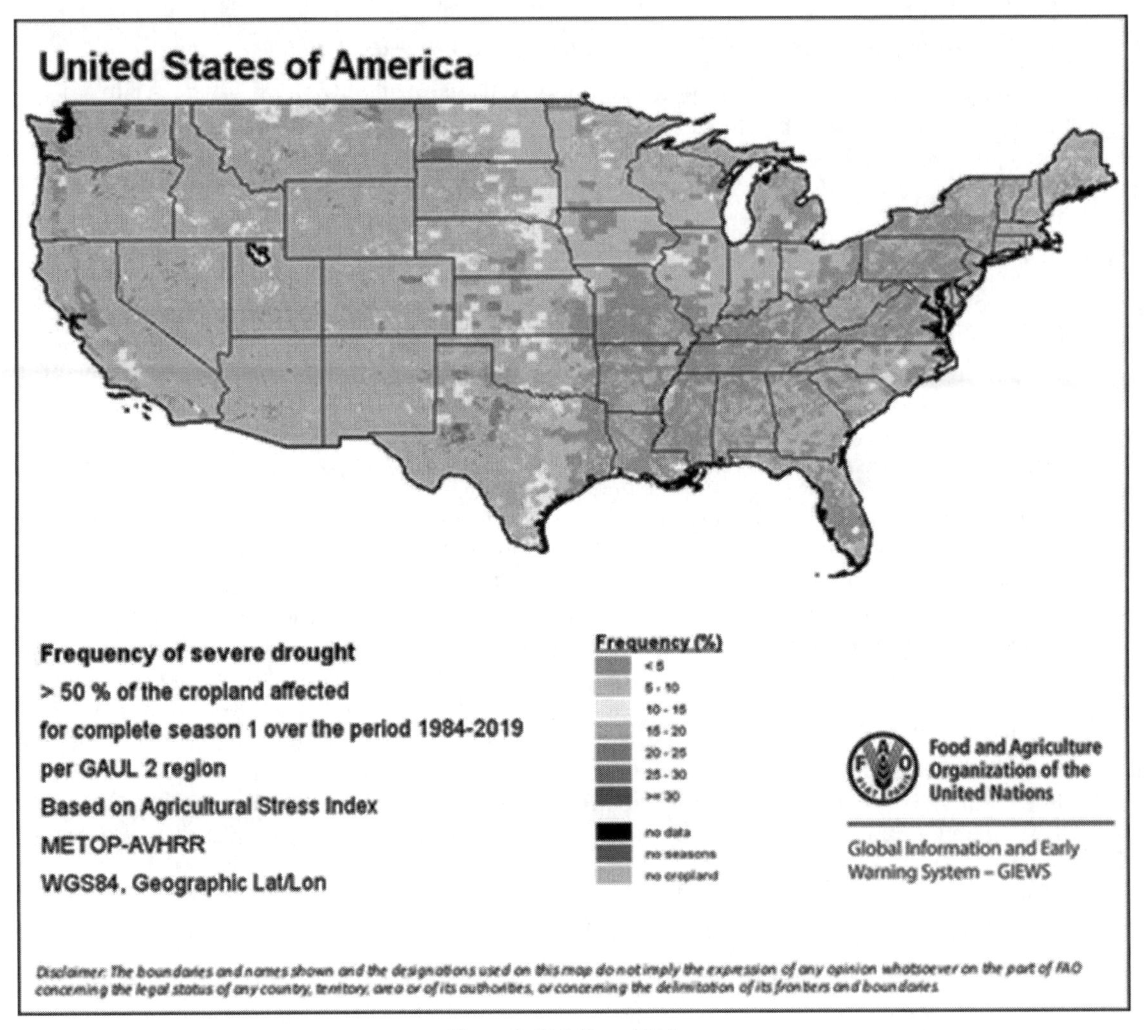

Figure 9: U.S. Drought Map.

Greater (>) than 30% of the crop land was affected for one complete season over the period 1984–2020 per GAUL 2 region. The calculations are based on the Agricultural Stress Index (ASI) using METOP—AVHRR (please see Figure 10).

Coupled Model Intercomparison Project Climate Projections (CMIP)

The CMIP is a standard experimental framework for studying the output of coupled atmosphere-ocean general circulation models. This facilitates assessment of the strengths and weaknesses of climate models which can enhance and focus on the development of future models [12]. A recent addition to CMIP is CMIP 6 extreme.

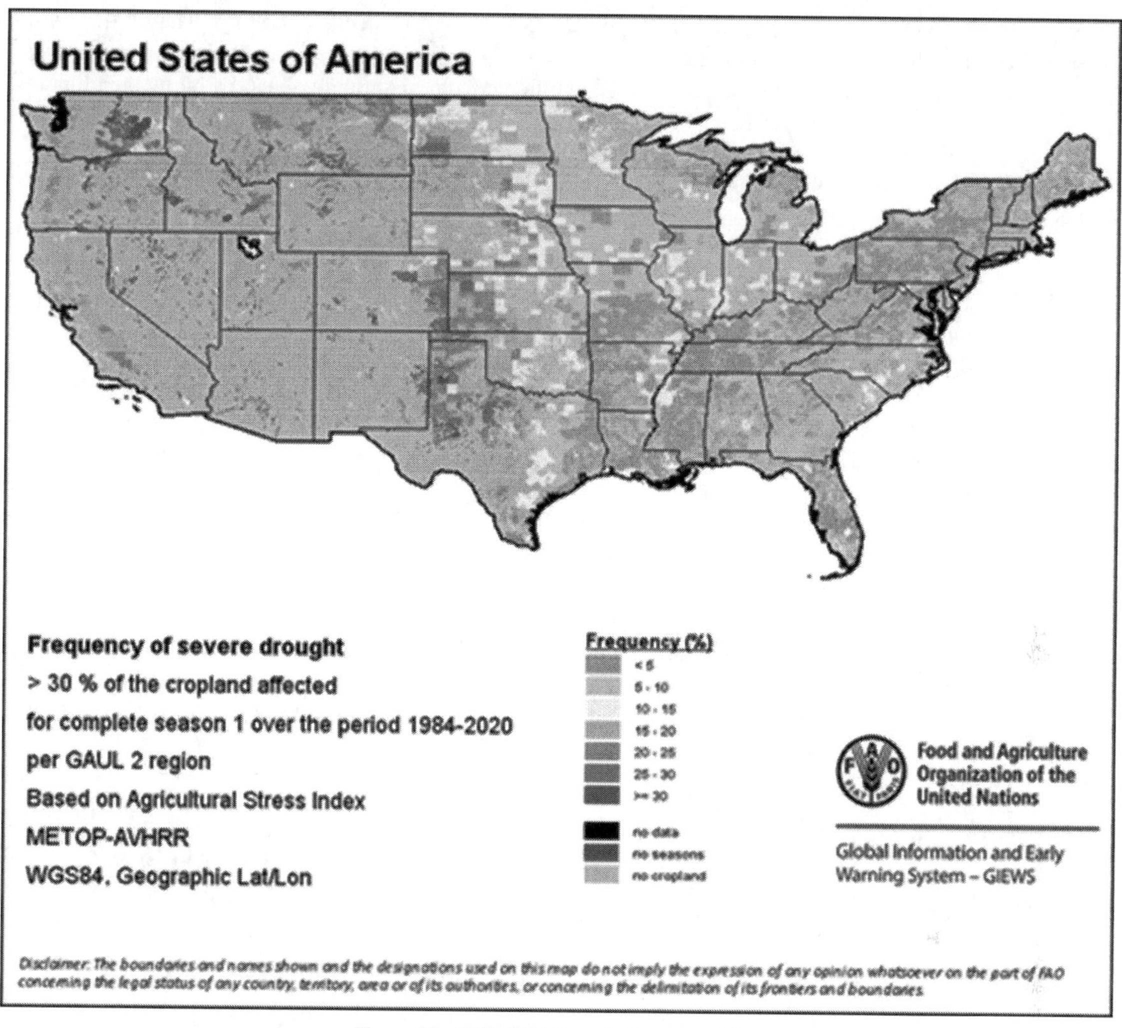

Figure 10: 1984–2020 Drought Frequency Map.

CMIP 5

This catalogue entry provides daily climate projections on single levels from many experiments, models, members, and time periods computed in the framework of the fifth phase of the Coupled Model Intercomparison Project (CMIP5).[461] The term "single levels" is used to express that the variables are computed at one vertical level which can be a surface (or a level close to the surface) or a dedicated pressure level in the atmosphere. Multiple vertical levels are excluded from this catalogue entry. CMIP5 data is used extensively in the Intergovernmental Panel on Climate Change Assessment Reports (the latest one is IPCC AR5, which was published in 2014).

CMIP6

CMIP6 represents a substantial expansion over CMIP5, in terms of the number of modelling groups participating, future scenarios examined, and different experiments conducted. The goal of the CMIP is to generate a set of standard simulations to ensure that each model runs. This catalogue entry provides daily and monthly global climate projections data from many experiments, models and time periods computed in the framework of the sixth phase of the Coupled Model Intercomparison Project (CMIP6).[462] The CMIP6 will consist of the "runs" from around 100 distinct climate models being produced across 49 different modelling groups. The purpose of Climate models is to model the climate from historical values, i.e., how the climate has changed in the past and may change in the future. These models [26] simulate the physics, chemistry and biology of the atmosphere, land, and oceans in detail, and require some of the largest supercomputers in the world to generate their climate projections.[463] For more on CMIP6 and SSPs, please refer to Artificial Intelligence and Heuristics for Enhanced Food Security [13].

CMIP 6 (Extreme)

Climate extreme indices and heat stress indicators are derived from CMIP6 global climate projections. The dataset provides climate extreme indices related to temperature and precipitation as defined by the Expert Team on Climate Change Detection and Indices (ETCCDI), as well as selected Heat Stress Indicators (HSI). The indices are provided for historical and future climate projections (SSP1-2.6, SSP2-4.5, SSP3-7.0, SSP5-8.5) included in the Coupled Model Intercomparison Project Phase 6 (CMIP6) and used in the 6th Assessment Report of the Intergovernmental Panel on Climate Change (IPCC). This dataset provides a comprehensive source of pre-calculated and consistent ETCCDI, and heat stress indicators commonly used by climate science and impact communities. Most models used in this catalogue entry are now available in the Climate Data Store though the indices offered in this entry additionally include ensemble members obtained from the Earth System Grid Federation [14].

Climate projection experiments following the combined pathways of Shared Socioeconomic Pathway (SSP) and Representative Concentration Pathway (RCP). SSPs (please figure 11) are an outcome of the work of an international team of climate scientists, economists and energy system modelers that have built a range

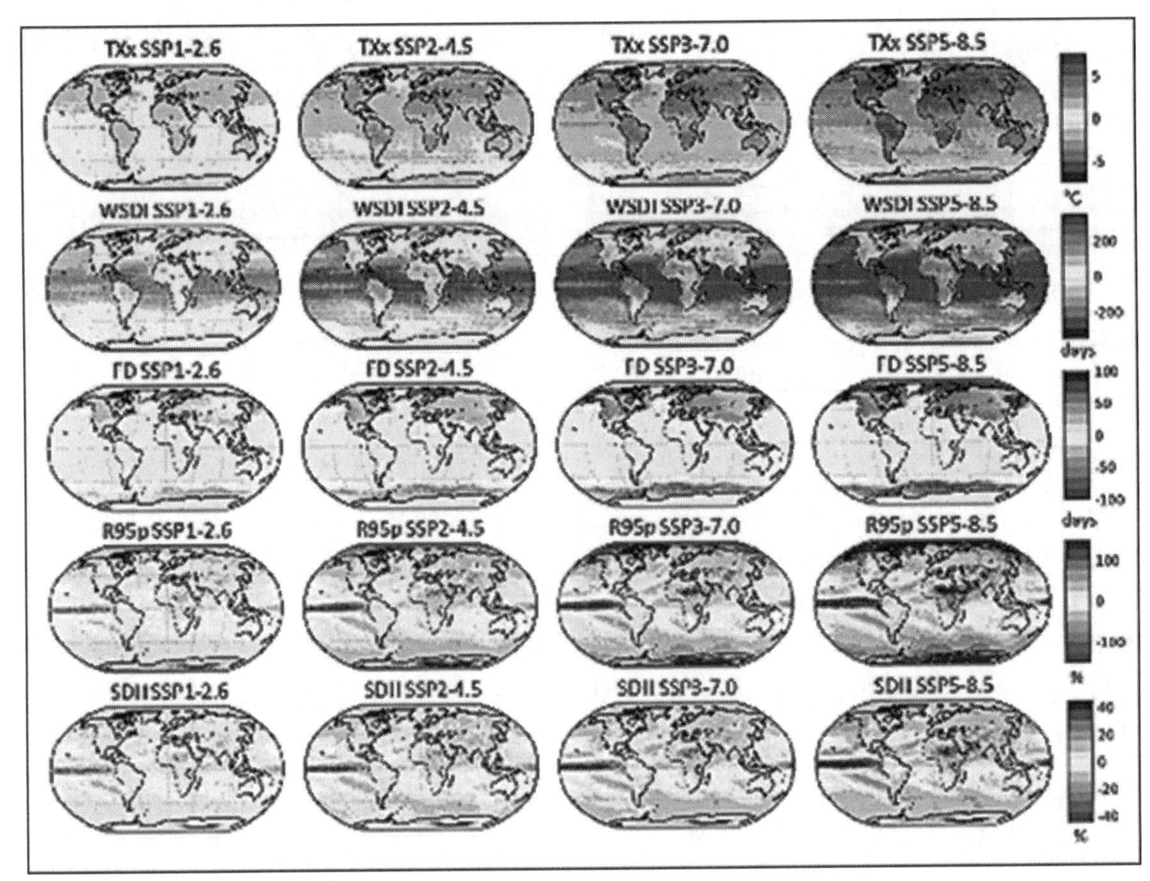

Figure 11: SSPs.

of new "pathways" that examine how global society, demographics and economics might change over the next century. In essence, the SSP scenarios provide different pathways of future climate forcing and provide a toolkit for the climate change research community to carry out integrated, multi-disciplinary analysis [13].

Shared Socioeconomic Pathway (SSP) Projection Models

The following 5 SSP models provide a temperature description of Uttar Pradesh, India for the purpose of growing wheat. For Shared Socioeconomic Pathways' Narratives for each Emission Scenario, please refer to the book Artificial Intelligence and Heuristics for Enhanced Food Security [13].

SSP 1-2.6 (SSP 1)

SSP1 promotes the world shifts gradually, but pervasively, toward a more sustainable path, emphasizing more inclusive development that respects perceived environmental boundaries. Management of the global commons slowly improves, educational and health investments accelerate the demographic transition, and the emphasis on economic growth shifts toward a broader emphasis on human well-being. Agriculture is at the forefront of the World producers with a heavy emphasis on the development of rural & marginalized communities. Driven by an increasing commitment to achieving development goals, inequality is reduced both across and within countries. Consumption is oriented toward low material growth by lowering resource and energy intensiveness. It visualizes the same socioeconomic shifts towards sustainability as SSP 1-1.9. However, temperatures stabilize around 1.8°C higher by the end of the century (please see Figure 12) [13].

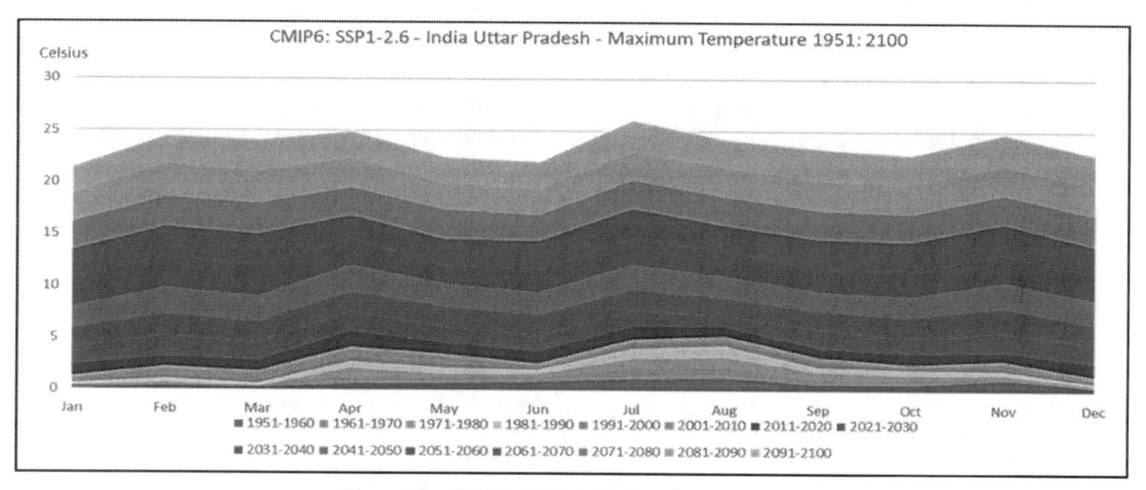

Figure 12: CMIP6 : SSP1-2.6 India UP (1951 : 2100).

SSP 2-4.5 (SSP2)

The world follows a path in which social, economic, and technological trends do not shift markedly from historical patterns. Development and income growth proceed unevenly, with some countries making relatively good progress (mostly energy exporters) while others fall short of expectations (generally countries that rely on a higher percentage of energy imports). Global and national institutions work toward but make slow progress in achieving sustainable development goals. Environmental systems experience degradation, although there are some improvements and overall, the intensity of resource and energy use declines. Global population growth is moderate and levels off in the second half of the century. Income inequality persists or improves only slowly and challenges to reducing vulnerability to societal and environmental changes remain [13]. Socioeconomic factors follow their historic trends, with no notable shifts. Progress toward sustainability is slow, with development and income growing unevenly. In this scenario, temperatures rise by 2.7°C by the end of the century (please see Figure 13).

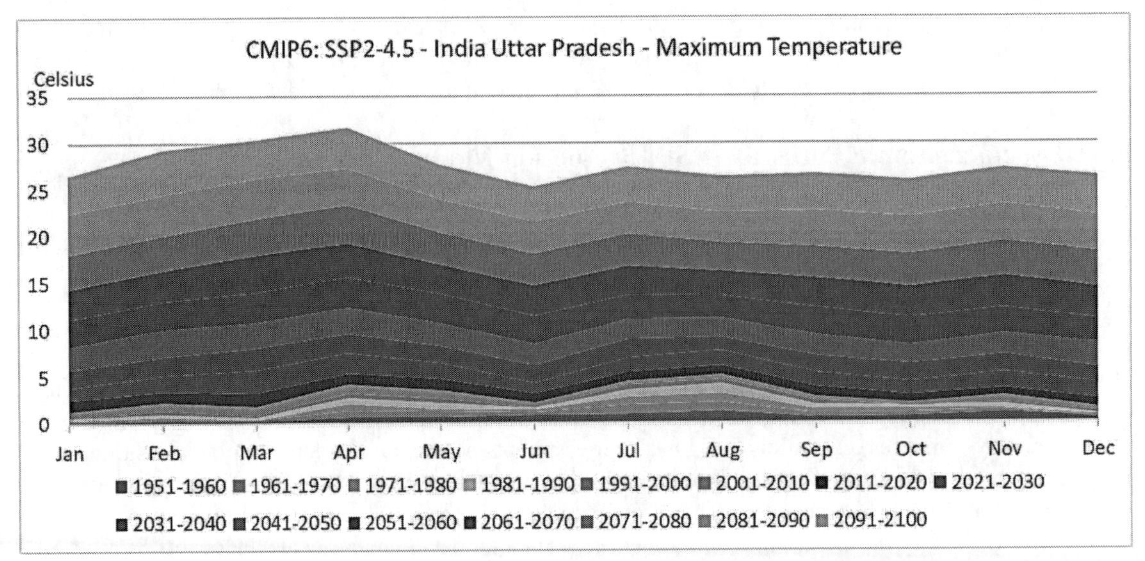

Figure 13: CMIP6 : SSP 2-4.5 India UP Maximum Temperature (1951–2100).

SSP 3-7.0 (SSP 3 Model)

A resurgent nationalism, concerns about competitiveness and security, and regional conflicts push countries to increasingly focus on domestic or, at most, regional issues. Policies shift over time to become increasingly oriented toward national and regional security issues [13]. Countries focus on achieving energy and food security goals within their own regions at the expense of broader-based development (a trend that many countries exhibited during Ukraine war). Investments in education and technological development decline. Economic development is slow, consumption is material-intensive, and inequalities persist or worsen over time. Population growth is low in industrialized and high in developing countries. By the end of the century, average temperatures have risen by 3.6°C (please see Figure 14).

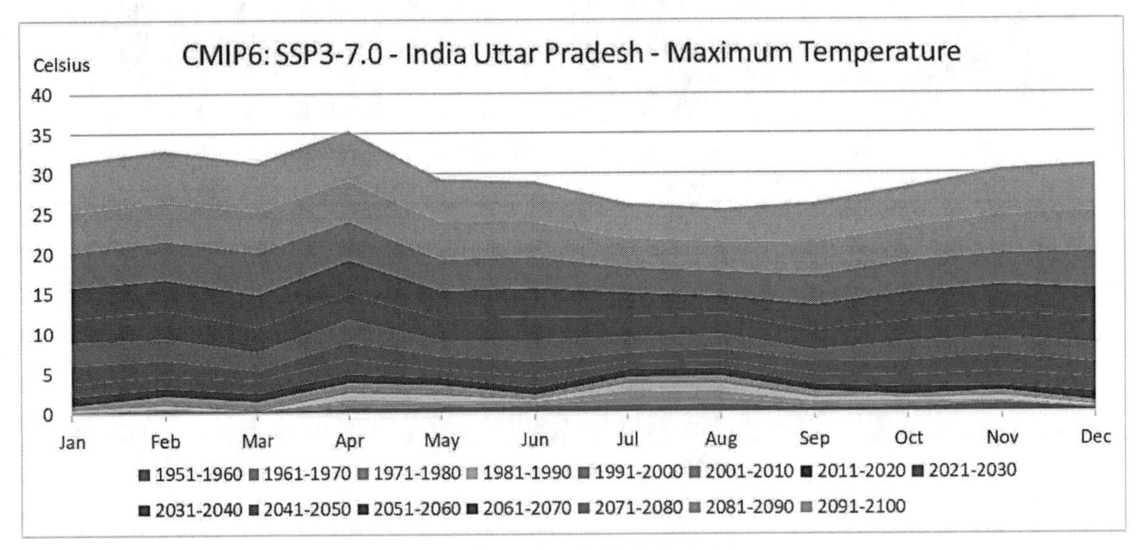

Figure 14: CMIP6: SSP 3-7.0 India UP.

SSP5-8.8 (SSP 5 Model)

This world places increasing faith in competitive markets, innovation, and participatory societies to achieve rapid technological progress and development of human capital as the path to sustainable development. Global markets are increasingly integrated. There are heavy investments in health, education, and institutions as well to enhance human and social capital. At the same time, the push for economic and social development is coupled with the exploitation of abundant fossil fuel resources and the adoption of resource and energy intensive lifestyles around the world (please see Figure 15).

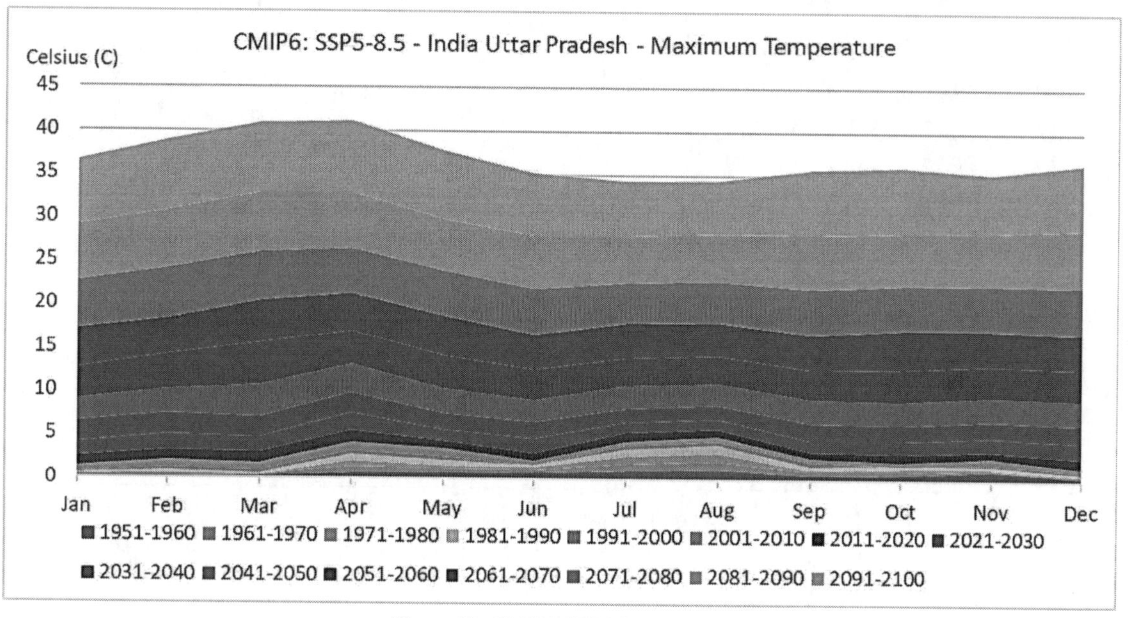

Figure 15: CMIP6: SSP5-8.5 India UP.

All these factors lead to the rapid growth of the global economy, while the global population peaks and declines in the 21st century. Local environmental problems like air pollution are successfully managed. There is faith in the ability to effectively manage social and ecological systems by geo-engineering if necessary. By 2100, the average global temperature is a scorching 4.4°C higher [13]. As can be seen as the SSPs move from SSP 1-2.6 to SSP 5-8.5, the temperature rise can be observed.

Kansas & Wheat Production

Kansas[464] grows winter wheat that is planted and sprouts in the fall, becomes dormant in the winter, grows again in the spring, and is harvested in early summer. All the wheat grown in Kansas in a single year would fit in a train stretching from western Kansas to the Atlantic Ocean. Kansas stores more wheat than any other state. On average, Kansas is the largest wheat producing state. Nearly one-fifth of all wheat grown in the United States is grown in Kansas. Therefore, it is called the "Wheat State" and "Breadbasket of the World." The net effect of a wheat yield drop is on food availability across the nation and to the world as Kansas is the breadbasket.

The following map (please see Figure 16) provides Kansas 2021 Winter Wheat Yield (Bushels Per Acre) County level data[465] & USDA National Agricultural Statistics Service (NASS) Query[466] can provide tabular data.

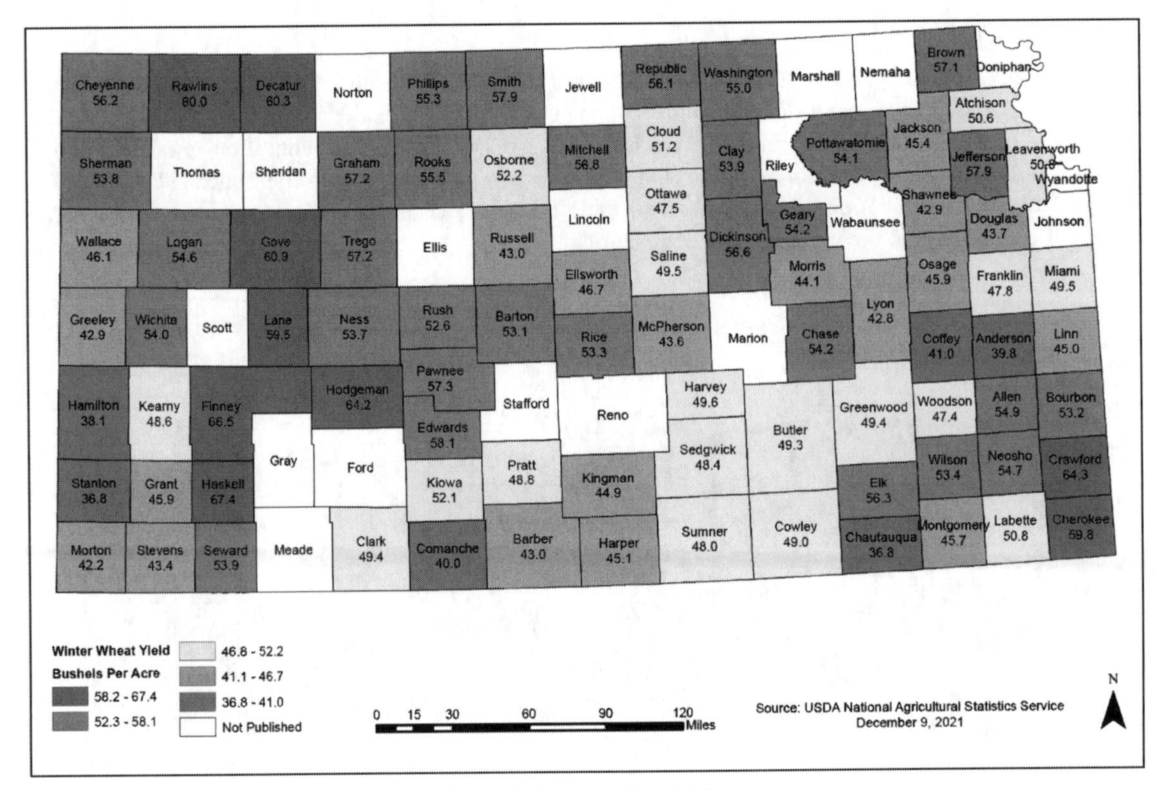

Figure 16: Kanasas - County Map.

Yield Loss & Opportunity Cost

Russia's war in Ukraine has disrupted global food supplies, driving up wheat demand and prices. However, after months of drought, many western Kansas farmers won't have crops to sell. Even with wheat selling for near-record-high prices as the war in Ukraine disrupts the world's food supplies, a lot of farmers in western Kansas won't have any to sell.[467] In addition, those who made it through the drought with enough crops to harvest will likely end up with far fewer bushels than they had last year, a downturn that limits the state's ability to help ease the global food crisis (please see Figure 17). The US Department of Agriculture estimates that wheat fields statewide will average roughly 39 bushels per acre this year,[468] down sharply from 52 bushels per acre last year. However, many farms in the western half of the state will produce far less than that [15].

Based on May 1 conditions, Kansas's 2022 winter wheat crop is forecast at *271 million bushels*, down *26%* from last year's crop, according to the USDA's National Agricultural Statistics Service. Average yield is forecast at *39 bushels per acre*, down 13 bushels from last year. Acreage to be harvested for grain is estimated at *6.95 million acres*, down 50,000 acres from last year. This would be 94% of the planted acres, below last year's 96% of harvested acreage.

Even land management could recover significantly more yields. As one farmer pointed out [15], his wheat field ended up higher than the 27-bushel average, something they credit to the way land rests between plantings. However, even with conservative land management strategies, fields might still only produce half of what they did last year—*all because of too much heat and not enough rain*. Not only food security concerns are there but farmers sustainability is also at the center of it. As a result of extreme weather and drought at least half the wheat fields in western Kansas won't produce enough for farmers to break even. We're losing money even with the highest price of wheat that we've probably ever seen in the past 50 or 100 years.

Figure 17: Kansas Drought fields (May 2022).

Opportunity Cost Conundrum

As a result of extreme weather and drought at least half the wheat fields in western Kansas won't produce enough for farmers to break even. We're losing money even with the highest price of wheat that we've probably ever seen in the past 50 or 100 years. [15]

Western Kansas wheat crops are failing just when the world needs them the most. Kansas's wheat production is getting extremely affected by climate change due to increased changes in temperature as well as the precipitation in that region (please see Figure 18). Plants, like humans, are organisms, requiring certain patterns in weather to provide a rich harvest. Yet because of the varying conditions of heat plants have a harder time adapting to the changes in temperature and have a harder time going through their chemical processes as such. This includes wheat as well—therefore, the quality of wheat products will inevitably be reduced. An increased precipitation that is unevenly spread throughout Kansas is also directly proportional to

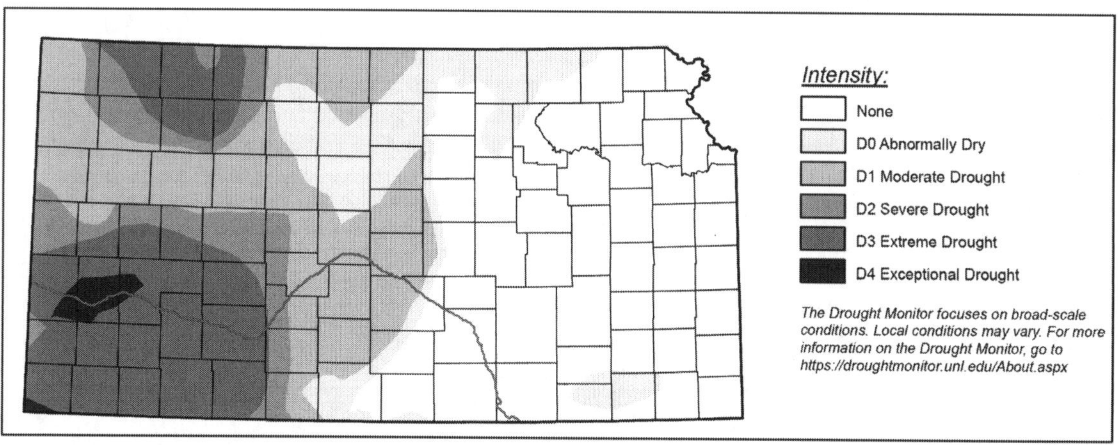

Figure 18: Drought monitor.

not enough water in certain areas to harvest wheat. Finally, climate change reduces the quality as well as the supply of wheat to the public resulting in a compromised food security to the US public.

New research from the Kansas State University shows that rising overnight temperatures hurt wheat harvests.[469] Rice, corn, and barley face similar problems [16].

An increase of only 1.8 degrees Fahrenheit in overnight temperatures caused a 5% reduction in wheat yield. The causality is simple—plants are like humans. As per Colleen Doherty, an associate professor of biochemistry at North Carolina State University, "plants are constantly regulating their biological processes—gearing up for photosynthesis just before dawn, winding that down in the late afternoon, determining precisely how and where to burn their energy resources" [16].

Wheat[470] yields of Kansas (1982:2020) experience a drastic drop for the years 1989, 1995, and 2014 [17]. The drastic drop has a real-life event associated with it. For instance, the 1989 drop was due to drought conditions [18]. The principal grain affected by the drought is hard red winter wheat, which is produced mainly in Kansas, Colorado, Montana, Nebraska, Oklahoma, and South Dakota.[471] The hardest-hit state is Kansas, where the Agriculture Department projected a harvest of 202.4 million bushels, compared with 366.3 million in 1987 and 323 million in 1988. Kansas is the leading wheat producer; its output normally accounts for about a third of the winter wheat and 16 to 17 percent of all wheat.

The implication was a direct hit on consumers and food security. Winter wheat is planted in the fall, lies dormant in the winter, and matures in the spring. It is normally about 75 percent of the total national wheat crop. The hard red variety, the major wheat used in bread and the kind produced in Kansas and other parts of the Central and Southern Plains, normally represents more than two-thirds of all winter wheat [18]. In 1995–1996, droughts[472] caused severe losses to the farmers of Kansas and the impact has been felt worldwide due to an increase in prices [19]. Droughts punctuated[473] by floods, and early rains, have contributed to the drop in wheat yields in 2014 [19]. The Kansas wheat harvest in 2014 may be one of the worst on record and the loss doesn't just hurt Kansas, according to a Kansas State University expert [20].

The net effect of a wheat yield drop is on the food availability across the nation and to the world as Kansas is the breadbasket (please see Figure 19). The Satellite of the ML model that would be developed

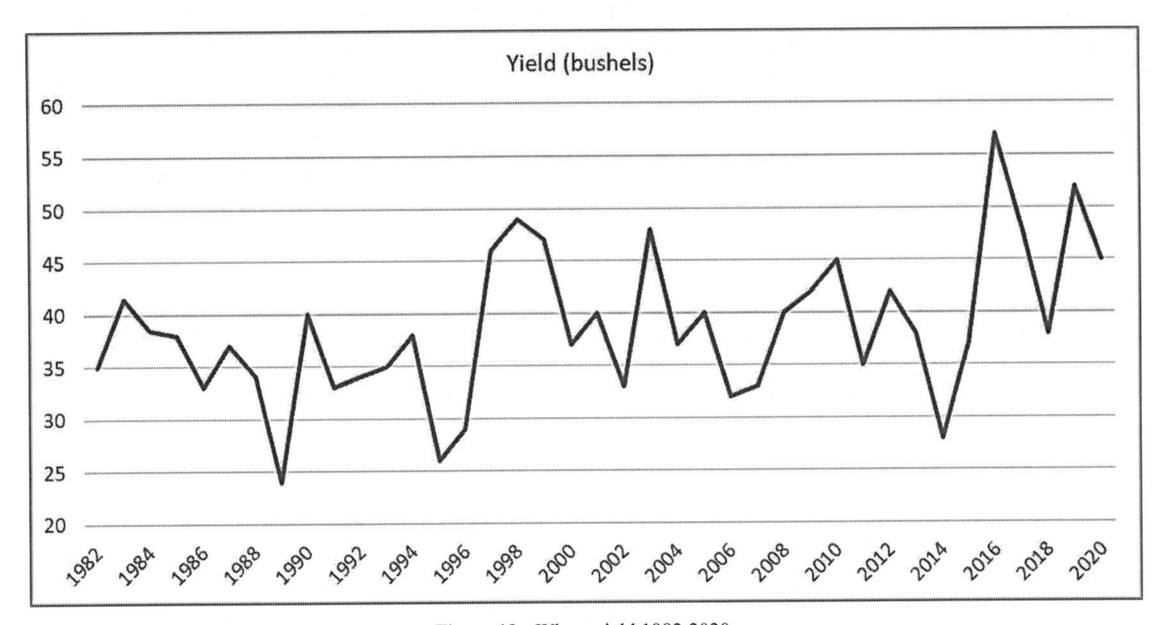

Figure 19: Wheat yield 1982:2020.

later is to tie together weather and satellite impacts[474] for imagery to predict future yield issues. The key parameters required to model the impacts of droughts and yield reductions are

- Vegetation Condition index (VCI)—VCI, based on the pre and post launch calibrated radiances converted to the no noise Normalized Difference Vegetation Index.[475]
- Temperature Condition index (TCI)—*TCI is a proxy for thermal conditions.*
- Vegetation Health index (VHI)—VHI is a proxy characterizing vegetative health or a combined estimation of *moisture and thermal conditions.*
- No noise (smoothed) Normalized Difference Vegetation Index (SMN)—The SMN is derived from the no noise NDVI, with components that were pre and post launch calibrated. SMN can be used to estimate the start and senescence of vegetation, start of the growing season and phenological phases.
- No noise (smoothed) Brightness Temperature (SMT)—The SMT is the BT with completely removed high frequency noise that can be used for the estimation of thermal conditions, cumulative degree days and other parameters of interest.

The purpose of satellite data[476] is to overlay SMN, SMT, VCI, TCI, and VHI for all years to predict the next occurrence of food insecurity (please see Figure 20).

Year	Yield (bushels)	FoodSecuri	TCI_WK1	VHI_WK1	TCI_WK2	VHI_WK2	TCI_WK3	VHI_WK3	TCI_WK4	VHI_WK4
1982	35	Yes	53.35	62.36	60.09	63.94	64.8	63.68	68.1	62.12
1983	41.5	Yes	43.86	39.34	52.47	40.18	62.01	41.98	69.16	42.94
1984	38.5	Yes	85.71	51.45	81.82	52.01	74.01	51.71	65.43	51.54
1985	38	Yes	-1	-1	-1	-1	-1	-1	-1	-1
1986	33	Yes	26.96	36.43	25.29	36.65	21.44	36.39	20.97	37.17
1987	37	Yes	29.93	38.23	33.86	40.36	35.46	41.95	33.7	42.77
1988	34	Yes	56.15	45.59	60.72	47.37	61.23	47.97	59.83	48.54
1989	24	No	8.75	41.97	11.36	42.37	14.18		17.39	41.68
1990	40	Yes	8.7	33.94	9.69	34.49	12.03	35.08	17.42	37.11
1991	33	Yes	36.22	44.23	38.75	45.77	37.01	46.14	32.57	45.45

Figure 20: Wheat Table.

Temperature Condition Index (TCI)—Global, 4 km, 7-day composite, validated

The TCI is based on AVHRR radiance measurements in the 10.3 to11.3 μm range converted to brightness temperatures (BTs), which were improved through the complete removal of high frequency noise. The BTs were expressed as an anomaly relative to the 25-year climatology estimates based on bio-physical and ecosystem laws (law-of-minimum, law-of- tolerance and carrying capacity). The TCI is a proxy for thermal conditions.[477]

$$Y = -0.3559\,x + 757.0745$$

As can be seen the TCI slope is negatively influenced over time. Put it another way, over the years, the temperature gradient has a negative drop of 35.59%. That is, a drop in the temperature health index (TCI) can be observed in future years (please see Figure 21).

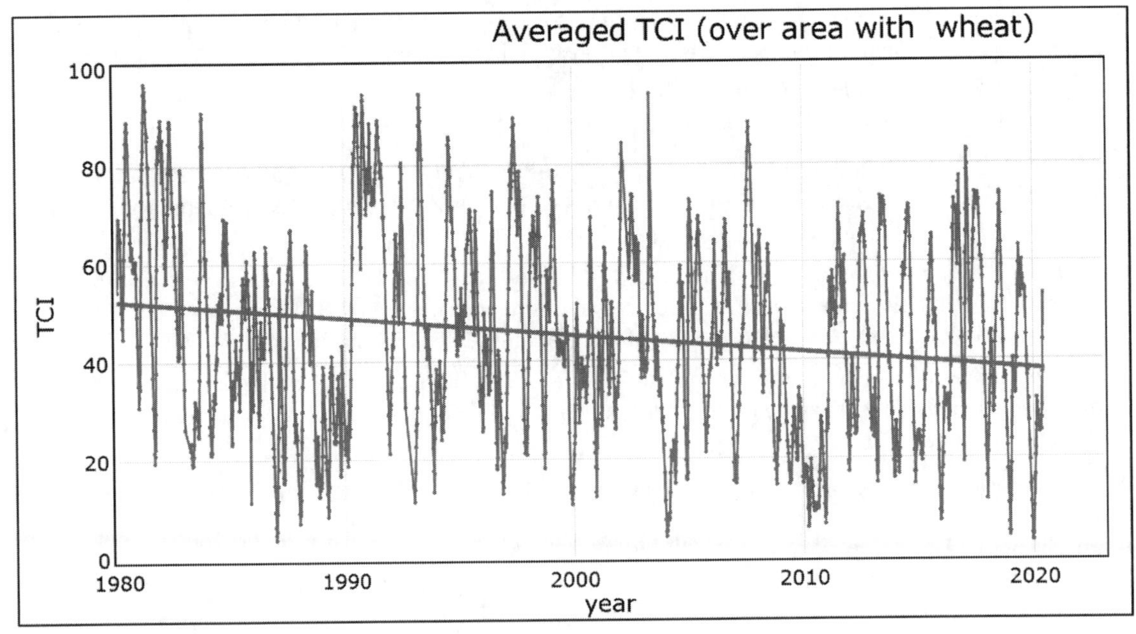

Figure 21: Kansas TCI (1980:2020).

Drought & TCI

Let's look at the Drought Area of Kansas during the drought and normal years. In the drought years Wheat[478] yield of Kansas (1982:2020) and you would see a drastic drop in the years 1989, 1995, and 2014 [17] since the heat stress or Temperature Condition index during these years is severe or extreme (please see Figure 22).

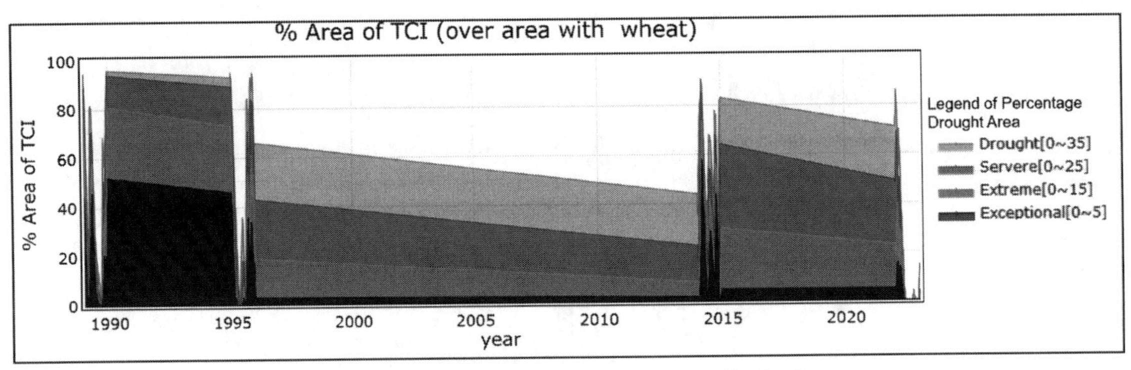

Figure 22: Kansas % Area of TCI (over area with wheat).

Drought & VHI

VHI during the drought years can be seen clearly: in 1989, February through June (1980.2–1989.6), Kansas has experienced severe, extreme, and exceptional droughts. For the year 1995,[479] June through December, Kansas has been under severe drought. The year 2014,[480] from February through June has registered severe droughts in Kansas.

As can be seen in 1989 droughts occurred during the planting season (February–June); in1995 droughts occurred during the harvest season; in 2014 droughts were severe in the planting season, and in 2022 droughts are passing though the planting season. The severity is high in 1989 and 2022. The prognosis for the 2022 yield is dire (please see Table 2).

Table 2: Area VHI (over area with wheat).

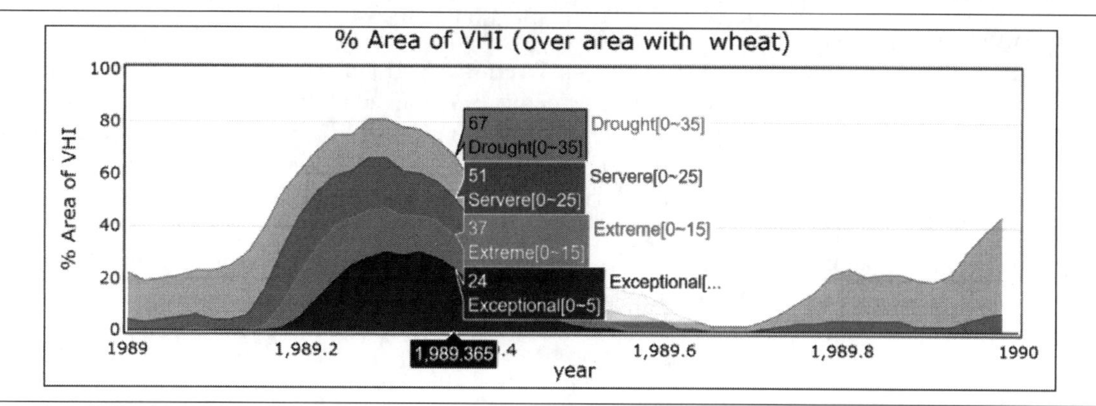

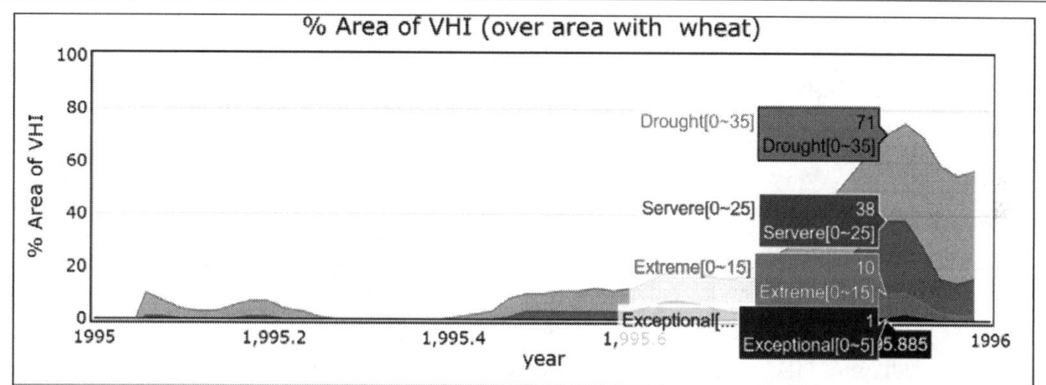

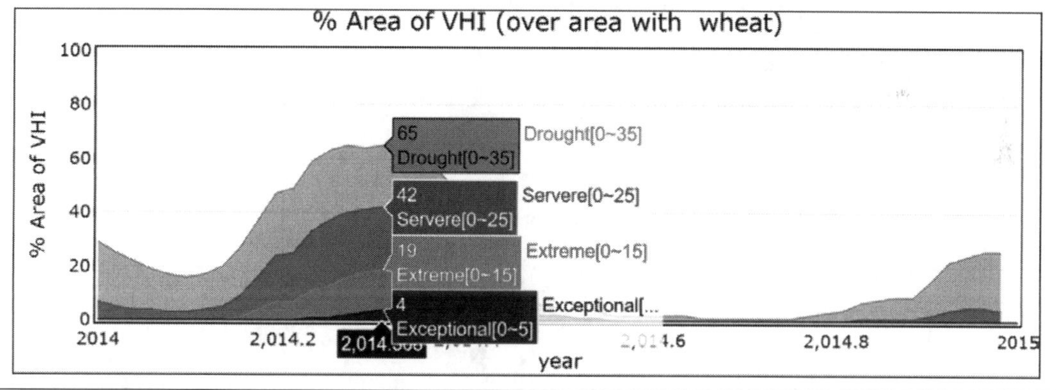

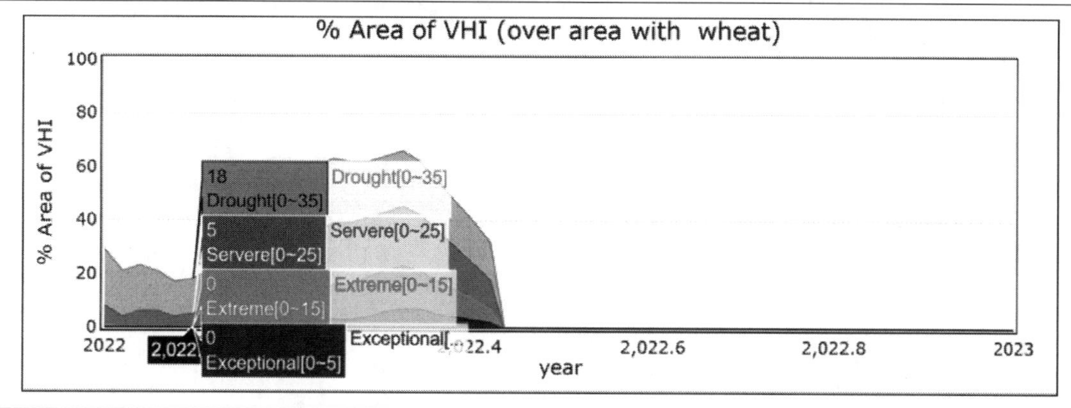

Mathematical Modeling

Wheat demand prediction and wheat futures can be analyzed with econometric models based on standard ordinary least squares (OLS) regression. In addition, the forecasting performance of the wheat futures market is analyzed and compared to out-of-sample forecasts derived from an additive AutoRegressive Integrated Moving Average (ARIMA)[481] model and the error-correction model [27]. In the following section, we have applied ensemble modeling and advanced machine learning to develop Kansas & Indian major wheat production models (please see Table 3 for the Kansas wheat model). Our unique contribution is the analysis of the signature between yield drops (below 1989, 1995, and 2014) due to extreme weather events (droughts & heat waves) and vegetation health index data that can hold patterns of water & temperature heat stressors. We have applied Linear regressive, ensemble, Vector Auto Regressive, and Prophet models.

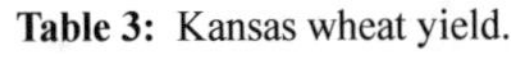

Table 3: Kansas wheat yield.

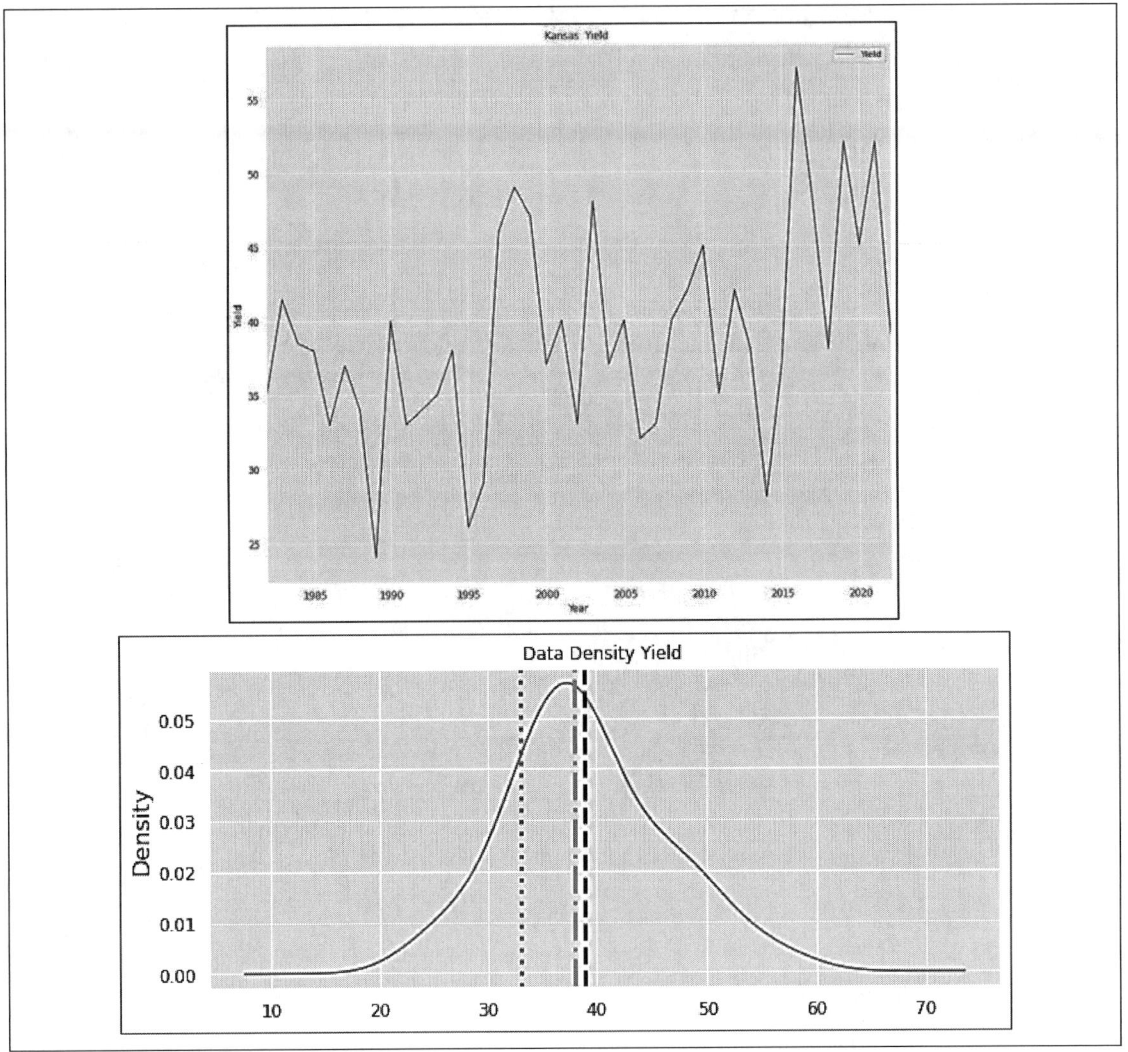

Machine Learning Model: Drought & Wheat Yield Production Linkage in Kansas

Kansas's wheat production is getting extremely affected by climate change: increased changes in temperatures and precipitation levels. Plants, like humans, are organisms that have certain weather pattern requirements to provide a rich harvest. Yet because of the varying heat conditions of plants have a harder time adapting to the changes in temperature and have a harder time going through their chemical processes as such. This includes wheat as well—therefore, the quality of wheat products will inevitably be inferior. An increased precipitation that is unevenly spread throughout Kansas is also directly proportional to inadequate water in certain areas to harvest wheat. Finally, climate change reduces the quality as well as the supply of wheat to the public resulting in a compromised food security to the US public. As part of the model, I would like to develop a Machine Learning model that learns drought impacts on the Wheat yield in Kansas.

Kansas Wheat Yield Production & Drought Lin kage	

Let's build a model that depicts the impact of drought on Kansas Wheat production:

	Software code for this model: FinalKansasWheatProduction.ipynb (Jupyter Notebook Code)

Data Sources

Kansas Wheat History[482]	Kansas Wheat Yield data (1982:2021) Important variables: Area Planted for all purposes Area Harvested for grains Yield per acre Production	USDA
NOAA STAR[483]—Global Vegetation Health Products: Province-Averaged VH	Mean data for USA Province = 17: Kansas, from 1982 to 2022, weekly for cropland[484] area only year, week, SMN, SMT, VCI, TCI, VHI	

Second, agricultural Wheat calendars for Kansas.[485]

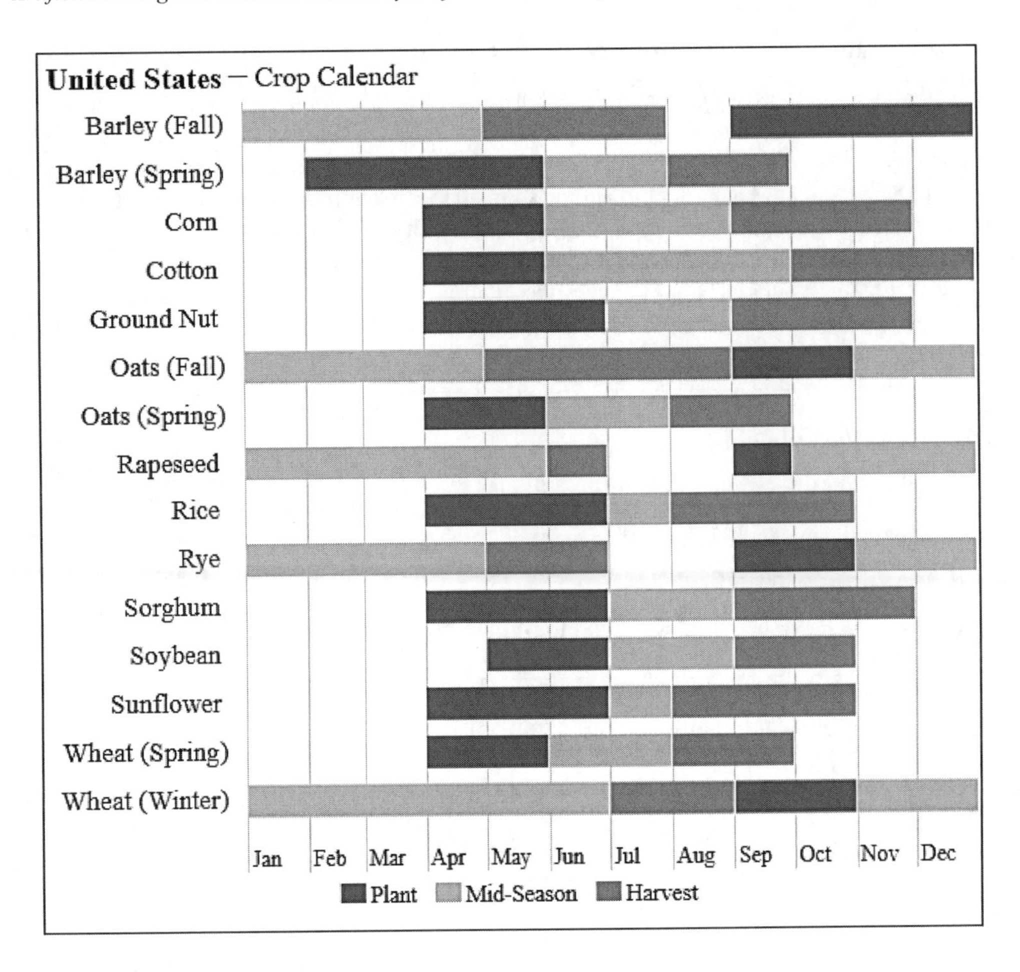

Step 1: Import Libraries

Import open source and machine learning libraries to process the data.

```
import numpy as np
import pandas as pd
import functools
from datetime import date
import plotly.graph_objects as go
import seaborn as sns; sns.set()
import statsmodels.api as sm
from statsmodels.tsa.arima_process import ArmaProcess
from scipy import stats
from scipy.stats import pearsonr
import matplotlib.pyplot as plt
from sklearn.preprocessing import MinMaxScaler
from sklearn.model_selection import train_test_split
from sklearn.linear_model import LinearRegression
from sklearn.metrics import r2_score
from sklearn import metrics
from sklearn.impute import SimpleImputer
import eli5
```

Step 2: Load Kansas Wheat Data Frame 1982:2020

Load Kansas Wheat Data.

```
KansasWheatdf = pd.read_csv("AverageKansasWheatProduction.csv")
KansasWheatdf = KansasWheatdf[1:]
KansasWheatdf["Year"] = KansasWheatdf["Year"].apply(lambda x : int(x[:4]))
KansasWheatdf.head(10)
```

Output:

The data frame consists of Area Harvested, Yield, and Production.

	Year	Area Planted	Area Harvested	Yield	Production
1	1982	14100	13100	35	458500
2	1983	13200	10800	41.5	448200
3	1984	13300	11200	38.5	431200
4	1985	12400	11400	38	433200
5	1986	11500	10200	33	336600
6	1987	10700	9900	37	366300
7	1988	10200	9500	34	323000
8	1989	12400	8900	24	213600
9	1990	12400	11800	40	472000
10	1991	11800	11000	33	363000

Step 3: Load NOAA STAR Mean data for USA Province 17 Kansas Data.

```
Satelite_data = pd.read_csv("MeandataforUSAProvince17Kansas.csv")
Satelite_data.head()
```

Output:

Satellite data consists of SMN, SMT, VCI, TCI, and VHI data.

	year	week	SMN	SMT	VCI	TCI	VHI
0	1982	1	0.193	276.46	71.30	54.53	62.91
1	1982	2	0.184	276.20	67.89	61.29	64.59
2	1982	3	0.175	276.35	62.83	66.07	64.45
3	1982	4	0.166	276.81	56.64	69.35	63.00
4	1982	5	0.161	278.08	53.47	68.64	61.06
5	1982	6	0.162	280.06	53.10	65.28	59.19
6	1982	7	0.166	282.20	49.07	65.65	57.36
7	1982	8	0.174	284.18	47.71	67.50	57.61
8	1982	9	0.187	286.30	49.62	64.85	57.23
9	1982	10	0.201	288.33	53.46	57.70	55.58

Step 4: Pivot Satellite data (Rows to Columns)

Satellite data is in a row format. We need to pivot the data to arrange the weekly data.

```
Satelite_data["Week"]= Satelite_data["week"].apply(lambda x: str(x) +'Wk' )
Statelite_df =Satelite_data.pivot_table(index = ["year"],values = [' SMN', 'SMT', 'VCI', 'TCI', ' VHI'],columns = "Week").reset_index()
Statelite_df.head()
Statelite_df.columns = list(map("_".join, Statelite_df.columns))
Statelite_df
```

Pivot the satellite rows into weekly column data.

Output:

	year_	SMN_10Wk	SMN_11Wk	SMN_12Wk	SMN_13Wk	SMN_14Wk	SMN_15Wk	SMN_16Wk	SMN_17Wk
0	1982	0.201	0.216	0.235	0.258	0.282	0.298	0.312	0.330
1	1983	0.167	0.180	0.194	0.211	0.233	0.257	0.282	0.304
2	1984	0.192	0.200	0.209	0.220	0.237	0.258	0.279	0.300
3	1985	0.225	0.242	0.262	0.286	0.311	0.335	0.355	0.373
4	1986	0.202	0.218	0.233	0.255	0.277	0.295	0.312	0.327
5	1987	0.210	0.224	0.239	0.256	0.275	0.296	0.318	0.338
6	1988	0.200	0.214	0.231	0.252	0.272	0.293	0.315	0.336
7	1989	0.155	0.162	0.170	0.179	0.195	0.212	0.228	0.247
8	1990	0.353	0.199	0.216	0.232	0.247	0.264	0.283	0.305
9	1991	0.205	0.217	0.231	0.247	0.262	0.279	0.295	0.310

Step 5: Combine Kansas Wheat and Satellite data

To model the drought yield drop, combine Satellite data with Kansas yield.

```
combined_df= KansasWheatdf.merge(Statelite_df , left_on ="Year", right_on = "year_" )
combined_df
```

Output:

	Year	Area Planted	Area Harvested	Yield	Production	year_	SMN_10Wk	SMN_11Wk	SMN_12Wk	SMN_13Wk	...	VCI_49Wk	VCI_4Wk	V
0	1982	14100	13100	35.0	458500	1982	0.201	0.216	0.235	0.258	...	49.19	56.64	
1	1983	13200	10800	41.5	448200	1983	0.167	0.180	0.194	0.211	...	22.25	16.04	
2	1984	13300	11200	38.5	431200	1984	0.192	0.200	0.209	0.220	...	60.44	38.64	
3	1985	12400	11400	38.0	433200	1985	0.225	0.242	0.262	0.286	...	36.36	-1.00	
4	1986	11500	10200	33.0	336600	1986	0.202	0.218	0.233	0.255	...	47.85	53.98	
5	1987	10700	9900	37.0	366300	1987	0.210	0.224	0.239	0.256	...	56.87	49.44	
6	1988	10200	9500	34.0	323000	1988	0.200	0.214	0.231	0.252	...	74.25	37.11	
7	1989	12400	8900	24.0	213600	1989	0.155	0.162	0.170	0.179	...	65.11	66.39	
8	1990	12400	11800	40.0	472000	1990	0.353	0.199	0.216	0.232	...	64.61	57.69	
9	1991	11800	11000	33.0	363000	1991	0.205	0.217	0.231	0.247	...	29.93	58.37	
10	1992	12000	10700	34.0	363800	1992	0.202	0.220	0.237	0.253	...	14.33	34.18	
11	1993	12100	11100	35.0	388500	1993	0.146	0.167	0.189	0.211	...	55.82	9.02	
12	1994	11900	11400	38.0	433200	1994	0.190	0.202	0.216	0.234	...	-1.00	47.31	

Step 6: Visualize Yield Data

Plot the model data to see the trends and seasonality in the data.

```
numerical_features=['Area Planted', 'Area Harvested', 'Yield', 'Production']

def show_density(var_name,var_data):
    from matplotlib import pyplot as plt
    print("\n" + var_name + "\n")
    rng = var_data.max() - var_data.min()
    var = var_data.var()
    std = var_data.std()
    print('\n{}:\n - Range: {:.2f}\n - Variance: {:.2f}\n - Std.Dev: {:.2f}\n'.format(var_name, rng, var, std))
    # Get statistics
    min_val = var_data.min()
    max_val = var_data.max()
    mean_val = var_data.mean()
    med_val = var_data.median()
    mod_val = var_data.mode()[0]
print('Minimum:{:.2f}\nMean:{:.2f}\nMedian:{:.2f}\nMode:{:.2f}\nMaximum:{:.2f}\n'.format(min_val,
                                                                                mean_val,
                                                                                med_val,
                                                                                mod_val,
                                                                                max_val))
    fig = plt.figure(figsize=(10,4))
    # Plot density
    var_data.plot.density()
    # Add titles and labels
    plt.title('Data Density ' + var_name, fontsize=12)
    plt.ylabel('Density', fontsize=16)
    plt.tick_params(axis = 'both', which = 'major', labelsize = 12)
    plt.tick_params(axis = 'both', which = 'minor', labelsize = 12)
    # Show the mean, median, and mode
    plt.axvline(x=var_data.mean(), color = 'black', linestyle='dashed', linewidth = 3)
    plt.axvline(x=var_data.median(), color = 'darkgray', linestyle='dashdot', linewidth = 3)
    plt.axvline(x=var_data.mode()[0], color = 'gray', linestyle='dotted', linewidth = 3)
    # Show the figure
    plt.show()

for col in numerical_features:
    show_density(col,combined_df[col])
```

Output:

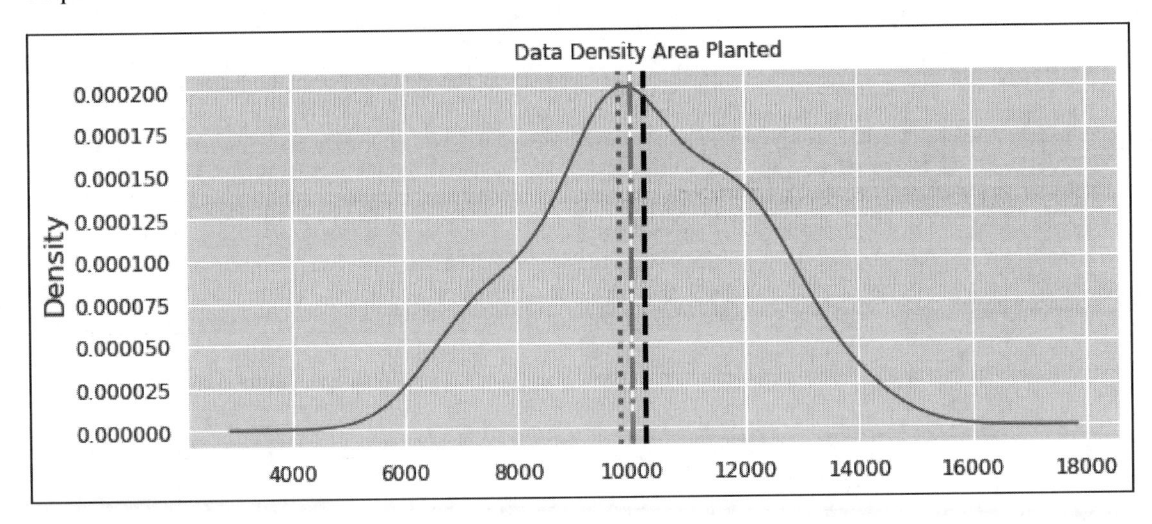

Data distribution:

Area Planted: - Range: 7500.00; - Variance: 3346304.88; - Std.Dev: 1829.29
Minimum:6600.00; Mean:10234.15 ; Median:10000.00; Mode:9800.00; and Maximum:14100.00

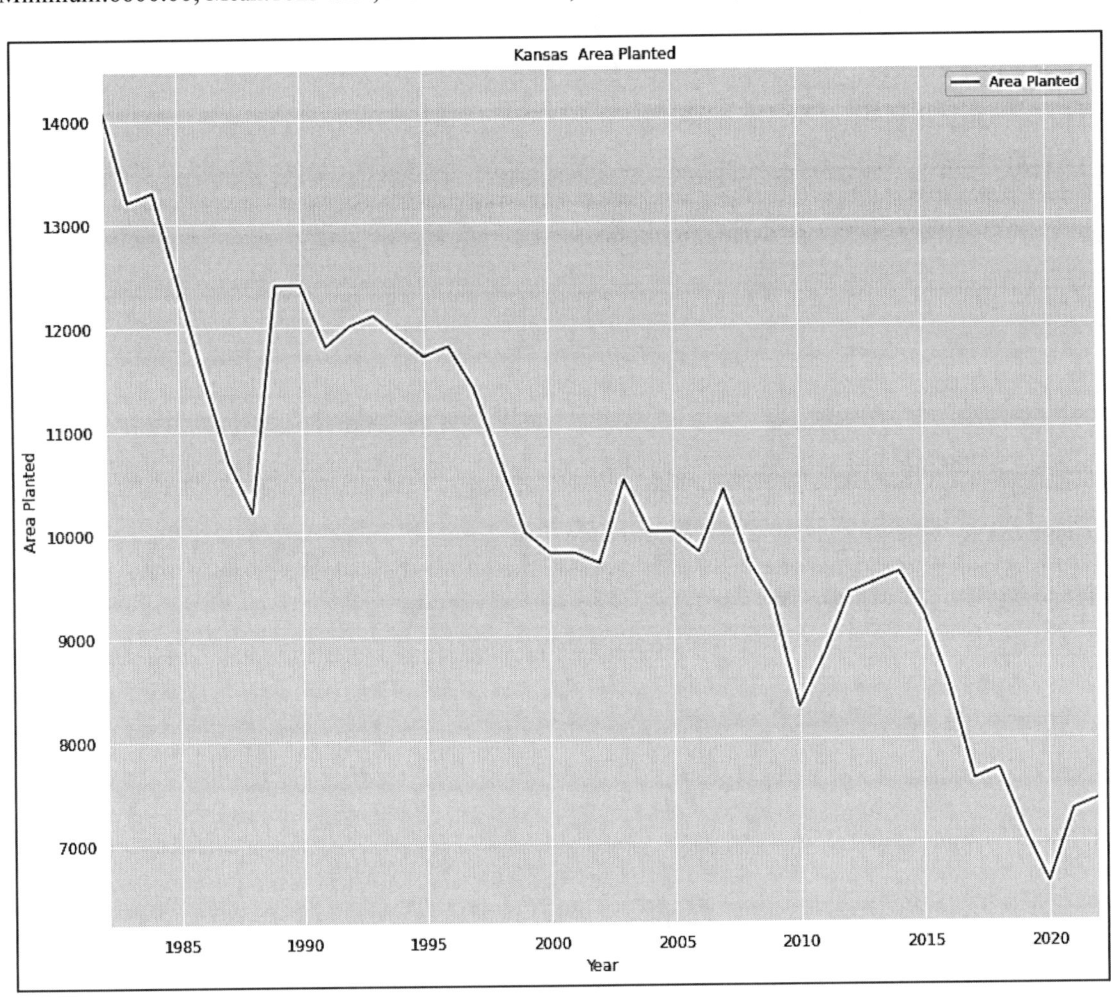

As it can be seen the area planted was 14,000 thousand acres in 1985 and was reduced to 7000 thousand acres in 2020.[486] **Area Harvested**: As it can be seen the area harvested was 13,000 thousand acres in 1985 and was reduced to 6600 thousand acres in 2020.[487]

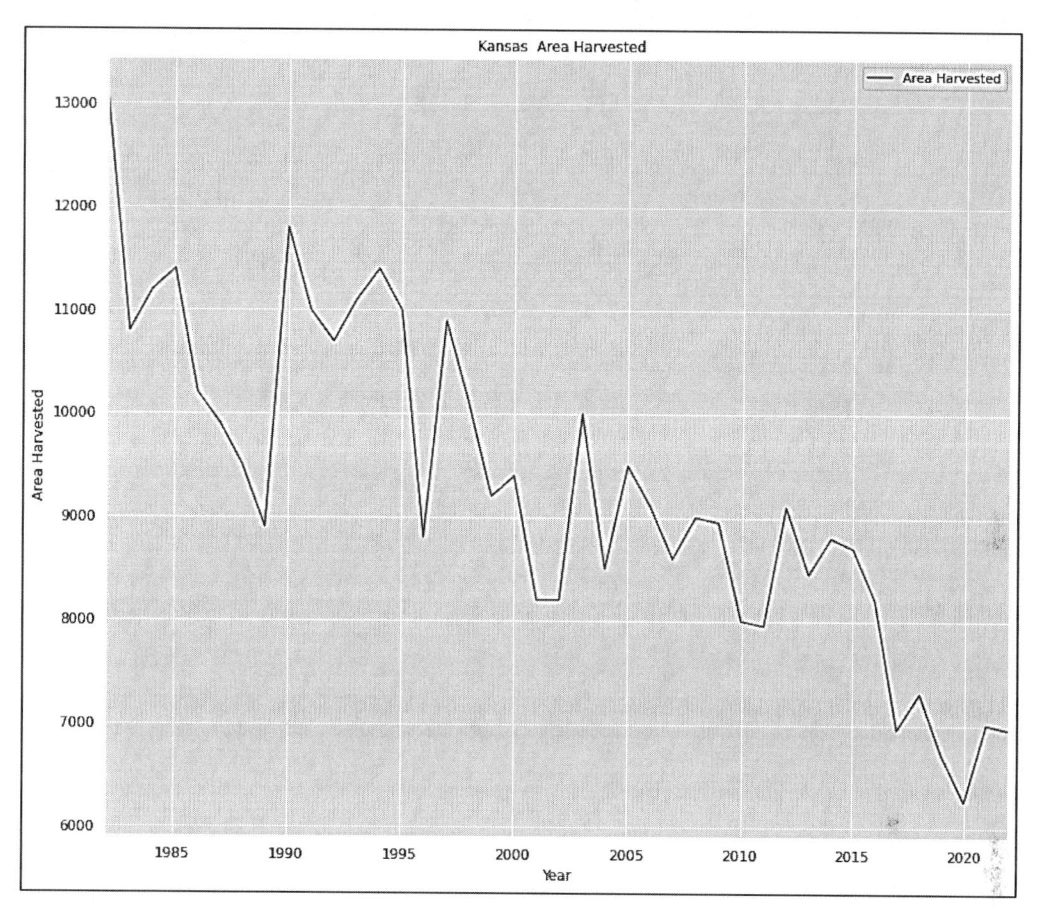

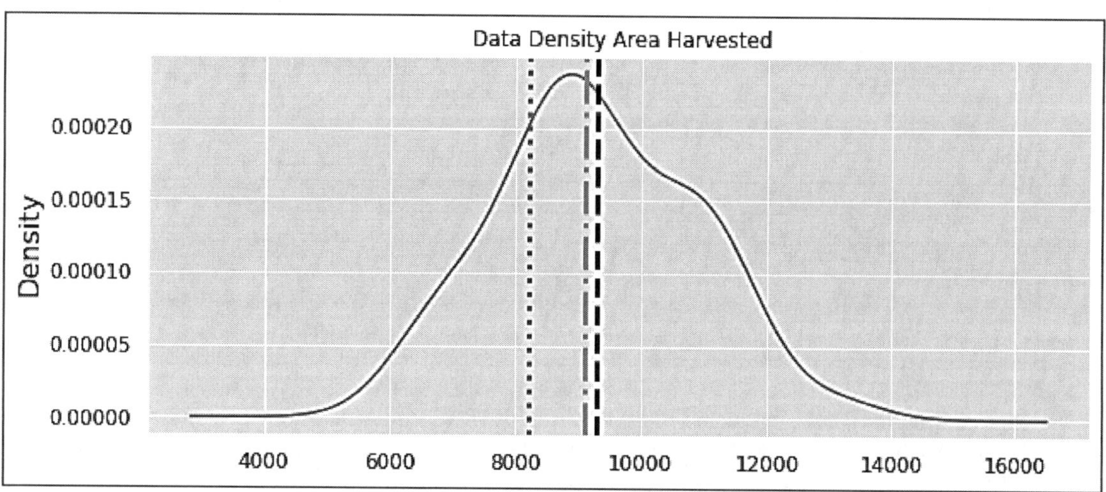

The data distribution is right tailed as there is a higher number of lower harvested area values. The data distribution has minimum:6250.00, mean:9287.80, median:9100.00, mode:8200.00, and maximum:13100.00. The Range: 6850.00, Variance: 2450222.56, and Std.Dev: 1565.32.

Kansan Wheat Production

Wheat production was 420,000 (1000 bushels) in 1980 and it was 281,250 (1000 bushels) in 2020. During drought years (1989–213,600 (1000 bushels), 286,000 (1000 bushels) in 1995, and 246,000 (1000 bushels), production was substantially lower than during regular crop years.

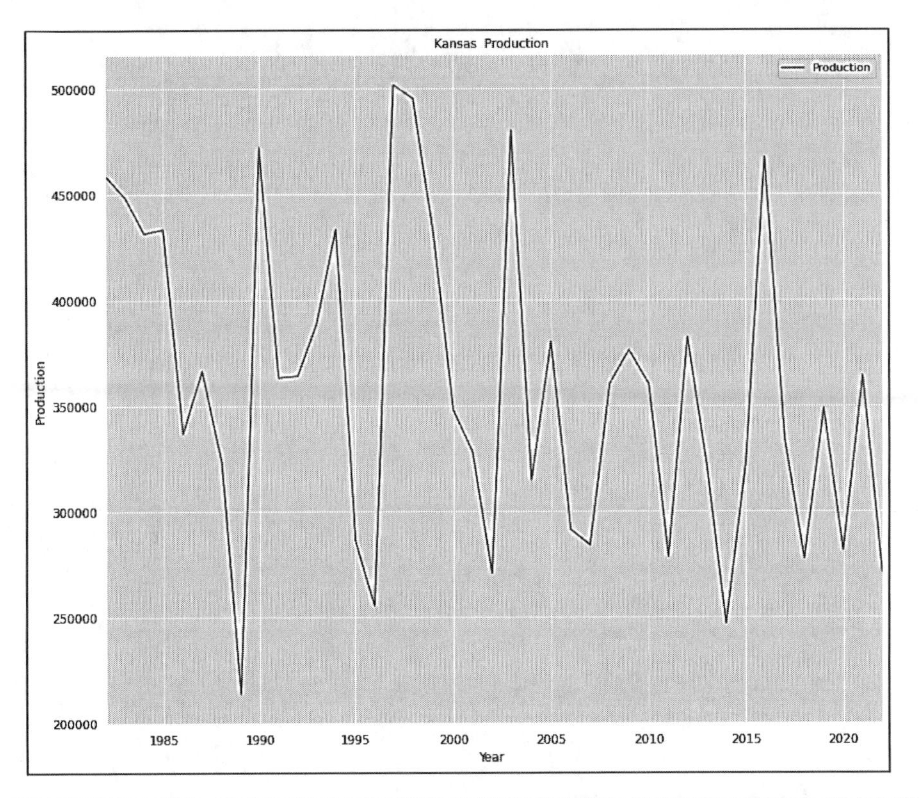

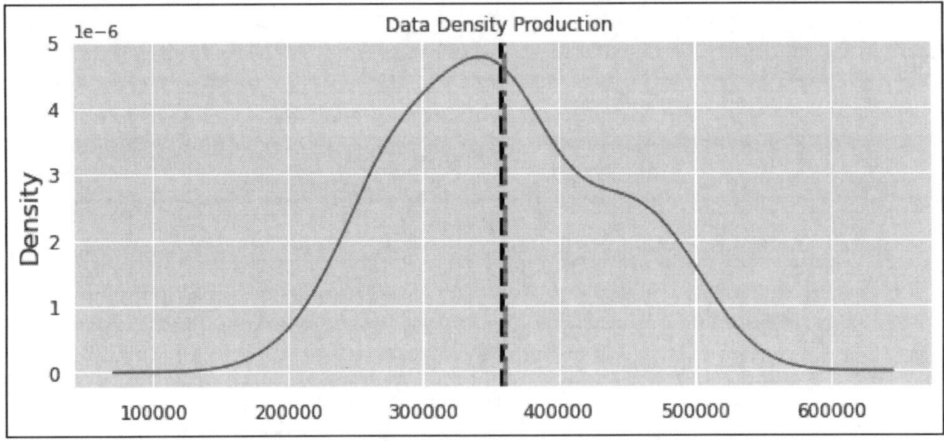

The data distribution is minimum:213600.00, mean:358189.02, median:360000.00, mode:360000.00, and maximum:501400.00. The 1989 drought has yielded minimum production. The data range was 287800.00, Variance: 5574678564.02, and Standard Deviation: 74663.77.

Step 7: Develop Linear Regressive Model (Winter Wheat)

Based on the Winter Wheat Harvest season, prepare input columns to regressors that take into consideration, July-August (Harvest) & September-October (Plant) months.

```
winter_X = combined_df[['Area Harvested',
 'Area Planted',
 'Production',
 ' SMN_1Wk', ' SMN_2Wk', ' SMN_3Wk', ' SMN_4Wk',
 ' SMN_5Wk', ' SMN_6Wk', ' SMN_7Wk', ' SMN_8Wk',
 ' SMN_9Wk', ' SMN_10Wk',' SMN_11Wk', ' SMN_12Wk',
 ' SMN_13Wk', ' SMN_14Wk', ' SMN_15Wk', ' SMN_16Wk',
 ' SMN_17Wk', ' SMN_18Wk', ' SMN_19Wk',' SMN_20Wk',
 ' SMN_21Wk', ' SMN_22Wk', ' SMN_23Wk',
 ' SMN_24Wk', ' SMN_25Wk', ' SMN_26Wk',
 ' VHI_1Wk', ' VHI_2Wk', ' VHI_3Wk', ' VHI_4Wk',
 ' VHI_5Wk', ' VHI_6Wk', ' VHI_7Wk', ' VHI_8Wk', ' VHI_9Wk',
 ' VHI_10Wk', ' VHI_11Wk', ' VHI_12Wk', ' VHI_13Wk',
 ' VHI_14Wk', ' VHI_15Wk', ' VHI_16Wk', ' VHI_17Wk',
 ' VHI_18Wk', ' VHI_19Wk', ' VHI_20Wk', ' VHI_21Wk',
 ' VHI_22Wk', ' VHI_23Wk',' VHI_24Wk', ' VHI_25Wk', ' VHI_26Wk',
 'SMT_1Wk', 'SMT_2Wk', 'SMT_3Wk', 'SMT_4Wk',
 'SMT_5Wk', 'SMT_6Wk', 'SMT_7Wk', 'SMT_8Wk', 'SMT_9Wk',
 'SMT_11Wk', 'SMT_12Wk', 'SMT_13Wk', 'SMT_14Wk', 'SMT_15Wk',
 'SMT_16Wk', 'SMT_17Wk', 'SMT_18Wk', 'SMT_19Wk', 'SMT_1Wk',
 'SMT_20Wk', 'SMT_21Wk', 'SMT_22Wk', 'SMT_23Wk', 'SMT_24Wk',
 'SMT_25Wk', 'SMT_26Wk',
 'TCI_1Wk','TCI_2Wk', 'TCI_3Wk','TCI_4Wk',
 'TCI_5Wk', 'TCI_6Wk','TCI_7Wk', 'TCI_8Wk', 'TCI_9Wk',
 'TCI_10Wk', 'TCI_11Wk', 'TCI_12Wk', 'TCI_13Wk', 'TCI_14Wk',
 'TCI_15Wk', 'TCI_16Wk', 'TCI_17Wk', 'TCI_18Wk', 'TCI_19Wk',
 'TCI_20Wk', 'TCI_21Wk', 'TCI_22Wk', 'TCI_23Wk', 'TCI_24Wk',
 'TCI_25Wk', 'TCI_26Wk',
 'VCI_1Wk', 'VCI_2Wk', 'VCI_3Wk', 'VCI_4Wk',
 'VCI_5Wk', 'VCI_6Wk', 'VCI_7Wk', 'VCI_8Wk', 'VCI_9Wk',
 'VCI_10Wk', 'VCI_11Wk', 'VCI_12Wk', 'VCI_13Wk', 'VCI_14Wk',
 'VCI_15Wk', 'VCI_16Wk', 'VCI_17Wk', 'VCI_18Wk', 'VCI_19Wk',
 'VCI_20Wk', 'VCI_21Wk', 'VCI_22Wk', 'VCI_23Wk', 'VCI_24Wk',
 'VCI_25Wk', 'VCI_26Wk'
]]
```

```
winter_y = combined_df['Yield']
```

Train and split the variables.

```
X_train_winter,X_test_winter, y_train_winter , y_test_winter = train_test_split(winter_X ,winter_y, test_size = 0.2,random_state = 0)
```

```
regressor_winter = LinearRegression()
regressor_winter.fit(X_train_winter,y_train_winter)
```

Output:

LinearRegression()

Step 8: Predict Model (Winter Wheat)

Predict the model.

```
y_pred_winter = regressor_winter.predict(X_test_winter)
y_pred_winter
```

Output:

```
array([35.71679137, 45.16248829, 35.88374313, 30.26330266, 31.14939575,
 37.54573852, 40.97354365, 32.96376814, 38.5914771 ])
```

Step 9: Evaluate the Model

Predict the model.

```
plt.scatter(y_test_winter,y_pred_winter)
plt.xlabel('Actual')
plt.ylabel("Predicted")
plt.title('Actual vs Predicted')
# overlay the regression line
z = np.polyfit(y_test_winter, y_pred_winter, 1)
p = np.poly1d(z)
plt.plot(y_test_winter,p(y_test_winter), color='b')
plt.show()
```

Output:

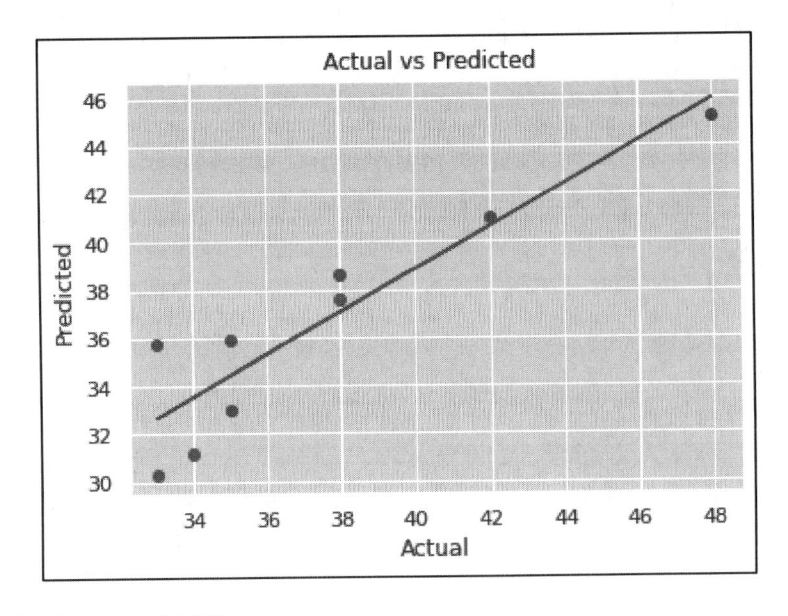

Calculate mean absolute error (MAE):

```
pred_y_df_winter = pd.DataFrame({'Actual Value':y_test_winter, 'Predicted value':y_pred_
winter,'Difference': y_test_winter - y_pred_winter})
pred_y_df_winter
```

```
from sklearn.metrics import mean_absolute_error
from matplotlib import pyplot
mae = mean_absolute_error(y_test_winter, y_pred_winter)
mae
```

Output:

	Actual Value	Predicted value	Difference
25	33.0	35.716791	-2.716791
35	48.0	45.162488	2.837512
29	35.0	35.883743	-0.883743
4	33.0	30.263303	2.736697
10	34.0	31.149396	2.850604
31	38.0	37.545739	0.454261
27	42.0	40.973544	1.026456
11	35.0	32.963768	2.036232
36	38.0	38.591477	-0.591477

MAE: 1.792

Regression Coefficient and R^2 Value:

```
regressor_winter.intercept_
r2_score(y_test_winter,y_pred_winter)
```

Output:

Regression Coefficient: 22.7247

R^2 Value: 0.8082 or 80.2%

The model accurately explains independent and dependent variables by 80%.

Step 10: Model Equation

Formulate Model equation.

```
mx=""
for ifeature in range(len(winter_X.columns)):
   if regressor_winter.coef_[ifeature] <0:
      # format & beautify the equation
      mx += " - " + "{:.2f}".format(abs(regressor_winter.coef_[ifeature])) + " * " + winter_X.
     columns[ifeature]
   else:
      if ifeature == 0:
         mx += "{:.2f}".format(regressor_winter.coef_[ifeature]) + " * " + winter_X.columns[ifeature]
      else:
         mx+=" + "+"{:.2f}".format(regressor_winter.coef_[ifeature])+" * "+winter_X.columns[ifeature]
print(mx)
```

```
# y=mx+c
if(regressor_winter.intercept_ <0):
  print("The formula for the " + y.name + " linear regression line (y=mx+c) is = " + " - {:.2f}".
format(abs(regressor_winter.intercept_)) + " + " + mx )
else:
  print("The formula for the "+y.name+" linear regression line (y=mx+c) is = "+"{:.2f}".format(regressor_
winter.intercept_) + " + " + mx )
```

Output:

```
The formula for the Yield linear regression line (y=mx+c) is = 22.72 + - 0.00 * Area Harvested - 0.00 * Area Planted + 0.00 * Production + 0.00 * SMN_1Wk - 0.00 * SMN_2Wk - 0.00 * SMN_3Wk - 0.00 * SMN_4Wk - 0.00 * SMN_5Wk + 0.00 * SMN_6Wk - 0.00 * SMN_7Wk + 0.00 * SMN_8Wk + 0.00 * SMN_9Wk - 0.00 * SMN_10Wk + 0.00 * SMN_11Wk + 0.00 * SMN_12Wk + 0.00 * SMN_13Wk + 0.00 * SMN_14Wk + 0.00 * SMN_15Wk + 0.00 * SMN_16Wk + 0.00 * SMN_17Wk + 0.00 * SMN_18Wk - 0.00 * SMN_19Wk + 0.00 * SMN_20Wk + 0.00 * SMN_21Wk - 0.00 * SMN_22Wk + 0.00 * SMN_23Wk - 0.00 * SMN_24Wk - 0.00 * SMN_25Wk - 0.00 * SMN_26Wk - 0.01 * VHI_1Wk - 0.01 * VHI_2Wk - 0.04 * VHI_3Wk - 0.01 * VHI_4Wk + 0.03 * VHI_5Wk + 0.06 * VHI_6Wk + 0.01 * VHI_7Wk - 0.01 * VHI_8Wk - 0.01 * VHI_9Wk - 0.01 * VHI_10Wk - 0.02 * VHI_11Wk - 0.01 * VHI_12Wk + 0.02 * VHI_13Wk + 0.03 * VHI_14Wk + 0.01 * VHI_15Wk - 0.03 * VHI_16Wk - 0.04 * VHI_17Wk - 0.02 * VHI_18Wk + 0.04 * VHI_19Wk - 0.01 * VHI_20Wk - 0.00 * VHI_21Wk + 0.01 * VHI_22Wk + 0.02 * VHI_23Wk - 0.00 * VHI_24Wk - 0.02 * VHI_25Wk + 0.00 * VHI_26Wk + 0.00 * SMT_1Wk - 0.00 * SMT_2Wk + 0.00 * SMT_3Wk + 0.00 * SMT_4Wk - 0.01 * SMT_5Wk - 0.02 * SMT_6Wk - 0.01 * SMT_7Wk - 0.00 * SMT_8Wk + 0.02 * SMT_9Wk + 0.01 * SMT_11Wk + 0.02 * SMT_12Wk + 0.01 * SMT_13Wk + 0.00 * SMT_14Wk + 0.00 * SMT_15Wk + 0.00 * SMT_16Wk + 0.01 * SMT_17Wk + 0.00 * SMT_18Wk - 0.00 * SMT_19Wk + 0.00 * SMT_1Wk + 0.01 * SMT_20Wk - 0.00 * SMT_21Wk - 0.00 * SMT_22Wk - 0.00 * SMT_23Wk + 0.00 * SMT_24Wk + 0.00 * SMT_25Wk + 0.00 * SMT_26Wk - 0.03 * TCI_1Wk - 0.01 * TCI_2Wk - 0.06 * TCI_3Wk - 0.01 * TCI_4Wk + 0.07 * TCI_5Wk + 0.11 * TCI_6Wk - 0.01 * TCI_7Wk - 0.07 * TCI_8Wk + 0.01 * TCI_9Wk + 0.07 * TCI_10Wk - 0.01 * TCI_11Wk - 0.08 * TCI_12Wk - 0.03 * TCI_13Wk + 0.04 * TCI_14Wk - 0.02 * TCI_15Wk - 0.02 * TCI_16Wk - 0.02 * TCI_17Wk + 0.01 * TCI_18Wk + 0.10 * TCI_19Wk - 0.05 * TCI_20Wk - 0.06 * TCI_21Wk + 0.01 * TCI_22Wk + 0.00 * TCI_23Wk + 0.00 * TCI_24Wk + 0.00 * TCI_25Wk + 0.03 * TCI_26Wk + 0.01 * VCI_1Wk - 0.01 * VCI_2Wk - 0.03 * VCI_3Wk - 0.00 * VCI_4Wk + 0.00 * VCI_5Wk + 0.01 * VCI_6Wk + 0.03 * VCI_7Wk + 0.06 * VCI_8Wk - 0.02 * VCI_9Wk - 0.09 * VCI_10Wk - 0.03 * VCI_11Wk + 0.05 * VCI_12Wk + 0.07 * VCI_13Wk + 0.01 * VCI_14Wk + 0.03 * VCI_15Wk - 0.03 * VCI_16Wk - 0.06 * VCI_17Wk - 0.05 * VCI_18Wk - 0.02 * VCI_19Wk + 0.02 * VCI_20Wk + 0.05 * VCI_21Wk + 0.02 * VCI_22Wk + 0.03 * VCI_23Wk - 0.01 * VCI_24Wk - 0.05 * VCI_25Wk - 0.03 * VCI_26Wk
```

Step 11: Model Explainability

Model Explainability would help analyze important model variables.

```
eli5.explain_weights(
  regressor_winter,
  feature_names = list(winter_X.columns) )
```

Output:

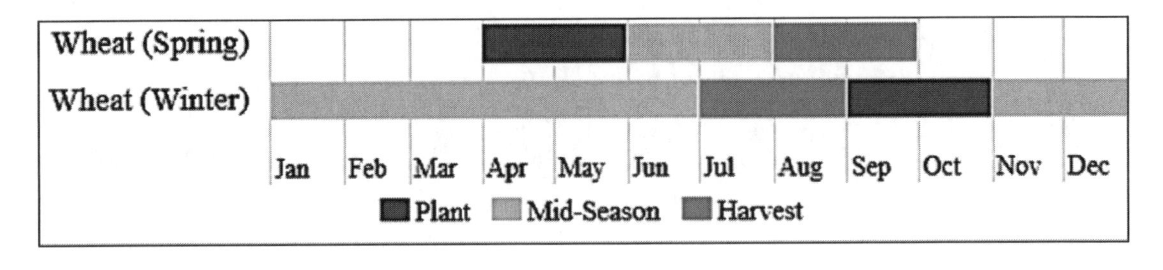

Temperature Condition index (TCI) of the 6th Week (2nd of February) positively correlates with wheat yield (11.3%). It tells us that stress during February has a negative wheat yield effect a fact that we can see as part of the 1989 drought.
Temperature Condition index (TCI) of the 19th Week (May-June) positively correlates with the wheat yield (10.2%). It tells us that temperature stress during the May to June period has a negative wheat yield effect a fact that we can see as part of the1989, 1995, & 2014 droughts.
Vegetation Condition Index (VCI_18WK) negatively correlates with Yield. The planting season early vegetation health index plays an important role in Wheat Yield.
Temperature Condition index (TCI) of the 12th Week (May) negatively correlates with wheat yield (–8%). It tells us that temperature stress during May has a negative wheat yield effect a fact that we can see as part of the 1989, 1995, & 2014 droughts.
Vegetation Condition Index (VCI_10WK) in March negatively correlates with Yield. The planting season early vegetation health index plays an important role in Wheat Yield.

y top features

Weight?	Feature
+22.725	<BIAS>
+0.113	TCI_6Wk
+0.102	TCI_19Wk
+0.074	TCI_10Wk
+0.069	TCI_5Wk
+0.065	VCI_13Wk
+0.062	VHI_6Wk
+0.055	VCI_8Wk
+0.054	VCI_21Wk
+0.051	VCI_12Wk
... 59 more positive ...	
... 55 more negative ...	
-0.045	VHI_3Wk
-0.048	VCI_18Wk
-0.050	VCI_25Wk
-0.052	TCI_20Wk
-0.056	TCI_21Wk
-0.059	TCI_3Wk
-0.059	VCI_17Wk
-0.073	TCI_8Wk
-0.080	TCI_12Wk
-0.093	VCI_10Wk

Step 12: Develop Linear Regressive Model (Spring Wheat Model)

Based on the Spring Wheat Harvest season, prepare input columns to regressors that take into consideration April-May (Plant) & August-September (Harvest) months.

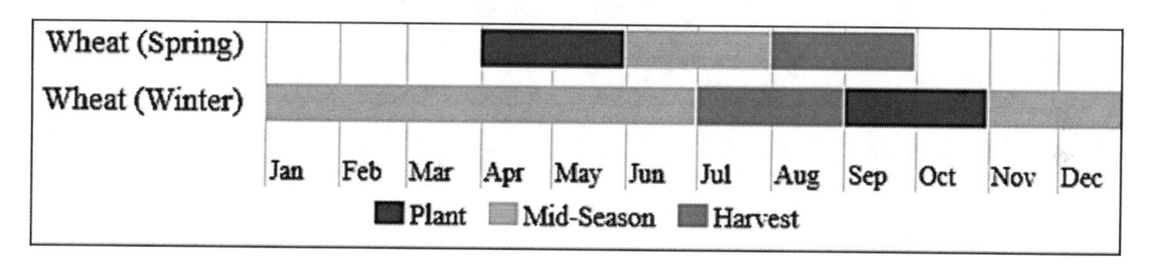

```
spring_X = combined_df[['Area Harvested',
 'Area Planted',
 'Production',
 ' SMN_14Wk', ' SMN_15Wk', ' SMN_16Wk',
 ' SMN_17Wk', ' SMN_18Wk', ' SMN_19Wk',' SMN_20Wk',
 ' SMN_21Wk', ' SMN_22Wk', ' SMN_23Wk',
 ' SMN_24Wk', ' SMN_25Wk', ' SMN_26Wk',
 ' SMN_27Wk', ' SMN_28Wk', ' SMN_29Wk',' SMN_30Wk',
 ' SMN_31Wk', ' SMN_32Wk', ' SMN_33Wk', ' SMN_34Wk',
 ' SMN_35Wk', ' SMN_36Wk', ' SMN_37Wk', ' SMN_38Wk', ' SMN_39Wk',
 'SMT_14Wk', 'SMT_15Wk', 'SMT_16Wk', 'SMT_17Wk', 'SMT_18Wk',
 'SMT_19Wk', 'SMT_20Wk', 'SMT_21Wk', 'SMT_22Wk', 'SMT_23Wk',
 'SMT_24Wk', 'SMT_25Wk', 'SMT_26Wk', 'SMT_27Wk', 'SMT_28Wk',
 'SMT_29Wk', 'SMT_30Wk', 'SMT_31Wk', 'SMT_32Wk', 'SMT_33Wk',
 'SMT_34Wk', 'SMT_35Wk', 'SMT_36Wk', 'SMT_37Wk', 'SMT_38Wk', 'SMT_39Wk',
 ' VHI_14Wk', ' VHI_15Wk', ' VHI_16Wk', ' VHI_17Wk', ' VHI_18Wk',
 ' VHI_19Wk', ' VHI_20Wk', ' VHI_21Wk', ' VHI_22Wk', ' VHI_23Wk',
 ' VHI_24Wk', ' VHI_25Wk', ' VHI_26Wk', ' VHI_27Wk', ' VHI_28Wk',
 ' VHI_29Wk', ' VHI_30Wk', ' VHI_31Wk', ' VHI_32Wk', ' VHI_33Wk',
 ' VHI_34Wk', ' VHI_35Wk', ' VHI_36Wk', ' VHI_37Wk', ' VHI_38Wk', ' VHI_39Wk',
 'TCI_1Wk','TCI_2Wk', 'TCI_3Wk','TCI_4Wk',
 'TCI_5Wk', 'TCI_6Wk','TCI_7Wk', 'TCI_8Wk', 'TCI_9Wk',
 'TCI_10Wk', 'TCI_11Wk', 'TCI_12Wk', 'TCI_13Wk', 'TCI_14Wk',
 'TCI_15Wk', 'TCI_16Wk', 'TCI_17Wk', 'TCI_18Wk', 'TCI_19Wk',
 'TCI_20Wk', 'TCI_21Wk', 'TCI_22Wk', 'TCI_23Wk', 'TCI_24Wk',
 'TCI_25Wk', 'TCI_26Wk','TCI_27Wk', 'TCI_28Wk', 'TCI_29Wk',
'TCI_30Wk', 'TCI_31Wk', 'TCI_32Wk', 'TCI_33Wk', 'TCI_34Wk',
'TCI_35Wk', 'TCI_36Wk', 'TCI_37Wk', 'TCI_38Wk', 'TCI_39Wk',
 'VCI_1Wk', 'VCI_2Wk', 'VCI_3Wk', 'VCI_4Wk',
 'VCI_5Wk', 'VCI_6Wk', 'VCI_7Wk', 'VCI_8Wk', 'VCI_9Wk',
 'VCI_10Wk', 'VCI_11Wk', 'VCI_12Wk', 'VCI_13Wk', 'VCI_14Wk',
 'VCI_15Wk', 'VCI_16Wk', 'VCI_17Wk', 'VCI_18Wk', 'VCI_19Wk',
 'VCI_20Wk', 'VCI_21Wk', 'VCI_22Wk', 'VCI_23Wk', 'VCI_24Wk',
 'VCI_25Wk', 'VCI_26Wk','VCI_27Wk','VCI_28Wk', 'VCI_29Wk', 'VCI_30Wk',
'VCI_31Wk', 'VCI_32Wk', 'VCI_33Wk', 'VCI_34Wk', 'VCI_35Wk', 'VCI_36Wk',
 'VCI_37Wk', 'VCI_38Wk', 'VCI_39Wk',
 ]]
```

```
spring_y = combined_df['Yield']
```

Train and split the variables.

```
X_train_spring,X_test_spring, y_train_spring , y_test_spring = train_test_split(spring_X ,spring_y, test_
size = 0.2,random_state = 0)
```

```
regressor_spring = LinearRegression()
regressor_spring.fit(X_train_spring,y_train_spring)
```

Output:
LinearRegression()

Step 13: Predict Model (Winter Wheat)

Predict the model.

```
# predicting the test
y_pred_spring = regressor_spring.predict(X_test_spring)
y_pred_spring
```

Output:

```
array([34.35541344, 46.36309057, 34.71074673, 30.57673788, 33.127054,
 36.20786958, 41.02262152, 32.43929707, 38.01605976])
```

Step 14: Evaluate the Model

Predict the model.

```
plt.scatter(y_test_spring,y_pred_spring)
plt.xlabel('Actual')
plt.ylabel("Predicted")
plt.title('Actual vs Predicted')
# overlay the regression line
z = np.polyfit(y_test_spring, y_pred_spring, 1)
p = np.poly1d(z)
plt.plot(y_test_spring,p(y_test_spring), color='b')
plt.show()
```

Output:

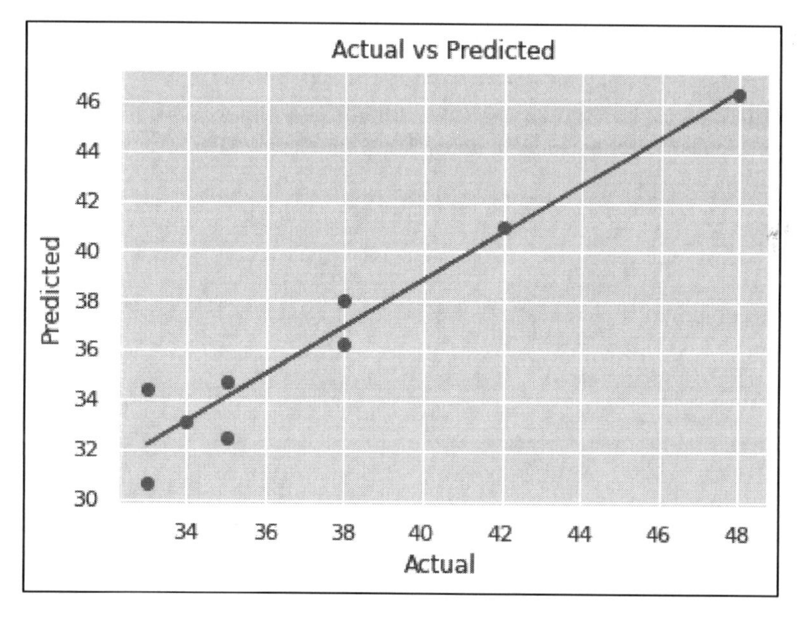

Calculate mean absolute error (MAE):

```
pred_y_df_spring = pd.DataFrame({'Actual Value':y_test_spring, 'Predicted value':y_pred_
spring,'Difference': y_test_spring - y_pred_spring})
pred_y_df_spring
```

```
from sklearn.metrics import mean_absolute_error
from matplotlib import pyplot
mae = mean_absolute_error(y_test_winter, y_pred_winter)
mae
```

Output:

	Actual Value	Predicted value	Difference
25	33.0	34.355413	-1.355413
35	48.0	46.363091	1.636909
29	35.0	34.710747	0.289253
4	33.0	30.576738	2.423262
10	34.0	33.127054	0.872946
31	38.0	36.207870	1.792130
27	42.0	41.022622	0.977378
11	35.0	32.439297	2.560703
36	38.0	38.016060	-0.016060

MAE: 1.324
Regression Coefficient and R^2 Value:

```
regressor_winter.intercept_
r2_score(y_test_spring, y_pred_spring)
```

Output:

Regression Coefficient: 22.7247

R^2 Value: 0.8879 or 88.79%

The model accurately explains independent and dependent variables by 80%.

Step 15: Model Equation

Formulate Model equation.

```
mx=""
for ifeature in range(len(spring_X.columns)):
   if regressor_spring.coef_[ifeature] <0:
      # format & beautify the equation
       mx += " - " + "{:.2f}".format(abs(regressor_spring.coef_[ifeature])) + " * " + spring_X.columns[ifeature]
   else:
      if ifeature == 0:
          mx += "{:.2f}".format(regressor_spring.coef_[ifeature]) + " * " + spring_X.columns[ifeature]
      else:
          mx+=" + "+"{:.2f}".format(regressor_spring.coef_[ifeature])+" * "+spring_X.columns[ifeature]
print(mx)
```

```
# y=mx+c
if(regressor_spring.intercept_ <0):
    print("The formula for the " + y.name + " linear regression line (y=mx+c) is = " + " - {:.2f}".format(abs(regressor_spring.intercept_)) + " + " + mx )
else:
    print("The formula for the " + y.name + " linear regression line (y=mx+c) is = " + "{:.2f}".format(regressor_spring.intercept_) + " + " + mx )
```

Output:

```
The formula for the Yield linear regression line (y=mx+c) is = 38.48 + - 0.01 * Area Harvested + 0.00 * Area Planted + 0.00 * Production + 0.00 * SMN_14Wk + 0.00 * SMN_15Wk + 0.00 * SMN_16Wk + 0.00 * SMN_17Wk - 0.00 * SMN_18Wk - 0.00 * SMN_19Wk - 0.00 * SMN_20Wk - 0.00 * SMN_21Wk + 0.00 * SMN_22Wk + 0.00 * SMN_23Wk + 0.00 * SMN_24Wk + 0.00 * SMN_25Wk - 0.00 * SMN_26Wk - 0.00 * SMN_27Wk - 0.00 * SMN_28Wk - 0.00 * SMN_29Wk + 0.00 * SMN_30Wk + 0.00 * SMN_31Wk + 0.00 * SMN_32Wk + 0.00 * SMN_33Wk + 0.00 * SMN_34Wk + 0.00 * SMN_35Wk + 0.00 * SMN_36Wk + 0.00 * SMN_37Wk + 0.00 * SMN_38Wk + 0.00 * SMN_39Wk + 0.00 * SMT_14Wk + 0.00 * SMT_15Wk + 0.00 * SMT_16Wk + 0.00 * SMT_17Wk + 0.00 * SMT_18Wk - 0.00 * SMT_19Wk + 0.00 * SMT_20Wk + 0.00 * SMT_21Wk - 0.00 * SMT_22Wk - 0.00 * SMT_23Wk - 0.00 * SMT_24Wk + 0.00 * SMT_25Wk + 0.00 * SMT_26Wk + 0.00 * SMT_27Wk - 0.00 * SMT_28Wk - 0.00 * SMT_29Wk - 0.00 * SMT_30Wk - 0.00 * SMT_31Wk - 0.00 * SMT_32Wk + 0.00 * SMT_33Wk - 0.00 * SMT_34Wk - 0.00 * SMT_35Wk + 0.00 * SMT_36Wk - 0.00 * SMT_37Wk - 0.00 * SMT_38Wk + 0.00 * SMT_39Wk + 0.01 * VHI_14Wk + 0.00 * VHI_15Wk - 0.01 * VHI_16Wk - 0.01 * VHI_17Wk - 0.00 * VHI_18Wk + 0.01 * VHI_19Wk - 0.03 * VHI_20Wk - 0.02 * VHI_21Wk + 0.02 * VHI_22Wk + 0.03 * VHI_23Wk + 0.02 * VHI_24Wk + 0.00 * VHI_25Wk - 0.00 * VHI_26Wk - 0.01 * VHI_27Wk - 0.01 * VHI_28Wk - 0.01 * VHI_29Wk - 0.00 * VHI_30Wk + 0.00 * VHI_31Wk + 0.00 * VHI_32Wk - 0.01 * VHI_33Wk + 0.01 * VHI_34Wk + 0.01 * VHI_35Wk + 0.00 * VHI_36Wk + 0.00 * VHI_37Wk + 0.01 * VHI_38Wk - 0.03 * VHI_39Wk - 0.03 * TCI_1Wk - 0.03 * TCI_2Wk - 0.02 * TCI_3Wk + 0.02 * TCI_4Wk + 0.06 * TCI_5Wk + 0.07 * TCI_6Wk - 0.01 * TCI_7Wk - 0.06 * TCI_8Wk + 0.01 * TCI_9Wk + 0.02 * TCI_10Wk - 0.00 * TCI_11Wk - 0.04 * TCI_12Wk - 0.00 * TCI_13Wk + 0.02 * TCI_14Wk - 0.02 * TCI_15Wk - 0.03 * TCI_16Wk - 0.03 * TCI_17Wk - 0.00 * TCI_18Wk + 0.04 * TCI_19Wk - 0.02 * TCI_20Wk - 0.04 * TCI_21Wk + 0.02 * TCI_22Wk + 0.03 * TCI_23Wk + 0.03 * TCI_24Wk + 0.01 * TCI_25Wk + 0.00 * TCI_26Wk + 0.01 * TCI_27Wk + 0.01 * TCI_28Wk + 0.00 * TCI_29Wk + 0.00 * TCI_30Wk + 0.01 * TCI_31Wk + 0.01 * TCI_32Wk - 0.03 * TCI_33Wk + 0.01 * TCI_34Wk + 0.03 * TCI_35Wk - 0.00 * TCI_36Wk - 0.01 * TCI_37Wk + 0.02 * TCI_38Wk - 0.03 * TCI_39Wk - 0.04 * VCI_1Wk - 0.01 * VCI_2Wk - 0.01 * VCI_3Wk + 0.02 * VCI_4Wk + 0.01 * VCI_5Wk + 0.00 * VCI_6Wk + 0.01 * VCI_7Wk + 0.01 * VCI_8Wk + 0.04 * VCI_9Wk + 0.02 * VCI_10Wk - 0.06 * VCI_11Wk - 0.01 * VCI_12Wk - 0.00 * VCI_13Wk + 0.01 * VCI_14Wk + 0.02 * VCI_15Wk + 0.01 * VCI_16Wk + 0.01 * VCI_17Wk - 0.01 * VCI_18Wk - 0.02 * VCI_19Wk - 0.03 * VCI_20Wk - 0.01 * VCI_21Wk + 0.02 * VCI_22Wk + 0.03 * VCI_23Wk + 0.01 * VCI_24Wk - 0.00 * VCI_25Wk - 0.01 * VCI_26Wk - 0.02 * VCI_27Wk - 0.02 * VCI_28Wk - 0.02 * VCI_29Wk - 0.01 * VCI_30Wk - 0.00 * VCI_31Wk - 0.00 * VCI_32Wk + 0.01 * VCI_33Wk + 0.00 * VCI_34Wk - 0.01 * VCI_35Wk + 0.00 * VCI_36Wk + 0.01 * VCI_37Wk + 0.00 * VCI_38Wk - 0.02 * VCI_3
```

Step 16: Model Explainability

Model Explainability would help analyze important model variables.

```
eli5.explain_weights(
   regressor_spring,
   feature_names = list(spring_X.columns) )
```

Output:

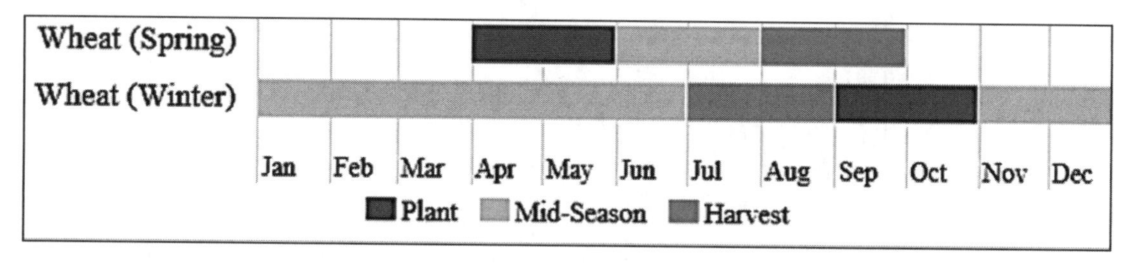

<table>
<tr><td>Temperature Condition index (TCI) of the 6th week (2nd of February) positively correlates with wheat yield (7.0%). It tells us that stress during February has a negative wheat yield effect a fact that we can see as part of the 1989 drought.</td><td rowspan="5">y top features

Weight? | Feature
+38.479 | <BIAS>
+0.070 | TCI_6Wk
+0.063 | TCI_5Wk
+0.045 | TCI_19Wk
+0.039 | VCI_9Wk
+0.035 | TCI_35Wk
+0.034 | VCI_23Wk
+0.030 | VHI_23Wk
+0.029 | TCI_24Wk
... 79 more positive ...
... 61 more negative ...
-0.027 | TCI_16Wk
-0.028 | TCI_2Wk
-0.028 | VCI_20Wk
-0.028 | TCI_17Wk
-0.032 | TCI_39Wk
-0.035 | TCI_1Wk
-0.039 | TCI_12Wk
-0.040 | VCI_1Wk
-0.042 | TCI_21Wk
-0.059 | VCI_11Wk
-0.065 | TCI_8Wk</td></tr>
<tr><td>Temperature Condition index (TCI) of the 19th Week (May–June) positively correlates with the wheat yield (4.5%). It tells us that the temperature stress during the May to June period has a negative wheat yield effect a fact that we can see as part of the 1989, 1995, & 2014 droughts.</td></tr>
<tr><td>Vegetation Condition Index (VCI_9WK) positively correlates with Yield. Planting season early vegetation health index plays an important role in the Wheat Yield.</td></tr>
<tr><td>Temperature Condition index (TCI) of the 16th Week (May) negatively correlates with the wheat yield (–2.7%). It tells us that temperature stress during May has a negative wheat yield effect a fact that we can see as part of the1989, 1995, & 2014 droughts.</td></tr>
<tr><td>Vegetation Condition Index (VCI_11WK) in March negatively correlates with yield. Planting season early vegetation health index plays an important role in Wheat Yield.</td></tr>
</table>

Step 17: Model summary

Wheat yield is highly influenced by Temperature Condition index (TCI), Vegetation Health index (VHI), and No noise (smoothed) Normalized Difference Vegetation Index (SMN). Drought during the planting months (Spring—April–May) & (Winter September–October) drastically influences wheat yield.

Kansas[488] grows winter wheat that is planted and sprouts in the fall, becomes dormant in the winter, grows again in the spring, and is harvested in early summer. Temperature Condition index (TCI), a proxy of thermal conditions, during February (mid-season) influences 11.3% of the wheat yield for winter planting. Weather is a driver of agricultural yield. The lack of moisture in the winter wheat and excessive moisture in the spring will affect yields and quality especially during the early spring planting season [10]. Hence VCI-9WK for Spring Wheat, VCI_13WK for Winter wheat, hence, play an important role in Wheat yield.

During the heading and flowering stages, excessively high or low temperatures and drought are harmful to wheat. Cloudy weather, with high humidity and low temperatures is conducive to rust attacks. Wheat plants require about 14–15°C (57.2–59°F) optimum average temperature at the time of ripening. The temperature conditions at the time of grain filling and development are very crucial for yield. Temperatures above 25°C (77°F) during this period tend to depress grain weight. When temperatures are high, too much energy is lost through the process of transpiration by the plants and the reduced residual energy results in

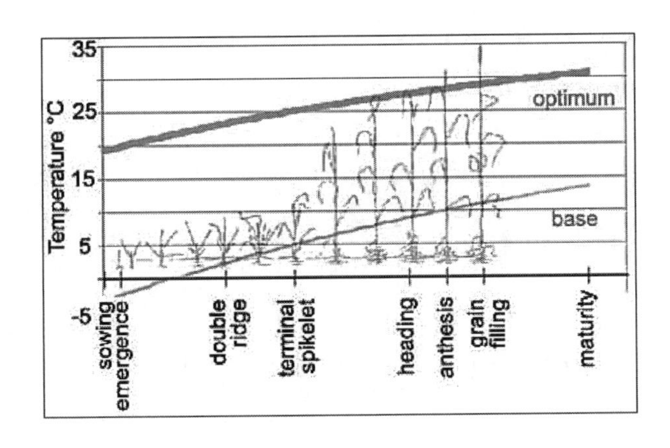

poorer grain formation and lower yields. Hence, for winter wheat {–0.056 - TCI_21Wk, –0.059-TCI_3Wk, –0.059-VCI_17Wk, –0.073-TCI_8Wk, –0.080-TCI_12Wk, and –0.093- VCI_10W} and Spring wheat {–0.039-TCI_12Wk, –0.040-VCI_1Wk, –0.042-TCI_21Wk, –0.059-VCI_11Wk, and –0.065-TCI_8Wk}.

India & Wheat Production

Wheat is the main cereal crop in India.[489] The total area under the crop plantation in 2018 is about 29.8 million hectares and has grown to 31.6 million hectares in 2021.[490] The production of wheat in the country has increased significantly from 75.81 million MT in 2006–07, 94.88 million MT in 2011–12 to a record high of 107.85 million MT in 2019–20.[491] The productivity of wheat which was 2602 kg/hectare in 2004–05, 3140 kg/hectare in 2011–12 has increased to 3440 kg/hectare in 2019–20. A major increase in the productivity of wheat has been observed in the states of Haryana, Punjab, and Uttar Pradesh. Higher area coverage is reported from Madhya Pradesh (MP) in recent years.

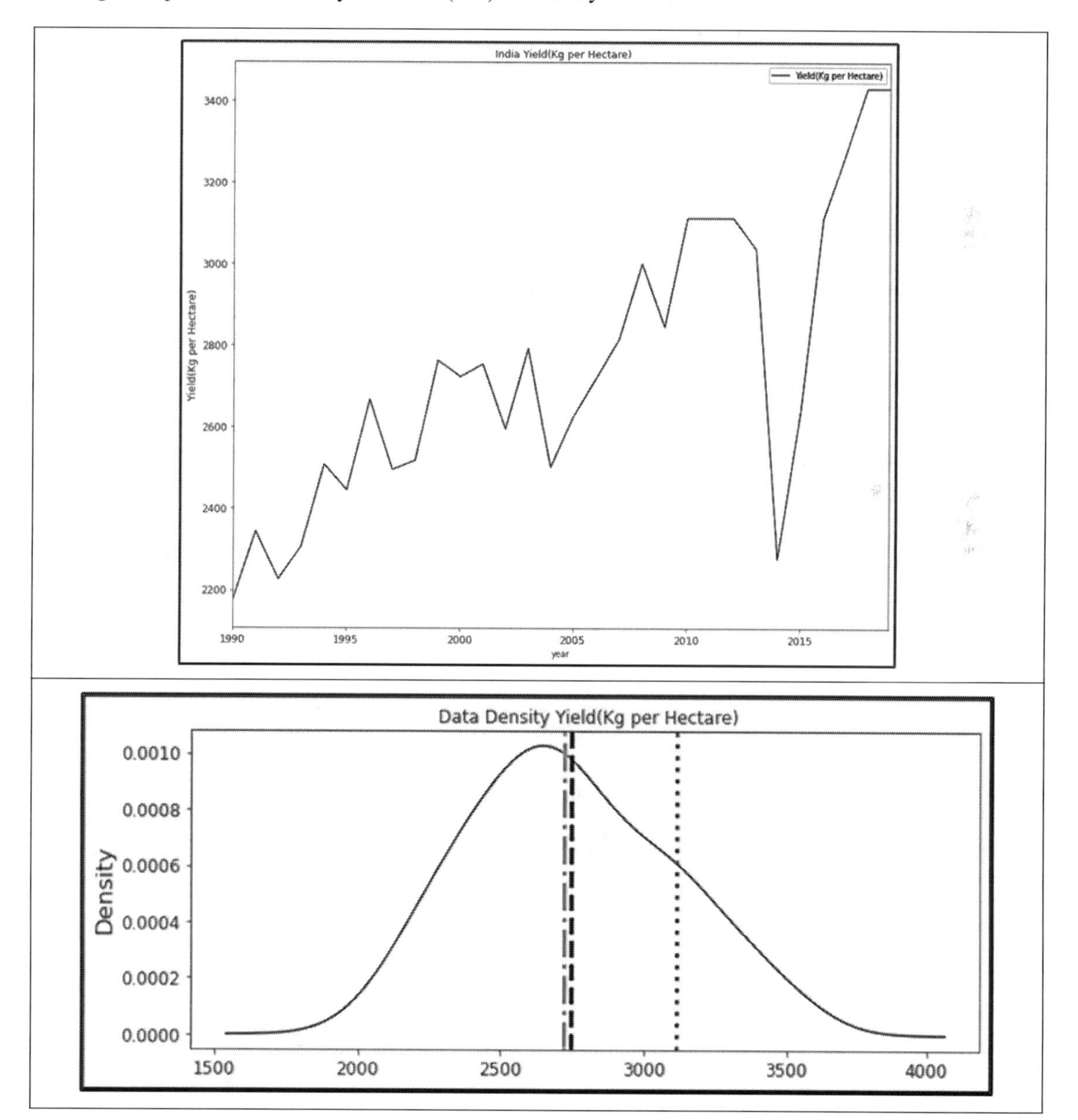

The data distribution of Yield (Kg per Hectare) is left skewed with minimum:2171.00 (Kg per Hectare), mean:2745.50 (Kg per Hectare), median:2722.50 (Kg per Hectare), mode:3113.00, and maximum:3432.00 (Kg per Hectare).

Indian wheat is largely a soft/medium hard, medium protein, white bread wheat, somewhat like U.S. hard white wheat. Wheat grown in central and western India is typically hard, with a high protein and gluten content. India also produces around 1.0–1.2 million tons of durum wheat, mostly in the state of Madhya Pradesh (MP). Most Indian durum is not marketed separately due to segregation problems in the market yards. However, some quantities are purchased by the private traders at a price premium, mainly for processing of higher value/branded products. The production and productivity of Wheat crop were quite low when India became independent in 1947. The production of Wheat was only 6.46 million tonnes and productivity was merely 663 kg per hectare during 1950–51, which was not sufficient to feed the Indian population. The Country used to import Wheat in large quantities for fulfilling the needs of our people from many countries like USA under PL-480.[492]

(a) The reasons of low production and productivity of Wheat at that time was the tall growing plant habit resulting in lodging, when grown under fertile soils,
(b) the poor tillering and low sink capacity of the varieties used,
(c) higher susceptibility to diseases,
(d) the higher sensitivity to thermo & photo variations resulting in poor adaptability, and
(e) longer crop durations resulting in a long exposure of plants to the climatic variations and insect pest/ disease attacks.

In the following Heatwave-Yield loss basic bending process (please see Figure 23), progression of food security due to heatwaves is outlined. Although the intermediary statistical steps could be performed in a non-sequential order, nonetheless, the inputs and outputs of the process are the start and the end of the process, respectively. The data is from Reserve Bank of India (RBI)—Handbook of statistics on Indian states.[493]

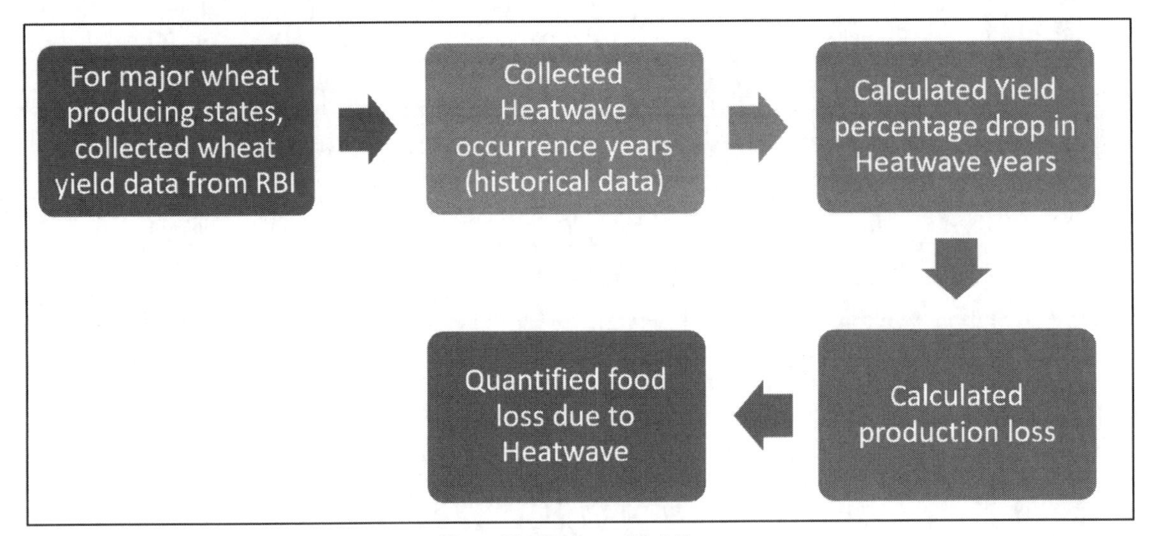

Figure 23: Heatwave Modeling.

Indian Wheat Growing Zones

The entire wheat growing areas of the country have been categorized into 6 major zones as follows (please see Table 4).[494]

Table 4: India - wheat growing zones.

Zones	States/Regions Covered	Approx. Area (million ha)
Northern Hill Zone(NHZ)	Hilly areas of J&K (except Jammu, Kathua and Samba districts), Himachal Pradesh (except Una & Paonta valley), Uttarakhand (excluding Tarai region) & Sikkim	0.8
North Western Plains Zone(NWPZ)	Punjab, Haryana, Western UP (except Jhansi Div), Rajasthan (excluding Kota & Udaipur div), Delhi, Tarai region of Uttarakhand, Una & Paonta valley of HP, Jammu, Samba & Kathua districts of J&K and Chandigarh	11.55
North Eastern Plains Zone(NEPZ)	Eastern UP (28 dist), Bihar, Jharkhand, West Bengal, Assam, Odisha and other NE states (except Sikkim)	10.5
Central Zone	MP, Gujarat, Chattisgarh, Kota & Udaipur Div of Rajasthan & Jhansi Div of UP	5.2
Peninsular Zone	Maharashtra, Tamil Nadu(except Nilgiris & Palani Hills),Karnataka & Andhra Pradesh	1.6
Southern Hill Zone(SHZ)	Nilgiris & Palani Hills of Tamil Nadu	0.1

Three major states are considered: Uttar Pradesh, Madya Pradesh, and Punjab. Both Uttar Pradesh (UP) and Punjab have been credited with being the Wheat breadbasket of India and Punjab as India's "grain bowl" and the government has encouraged cultivation of wheat and rice here since the 1960s. It is typically the biggest contributor to India's national reserves and the government had hoped to buy about a third of this year's stock from the region. However, government assessments predict lower yields this year, 2022 (please see Figure 24), of about 25% less. The story is the same in other major wheat-producing states like Uttar Pradesh and Madhya Pradesh.

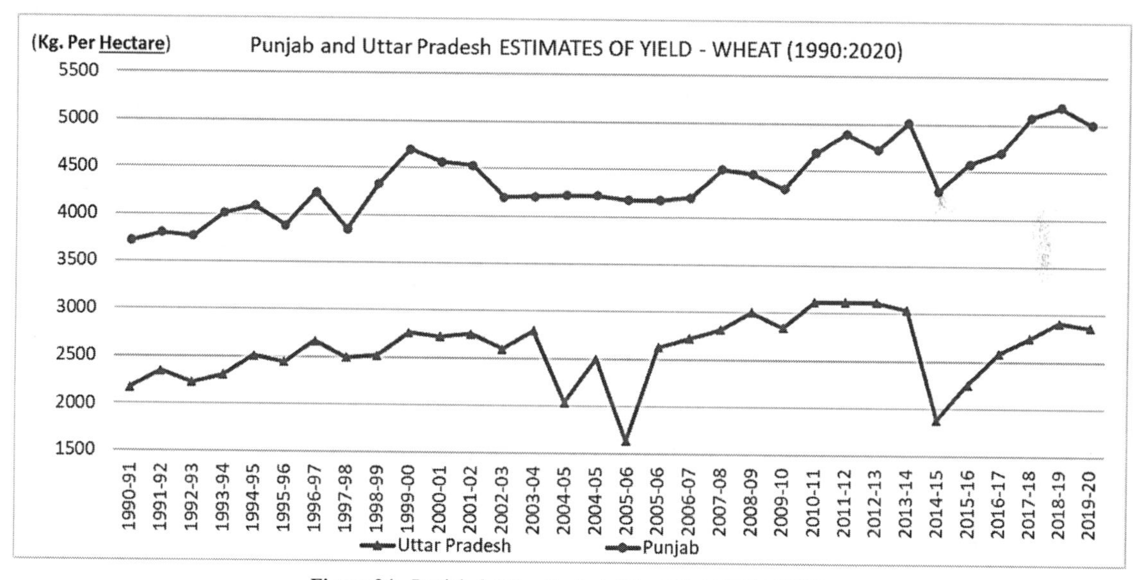

Figure 24: Punjab & Uttar Pradesh Wheat Yield (1990:2020).

Beyond India, other countries are also grappling with poor harvests that hinder their ability to help offset the potential shortfall of supplies from Russia and Ukraine, normally the world's largest and fifth-largest exporters of wheat. According to China's agriculture minister, the 2022 winter wheat harvest was likely to be poor, hindered by flooding and delays in planting (please see Figure 25).

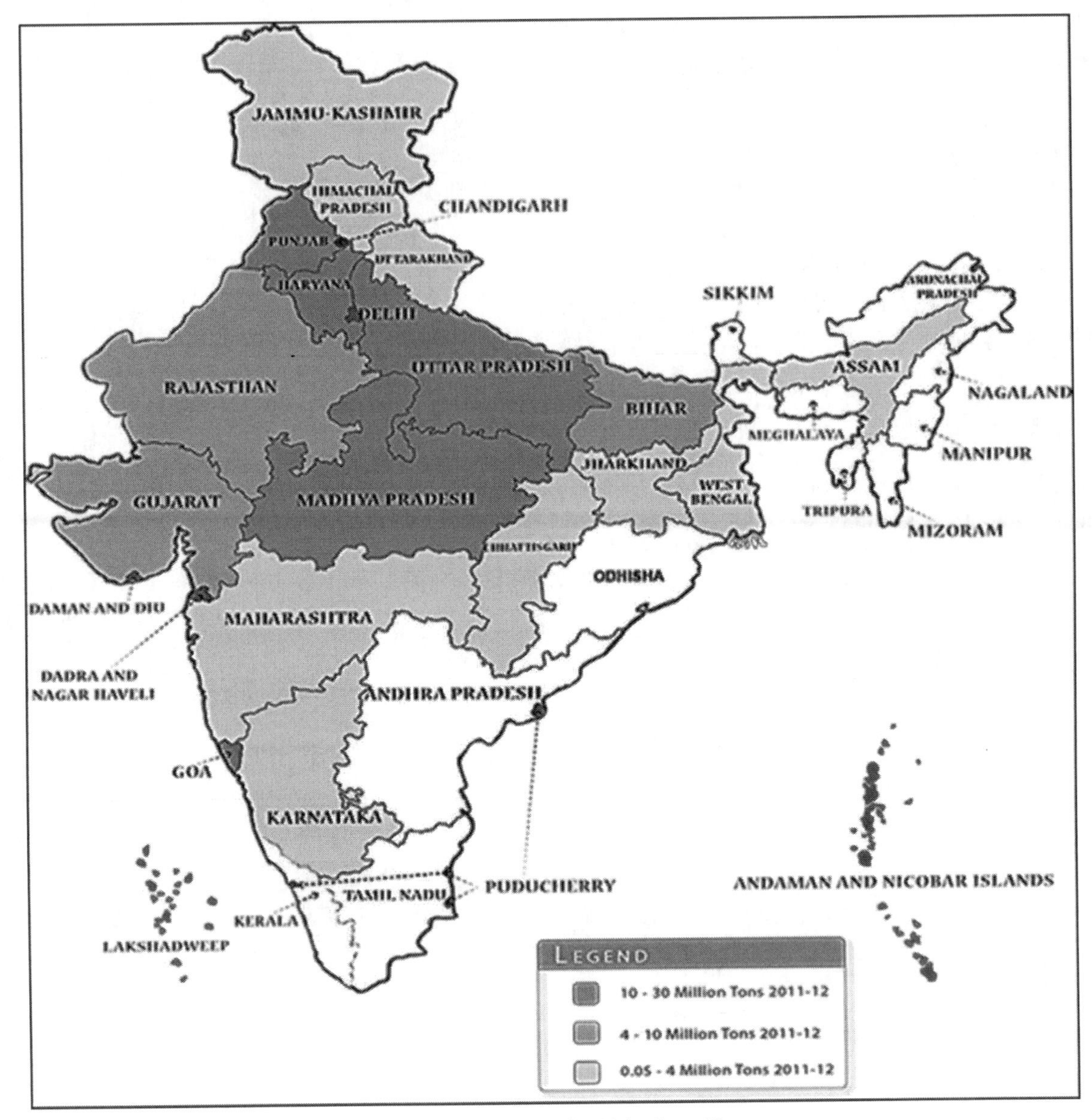

Figure 25: Wheat Producing Major States Map.

Map: curtesy of Farmer-India.gov website: https://farmer.gov.in/M_cropstaticswheat.aspx (access date: June 11, 2022).

Climate change and Heat waves are attributed to recent hikes in productivity and yields in major wheat producing areas of India. To assess the impact, the following statistical process is used:

India & heat waves—wheat yield drop

India's vulnerability to extreme heat increased by 15% from 1990 to 2019, according to a 2021 [21]. It is among the top five countries where vulnerable people, like the old and the poor, have the highest exposure to heat. India and Brazil have the highest heat-related mortality in the world. An unusually early, record-shattering heat wave (daytime temperatures peaking at 46.5 degrees Celsius (114.08 Fahrenheit) was experienced. Temperatures breached the 45°C (113°F) mark in nine other cities in India reducing wheat yields, raising questions about how the country will balance its domestic needs with ambitions to increase

Table 5: India-Heatwave Historical Table.

Year	Duration	Maximum Temperature	Agricultural Impact
1956		50.6°C (123.1°F)[495]	
1995	June	45.5°C (113.9°F)	A deadly heat wave has left millions of Indians longing for the monsoon, a yearly deluge that turns down the temperature, puts out fires and fills the rivers.[496]
1998	May–June	49.5°C (121.1°F)	
2002	April–May	49°C (120°F)	
2015	May–June	49.4°C (120.9°F)	
2016	April–May	51.0°C (123.8°F)	
2019	May–June	50.8°C (123.4°F)	
2022	March–May	50.8°C (123.4°F)	

exports and make up for shortfalls due to Russia's war in Ukraine [22]. It was the heat in March 2022—the hottest in India since records first started being kept in 1901—that stunted crops (please see Table 5). Wheat is very sensitive to heat, especially during the *final stage when its kernels mature and ripen*. Additionally, maximum temperatures are again likely to reach 50°C (122°F) in some spots later in the week or into the weekend, with continuously high overnight temperatures [23]. Indian farmers time their planting so that this stage coincides with India's usually cooler spring. The following table has heat wave occurrences in India, especially in major heat producing states.

As the intensity of heat waves increased, both maximum temperatures and durations increased and the drop in wheat yield was pronounced. One can see drops in Yields during Heatwave years. The drop is proportional to the time of the year the heat waves occurred (May–June), days of heat waves, and winds that contribute to the propagation of heat waves to wider areas (please see Figure 26). The grey areas indicate years of heat waves. Allahabad, a part of Uttar Pradesh, experienced a peak in 2015 (30th June) and it had disrupted the harvest season.

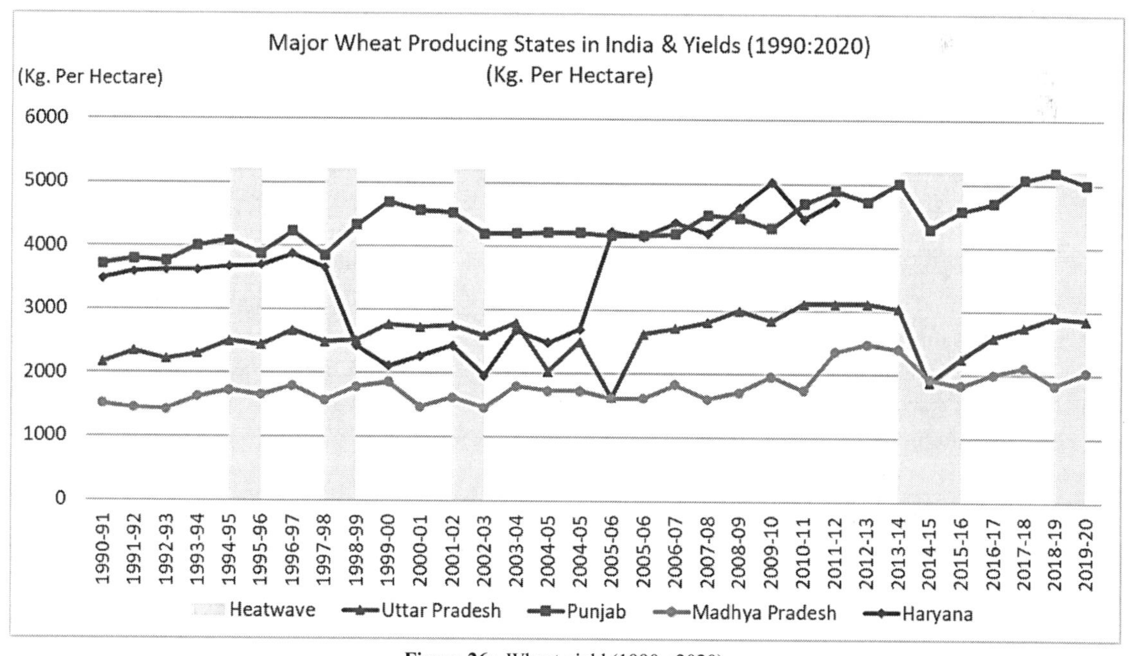

Figure 26: Wheat yield (1990 : 2020).

2015 heat wave in India and Pakistan

What yields have seen a drastic reduction in 2015, especially in Uttar Pradesh (UP). For much of May, parts of India were gripped in an intense heat wave that has seen the mercury rising above 110°F (43°C). Temperatures have been hot enough (over 111°F, or 44°C) to melt pavements in the capital city, New Delhi. The searing heat across the country—which has seen the capital observe five consecutive days with high temperatures over 110°F (43°C)—is the worst heat event in a decade according to the Indian Meteorological Department[497] a fact we can see in the production of wheat (please see Figure 27). In the figure below, as intensity of heat wave increases the drop in yield can be seen.

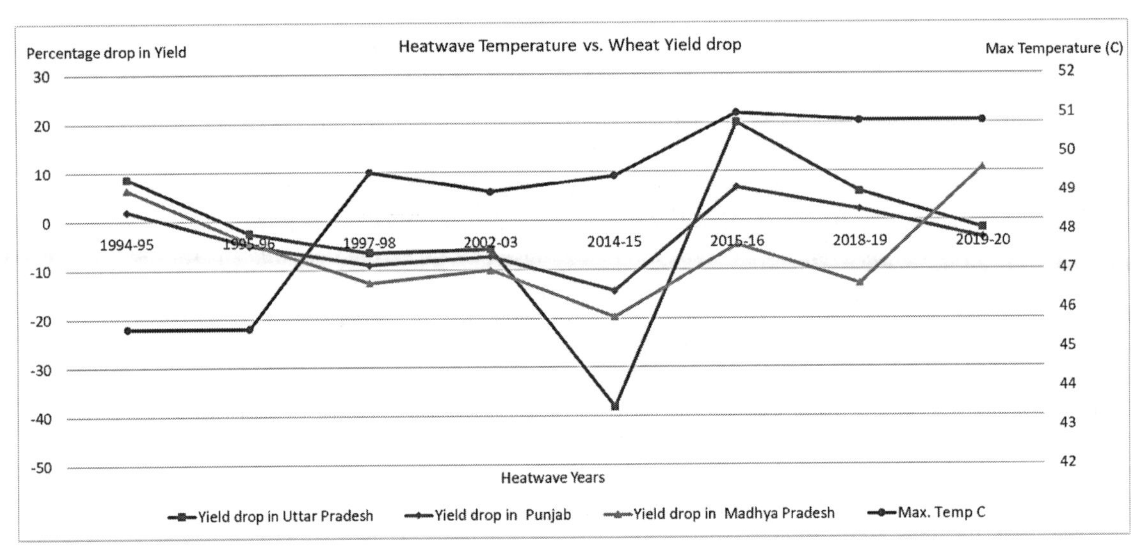

Figure 27: Heatwave & Yield Drop.

Hit even harder, the Odisha, Andhra Pradesh, and Telangana states of southeast India have observed temperatures soaring past 113°F (45°C). In the city of Titlagarh in the state of Odisha, temperatures reached 117.7°F (47.6°C)! Near the coast, the city of Ongole, in the state of Andhra Pradesh, higher temperatures averaged 110°F (43.47°C) from May 24–30. Farther inland, in the state of Jharkhand, the city of Daltonganj averaged 115.6°F for the entire week of May 24–30, peaking at 117° (47.2°C) on May 27. For coastal cities, this was not "dry heat:" humid air made the high temperatures even more unbearable and life-threatening. Even for India, this is extreme heat (please see Figure 28).

In the future, though, according to the Intergovernmental Panel on Climate Change (IPCC), more frequent and intense heat waves in Asia (including India) will negatively impact vulnerable communities and increase mortality. In fact, it is likely that heat waves already occur more often now in Asia than they did in 1950. In a future with high carbon emissions, it is likely that the maximum temperature that occurs once every 20 years will at least double in frequency (to a 1-in-10-year event) by the end of the 21st century. Research focusing solely on India also concludes that heat waves will last longer, will be more intense and occur more often in the future. Since major producing states were affected by heat waves, the 2015 wheat yield has dropped considerably.

2019 heat wave in India and Pakistan

In early June 2019,[498] an intense heatwave scorched northern India. Some regions experienced temperatures surpassing 45°C (113°F) for the better part of three weeks. On June 10, Delhi (please see Figure 29) reached its hottest day on record for the month, reaching 48°C (118°F). The map above shows temperatures on June 10 in India and Pakistan, which has also been experiencing hot and dry conditions in the past two months. The map was derived from the Goddard Earth Observing System (GEOS) model and represents air temperatures

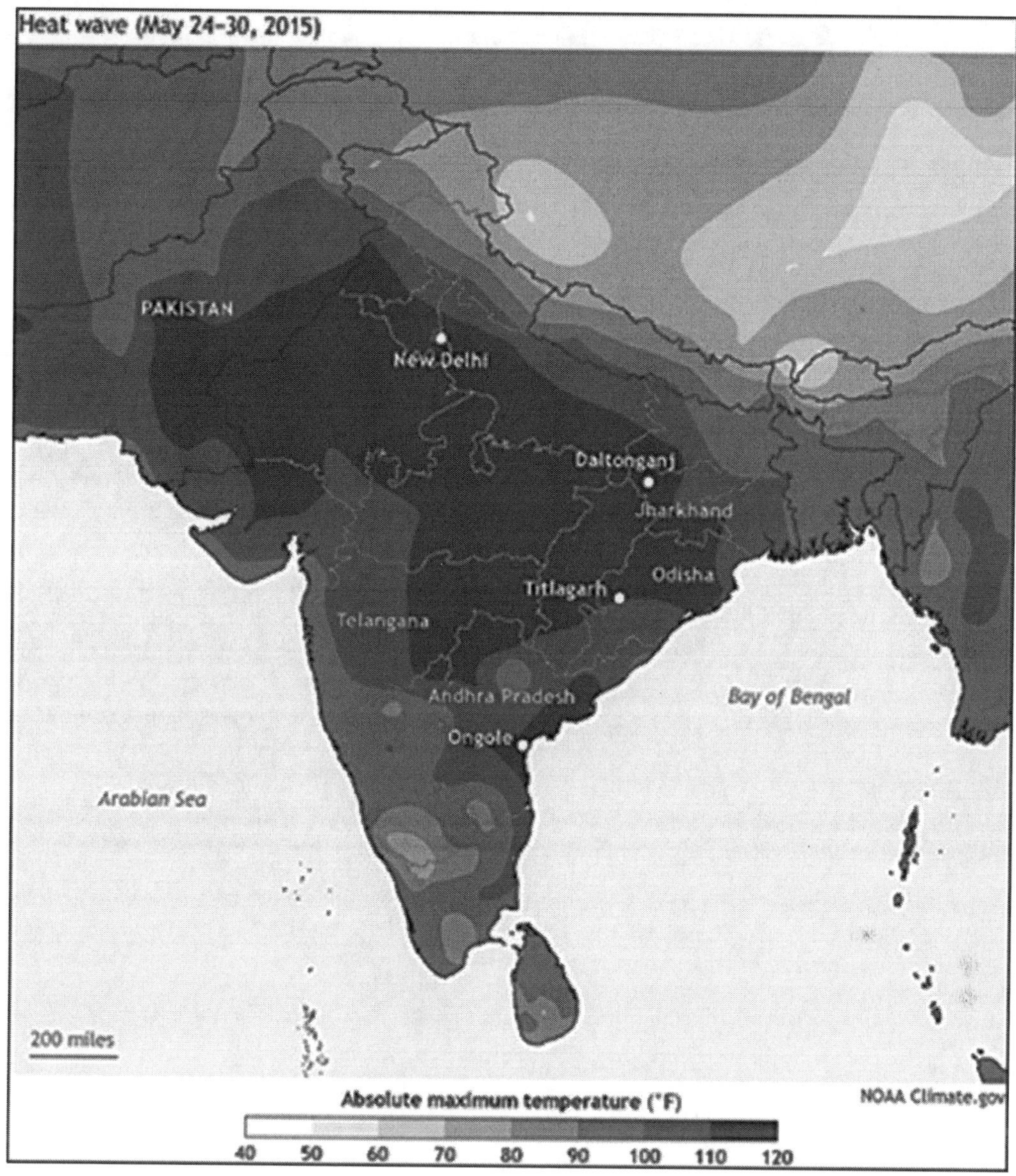

Figure 28: Heat wave May 24–30, 2015.

at 2 meters (about 6.5 feet) above the ground. The GEOS-5 model, like all weather and climate models, uses mathematical equations that represent physical processes (like precipitation and cloud processes) to calculate what the atmosphere will do. Actual measurements of physical properties, like temperature, moisture, and winds, are routinely folded into the model to keep the simulation as close to measured reality as possible. Heatwaves were predominately active in Punjab and Haryana and a decline can be observed.

Record-breaking heatwaves in north-west India and Pakistan have been made 100 times more likely by the climate crisis, according to scientists. The analysis means scorching weather once expected every three centuries is now likely to happen every three years [23]. The models developed take into account the factors, Yield (Hg/Ha), Area Harvested (Ha), Production (Mt) [21], and climate parameters need to be run, ironically, more frequently to predict food security assessment impacts as heatwave related yield assessments are the need of the day.

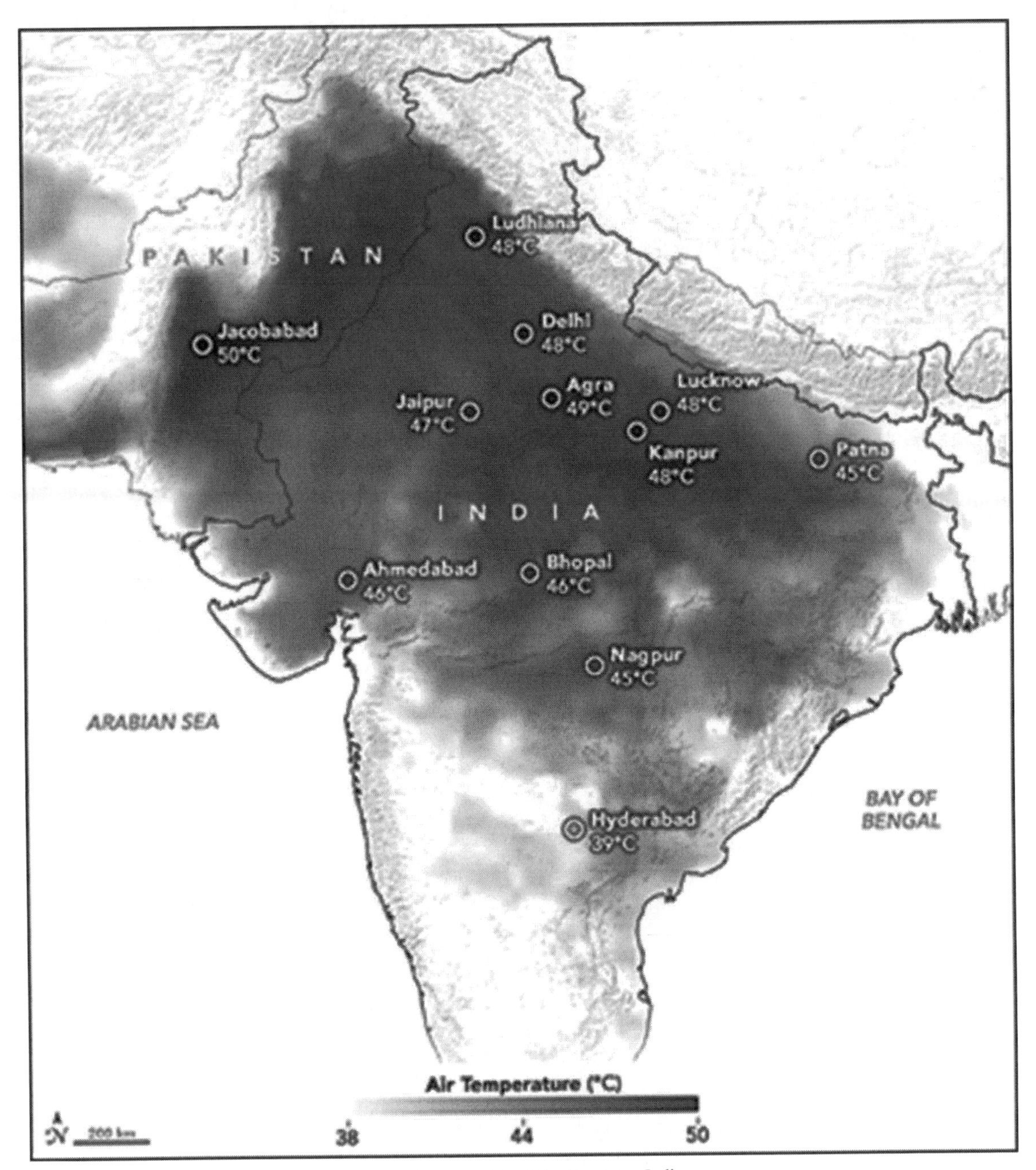

Figure 29: June 10, 2019 - Heatwave India.

The following table has the wheat data[499] for state wise production, area harvested, and yield details: the heat wave years and production yield drops are from Reserve Bank of India—Handbook of statistics on Indian states.[500] The percentage drop is computed from the Yield data and Maximum heatwave temperature registered during the heatwave episode - Indian Meteorological Department (IMD).[501]

Table 6: Wheat yield & percentage drop.

Yield (Kg. Per Hectare)					Percentage of Yield Drop			Heatwave
Heatwave Year	Uttar Pradesh	Punjab	Madhya Pradesh	Haryana	Uttar Pradesh	Punjab	Madhya Pradesh	Max. Temp C
1995–96	2445	3884	1658	3697	–2.51196	–5.03667	–4.49309	45.5
1997–98	2495	3853	1573	3660	–6.48426	–8.99858	–12.6596	49.5
2002–03	2596	4200	1456	1966	–5.77132	–7.32568	–10.1235	49
2014–15	1881	4294	1931		–38.0843	–14.411	–19.7089	49.4
2015–16	2258	4583	1835		20.04253	6.730321	–4.97152	51
2018–19	2910	5188	1847		5.856675	2.186331	–12.9184	50.8
2019–20	2861	5003	2046		–1.68385	–3.56592	10.77423	50.8

The average drop in Yield due to heatwave temperatures can be computed with a 0.8°C increase in temperature that has resulted in a 6.59% drop in yield. That is, if we take 2020 yields in Uttar Pradesh (2230 Kg. Per Hectare), Madya Pradesh (2046 Kg. Per Hectare), and Punjab (5003 Kg. Per Hectare) 2020, the heatwave in 2022 could result in a yield of 203.82 (Kg. Per Hectare). The total area harvested in Uttar Pradesh (9853 Thousand Hectares), Madya Pradesh (6551 Thousand Hectares), and Punjab (3521 Thousand Hectares) is 19, 925 thousand hectares or 19.9 million hectares and the heatwave resulted in 4.061 billion kgs or 8952972467.328 pounds (8.95 billion pounds) of wheat production.

Year	Average Drop	Temperature Increase
1995–96	–4.01391	4°C
1997–98	–9.38082	–0.5°C
2002–03	–7.74016	0.4°C
2014–15	–24.0681	1.6°C
2015–16	7.267112	–0.2°C
2018–19	–1.62514	0°C

In summary, heatwaves are expensive, needless to mention the human toll they induce is unimaginable in terms of life loss and property losses due to wildfires. Food security issues are humongous with yield loss resulting into wheat loss that can feed 49.7 million in the world for year, assuming 180 pounds[502] (81.6 kg) of wheat per person. With an increase in frequency and temperature intensity, the drop in yield will result in severe food insecurity.

"Climate change has made India's heat waves hotter. Earlier human activities increased global temperatures, heat waves like in the year 2022 would have struck India once in about half a century. However, now it is a more common occurrence—we can expect such high temperatures about once every four years."

Friederike Otto,
a climate scientist at the Imperial College of London

The frequency of heatwaves due to climate change has increased across the world. As of the writing of this book in June 2022, India just exited its worst heatwave in decades and Spain is witnessing its worst heatwave—Spain melts under the earliest heat wave in over 40 years.[503]. Spain is in the grip of its first heatwave of the year in June 2022, with temperatures in parts of the west and south expected to reach 44ºC (111.2ºF).[504] Climate scientists say the prolonged heatwave is undoubtedly the result of global warming. The heatwave has prompted India to take a strong stand[505] in Bonn, Germany, where officials are meeting to prepare for the next UN climate conference in Egypt in the month of November.

"The main input for growing the crop is rain and You don't have any rain, you don't have any grain basically"[506] [26].

Breadbasket to the World

Wheat is one of the essential and staple food grains. The world needs Wheat to maintain a healthy life balance and to maintain the required nutrition. The breadbasket of the world, in terms of production, includes USA, Brazil, EU, Argentina, Russia & Ukraine, and others. Ukraine also holds most of the wheat produced by the world and is a crucial country in the production of barley—winter barley—a type of wheat that is more accustomed to colder temperatures. Yet weather patterns show an increase in temperature due to the greenhouse gas emissions that are constantly being released because of the on-going war between Ukraine and Russia. Which means that the Winter Barley plant is not being harvested because of the horrible conditions of the area.[507] Alongside the need for increased irrigation that is fundamental to the growth of crops such as oats, wheat, corn, and legumes, which is being hindered by the increased temperatures and decreased precipitation levels in Ukraine. Ukraine is suffering from increased Greenhouse gas emissions as well as food and money shortages that come from the war that they have with Russia, which directly takes a toll on the wheat being produced.

Though the rest of the world might be suffering from climate change, Russia is benefiting from it since increased temperatures of the country are making it easier to grow crops [24]. Because of this, Russia has another profitable way to make money and grow from it.[508] As many counties around Russia are being affected badly by climate change, they are importing wheat and other resources from it. Russia is benefiting from the agricultural production increase that climate change is providing it with [22].

Production of 2021 Wheat Crop Officially Forecast at Record Level

Harvesting of the mostly irrigated, main 2021 "Rabi" wheat crop is about to start, and early official forecasts indicate a record production of 109.2 million tonnes. The area planted is officially estimated at an all-time high of 34.6 million hectares. Farmers have been encouraged to increase the planted area by the remunerative producer prices guaranteed by the Government, coupled with optimal soil moisture conditions at planting time and the timely harvest of previously grown summer crops that made land available. Overall, yields are forecasted close to average as weather conditions were generally favorable throughout the cropping season and supplies of agricultural inputs, including irrigation water and fertilizer, were adequate.[509]

No noise (smoothed) Brightness Temperature (SMT)—Uttar Pradesh

The SMT is the Brightness Temperature (BT) with complete[510] removal of high frequency noise, that can be used for estimating of thermal conditions, cumulative degree days and other important parameters.

A glimpse of heatwave temperatures in March 2022 (please see Figure 30) can be visualized using the MODIS[511] data. The images below were created from the MODIS product: 'Land Surface Temperature 8-Day composed, Global 1000m' (MYD11A2.005) data source: ftp://e4ftl01.cr.usgs.gov/MOLA/MYD11A2.005/

MODIS Land Surface Temperature of 2022 Period=010: composed by MODIS data from 3/14/2022 to 3/21/2022 (each period has 8 days)—8-days composed LST of 2022. As India is under extreme heat throughout!

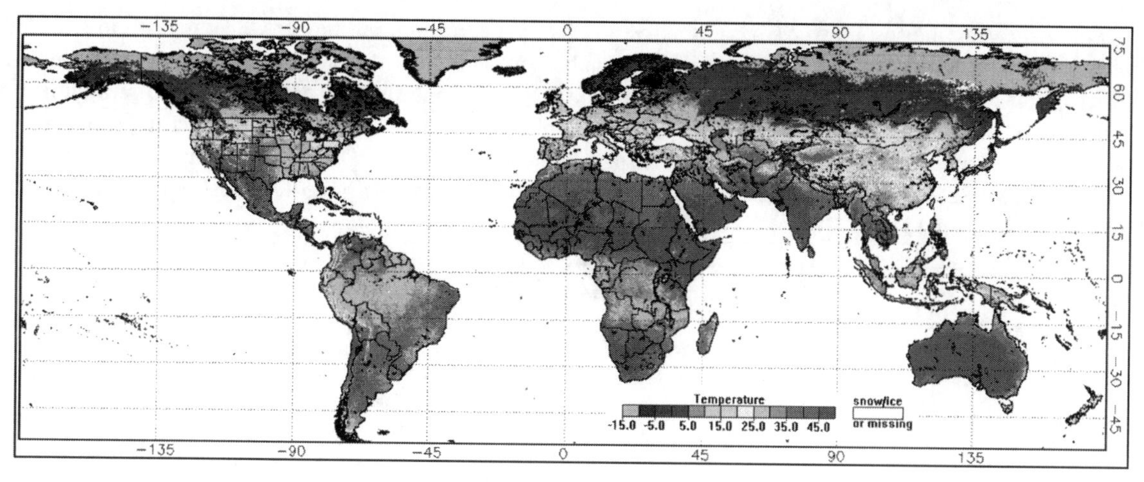

Figure 30: Heat wave.

VH Time Series by administrative regions for the wheat crop specifically – India: Uttar Pradesh

Collect Vegetation Health indexes using the VH Dataset for the Wheat crop[512] for the years 1980 to 2022. The trendline of the Vegetation Health of Uttar Pradesh's wheat crop is (please see Figure 31):

$$Y = 0.0007x + (-1.1697)$$

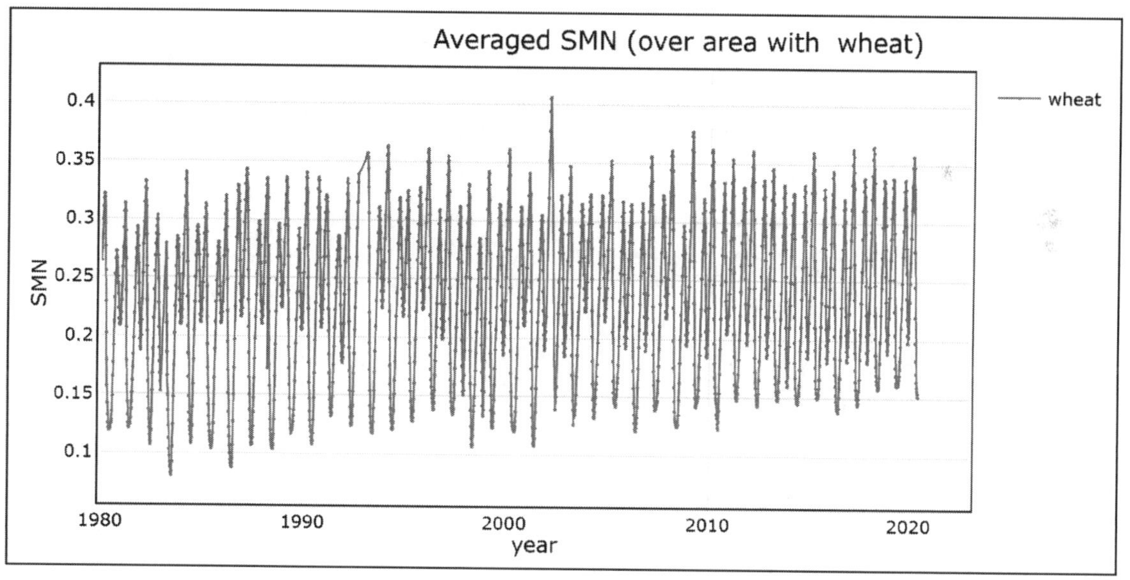

Figure 31: Average SMN Over Wheat Areas.

Wheat moisture condition

Vegetation Condition index (VCI) is a proxy for moisture conditions. The target yield is the expected crop yield based on the amount of spring soil moisture and the growing season precipitation levels. Heat determines whether a crop will mature but moisture establishes its yield potential. Each crop requires a minimum amount of moisture to produce the first bushel.

$$Y = 0.4854\,x + (-918.8514)$$

Year over year, the soil moisture has a positive trend and a contributor to higher wheat growth in Uttar Pradesh. For wheat, this amount is about 4 inches. Each additional inch of water increases the target yield by about 4 bushels per acre. The amount of water available to a crop during the growing season is equal to the amount of soil moisture available at the seeding time plus the amount of precipitation over the growing season. Soil moisture can be assessed easily in the spring by using the "feel and appearance" method or a soil moisture probe. The expected growing season rainfall is determined by referring to historical rainfall data for the area[513] (please see Figure 32).

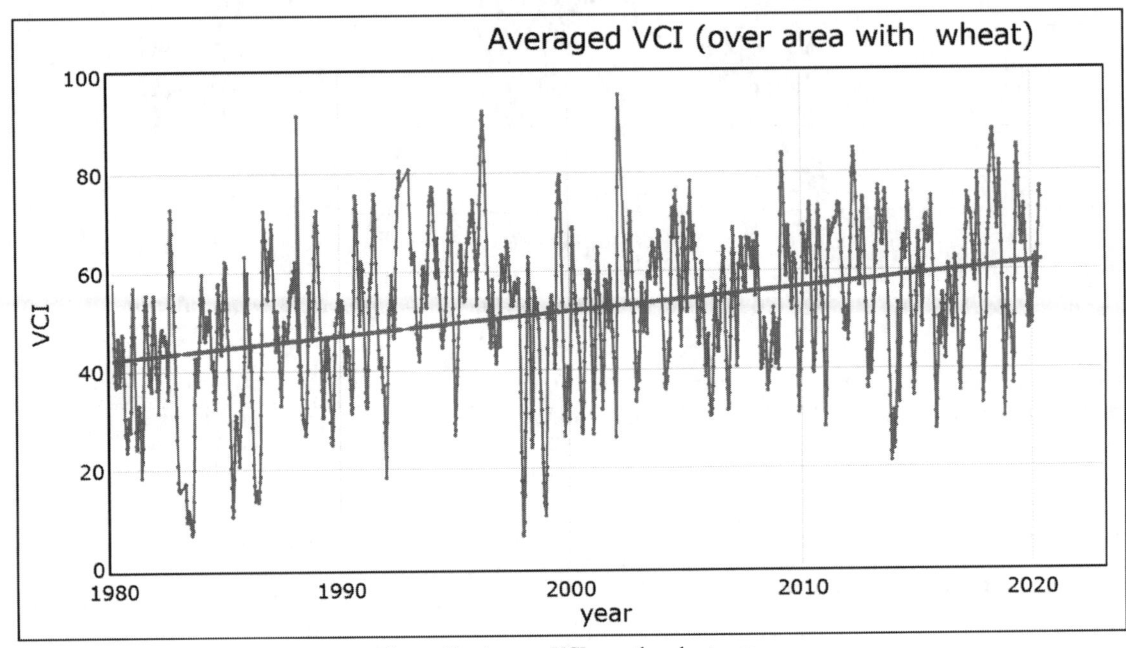

Figure 32: Average VCI over the wheat area.

Province-Averaged VH data for 'wheat'

The province-Averaged VH data for 'wheat' has the following structure: Mean data for IND Province= 33: Uttar Pradesh, from 1982 to 2022, weekly[514] for area with 'WHEA'

```
year,week, SMN,SMT,VCI,TCI, VHI
1982, 1, 0.264,294.90, 57.65, 45.68, 51.67,
1982, 2, 0.274,294.65, 52.40, 51.91, 52.15,
1982, 3, 0.286,294.41, 48.10, 63.22, 55.66,
1982, 4, 0.299,294.35, 45.10, 73.90, 59.50,
1982, 5, 0.310,294.64, 42.24, 80.51, 61.37,
1982, 6, 0.317,295.03, 39.42, 85.62, 62.52,
1982, 7, 0.321,295.71, 37.97, 88.56, 63.26,
1982, 8, 0.318,296.82, 37.20, 89.07, 63.14,
1982, 9, 0.310,298.03, 36.80, 88.89, 62.84,
1982,10, 0.295,299.41, 37.73, 87.54, 62.64,
1982,11, 0.276,300.97, 40.50, 85.04, 62.77,
```

Maximum, Minimum, Average Temperature & Precipitation (1901:2020)

India's temperature from 1901to 2020 can be downloaded from the Climate Change Knowledge Portal (CCKP).

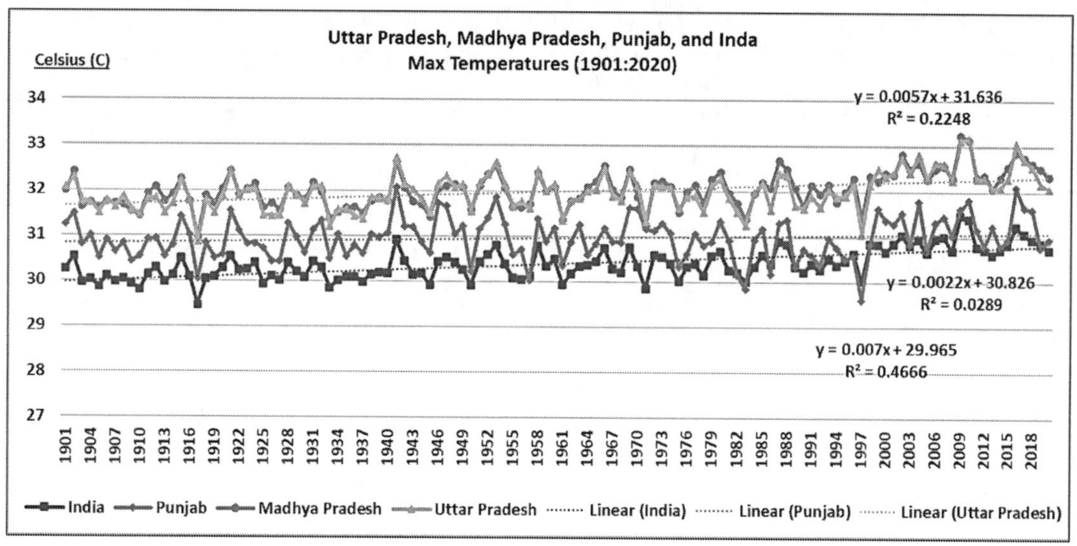

Figure 33: Max temperatures (1901 : 2020).

Maximum Temperatures in wheat growing states and India: as can be seen the temperatures are trending upwards year over year. The future temperatures will be much higher ($R^2_{UP} = 0.2248$ & $R^2_{India} = 0.4666$) as the prevailing equation predicates (please see Figure 33):

$$Y_{UP} = 0.0057x + 31.636$$

$$Y_{India} = 0.007x + 29.965$$

Similar observations can be made for minimum and mean or average temperatures.

Average temperatures play an important role for wheat crops. The average Temperatures in wheat growing states and India: as can be seen the temperatures are trending upwards year over year. The future temperatures will be much higher ($R^2_{UP} = 0.3257$ & $R^2_{India} = 0.5111$) as the prevailing equation predicates (please see Figure 34):

$$Y_{India} = 0.0076x + 23.935$$

$$Y_{UP} = 0.0072x + 24.981$$

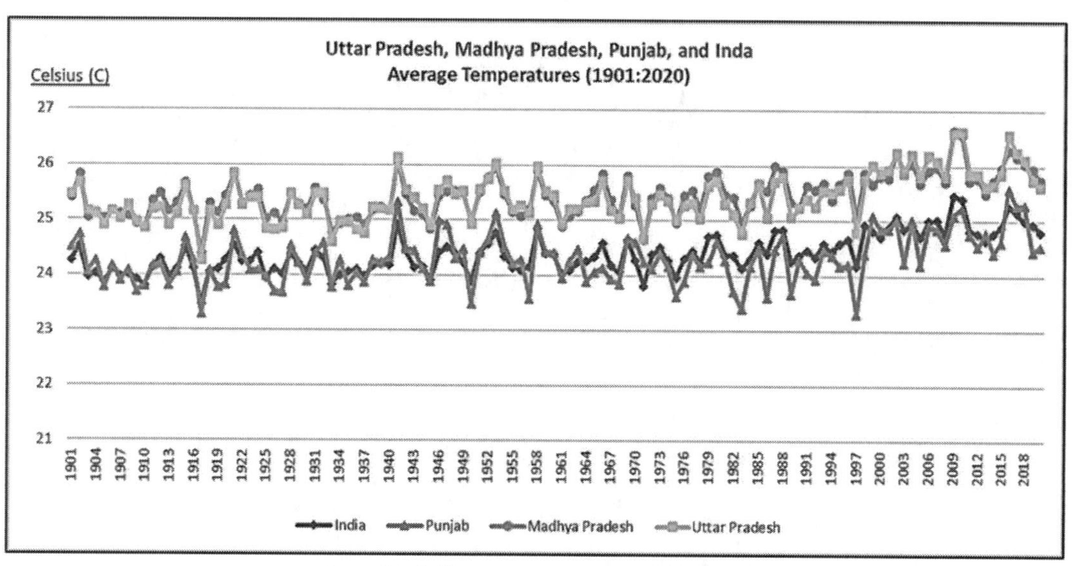

Figure 34: India average temperatures (1901:2020).

A clear indication from the temperature movement is that higher temperatures are something slated for the next century. The above data from CCKP is downloaded using the following options (please see Figure 35):

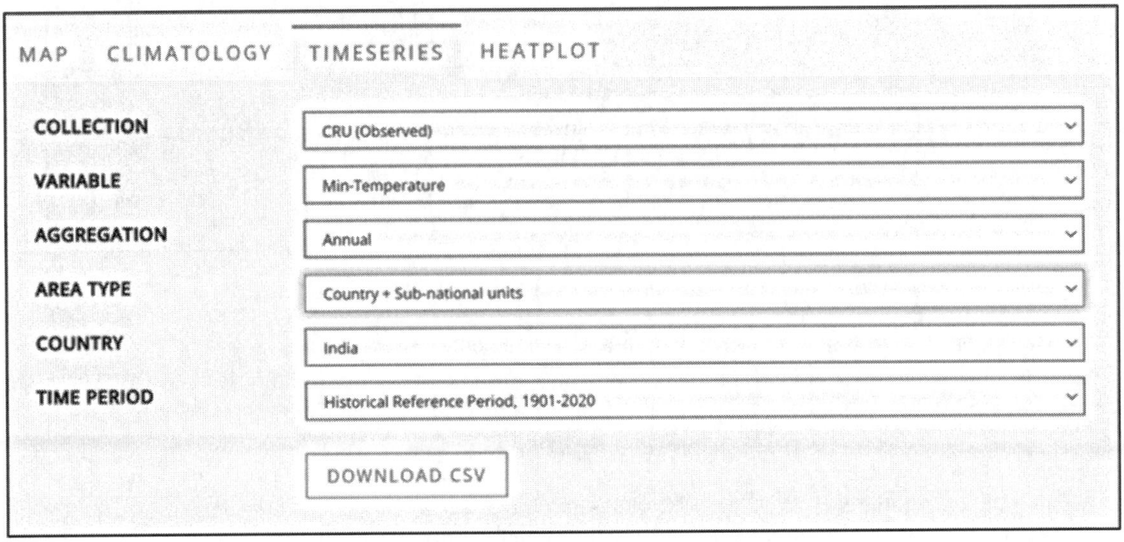

MAP
CLIMATOLOGY
TIMESERIES
HEATPLOT
COLLECTION
CRU (Observed)
VARIABLE
Min-Temperature
AGGREGATION
Annual
AREA TYPE
Country + Sub-national units
COUNTRY
India
TIME PERIOD
Historical Reference Period, 1901-2020
DOWNLOAD CSV

Figure 35: CCKP Portal.

Climate Modeling: CMIP6-Scenario SSP5-85, and 2040:2059

Subject the model to CMIP6, SSP5-8.5 (2040:2059) please see Figure 36.

Heat Waves & Wheat Yield Production Linkage in India

India - major wheat producing steates
CMIP6, 2040-2059, SSP 5-8.5, Multi-Model Ensemble
50
45
40
35
30
25
20
15
10
5
0
Jan
Feb
Mar
Apr
May
Jun
Jul
Aug
Sep
Oct
Nov
Dec
India
Madhya Pradesh
Uttar Pradesh

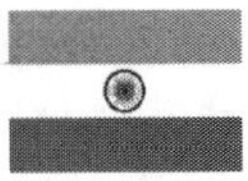

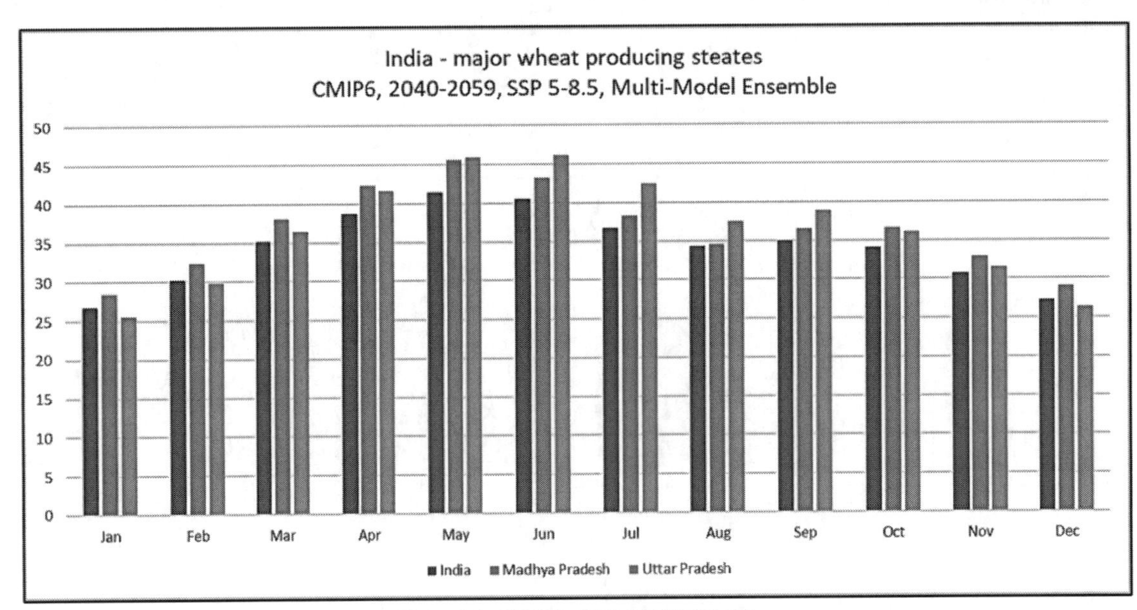

Figure 36: CMIP 6 - 2040-2059 SSP5-8.5.

Machine Learning Model: Heat Waves & Wheat Yield Production Linkage in India

India is the world's second most populous country and one of the world's fastest major growing economies since 2000.[515] India is also among the world's largest producers and consumers of a range of crop and livestock commodities. While still classified as a lower middle-income country, rapid expanding economic growth and diversifying food demand are pressuring the farm sector to boost currently low levels of productivity, and challenging agricultural policies traditionally focused on self-sufficiency [25].

India's behavior in world markets for commodities is important to U.S. agriculture and the world in general. India is the world's leading importer of soybean oil, ahead of the European Union and China. India is the world's largest producer and consumer of milk. Growth in milk supply and demand has been robust, but projections indicate that production targets will be difficult to reach without stronger gains in productivity. Rising incomes, urbanization, and India's youthful demographic profile are expected to shape continued growth and increased diversity in consumer food demand. While per capita consumption of staple grains, such as rice and wheat, has been relatively stable, demand for higher valued foods—fruits, vegetables, edible oils, dairy products, and poultry meat—has been robust. Alongside this growth story, however, Government estimates indicate that about 22 percent of India's population—or about 260 million people—remain in poverty. According to USDA estimates, India still accounts for the largest share of the global population classified as food insecure. In addition, because India accounts for the largest share of the global population classified as food insecure [25], productivity enhancement technologies and climate smart sensors are needed to feed future generations.

	Reforms that improve the efficiency of India's agricultural markets could generate economy-wide gains in output and wages. These reforms could raise agricultural producer prices, reduce consumer food prices—and increase consumption, particularly by low-income households.

From a food security enhancement point of view, India should focus on two important aspects: (a) climate change and (b) Data driven approaches to improve agricultural practices. Although India has made strides in agricultural productivity, there is still a huge enhancement to be made to provide food security to the nation. Agricultural productivity enhancement is pivotal to India to enhance food security. Given temperature sensitive aspects, and global climate change, India needs to not only improve production technologies but also an infusion of data science and data driven approaches to tackle productivity improvement practices.

Heatwaves have a negative effect on agricultural outputs, especially for staple cereals. The area planted is officially estimated at 34.3 million hectares, close to the last 2021 level, supported by the remunerative Minimum Support Price (MSP) by the government and optimal soil moisture conditions at the time of planting. Weather conditions were favorable between October 2021 and February 2022 in the main wheat producing areas, raising expectations for record yields. Unfortunately, unseasonal high temperatures and a well below-average precipitation from mid-March to April over northwestern parts of the country affected crops just before the harvest. Wheat crops at the grain filling stages of development were the most affected by the high temperatures, given that the crops experienced increased susceptibility to heat at this phenological stage. As a result, yields are now expected to fall below record levels that had been forecast earlier. Wheat cultivation in the states of Punjab, Haryana and Uttar Pradesh were the most affected. Based on current official estimates, 2022 wheat production is expected at about 106.4 million tons, which is above the five-year average. However, it is likely that production may be lower as the extent of the heatwave damage is still to be fully assessed. Let's build the model:

 	Software code for this model: Final_Indian_wheat_model_Punjab.ipynb (Jupyter Notebook Code) Final_India_UttarPradesh (UP)_WheatYield_and_HeatWave_NDVIdata-MLModel.ipynb (Jupyter Notebook Code)

Time Series for Terra MODIS-8-day Vegetation Index

There are many standard MODIS[516] data products that scientists are using to study global change. These products are being used by scientists from a variety of disciplines, including oceanography, biology, and atmospheric science. This section provides some detail for each product individually, introducing products, explaining the science behind them, and alerting readers to the known areas of concern with products' data. Additional information about these products can be obtained by going to the appropriate URL's noted below. Please find Vegetation Health (VH) Index for India[517] and Province-Averaged VH data for Crop Land data[518] (please see Table 7).

Table 7: India VH Data.

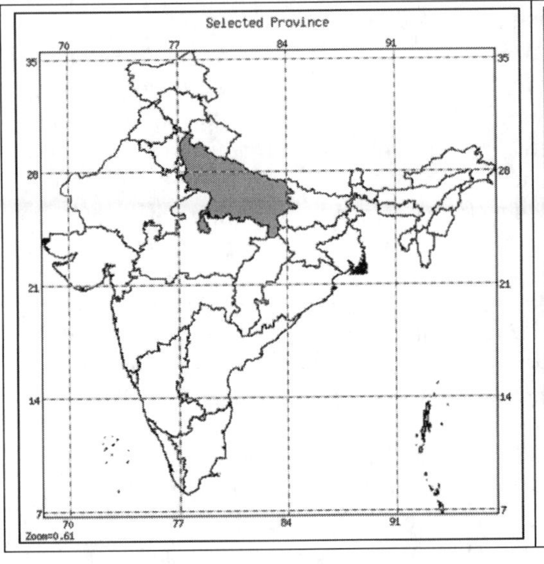


```
Mean data for IND Province= 33: Uttar Pradesh, from 1982 to 2022, weekly
for area with 'WHEA'
year,week, SMN,SMT,VCI,TCI, VHI

1982, 1, 0.264,294.90, 57.65, 45.68, 51.67,
1982, 2, 0.274,294.65, 52.40, 51.91, 52.15,
1982, 3, 0.286,294.41, 48.10, 63.22, 55.66,
1982, 4, 0.299,294.35, 45.10, 73.90, 59.50,
1982, 5, 0.310,294.64, 42.24, 80.51, 61.37,
1982, 6, 0.317,295.03, 39.42, 85.62, 62.52,
1982, 7, 0.321,295.71, 37.97, 88.56, 63.26,
1982, 8, 0.318,296.82, 37.20, 89.07, 63.14,
1982, 9, 0.310,298.03, 36.80, 88.89, 62.84,
1982,10, 0.295,299.41, 37.73, 87.54, 62.64,
1982,11, 0.276,300.97, 40.50, 85.04, 62.77,
1982,12, 0.255,302.61, 44.58, 81.33, 62.96,
1982,13, 0.229,304.27, 46.75, 78.20, 62.48,
1982,14, 0.202,305.72, 46.06, 77.39, 61.73,
1982,15, 0.179,306.79, 46.11, 78.63, 62.37,
1982,16, 0.157,307.39, 40.46, 81.20, 60.83,
1982,17, 0.142,307.66, 37.18, 82.46, 59.82,
1982,18, 0.132,308.05, 38.99, 79.15, 59.07,
1982,19, 0.125,308.23, 39.90, 73.99, 56.94,
1982,20, 0.121,308.17, 41.36, 68.98, 55.17,
1982,21, 0.119,307.74, 43.58, 62.65, 53.11,
```

Data Sources

The data is from Reserve Bank of India (RBI) – Handbook of statistics on Indian states.[519]	Table 52: State-wise Production of Foodgrains - Wheat (1982:2021) Table 61: State-wise Area for Foodgrains – Wheat Table 74: State-wise Estimates of Wheat Yields	RESERVE BANK OF INDIA
NOAA STAR[520]—Global Vegetation Health Products: Province-Averaged VH	Mean data for USA Province= 17: Kansas, from 1982 to 2022, weekly for cropland[521] area only year, week, SMN, SMT, VCI, TCI, VHI	NATIONAL OCEANIC AND ATMOSPHERIC ADMINISTRATION NOAA U.S. DEPARTMENT OF COMMERCE
The Climate Change Knowledge Portal (CCKP)[522]	India temperatures[523] Mean Temperature Minimum Temperature Maximum Temperature Precipitation	THE WORLD BANK

Step 1: Load Libraries

Load libraries to process major Indian wheat producing states' data.

```
import pandas as pd
import numpy as np
import plotly.graph_objects as go
import seaborn as sns
from sklearn.linear_model import LinearRegression
from sklearn.model_selection import train_test_split
import eli5
from sklearn.preprocessing import MinMaxScaler
import matplotlib.pyplot as plt
import pickle
coffeeAreaHarvested_Original = pd.read_csv("FAOSTAT_Vietnam_CoffeeareaHarvested.csv")
```

Step 2: Load Punjab Wheat data (1961:2020)

Load wheat data of Punjab.

```
data = pd.read_csv('wheat_india.csv')
data.tail()
```

Output:

	Area_harvested_Value	Yield_Value	Production_Value	Nov_temp	Dec_temp	Jan_temp	Feb_temp	Mar_temp	Apr_temp	May_temp	...
Year											
1961	12927000	8507	10997000	17.64	12.56	13.45	13.61	21.36	26.29	31.63	...
1962	13570000	8896	12072000	18.77	13.53	12.32	16.19	20.68	27.74	31.45	...
1963	13590000	7929	10776000	19.99	14.17	12.52	17.15	20.52	26.38	30.04	...
1964	13499000	7299	9853000	18.03	13.90	11.06	14.99	22.28	27.67	30.13	...
1965	13422000	9132	12257000	20.24	13.31	14.47	15.21	20.38	24.47	29.98	...
1966	12572000	8268	10394000	18.41	13.23	13.53	17.53	20.71	26.43	31.18	...
1967	12838000	8874	11393000	18.73	13.92	11.74	17.19	19.63	25.97	30.44	...
1968	14998200	11028	16540100	18.96	14.06	11.56	13.12	20.65	26.69	30.17	...
1969	15958100	11688	18651600	20.23	14.76	12.74	15.49	23.24	26.97	30.13	...
1970	16625500	12086	20093296	18.63	14.47	13.40	15.22	20.46	28.55	32.76	...

Step 3: Load Minimum, Maximum, and Average Temperature & Precipitation Data - Punjab (1901 to 2020)

Load Minimum, Maximum, and Average temperatures data (1901 to 2020).

```
data_temp = pd.read_csv('tas_timeseries_monthly_cru_1901-2020_IND_1505.csv')
data_temp.head()
```

```
data_temp_min = pd.read_csv('tasmin_timeseries_monthly_cru_1901-2020_IND_1505.csv')
data_temp_min.head()
```

```
data_temp_max = pd.read_csv('tasmax_timeseries_monthly_cru_1901-2020_IND_1505.csv')
data_temp_max.head()
```

```
data_precp = pd.read_csv('pr_timeseries_monthly_cru_1901-2020_IND_1505.csv')
data_precp.head()
```

Output:

Step 4: Apply Imputation Strategies to fix nulls

Check for nulls in the data.

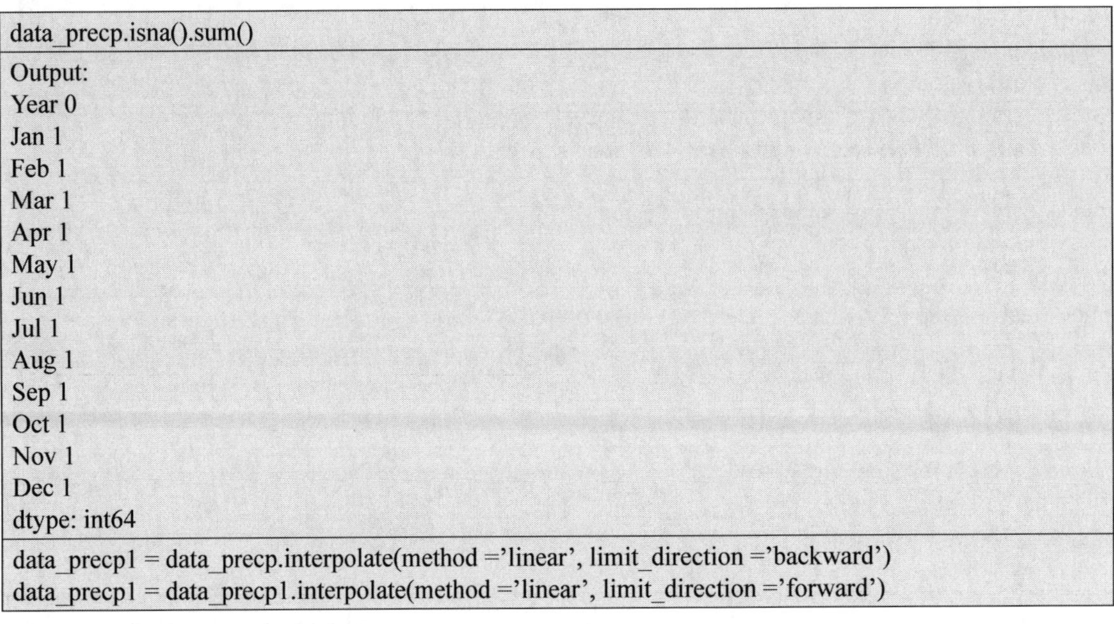

```
data_precp.isna().sum()
```

```
Output:
Year 0
Jan 1
Feb 1
Mar 1
Apr 1
May 1
Jun 1
Jul 1
Aug 1
Sep 1
Oct 1
Nov 1
Dec 1
dtype: int64
```

```
data_precp1 = data_precp.interpolate(method ='linear', limit_direction ='backward')
data_precp1 = data_precp1.interpolate(method ='linear', limit_direction ='forward')
```

This would resolve data null issues.

Step 5: Merge Temperature and Production Yield Data

Combine all datasets into one master to perform regression analysis.

```
df = data[[ 'Year', 'Area_harvested_Value', 'Yield_Value', 'Production_Value']]
```

```
result = df.set_index('Year').join(dt1.set_index('Year_temp'))
result = result.join(dp1.set_index('Year_precp'))
result = result.join(dt1_max1.set_index('Year_Max_temp'))
result = result.join(dt1_min1.set_index('Year_min_temp'))
```

```
Result
```

Output:

Year	Area_harvested_Value	Yield_Value	Production_Value	Nov_temp	Dec_temp	Jan_temp	Feb_temp	Mar_temp	Apr_temp	May_temp	...
1961	12927000	8507	10997000	17.64	12.56	13.45	13.61	21.36	26.29	31.63	...
1962	13570000	8896	12072000	18.77	13.53	12.32	16.19	20.68	27.74	31.45	...
1963	13590000	7929	10776000	19.99	14.17	12.52	17.15	20.52	26.38	30.04	...
1964	13499000	7299	9853000	18.03	13.90	11.06	14.99	22.28	27.67	30.13	...
1965	13422000	9132	12257000	20.24	13.31	14.47	15.21	20.38	24.47	29.98	...
1966	12572000	8268	10394000	18.41	13.23	13.53	17.53	20.71	26.43	31.18	...
1967	12838000	8874	11393000	18.73	13.92	11.74	17.19	19.63	25.97	30.44	...
1968	14998200	11028	16540100	18.96	14.06	11.56	13.12	20.65	26.69	30.17	...
1969	15958100	11688	18651600	20.23	14.76	12.74	15.49	23.24	26.97	30.13	...
1970	16625500	12086	20093296	18.63	14.47	13.40	15.22	20.46	28.55	32.76	...

Step 6: Correlation between Yield and Temperatures/Precipitation

To witness the correlation between yield and weather, let's perform data frame correction.

```
fig = plt.figure(figsize=(30, 20))
corrMatrix = fr.corr()
sn.heatmap(corrMatrix, annot=True)
plt.show()
```

Output:

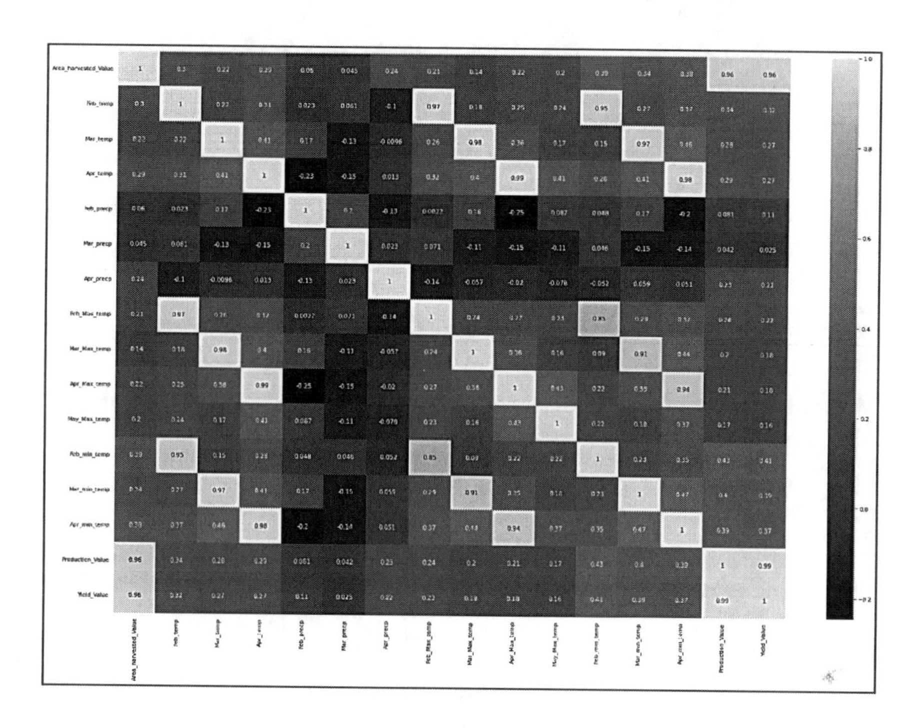

For clarity, extract only the relationship between yield value and other independent time series variables:

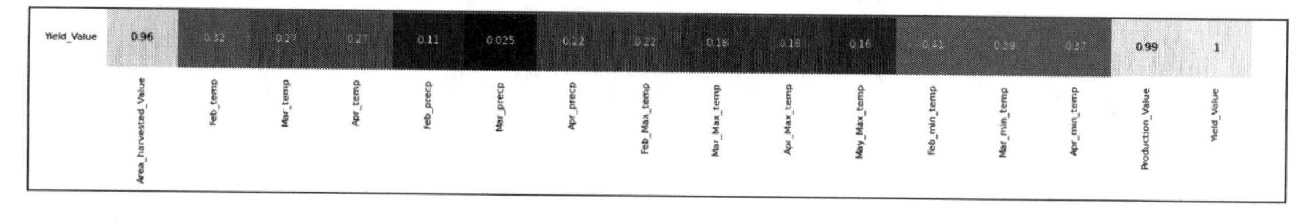

Yield value is highly correlated with Area Harvested (96%), Production Value (99%), February minimum temperature (41%), March Minimum Temperature (39%), April Minimum Temperature (37%), Average Temperature February (32%), March average temperature (27%), April average temperature (27%), February Precipitation (11%), April precipitation (22%), March Maximum temperature (18%), April maximum temperature (18%), and May maximum temperature (16%).

Very similar to that of the Kansas[524] agriculture pattern, Wheat is a winter Rabi crop in India. The wheat plants are planted in October, November, and December, become dormant in the winter, grow again in the spring (February through June), and are harvested in early summer.

So, temperatures in February through June play an important role in the successful yield of crops.

Wheat plants behave very similarly across geographical regions. All they need is sufficient moisture, temperature, and light. The rest is the magic of nature, and we can see a golden harvest.

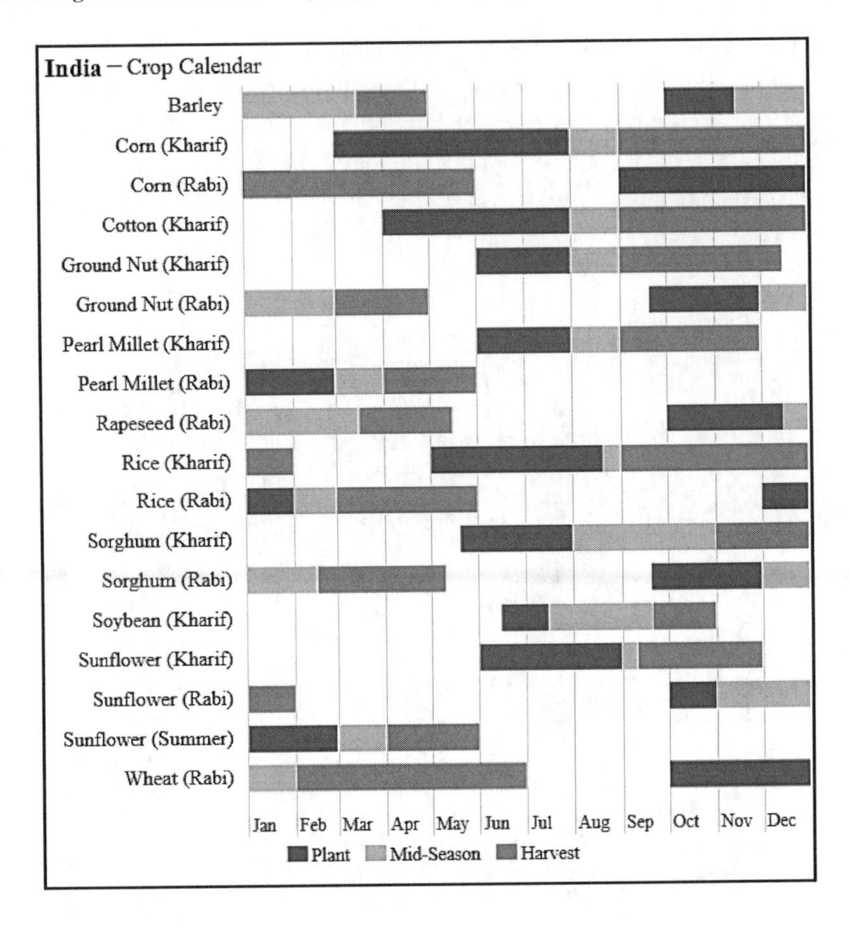

Step 7: Train Regressive Model

Prepare regressive model to see the influence of temperature and precipitation coefficients on the wheat yield.

```
X = fr[['Area_harvested_Value', 'Feb_temp', 'Mar_temp','Apr_temp',
    'Feb_precp', 'Mar_precp', 'Apr_precp',
    'Feb_Max_temp','Mar_Max_temp', 'Apr_Max_temp','May_Max_temp',
    'Feb_min_temp', 'Mar_min_temp', 'Apr_min_temp','Production_Value' ]]
y = fr['Yield_Value']
from sklearn.model_selection import train_test_split
X_train, X_test, y_train, y_test = train_test_split(X, y, test_size = 0.3, random_state = 0)
```

```
# Train the model
from sklearn.linear_model import LinearRegression
# Fit a linear regression model on the training set
model = LinearRegression(normalize=False).fit(X_train, y_train)
print (model)
```

Output:

```
LinearRegression()
```

```
import matplotlib.pyplot as plt
%matplotlib inline
plt.scatter(y_test, predictions)
plt.xlabel('Actual Labels')
plt.ylabel('Predicted Labels')
plt.title('Predictions')
# overlay the regression line
z = np.polyfit(y_test, predictions, 1)
p = np.poly1d(z)
plt.plot(y_test,p(y_test), color='magenta')
plt.show()
```

Output:

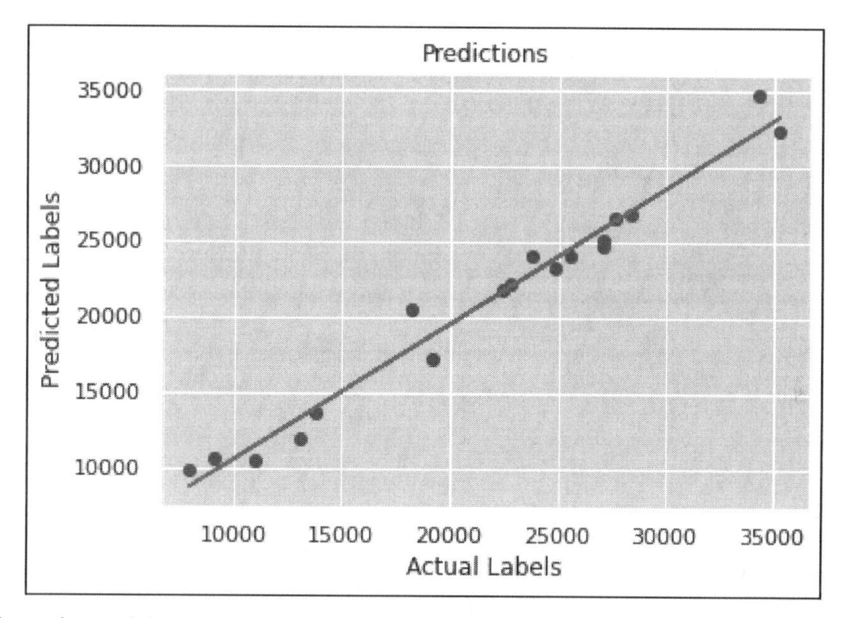

Step 8: Evaluate the model

Let's calculate the model metrics MSE, RMSE and MAE.

```
from sklearn.metrics import mean_squared_error, r2_score
mse = mean_squared_error(y_test, predictions)
print("MSE:", mse)
rmse = np.sqrt(mse)
print("RMSE:", rmse)
r2 = r2_score(y_test, predictions)
print("R2:", r2)
```

Output:

```
MSE: 2407545.833409552
RMSE: 1551.6268344578061
R2: 0.9617092959089765
```

Step 9: Model CMIP6, SSP Models

Under CMIP6, SSP5-8.5, for the year 2040–2059, the model yielded a lower value as baselined to 34311 (hg/ha).

```
# baseline year 2020
# ha      Area_harvested_Value      Yield_Unit      Yield_Value      Production_Unit
          Production_Value
#ha       31357000                  hg/ha           34311            tonnes          107590000
td = [[ 31357000 ,
    18.31,24.28,31.09,
    25.12,24.46,24.95,
    25.65,31.94,38.01,43.19,
    11.85,17.09,22.74, 107590000
 ]]
result = loaded_model.predict(td)
print(result)
```

Output:

18865.6647 (hg/ha).

Under CMIP6, SSP5-8.5, for the year 2060–2079, the model yielded a lower value as baselined to 34311 (hg/ha).

```
# baseline year 2020
# ha      Area_harvested_Value      Yield_Unit      Yield_Value      Production_Unit
          Production_Value
#ha       31357000                  hg/ha                  34311            tonnes    107590000
td = [[ 31357000 ,
    20,26.15,32.42,
    31.35,24.29,26.01,
    27.65,33.94,40.01,45.19,
    13.94,18.6,24.66,107590000
 ]]
result = loaded_model.predict(td)
print(result)
```

Output:

20297.9193 (hg/ha).

Under CMIP6, SSP5-8.5, for the year 2080–2099, the model yielded a lower value as baselined to 34311 (hg/ha).

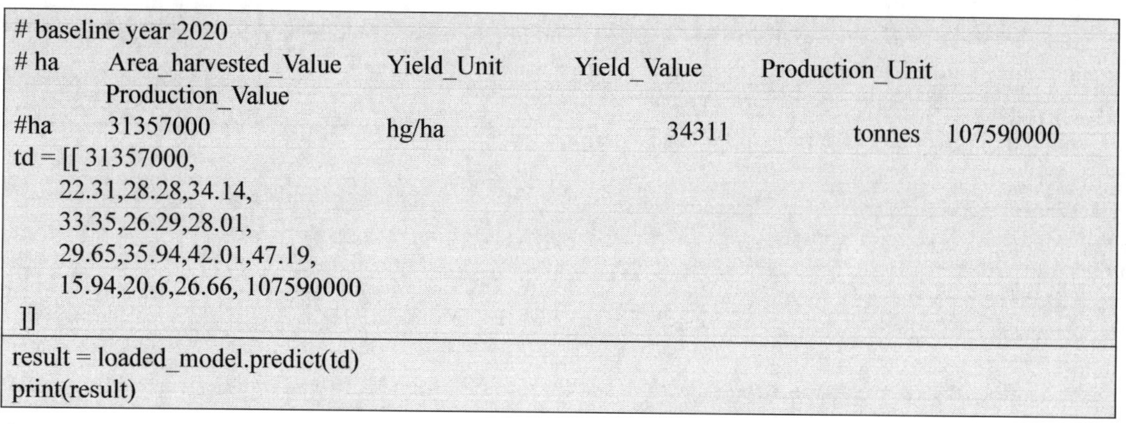

```
# baseline year 2020
# ha      Area_harvested_Value      Yield_Unit      Yield_Value      Production_Unit
          Production_Value
#ha       31357000                  hg/ha                34311            tonnes   107590000
td = [[ 31357000,
     22.31,28.28,34.14,
     33.35,26.29,28.01,
     29.65,35.94,42.01,47.19,
     15.94,20.6,26.66, 107590000
 ]]
result = loaded_model.predict(td)
print(result)
```

Output:
27253.3290 (hg/ha).

Step 10: Model summary

Climate change imposes a huge tax on the wheat yield for major wheat producing states (Uttar Pradesh). There is a drop of –45.01% in 2040–2059, –40.84% in 2060–2079, and –20.56% in 2080–2099. In terms of yield (hg/ha), the drop in yield imposes huge food insecurity (please see Figure 37).

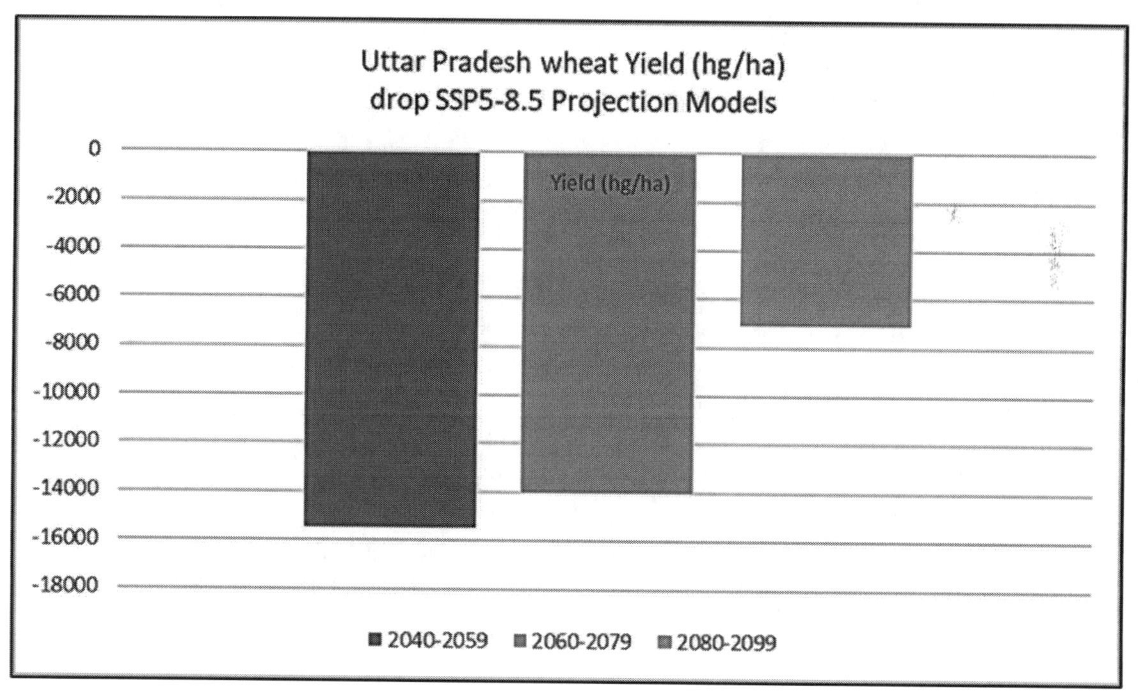

Figure 37: Wheat Yield.

Machine Learning Forecasting Model: Heat Waves & Wheat Yield Production in India

The model uses data from Punjab from 1961 to 2020 to build a time series model and forecast the impact of climate change induced by heatwaves on overall food security.

Step 1: Prepare Time Series Prophet Model

Load Python libraries, specifically the Prophet time series model (bolded).

```
import numpy as np
import pandas as pd
import functools
from datetime import date
import plotly.graph_objects as go
import seaborn as sns; sns.set()
from sklearn.impute import SimpleImputer
from scipy import stats
from scipy.stats import pearsonr
import matplotlib.pyplot as plt
from sklearn.preprocessing import MinMaxScaler
from sklearn.model_selection import train_test_split
from sklearn.linear_model import LinearRegression
from sklearn.metrics import r2_score
from sklearn import metrics
from sklearn.impute import SimpleImputer
from fbprophet import Prophet
from fbprophet.diagnostics import cross_validation, performance_metrics
```

Step 2: Prepare Prophet Regressors Time Series Variables

Prepare Prophet Data Frame to load key time series attributes that are validated in earlier regressive model.

```
dfProphet = fr[['Area_harvested_Value', 'Feb_temp', 'Mar_temp','Apr_temp',
   'Feb_precp', 'Mar_precp', 'Apr_precp',
   'Feb_Max_temp','Mar_Max_temp', 'Apr_Max_temp','May_Max_temp',
   'Feb_min_temp', 'Mar_min_temp', 'Apr_min_temp', 'Production_Value','Yield_Value' ]].copy()
```

```
Prophet_cols = ['Area_harvested_Value', 'Feb_temp', 'Mar_temp','Apr_temp',
   'Feb_precp', 'Mar_precp', 'Apr_precp',
   'Feb_Max_temp','Mar_Max_temp', 'Apr_Max_temp','May_Max_temp',
   'Feb_min_temp', 'Mar_min_temp', 'Apr_min_temp','Production_Value']
```

```
dfProphet = dfProphet.reset_index()
```

```
dfProphet.rename(columns = {"Year":'ds','Yield_Value' : 'y'},inplace = True)
```

```
dfProphet
```

Prophet requires two key columns: Date Time "ds" and Y variable " Yield _Value".

Output:

	ds	Area_harvested_Value	Feb_temp	Mar_temp	Apr_temp	Feb_precp	Mar_precp	Apr_precp	Feb_Max_temp	Mar_Max_temp	Apr_Max_temp
0	1961	12927000	13.61	21.36	26.29	27.390000	24.25000	4.180000	20.14	28.38	34.08
1	1962	13570000	16.19	20.68	27.74	14.270000	33.92000	13.000000	22.93	27.62	35.48
2	1963	13590000	17.15	20.52	26.38	6.780000	8.27000	13.480000	24.79	27.69	33.69
3	1964	13499000	14.99	22.28	27.67	26.880000	16.10000	34.210000	21.83	29.80	35.30
4	1965	13422000	15.21	20.38	24.47	45.220000	16.66000	10.000000	21.99	27.68	31.94
5	1966	12572000	17.53	20.71	26.43	16.970000	67.25000	5.960000	24.65	28.09	33.95
6	1967	12838000	17.19	19.63	25.97	27.220000	18.65000	4.180000	25.07	26.74	33.74
7	1968	14998200	13.12	20.65	26.69	19.740000	26.97000	20.430000	20.16	28.31	34.56
8	1969	15958100	15.49	23.24	26.97	26.060000	22.27000	2.230000	22.23	31.43	34.48
9	1970	16625500	15.22	20.46	28.55	25.180000	3.89000	13.040000	22.30	27.37	36.77
10	1971	18240496	15.73	21.48	28.42	20.530000	13.11000	13.440000	23.30	29.05	36.55
11	1972	19138896	13.29	21.42	25.98	15.860000	16.79000	2.620000	20.20	29.08	33.99
12	1973	19463600	16.66	20.33	28.42	7.250000	6.13000	4.030000	23.96	27.65	36.92

Step 3: Convert DS column to Date time Format

```
from datetime import datetime
dfProphet[“ds”] = pd.to_datetime(dfProphet[“ds”],format=’%Y’)
```

Step 4: Split Data to Model Time Series

Based on the number of observations, split the data into train and test data frames.

```
n_obs=3
fbProphet_train, fbProphet_test = dfProphet[0:-n_obs],dfProphet[-n_obs:]
print(fbProphet_train.shape,fbProphet_test.shape)
```

Output:

```
(57, 17) (3, 17)
```

Step 5: Prepare Prophet Time Series Model

By default, Prophet will return uncertainty intervals for the forecast $\bar{Y}$. There are several important assumptions behind these uncertainty intervals.[525]

Uncertainty in the trend: The biggest source of uncertainty in the forecast is the potential for future trend changes. The Prophet can detect and fit these, but what trend changes should we expect moving forward? It's impossible to know for sure, so we do the most reasonable thing we can, and we assume that the future will see similar trend changes as history. We assume that the average frequency and magnitude of trend changes in the future will be the same as that we have observed in the past. We project these trend changes forward and by computing their distribution we obtain the uncertainty intervals. The width of the uncertainty intervals (by default 80%) can be set using the parameter interval_width:

```
multi_model = Prophet(interval_width = 0.95)
```

Uncertainty in seasonality: By default, the Prophet will only return uncertainty in the trend and the noise observations. To get the uncertainty in seasonality, full Bayesian sampling must be done. This is done using the parameter mcmc.samples (which defaults to 0).

```
multi_model.fit(fbProphet_train)
```

Next, fit the model.

We make a DataFrame for future predictions. Here we keep capacity constant at the same value as in history, and forecast 5 years into the future:

```
future = multi_model.make_future_dataframe(periods = 3,freq='YS', include_history=True)
future.tail()
```

```
fbProphet_test['ds']
```

```
# filling all columns data in future frame except for y
for col in Prophet_cols:
  future[col] = dfProphet[col]
```

The above code created data frames and filled all data columns in future frames except for y.

Step: Forecast Model

```
forecastProphet = multi_model.predict(future)
forecastProphet
```

```
fig1 = multi_model.plot(forecastProphet)
```

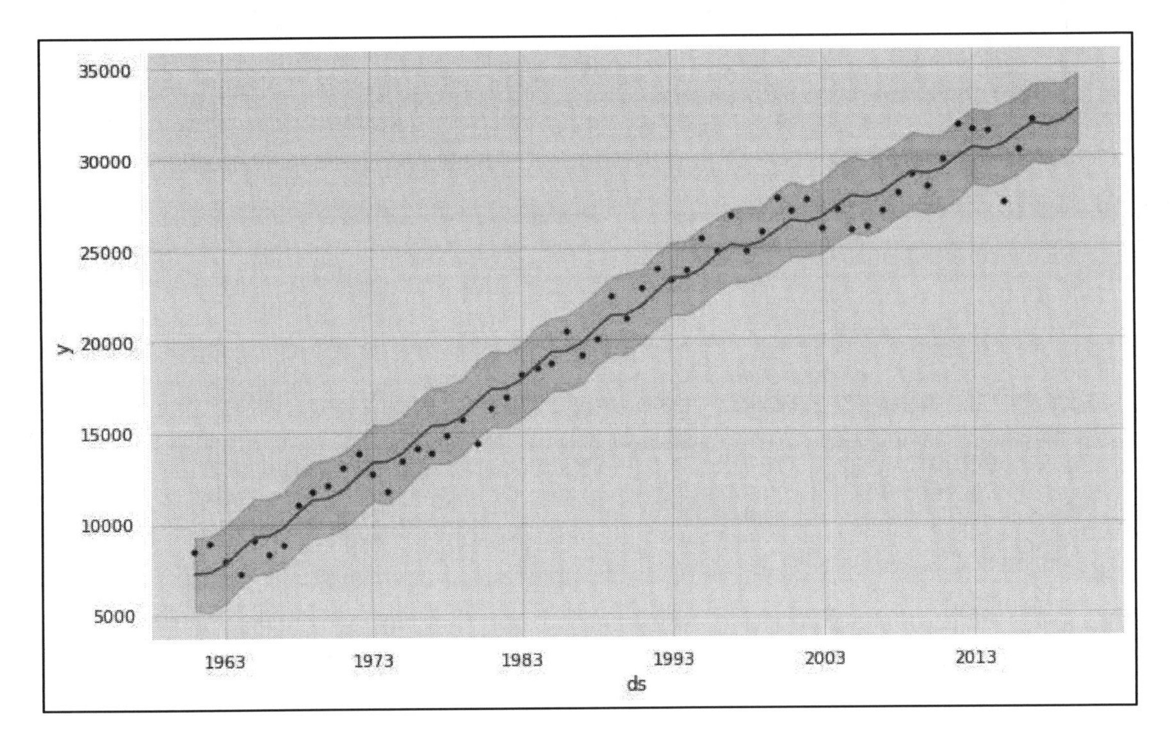

The above graph closely predicts the values from actuals with a high degree of accuracy.

```
multi_model.plot_components(forecastProphet)
```

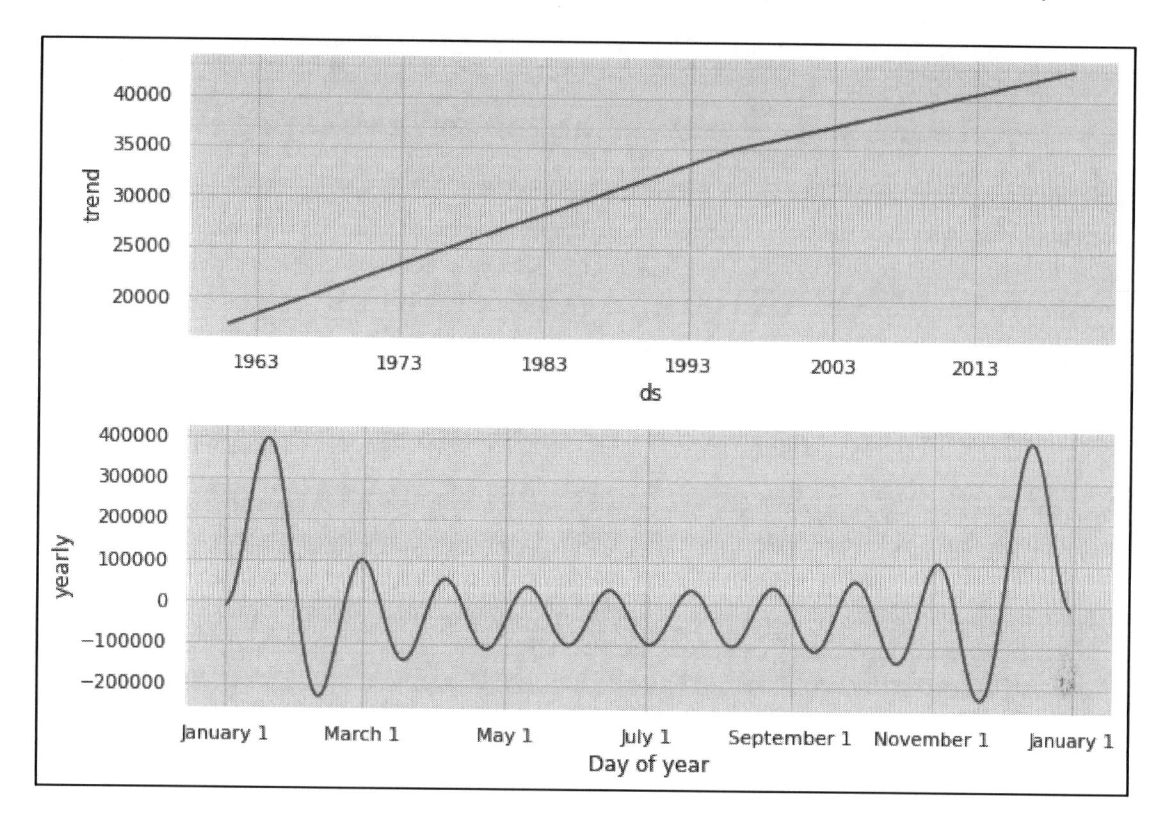

Step 6: Evaluate the model

```
#combining predicted and real data set
combineProphetUpper = pd.concat([forecastProphet['yhat'], dfProphet['y']], axis=1)
combineProphetUpper['accuracy'] = round(combineProphetUpper.apply(lambda row: row.yhat /row.y * 100, axis = 1),2)
combineProphetUpper['accuracy'] = pd.Series(["{0:.2f}%".format(val) for val in combineProphetUpper['accuracy']],index = combineProphetUpper.index)
combineProphetUpper = combineProphetUpper.round(decimals=2)
combineProphetUpper
```

Output:

	yhat	y	accuracy
0	7258.95	8507	85.33%
1	7322.74	8896	82.31%
2	7749.10	7929	97.73%
3	8538.16	7299	116.98%
4	9269.39	9132	101.50%
5	9333.18	8268	112.88%
6	9759.55	8874	109.98%
7	10548.60	11028	95.65%
8	11279.84	11688	96.51%
9	11343.63	12086	93.86%
10	11770.00	13066	90.08%

```
from sklearn.metrics import mean_absolute_error
from matplotlib import pyplot
plt.figure(figsize=(18, 9))
mae = mean_absolute_error(y_true, y_pred)
print('MAE: %.3f' % mae)
# plot expected vs actual
pyplot.plot(y_true, label='Actual')
pyplot.plot(y_pred, label='Predicted')
pyplot.legend()
pyplot.show()
```

Output:
MAE: 951.568

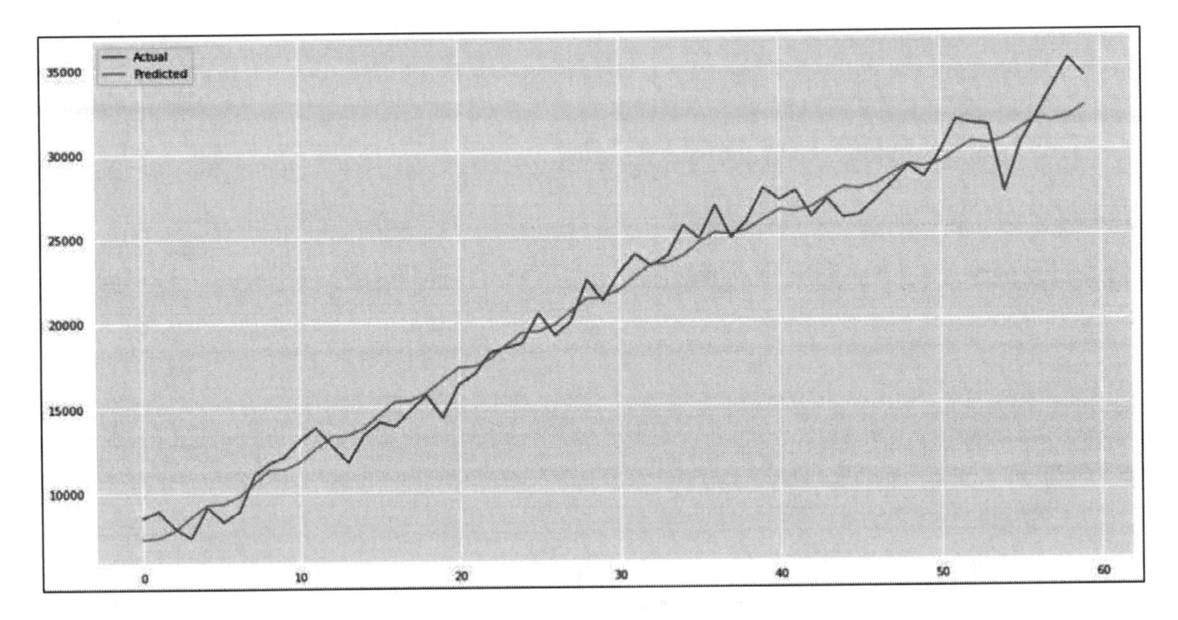

Machine Learning (Mid-century 2050) Projection Model: Heat Waves with Increased Frequencies and Intensities & Wheat Yield Production in India

We can use the above Forecasting model to assess the impact of increased frequencies of heatwaves on the wheat yield. The following table has years of future when heatwave could occur.

Year	Area_harvested_Value (hg/ha)	Yield_Value	Production_Value (tonnes)
2024	31470000	27496	86530000
2027	31470000	27496	86530000
2030	31470000	27496	86530000
2034	31470000	27496	86530000
2037	31470000	27496	86530000
2041	31470000	27496	86530000
2045	31470000	27496	86530000
2049	31470000	27496	86530000

The frequency of Heatwaves for 3 years has been increased as per the climate change models. In the future, though, according to the Intergovernmental Panel on Climate Change (IPCC), more frequent and intense heat waves in Asia (including India) will negatively impact vulnerable communities and increase mortality. In fact, it is likely that heat waves already occur more often now in Asia than they did in 1950. In a future with high carbon emissions, it is likely that the maximum temperature that occurs once every 20 years will at least double in frequency (to a 1-in-10-year event) by the end of the 21st century. Research focusing solely on India also concludes that heat waves will last longer, will be more intense and occur more often in the future Since major producing states were affected by heat waves, 2015 wheat yields have dropped considerably.

Step 1: Load Heatwave specific data

```
# 2050 Heatwave Projections
fileWheatProdIndiaPunjab='wheat_india_2050_future.csv'
fileaverageTemperaturesIndiaPunjab='tas_timeseries_monthly_cru_1901-2020_IND_1505_future_2050.csv'
fileminTemperatureIndiaPunjab='tasmin_timeseries_monthly_cru_1901-2020_IND_1505_future_2050.csv'
filemaxTemperatureIndiaPunjab='tasmax_timeseries_monthly_cru_1901-2020_IND_1505_future_2050.csv'
filePrecipitationIndiaPunjab='pr_timeseries_monthly_cru_1901-2020_IND_1505_future_2050.csv'
```

```
data = pd.read_csv(fileWheatProdIndiaPunjab)
data.tail()
```

Output:

	Domain	Item	Year	Area_harvested_Unit	Area_harvested_Value	Yield_Unit	Yield_Value	Production_Unit	Production_Value
0	Crops and livestock products	Wheat	1961	ha	12927000.0	hg/ha	8507.0	tonnes	10997000.0
1	Crops and livestock products	Wheat	1962	ha	13570000.0	hg/ha	8896.0	tonnes	12072000.0
2	Crops and livestock products	Wheat	1963	ha	13590000.0	hg/ha	7929.0	tonnes	10776000.0
3	Crops and livestock products	Wheat	1964	ha	13499000.0	hg/ha	7299.0	tonnes	9853000.0
4	Crops and livestock products	Wheat	1965	ha	13422000.0	hg/ha	9132.0	tonnes	12257000.0
...	...	...	...	...	...	...	...	...	...
85	Crops and livestock products	Wheat	2046	ha	NaN	hg/ha	NaN	tonnes	NaN
86	Crops and livestock products	Wheat	2047	ha	NaN	hg/ha	NaN	tonnes	NaN
87	Crops and livestock products	Wheat	2048	ha	NaN	hg/ha	NaN	tonnes	NaN
88	Crops and livestock products	Wheat	2049	ha	31470000.0	hg/ha	27496.0	tonnes	86530000.0
89	Crops and livestock products	Wheat	2050	ha	NaN	hg/ha	NaN	tonnes	NaN

Step 2: Check for Nulls

```
data.isna().sum()
```

```
data_temp.isna().sum()
```

Output:

```
Domain 0
Item 0
Year 0
Area_harvested_Unit 0
Area_harvested_Value 22
Yield_Unit 0
Yield_Value 22
Production_Unit 0
Production_Value 22
dtype: int64
```

```
Year 0
Jan 22
Feb 22
Mar 22
Apr 22
May 22
Jun 22
Jul 22
Aug 22
Sep 22
Oct 22
Nov 22
Dec 22
dtype: int64
```

As can be seen for the years 2023–2050, we will not have Harvest, Production, and Yield data. It is the future!

Step 3: Impute Data using interpolate – Spline

Since we need to impute time series data, apply Spline and linear interpolate strategies.

```
data_precp1 = data_precp.interpolate('spline', order=1)
data_precp1
```

```
data_temp1 = data_temp.interpolate('spline', order=1)
data_temp1
```

```
data_temp_min1 = data_temp_min.interpolate('spline', order=1)
data_temp_min1
```

```
data_temp_max1 = data_temp_max.interpolate('spline', order=1)
data_temp_max1
```

```
data = data.interpolate(method ='linear', limit_direction ='backward')
data = data.interpolate(method ='linear', limit_direction ='forward')
```

- Impute Precipitation data
- Average Temperature Data
- Minimum temperature data
- And maximum temperature data.

Output:

	Year	Jan	Feb	Mar	Apr	May	Jun	Jul	Aug	Sep	Oct	Nov	Dec
0	1901	18.860000	20.860000	28.450000	34.320000	39.470000	41.360000	36.920000	35.190000	35.000000	34.410000	27.860000	22.170000
1	1902	21.410000	23.520000	29.610000	35.330000	40.190000	40.130000	36.440000	35.430000	34.640000	32.870000	27.150000	21.040000
2	1903	19.800000	21.640000	26.030000	33.250000	39.050000	41.680000	37.110000	34.660000	35.560000	33.270000	26.570000	21.020000
3	1904	19.170000	22.760000	27.240000	35.680000	39.500000	40.980000	35.970000	34.720000	34.550000	32.870000	26.800000	21.750000
4	1905	18.150000	18.220000	24.320000	32.620000	40.850000	41.240000	36.700000	35.770000	35.170000	33.400000	28.170000	21.480000
...	...	...	...	...	...	...	...	...	...	...	...	...	...
145	2046	19.811668	24.245959	27.115879	35.016957	39.712014	39.154236	35.217912	34.129413	34.625533	33.193367	27.762933	22.186922
146	2047	19.811997	24.244838	27.119547	35.025457	39.713614	39.141481	35.208603	34.124940	34.625650	33.196262	27.769884	22.193675
147	2048	19.812326	24.243718	27.123214	35.033957	39.715213	39.128727	35.199294	34.120467	34.625767	33.199156	27.776834	22.200429
148	2049	19.570000	24.250000	27.100000	34.980000	39.750000	38.500000	34.630000	34.030000	34.920000	33.600000	27.850000	22.130000
149	2050	19.812984	24.241478	27.130549	35.050958	39.718412	39.103217	35.180675	34.111521	34.626000	33.204945	27.790736	22.213935

150 rows × 13 columns

The interpolate imputation has filled nulls with temperature, precipitation, and harvest data.

Step 4: Run the time series model and evaluate

```
dfProphet = fr[['Area_harvested_Value', 'Feb_temp', 'Mar_temp','Apr_temp',
'Feb_precp', 'Mar_precp', 'Apr_precp',
'Feb_Max_temp','Mar_Max_temp', 'Apr_Max_temp','May_Max_temp',
'Feb_min_temp', 'Mar_min_temp', 'Apr_min_temp', 'Production_Value','Yield_Value' ]].copy()
```

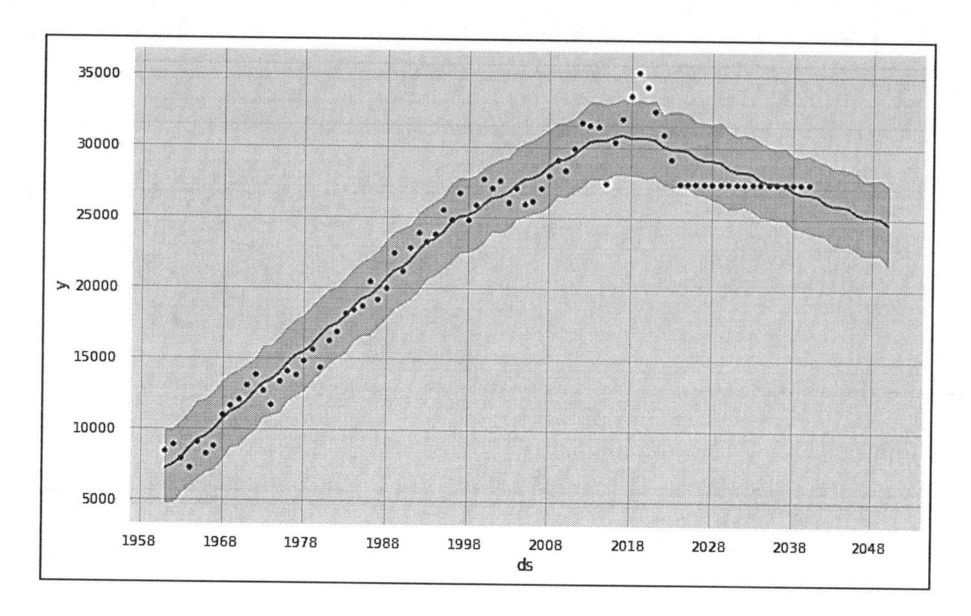

Forecast model:

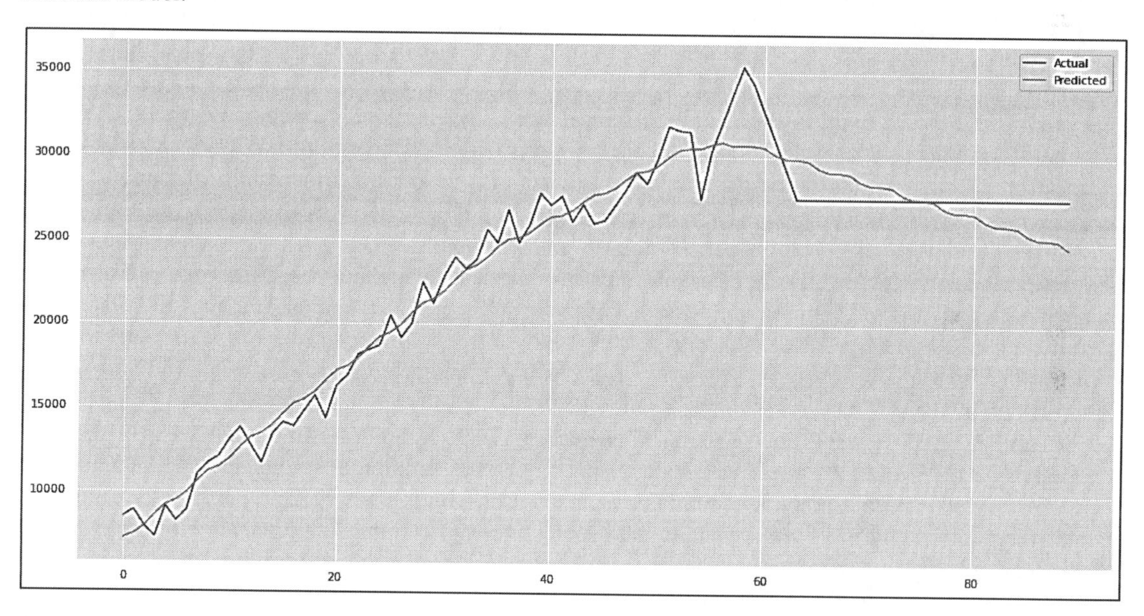

As can be seen, future heatwaves will result in reduced yield values.

	Reforms that improve the efficiency of India's agricultural markets could generate economy-wide gains in output and wages. These reforms could raise agricultural producer prices, reduce consumer food prices—and increase consumption, particularly in low-income households.

In summary, the impact of heatwaves is not only deadly with respect to human tolls but also on the food security as it inhibits marginalized communities to achieve full food nutrition as heatwaves, historically have resulted in a minimum increase of 10%–16% in wheat prices (NCDEX NWTc1[526]) especially during the March to June time period (please see in Figure 38). In the figure below, the wheat futures prices in the National Commodity & Derivatives Exchange Limited (NCDEX), an Indian online commodity and derivative exchange under the ownership of Ministry of Finance, Government of India, were captured during the heatwave years. The increase in futures is very consistent during the heatwave years as compared to the years with normal weather.

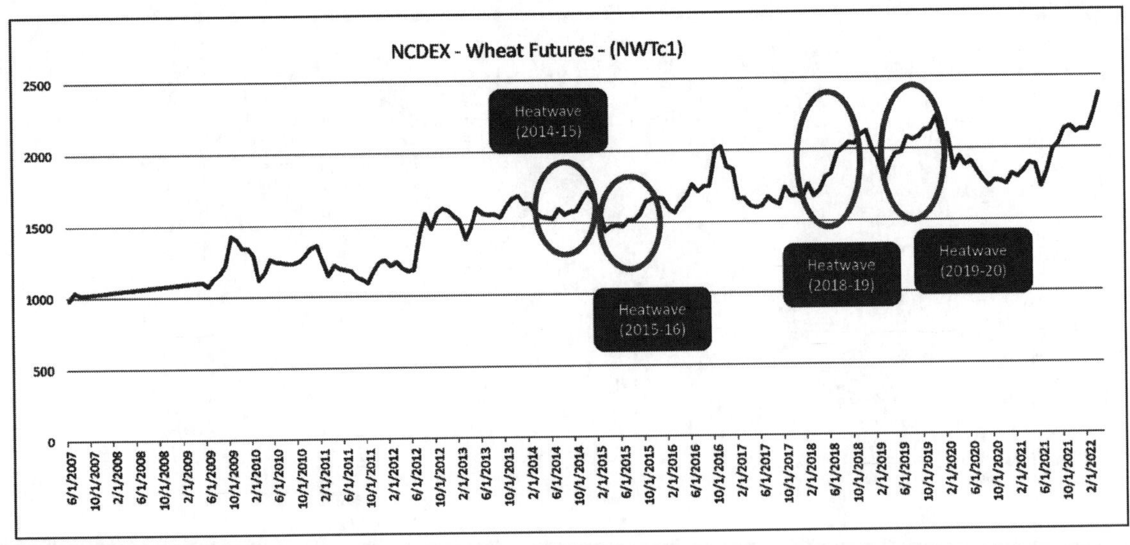

Figure 38: NCDEX Wheat Futures.

The price & shock transmission could be observed in other parts of the world. In 2018 Global Heat Wave Toasts Wheat and Prices Soared. Global wheat prices have soared to multiyear highs as a heat wave sweeping across Europe and Asia slashes forecasts for the 2018 harvest [27,28]. Chicago wheat futures hit three-year highs on Thursday, Aug. 4, 2018, while a key European benchmark topped a four-year high.[527] Wheat prices hit a record high, in 2022, as India's heat wave-driven export ban compounds the Ukraine war supply woes. India's recent decision to severely restrict wheat exports amid a devastating heatwave has driven up global prices for the basic commodity to record levels and drawn warnings of looming food shortages around the globe. The price of wheat futures rose by 5.9% on Monday to touch an all-time high of $12.68 per half bushel in Chicago. In the European market, the price rose to 436.25 euros per ton—up 4.68% during the day's trading. The prolonged heat wave has sent temperatures soaring over 120 degrees Fahrenheit, killed dozens of people in India and Pakistan since March 2022, and taken a huge toll on crops. As a result, the government in India, the world's second-largest wheat producer after China, banned private exports of wheat in May 2022, saying the move was necessary to manage national food security[528] amid the threat presented by the severe heat hitting the country [29]. Climate change is a costly business proposition, and it has an unprecedented impact on human food security and human survival. "Food security will be increasingly affected by projected future climate changes", the U.N.'s Intergovernmental Panel on Climate Change (IPCC) said in its latest report, adding: "Low-income consumers are particularly at risk, with models projecting increases of up to 183 million additional people at the risk of hunger… compared to a no climate change scenario" [30].

Food Security and Climate Resilient Economy: Heatwaves and Dairy Productivity Signal Mining to create a Smart Climate Sensor for Enhanced Food Security

In many parts of the United States, climate change is likely to result in higher average temperatures, hotter daily maximum temperatures, and more frequent heat waves, which could increase heat stress for livestock [31]. Heat stress can reduce meat and milk production and lower animal reproduction rates. Livestock producers can mitigate heat stress with shade structures, cooling systems, or altered feed mixes, but these methods increase production and capital costs. Dairy cows are particularly sensitive to heat stress, higher temperatures lower milk output and milk quality. Heat stress (HS) causes cows to produce less milk with the same nutritional inputs, which effectively increases farmers' production costs. Across the world, heatwaves would have a detrimental effect on cattle, dairy farms, and animal husbandry.

How cows produce milk all boils down to their anatomy as well as their ability to maintain homeostasis. Homeostasis is the ability of an organism to maintain an average internal temperature despite changes in external factors. For a cow to produce the best quality of milk and keep up an active well-being, the internal temperature of the said cow needs to be between 25 and 65 degrees Fahrenheit (–3.89°C to 18.33°C respectively). Due to the heat and global warming in general, this temperature is continuously becoming harder to manage leading to the cow not being able to produce good milk. The organism's well-being is also being affected because of the drastic change in weather which results in the cow feeling less hungry and more lethargic, completely losing the ability to produce milk.

The economic toll due to higher-temperature, heat stress is a $1 billion annual problem. Not only in the United States, but also around the globe heat stress causes an adverse impact on dairy productivity. The opportunity, however, for the dairy industry is to electronically monitor cattle temperature and implement appropriate measures so that the impact of HS can be minimized. The U.S. Department of Agriculture estimates nearly $2.4 billion a year in losses from animal illnesses that lead to death that can be prevented by electronically checking on the cattle's vital signs [32]. There is a higher risk of deaths due to heat stress. It affects cattle and lactating cows-results in lower birth weights and less milk production.

To predict how much climate change-related heat stress will affect dairy cows, ERS researchers used climate forecasts from four different climate change models. These models estimate patterns of temperature and precipitation based on assumptions about future carbon emissions levels. Using an emissions scenario that incorporates midrange assumptions about future population growth, technological change, economic growth, and international political cooperation, the predicted change in regional THI loads in 2030 can vary considerably among the four models. For all models, however, the largest increases are forecast in the South. For the sample of dairies surveyed in 2010, the average annual temperature is forecast to increase between 1.45- and 2.37-degrees Fahrenheit by 2030, depending on which climate model is used. The THI load is predicted to increase by an average of 1,670–3,940 humidity-adjusted degree hours [31].

When temperatures began to soar in Uttar Pradesh state several weeks ago, not only did wheat crops begin to shrivel, but dairy cows also provided less milk [33]. When the mercury hit 117 degrees Fahrenheit (47.22°C) in May 2022, a record high in small towns in Uttar Pradesh, it became punishing for humans, too: Seven of 25 farmhands came down with diarrhea, a symptom of heat stroke. Others refused to stay outside past 10 a.m. [33].

Heat waves can also be the main cause for small dairy companies losing their business and closing down due to less money powering their organization. Farmers back in 2014 came up with the idea of offsetting the rise in temperature by installing cooling systems in farms. It did seem like a good solution, yet it only resulted in much higher production costs to keep farms running. With farmers not having enough money to keep their businesses afloat many of the small farms closed due to the lack of money needed to maintain them. Ultimately, the world's need for dairy farming is only increasing and more farms closing shops can create more demand yet not enough supply.

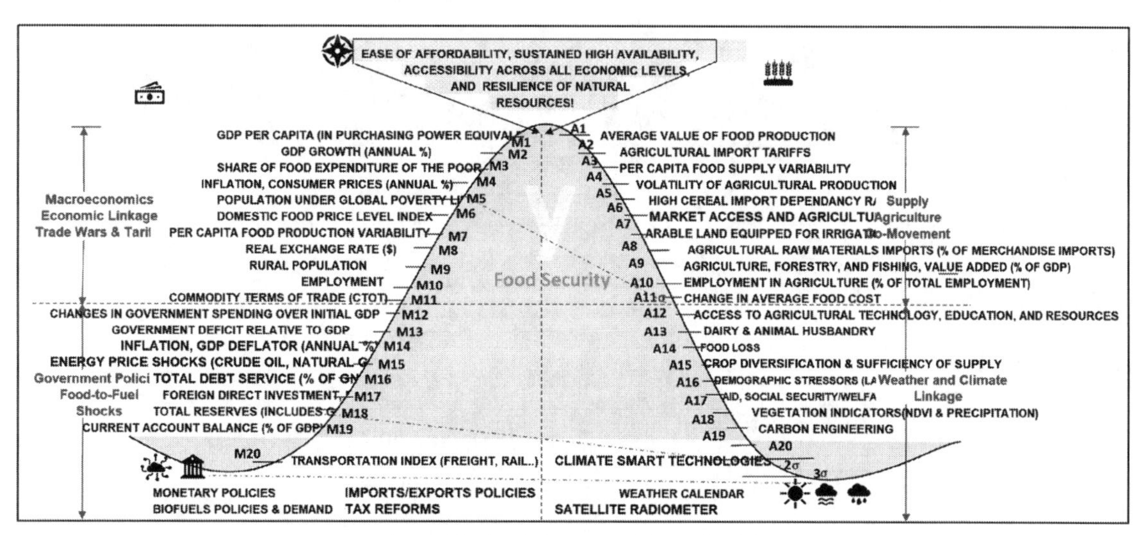

Figure 39: Food Security Bell Curve Model.

Dairy and Animal Husbandry offer off-farm employment & cash to farmers (A13 in Figure 39). The income from milk keeps small farmers afloat and keeps their agricultural farms afloat. Heatwaves are double jeopardy to small farmers, pushing them into economic abysses—a classical case for the prevalence of undernourishment and higher food insecurity. And for consumers, higher milk prices only mean lack of enough dairy food as part of day to day food intake—a solid case for undernutrition and reduced overall caloric intake.

	Software code for this model: Final_Indian_wheat_model_Punjab.ipynb (Jupyter Notebook Code) Final_India_UttarPradesh (UP)_WheatYield_and_HeatWave_NDVIdata-MLModel.ipynb (Jupyter Notebook Code)

Machine Learning Model: Drought & Heatwave Signature Mining through the Application of Sensor, Satellite Data to reduce overall Food Insecurity

Data Sources

NOAA STAR[529]—Global Vegetation Health Products: Province-Averaged VH	Mean data for USA Province= 17: Kansas, from 1982 to 2022, weekly for cropland[530] area only year, week, SMN, SMT, VCI, TCI, VHI	NOAA
The Climate Change Knowledge Portal (CCKP)[531]	India temperatures[532] • Mean Temperature • Minimum Temperature • Maximum Temperature • Precipitation	THE WORLD BANK
Dairy Sensor Data	Cow Necklace[533] Datasets from California and Punjab	

The Data for the Sensors are in Dairy Farms in California—so both land and crop land NOAA data are required for developing signature mapping for extreme temperatures. Data from NOAA STAR- California.[534]

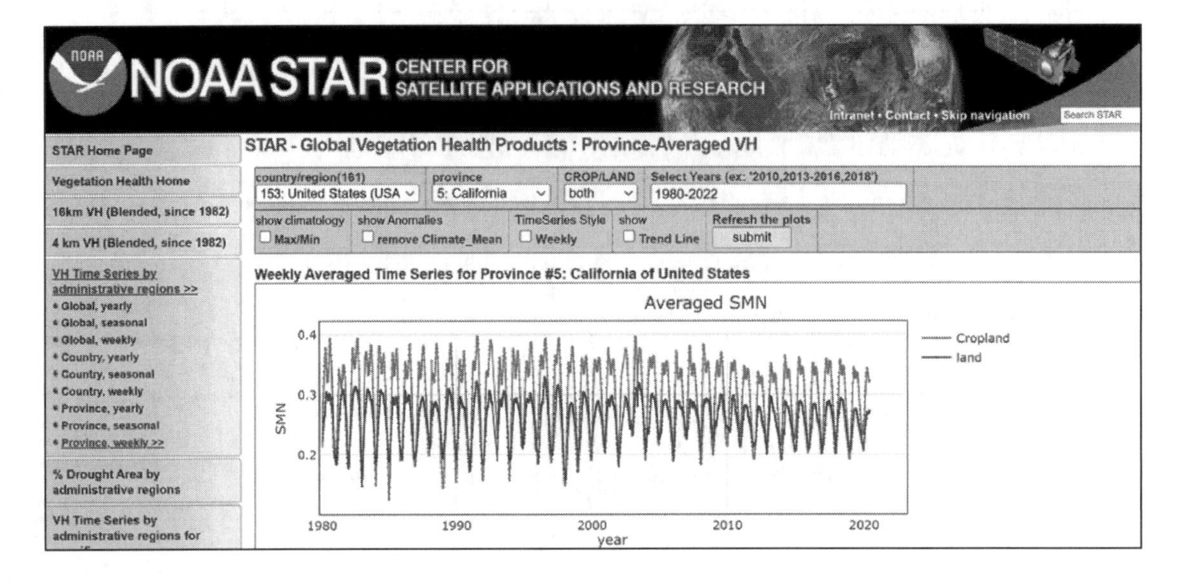

Temperature Condition index (TCI): TCI is based on a 10.3–11.3 μm range of AVHRR radiance measurements converted to a brightness temperature (BT), which is a proxy for thermal conditions.[535]

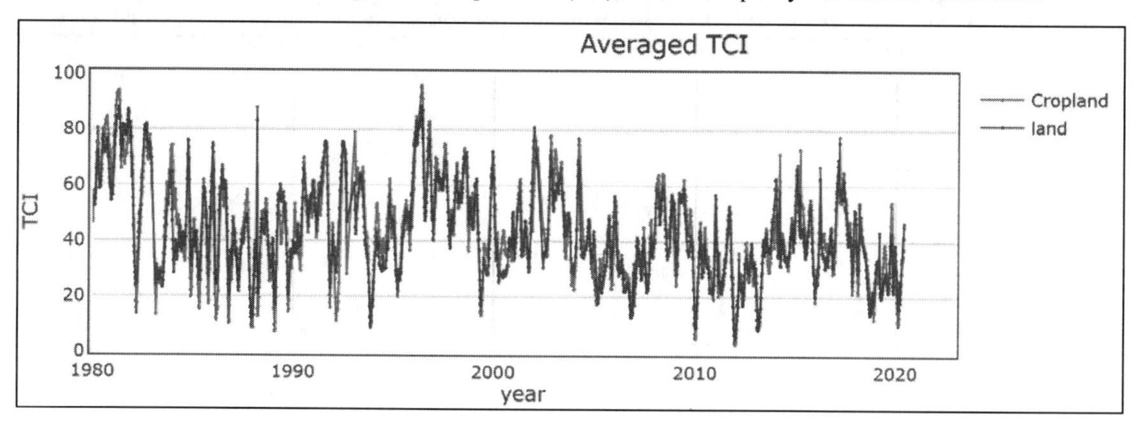

No noise (smoothed) Brightness Temperature (SMT)—The SMT is the BT with a complete removal of high frequency noise, that can be used for the estimation of *thermal conditions, cumulative degree days and other significant parameters.*

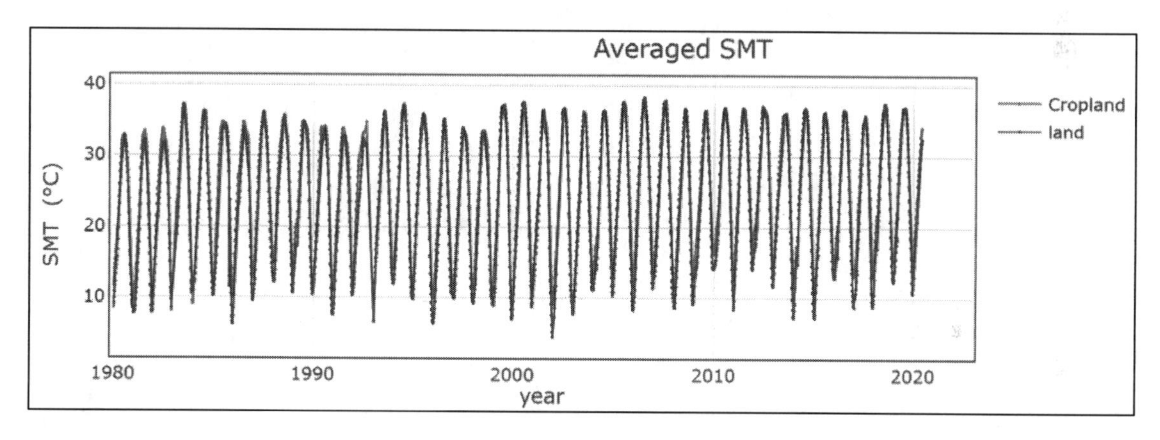

Step 1: Load Libraries

Load libraries to process Sensor and California Satellite Data

```
import pandas as pd
import numpy as np
import seaborn as sns
import scipy
import matplotlib.pyplot as plt
import plotly.graph_objects as go
```

Step 2: Load Sensor Data

```
Sensor_12df = pd.read_csv("COWNeckLaceSensorCA12UUID_2dwf3_7783.csv")
sensor_12df
```

```
df["year"] = pd.date_range(start='2020-08-01', end='2020-12-31')
```

```
weekly_tempdf = combined_df.resample("W" ).mean()
```

Output:

	TimeStamp_T	Cattle_temperature	Cattle_humidity	Internal_sensor_temperature	Internal_sensor_humidity	BatteryCharge	AbnormalTemperature
0	2020-09-06 00:33:08	27.1	60.3	27.0	59.6	100	False
1	2020-09-06 01:33:10	27.2	60.8	27.2	61.3	100	False
2	2020-09-06 02:33:12	26.7	60.6	26.4	58.0	100	False
3	2020-09-06 03:33:14	26.7	60.5	26.3	58.5	100	False
4	2020-09-06 04:33:16	26.6	60.4	26.2	58.4	100	False
...	...	...	...	...	...	...	...
707	2020-08-31 19:28:58	24.7	59.8	24.3	59.9	100	False
708	2020-08-31 20:29:00	24.7	59.8	24.4	59.6	100	False
709	2020-08-31 21:29:02	24.9	60.8	24.5	61.7	100	False
710	2020-08-31 22:29:04	25.1	62.2	24.7	62.6	100	False

Sensor offers the following key attributes (please see Figure 40):

- Cattle Temperature
- Case Temperature
- Internal Temperature
- Internal Sensor Humidity, and

Signal mappings of interest with climate change drivers include Internal Sensor Temperature and Internal Sensor Humidity.

Step 3: Load NOAA California Data

```
Satelite_df = pd.read_excel("CaliforniaSateliteData.xlsx")
Satelite_df
```

Output:

	year	week	SMN	SMT	VCI	TCI	VHI
0	1982	1	0.215	281.70	53.42	55.53	54.48
1	1982	2	0.219	281.85	54.43	55.13	54.78
2	1982	3	0.224	282.25	55.33	55.38	55.35
3	1982	4	0.229	282.95	55.92	54.84	55.38
4	1982	5	0.236	283.82	56.73	52.53	54.63
...	...	...	...	...	...	...	...
2127	2022	48	-1.000	-1.00	-1.00	-1.00	-1.00
2128	2022	49	-1.000	-1.00	-1.00	-1.00	-1.00
2129	2022	50	-1.000	-1.00	-1.00	-1.00	-1.00

If values are captured, they will be reported as –1.

Step 4: Exploratory analysis on Satellite data—add Week and Year Labels

Hanumayamma' s Cow Necklace (Figure 40) sensor provides data frequencies on an hourly basis and Satellite data on a weekly basis. Before resampling all into one frequency, we need to synthetically add week and year labels.

```
atelite_df["Week"]= Satelite_df["week"].apply(lambda x: str(x) +'Wk' )
```

```
Satelite_df["Year_wk"] = Satelite_df["year"].astype(str) + "_" + Satelite_df["Week"]
Satelite_df
```

Output:

	year	week	SMN	SMT	VCI	TCI	VHI	Column1	Week	Year_wk
0	1982	1	0.215	281.70	53.42	55.53	54.48	NaN	1Wk	1982_1Wk
1	1982	2	0.219	281.85	54.43	55.13	54.78	NaN	2Wk	1982_2Wk
2	1982	3	0.224	282.25	55.33	55.38	55.35	NaN	3Wk	1982_3Wk
3	1982	4	0.229	282.95	55.92	54.84	55.38	NaN	4Wk	1982_4Wk
4	1982	5	0.236	283.82	56.73	52.53	54.63	NaN	5Wk	1982_5Wk

Step 5: Merge Sensor and Satellite Data

```
sensor_satelite_df = pd.merge(Satelite_df,weekly_tempdf2, on = "Year_wk" )
sensor_satelite_df
```

```
sensor_satelite_df1 = sensor_satelite_df.set_index("Year_wk")
sensor_satelite_df1
```

Output:

	year	week	SMN	SMT	VCI	TCI	VHI	Column1	Week	Year_wk	Internal_sensor_temperature
0	2020	31	0.269	310.60	37.71	27.24	32.52	NaN	31Wk	2020_31Wk	24.545833
1	2020	32	0.264	310.33	36.13	26.95	31.59	NaN	32Wk	2020_32Wk	24.918750
2	2020	33	0.261	310.18	35.70	24.06	29.91	NaN	33Wk	2020_33Wk	23.368517
3	2020	34	0.256	309.87	33.84	21.25	27.57	NaN	34Wk	2020_34Wk	23.368517
4	2020	35	0.250	309.23	31.94	19.84	25.90	NaN	35Wk	2020_35Wk	23.368517
5	2020	36	0.247	308.56	31.59	18.08	24.85	NaN	36Wk	2020_36Wk	24.593421
6	2020	37	0.243	307.80	31.52	16.38	23.96	NaN	37Wk	2020_37Wk	24.830208
7	2020	38	0.240	306.81	31.97	14.79	23.39	NaN	38Wk	2020_38Wk	23.368517
8	2020	39	0.238	305.58	33.08	13.54	23.32	NaN	39Wk	2020_39Wk	23.368517

And set the index to Week.

Step 6: Visualize the Data

```
import plotly.express as px
fig = go.Figure()
fig.add_trace(go.Scatter(x = sensor_satelite_df.Year_wk, y=Satelite_df['SMT'],mode = 'lines',name='SMT'))
```

Plot SMT. Output:

As can be seen, the SMT values increased with a trend in California – especially at the locations near the bay area.

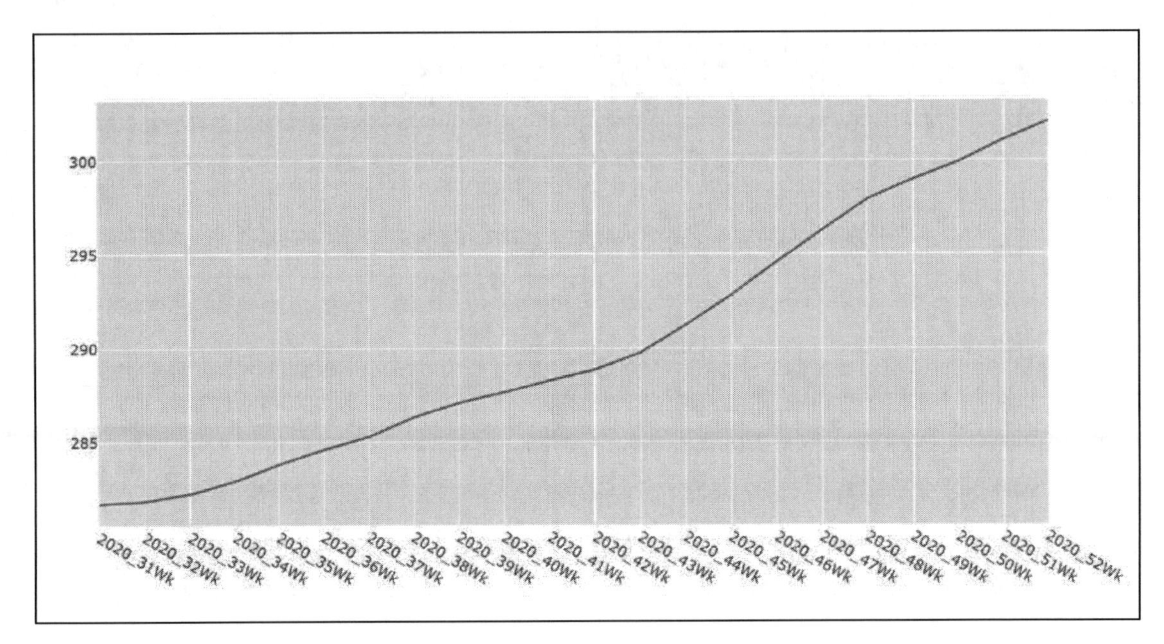

It tells us about higher thermal conditions and higher cumulative degree days.

Sensor Internal Temperature

```
fig = go.Figure()
fig.add_trace(go.Scatter(x = weekly_tempdf2.index, y=weekly_tempdf2["Internal_sensor_temperature"],mode = 'lines',name='Internal_sensor_temperature'))
```

Output:

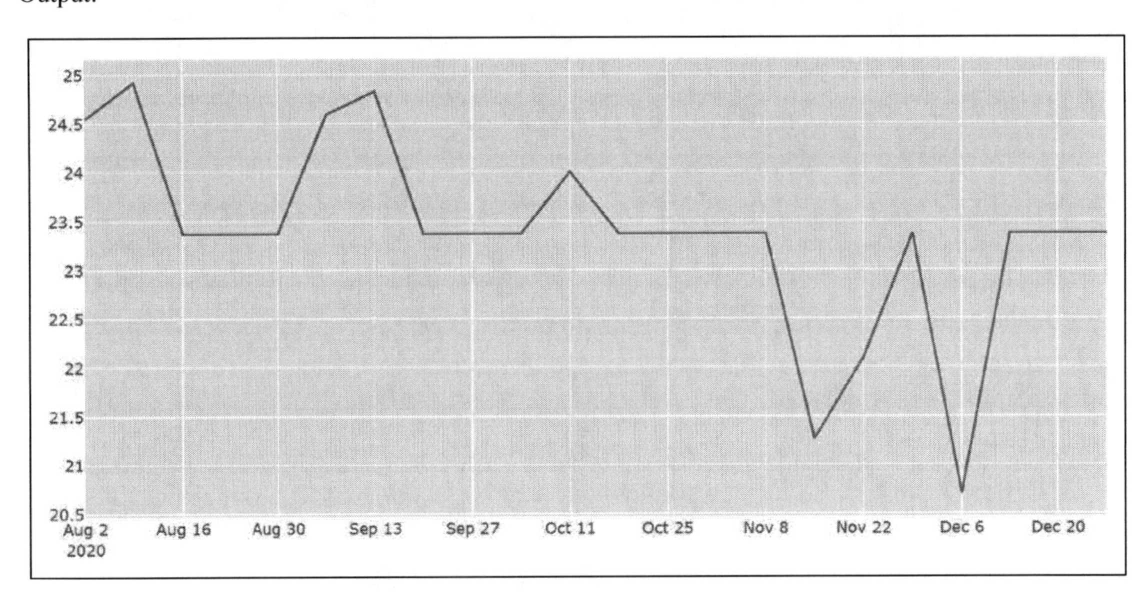

During the August 2020 time frame, the internal values registered by the sensor in Fremont varied from 21.4ºC to 24.5ºC. Sensor internal temperature is not accurately covariant with the ambient temperature due to the internal packing and other materials properties.

Step 7: Perform standardization to mine the Signal

```
from sklearn import preprocessing
scaler = preprocessing.StandardScaler().fit(sensor_satelite_df2)
scaler
```

```
scaled_df = scaler.transform(sensor_satelite_df2)
```

```
sensor_satelite_df3 = pd.DataFrame(scaled_df , columns =sensor_satelite_df2.columns, index = sensor_satelite_df2.index )
sensor_satelite_df3
```

Output:

	Internal_sensor_temperature	SMN	SMT	VCI	TCI	VHI
Year_wk						
2020_31Wk	1.205640	1.890705	1.219641	-0.331338	1.838389	0.609569
2020_32Wk	1.587528	1.634575	1.189568	-0.575297	1.766740	0.398290
2020_33Wk	0.000000	1.480897	1.172860	-0.641691	1.052722	0.016626
2020_34Wk	0.000000	1.224767	1.138332	-0.928883	0.358469	-0.514979
2020_35Wk	0.000000	0.917411	1.067047	-1.222251	0.010107	-0.894372
2020_36Wk	1.254372	0.763733	0.992420	-1.276293	-0.424727	-1.132912
2020_37Wk	1.496856	0.558829	0.907769	-1.287101	-0.844738	-1.335104
2020_38Wk	0.000000	0.405151	0.797501	-1.217619	-1.237571	-1.464597
2020_39Wk	0.000000	0.302699	0.660500	-1.046230	-1.546403	-1.480500
2020_40Wk	0.000000	0.149021	0.485629	-0.826976	-1.440165	-1.271493
2020_41Wk	0.645233	0.046569	0.300733	-0.583017	-1.306750	-1.030681

Step 8: Signal Mapping of Sensor & NOAA Satellite data

Let's overlap the sensor and Satellite data to see any signal intuition.

```
fig = go.Figure()
fig.add_trace(go.Scatter(x = sensor_satelite_df3.index, y=sensor_satelite_df3['SMT'],mode = 'lines',name='SMT', line=dict(color='firebrick', width=4, dash='dash')))
fig.add_trace(go.Scatter(x = sensor_satelite_df3.index, y=sensor_satelite_df3['Internal_sensor_temperature'],mode='lines',name='Internal_sensor_temperature',line=dict(color='royalblue', width=4)))
# Edit the layout
fig.update_layout(title='Average Weekly Temperature sensor and satelite data',
        xaxis_title='weeks',
        yaxis_title='scaled Temperature ')
fig.update_xaxes(
    tickangle = 45,
    title_text = "Weeks",
    title_font = {"size": 20},
    title_standoff = 25)
```

Output:

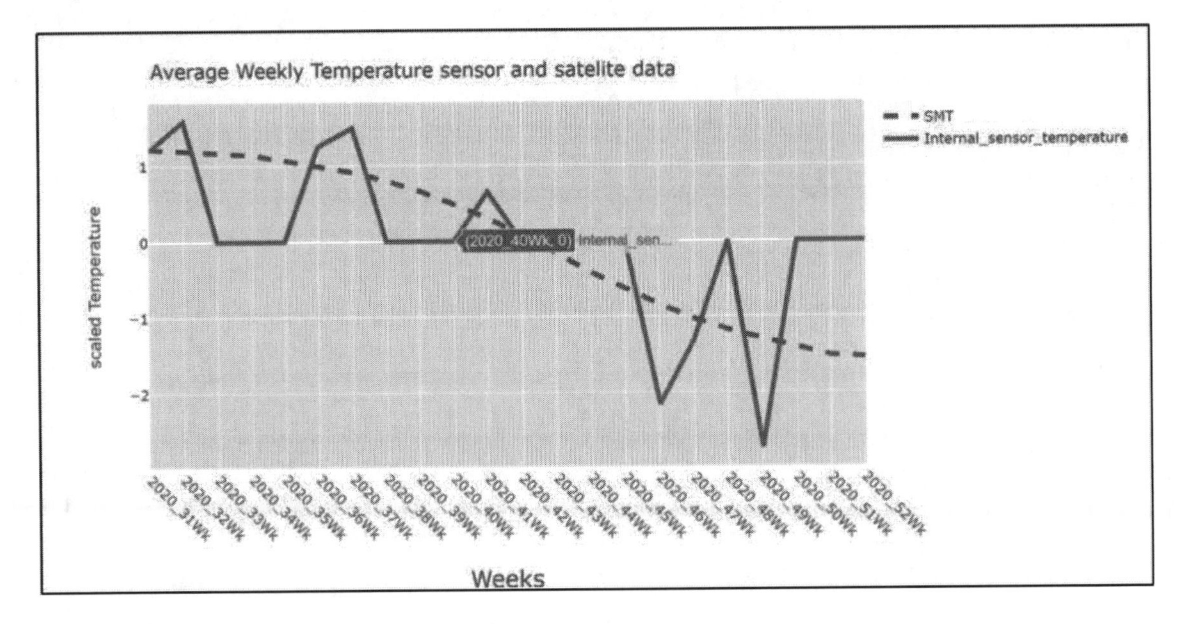

The SMTs of Satellite data vary from the peak to lower values of the normalized sensor. What it tells is that the sensor is exposed to overall sunlight (as correlated with SMT) and temperature values proportionately cut the sensor internal temperature.

The advantage of capturing the data is when a new sensor is deployed in a new location, given cold start issues, to fine tune the sensor to local thermal conditions, the data mined sensor to SMT signatures could be clustered to deploy proximity sensors with local temperatures and ambient properties. The real value it delivers to farmers is in terms of productivity improvement & overall increase in milk output for better economic opportunities.

Given the rate at which climate changes are taking place, it is essential that we apply data driven techniques to understand and improve our long-term sustainable development & and gain an opportunity cost that is forcefully induced due to climate change—build a climate resilient economy.[536] A real adaptation strategy to be future ready!

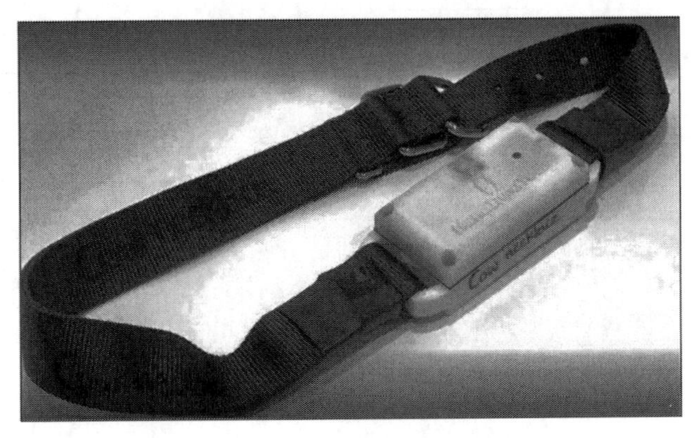

Figure 40: Climate Smart Veterinarian Cow Sensor.

After reading this chapter, you should be able to answer queries on the impact of climate change on agricultural production. You should be able to apply time series analysis on Phenological Stages of wheat and develop models to understand the impact of climate change. You should be able to answer questions on sNOAA Star Global Vegetation Health, Vegetation Condition index (VCI), Temperature Condition index (TCI), Vegetation Health index (VHI), No noise (smoothed) Normalized Difference Vegetation Index (SMN). Next, Coupled Model Intercomparison Project climate projections (CMIP) and Shared Socioeconomic Pathway (SSP) Projection Models have been introduced. Finally, you should be able to apply Machine Learning models for Animal Husbandry, Kansas wheat, and major Indian wheat producing states with the application of climate change impacts such as droughts and heatwaves.

References

[1] Growth and physiology, FAO 2002, https://www.fao.org/3/Y4011E/y4011e00.htm#Contents, Access Date: May 03, 2022.

[2] Howard, M. Rawson and Helena Gómez Macpherson. IRRIGATED WHEAT, 2000, https://www.fao.org/3/x8234e/x8234e00.htm#Contents, Access Date: June 12, 2022.

[3] Patrick Galey. Heat wave in India threatens residents and crucial wheat harvest, PUBLISHED TUE, APR 26 20229:26 PM EDTUPDATED TUE, APR 26 202211:01 PM EDT, https://www.cnbc.com/2022/04/27/heat-wave-in-india-threatens-residents-and-crucial-wheat-harvest.html, Access Date: May 13, 2022.

[4] FAO, note on the impact of the war on food security in Ukraine, 25 March 2022, https://reliefweb.int/sites/reliefweb.int/files/resources/cb9171en.pdf, Access Date: May 15, 2022.

[5] FAO, Information Note: The importance of Ukraine and the Russian Federation for global agricultural markets and the risks associated with the current conflict, 25 March 2022 Update, https://www.fao.org/3/cb9236en/cb9236en.pdf, Access Date: May 13, 2022.

[6] Denise Chow, Fast changes, food woes and who's vulnerable: 7 big takeaways from the U.N. climate report, Feb. 28, 2022, 2:14 PM PST, https://www.nbcnews.com/science/environment/fast-changes-food-woes-s-vulnerable-7-big-takeaways-un-climate-report-rcna17932, Access Date: May 13, 2022.

[7] Jenny Dewey Rohrich, WHEAT GROWTH STAGES, JUNE 24, 2015, https://prairiecalifornian.com/wheat-growth-stages/, Access Date: May 03, 2022.

[8] Patrick, Thomas and Kirk Maltais. 2022. Farmers Are Racing Against Poor Weather to Plant Crops, May. 11, 2022 1:29 pm ET, https://www.wsj.com/articles/poor-planting-weather-puts-squeeze-on-u-s-farmers-11652290167, Access Date: May 12, 2022.

[9] Jacob, BungeFollow and Kirk Maltais. Drenched Land Confront Tough Choice on Planting, Updated Jun. 5, 2019 12:24 pm ET, https://www.wsj.com/articles/farmers-on-drenched-land-confront-tough-choice-on-planting-11559727000, Access Date: May 12, 2022.

[10] National Center for Atmospheric Research Staff (Eds). Last modified 25 Apr 2022. "The Climate Data Guide: CMIP (Climate Model Intercomparison Project) Overview." Retrieved from https://climatedataguide.ucar.edu/climate-model-evaluation/cmip-climate-model-intercomparison-project-overview.

[11] Chandrasekar Vuppalapati. 2022. Artificial Intelligence and Heuristics for Enhanced Food Security, Publisher : Springer; 1st ed. 2022 edition (August 31, 2022), ISBN-13 : 978-3031087424.

[12] Sandstad, M., Schwingshackl, C., Iles, C.E. and Sillmann, J. 2022. Climate extreme indices and heat stress indicators derived from CMIP6 global climate projections, [specify product used], v1, Copernicus Climate Change Service (C3S) Climate Data Store (CDS). (Accessed on [02-JUN-2022]), https://doi.org/10.24381/cds.776e08bd.

[13] David Condos. 2022. Western Kansas wheat crops are failing just when the world needs them most, June 9, 2022 at 3:00 AM CDT, https://www.kcur.org/news/2022-06-09/western-kansas-wheat-crops-are-failing-just-when-the-world-needs-them-most, Access Date: June 12, 2022.

[14] Brian Grimmett. Climate change means Kansas farmers are dealing with hotter nights and change in rainfall and freezing patterns, July 2021, https://www.cjonline.com/story/news/2021/07/04/such-climate-changes-temperature-rainfall-stresses-kansas-farmers-wheat/7837492002/, Access Date: June 12, 2022.

[15] Zack Adkins. Satellite Imagery Resources and Usage for the Farm Service Agency, April 2014, https://www.fsa.usda.gov/Internet/FSA_File/satellite_imageryresources.pdf, Access Date: May 26, 2022.

[16] William Robbins, Lingering Drought Stunts Wheat Crop, May 12, 1989, https://www.nytimes.com/1989/05/12/us/lingering-drought-stunts-wheat-crop.html, Access Date: May 26, 2022.

[17] CNN. Wheat Belt drought taking heavy toll, April 26, 1996, http://www.cnn.com/US/9604/26/kansas.wheat.woes/index.html, Access Date: May 26, 2022.

[18] Kansas State University. Drought, poor wheat harvest in Kansas has effects on national economy, says climatologist. ScienceDaily. ScienceDaily, 10 July 2014, Access Date: May 26, 2022.

[19] FAO, Trade Reforms and Food Security, 2003, https://www.fao.org/3/y4671e/y4671e.pdf, Access Date: June 10, 2022.

[20] Aniruddha Ghosal Ap Science Writer, Heat wave scorches India's wheat crop, snags export plans, April 29, 2022, 10:11 AM, https://abcnews.go.com/International/wireStory/heat-wave-scorches-indias-wheat-crop-snags-export-84389631, Access Date: June 11, 2022.

[21] Damian Carrington. Climate crisis makes extreme Indian heatwaves 100 times more likely – study, Wed 18 May 2022 08.29 EDT, https://www.theguardian.com/environment/2022/may/18/climate-crisis-makes-extreme-indian-heatwaves-100-times-more-likely-study, Access Date: June 11, 2022.

[22] Leah Emanuel. Climate Change in Russia and the Weaponization of Wheat, July 2021, https://climateandsecurity.org/2020/08/goodman-and-bergenas-why-the-us-needs-to-restructure-its-national-security-strategy/, Access Date: June 12, 2022.

[23] Kayode Ajewole. India, Thursday, Last updated: Thursday, October 08, 2020, https://www.ers.usda.gov/topics/international-markets-u-s-trade/countries-regions/india/, Access Date: June 09, 2022.

[24] David. Hodari and Benjamin Parkin. Global Heat Wave Toasts Wheat and Prices Soar, Aug. 4, 2018 10:00 am ET, https://www.wsj.com/articles/global-heat-wave-toasts-wheat-and-prices-soar-1533391201, Access Date: June 15, 2022.

[25] Arshad, R. Zargar. MAY 17, 2022/10:47 AM, Wheat prices hit record high as India's heat wave-driven export ban compounds Ukraine war supply woes, https://www.cbsnews.com/news/india-heat-wave-wheat-prices-soar-climate-change-ukraine-war-supplies/, Access Date: June 15, 2022.

[26] Sybilla Gross. 2022. Bumper Wheat Crop Looms Again in Australia on Ample Rains, April 5, 2022, 2:00 PM PDT Updated on April 6, 2022, 12:40 AM PDT, https://www.bloomberg.com/news/articles/2022-04-05/bumper-wheat-crop-looms-again-in-australia-on-high-prices-rains, Access Date: April 06, 2022.

[27] Andrew, M. McKenzie, Bingrong Jiang, Harjanto Djunaidi, Linwood A. Hoffman and Eric J. Wailes. 2022. Unbiasedness and Market Efficiency Tests of the U.S. Rice Futures Market, Vol. 24, No. 2 (Autumn - Winter, 2002), pp. 474–493, https://www.jstor.org/stable/1349773, Access Date: October 07, 2022.

[28] IPCC. 2019. Summary for Policymakers. In: Climate Change and Land: an IPCC special report on climate change, desertification, land degradation, sustainable land management, food security, and greenhouse gas fluxes in terrestrial ecosystems [Shukla, P.R., Skea J., Calvo Buendia, E., Masson-Delmotte, V., Pörtner, H.-O., Roberts, D.C., Zhai, P., Slade, R., Connors, S., van Diemen, R., Ferrat, M., Haughey, E., Luz, S., Neogi, S., Pathak, M., Petzold, J., Portugal Pereira, J., Vyas, P., Huntley, E., Kissick, K., Belkacemi, M., Malley, J. (eds.)]. In press.

[29] Ilapakurti, A. and Vuppalapati, C.. 2015. Building an IoT Framework for Connected Dairy. 2015 IEEE First International Conference on Big Data Computing Service and Applications, 2015, pp. 275–285, doi: 10.1109/BigDataService.2015.39.

[30] Chandrasekar Vuppalapati. 2019. Building Enterprise IoT Applications, Publisher : CRC Press; 1st edition (December 17, 2019), ISBN-10 : 0367173859.

[31] Chandrasekar Vuppalapati. Democratization of Artificial Intelligence for the Future of Humanity, Publisher : CRC Press; 1st edition (January 18, 2021), ISBN-13 : 978-0367524128.

[32] Nigel Key and Stacy Sneeringer, Greater Heat Stress From Climate Change Could Lower Dairy Productivity, November 03, 2014, https://www.ers.usda.gov/amber-waves/2014/november/greater-heat-stress-from-climate-change-could-lower-dairy-productivity/, Access Date: June 15, 2022.

[33] Gerry Shih and Kasha Patel. 2022. India tries to adapt to extreme heat but is paying a heavy price, May 9, 2022 at 2:00 a.m. EDT, https://www.washingtonpost.com/world/2022/05/09/india-heat-wave-climate-change/, Access Date: June 15, 2022.

---

[415] Issues and Challenges of Inclusive Development: Essays in Honor of Prof. R. Radhakrishna by R. Maria Saleth, S. Galab and E. Revathi, Publisher : Springer; 1st ed. 2020 edition (June 19, 2020), ISBN-13 : 978-9811522284

[416] Food Security - https://www.fao.org/fileadmin/templates/faoitaly/documents/pdf/pdf_Food_Security_Cocept_Note.pdf

[417] Food Security and Nutrition Assistance - https://www.ers.usda.gov/data-products/ag-and-food-statistics-charting-the-essentials/food-security-and-nutrition-assistance/

[418] The prevalence of food insecurity in 2020 is unchanged from 2019 - https://www.ers.usda.gov/data-products/chart-gallery/gallery/chart-detail/?chartId=58378

[419] Prevalence of food insecurity is not uniform across the country - https://www.ers.usda.gov/data-products/chart-gallery/gallery/chart-detail/?chartId=58392

[420] Extreme heat impacting millions across India and Pakistan - https://news.un.org/en/story/2022/04/1117272

[421] The Economic Effects of Energy Price Shocks - https://www.aeaweb.org/conference/2009/retrieve.php?pdfid=145

[422] Wheat growth and physiology - https://www.fao.org/3/Y4011E/y4011e00.htm#Contents

[423] The difference between C3 and C4 plants - https://ripe.illinois.edu/blog/difference-between-c3-and-c4-plants

[424] Wheat – India - https://farmer.gov.in/M_cropstaticswheat.aspx

[425] Calvin Cycle - https://www.nationalgeographic.org/media/calvincycle/

[426] About wheat - https://farmer.gov.in/M_cropstaticswheat.aspx

[427] STAR - Global Vegetation Health Products : Province-Averaged VH (Wheat)- https://www.star.nesdis.noaa.gov/smcd/emb/vci/VH/vh_adminMeanByCrop.php?type=Province_Weekly_MeanPlot

[428] WHEAT GROWTH STAGES - https://prairiecalifornian.com/wheat-growth-stages/

[429] Crop Calendars for United States- https://ipad.fas.usda.gov/rssiws/al/crop_calendar/us.aspx

[430] Section 6: Explanations of plant development - https://www.fao.org/3/x8234e/x8234e09.htm

[431] Heat wave in India threatens residents and crucial wheat harvest - https://www.cnbc.com/2022/04/27/heat-wave-in-india-threatens-residents-and-crucial-wheat-harvest.html

[432] Agriculture Calendar of India - https://ipad.fas.usda.gov/rssiws/al/crop_calendar/sasia.aspx

[433] Crop Calendar Charts - https://ipad.fas.usda.gov/ogamaps/cropcalendar.aspx

[434] India - Country Brief - https://www.fao.org/giews/countrybrief/country.jsp?lang=en&code=IND

[435] Note on the impact of the war on food security in Ukraine 25 March 2022 - https://reliefweb.int/sites/reliefweb.int/files/resources/cb9171en.pdf

[436] UN climate report: It's 'now or never' to limit global warming to 1.5 degrees - https://news.un.org/en/story/2022/04/1115452

[437] Climate Change 2022: Mitigation of Climate Change - https://www.ipcc.ch/report/ar6/wg3/

[438] Climate Change 2022 – Mitigation of Climate Change Full Report - https://report.ipcc.ch/ar6wg3/pdf/IPCC_AR6_WGIII_FinalDraft_FullReport.pdf

[439] Climate Change 2022: Impacts, Adaptation and Vulnerability - https://www.ipcc.ch/report/ar6/wg2/

[440] Climate Change 2022: Impacts, Adaptation and Vulnerability - https://www.ipcc.ch/report/ar6/wg2/downloads/report/IPCC_AR6_WGII_FinalDraft_FullReport.pdf

[441] Climate change is the number one challenge of our time, EBRD president says - https://www.cnbc.com/video/2022/04/26/climate-change-is-the-number-one-challenge-of-our-time-ebrd-president-says.html

[442] Wheat - https://farmer.gov.in/M_cropstaticswheat.aspx

[443] Farmers Are Racing Against Poor Weather to Plant Crops - https://www.wsj.com/articles/poor-planting-weather-puts-squeeze-on-u-s-farmers-11652290167

[444] Drenched Land Confront Tough Choice on Planting - https://www.wsj.com/articles/farmers-on-drenched-land-confront-tough-choice-on-planting-11559727000

[445] Agriculture and Climate - https://www.epa.gov/agriculture/agriculture-and-climate

[446] Global Climate Change Impact on Crops Expected Within 10 Years, NASA Study Finds - https://climate.nasa.gov/news/3124/global-climate-change-impact-on-crops-expected-within-10-years-nasa-study-finds/

[447] Orbital and Millennial Antarctic Climate Variability over the Past 800,000 Years - https://www.ncei.noaa.gov/access/paleo-search/study/6080

[448] 12 Things to Know: Food Security in Asia and the Pacific - https://www.adb.org/news/features/12-things-know-food-security-asia-and-pacific

[449] Deconstructing Wheat Price Spikes: A Model of Supply and Demand, Financial Speculation, and Commodity Price Comovement - https://www.ers.usda.gov/webdocs/publications/45199/46438_err165_summary.pdf?v=0

[450] Agricultural Commodity Price Spikes in the 1970s and 1990s: Valuable Lessons for Today - https://www.ers.usda.gov/amber-waves/2009/march/agricultural-commodity-price-spikes-in-the-1970s-and-1990s-valuable-lessons-for-today/

[451] Wheat Futures - https://tradingeconomics.com/commodity/wheat

[452] Global Food Prices Drop to a Five-Year Low - https://www.worldbank.org/en/news/press-release/2015/07/01/global-food-prices-drop-to-a-five-year-low

[453] Grain Price Shock - https://www.bls.gov/opub/mlr/1998/08/art1full.pdf

[454] El Niño & La Niña (El Niño-Southern Oscillation) - https://www.climate.gov/enso

[455] Vegetation Index - https://www.star.nesdis.noaa.gov/smcd/emb/vci/VH/VH-Syst_10ap30.php

[456] TCI - https://www.star.nesdis.noaa.gov/smcd/emb/vci/VH/VH-Syst_10ap30.php

[457] Vegetation Index - https://www.fao.org/giews/earthobservation/country/index.jsp?type=21&code=GMB

[458] USA Earth Observation - https://www.fao.org/giews/earthobservation/country/index.jsp?lang=en&type=1141&code=USA#

[459] USA - https://www.fao.org/giews/earthobservation/country/show_img.jsp?img=/giews/earthobservation/asis/data/country/USA/PHE/HIS/PEy_ASy_c1_s1_g2_t30.png

[460] Historical Drought - https://www.fao.org/giews/earthobservation/country/index.jsp?lang=en&type=1141&code=USA#

[461] CMIP5 daily data on single levels - https://cds.climate.copernicus.eu/cdsapp#!/dataset/projections-cmip5-daily-single-levels?tab=overview

[462] CMIP6 climate projections - https://cds.climate.copernicus.eu/cdsapp#!/dataset/projections-cmip6?tab=overview

[463] CMIP6: the next generation of climate models explained - https://www.carbonbrief.org/cmip6-the-next-generation-of-climate-models-explained

[464] Kansas Wheat - https://nationalfestivalofbreads.com/nutrition-education/wheat-facts

[465] 2021 Winter Wheat Yield- https://www.nass.usda.gov/Statistics_by_State/Kansas/Publications/County_Estimates/21KSww.pdf

[466] NASS Query - https://quickstats.nass.usda.gov/results/FECB7627-0F43-3346-9623-B01D1AF156F5#C6B7BB28-49D3-37E4-AE98-351BD13E80E7

[467] Western Kansas wheat crops are failing just when the world needs them most - https://www.kcur.org/news/2022-06-09/western-kansas-wheat-crops-are-failing-just-when-the-world-needs-them-most

[468] KANSAS CROP PRODUCTION REPORT - https://www.nass.usda.gov/Statistics_by_State/Kansas/Publications/Crops_Releases/Crop_Production/2022/KS-croppr2205.pdf

[469] Climate change means Kansas farmers are dealing with hotter nights and change in rainfall and freezing patterns - https://www.cjonline.com/story/news/2021/07/04/such-climate-changes-temperature-rainfall-stresses-kansas-farmers-wheat/7837492002/

[470] BREAD WHEAT - https://www.fao.org/3/y4011e/y4011e00.htm#Contents

[471] Lingering Drought Stunts Wheat Crop - https://www.nytimes.com/1989/05/12/us/lingering-drought-stunts-wheat-crop.html

[472] Wheat Belt drought taking heavy toll - http://www.cnn.com/US/9604/26/kansas.wheat.woes/index.html

[473] Drought, poor wheat harvest in Kansas has effects on national economy, says climatologist - https://www.sciencedaily.com/releases/2014/07/140710094340.htm

[474] Satellite Imagery Resources and Usage for the Farm Service Agency, April 2014, https://www.fsa.usda.gov/Internet/FSA_File/satellite_imageryresources.pdf

[475] STAR - Global Vegetation Health Products : Background and Explanation - https://www.star.nesdis.noaa.gov/smcd/emb/vci/VH/VH-Syst_10ap30.php

[476] Satellite Imagery Resources and Usage for the Farm Service Agency - https://www.fsa.usda.gov/Internet/FSA_File/satellite_imageryresources.pdf

[477] TCI - https://www.star.nesdis.noaa.gov/smcd/emb/vci/VH/VH-Syst_10ap30.php

[478] BREAD WHEAT - https://www.fao.org/3/y4011e/y4011e00.htm#Contents

[479] 1995 - https://www.star.nesdis.noaa.gov/smcd/emb/vci/VH/vh_adminMeanByCrop.php?type=Province_Weekly_PAreaPlot

[480] Weekly Percentage of Drought Area for Province #17: Kansas of United States - https://www.star.nesdis.noaa.gov/smcd/emb/vci/VH/vh_adminMeanByCrop.php?type=Province_Weekly_PAreaPlot

[481] Unbiasedness and Market Efficiency Tests of the U.S. Rice Futures Market - https://www.jstor.org/stable/1349773

[482] Kansas Wheat History - nass.usda.gov/Statistics_by_State/Kansas/Publications/Cooperative_Projects/KS-wheat-history21.pdf

[483] NOAA STAR - https://www.star.nesdis.noaa.gov/smcd/emb/vci/VH/vh_adminMean.php?type=Province_Weekly_MeanPlot

[484] Weekly Crop Land - https://www.star.nesdis.noaa.gov/smcd/emb/vci/VH/get_TS_admin.php?provinceID=17&country=USA&yearlyTag=Weekly&type=Mean&TagCropland=land&year1=1982&year2=2022

[485] Crop Calendars for United States- https://ipad.fas.usda.gov/rssiws/al/crop_calendar/us.aspx

[486] Kansas Wheat - https://www.nass.usda.gov/Statistics_by_State/Kansas/Publications/Cooperative_Projects/KS-wheat-history21.pdf

[487] Kansas Wheat - https://www.nass.usda.gov/Statistics_by_State/Kansas/Publications/Cooperative_Projects/KS-wheat-history21.pdf

[488] Kansas Wheat - https://nationalfestivalofbreads.com/nutrition-education/wheat-facts

[489] Farmer Portal - https://farmer.gov.in/M_cropstaticswheat.aspx

[490] What Area under cultivation - https://www.statista.com/statistics/765704/india-area-of-cultivation-for-wheat/

[491] Hand Book of Statistics on Indian States - https://www.rbi.org.in/Scripts/AnnualPublications.aspx?head=Handbook+of+Statistics+on+Indian+States

[492] USAID and PL–480, 1961–1969 - https://history.state.gov/milestones/1961-1968/pl-480

[493] Handbook of statistics on India states - https://www.rbi.org.in/Scripts/AnnualPublications.aspx?head=Handbook+of+Statistics+on+Indian+States

[494] India Wheat - https://farmer.gov.in/M_cropstaticswheat.aspx

[495] Heatwave in India: 1956 - https://news.google.com/newspapers?nid=1241&dat=19980612&id=FI5TAAAAIBAJ&sjid=L4YDAAAAIBAJ&pg=5859,2125235

[496] Heatwave in India – 1995 - https://news.google.com/newspapers?nid=1980&dat=19950617&id=yDtAAAAAIBAJ&sjid=_K0FAAAAIBAJ&pg=5707,3954685

[497] India heat wave kills thousands - https://www.climate.gov/news-features/event-tracker/india-heat-wave-kills-thousands

[498] NASA Earth Observatory (EO) June 10th, 2019 - https://earthobservatory.nasa.gov/images/145167/heatwave-in-india

[499] State-wise Estimates of Yield – Wheat - https://www.rbi.org.in/Scripts/AnnualPublications.aspx?head=Handbook+of+Statistics+on+Indian+States

[500] Handbook of Statistics – India states - https://www.rbi.org.in/Scripts/AnnualPublications.aspx?head=Handbook+of+Statistics+on+Indian+States

[501] IMD - https://mausam.imd.gov.in/imd_latest/contents/pdf/pubbrochures/Heat%20Wave%20Warning%20Services.pdf

[502] Per capita wheat flour consumption declines along with other starches - https://www.ers.usda.gov/data-products/chart-gallery/gallery/chart-detail/?chartId=81227

[503] Spain heatwave - https://www.reuters.com/world/europe/spain-melts-under-earliest-heat-wave-over-40-years-2022-06-13/

[504] Spain in grip of heatwave with temperatures forecast to hit 44C - https://www.theguardian.com/world/2022/jun/10/spain-heatwave-temperatures-forecast-hit-44c

[505] India takes tough stand at climate talks as Delhi endures brutal heatwave - https://www.theguardian.com/world/2022/jun/14/india-takes-tough-stand-at-climate-talks-as-delhi-endures-brutal-heatwave

[506] Bumper Wheat Crop Looms Again in Australia on Ample Rains - https://www.bloomberg.com/news/articles/2022-04-05/bumper-wheat-crop-looms-again-in-australia-on-high-prices-rains

[507] Agriculture and Horticulture Ukraine - https://www.climatechangepost.com/ukraine/agriculture-and-horticulture/

[508] Climate Change in Russia and the Weaponization of Wheat - https://climateandsecurity.org/2020/08/climate-change-in-russia-and-the-weaponization-of-wheat/
[509] GIEWS - Global Information and Early Warning System - https://www.fao.org/giews/countrybrief/country.jsp?lang=en&code=IND
[510] SMT - https://www.star.nesdis.noaa.gov/smcd/emb/vci/VH/VH-Syst_10ap30.php
[511] MODIS VH - https://www.star.nesdis.noaa.gov/smcd/emb/vci/VH/modis_browse8daysLST.php
[512] STAR - Global Vegetation Health Products : Province-Averaged VH - https://www.star.nesdis.noaa.gov/smcd/emb/vci/VH/vh_adminMeanByCrop.php?type=Province_Weekly_MeanPlot
[513] Moisture and Target Yields - https://www.gov.mb.ca/agriculture/environment/soil-management/moisture-and-target-yields.html
[514] UP Wheat Data - https://www.star.nesdis.noaa.gov/smcd/emb/vci/VH/get_TS_admin.php?provinceID=33&country=IND&yearlyTag=Weekly&type=Mean&TagCropland=WHEA&year1=1982&year2=2022
[515] India - https://www.ers.usda.gov/topics/international-markets-u-s-trade/countries-regions/india/
[516] MODIS - https://modis.gsfc.nasa.gov/data/dataprod/index.php#atmosphere
[517] India State UP - https://www.star.nesdis.noaa.gov/smcd/emb/vci/VH/vh_adminMeanByCrop.php?type=Province_Weekly_MeanPlot
[518] Province-Averaged VH data for CropLand Data - https://www.star.nesdis.noaa.gov/smcd/emb/vci/VH/get_TS_admin.php?provinceID=33&country=IND&yearlyTag=Weekly&type=Mean&TagCropland=WHEA&year1=1982&year2=2022
[519] Handbook of statistics on India states - https://www.rbi.org.in/Scripts/AnnualPublications.aspx?head=Handbook+of+Statistics+on+Indian+States
[520] NOAA STAR - https://www.star.nesdis.noaa.gov/smcd/emb/vci/VH/vh_adminMean.php?type=Province_Weekly_MeanPlot
[521] Weekly Crop Land - https://www.star.nesdis.noaa.gov/smcd/emb/vci/VH/get_TS_admin.php?provinceID=17&country=USA&yearlyTag=Weekly&type=Mean&TagCropland=land&year1=1982&year2=2022
[522] The Climate Change Knowledge Portal - https://climateknowledgeportal.worldbank.org/
[523] The CCKP Time Series - https://climateknowledgeportal.worldbank.org/download-data
[524] Kansas Wheat - https://nationalfestivalofbreads.com/nutrition-education/wheat-facts
[525] Uncertainty Intervals - https://facebook.github.io/prophet/docs/uncertainty_intervals.html
[526] Wheat futures - https://www.investing.com/commodities/ncdex-wheat-futures-historical-data
[527] Global Heat Wave Toasts Wheat and Prices Soar - https://www.wsj.com/articles/global-heat-wave-toasts-wheat-and-prices-soar-1533391201
[528] Wheat prices hit record high as India's heat wave-driven export ban compounds Ukraine war supply woes- https://www.cbsnews.com/news/india-heat-wave-wheat-prices-soar-climate-change-ukraine-war-supplies/
[529] NOAA STAR - https://www.star.nesdis.noaa.gov/smcd/emb/vci/VH/vh_adminMean.php?type=Province_Weekly_MeanPlot
[530] Weekly Crop Land - https://www.star.nesdis.noaa.gov/smcd/emb/vci/VH/get_TS_admin.php?provinceID=17&country=USA&yearlyTag=Weekly&type=Mean&TagCropland=land&year1=1982&year2=2022
[531] The Climate Change Knowledge Portal - https://climateknowledgeportal.worldbank.org/
[532] The CCKP Time Series - https://climateknowledgeportal.worldbank.org/download-data
[533] Cow Necklace – trademark of Hanumayamma Innovations and Technologies, inc., https://www.hanuinnotech.com
[534] California - https://www.star.nesdis.noaa.gov/smcd/emb/vci/VH/vh_adminMean.php?type=Province_Weekly_MeanPlot
[535] NOAA TCI - https://www.star.nesdis.noaa.gov/smcd/emb/vci/VH/VH-Syst_10ap30.php
[536] California Climate Adaptation Strategy - https://resources.ca.gov/Initiatives/Building-Climate-Resilience/2021-State-Adaptation-Strategy-Update

Chapter 7

Energy Shocks and Macroeconomic Linkage Analytics

The chapter introduces the impact of energy prices on macroeconomic drivers. The chapter starts with changes in energy prices from the U.S. economic point of view and analyzes the impact of energy price fluctuations on agricultural commodities. Next, the chapter looks at agricultural farm inputs, fertilizers, and develops a linkage model. Next it develops a Machine Learning model using a bell curve food security model. Finally, the chapter concludes with the development of Machine Learning Models for Energy Prices & Urea and Phosphate Fertilizer Costs and Commodities Demand and Energy Shocks on the Phosphate Model.

Large fluctuations in energy prices, recently, have been a distinguishing characteristic of the World economy. The drivers of price fluctuation could be due to increased global conflicts, war in Ukraine, climatic events, heatwaves, droughts, and the COVID-19 pandemic. The U.S. economy has experienced price fluctuations since the 1970s. Turmoil in the Middle East, rising energy prices in the U.S. and evidence of global warming recently have reignited interest[537] in the link between energy prices and economic performance [1]. Energy prices (oil price) have a significant impact on food prices. According to the results of the model that was developed as part of this chapter, agricultural food prices respond positively to any shock from oil prices. There is a linkage between energy and food security through price volatility. Since inflation in oil price is harmful for food security, it would be necessary to diversify the energy consumption in this sector and optimize agricultural productivity that will be in favor of not only energy security but also food security.

In this chapter, I would like to model events that are drivers of large fluctuations and their events on food security. Specifically, events that I would like to focus are those that have a considerable impact on the global economy: extreme weather, climate change, and fertilizer price spikes due to wars. It is widely accepted that energy prices in general and crude oil & natural gas prices have been endogenous with respect to European, U.S. and the World macroeconomic conditions. *Endogeneity* here refers to the fact that not only do energy prices affect the U.S. economy, but that there is reverse causality from the U.S. and more generally from the global macroeconomic aggregates to the price of energy [1]. Clearly, both the supply and demand for energy depend on global macroeconomic aggregates such as real global economic activity and interest rates and could be affected by conflicts and wars.

Extremely hot weather is a new normal!

The extreme heat is impacting hundreds of millions of people in one of the most densely populated parts of the world, threatening to damage whole ecosystems.[538]

There is a cascading effect. Extreme heat has multiple and cascading impacts not just on human health, but also on ecosystems, agriculture, water and energy supplies and key sectors of the economy.

Climate Change and Energy Shocks Linkage

Extreme weather has a considerable impact on agricultural outcomes. For instance, consider heat spells. The heat spell occurs very fast and the crop matures at a faster pace, which shrivels the grain size. This also results in a yield drop. "Wheat prices will be driven up, and if you look at what is happening in Ukraine, with many countries relying on wheat from main producers such as India to compensate, the impact will be felt well beyond India [2]. Another impact of extreme weather is the one on a country's foreign reserves and gross domestic product (GDP), for instance when Early Season Heat Waves Strike India[539] these macroeconomic variables get affected. A heat wave[540] is generally defined as a prolonged period of excessively hot weather. Prolonged and intense heat waves have become more frequent in many parts of the globe. The figure shows the annual values of the U.S. Heat Wave Index[541] from 1895 to 2020 (please see Figure 1). This data covers the contiguous 48 states. An index value of 0.2 (for example) could mean that 20 percent of the country experienced one heat wave, 10 percent of the country experienced two heat waves, or some other combination of frequency and area resulted in this value.[542]

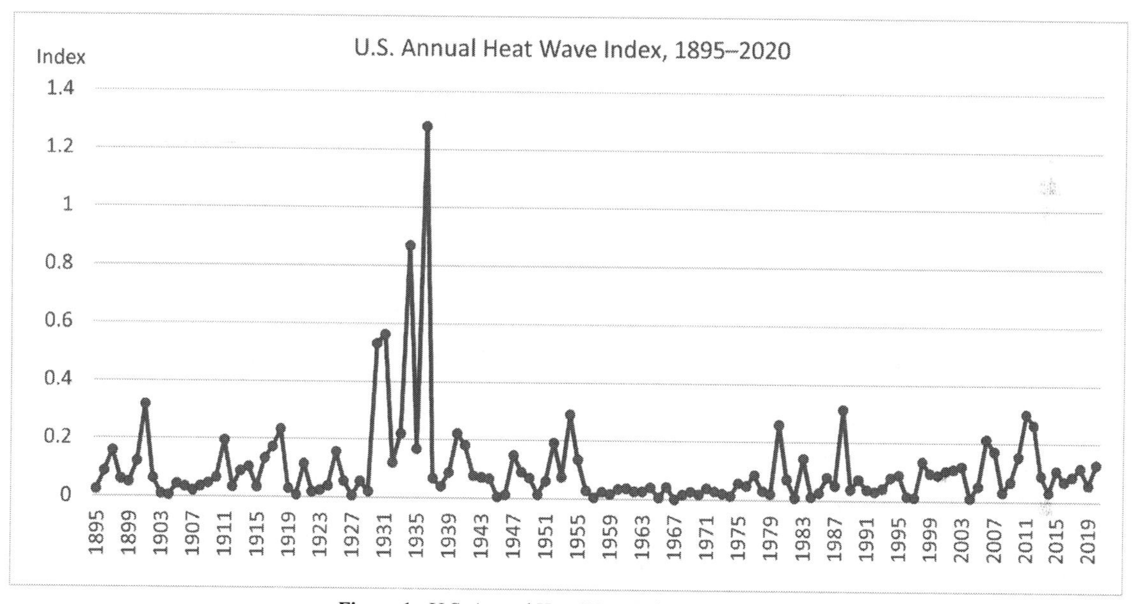

Figure 1: U.S. Annual Heat Wave Index - 1895–2020.

Heat waves lead to short-term increases in mortality and negative impacts on infrastructure and on biophysical systems [3]. The effects of the heat wave include heat-related illnesses, poor air quality, little rainfall, and reduced crop yields. Additionally, power demand has spiked, and coal inventories have dropped, leaving the country with its worst electricity shortage [4] in more than six years [5]. As a result of heat spell events India has been led to buy expensive foreign gas to ease its power crisis.[543] Sweltering heat and ongoing blackouts are forcing India's liquefied natural gas importers to top up with expensive shipments. The purchases are unusual for India's cost-sensitive power generators that are heavily reliant on Coal (72.08% of total electricity generated), which tends to avoid buying Liquefied Natural Gas (LNG) at such high rates.[544] This illustrates how a domestic coal shortage is forcing the South Asian nation to look for alternative fuels no matter what the price is, further elevating international demand [6].

During heat wave episodes, there is an unprecedented increase in energy consumption. According to an Indian government report by the Ministry of Earth Sciences, the average frequency of summer heat waves will increase to about 2.5 events per season by the mid-21st century with a further rise to about 3 events by the end of the century [7]. The average duration of heat waves is also expected to increase to 18 days per season[545] toward the end of the century.

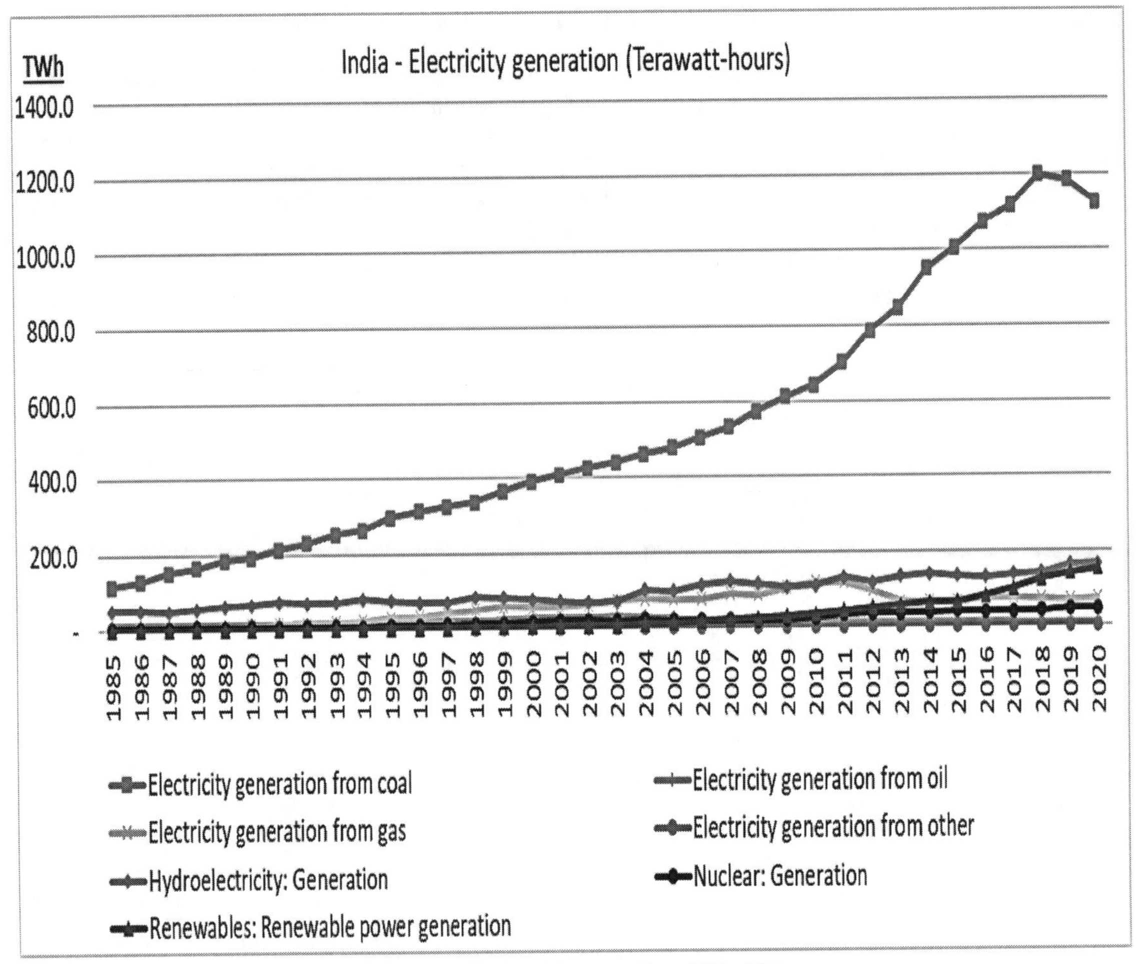

Figure 2: India electricity generation - 1985 : 2020.

Data source: bp Statistical Review of World Energy July 2021[546]

The U. N.[547]

In Indian Heat waves typically occur from March to June, and in some rare cases, even extend till July (please see Figure 2). On average, five to six heat wave events occur every year over the northern parts of the country. Single events can last weeks, occur consecutively, and can impact large populations. Additionally, more than 300 large wildfires were burning around the country on April 27, according to the Forest Survey of India. Nearly a third of those were in Uttarakhand [4]. The Intergovernmental Panel on Climate Change has predicted a similar scenario for India. Consulting company McKinsey & Company estimates that by the end of the decade, the country could lose $250 billion or 4.5% of its gross domestic product to work hours lost to heat waves. As we can see, the economy became more vulnerable to supply shocks with its increasing dependence on coal. However, the (please see Figure 3) transition from coal to importing oil & LNG has increased the negative impacts of demand shocks for economies [8].

Under the Representative Concentration Pathway (RCP) 4.5, the risk of heat waves is projected to increase tenfold during the twenty-first century. More than ~70% of the land areas in India is projected to be influenced by heat waves with magnitudes greater than nine.

The climate change impact leads to foreseeable food security issues for the country's poor & vulnerable as the government could divert resources from the public SafetyNet programs to pay foreign reserves' deficits caused by energy imports. The impact could be cyclic and systematic. Purchase of high cost energy by a country would lead to depletion of foreign reserves that directly impact the USD exchange rate. The

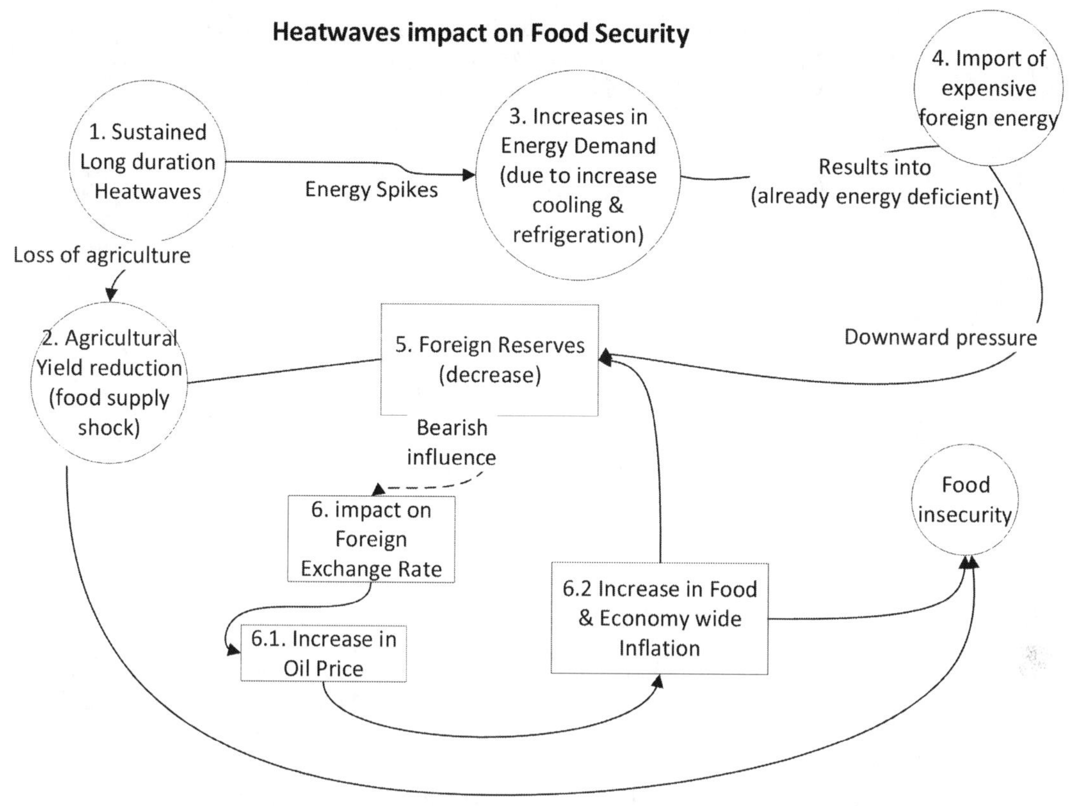

Figure 3: Heatwave and Food Security Linkage.

downward pressure of the exchange rate would increase overall inflation & thus impact the overall food insecurity [3].

Unexpected increases in energy prices (also called energy price shocks) can adversely affect food security;[548] purchasing food and gasoline and paying utility bills compete for the same limited resources of low-income households. For poor families, rising energy prices create a difficult tradeoff between buying enough food, staying warm, or having enough gas for transportation. *Energy price shocks* may be particularly detrimental to low-income households and vulnerable communities because they have fewer resources available to absorb the higher unplanned expenses. For instance, a 41-percent increase in natural gas prices led to the prevalence of food insecurity among low-income households rising from 12.4 to 14.7 percent [9,10]. Three important energy sources that are primary candidates for energy price shocks include gasoline, natural gas, and electricity [9].

Unexpected Price Shocks—Gasoline, Natural Gas, and Electricity

Price shocks are defined as a large deviation from prices that households expect to bear [10]. Unexpected price increases[549] for each energy source (gasoline, natural gas, and electricity) adversely affect at least one food-security indicator (availability, access, and others) for households across the globe and, in addition, increase food stress and the money required for food. Food stress is generally the case for households that had enough food but not always the kinds of food that they want to eat, sometimes not enough to eat, or often not enough to eat. One of the salient, unfortunately, indicators of energy price shocks is the "more money for food" indicator, needing to spend more than a family does to maintain just enough food to meet the household's food needs [9]. *Natural gas price shocks* had the most profound effects on all households, increasing the likelihood of food insecurity, food stress, & more money required, especially the impact could be perpetual for low-income households. On a national level, natural gas price shocks increased overall agricultural farm input costs that has increased overall food insecurity.

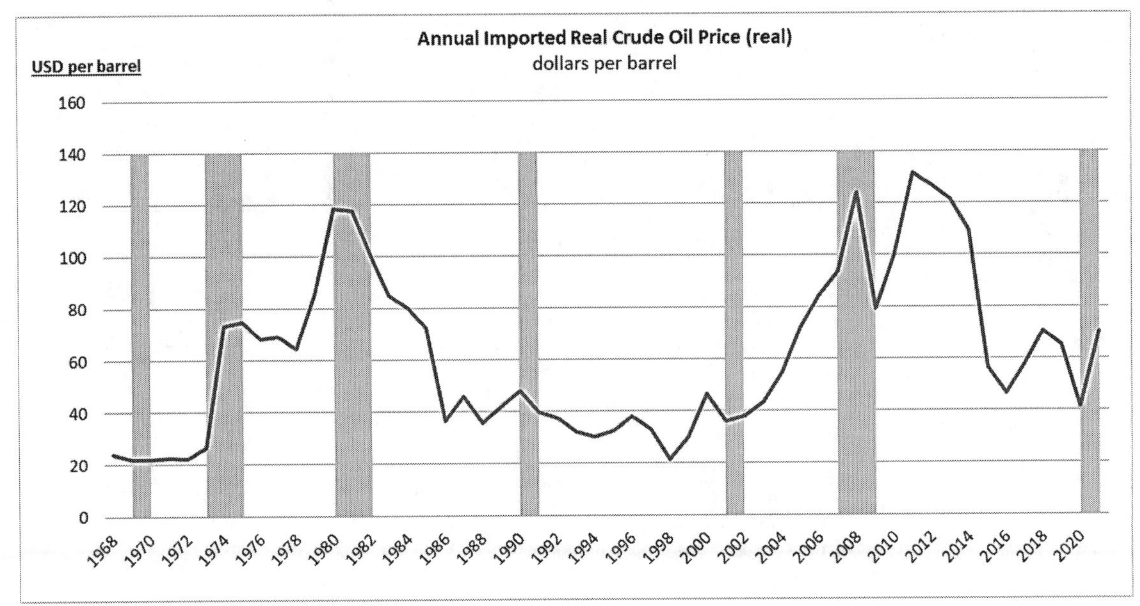

Figure 4: Annual Imported Crude Oil Prices.

Figure 4 shows the history of the price of oil since the early 1960s (U.S. Energy Information Administration Real Prices Viewer (real_prices.xlsx).[550] The price shown is the monthly average spot price of a barrel of West Texas intermediate crude oil, measured in U.S. dollars. The gray bars[551] in this and all the following figures represent recessions, as defined by the National Bureau of Economic Research [11,12].

As you can see from Figure 4, a long period of oil price stability was interrupted in 1973. In fact, the 1970s show two distinct jumps in oil prices: one was triggered by the Yom Kippur War in 1973 [11], and one was prompted by the Iranian Revolution of 1979. Since then, oil prices have regularly displayed volatility relative to the 1950s and 1960s [11]. The price jumps have arrived post the 2002 recession. During the great recession (December 2007–June 2009), price jumps became severe, a fact of the great recession.

Real Petroleum Prices[552]

Real Petroleum Prices are computed by dividing the nominal price for each month by the ratio of the Consumer Price Index (CPI) in that month to the CPI in some "base" period. The Real Petroleum Prices spreadsheet and charts are updated every month so that the current month is the base period in the monthly price series. Consequently, all real prices are expressed in "current" dollars and any current month price may be compared directly with any past or projected real prices.

Impact of Higher Gas prices on Macroeconomic Level

As a consumer, you may already understand the microeconomic implications of higher oil prices. When observing higher oil prices, most of us are likely to think about the price of gasoline as well, since gasoline purchases are necessary for most households. When gasoline prices increase, a larger share of household budgets is likely to be spent on it, which leaves less to spend on other goods and services. The same goes for businesses whose goods must be shipped from place to place or that use fuel as a major input (such as the airline industry). Higher oil prices tend to make production more expensive for businesses, just as they make it more expensive for households to do the things they normally do. This would directly affect food security for low-income and vulnerable communities.

It turns out that oil and gasoline prices are indeed very closely related. Figure 5 plots yearly oil prices from 1979 through early 2021, using the spot oil price for West Texas intermediate Short-Term Energy Outlook Real and Nominal Prices. The following series track each other very closely over time:

- increases in motor gasoline prices ($/gallon) are accompanied by increases in imported crude oil prices.
- increases in heating oil prices ($/gallon) are accompanied by increases in imported crude oil prices ($/barrel).
- increases in diesel prices ($/gallon) are accompanied by increases in imported crude oil prices ($/barrel).

These are facts we can witness through the lens of correlation relationships. As shown in the below table, the correlation coefficients for the series are as follows:

- correlation relationship of 0.99 between motor gasoline oil and imported crude oil prices.
- correlation relationship of 0.95 between heating oil price and imported crude oil prices.
- correlation relationship of 0.94 between diesel price and imported crude oil prices.

We can fairly say heating oil, diesel, and motor gasoline prices are highly correlated with imported crude oil prices. So, when imported crude oil prices spike, one can expect gasoline prices, diesel prices, and heating oil prices to spike as well, and that affects the costs faced by most households and businesses. Oil price increases are generally thought to increase inflation and reduce economic growth. In terms of inflation, oil prices directly affect the prices of goods made with petroleum products.

Diesel prices have been rising around the world as economies have rebounded from the Covid-19 pandemic, several refineries have closed, and Western countries have attempted to curtail imports of Russian energy.[553] Crude and gasoline prices have been on the rise, too, but diesel has outpaced them because of refinery closures and because Russia was such a big supplier of refined fuels to Europe, causing ripple effects world-wide [13]. The Biden administration is exploring the release of diesel fuel reserves amid high prices to act quickly if needed to address supply outages on the East Coast to alleviate inflationary trends on the trucking industry and economy. Another industry feeling the heat is agriculture, which relies on diesel to fuel tractors and dry crops and is also facing a historic rise in fertilizer prices. Fertilizer prices are most sensitive to oil prices [13]. Given the high inflation and the Ukraine war, May 2022, some suppliers had told farms that they could order diesel but would be quoted a price only when the fuel is delivered. In a lot of cases, farmers have ordered diesel but have no idea what they are going to be paying for it. We've never seen such concern about pricing and availability [11]. As mentioned above, oil prices indirectly affect costs such as transportation, manufacturing, and heating. The increase in these costs can in turn affect the prices of a variety of goods and services, as producers may pass production costs on to consumers. The extent to which oil price increases lead to consumption price increases depends on how important oil is to produce a given type of good or service [14].

One direct impact of the increase in oil prices can be felt both by individual consumers and businesses is in terms of increased shipping costs for goods and materials. With increased imports of crude oil, prices for the fuel powering big rigs (no wonder a stronger correlation relationship –0.94—between crude oil and diesel prices) have hit record levels, Spring 2022, adding to inflationary pressures in the U.S. economy.[554] Diesel costs are reaching new highs (given diesel is produced from the same slice of the crude barrel [15] that can be witnessed from the 0.96 correlation relationship with crude oil prices) across the U.S., straining the operations of trucking companies and wrecking the transportation budgets of businesses that need to ship goods [13]. The price of the fuel that powers heavy-duty trucks has increased by more than $1.50 a gallon in roughly two months, according to the U.S. Energy Information Administration[555] (please see Figure 6). The national average price has climbed to $5.62 a gallon, setting a record for the second week in a row, as prices at the pump surpassed $6 in some markets. Many small-trucker shipping costs within the U.S. have risen by 15% to 20% from 2021, pushing them to make changes to their distribution operations as some customers reconsider projects because of rising costs. The spillover effects many retail prices of perishable commodities; stock units, and essentials for day-to-day life would go-up, fueling inflation as Diesel is used in the U.S. mostly in trucks, which means higher prices add to shipping and delivery costs. Inventories of distillates, which also include heating oil, fell recently to a 17-year low during lower refining activity

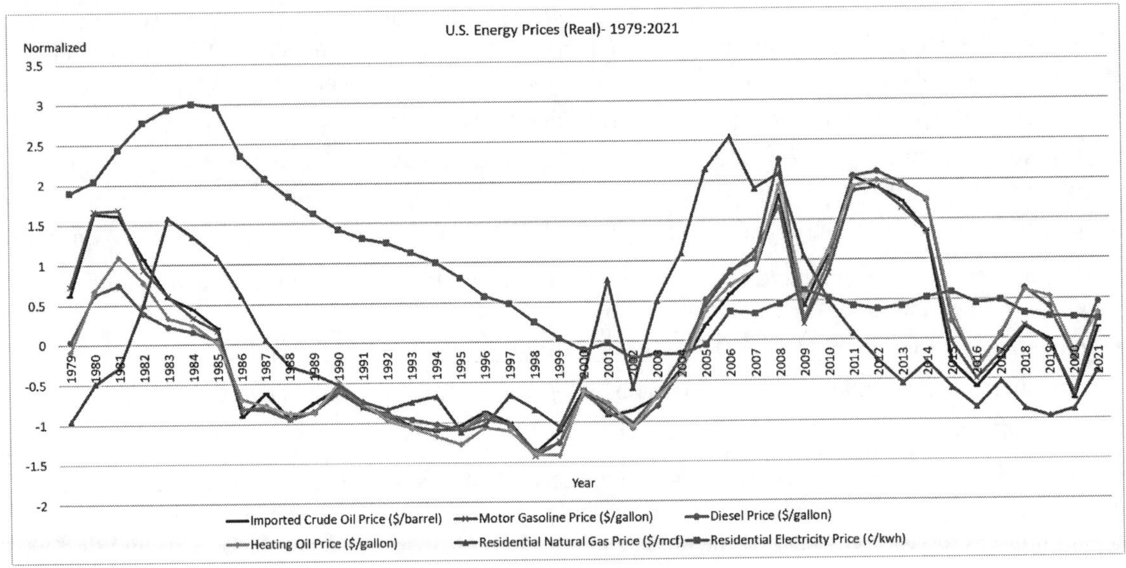

Figure 5: U.S. Energy Prices (1970–2021).

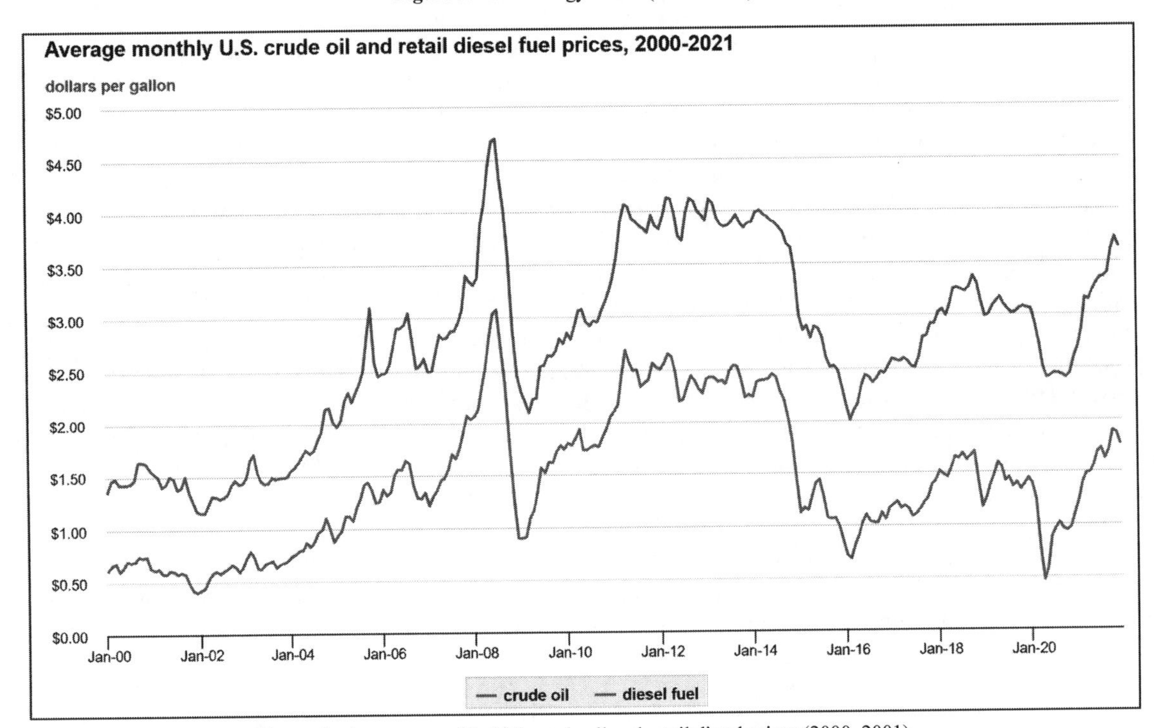

Figure 6: Average monthly U.S. crude oil and retail diesel prices (2000–2001).

and higher demand domestically and abroad,[556] according to the U.S. Energy Information Administration. Supplies are particularly tight along the East Coast, where inventories have dropped to their lowest level since at least 1990 [15]. Rising energy prices are a major factor contributing to the persistence of inflation, which has sparked steep declines in the stock and bond markets. The S&P 500 fell by1.6% on Wednesday after a gauge of U.S. consumer prices came in higher than what Wall Street expected [15].

Trucking companies generally can cover rising diesel prices through fuel surcharges that are built into contracts. However, the thousands of smaller fleets and independent owner-operators that make up the bulk of the highly fragmented truck market have a harder time passing along the added expenses. The rising operating costs are hitting those operators just as the base shipping prices on trucking spot markets are dropping on wavering freight demand.

Oil price increases can also stifle the growth of the economy through their effect on the supply and demand for goods other than oil. Increases in oil prices can depress the supply of other goods because they increase the costs of producing them. In economics terminology, high oil prices can shift down the supply curve for the goods and services for which oil is an input [14].

High oil prices also can reduce demand for other goods because they reduce wealth, as well as induce uncertainty about the future [16]. One way to analyze the effects[557] of higher oil prices is to think about the higher prices as a tax on consumers. The simplest example occurs in the case of imported oil. The extra payment that U.S. consumers make to foreign oil producers can now no longer be spent on other kinds of consumption goods [17].

The Macroeconomics of Oil Shocks[558]

For various reasons, oil-price increases may lead to significant slowdowns in economic growth. Five of the last seven U.S. recessions were preceded by significant increases in the price of oil [16].

Please see U.S. energy prices from 1979 to 2021 (source: U.S. Energy Information Administration (eia) Real Price Viewer[559]—all real and nominal price series).[560]

	Imported Crude Oil Price ($/barrel)
Imported Crude Oil Price ($/barrel)	1
Motor Gasoline Price ($/gallon)	0.992082433
Diesel Price ($/gallon)	0.94245507
Heating Oil Price ($/gallon)	0.951778372
Residential Natural Gas Price ($/mcf)	0.377119369
Residential Electricity Price (¢/kwh)	0.154742135

Coming to natural gas and residential electricity prices, both are less correlated with imported crude oil. The reason is straight forward: electricity generation can be spawned from different sources.

Gasoline price shocks, however, had a less consistent effect, increasing the likelihood of needing more money for food in each sample but not increasing the other measures of food distress. However, gas price shocks will have an adverse impact on farm inputs.

Figure 7 shows the "real" oil price (downloaded from U.S. Energy Information Administration).[561] This removes the effect of inflation and thus gives a more accurate sense of what is happening to the price of the commodity itself. In essence, the "real" measure allows you to compare oil prices over time in a way that you can't when inflation is also part of the change in price. You can see that real oil prices have varied a lot over time, and large fluctuations tend to be concentrated over somewhat short periods. You can also see that by the spring of 2008 [14] the real price of oil has easily exceeded that of the late 1970s. Without no exception, the next wave of higher real crude prices has arrived in 2012. Gasoline prices have averaged the highest ever in 2012,[562] despite never topping the single day spikes seen in 2008 due to refinery outages and tensions in the Middle East for the high prices [18].

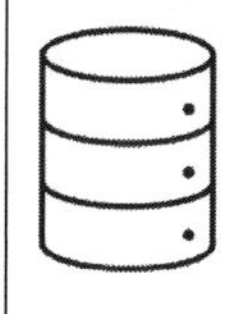

Data systems capture the pulse of a nation in terms of the gross national economic activities namely, entertainment & recreational activities and social citizen activities, national health & examination survey data, gross domestic output & commercial transactions, agricultural and animal husbandry data, food tastes & consumption patterns, citizen food choices & preferences, weather & climatology, science & engineering data, and datasets that depict just a basic way-of-life are the assets of national prominence and will have to be secured just like other national security systems!

In simple single liner, "nations that take care of their data & data systems are those that prosper"!

Author

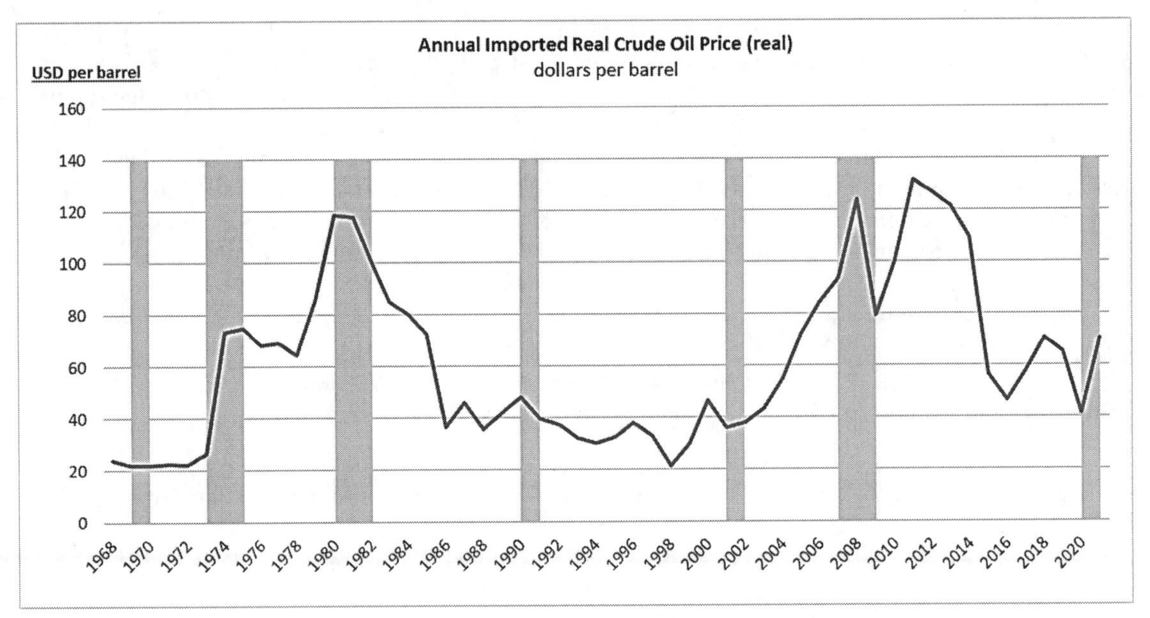

Figure 7: Annual Imported Real Crude Oil Price (1968 : 2022).

Fact #915: March 7, 2016, Average Historical Annual Gasoline Pump Price, 1929-2015

When adjusted for inflation, the average annual price of gasoline has fluctuated significantly, and has recently experienced sharp increases and decreases. The effect of the U.S. embargo on oil from Iran can be seen in the early 1980's with the price of gasoline peaking in 1982. From 2002 to 2008 the price of gasoline rose substantially, but fell sharply in 2009 during the economic recession. In 2012, prices reached the highest level in the eighty-year series in both current and constant dollars but began a steep decline thereafter. In constant dollar terms, the price of gasoline in 2015 was only seven cents higher than in 1929. Fact #915 Dataset (fotw#915_web.xlsx):[563] https://www.energy.gov/eere/vehicles/downloads/fact-915-march-7-2016-average-historical-annual-gasoline-pump-price-1929

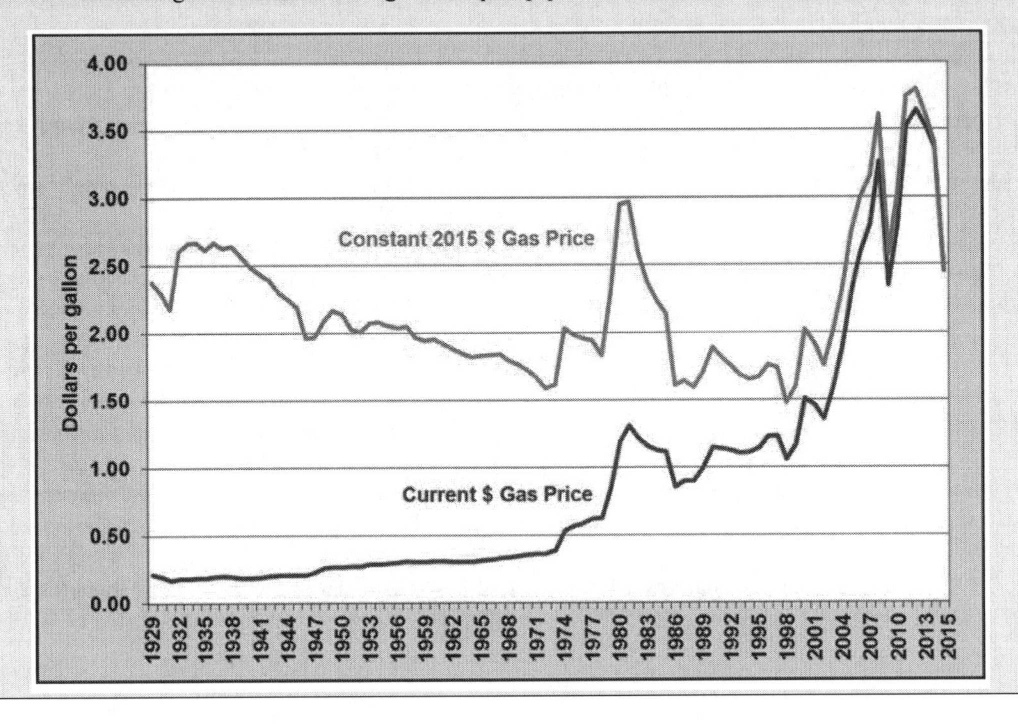

Electricity is the primary source of energy across the world. Electricity price shocks could lead to depressive economic outcomes and increase input costs across the economies, resulting in lower demand & induce a country's economy into recession.

The strongest responses are observed for gasoline and for heating oil and coal. Contrary to the conventional wisdom, gasoline consumption responds immediately to unanticipated energy price increases reaching an elasticity of –0.48 after one year. The strikingly large response of –1.47 for heating oil and coal is likely due to households' ability to store heating oil in tanks. This storage feature allows households to delay purchases of new heating oil when the price of heating oil is high and to fill the tank completely when prices are low. In contrast, electricity and natural gas are inherently un-storable, and gasoline may not be stored for safety reasons beyond the tank capacity of a vehicle. Indeed, the declines in electricity consumption and in natural gas use are smaller and not statistically significant [1].

The Channels of Transmission & Behavioral Economy

The literature has focused on four complementary mechanisms by which consumption expenditures may be directly affected by energy price changes [1].

- First, higher energy prices are expected to reduce discretionary income, as consumers have less money to spend after paying their energy bills. All else equal, this discretionary income effect will be the largest, the less elastic the demand for energy, but even with perfectly inelastic energy demand the magnitude of the effect of a unit change in energy prices is bounded by the energy share in consumption.
- Second, changing energy prices may create uncertainty about the future path of the price of energy, causing consumers to postpone irreversible purchases [19,20]. Unlike the first effect, this uncertainty effect is limited to consumer durables.
- Third, even when purchase decisions are reversible, consumption may fall in response to energy price shocks, as consumers increase their precautionary savings. This response may arise if consumers smoothen their consumption because they perceive a greater likelihood of future unemployment and hence future income losses. By construction, this effect will embody general equilibrium effects on employment and real income otherwise ignored by the demand channel of transmission. In addition, the precautionary savings effect may also reflect greater uncertainty about the prospects of remaining gainfully employed, in which case any unexpected change in energy prices would lower consumption.
- Fourth, consumption of durables that are complementary in use with energy (in that their operation requires energy) will tend to decline even more, as households delay or forego purchases of energy-using durables. This operating cost effect is more limited in scope than the uncertainty effect in that it only affects specific consumer durables. In addition, there may be indirect effects related to the changing patterns of consumption expenditures.
- Finally, for the low-income and marginalized communities, in addition to the above four, there is a direct decline in food security. To offset higher energy prices and shocks, they tend to pivot, no choice, to high calorie & low healthy food. Additionally, this not only increases public health concerns but also increases severe food insecurity.

ERS research suggests that energy price shocks can lead to higher rates of food insecurity and food hardship among low-income households. A 38-percent increase in gasoline prices was found to increase the likelihood of low-income households needing more money for food by 7.3 percentage points. Similarly sized price shocks for natural gas and electricity also affected food distress measures. As examples, an unexpected increase in the price of electricity raised the likelihood of low-income households needing more money for food by around 2 percentage points. And natural gas price shocks were found to increase the likelihood of low-income households experiencing food insecurity by 2.3 percentage points [10].

Fertilizer Use and Price[564]

The purpose of the model is to analyze fertilizer consumption in the United States in the form of plant nutrients and major fertilizer products—as well as consumption of mixed fertilizers, secondary nutrients, and micronutrients—for 1960 through the latest year for which statistics are available. The share of planted crop acreage receiving fertilizers, and fertilizer applications per receiving acre (by nutrient), are presented for major corn, cotton, soybeans, and wheat producing (nutrient consumption by crop data starts in 1964) states. Fertilizer farm prices and wholesale indices of fertilizer prices are also available. The goal of our time series is to develop movements in fertilizer prices based on the endogenous and exogenous variables.

Please find Primary nutrient content (Nitrogen (N), Phosphate (P_2O_5), and Potash (K_2O)) U.S. consumption of plant nutrients (1960:2015):[565] USDA fertilizeruse.xlsx[566] (All fertilizer uses, and price tables are in one workbook) [21].

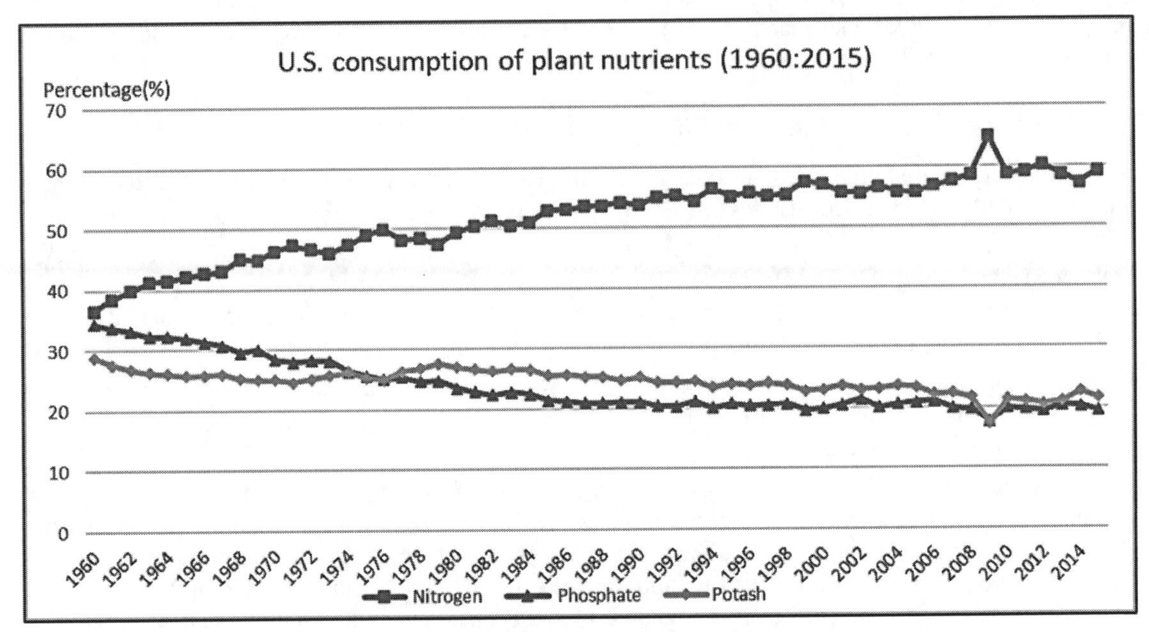

Figure 8: U.S. Consumption of plant nutrients.

As it can be seen from the above graph (please see Figure8) Nitrogen use has steadily increased whereas the consumption of Phosphate and Potash have reduced. Nitrogen is the most abundant gas in the atmosphere and makes up approximately 78% of the atmosphere. One of the most important scientific discoveries of the 20th century is the Haber-Bosch process, which transforms atmospheric nitrogen into synthetic nitrogen for crop fertilization. The discovery of the Haber-Bosch [22] process allowed for the widespread fertilization of crops, and together with other agricultural technology advancements, helped revolutionize food production for a growing world population.[567] The Haber-Bosch process utilizes hydrogen and atmospheric nitrogen under extremely high pressure and temperature in combination with a metal catalyst such as iron to produce anhydrous ammonia gas, which is condensed with the help of cold water, forming liquid ammonia. In this process, natural gas is typically used as a hydrogen feedstock and as a source of energy to obtain the high pressure and temperature required for the reaction, and for this reason, natural gas and nitrogen fertilizer prices are closely related [22].

Bayer's Monsanto Division Rebounds

Ukraine is a top exporter of grains, accounting for 13% of global corn and 12% of wheat exports, but the Russian invasion is preventing it from shipping much of those volumes to world markets.[568] The resulting threat of food shortages underscores the demand for Bayer's genetically modified crop seeds, as well as its weedkiller and pest-control products that are designed to boost yields. The food shortages stoked by Russia's invasion of Ukraine drives the demand for seeds and pesticides to boost global crop production [23].

The U.S. is the world's fourth-largest[569] producer [24] of nitrogen fertilizers with China (36,957,467.98 metric tons) leading the top five producer pack, followed by India (36,957,467.98 metric tons), Russia (36,957,467.98 metric tons)[570] and Canada (3,948,980.00 metric tons). In 2019, ammonia was produced in the U.S. by 16 companies at 35 plants in 16 states [24]. Unlike elements mined from the Earth, such as *potassium and phosphorus*,[571] where the mine location is determined by the presence of rocks containing significant amounts of these elements, ammonia is produced from the atmosphere so theoretically production facilities could be located anywhere [25]. Most ammonia production in the U.S. occurs near large reserves of natural gas in Louisiana, Oklahoma, and Texas due to the use of natural gas as a hydrogen feedstock, and to fuel the high temperature and pressure needed to produce ammonia (Apodaca 2020a). Most ammonia in the U.S. is produced by international companies and used for domestic consumption, with some imports from production facilities located in Trinidad and Tobago and Canada. A small amount of ammonia produced in the U.S. is exported. Although U.S. imports and exports of ammonia appear small compared to domestic production and consumption, the numbers are significant when compared with other ammonia importing and exporting countries. In 2019, the U.S. was the 9th largest exporter and 2nd largest importer of ammonia in the world. The U.S. also imports urea, and in 2019 they were the 2nd largest importer of urea in the world. Figure 9 shows total production, imports, exports, and apparent consumption of ammonia from 2015 to 2019. Please note short ton (symbol tn) is a mass measurement unit equal to 2,000 pounds (lb) (907.18474 kg). It is commonly used in the United States, where it is known simply as a ton (please see Figure 9).

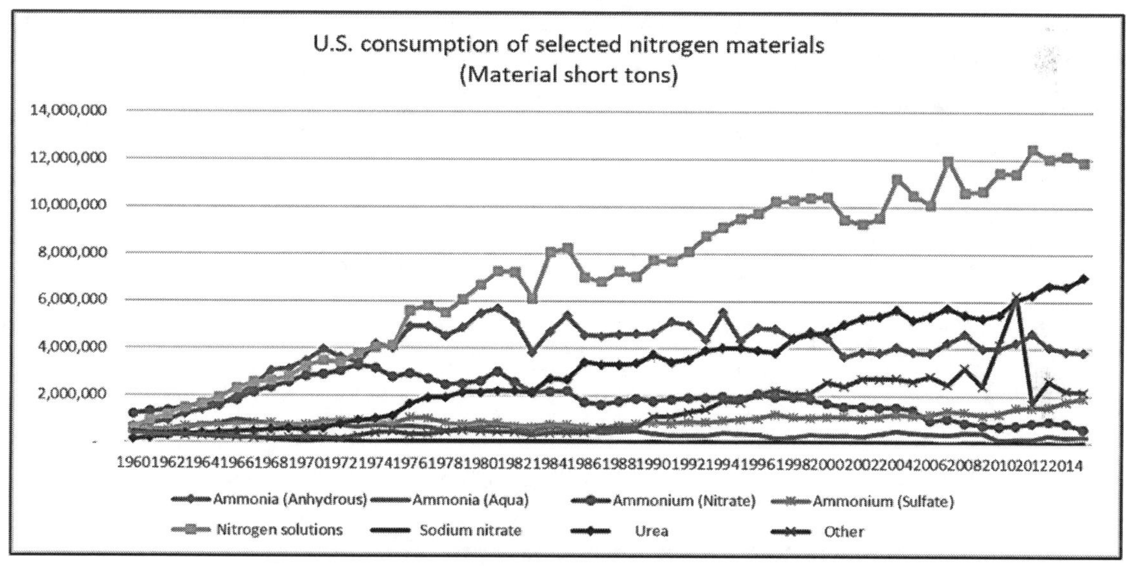

Figure 9: U.S. consumption of selected nitrogen materials.

The war in Ukraine has increased unprecedented global risks, especially for the agriculture and allied sectors. The Russian invasion of Ukraine substantially elevates the risk of disruptions in the global fertilizer trade. Russia is the world's largest exporter of fertilizers, accounting for 23% of ammonia exports, 14% of urea exports, 10% of processed phosphate exports, and 21% of potash exports, according to data from The Fertilizer Institute. The primary destinations of fertilizers from Russia are Brazil (21%), China (10%), the US (9%), and India (4%). The Ukraine war has not only increased the risk for major fertilizer producers but also has put other countries that depend on Russian imports at risk. In this regard, compared to the US, Brazil will be affected more directly [26] as it imports 85% of its fertilizers.[572] Furthermore, *gas is a key input for fertilizer production*. High gas prices have resulted in a curtailment of production in regions such as Europe, further constricting an already tight market. Meanwhile, sanctions on Russian ally Belarus have substantial implications for the *potash*[573] *market*, with Russia and Belarus contributing a combined 40% of annual traded volumes. Given the asymmetrical impact, overall, the current commodity wars will lead to higher and more persistent inflation around the world, including in the U.S. [27].

Nitrogen (N) [28], Phosphorous (P) [29], and Potassium (K) [29] are essential to life.

The usage of fertilizer is crucial for staple and essential agricultural crops. For instance, consider the average percentage of nitrogen fertilizers received by acreage by the following commodities in the U.S: 25% of soyabeans acreage, 98% of corn acreage, 100% of cotton acreage, and above 80% of wheat acreage receives nitrogen fertilizers. As can be seen corn (please see Figure 10), wheat, and cotton use high percentages of Nitrogen fertilizer. It is no coincidence as Nitrogen is key to life.[574]

Nitrogen, the most abundant element in our atmosphere, is crucial to life. Nitrogen is found in soils and plants, in the water we drink, and in the air we breathe. It is also essential to life: a key building block of DNA, which determines our genetics, is essential to plant growth, and therefore necessary for the food we grow. However, as with everything else, balance is the key: with inadequate nitrogen plants cannot thrive, leading to low crop yields; but too much nitrogen can be toxic to plants, and can also harm our environment. Plants that do not have enough nitrogen become yellowish and do not grow well and can have smaller flowers and fruits. Farmers can add nitrogen fertilizer to produce better crops, but too much can hurt plants and animals, and pollute our aquatic systems. Understanding the Nitrogen Cycle—how nitrogen moves from the atmosphere to earth, through soils and back to the atmosphere in an endless Cycle—can help us grow healthy crops and protect our environment. Agricultural harvest depletes Nitrogen that is naturally found in the soil and hence Nitrogen, in terms of Ammonia, is applied to replenish the lost Nitrogen.[575] And this process repeats itself and is cyclic.

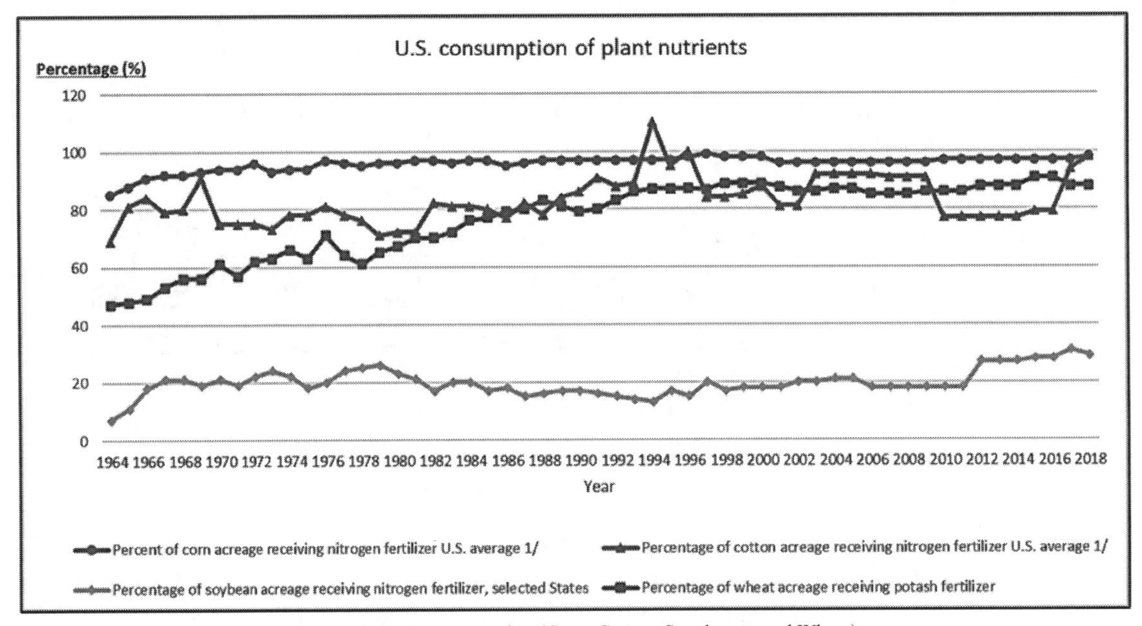

Figure 10: U.S. consumption (Corn, Cotton, Soyabeans, and Wheat).

It is intuitively obvious fertilizers are a national essential for crop life and clearly this is true for producing agricultural commodities. Given the crucial and central role that fertilizers play, the U.S. Department of Agriculture (USDA) is announcing its support with additional fertilizer production for American farmers to address rising costs, including the impact of Putin's price hike on farmers, and spur competition. USDA will make available $250 million[576] through a new grant program this summer to support independent, innovative, and sustainable American fertilizer production to supply American farmers with them [30].

The USDA announces plans for a $250 Million investment to support innovative American-made fertilizers to give US farmers more choices in the marketplace [30].

Recent supply chain disruptions due to the global pandemic and Putin's unprovoked war against Ukraine have shown just how important it is to invest in this crucial link in the agricultural supply chain here at home. The planned investment is one example of many Biden-Harris Administration initiatives to bring production and jobs back to the United States, promote competition, and support American goods and services.

U.S. Agriculture Secretary

The U.S. Dollar Index

The U.S. dollar index[577] is a measurement of the dollar's value relative to six foreign currencies as measured by their exchange rates [31]. Over half the index's value is represented by the dollar's value measured against the euro. The other five currencies include the Japanese yen, the British pound, the Canadian dollar, the Swedish krona, and the Swiss franc. The elimination of the gold standard fostered the birth of the U.S. Dollar Index.[578] Economic conditions in the U.S, and abroad can affect the dollar's index value. The competitiveness of agriculture [32] depends on the value of the dollar and I will cover more in the next section.

The dollar index began at 100 (please see Figure 11). The index has measured the percent change in the dollar's value since the establishment of its base value. Its all-time high was 163.83 on March 5, 1985. Its all-time low was 71.58 on April 22, 2008, 28.4% lower than at its inception. The distribution is bimodal with two recent peaks (2016 and 2020). Data distributions in statistics can have one peak, or they can have several peaks.[579] The type of distribution you might be familiar with is the normal distribution, or the bell curve, which has one peak [33]. The bimodal distribution has two peaks. The Dollar Index has two recent peaks (2006:2022). There are two peaks in the data, which usually indicates there are two different groups.

1. In 2016, two significant developments have caused it—the surprise win of the Republican candidate Donald Trump[580] in the US presidential elections and the rate hike by the Federal Reserve in its December 2016 policy meet left the currency soaring. The US dollar hit an 11-month peak on November 14, 2016, in the hope that Donald Trump will go on a US spending spree as he had said in the election campaign [34]. The dollar has surged to the highest level[581] since March 2003 according to the US economic data [35].
2. In 2020, the dollar rose until March 19 when it peaked at 102.82. Investors flocked to the safe-haven dollar in response to the COVID-19 pandemic. The Fed lowered the rate to zero in March. The USDX fell to its lowest of 89.63 for the year at closing on Dec. 30 as U.S. outbreaks grew worse. It ended the year slightly higher at 89.94 [31].

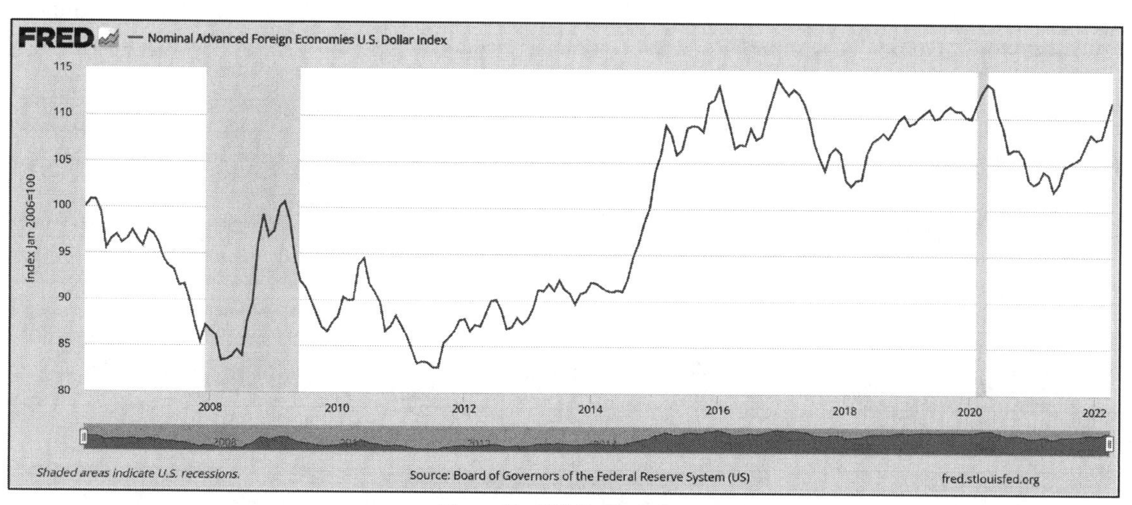

Figure 11: U.S. Dollar Index.

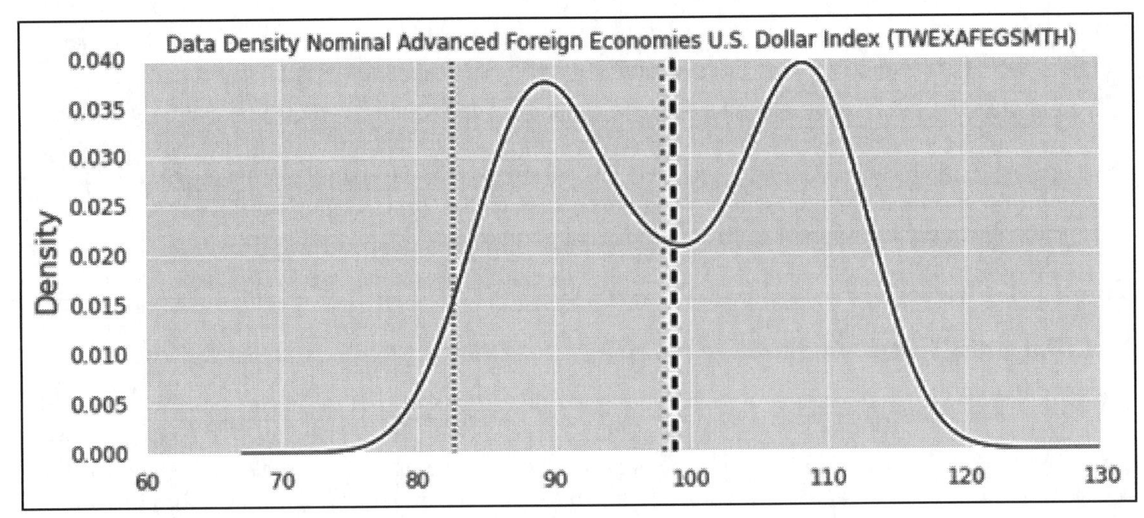

Figure 12: Data Distribution : Dollar Index.

The series is not stationary. Data Distribution statistics (please see Figure 12): Minimum:82.68; Mean:98.66 (dashed line); Median:97.97 (dash dot line); Mode:82.68 (dotted line); Maximum:114.01 (based on Linkage ML Model below). To summarize, the distribution of the data is skewed to the right—statistically, if the mode is less than the median, which is less than the mean. This kind of distribution is called *right skewed*. The mass of the data is on the left side of the distribution, creating a long tail to the right because of the values at the extreme high end, which pull the mean to the right. This is a fact that underscores the dollar's strength.

U.S. Dollar and Global Commodity Prices Linkage

There's normally an inverse[582] relationship between the value of the dollar and prices of commodities [36]. Historically, the prices of commodities have tended to drop when the dollar strengthens against other major currencies (please see the years 2016 and 2020 for commodity prices in the figure below), and when the value of the dollar weakens against other major currencies, the prices of commodities generally move higher (please see the years 2008 and 2011, for commodity prices in Figure 13 below). This is a general rule, and the correlation isn't perfect, but there's often a significant inverse relationship over time.

The primary reason the value of the dollar influences commodities prices is that the dollar is the benchmark pricing mechanism for most commodities and commodities are global assets. Commodities trade all over the world. Foreign buyers purchase U.S. commodities such as corn, soyabeans, wheat, and oil with

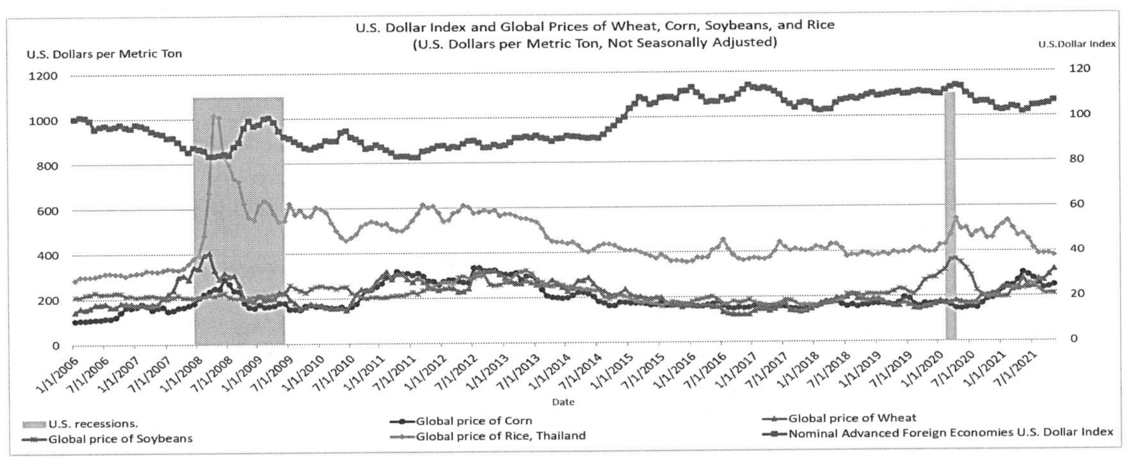

Figure 13: U.S. Dollar Index & Commodities.

dollars. When the value of the dollar drops, they have more buying power, because it requires lower amounts of their currencies to purchase each dollar [36]. Classic economics teaches that demand typically increases as prices drop.

Putting it simply, agricultural commodities are international trade assets and denominated in U.S. dollars. The principle goes as follows: a strong dollar implies that more of the other currencies must be spent to buy commodities. Of course, commodity demand goes down due to lower purchasing power. On the other hand, when the dollar weakens, more commodities can be bought with other currencies increasing the demand for and prices of commodities. The only concern is that huge swings in the dollar exchange rates affect food security adversely. Please see IMF Primary Commodity Prices Index[583] vs. U.S. dollar index (please see Figure 14).

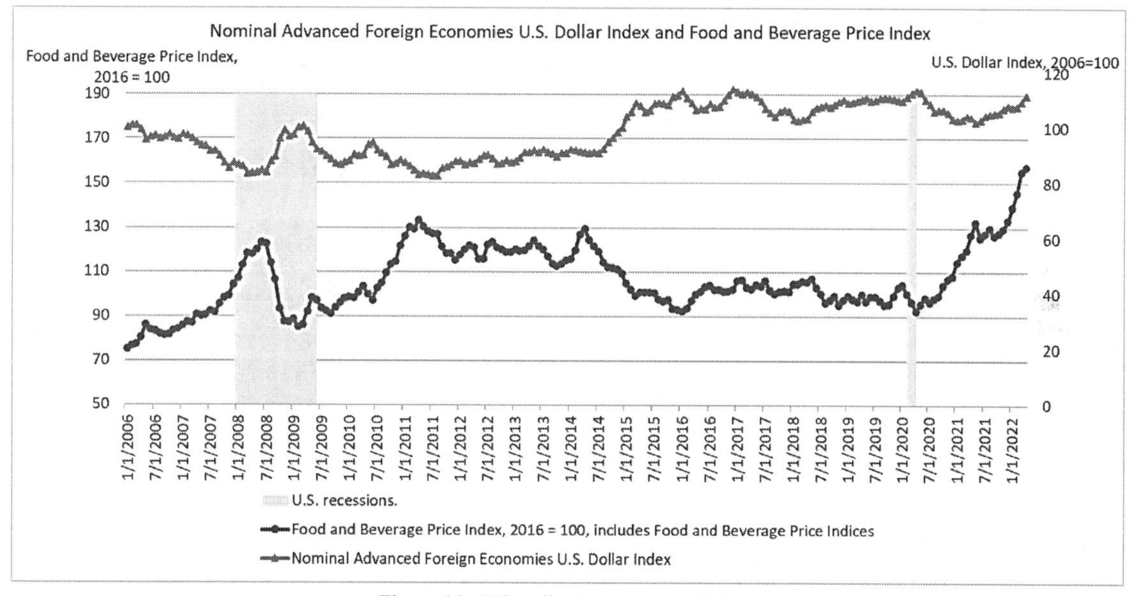

Figure 14: U.S Dollar Index vs. Food Prices.

Declining Exports with Dollar Appreciation

Agriculture plays an important role in the U.S. economy and as a matter of fact in the world economy as well. Agricultural exports support more than one million American jobs, with roughly 70 percent of these jobs in non-farming sectors, such as processing and agricultural based manufacturing.[584] Overall, U.S. farmers and ranchers export more than 20 percent of what they produce. In 2018, domestic agricultural exports reached nearly $145 billion, an increase of 1.4 percent over 2017. The U.S Administration will continue working to open new markets for safe, wholesome U.S. food and agricultural products to be enjoyed by consumers around the world. As mentioned, a major portion of U.S. agricultural production is exported to other countries [32]. So, our competitiveness in export markets is crucial to the stability and growth of U.S. agriculture. One of the fundamental factors in our competitiveness in export markets is the *currency exchange rate*. The currency exchange rate is the ratio of the value of a nation's currency to the value of another nation's currency. Many factors affect exchange rates, including the countries' macroeconomic policies, fiscal situation, and expected economic growth. Changes in the exchange rate affect our agricultural trade competitiveness because they indicate relative changes in the prices for traded goods in other countries. Nearly half of the change in the real value of U.S. agricultural exports can be attributed to changes in exchange rates.

Farm Inputs & Fertilizer Linkage Model

The Russian invasion of Ukraine substantially elevates the risk of disruptions in the global fertilizer trade. Russia is the world's largest exporter of fertilizers, accounting for 23% of ammonia exports, 14% of urea exports, 10% of processed phosphate exports, and 21% of potash exports, according to data from The

Fertilizer Institute. The primary destinations of fertilizers from Russia are Brazil (21%), China (10%), the US (9%), and India (4%). Compared to the US, Brazil will be affected more directly as Brazil imports 85% of its fertilizers. Supply in the US should be less of an issue as the US has robust domestic production. However, US farmers are likely to face higher prices because of the global interconnectedness of the global fertilizer industry.

Too many factors are influencing prices of the fertilizers, and this is the most important factor for farmers as fertilizer input costs[585] vary based on the crop. For instance, the 2021 Corn Crop fertilizer cost per acre is 35% ($ 116.57) of the total operating cost [37]. The cost of fertilizer inputs includes the cost of commercial fertilizers, soil conditioners, and manure. The gross value of production includes grain (primary product) and silage (secondary product). Corn recent cost return[586] data (CornCostReturm.xlsx)[587] provides cost variations since 1997. A notable observation, as per Figure 15, is that the fertilizer cost as a % of total operating cost peaked in 2008 due to the great recession.

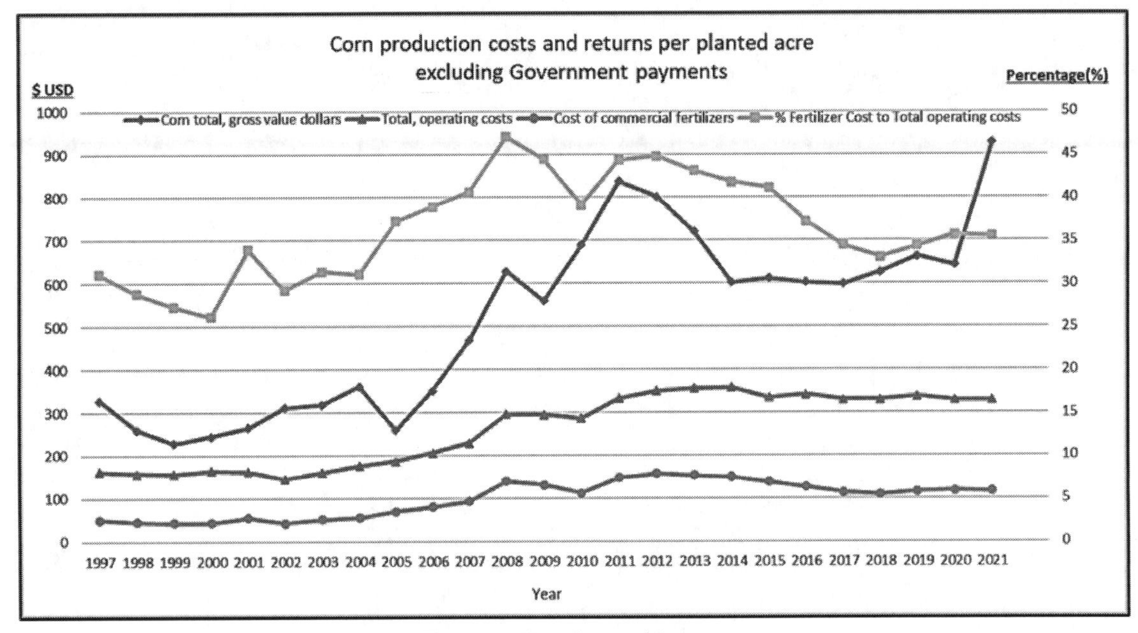

Figure 15: Corn Costs and Returns.

Similarly, the 2021 Cotton Crop fertilizer cost per acre is 16.04% ($ 68.96) of the total operating cost. [37]. The cost of fertilizer inputs includes the cost of commercial fertilizers, soil conditioners, and manure. The gross value of production includes cotton lint (primary product) and cottonseed (secondary product). Cotton recent cost return [38] data (CottonCostReturm.xlsx)[588] provides cost variations since 1997. A notable observation, as per Figure 16, is that the fertilizer cost as a % of total operating cost peaked in 2008 due to the great recession.

The wheat crop fertilizer cost per acre in 2021 is 34.48% ($ 43.63) of total operating cost [37]. The cost of fertilizer inputs includes the cost of commercial fertilizers, soil conditioners, and manure. The gross value of production includes grain (primary product) and silage/straw/grazing (secondary product). Wheat recent cost return [38] data (WheatCostReturm.xlsx)[589] provides cost variations since 1997. A notable observation, as per Figure 17, is that the fertilizer cost as a % of the total operating cost peaked in 2008 due to the great recession.

Finally, Soyabeans and Rice fertilizer cost per acre is approximately 20.00% of the total operating cost [37]. Net-net, the cost of fertilizer is a major cost component of farm inputs and significantly affects the profitability of farmers. Especially, for high input cost crops such as Corn and Wheat, the impact of

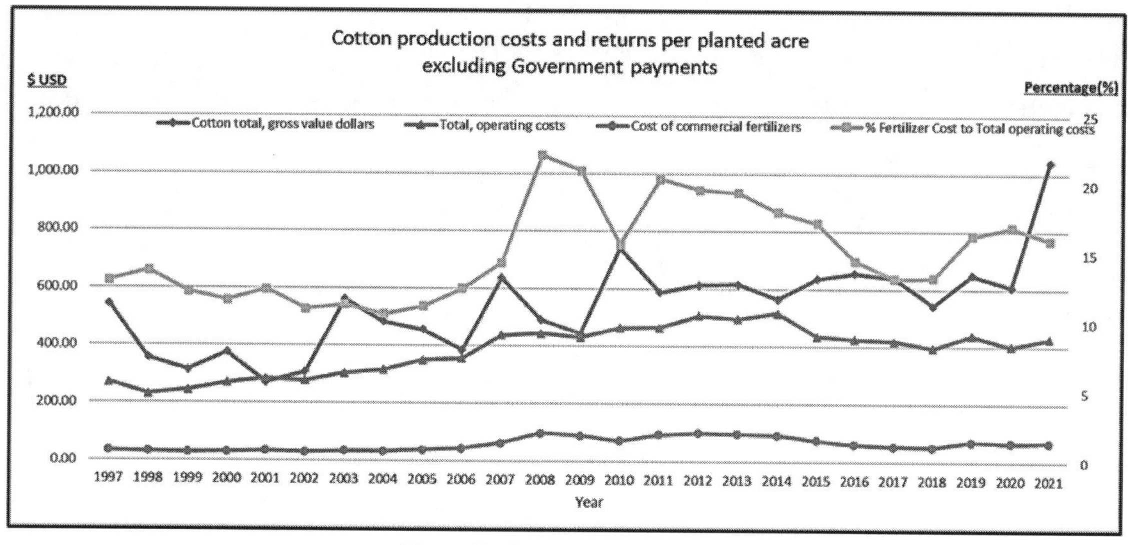

Figure 16: Cotton - Costs and returns.

Corn production costs and returns per planted acre, excluding Government payments
(dollars per planted acre, except where indicated)

U.S. total | Southern Seaboard | Prairie Gateway | Northern Great Plains | Northern Crescent | Heartland | Eastern Uplands

	Base survey of 2016						Base survey of 2010						
	2021	2020	2019	2018	2017	2016	2015	2014	2013	2012	2011	2010	2009
Gross value of production													
Primary product, grain	927.36	642.58	662.59	625.86	598.00	602.07	611.22	601.80	719.16	801.22	836.58	688.47	560.04
Secondary product, silage	2.34	2.13	2.35	2.25	2.03	1.85	1.38	1.38	1.35	1.33	1.19	0.92	1.18
Total, gross value of production	929.70	644.71	664.94	628.11	600.03	603.92	612.60	603.18	720.51	802.55	837.77	689.39	561.22
Operating costs													
Seed	91.42	91.84	93.48	95.96	97.07	98.36	101.62	101.04	97.59	92.04	84.37	81.58	78.92
Fertilizer [1]	116.57	116.93	115.86	108.97	113.46	126.53	137.33	149.23	153.33	156.51	147.36	112.03	131.11
Chemicals	31.48	32.63	34.01	34.02	34.77	35.65	27.95	29.20	28.57	27.52	26.35	26.29	27.83
Custom services [2]	23.49	22.94	22.74	22.48	22.05	22.69	19.04	18.24	17.77	17.07	16.77	16.36	11.76
Fuel, lube, and electricity	26.96	27.19	32.41	30.71	27.21	24.08	21.28	32.80	32.27	30.63	32.42	25.80	29.13
Repairs	37.96	35.56	35.13	33.91	32.75	32.20	26.18	26.17	25.79	25.48	24.79	23.96	15.69
Purchased irrigation water	0.28	0.28	0.29	0.27	0.26	0.26	0.12	0.12	0.12	0.11	0.10	0.11	0.14
Interest on operating capital	0.10	0.69	3.46	3.39	1.72	0.78	0.28	0.12	0.16	0.23	0.17	0.28	0.43
Total, operating costs	328.26	328.06	337.38	329.71	329.29	340.55	333.80	356.92	355.60	349.59	332.33	286.41	295.01
Allocated overhead													
Hired labor	5.64	5.33	5.25	4.85	4.58	4.49	3.28	3.16	3.12	3.02	2.92	2.96	2.41

Figure 17: Corn production costs.

high gas prices is extremely critical! All major nutrients used in the production of primary row crops in the U.S., nitrogen (in the forms of anhydrous ammonia, urea, or liquid nitrogen), phosphorus (diammonium phosphate—DAP and monoammonium phosphate—MAP) and potassium (potash), have experienced varying degrees of upward price pressures (please see Figure 18).

Please see the top 10 countries consuming Urea for agricultural use (FAO Stat → Land, Input, and Sustainability → Inputs → Fertilizers By Product → Select Urea)[590] (please see Figure 19).

Fertilizers are global commodities possibly influenced by multiple market factors beyond the control of U.S. producers. Like globally traded commodities, 44% of all fertilizer materials are exported to different countries. This factor has an exaggerated impact on fertilizer prices because production is not only influenced by events occurring in the producing country, but also by the demand in numerous other countries for fertilizer products and reasonable transportation rates till the desired destinations [21,37].

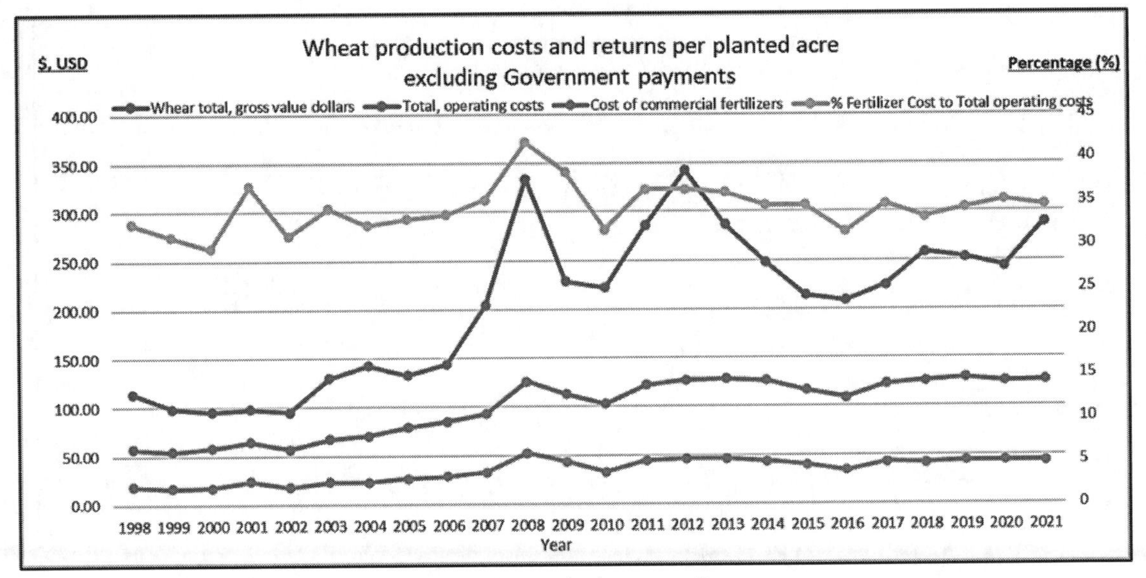

Figure 18: Wheat production costs and returns.

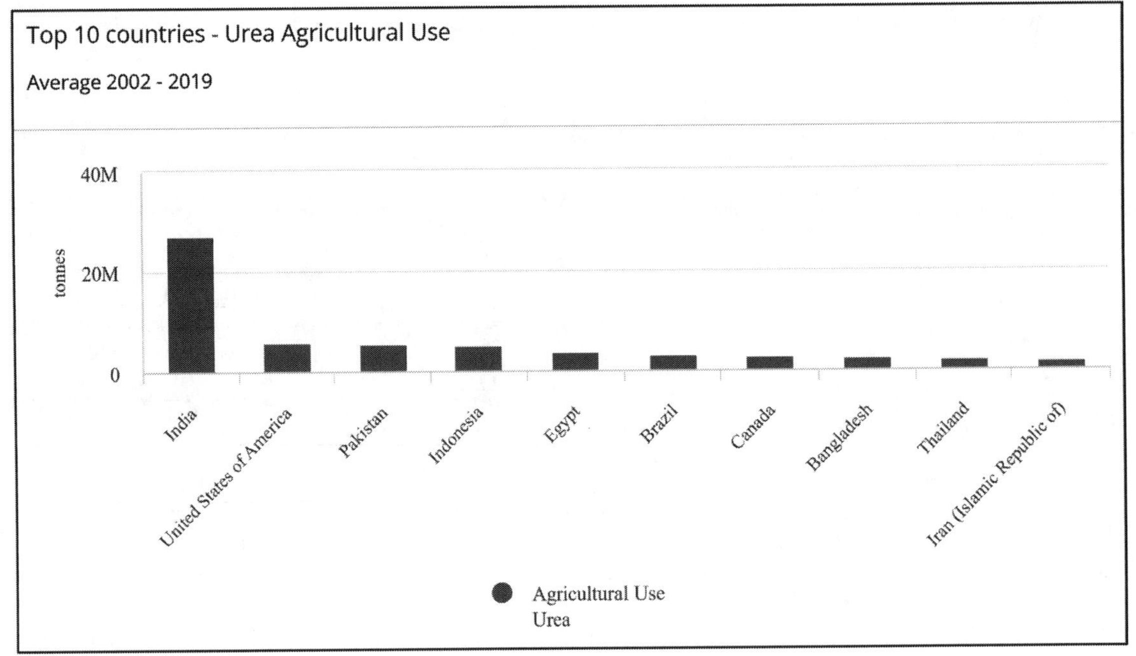

Figure 19: Top 10 countries - Urea Agricultural Use.

Urea

For instance, consider Urea distribution (please see Figure 20):

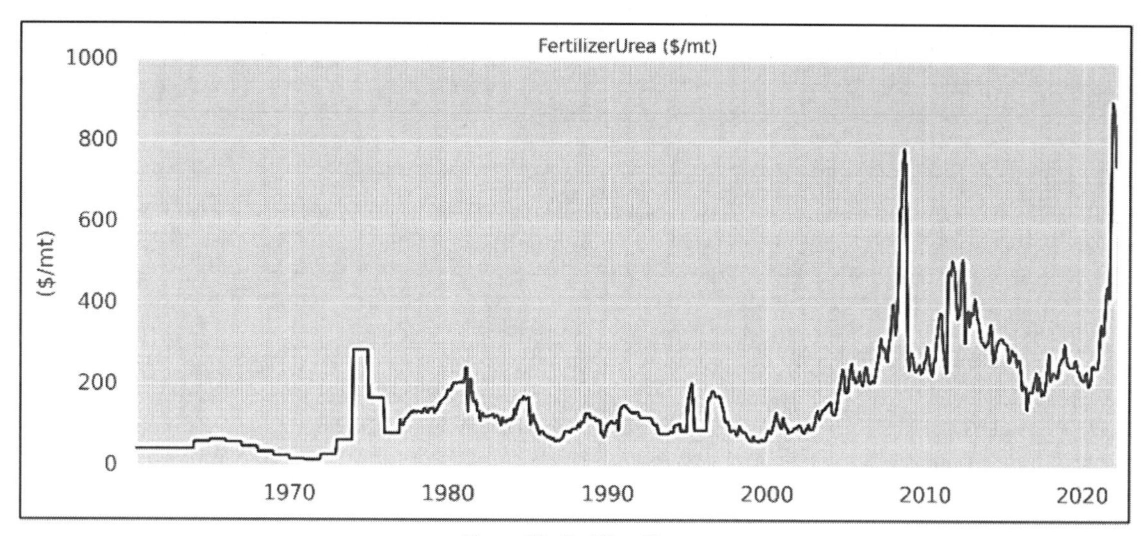

Figure 20: Fertilizer Urea.

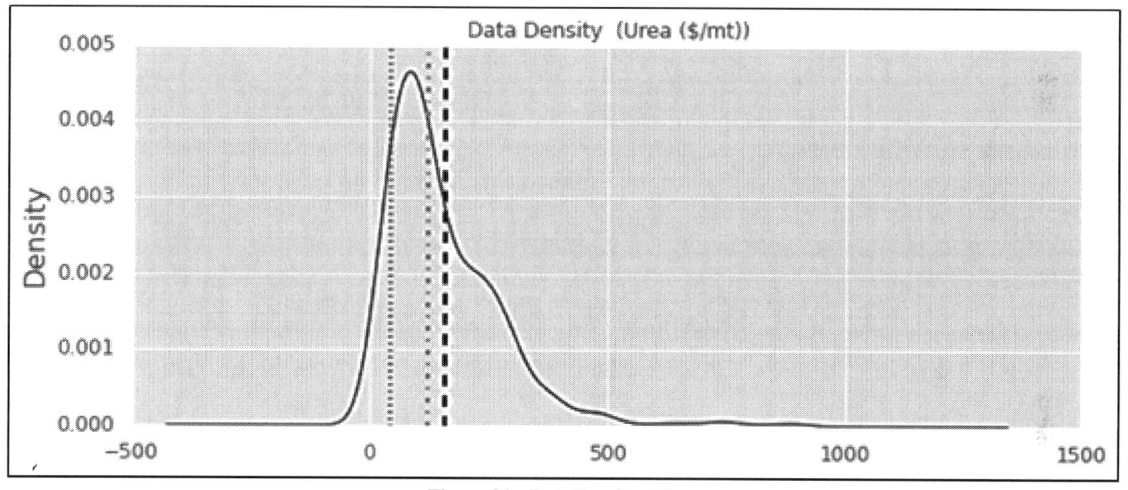

Figure 21: Data Density - Urea.

To summarize, the distribution of data is skewed to the right—statistically, if the mode 42.25 ($/mt) is less than the median 121.00 ($/mt), which is less than the mean 157.60 ($/mt). This kind of distribution is called *right skewed*. The mass of the data is on the left side of the distribution, creating a long tail to the right because of the values at the extreme high end, which pull the mean to the right. A fact that underscores the spike in Urea prices (especially after the great recession 2008) as can be seen in Figure 21 above.

Data Distribution statistics: Minimum:16.00 ($/mt); Mean:157.60 ($/mt) (dashed line); Median:121.00 ($/mt) (dash dot line); Mode:42.25 ($/mt) (dotted line); Maximum:900.50 ($/mt) (based on Linkage ML Model below). Statistical significance of the mean shifting to the right implies ~~more~~ higher food prices and more food insecurity. For instance, In the U.S., some corn growers are seeing prices that are more than double of what they paid last year. Across Brazil, about a third of the nation's coffee farmers don't have enough fertilizers.[591] In Thailand, some rice farmers are calling on the government to intervene in the spiraling market. With fertilizer markets now seeing unprecedented supply shocks and record prices, it means even more food inflation across the world [39].

Rising fertilizer prices threaten to exacerbate food inflation.

Phosphate Rock

Phosphate Rock which is mined demonstrates different signatures unlike Urea that is Nitrogen based and manufactured across the world, hence prices of Urea demonstrate co-movement with Oil & gas prices. Hence, we tend to observe less fluctuations in the Phosphate production data movement. Production refers to the beneficiated phosphate rock product of phosphate ore ranging in a 26 percent to about 34 percent P2O5 grade. This production is commonly referred to as marketable phosphate rock. Production data for 1900–14, 1932, and 1953 is equal to sold or used data. Data is from the Mineral Resources of the United States (MR) and the Minerals Yearbook (MYB).[592] Worldwide, more than 85 percent of the phosphate rock mined is used to manufacture phosphate fertilizers. The remaining 15 percent is used to make elemental phosphorus and animal feed supplements, or is applied directly to soils. Elemental phosphorus is used to manufacture a wide range of chemical compounds.

Here are the facts[593] [40]:

- The U.S. production of phosphate rock in 2012 was 30.1 million metric tons, valued at $3.08 billion.
- Total world production of phosphate rock in 2012 was 233 million metric tons (please see Figure 22). China was the leading producer, with 41 percent of world production, followed by the United States, Morocco and Western Sahara.
- China, the United States, Morocco, Russia and India are the leading consumers of phosphate rock.
- Morocco has 50 billion metric tons of phosphate rock reserves, which is about 75 percent of the world's total.

As it can be seen from the above figure, the randomness of price changes for Phosphate rock became evident only after the great recession of 2008. Prior to the recession it was a very stable commodity.

So, from the statistical distribution point of view, the spread (standard deviation) should demonstrate lesser variance as compared to Urea.

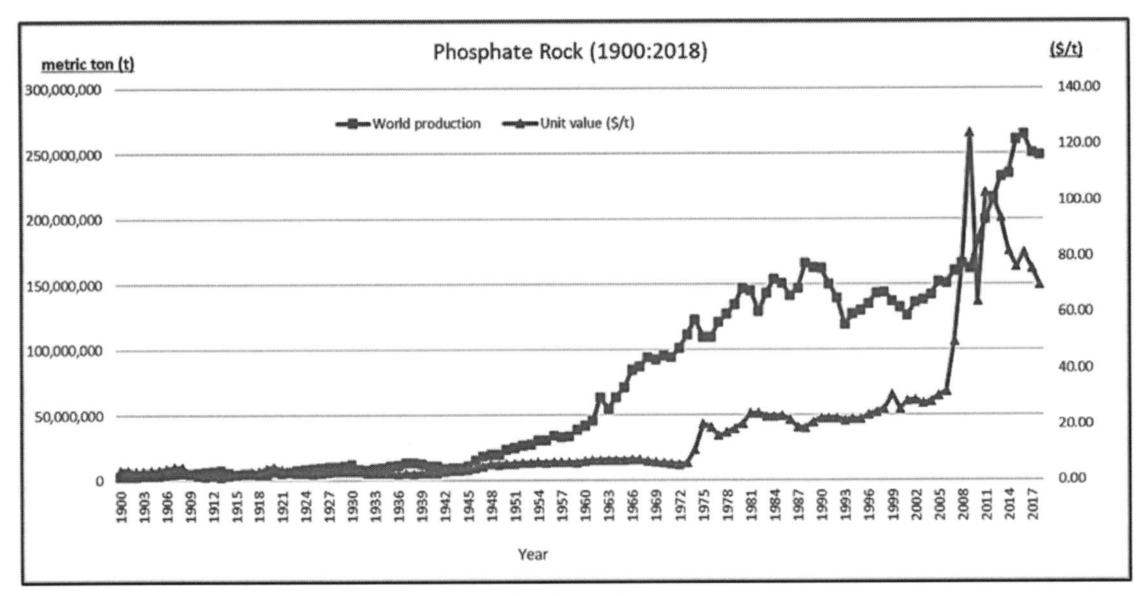

Figure 22: Phosphate Rock (1900 :2018).

The series is stationary (please see Figure 23). Data Distribution statistics: Minimum:11.00($/mt); Mean:54.97 ($/mt) (dashed line); Median:36.00 ($/mt) (dash dot line); Mode:44.00 ($/mt) (dotted line); Maximum:450.0 ($/mt) (based on Linkage ML Model below).

- Increased Global Fertilizer Demand
- A multitude of Supply Factors
  - o Domestic Production vs. Imports
  - o Energy and Other Variable Costs Rising
  - o Pivots in Fertilizer Demand Outlook Impacting Production
  - o Distribution and Supply Chain Disruptions
  - o Trade Duties
  - o Other Geopolitics

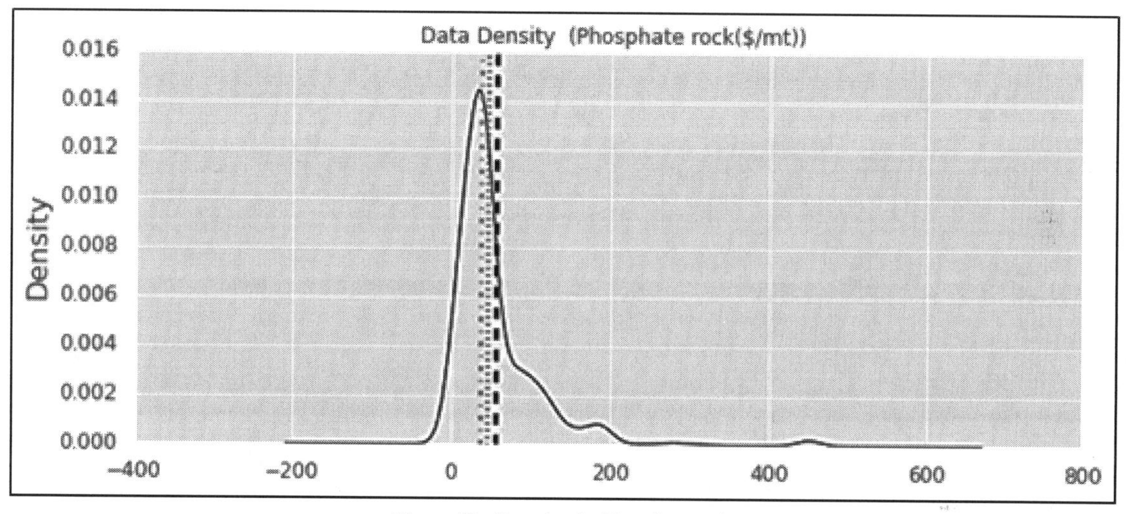

Figure 23: Data density Phosphate rock.

Machine Learning Model Energy Prices & Fertilizer Costs—Urea

The purpose of our model is to develop a linkage between Urea & Inputs.

	UREA_EE_BULK Fertilizer Time Series.ipynb (Jupyter Notebook Code)

Data Sources

Source	Time Frequency	Source
Pink Sheet Commodities Data	Monthly Commodity Prices[594] Commodities Data from World Bank Pink sheet. Commodity Prices - Crude Oil, Commodities (Corn, Wheat, Cotton, Soybeans, Rice, Soybean Oil), Exchange Rate (FX Rate), Crude oil, average, Urea, TSP, DAP, Phosphate rock, Corn, Wheat, Cotton, Soybeans, Rice, Soybean Oil, Vegetable Oil, and Exchange Rate (FX Rate)	THE WORLD BANK IBRD · IDA
St. Louis FRED U.S. Dollar Index	Nominal Advanced Foreign Economies U.S. Dollar Index (TWEXAFEGSMTH)[595]	FRED ECONOMIC DATA ST. LOUIS FED

Step 1: Load Libraries for performing Time Series & ML Modeling

Load Scikit Learn and Matplot Libraries.

```
import pandas as pd
import numpy as np
import plotly.graph_objects as go
import seaborn as sns; sns.set()
import matplotlib.pyplot as plt
from scipy import stats
from scipy.stats import pearsonr
import matplotlib.pyplot as plt
from sklearn.preprocessing import MinMaxScaler
from sklearn.model_selection import train_test_split
from sklearn.linear_model import LinearRegression
from sklearn.ensemble import RandomForestClassifier
from sklearn.metrics import r2_score
from sklearn import metrics
import eli5
import matplotlib.pyplot as plt
import matplotlib as mpl
import numpy as np
import scipy.stats as spstats
import seaborn as sns
import utils
%matplotlib inline
mpl.style.reload_library()
mpl.style.use('classic')
mpl.rcParams['figure.facecolor'] = (1, 1, 1, 0)
mpl.rcParams['figure.figsize'] = [6.0, 4.0]
mpl.rcParams['figure.dpi'] = 100
```

Step 2: Load Commodities Pricing Data

Load Commodities prices data.[596] Commodity markets are integral to the global economy. Understanding what drives developments of these markets is critical to the design of policy frameworks that facilitate the economic objectives of sustainable growth, inflation stability, poverty reduction, food security, and the mitigation of climate change.

```
CMOHistoricalDataDF = pd.read_csv("CMO_historical_data.csv")
CMOHistoricalDataDF
```

Output:
The focus is on natural gas, crude petroleum, maize, sorghum, rice, wheat, and other commodities.

	Year	NGAS_US US ($/mmbtu)	SOYBEANS ($/mt)	CRUDE_PETRO ($/bbl)	MAIZE ($/mt)	SORGHUM ($/mt)	RICE_05 ($/mt)	WHEAT_US_SRW ($/mt)	WHEAT_US_HRW ($/mt)	UREA_EE_BULK ($/mt)
0	197901	1.0200	284.00	17.45	105.410000	97.06	278.52	140.360000	137.79	143.00
1	197902	1.0500	298.00	20.75	107.010000	97.83	279.50	144.770000	140.36	143.00
2	197903	1.1000	310.00	22.02	109.520000	99.21	293.00	144.040000	139.99	131.88
3	197904	1.1100	300.00	22.43	111.490000	98.27	295.46	142.200000	139.99	135.00
4	197905	1.1500	300.00	33.50	112.570000	100.49	296.78	141.430000	143.30	141.75
...	...	...	...	...	...	...	...	...	...	...
515	202112	3.7327	554.14	71.53	264.537213	189.49	400.00	327.819675	376.81	890.00
516	202201	4.3325	606.22	83.12	276.623189	189.49	427.00	332.059894	374.24	846.38
517	202202	4.6577	661.63	91.74	292.622344	189.49	427.00	533.121257	390.50	744.17
518	202203	4.8839	720.60	108.49	335.529527	189.49	422.00	533.121257	486.30	872.50
519	202204	6.5306	720.79	101.78	348.166655	189.49	431.00	672.456928	495.28	925.00

520 rows × 10 columns

The following commodities are loaded:
Energy commodities:

- Crude oil, US, West Texas Intermediate (WTI) 40` API
- Natural Gas (U.S.), spot price at Henry Hub, Louisiana

Grains

- Maize (U.S.), no. 2, yellow, f.o.b. US Gulf ports
- Rice (Thailand), 5% broken, white rice (WR), milled, indicative price based on weekly surveys of export transactions, government standard, F.O.B. Bangkok
- Wheat (U.S.), no. 2 hard red winter Gulf export price; June 2020 backwards, no. 1, hard red winter, ordinary protein, export price delivered at the US Gulf port for prompt or 30 days shipment
- Wheat (U.S.), no. 2, soft red winter, export price delivered at the US Gulf port for prompt or 30 days shipment

Other Raw Materials

- Cotton (US), Memphis/Eastern, middling 1-3/32 inch, Far East , C/F beginning October 2008; previously C.I.F. Northern Europe

Fertilizers

- DAP (diammonium phosphate), spot, F.O.B. US Gulf
- Urea, (Ukraine), prill spot f.o.b. Middle East, beginning March 2022; previously, F.O.B. Black Sea.
- Phosphate rock, F.O.B. North Africa
- Potassium chloride (muriate of potash), F.O.B. Vancouver

Step 3: Load U.S. Dollar Index (TWEXAFEGSANL)

Load Nominal Advanced Foreign Economies U.S. Dollar Index Nominal Advanced Foreign Economies U.S. Dollar Index (TWEXAFEGSMTH).[597]

```
CMOHistoricalDataDF = pd.read_csv("CMO_historical_data.csv")
CMOHistoricalDataDF
```

Output:

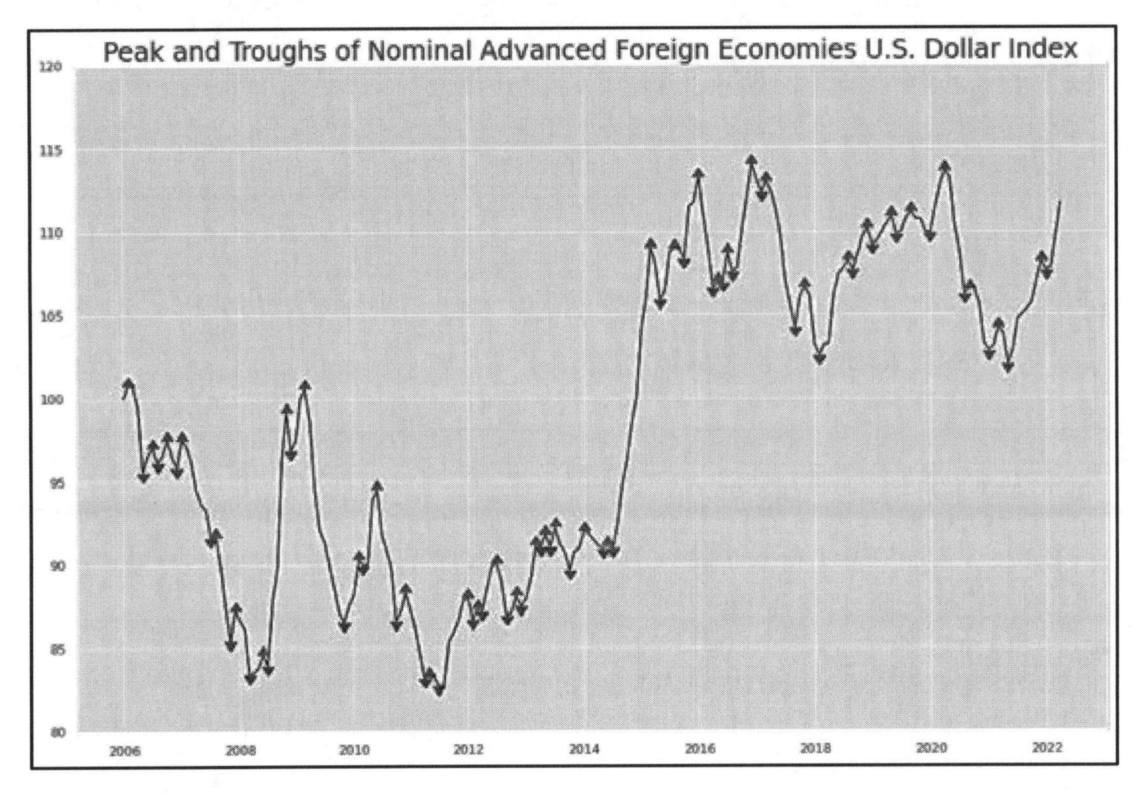

Step 4: Load Fertilizer Data (TWEXAFEGSANL)

Load Fertilizers data.[598]

```
def plottimeseries(column_name,var_ts, var_data):
   fig = go.Figure()
   # Add traces
   fig.add_trace(go.Scatter(x=var_ts, y=var_data,
      mode='lines',
      name= column_name))
   fig.update_layout( title=column_name)
   fig.show()
```

```
df_fert = pd.read_csv("Fertilizers.csv")
df_fert
fertilizers_numeric_features=['Phosphate rock($/mt)','DAP($/mt)','TSP($/mt)','Urea ($/mt)']
for col in fertilizers_numeric_features:
  print(col)
  # Call the function
  plottimeseries(col, df_fert.Date,df_fert[col])
```

Output:
Distribution diagrams of all fertilizers used in the model.

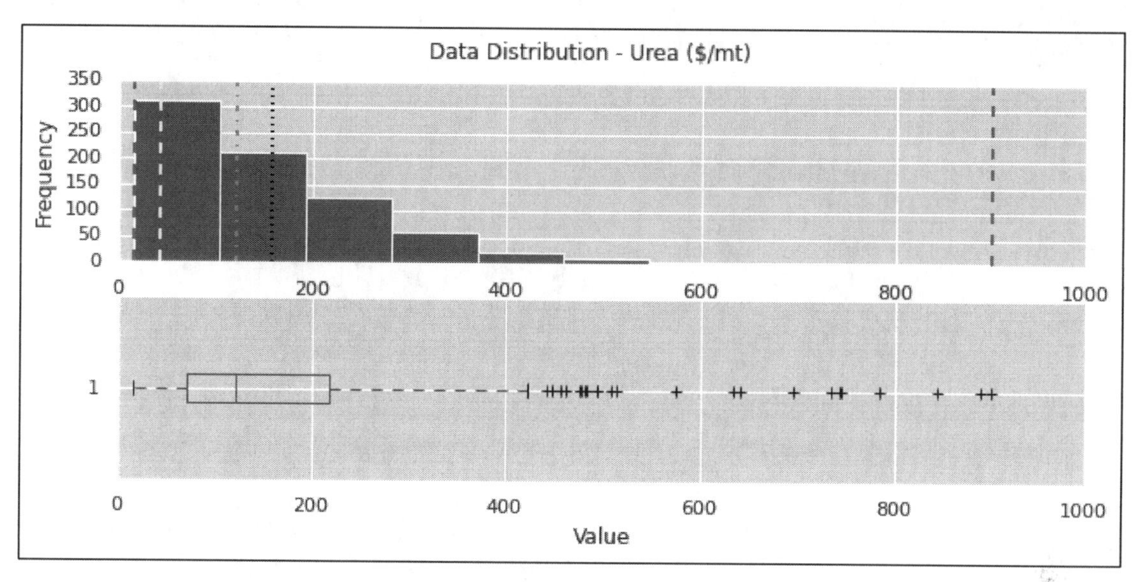

Step 5: Combine Data Frames (Pink Sheet, Fertilizers, USD Dollar Index, and Oil)
Merge data frames. Combine Pink Sheet data and USD Index.

```
df2= CMOHistoricalDataDF.merge(dollar_index , on = "Date")
df2
df3 = df2.merge(pinkdf_oil, on = "Date")
df3
df4 = df3.merge(df_fert, on ="Date")
df4
df4.drop(["Unnamed: 0","Urea ($/mt)"], axis = 1, inplace = True)
CMOHistoricalDataDF1 = df4
CMOHistoricalDataDF.isnull().sum() ## missing values
```

Output:

```
NGAS_US US ($/mmbtu) 0
SOYBEANS ($/mt) 0
CRUDE_PETRO ($/bbl) 0
MAIZE ($/mt) 0
SORGHUM ($/mt) 0
RICE_05 ($/mt) 0
WHEAT_US_SRW ($/mt) 0
WHEAT_US_HRW ($/mt) 0
UREA_EE_BULK ($/mt) 0
dtype: int64
```

Step 6: Correlation Analysis
Perform correlation analysis to see the relationship between commodities, USD Index, and Oil.

Merge data frames. Combine Pink Sheet data and USD Index.

```
plt.figure(figsize=(12, 9))
sns.heatmap(CMOHistoricalDataDF1.corr())
```

Output:

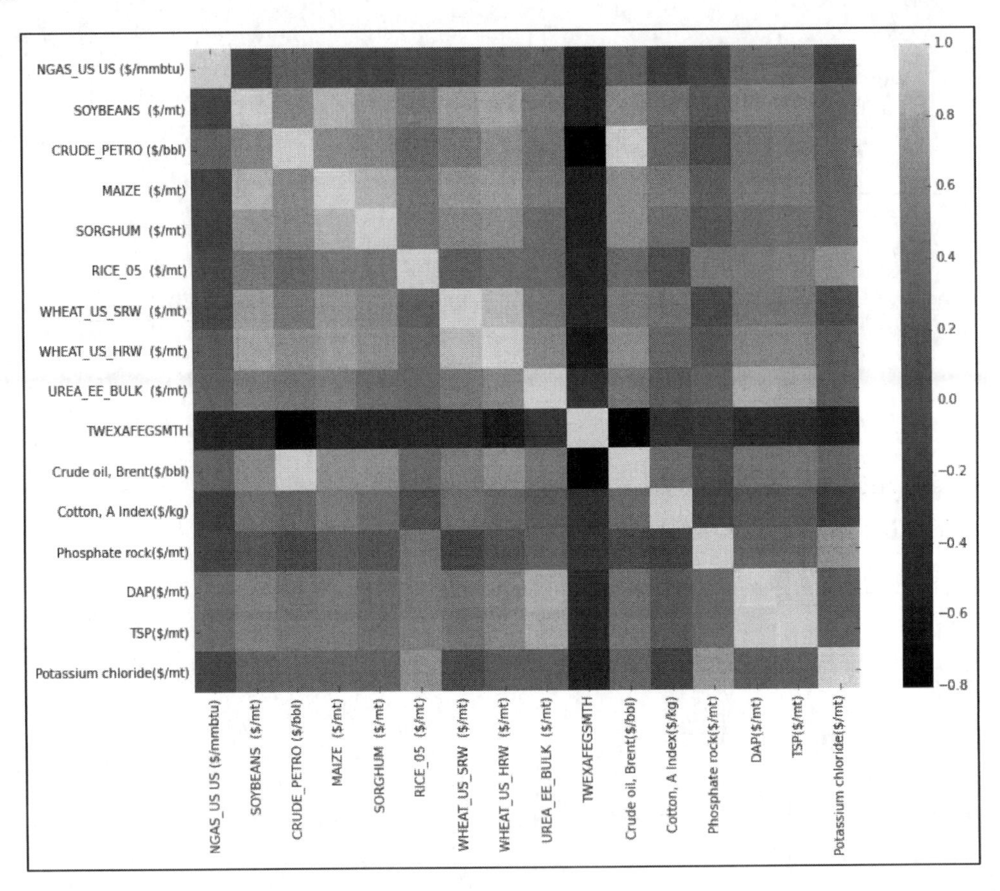

Price of UREEA_EE_BULK ($/mt) is correlated as follows: 0.366026 (GAS_US US ($/mmbtu)), **0.604525(SOYBEANS ($/mt))**, 0.551770 (CRUDE_PETRO ($/bbl)), 0.604408 (MAIZE ($/mt)), 0.466162 (SORGHUM ($/mt)), 0.440983 (RICE_05 ($/mt)), 0.560462 (WHEAT_US_SRW ($/mt)), **0.664282 (WHEAT_US_HRW ($/mt))**, **-0.341174 (TWEXAFEGSMTH)**, 0.546572 (Crude oil, Brent($/bbl)), 0.309936 (Cotton, A Index($/kg)), 0.391608 (Phosphate rock($/mt)), **0.831219 (DAP($/mt))**, **0.810773 (TSP($/mt))** and 0.324036 (Potassium chloride($/mt)).

- Phosphate Rock is highly correlated (0.8312) with Diammonium phosphate (DAP). It is no wonder that the DAP is based on the Phosphate component, the relationship justifies. DAP is the most concentrated phosphate-based fertilizer. It is perfect for any agriculture crop to provide full phosphorus nutrition throughout crop growth and development, as well as a starter dose of nitrogen and low Sulphur.
- Phosphate Rock is highly correlated (0.8107) with Triple superphosphate (TSP) and is one of the first high-analysis phosphorus (P) fertilizers that became widely used in the 20th century. Technically, it is known as calcium dihydrogen phosphate and as monocalcium phosphate [$Ca(H_2PO_4)_2$. H_2O].
- Demand of Soybeans (0.6045) and Hard Red Winter Wheat (0.6442) correlate highly with Phosphate.
- Finally, Phosphate Rock negatively correlates with the U.S. Dollar Index.

Demand plays an important role in fertilizer prices. The price relationship between Phosphorous fertilizer and price is as follows: fertilizer market policies, doubled the import of P-fertilizer, in India, the largest global importer of phosphorus fertilizers and phosphate rock and turned out to be a major contributor to the global price spike.[599] Second, the price spike was magnified on the one hand by protective trade measures of fertilizer suppliers leading to a 19% drop in global phosphate fertilizer exports. On the other hand, the Indian fertilizer subsidy scheme led to farmers not adjusting their demand for fertilizer. The triggering mechanism appeared to be the Indian production outage of P-fertilizer resulting in the additional import demand for DAP in a quantity of about 20% of the annual global supply [41].

Step 7: Exploratory Data Analysis (WDA)

Check for nulls or data issues.

```
dataset.isnull().sum() ## missing values
```

Output:

```
NGAS_US US ($/mmbtu)   0
SOYBEANS ($/mt)   0
CRUDE_PETRO ($/bbl)   0
MAIZE ($/mt)   0
SORGHUM ($/mt)   0
RICE_05 ($/mt)   0
WHEAT_US_SRW ($/mt)   0
WHEAT_US_HRW ($/mt)   0
UREA_EE_BULK ($/mt)   0
TWEXAFEGSMTH   0
Crude oil, Brent($/bbl)   0
Cotton, A Index($/kg)   0
Phosphate rock($/mt)   0
DAP($/mt)   0
TSP($/mt)   0
Potassium chloride($/mt)   0
dtype: int64
```

Step 8: Plot dataset

Plot dataset.

```
# Plot
fig, axes = plt.subplots(nrows=4, ncols=4, dpi=120, figsize=(10,16))
for i, ax in enumerate(axes.flatten()):
  data = dataset[dataset.columns[i]]
  ax.plot(data, color='red', linewidth=1)
  ax.set_title(dataset.columns[i])
  ax.xaxis.set_ticks_position('none')
  ax.yaxis.set_ticks_position('none')
  ax.spines["top"].set_alpha(0)
  ax.tick_params(labelsize=6)
plt.tight_layout();
```

Output:

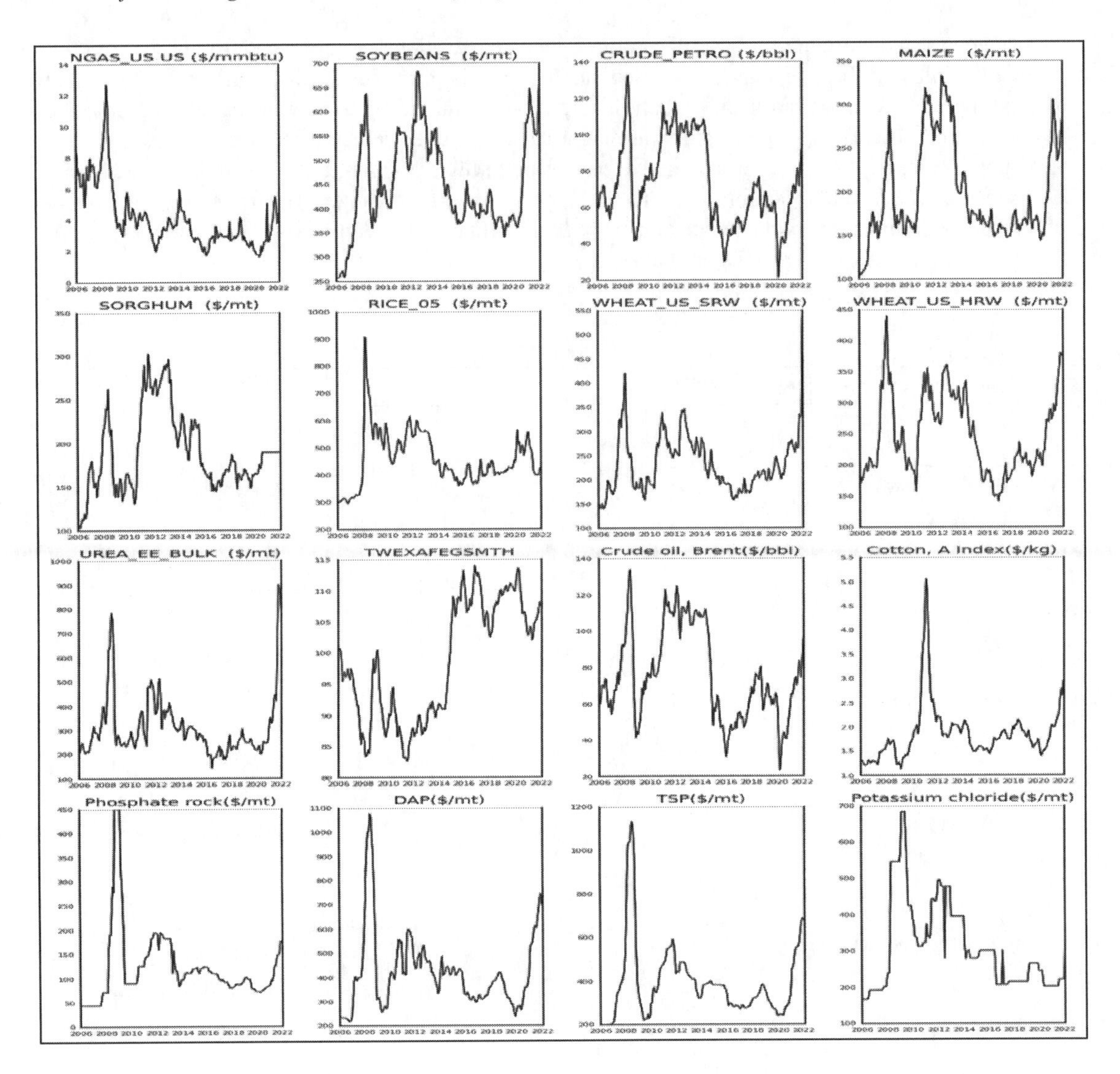

Step 9: Normality Test

To extract the maximum information from our data, it is important to have a normal or Gaussian distribution of the data. To check for that, we have done a normality test based on the Null and Alternate Hypothesis intuition. Augmented Dickey-Fuller tests[600] make strong assumptions about the data. They can only be used to inform the degree to which a null hypothesis can be rejected or fail to be rejected. The result must be interpreted for a given problem to be meaningful [42]. Nevertheless, they can provide a quick check and confirmatory evidence that your time series is stationary or non-stationary. The Augmented Dickey-Fuller test is a type of statistical test called a unit root test. The intuition behind a unit root test is that it determines how strongly a time series is defined by a trend. There are several unit roots tests and the Augmented Dickey-Fuller may be one of the more widely used. It uses an autoregressive model and optimizes information criterion across different multiple lag values.

```
from statsmodels.tsa.stattools import adfuller
def stationary_test(df,col):
   print("Observations of Dickey-fuller test")
   dftest = adfuller(df[col],autolag='AIC')
   dfoutput=pd.Series(dftest[0:4],index=['Test Statistic','p-value','#lags used','number of observations used'])
   for key,value in dftest[4].items():
      dfoutput['critical value (%s)'%key]= value
   print(dfoutput)
   if dfoutput[0] < dfoutput[4] or dfoutput[0] < dfoutput[5] or dfoutput[0] < dfoutput[6]:
      print(f'{col}: The series is stationary ")
   else :
      print(f'{col}: The series is not stationary ")
```

```
for i in dataset.columns:
  stationary_test(dataset,i )
```

- p-value > 0.05: Fail to reject the null hypothesis (H0), the data has a unit root and is non-stationary.
- p-value <= 0.05: Reject the null hypothesis (H0), the data does not have a unit root and is stationary.

Output:

```
Observations of Dickey-fuller test
Test Statistic                    –2.852234
p-value                            0.051191
#lags used                         0.000000
number of observations used      193.000000
critical value (1%)               –3.464694
critical value (5%)               –2.876635
critical value (10%)              –2.574816
dtype: float64
NGAS_US US ($/mmbtu): The series is stationary
```

```
Observations of Dickey-fuller test
Test Statistic                    –2.920258
p-value                            0.043042
#lags used                         1.000000
number of observations used      192.000000
critical value (1%)               –3.464875
critical value (5%)               –2.876714
critical value (10%)              –2.574859
dtype: float64
CRUDE_PETRO ($/bbl): The series is stationary
```

In summary, the following is the stationarity of the variables:

- GAS_US US ($/mmbtu): The series is stationary
- SOYBEANS ($/mt): The series is **not stationary**
- CRUDE_PETRO ($/bbl): The series is stationary
- MAIZE ($/mt): The series is **not stationary**
- SORGHUM ($/mt): The series is **not stationary**
- RICE_05 ($/mt): The series is stationary
- WHEAT_US_SRW ($/mt): The series is **not stationary**
- WHEAT_US_HRW ($/mt): The series is **not stationary**
- TWEXAFEGSMTH: The series is **not stationary**
- Crude oil, Brent($/bbl): The series is stationary
- Cotton, A Index($/kg): The series is stationary
- Phosphate rock($/mt): The series is stationary
- DAP($/mt): The series is stationary
- TSP($/mt): The series is stationary
- Potassium chloride($/mt): The series is **not stationary**

Perform Kurtosis analysis:

```
import matplotlib.pyplot as plt
import scipy.stats as stats
from scipy.stats import kurtosis
dataset['WHEAT_US_HRW ($/mt)'].plot(kind =
'density')
print('WHEAT_US_HRW: Kurtosis of
normal distribution: {}'.format(stats.
kurtosis(dataset['WHEAT_US_HRW ($/mt)'])))
print('WHEAT_US_HRW:Skewness
of normal distribution: {}'.format(stats.
skew(dataset['WHEAT_US_HRW ($/mt)'])))
```

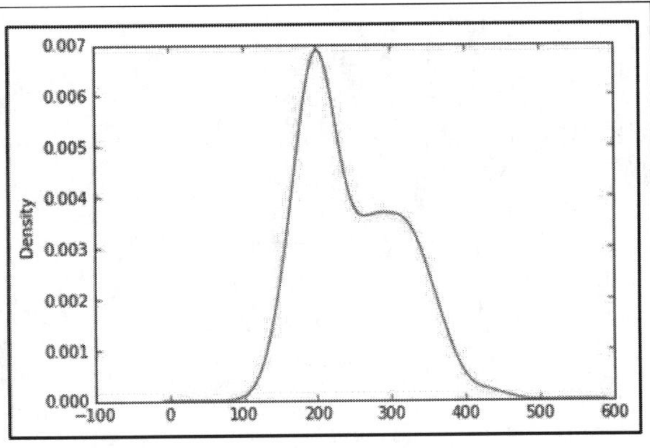

Output:
WHEAT_US_HRW: Kurtosis of normal distribution: -0.6584650413709592
WHEAT_US_HRW:Skewness of normal distribution: 0.5345253691699535

```
import matplotlib.pyplot as plt
import scipy.stats as stats
from scipy.stats import kurtosis
dataset['SORGHUM ($/mt)'].plot(kind =
'density')
print('SORGHUM: Kurtosis of
normal distribution: {}'.format(stats.
kurtosis(dataset['SORGHUM ($/mt)'])))
print('SORGHUM:Skewness of
normal distribution: {}'.format(stats.
skew(dataset['SORGHUM ($/mt)'])))
```

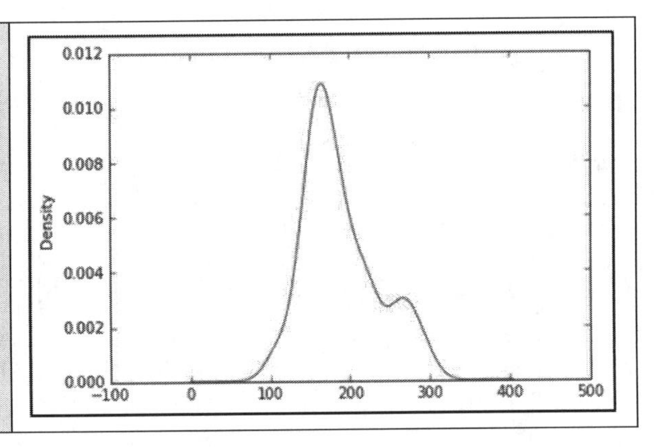

Output:
SORGHUM: Kurtosis of normal distribution: -0.24615643358193573
SORGHUM:Skewness of normal distribution: 0.676617922840901

These WHEAT_US_HRW, SORGHUM, CRUDE_PETRO, and SOYBEANS distributions give us some intuition about the distribution of our data. The kurtosis of this dataset is less than 0, it is a light-tailed dataset. It has as much data in each tail as it does in the peak. Moderate skewness refers to the values between –1 and –0.5 or 0.5 and 1.

UREA_EE_BULK with kurtosis (6.1235), WHEAT_US_SRW with 2.96 Kurtosis, RICE_05 with 2.76 Kurtosis, NGAS_US with 2.69 Kurtosis indicates a large departure from normality. Very small values of kurtosis also indicate a deviation from normality,[601] but it is a very benign deviation [43].

Step 10: Normality Probability Plot

Plot Normal probability to see how the data looks:

```
plt.figure(figsize=(14,6))
plt.subplot(1,2,1)
dataset['UREA_EE_BULK ($/mt)'].hist(bins=30)
plt.title('UREA_EE_BULK ($/mt)')
plt.subplot(1,2,2)
stats.probplot(dataset['UREA_EE_BULK ($/mt)'], plot=plt);
dataset['UREA_EE_BULK ($/mt)'].describe().T
```

Output:

```
count  194.000000
mean   312.700876
std    131.812786
min    142.630000
25%    234.220000
50%    269.000000
75%    334.375000
max    900.500000
```

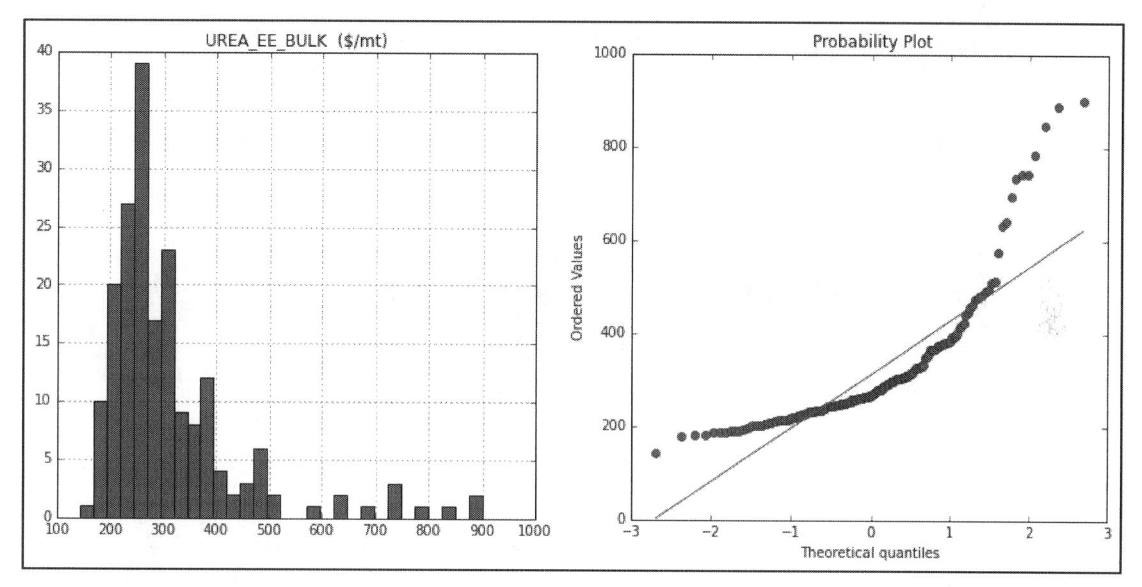

Normal probability plot also shows the data is far from normally distributed.

Step 11: Model Urea and other key drivers

Keep only UREA specific influencers.

```
dataset_all_fertilizers=dataset.copy()
dataset=dataset_all_fertilizers.copy()
# only UREA
dataset.drop(['Phosphate rock($/mt)', 'DAP($/mt)', 'TSP($/mt)','Potassium chloride($/mt)'], axis = 1,
inplace = True)
```

```
import matplotlib.pyplot as plt
import pandas as pd
import numpy as np
import seaborn as sns; sns.set()
corr = dataset.corr()
fig, ax = plt.subplots()
sns.heatmap(corr,xticklabels=corr.columns.values, yticklabels=corr.columns.values, annot=True,annot_
kws={'size':12})
heat_map=plt.gcf()
heat_map.set_size_inches(18,10)
plt.xticks(fontsize=10)
plt.yticks(fontsize=10)
plt.show()
```

Output:

Urea is highly correlated with Soybeans, Maize, and Wheat HRW & negatively correlated with the U.S. Dollar Index.

	NGAS_US US ($/mmbtu)	SOYBEANS ($/mt)	CRUDE_PETRO ($/bbl)	MAIZE ($/mt)	SORGHUM ($/mt)	RICE_05 ($/mt)	WHEAT_US_SRW ($/mt)	WHEAT_US_HRW ($/mt)	UREA_EE_BULK ($/mt)	TWEXAFEGSMTH	Crude oil, Brent($/bbl)	Cotton, A Index($/kg)
NGAS_US US ($/mmbtu)	1	0.025	0.38	0.0076	-0.052	0.13	0.13	0.31	0.37	-0.5	0.35	-0.13
SOYBEANS ($/mt)	0.025	1	0.67	0.89	0.76	0.59	0.8	0.82	0.6	-0.49	0.68	0.53
CRUDE_PETRO ($/bbl)	0.38	0.67	1	0.69	0.68	0.46	0.59	0.72	0.55	-0.81	1	0.46
MAIZE ($/mt)	0.0076	0.89	0.69	1	0.87	0.57	0.76	0.79	0.6	-0.51	0.72	0.6
SORGHUM ($/mt)	-0.052	0.76	0.68	0.87	1	0.54	0.66	0.7	0.47	-0.52	0.71	0.5
RICE_05 ($/mt)	0.13	0.59	0.46	0.57	0.54	1	0.35	0.45	0.44	-0.46	0.45	0.2
WHEAT_US_SRW ($/mt)	0.13	0.8	0.59	0.76	0.66	0.35	1	0.91	0.56	-0.44	0.6	0.52
WHEAT_US_HRW ($/mt)	0.31	0.82	0.72	0.79	0.7	0.45	0.91	1	0.66	-0.6	0.72	0.45
UREA_EE_BULK ($/mt)	0.37	0.6	0.55	0.6	0.47	0.44	0.56	0.66	1	-0.34	0.55	0.31
TWEXAFEGSMTH	-0.5	-0.49	-0.81	-0.51	-0.52	-0.46	-0.44	-0.6	-0.34	1	-0.79	-0.28
Crude oil, Brent($/bbl)	0.35	0.68	1	0.72	0.71	0.45	0.6	0.72	0.55	-0.79	1	0.47
Cotton, A Index($/kg)	-0.13	0.53	0.46	0.6	0.5	0.2	0.52	0.45	0.31	-0.28	0.47	1

Step 12: Check for Autocorrelation

It is essential to check for serial correlation within the historical data.

```
import pandas as pd
import matplotlib.pyplot as plt
import statsmodels.api as sm
MAX_LAGS=30
# plots the autocorrelation plots for each stock's price at 150 lags
for i in dataset:
    sm.graphics.tsa.plot_acf(dataset[i])
    plt.title('ACF for %s' % i)
plt.show()
```

Output:

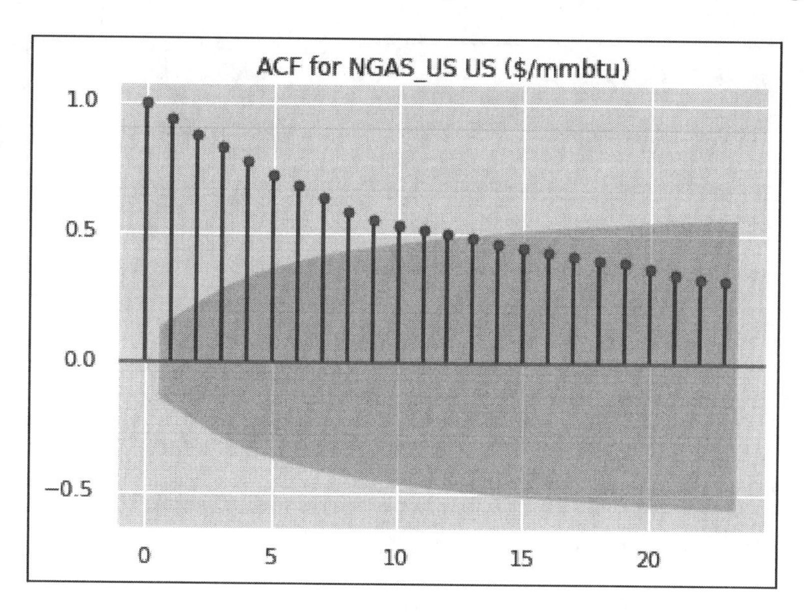

Other series also show an autocorrelation of +1 which represents a perfect positive correlation which means, an increase seen in one time series (CRUDE_PETRO ($/bbl)) leads to a proportionate increase in the other time series (UREA_EE_BULK ($/mt)). We need to apply transformations and neutralize this to make the series stationary. It measures linear relationships; even if the autocorrelation is minuscule, there may still be a nonlinear relationship between a time series and a lagged version of itself.

Step 13: Train and Test Data

The VAR model will be fitted on X_train and then used to forecast the next N observations. These forecasts will be compared against the actual present in test data. Observation could vary based on the frequency of the data.

```
n_obs=12
X_train, X_test = dataset[0:-n_obs], dataset[-n_obs:]
print(X_train.shape, X_test.shape)
```

Output:

(182, 12) (12, 12)

Step 14: Transformation

Applying first differencing on the training set to make all the series stationary. However, this is an iterative process where after the first differencing, the series may still be non-stationary. We shall have to apply a second difference or log transformation to standardize the series in such cases.

```
transform_data = X_train.diff().dropna()
transform_data.head()
```

Output:

Date	NGAS_US US ($/mmbtu)	SOYBEANS ($/mt)	CRUDE_PETRO ($/bbl)	MAIZE ($/mt)	SORGHUM ($/mt)	RICE_05 ($/mt)	WHEAT_US_SRW ($/mt)	WHEAT_US_HRW ($/mt)	UREA_EE_BULK ($/mt)	TWEXAFEGSMTH
2006-02-01	-1.17	0.0	-2.76	4.48	5.43	10.25	5.09	12.68	11.58	0.7907
2006-03-01	-0.59	0.0	1.23	-1.88	-2.41	2.00	-6.59	-5.40	29.25	-0.0340
2006-04-01	0.19	1.0	7.04	2.47	5.69	-1.25	-1.84	5.91	4.37	-1.3449
2006-05-01	-0.89	8.0	0.71	2.93	4.41	5.75	10.10	12.82	-18.00	-3.8549
2006-06-01	-0.01	1.0	-0.39	-1.25	-1.88	4.50	-10.78	1.99	-19.50	1.0048

Step 15: Stationarity check:

Re-run Augmented Dickey Fuller test.

```
Iimport statsmodels.tsa.stattools as sm
def augmented_dickey_fuller_statistics(time_series):
    result = sm.adfuller(time_series.values, autolag='AIC')
    print('ADF Statistic: %f' % result[0])
    print('p-value: %f' % result[1])
    print('Critical Values:')
    for key, value in result[4].items():
    print('\t%s: %.3f' % (key, value))
```

```
for i in dataset.columns:
    stationary_test(transform_data,i )
```

Output:

```
+++++++++++++++++++++++++++++++
Observations of Dickey-fuller test
Test Statistic                 -1.383109e+01
p-value                         7.590267e-26
#lags used                      0.000000e+00
number of observations used     1.860000e+02
critical value (1%)            -3.466005e+00
critical value (5%)            -2.877208e+00
critical value (10%)           -2.575122e+00
dtype: float64
NGAS_US US ($/mmbtu): The series is stationary
```

```
+++++++++++++++++++++++++++++++
Observations of Dickey-fuller test
Test Statistic                  -4.335102
p-value                          0.000386
#lags used                      10.000000
number of observations used    176.000000
critical value (1%)             -3.468062
critical value (5%)             -2.878106
critical value (10%)            -2.575602
dtype: float64
SOYBEANS ($/mt): The series is stationary
```

In summary, the following is the stationarity of the variables:

- NGAS_US US ($/mmbtu): The series is stationary.
- SOYBEANS ($/mt): The series is stationary.
- CRUDE_PETRO ($/bbl): The series is stationary.
- MAIZE ($/mt): The series is stationary.
- SORGHUM ($/mt): The series is stationary.
- RICE_05 ($/mt): The series is stationary.
- WHEAT_US_SRW ($/mt): The series is stationary.
- WHEAT_US_HRW ($/mt): The series is stationary.
- TWEXAFEGSMTH: The series is stationary.

- Crude oil, Brent($/bbl): The series is stationary.
- Cotton, A Index($/kg): The series is stationary.

Step 16: Plot Visualization to see variables stationarity

```
fig, axes = plt.subplots(nrows=4, ncols=4, dpi=120, figsize=(10,6))
for i, ax in enumerate(axes.flatten()):
    d = transform_data[transform_data.columns[i]]
    ax.plot(d, color='red', linewidth=1)
    # Decorations
    ax.set_title(dataset.columns[i])
    ax.xaxis.set_ticks_position('none')
    ax.yaxis.set_ticks_position('none')
    ax.spines['top'].set_alpha(0)
    ax.tick_params(labelsize=6)
plt.tight_layout();
```

Output:

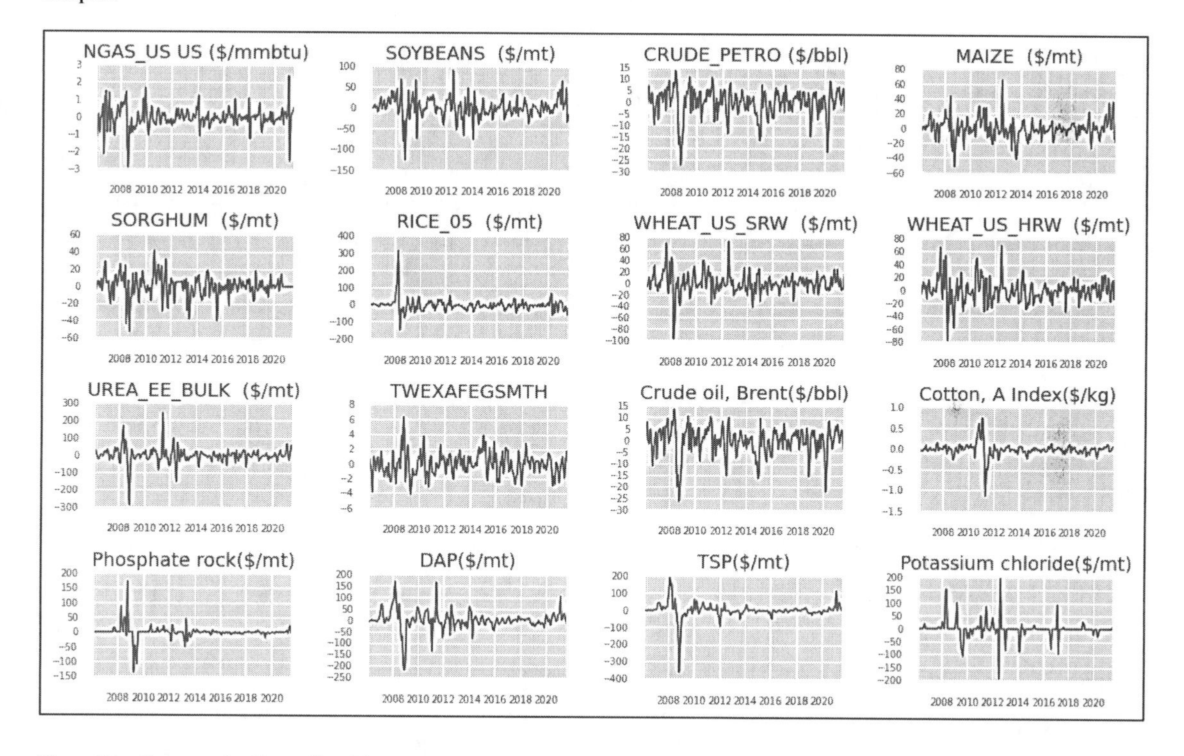

Step 17: Granger's Causality Test

Prof. Clive W.J. Granger, recipient of the 2003 Nobel Prize in Economics developed the concept of causality to improve the performance of forecasting. It is basically a hypothetical econometric test for verifying the usage of one variable in forecasting another in multivariate time series data with a particular lag.

A prerequisite for performing the Granger Causality test[602] is that the data needs to be stationary, i.e., it should have a constant mean, constant variance, and no seasonal component. Transform the non-stationary data to stationary data by differencing it, either first-order or second-order differencing. Do not proceed with the Granger causality test if the data is not stationary after second-order differencing.

The formal definition of Granger causality can be explained as whether past values of x aid in the prediction of y(t), conditional on having already accounted for the effects on y(t) of past values of y (and perhaps of past values of other variables). If they do, the x is said to "Granger cause" y. So, the basis behind the VAR is that each of the time series in the system influences each other.

Granger's causality Tests the null hypothesis that the coefficients of past values in the regression equation is zero. So, if the p-value obtained from the test is lesser than the significance level of 0.05, then, you can safely reject the null hypothesis. This has been performed on the original dataset.

```
import statsmodels.tsa.stattools as sm
maxlag=MAX_LAGS
test = 'ssr-chi2test'
def grangers_causality_matrix(X_train, variables, test = 'ssr_chi2test', verbose=False):
  dataset = pd.DataFrame(np.zeros((len(variables), len(variables))), columns=variables, index=variables)
  for c in dataset.columns:
    for r in dataset.index:
      test_result = sm.grangercausalitytests(X_train[[r,c]], maxlag=maxlag, verbose=False)
      p_values = [round(test_result[i+1][0][test][1],4) for i in range(maxlag)]
      if verbose: print(f'Y = {r}, X = {c}, P Values = {p_values}')
      min_p_value = np.min(p_values)
      dataset.loc[r,c] = min_p_value
  dataset.columns = [var + '_x' for var in variables]
  dataset.index = [var + '_y' for var in variables]
  return dataset
```

```
grangers_causality_matrix(dataset, variables = dataset.columns)
```

Output:

	NGAS_US US ($/mmbtu)_x	SOYBEANS ($/mt)_x	CRUDE_PETRO ($/bbl)_x	MAIZE ($/mt)_x	SORGHUM ($/mt)_x	RICE_05 ($/mt)_x	WHEAT_US_SRW ($/mt)_x	WHEAT_US_HRW ($/mt)_x	UREA_EE_BULK ($/mt)_x	TWEX
NGAS_US US ($/mmbtu)_y	1.0000	0.0111	0.0000	0.0992	0.0335	0.0000	0.0000	0.0000	0.1788	
SOYBEANS ($/mt)_y	0.0000	1.0000	0.2207	0.0251	0.0012	0.0000	0.0006	0.0069	0.0001	
CRUDE_PETRO ($/bbl)_y	0.0000	0.0012	1.0000	0.0000	0.0001	0.0000	0.0000	0.0000	0.0015	
MAIZE ($/mt)_y	0.0001	0.0038	0.4742	1.0000	0.0035	0.0000	0.0000	0.0000	0.0011	
SORGHUM ($/mt)_y	0.0000	0.0000	0.2989	0.0000	1.0000	0.0000	0.0000	0.0000	0.0000	
RICE_05 ($/mt)_y	0.0000	0.0009	0.0040	0.0125	0.0105	1.0000	0.0000	0.0000	0.0268	
WHEAT_US_SRW ($/mt)_y	0.0000	0.0001	0.0000	0.0000	0.0247	0.0000	1.0000	0.0000	0.0000	
WHEAT_US_HRW ($/mt)_y	0.0000	0.0040	0.0005	0.0002	0.0041	0.0000	0.0759	1.0000	0.0005	
UREA_EE_BULK ($/mt)_y	0.0000	0.0000	0.0000	0.0001	0.0159	0.0000	0.0000	0.0000	1.0000	
TWEXAFEGSMTH_y	0.0000	0.0849	0.0006	0.0000	0.0000	0.0000	0.0000	0.0001	0.0000	
Crude oil, Brent($/bbl)_y	0.0000	0.0002	0.0691	0.0000	0.0001	0.0000	0.0000	0.0000	0.0085	
Cotton, A Index($/kg)_y	0.0000	0.0000	0.0003	0.0559	0.0000	0.0027	0.0000	0.0000	0.0000	

The rows are the responses (y) and the columns are the predictor series (x).

- If we take the value 0.0000 in (row 9, column 2), it refers to the p-value of the Granger's Causality test for SOYBEANS ($/mt) _x causing UREA_EE_BULK ($/mt) _y. The 0.0001 in (row 2, column 9) refers to the p-value of SOYBEANS ($/mt) _y causing UREA_EE_BULK ($/mt) _x and so on.
- Our variables of interest is causation of UREA_EE_BULK on agricultural commodities - SOYBEANS, MAIZE, SORGHUM, RICE_05, WHEAT_US_SRW , WHEAT_US_HRW, and COTTON.

Step 18: VAR Model

VAR requires stationarity of the series which means the meaning to the series does not change over time (we can find this out from the plot drawn next to Augmented Dickey-Fuller Test). So, I will fit the VAR model on the training set and then use the fitted model to forecast the next 6 observations. These forecasts will be compared against the actual present in test data. I have taken the maximum lag (6) to identify the required lags for VAR model.

```
import numpy as np
import pandas
import statsmodels.api as sm
from statsmodels.tsa.api import VAR
mod = VAR(transform_data)
results = mod.fit(maxlags=n_obs, ic='aic')
print(results.summary())
```

Output:

Correlation matrix of residuals	NGAS_US US ($/mmbtu)	SOYBEANS ($/mt)	CRUDE_PETRO ($/bbl)	MAIZE ($/mt
NGAS_US US ($/mmbtu)	1.000000	0.140488	0.223285	0.03517
SOYBEANS ($/mt)	0.140488	1.000000	0.007040	0.52525
CRUDE_PETRO ($/bbl)	0.223285	0.007040	1.000000	-0.22359
MAIZE ($/mt)	0.035176	0.525255	-0.223592	1.00000
SORGHUM ($/mt)	0.230324	0.740643	-0.039025	0.54449
RICE_05 ($/mt)	-0.065169	-0.483597	0.063443	-0.29968
WHEAT_US_SRW ($/mt)	0.018966	0.529707	-0.046670	0.49201
WHEAT_US_HRW ($/mt)	0.193889	0.528638	-0.018617	0.43027
UREA_EE_BULK ($/mt)	-0.055807	0.006437	-0.077103	-0.01045
TWEXAFEGSMTH	-0.442855	-0.208540	-0.512725	0.10259
Crude oil, Brent($/bbl)	0.230640	0.021937	0.984345	-0.16927
Cotton, A Index($/kg)	-0.151446	-0.105022	0.238652	-0.02362

The correlation matrix of residuals: with the highest for Cotton (42,84%), Rice (35.92%), Sorghum (15.44%) and negative correlations of –29.29% for Wheat HRW and –26.27% wheat SRW.

SORGHUM ($/mt)	RICE 05 ($/mt)	WHEAT US SRW ($/mt)	WHEAT US HRW ($/mt)	UREA EE BULK ($/mt)	TWEXAFEGSMTH	C
0.230324	-0.065169	0.018966	0.193889	-0.055807	-0.442855	
0.740643	-0.483597	0.529707	0.528638	0.006437	-0.208540	
-0.039025	0.063443	-0.046670	-0.018617	-0.077103	-0.512725	
0.544498	-0.299685	0.492011	0.430277	-0.010454	0.102592	
1.000000	-0.492177	0.215881	0.287002	0.154462	-0.187942	
-0.492177	1.000000	-0.200462	-0.133405	0.359268	0.020907	
0.215881	-0.200462	1.000000	0.928458	-0.262733	-0.007179	
0.287002	-0.133405	0.928458	1.000000	-0.292913	-0.210031	
0.154462	0.359268	-0.262733	-0.292913	1.000000	0.216855	
-0.187942	0.020907	-0.007179	-0.210031	0.216855	1.000000	
0.011428	0.093266	-0.033694	0.001863	-0.001850	-0.497571	
0.048924	0.319575	-0.283737	-0.390145	0.428491	0.152343	

Step 19: Residual Plot

Plot the residuals.

```
y_fitted = results.fittedvalues
y_fitted
y_fitted = results.fittedvalues
residuals = results.resid
plt.figure(figsize = (15,5))
plt.plot(residuals, label='resid')
plt.plot(y_fitted, label='VAR prediction')
plt.xlabel('Date')
plt.xticks(rotation=45)
plt.ylabel('Residuals')
plt.grid(True)
```

Output:

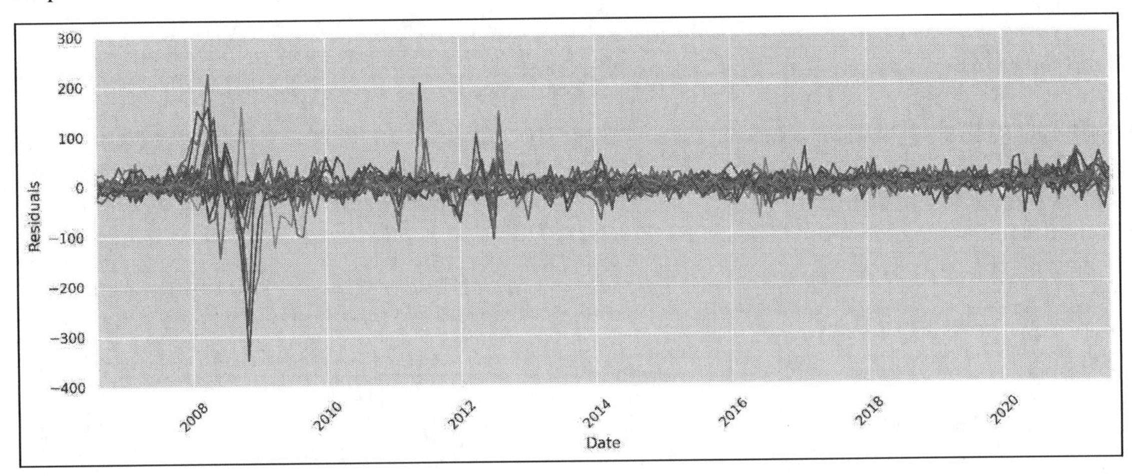

As it can be seen, the spikes of prices during the great recession.

Step 20: Durbin-Watson Statistic

The Durbin-Watson statistic will always have a value between 0 and 4. A value of 2.0 means that there is no autocorrelation detected in the sample. Values from 0 to less than 2 indicate a positive autocorrelation and values from 2 to 4 indicate a negative autocorrelation. A rule of thumb is that test statistic values in the range of 1.5 to 2.5 are relatively normal. Any value outside this range could be a cause for concern.

```
from statsmodels.stats.stattools import durbin_watson
out = durbin_watson(results.resid)
for col,val in zip(transform_data.columns, out):
  print((col), ":", round(val,2))
```

Output:

```
NGAS_US US ($/mmbtu) : 1.96
SOYBEANS ($/mt) : 2.46
CRUDE_PETRO ($/bbl) : 2.39
MAIZE ($/mt) : 2.23
SORGHUM ($/mt) : 2.17
RICE_05 ($/mt) : 2.23
WHEAT_US_SRW ($/mt) : 2.02
WHEAT_US_HRW ($/mt) : 2.02
UREA_EE_BULK ($/mt) : 2.2
TWEXAFEGSMTH : 2.5
Crude oil, Brent($/bbl) : 2.4
Cotton, A Index($/kg) : 1.84
```

All the values of DW is between 1.5 and 2.5.

The Durbin-Watson statistic will always have a value between 0 and 4. A value of 2.0 means that there is no autocorrelation.

Step 21: Prediction or Forecast VAR

To forecast, the VAR model expects up to the lag order number of *observations* from the past data. This is because the terms in the VAR model are essentially the lags of the various time series in the dataset, so we need to provide as many of the previous values as indicated by the lag order used by the model.

```
# Get the lag order
lag_order = results.k_ar
print(lag_order)
# Input data for forecasting
input_data = transform_data.values[-lag_order:]
print(input_data)
# forecasting
pred = results.forecast(y=input_data, steps=n_obs)
pred = (pd.DataFrame(pred, index=X_test.index, columns=X_test.columns + '_pred'))
print(pred)
```

Output:

```
12
CRUDE_PETRO ($/bbl)_pred MAIZE ($/mt)_pred \ ...
Date
2021-03-01      4.560647     -15.359362
2021-04-01     -0.602990     -22.412340
2021-05-01    -15.247281      -2.387266
2021-06-01      4.413274      -1.859596
2021-07-01     16.875809     -11.824071
2021-08-01      4.261913     -45.877150
2021-09-01      2.148813      -6.160637
2021-10-01     -0.179986     -17.419243
2021-11-01     -2.449792     -11.380505
2021-12-01      0.837611     -15.686079
2022-01-01     -5.209124     -18.768586
2022-02-01    -13.183669     -29.005563
```

Step 22: Invert the Transformation

The forecasts are generated but it is on the scale of the training data used by the model. So, to bring it back up to its original scale, we need to de-difference it. The way to convert the differencing is to add these differences consecutively to the base number. An easy way to do it is to first determine the cumulative sum at index and then add it to the base number. This process can be reversed by adding the observation at the prior time step to the difference value. inverted(ts) = differenced(ts) + observation(ts-1)

```
# inverting transformation
def invert_transformation(X_train, pred):
    forecast12 = pred.copy()
    columns = X_train.columns
    #col='CPI_AllUrban_FoodandBeverages'
    #forecast[str(col)+'_pred'] = dataset[col].iloc[-1] + forecast[str(col)+'_pred'].cumsum()
    for col in columns:
        print('++++++++++++++')
        print(col)
        forecast12[str(col)+'_pred'] = X_train[col].iloc[-1] + forecast12[str(col)+'_pred'].cumsum()
        print('++++++++++++++')
    return forecast12
output = invert_transformation(X_train, pred)
output
```

Output:

Date	NGAS_US US ($/mmbtu)_pred	SOYBEANS ($/mt)_pred	CRUDE_PETRO ($/bbl)_pred	MAIZE ($/mt)_pred	SORGHUM ($/mt)_pred	RICE_05 ($/mt)_pred	WHEAT_US_SRW ($/mt)_pred	WHEAT_US_HRW ($/mt)_pred	UREA_EE_BULK ($/mt)_pred
2021-03-01	7.728015	614.595126	65.020647	229.880638	171.102782	517.187877	279.477986	331.880876	323.722949
2021-04-01	7.093524	575.588895	64.417658	207.468298	141.511508	483.470732	270.721554	322.429445	370.435144
2021-05-01	5.867328	546.828473	49.170377	205.081031	145.453940	467.356046	271.142523	345.897096	351.848791
2021-06-01	5.154814	549.562924	53.583650	203.221436	160.298889	341.408336	285.986744	356.306786	293.349003
2021-07-01	5.051676	557.708560	70.459460	191.397365	159.838101	324.373557	336.767322	386.304644	123.412107
2021-08-01	5.641775	526.477556	74.721373	145.520215	118.485780	419.254754	306.789413	349.392720	152.135845
2021-09-01	8.433454	496.656262	76.870186	139.359578	100.760978	522.736118	272.031503	326.091774	237.322343
2021-10-01	9.144438	473.451791	76.690200	121.940335	96.919953	498.642069	203.791039	283.322885	362.081159
2021-11-01	8.607029	412.115113	74.240409	110.559830	82.698311	396.280313	178.714525	281.365782	313.339395
2021-12-01	7.352646	411.669489	75.078020	94.873751	72.354407	342.596662	200.697258	312.085771	269.774786
2022-01-01	5.946548	350.244204	69.868897	76.105164	84.253050	423.070148	215.259538	313.825704	301.662289

```
#combining predicted and real data set
combine = pd.concat([output['UREA_EE_BULK ($/mt)_pred'], X_test['UREA_EE_BULK ($/mt)']], axis=1)
combine['accuracy'] = round(combine.apply(lambda row: row['UREA_EE_BULK ($/mt)_pred'] / row['UREA_EE_BULK ($/mt)'] * 100, axis = 1),2)
combine['accuracy'] = pd.Series(["{0:.2f}%".format(val) for val in combine['accuracy']],index = combine.index)
combine = combine.round(decimals=2)
combine
```

Output:

Date	UREA_EE_BULK ($/mt)_pred	UREA_EE_BULK ($/mt)	accuracy
2021-03-01	323.72	352.88	91.74%
2021-04-01	370.44	328.10	112.90%
2021-05-01	351.85	331.63	106.10%
2021-06-01	293.35	393.25	74.60%
2021-07-01	123.41	441.50	27.95%
2021-08-01	152.14	446.88	34.04%
2021-09-01	237.32	418.75	56.67%
2021-10-01	362.08	695.00	52.10%
2021-11-01	313.34	900.50	34.80%
2021-12-01	269.77	890.00	30.31%
2022-01-01	301.66	846.38	35.64%
2022-02-01	283.17	744.17	38.05%

Step 23: Evaluation

To evaluate the forecasts, a comprehensive set of metrics, such as MAPE, ME, MAE, MPE and RMSE can be computed. We have computed some of these below.

```
from sklearn.metrics import mean_absolute_error
from sklearn.metrics import mean_squared_error
import math
#Forecast bias
forecast_errors = [combine['UREA_EE_BULK ($/mt)'][i]- combine['UREA_EE_BULK ($/mt)_pred'][i]
for i in range(len(combine['UREA_EE_BULK ($/mt)']))]
bias = sum(forecast_errors) * 1.0/len(combine['UREA_EE_BULK ($/mt)'])
print('Bias: %f' % bias)
print('Mean absolute error:', mean_absolute_error(combine['UREA_EE_BULK ($/mt)'].values,
combine['UREA_EE_BULK ($/mt)_pred'].values))
print('Mean squared error:', mean_squared_error(combine['UREA_EE_BULK ($/mt)'].values,
combine['UREA_EE_BULK ($/mt)_pred'].values))
print('Root mean squared error:', math.sqrt(mean_squared_error(combine['UREA_EE_BULK ($/mt)'].
values, combine['UREA_EE_BULK ($/mt)_pred'].values)))
```

Output:

Date	UREA_EE_BULK ($/mt)_pred	UREA_EE_BULK ($/mt)	accuracy
2021-03-01	323.72	352.88	91.74%
2021-04-01	370.44	328.10	112.90%
2021-05-01	351.85	331.63	106.10%
2021-06-01	293.35	393.25	74.60%
2021-07-01	123.41	441.50	27.95%
2021-08-01	152.14	446.88	34.04%
2021-09-01	237.32	418.75	56.67%
2021-10-01	362.08	695.00	52.10%
2021-11-01	313.34	900.50	34.80%
2021-12-01	269.77	890.00	30.31%
2022-01-01	301.66	846.38	35.64%
2022-02-01	283.17	744.17	38.05%

Step 24: Model summary

As it is clear from the time series analysis, the Urea fertilizer is sensitive to the demand of key agricultural commodities—Cotton (42,84%), Rice (35.92%), Sorghum (15.44%) and negative correlations of –29.29% for Wheat HRW and –26.27% wheat SRW were observed. The positive relationship suggests that the increase in demand of Cotton, Rice and Sorghum will lead to increases in the price of Urea. Similarly, a negative correlation of Wheat Hard Winter and Soft Winter suggests that the increase in Urea prices would increase the farm costs of Wheat.

Machine Learning Model Energy Prices and Fertilizer Costs—Phosphate

Data Sources are the same as for the Urea machine learning model except respective Urea columns are dropped to model the influence of Phosphate.

NGAS_US US ($/mmbtu)', 'SOYBEANS ($/mt)', 'CRUDE_PETRO ($/bbl)', 'MAIZE ($/mt)', 'SORGHUM ($/mt)', 'RICE_05 ($/mt)', 'WHEAT_US_SRW ($/mt)', 'WHEAT_US_HRW ($/mt)', 'UREA_EE_BULK ($/mt), 'TWEXAFEGSMTH', 'Crude oil, Brent($/bbl)', 'Cotton, A Index($/kg)', 'Phosphate rock($/mt)', 'DAP($/mt)', 'TSP($/mt)', and 'Potassium chloride($/mt)'

Step 1: Prepare Data Sources to Model Phosphate

Drop the UREA column

```
# For Phosphate
dataset.drop(['UREA_EE_BULK ($/mt)'], axis = 1, inplace = True)
```

```
import matplotlib.pyplot as plt
import pandas as pd
import numpy as np
import seaborn as sns; sns.set()
corr = dataset.corr()
fig, ax = plt.subplots()
sns.heatmap(corr,xticklabels=corr.columns.values, yticklabels=corr.columns.values, annot=True,annot_
kws={'size':12})
heat_map=plt.gcf()
heat_map.set_size_inches(18,10)
plt.xticks(fontsize=10)
plt.yticks(fontsize=10)
plt.show()
```

Output:

	NGAS_US US ($/mmbtu)	SOYBEANS ($/mt)	CRUDE_PETRO ($/bbl)	MAIZE ($/mt)	SORGHUM ($/mt)	RICE_05 ($/mt)	WHEAT_US_SRW ($/mt)	WHEAT_US_HRW ($/mt)	TWEXAFEGSMTH	Crude oil, Brent($/bbl)	Cotton, A Index($/kg)	Phosphate rock($/mt)	DAP($/mt)	TSP($/mt)	Potassium chloride($/mt)
NGAS_US US ($/mmbtu)	1	0.025	0.38	0.0076	-0.052	0.13	0.13	0.31	-0.5	0.35	-0.13	0.072	0.48	0.48	0.077
SOYBEANS ($/mt)	0.025	1	0.67	0.89	0.76	0.59	0.8	0.82	-0.49	0.68	0.53	0.29	0.64	0.63	0.39
CRUDE_PETRO ($/bbl)	0.38	0.67	1	0.69	0.68	0.46	0.59	0.72	-0.81	1	0.46	0.17	0.55	0.58	0.42
MAIZE ($/mt)	0.0076	0.89	0.69	1	0.87	0.57	0.76	0.79	-0.51	0.72	0.6	0.35	0.59	0.6	0.41
SORGHUM ($/mt)	-0.052	0.76	0.68	0.87	1	0.54	0.66	0.7	-0.52	0.71	0.5	0.27	0.48	0.51	0.4
RICE_05 ($/mt)	0.13	0.59	0.46	0.57	0.54	1	0.35	0.45	-0.46	0.45	0.2	0.57	0.59	0.62	0.72
WHEAT_US_SRW ($/mt)	0.13	0.8	0.59	0.76	0.66	0.35	1	0.91	-0.44	0.6	0.52	0.14	0.56	0.56	0.17
WHEAT_US_HRW ($/mt)	0.31	0.82	0.72	0.79	0.7	0.45	0.91	1	-0.6	0.72	0.45	0.26	0.66	0.66	0.3
TWEXAFEGSMTH	-0.5	-0.49	-0.81	-0.51	-0.52	-0.46	-0.44	-0.6	1	-0.79	-0.28	-0.22	-0.39	-0.41	-0.52
Crude oil, Brent($/bbl)	0.35	0.68	1	0.72	0.71	0.45	0.6	0.72	-0.79	1	0.47	0.17	0.54	0.56	0.42
Cotton, A Index($/kg)	-0.13	0.53	0.46	0.6	0.5	0.2	0.52	0.45	-0.28	0.47	1	0.055	0.29	0.31	0.055
Phosphate rock($/mt)	0.072	0.29	0.17	0.35	0.27	0.57	0.14	0.26	-0.22	0.17	0.055	1	0.44	0.47	0.75
DAP($/mt)	0.48	0.64	0.55	0.59	0.48	0.59	0.56	0.66	-0.39	0.54	0.29	0.44	1	0.98	0.33
TSP($/mt)	0.48	0.63	0.58	0.6	0.51	0.62	0.56	0.66	-0.41	0.56	0.31	0.47	0.98	1	0.37
Potassium chloride($/mt)	0.077	0.39	0.42	0.41	0.4	0.72	0.17	0.3	-0.52	0.42	0.055	0.75	0.33	0.37	1

1.0 0.8 0.6 0.4 0.2 0.0 −0.2 −0.4 −0.6 −0.8

Phosphate mineral highly correlated with DAP (44%), TSP (47%), Potassium Chloride (75%), and demand of Soybeans (29%), Maize (35%), Rice (57%), and Wheat HRW (26%) & negatively correlated with U.S. Dollar Index (–22%).

Step 2: Check for Autocorrelation

It is essential to check for serial correlation within the historical data.

```
import pandas as pd
import matplotlib.pyplot as plt
import statsmodels.api as sm
MAX_LAGS=30
# plots the autocorrelation plots for each stock's price at 150 lags
for i in dataset:
  sm.graphics.tsa.plot_acf(dataset[i])
  plt.title('ACF for %s' % i)
plt.show()
```

Output:

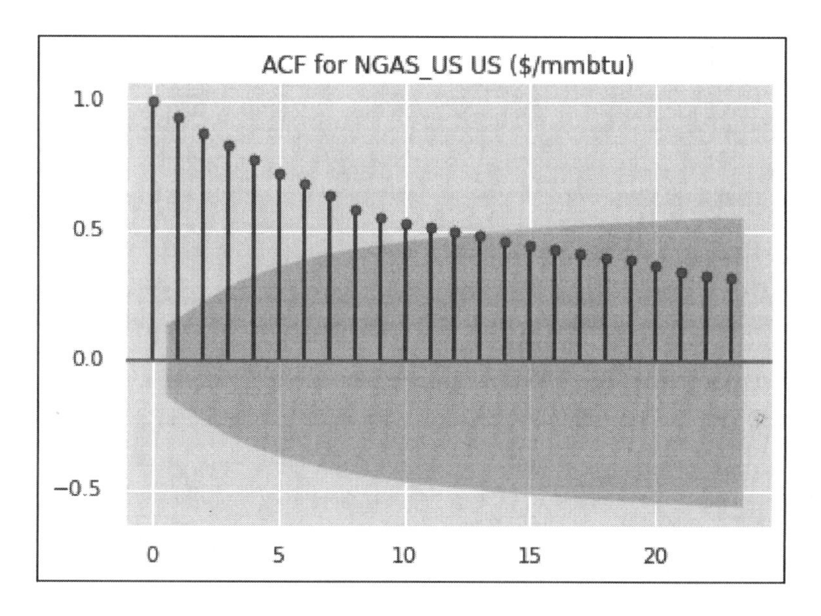

Other series also show auto-correlation of +1 which represents a perfect positive correlation which means, an increase seen in one time series (CRUDE_PETRO ($/bbl)) leads to a proportionate increase in the other time series (UREA_EE_BULK ($/mt)). We need to apply transformation and neutralize this to make the series stationary. It measures linear relationships; even if the autocorrelation is minuscule, there may still be a nonlinear relationship between a time series and a lagged version of itself.

Step 3: Train and Test Data

The VAR model will be fitted on X_train and then used to forecast the next N observations. These forecasts will be compared against the actual present in test data. Observation could vary based on the frequency of the data.

```
n_obs=12
X_train, X_test = dataset[0:-n_obs], dataset[-n_obs:]
print(X_train.shape, X_test.shape)
```

Output:

(182, 15) (12, 15)

Step 4: Transformation

Applying first differencing on training set to make all the series stationary. However, this is an iterative process where we after first differencing, the series may still be non-stationary. We shall have to apply a second difference or log transformation to standardize the series in such cases.

```
transform_data = X_train.diff().dropna()
transform_data.head()
```

Output:

Date	NGAS_US US ($/mmbtu)	SOYBEANS ($/mt)	CRUDE_PETRO ($/bbl)	MAIZE ($/mt)	SORGHUM ($/mt)	RICE_05 ($/mt)	WHEAT_US_SRW ($/mt)	WHEAT_US_HRW ($/mt)	TWEXAFEGSMTH	Crude oil, Brent($/bbl)
2006-02-01	-1.17	0.0	-2.76	4.48	5.43	10.25	5.09	12.68	0.7907	-3.651286
2006-03-01	-0.59	0.0	1.23	-1.88	-2.41	2.00	-6.59	-5.40	-0.0340	2.330043
2006-04-01	0.19	1.0	7.04	2.47	5.69	-1.25	-1.84	5.91	-1.3449	8.189062
2006-05-01	-0.89	8.0	0.71	2.93	4.41	5.75	10.10	12.82	-3.8549	-0.254833
2006-06-01	-0.01	1.0	-0.39	-1.25	-1.88	4.50	-10.78	1.99	1.0048	-1.329545

Step 5: Stationarity check:

Re-run Augmented Dick Fuller test.

```
Iimport statsmodels.tsa.stattools as sm
def augmented_dickey_fuller_statistics(time_series):
   result = sm.adfuller(time_series.values, autolag='AIC')
   print('ADF Statistic: %f' % result[0])
   print('p-value: %f' % result[1])
   print('Critical Values:')
   for key, value in result[4].items():
     print('\t%s: %.3f' % (key, value))
```

```
for i in dataset.columns:
   stationary_test(transform_data,i )
```

Output:

```
++++++++++++++++++++++++++++++++
Observations of Dickey-fuller test
Test Statistic                  -1.383109e+01
p-value                          7.590267e-26
#lags used                       0.000000e+00
number of observations used      1.860000e+02
critical value (1%)             -3.466005e+00
critical value (5%)             -2.877208e+00
critical value (10%)            -2.575122e+00
dtype: float64
NGAS_US US ($/mmbtu): The series is stationary
```

```
++++++++++++++++++++++++++++++++
Observations of Dickey-fuller test
Test Statistic                  -4.335102
p-value                          0.000386
#lags used                       10.000000
number of observations used      176.000000
critical value (1%)             -3.468062
critical value (5%)             -2.878106
critical value (10%)            -2.575602
dtype: float64
SOYBEANS ($/mt): The series is stationary
```

In summary, the following is the stationarity of the variables:

- NGAS_US US ($/mmbtu): The series is stationary.
- SOYBEANS ($/mt): The series is stationary.

- CRUDE_PETRO ($/bbl): The series is stationary.
- MAIZE ($/mt): The series is stationary.
- SORGHUM ($/mt): The series is stationary.
- RICE_05 ($/mt): The series is stationary.
- WHEAT_US_SRW ($/mt): The series is stationary.
- WHEAT_US_HRW ($/mt): The series is stationary.
- TWEXAFEGSMTH: The series is stationary.
- Crude oil, Brent($/bbl): The series is stationary.
- Cotton, A Index($/kg): The series is stationary.

Step 6: Plot Visualization to see variables stationarity

```
fig, axes = plt.subplots(nrows=4, ncols=4, dpi=120, figsize=(10,6))
for i, ax in enumerate(axes.flatten()):
   d = transform_data[transform_data.columns[i]]
   ax.plot(d, color='red', linewidth=1)
   # Decorations
   ax.set_title(dataset.columns[i])
   ax.xaxis.set_ticks_position('none')
   ax.yaxis.set_ticks_position('none')
   ax.spines['top'].set_alpha(0)
   ax.tick_params(labelsize=6)
plt.tight_layout();
```

Output:

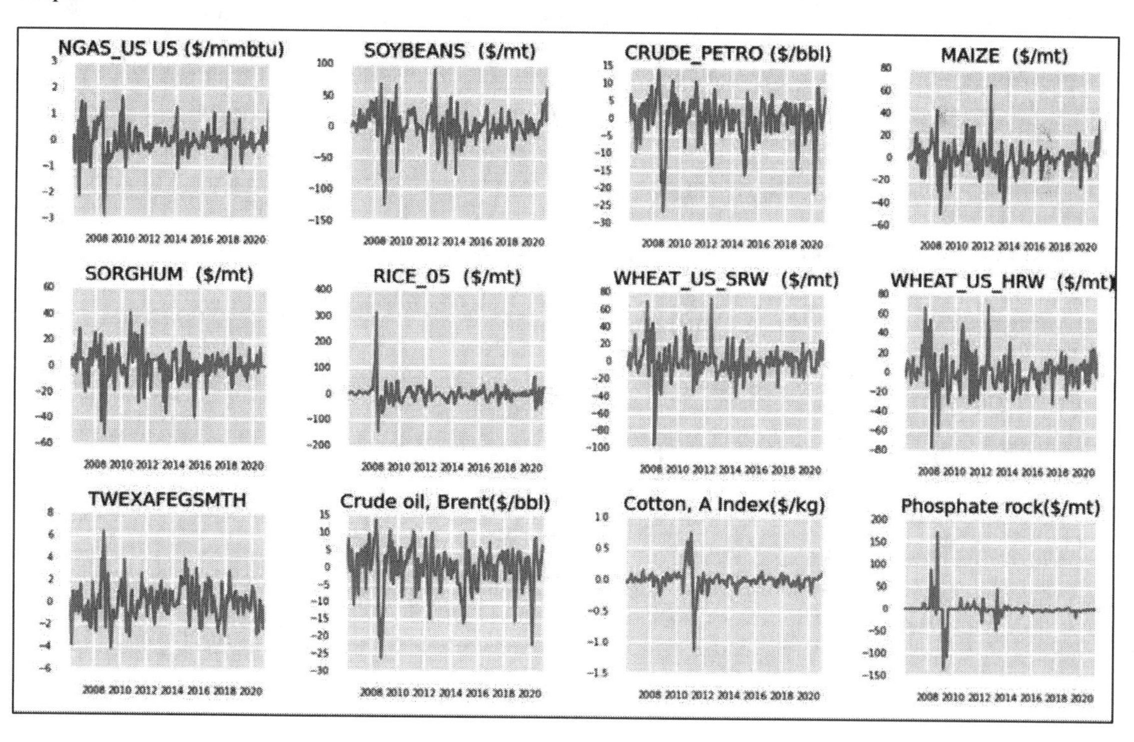

Step 7: Granger's Causality Test

Granger's causality Tests the null hypothesis that the coefficients of past values in the regression equation are zero. So, if the p-value obtained from the test is lesser than the significance level of 0.05, then, you can safely reject the null hypothesis. This has been performed on the original dataset.

```
import statsmodels.tsa.stattools as sm
maxlag=MAX_LAGS
test = 'ssr-chi2test'
def grangers_causality_matrix(X_train, variables, test = 'ssr_chi2test', verbose=False):
  dataset = pd.DataFrame(np.zeros((len(variables), len(variables))), columns=variables, index=variables)
  for c in dataset.columns:
    for r in dataset.index:
      test_result = sm.grangercausalitytests(X_train[[r,c]], maxlag=maxlag, verbose=False)
      p_values = [round(test_result[i+1][0][test][1],4) for i in range(maxlag)]
      if verbose: print(f'Y = {r}, X = {c}, P Values = {p_values}')
      min_p_value = np.min(p_values)
      dataset.loc[r,c] = min_p_value
  dataset.columns = [var + '_x' for var in variables]
  dataset.index = [var + '_y' for var in variables]
  return dataset
```

```
grangers_causality_matrix(dataset, variables = dataset.columns)
```

Output:

	NGAS_US US ($/mmbtu)_x	SOYBEANS ($/mt)_x	CRUDE_PETRO ($/bbl)_x	MAIZE ($/mt)_x	SORGHUM ($/mt)_x	RICE_05 ($/mt)_x	WHEAT_US_SRW ($/mt)_x	WHEAT_US_HRW ($/mt)_x	TWEXAFEGSMTH_x
NGAS_US US ($/mmbtu)_y	1.0000	0.0111	0.0000	0.0992	0.0335	0.0000	0.0000	0.0000	0.0015
SOYBEANS ($/mt)_y	0.0000	1.0000	0.2207	0.0251	0.0012	0.0000	0.0006	0.0069	0.0651
CRUDE_PETRO ($/bbl)_y	0.0000	0.0012	1.0000	0.0000	0.0001	0.0000	0.0000	0.0000	0.0030
MAIZE ($/mt)_y	0.0001	0.0038	0.4742	1.0000	0.0035	0.0000	0.0000	0.0000	0.0002
SORGHUM ($/mt)_y	0.0000	0.0000	0.2989	0.0000	1.0000	0.0000	0.0000	0.0000	0.0107
RICE_05 ($/mt)_y	0.0000	0.0009	0.0040	0.0125	0.0105	1.0000	0.0000	0.0000	0.0000
WHEAT_US_SRW ($/mt)_y	0.0000	0.0001	0.0000	0.0000	0.0247	0.0000	1.0000	0.0000	0.0125
WHEAT_US_HRW ($/mt)_y	0.0000	0.0040	0.0005	0.0002	0.0041	0.0000	0.0759	1.0000	0.0372
TWEXAFEGSMTH_y	0.0000	0.0849	0.0006	0.0000	0.0000	0.0000	0.0000	0.0001	1.0000
Crude oil, Brent($/bbl)_y	0.0000	0.0002	0.0691	0.0000	0.0001	0.0000	0.0000	0.0000	0.0057
Cotton, A Index($/kg)_y	0.0000	0.0000	0.0003	0.0559	0.0000	0.0027	0.0000	0.0000	0.0000
Phosphate rock($/mt)_y	0.0000	0.0000	0.0000	0.0000	0.0000	0.0000	0.0000	0.0000	0.0001
DAP($/mt)_y	0.0000	0.0000	0.0000	0.0000	0.0001	0.0000	0.0000	0.0000	0.0021
TSP($/mt)_y	0.0000	0.0000	0.0003	0.0000	0.0000	0.0000	0.0000	0.0000	0.0001
Potassium chloride($/mt)_y	0.0000	0.0000	0.0000	0.0000	0.0208	0.0000	0.0000	0.0000	0.0000

The rows are the response (y) and the columns are the predictor series (x).

- If we take the value 0.0000 in (row 12, column 2), it refers to the p-value of the Granger's Causality test for SOYBEANS ($/mt) _x causing Phosphate rock ($/mt) _y. The 0.0026 in (row 2, column 12) refers to the p-value of SOYBEANS ($/mt) _y causing Phosphate rock ($/mt) _x and so on.
- Our variables of interest are causation of Phosphate rock on agricultural commodities—SOYBEANS, MAIZE, SORGHUM, RICE_05, WHEAT_US_SRW, WHEAT_US_HRW, COTTON, and TSP & DSP. Of course, U.S. Dollar Index.
- Energy prices (CRUDE and NGAS) are granger caution with Phosphate & DAP/TSP.
- DAP exhibits bi-directional causality with agricultural commodities in the model and energy commodities [44].

Step 8: VAR Model

VAR requires stationarity of the series which means the meaning to the series does not change over time (we can find this out from the plot drawn next to Augmented Dickey-Fuller Test). So, I will fit the VAR model on training set and then use the fitted model to forecast the next 12 observations. These forecasts will be compared against the actual present in test data. I have taken the maximum lag (12) to identify the required lags for VAR model.

```
import numpy as np
import pandas
import statsmodels.api as sm
from statsmodels.tsa.api import VAR
mod = VAR(transform_data)
results = mod.fit(maxlags=6, ic='aic')
print(results.summary())
```

Output:

Summary of Regression Results

```
==============================================================================
Model:                       VAR
Method:                      OLS
Date:          Sun, 22, May, 2022
Time:                   09:21:39
------------------------------------------------------------------------------
No. of Equations:        15.0000     BIC:                          80.2931
Nobs:                    175.000     HQIC:                         65.6208
Log likelihood:         –7225.39     FPE:                      7.65922e+24
AIC:                     55.6078     Det(Omega_mle):           1.43394e+22
------------------------------------------------------------------------------
```

The Akaike information criterion (AIC) is an estimator of out-of-sample prediction error and thereby relative quality of statistical models for a given set of data. Given a collection of models for the data, AIC estimates the quality of each model, relative to each of the other models. Thus, AIC provides a means for model selection.[603]

Correlation matrix of residuals: with the highest for Potassium Chloride (29.61%), DAP (23.6%) Cotton (22.16%), and Soybeans (-30.01%). A fact that can verified by percentage of fertilizer received[604] on cotton acreage in U.S.

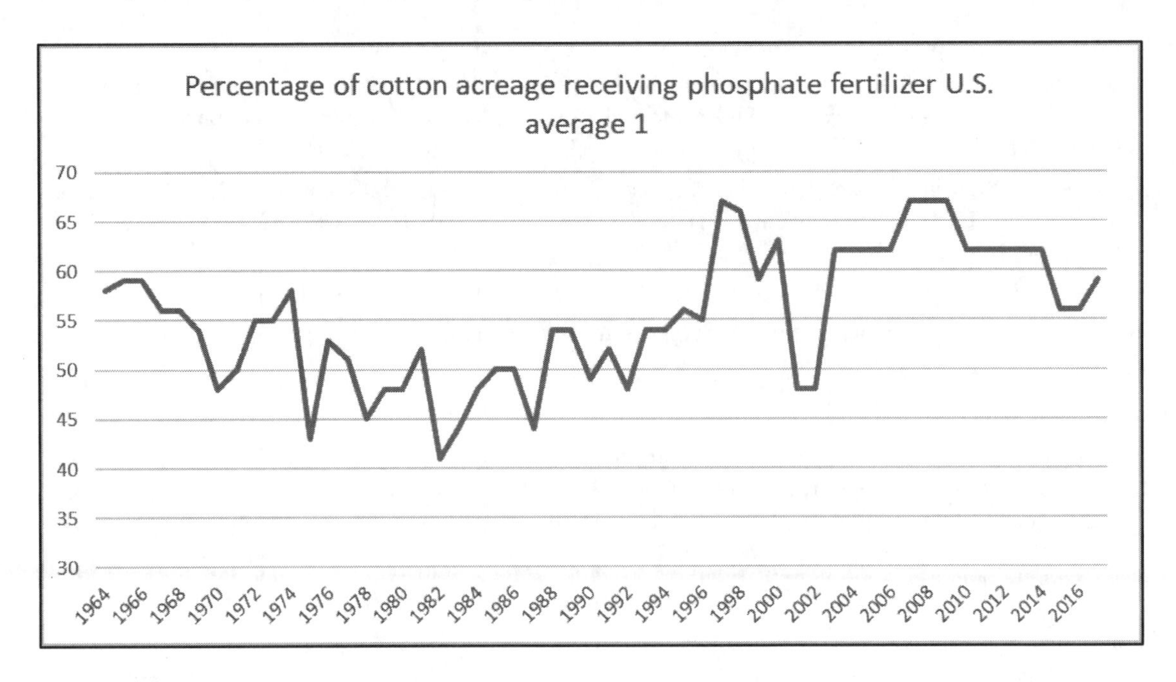

Additionally, energy prices, Crude Pero and Natural Gas heavily influence the price movements of TSP & DAP. DAP is highly correlated with PETRO (12.90%) and Natural Gas (17.55%) whereas TSP correlated with PETRO (4.8%) and Natural Gas (23.4%). By the by, mineral Potash is low correlated with Gas (–14.49%) and Petro (–11.6%) as expected given it is mined.

Correlation matrix of residuals

	NGAS_US US ($/mmbtu)	SOYBEANS ($/mt)	CRUDE_PETRO ($/bbl)	MAIZE ($/mt)
NGAS_US US ($/mmbtu)	1.000000	0.140260	0.217635	0.087444
SOYBEANS ($/mt)	0.140260	1.000000	0.275561	0.484709
CRUDE_PETRO ($/bbl)	0.217635	0.275561	1.000000	0.116323
MAIZE ($/mt)	0.087444	0.484709	0.116323	1.000000
SORGHUM ($/mt)	0.038929	0.300681	0.085345	0.565010
RICE_05 ($/mt)	0.027754	-0.123702	-0.126833	-0.064216
WHEAT_US_SRW ($/mt)	-0.072312	0.459330	0.121934	0.525732
WHEAT_US_HRW ($/mt)	-0.032203	0.450267	0.129270	0.473501
TWEXAFEGSMTH	-0.226946	-0.182711	-0.436104	-0.136250
Cotton, A Index($/kg)	0.002628	0.015693	0.231261	0.122670
Phosphate rock($/mt)	-0.123764	-0.248596	-0.120878	-0.027791
DAP($/mt)	0.175583	0.098230	0.129631	-0.114393
TSP($/mt)	0.234953	0.126257	0.048754	0.139987
Potassium chloride($/mt)	-0.095116	-0.110769	-0.008416	-0.168178

PETRO and Natural Gas are the key drivers of prices of DAP & TSP. Net-net, 30% price influence of DAP & TSP caused by Energy prices.[605]

Step 9: Residual Plot

Plot the residuals.

```
y_fitted = results.fittedvalues
y_fitted

y_fitted = results.fittedvalues
residuals = results.resid
plt.figure(figsize = (15,5))
plt.plot(residuals, label='resid')
plt.plot(y_fitted, label='VAR prediction')
plt.xlabel('Date')
plt.xticks(rotation=45)
plt.ylabel('Residuals')
plt.grid(True)
```

Output:

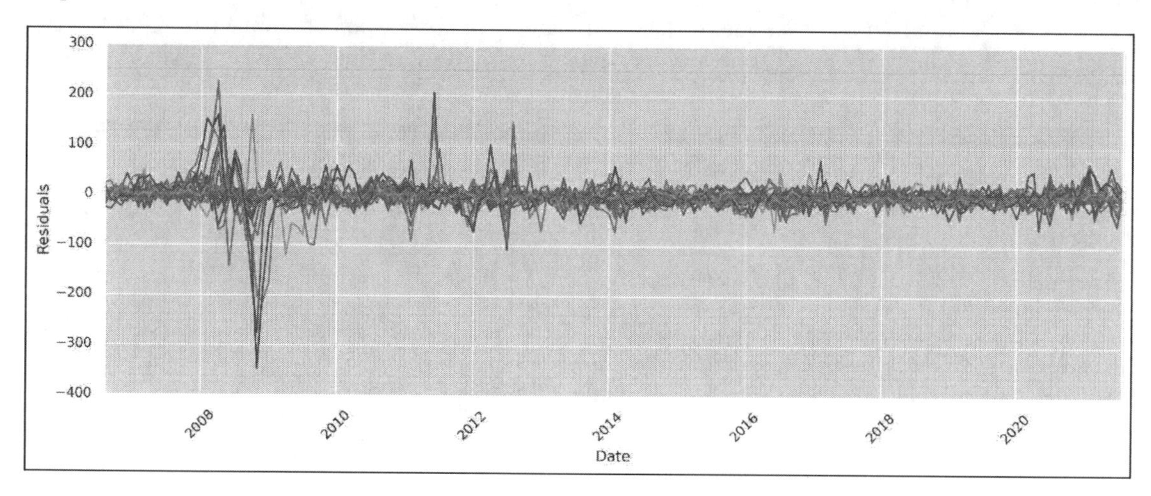

As it can be seen, the spikes of prices during the great recession.

Step 10: Durbin-Watson Statistic

The Durbin-Watson statistic will always have a value between 0 and 4. A value of 2.0 means that there is no autocorrelation detected in the sample. Values from 0 to less than 2 indicate positive autocorrelation and values from 2 to 4 indicate negative autocorrelation. A rule of thumb is that test statistic values in the range of 1.5 to 2.5 are relatively normal. Any value outside this range could be a cause for concern.

```
from statsmodels.stats.stattools import durbin_watson
out = durbin_watson(results.resid)
for col,val in zip(transform_data.columns, out):
  print((col), ":", round(val,2))
```

Output:

```
NGAS_US US ($/mmbtu) : 2.07
SOYBEANS ($/mt) : 1.91
CRUDE_PETRO ($/bbl) : 1.97
MAIZE ($/mt) : 1.88
SORGHUM ($/mt) : 2.1
RICE_05 ($/mt) : 2.07
WHEAT_US_SRW ($/mt) : 2.07
WHEAT_US_HRW ($/mt) : 1.93
TWEXAFEGSMTH : 1.94
Crude oil, Brent($/bbl) : 1.99
Cotton, A Index($/kg) : 1.96
Phosphate rock($/mt) : 2.07
DAP($/mt) : 1.99
TSP($/mt) : 1.93
Potassium chloride($/mt) : 2.02
```

All the values of DW is between 1.5 and 2.5.

The Durbin-Watson statistic will always have a value between 0 and 4. A value of 2.0 means that there is no autocorrelation

Step 11: Prediction or Forecast VAR

To forecast, the VAR model expects up to the lag order number of *observations* from the past data. This is because the terms in the VAR model are essentially the lags of the various time series in the dataset, so we need to provide as many of the previous values as indicated by the lag order used by the model.

```
# Get the lag order
lag_order = results.k_ar
print(lag_order)
# Input data for forecasting
input_data = transform_data.values[-lag_order:]
print(input_data)
# forecasting
pred = results.forecast(y=input_data, steps=n_obs)
pred = (pd.DataFrame(pred, index=X_test.index, columns=X_test.columns + '_pred'))
print(pred)
```

Output:

```
6
         DAP($/mt)_pred TSP($/mt)_pred Potassium chloride($/mt)_pred
Date
2021-03-01     83.643096      77.590662      6.147485
2021-04-01     31.450399      59.383043     11.799045
2021-05-01     10.473914      43.148238     28.333356
2021-06-01     15.694743      28.404281     22.793870
2021-07-01    –10.456153     –13.131657    –33.553211
2021-08-01     –3.340081      –2.153446     34.968971
2021-09-01    –13.745556      –1.876392      1.874642
2021-10-01     –2.053123     –16.069147     15.167234
2021-11-01    –11.880013     –22.694586     16.025429
2021-12-01      4.881489     –23.766734     37.291947
2022-01-01    –21.390670      –3.327082     –4.214219
2022-02-01     –2.116329     –18.581897     26.880335
```

Step 12: Invert the Transformation

The forecasts are generated but it is on the scale of the training data used by the model. So, to bring it back up to its original scale, we need to de-difference it. The way to convert the differencing is to add these differences consecutively to the base number. An easy way to do it is to first determine the cumulative sum at index and then add it to the base number. This process can be reversed by adding the observation at the prior time step to the difference value. inverted(ts) = differenced(ts) + observation(ts-1)

```
# inverting transformation
def invert_transformation(X_train, pred):
    forecast12 = pred.copy()
    columns = X_train.columns
    #col='CPI_AllUrban_FoodandBeverages'
    #forecast[str(col)+'_pred'] = dataset[col].iloc[-1] + forecast[str(col)+'_pred'].cumsum()
    for col in columns:
        print('++++++++++++++++')
        print(col)
        forecast12[str(col)+'_pred'] = X_train[col].iloc[-1] + forecast12[str(col)+'_pred'].cumsum()
        print('++++++++++++++++')
    return forecast12
output = invert_transformation(X_train, pred)
output
```

Output:

Date	NGAS_US US ($/mmbtu)_pred	SOYBEANS ($/mt)_pred	CRUDE_PETRO ($/bbl)_pred	MAIZE ($/mt)_pred	SORGHUM ($/mt)_pred	RICE_05 ($/mt)_pred	WHEAT_US_SRW ($/mt)_pred	WHEAT_US_HRW ($/mt)_pred	TWEXAFEGSMTH_pred
2021-03-01	5.314733	585.313568	65.002505	241.949756	194.688645	598.440847	294.224219	311.125722	101.947065
2021-04-01	5.212779	596.122901	63.066306	253.262961	199.876434	641.393884	309.081341	330.824647	102.161687
2021-05-01	5.365581	613.184032	65.413145	274.242086	215.484400	647.742961	317.546105	351.234970	102.485150
2021-06-01	5.774088	624.270167	70.577265	270.883702	213.115544	606.681340	299.174487	340.599394	103.090387
2021-07-01	5.352404	609.711081	73.864043	261.092162	204.318993	610.736440	296.046709	341.358623	102.826401
2021-08-01	5.199349	587.616494	70.840420	243.266443	192.676556	611.870978	285.779908	325.660874	103.694832
2021-09-01	5.243049	570.199083	67.964612	254.332386	203.149266	605.382618	283.240405	325.813155	104.508987
2021-10-01	5.346212	560.519269	64.890419	251.620524	194.990283	596.610858	282.368351	322.768948	105.451231
2021-11-01	5.198750	545.704569	61.792477	250.818728	196.334205	611.729315	278.786816	316.460062	106.265811

```
#combining predicted and real data set
combine = pd.concat([output['Phosphate rock($/mt)_pred'], X_test['Phosphate rock($/mt)']], axis=1)
combine['accuracy'] = round(combine.apply(lambda row: row['Phosphate rock($/mt)_pred'] / row['Phosphate rock($/mt)'] * 100, axis = 1),2)
combine['accuracy'] = pd.Series(["{0:.2f}%".format(val) for val in combine['accuracy']],index = combine.index)
combine = combine.round(decimals=2)
```

Output: a high accuracy can be observed.

Date	Phosphate rock($/mt)_pred	Phosphate rock($/mt)	accuracy
2021-03-01	98.40	96.25	102.23%
2021-04-01	123.75	95.00	130.26%
2021-05-01	128.60	102.50	125.47%
2021-06-01	145.31	125.00	116.25%
2021-07-01	167.19	125.00	133.75%
2021-08-01	182.18	136.88	133.09%
2021-09-01	200.86	147.50	136.18%
2021-10-01	231.92	147.50	157.24%
2021-11-01	257.62	153.13	168.24%
2021-12-01	254.88	176.67	144.27%
2022-01-01	257.10	173.13	148.50%
2022-02-01	261.16	172.50	151.40%

Step 13: Evaluation

To evaluate the forecasts, a comprehensive set of metrics, such as MAPE, ME, MAE, MPE and RMSE can be computed. We have computed some of these below.

```
from sklearn.metrics import mean_absolute_error
from sklearn.metrics import mean_squared_error
import math
#Forecast bias
forecast_errors = [combine['Phosphate rock($/mt)'][i]- combine['Phosphate rock($/mt)_pred'][i] for i in range(len(combine['Phosphate rock($/mt)']))]
bias = sum(forecast_errors) * 1.0/len(combine['Phosphate rock($/mt)'])
print('Bias: %f' % bias)
print('Mean absolute error:', mean_absolute_error(combine['Phosphate rock($/mt)'].values, combine['Phosphate rock($/mt)_pred'].values))
print('Mean squared error:', mean_squared_error(combine['Phosphate rock($/mt)'].values, combine['Phosphate rock($/mt)_pred'].values))
print('Root mean squared error:', math.sqrt(mean_squared_error(combine['Phosphate rock($/mt)'].values, combine['Phosphate rock($/mt)_pred'].values)))
```

Output:

Bias: –54.825833

Mean absolute error: 54.82583333333334

Mean squared error: 3973.1269916666674

Root mean squared error: 63.03274539211082

Step 14: Model Summary

Energy prices are drivers of inflation and food inflation. Farm input costs driven by high energy prices holistically influence the entire agricultural commodity value chain. As the model clearly shows, CRUDE and Natural Gas drive around 30% prices of the DAP & TSP fertilizers that are key farm input supplies of agriculture.[606]

Machine Learning Model—Commodities Demand and Energy Shocks on the Phosphate Model[607]

World prices for phosphorus and nitrogen fertilizers reached low levels in early 2020 and have been increasing sharply to reach levels not seen since 2008.[608] Futures' prices for these fertilizers for November delivery are even higher, above $700/mt. Potassium fertilizer prices remain low internationally but have reached much higher levels in the Corn Belt region and Brazil. The World Bank reports a 66% increase in fertilizer prices in 2021 in its commodity outlook and anticipates potassium fertilizer prices to sharply increase in 2022. What explains these steep increases in most fertilizer prices? In short, a perfect storm of supply shocks, weather events, plant closures, imperfectly competitive and shallow market structures, and a series of brewing policy changes are culminating into price volatility concerns and deterioration in fertilizer affordability. We review these varied elements in turn.

Here I would like to apply regression to develop a model that provides predictive capabilities to the following:

- Fertilizer markets are characterized by the outsized presence of a few very large players having an influence on these markets and prices.

- World supply of fertilizers has been curtailed with output and supply decreases in Europe, China and even Russia caused by sharp increases in production costs. Specifically, prices of natural gas and ammonia in Europe and Russia, and in China, high prices for coal have created electricity shortfalls and disruptions in the energy intensive fertilizer industry. These supply difficulties have pushed the Chinese government to put export restrictions in place, likely to last well into 2022. Russia has also placed restrictions on exports to keep domestic fertilizer costs down as higher fossil energy prices increase the cost of fertilizer production.
- In the U.S., hurricane Ida induced some plant closings in the southeast. These U.S. supply disruptions are temporary, but they have exacerbated existing logistic issues in markets.
- Fertilizers are differentiated by type of application and timing in the growing season and availability is a key attribute of supply which can induce scarcity.
- The increase in fossil fuel prices induced by the 2021 economic recovery and anti-fossil policies in many OECD countries is a fundamental factor which feeds into ammonium fertilizer prices and energy-intensive fertilizer production.

Step 1: Prepare the Dataset

Use the dataset as developed in the above ML models.

```
dataset.columns
```

Output:

```
Index(['NGAS_US US ($/mmbtu)', 'SOYBEANS ($/mt)', 'CRUDE_PETRO ($/bbl)',
   'MAIZE ($/mt)', 'SORGHUM ($/mt)', 'RICE_05 ($/mt)',
   'WHEAT_US_SRW ($/mt)', 'WHEAT_US_HRW ($/mt)', 'TWEXAFEGSMTH',
   'Cotton, A Index($/kg)', 'Phosphate rock($/mt)', 'DAP($/mt)',
   'TSP($/mt)', 'Potassium chloride($/mt)'],
   dtype='object')
```

Step 2: Train and Split the Dataset

Prepare the X and Y variables.

```
X = dataset[['NGAS_US US ($/mmbtu)', 'SOYBEANS ($/mt)', 'CRUDE_PETRO ($/bbl)',
   'MAIZE ($/mt)', 'SORGHUM ($/mt)', 'RICE_05 ($/mt)',
   'WHEAT_US_SRW ($/mt)', 'WHEAT_US_HRW ($/mt)',
   'TWEXAFEGSMTH', 'Cotton, A Index($/kg)',
   'Phosphate rock($/mt)']]
y = dataset['DAP($/mt)']
from sklearn.model_selection import train_test_split
X_train, X_test, y_train, y_test = train_test_split(X, y, test_size = 0.3, random_state = 0)
```

Step 3: Train the Model

```
# Train the model
from sklearn.linear_model import LinearRegression
# Fit a linear regression model on the training set
model = LinearRegression(normalize=False).fit(X_train, y_train)
print (model)
```

Output:

LinearRegression()

Step 4: Evaluate the Model

```
import numpy as np
predictions = model.predict(X_test)
np.set_printoptions(suppress=True)
print('Predicted labels: ', np.round(predictions))
print('Actual labels : ' ,y_test)
```

Output:

```
Predicted labels: [351. 496. 383. 456. 294. 497. 305. 360. 294. 271. 227. 251. 329. 313.
 438. 650. 317. 465. 488. 451. 380. 412. 341. 359. 426. 403. 397. 270.
 439. 418. 512. 218. 240. 219. 613. 792. 456. 315. 729. 362. 303. 349.
 153. 344. 320. 518. 422. 598. 332. 583. 554. 995. 386. 383. 610. 439.
 288. 316. 299.]
Actual labels :  Date
2015-06-01    421.375
2021-05-01    574.630
2010-09-01    506.000
2013-06-01    424.125
2007-01-01    249.200
2009-02-01    313.750
2007-07-01    398.500
2016-06-01    306.625
2019-08-01    292.900
2006-08-01    228.250
2006-06-01    232.375
2016-04-01    330.500
```

Step 5: Plot the Model

Plot the regression line.

```
import matplotlib.pyplot as plt
%matplotlib inline
plt.scatter(y_test, predictions)
plt.xlabel('Actual Labels')
plt.ylabel('Predicted Labels')
plt.title('Predictions')
# overlay the regression line
z = np.polyfit(y_test, predictions, 1)
p = np.poly1d(z)
plt.plot(y_test,p(y_test), color='magenta')
plt.show()
```

Output:

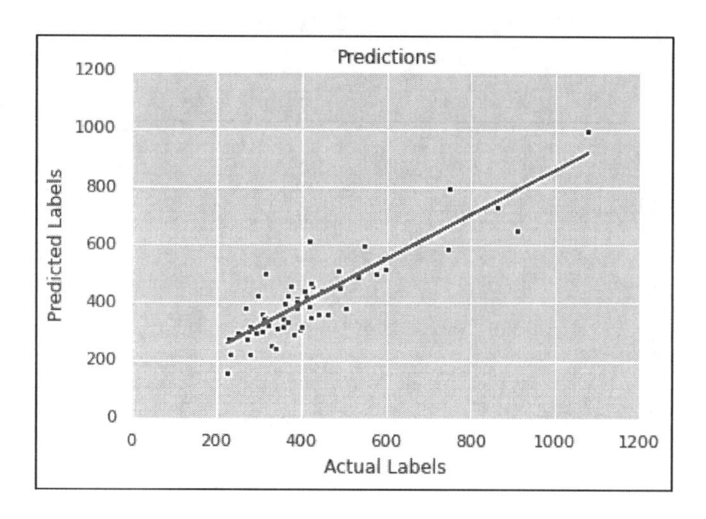

Step 6: Model evaluation

Evaluate the model metrics in terms of R^2 and others.

```
from sklearn.metrics import mean_squared_error, r2_score
mse = mean_squared_error(y_test, predictions)
print("MSE:", mse)
rmse = np.sqrt(mse)
print("RMSE:", rmse)
r2 = r2_score(y_test, predictions)
print("R2:", r2)
```

Output:

MSE: 6776.037218362257

RMSE: 82.31668857748261

R2: 0.7575290391850149

There is a high probability of price prediction with upper and lower bounds, +/– $82.31 dollars when predicting DAP ($/mt).

Step 7: Model Equation (y=mx+c)

Let's construct the model with coefficients,

```
mx=""
for ifeature in range(len(X.columns)):
    if model.coef_[ifeature] <0:
      # format & beautify the equation
      mx += " - " + "{:.2f}".format(abs(model.coef_[ifeature])) + " * " + X.columns[ifeature]
    else:
      if ifeature == 0:
         mx += "{:.2f}".format(model.coef_[ifeature]) + " * " + X.columns[ifeature]
      else:
         mx += " + " + "{:.2f}".format(model.coef_[ifeature]) + " * " + X.columns[ifeature]
print(mx)
```

Output:

```
51.20 * NGAS_US US ($/mmbtu) + 0.72 * SOYBEANS ($/mt) + 1.24 * CRUDE_PETRO ($/bbl) - 0.39
* MAIZE ($/mt) + 0.61 * SORGHUM ($/mt) + 0.38 * RICE_05 ($/mt) + 0.23 * WHEAT_US_SRW ($/
mt) - 0.11 * WHEAT_US_HRW ($/mt) + 8.81 * TWEXAFEGSMTH + 49.11 * Cotton, A Index($/kg) +
0.42 * Phosphate rock($/mt)
```

```
# y=mx+c
if(model.intercept_ <0):
 print("The formula for the " + y.name + " linear regression line (y=mx+c) is = " + " - {:.2f}".
format(abs(model.intercept_)) + " + " + mx )
else:
 print("The formula for the " + y.name + " linear regression line (y=mx+c) is = " + "{:.2f}".format(model.
intercept_) + " + " + mx )
```

Output:

```
The formula for the DAP($/mt) linear regression line (y=mx+c) is = - 1451.06 + 51.20 * NGAS_US US
($/mmbtu) + 0.72 * SOYBEANS ($/mt) + 1.24 * CRUDE_PETRO ($/bbl) - 0.39 * MAIZE ($/mt) + 0.61
* SORGHUM ($/mt) + 0.38 * RICE_05 ($/mt) + 0.23 * WHEAT_US_SRW ($/mt) - 0.11 * WHEAT_US_
HRW ($/mt) + 8.81 * TWEXAFEGSMTH + 49.11 * Cotton, A Index($/kg) + 0.42 * Phosphate rock($/mt)
```

Step 8: Conclusion

Fertilizers and agricultural staple commodities (wheat, rice, sorghum, maize) have a special relationship. In terms of statistical causation, more precisely Granger's causation, they exhibit a bi-directional influence. The prices of fertilizers increase as prices of commodities do. The vice versa is also true: the demand for agricultural commodities rises so fertilizer prices increase [45].[609]

Nitrogen fertilizers are a key component in the production of field crops. Fertilizers constitute an average of 36 percent[610] of a farmer's operating costs for corn, 35 percent for wheat, and 30 percent for sorghum, according to estimates in the USDA [46]. Global fertilizer prices soared to multi-year highs in the past few months following surges in prices of key feedstocks, natural gas and coal, and certain export restrictions were put in place by supplying countries. This causes an upward swing in the prices of commodities.

In addition, around the globe, natural gas is used as a raw material as well as fuel for nitrogen fertilizer production. Nitrogen fertilizers are the most used fertilizers in the world. Ammonia, phosphorous and potash are the other important fertilizer components.[611] In late 2021, fertilizer prices began to spike *alongside rising prices of natural gas*—a primary input in nitrogen fertilizer production. By December 2021, average monthly spot prices of natural gas at the Henry Hub, a distribution hub in Louisiana, as published by the U.S. Energy Information Administration, were 45 percent higher than in December 2020.[612] The key driver of the prices of agricultural commodities, energy, and fertilizers is an explosive price territory. High natural gas prices, if sustained through early 2022, will translate into higher food costs and reinforce an inflationary trend that is already being driven by supply chain disruptions, labor shortages and increased demand from the biofuels sector, thus one can expect high inflation till agricultural inputs normalize to stable prices. The same inference can be made from the coefficients when predicting the DAP ($/mt) [46].

Model Variable	Coefficients	Description
NGAS_US US ($/mmbtu)	51.20	All held constant, a dollar increase in the natural gas ($/mmbtu) would increase[613] the DAP price by 51.20 dollars per metric ton ($/mt) [47].
SOYBEANS ($/mt)	0.72 {exhibits bi-directional[614] causality, as described in the VAR model}	All held constant, a demand increase in the Soybeans would increase DAP price by 0.72 dollar per metric ton ($/mt) [44]
CRUDE_PETRO ($/bbl)	1.24	All held constant, a dollar increase in the crude petro ($/bbl) would increase[615] the DAP price [47].
MAIZE ($/mt)	-0.39	Both MAP and DAP are effective providers of P and N for corn when applied prior to planting.[616] But when banded at planting time, MAP has the advantage. Growers prefer MAP based largely upon two soil conditions during banding: the possibility of release of free ammonia and the change in pH of the soil solution surrounding the fertilizer particle. The negative regression coefficient justifies as MAP is preferred over DAP.
SORGHUM ($/mt)	0.61	Price hikes in gases would increase DAP Prices same goes with demand for Sorghum, Rice, Cotton, and Wheat.
RICE	0.38	
* WHEAT_US_ SRW ($/mt)	0.23	
Cotton	49.11	

MAP AS A STARTER FERTILIZER FOR CORN[617]

Monoammonium phosphate (MAP) and diammonium phosphate (DAP) are excellent sources of phosphorus (P) and nitrogen (N) for high-yield, high-quality crop production. Research trials at 42 field sites in seven Corn Belt states showed an average corn yield of 162 bushels per acre with MAP and 159 with DAP. MAP (11-52-0) and DAP (18-46-0) contain about 90 percent water-soluble P, which is well above the 60 percent needed for optimum crop growth.

Growers who plan to apply P and part of the N as a starter fertilizer for corn should be aware of the qualities that make MAP a preferred P source.

Both MAP and DAP are effective providers of P and N for corn when applied prior to planting. But when banded at planting time, MAP has the advantage. Growers prefer MAP based largely upon two soil conditions during banding: the possibility of release of free ammonia and the change in pH of the soil solution surrounding the fertilizer particle.

Since DAP contains a higher content of ammonium nitrogen, its granules can release free ammonia as they dissolve in soil solution. In acidic soils, this release of free ammonia can injure seeds if DAP is placed with or near germinating seeds. Also, the release of free ammonia results in a temporary elevation of soil pH near the fertilizer granules, which can lower the availability of certain micronutrients such as zinc and manganese. Ultimately, both MAP and DAP have an acidic effect on the soil due to the acidity created when ammonium ions convert to nitrates in the soil, but since the soil solution surrounding MAP granules remains acidic, there's no significant release of free ammonia to injure seedlings.

While 80% of the gas is used as feedstock for fertilizer, 20% is used for heating the process and producing electricity.

Yara Fertilizers

Machine Learning Model—Prophet Time Series Commodities Demand and Energy Shocks on Phosphate Model

The purpose of the model is to establish linkage drivers of Phosphate fertilizers. World prices for phosphorus and nitrogen fertilizers reached low levels in early 2020 and have been increasing sharply levels not seen since 2008.[618] Futures prices for these fertilizers are even higher.

Step 1: Load dataset

We can re-use the same dataset that we developed above.

```
Dataset
```

Output:

Date	NGAS_US US ($/mmbtu)	SOYBEANS ($/mt)	CRUDE_PETRO ($/bbl)	MAIZE ($/mt)	SORGHUM ($/mt)	RICE_05 ($/mt)	WHEAT_US_SRW ($/mt)	WHEAT_US_HRW ($/mt)	TWEXAFEGSMTH	Cotton, A Index($/kg)
2006-01-01	8.6600	257.00	62.46	102.650000	100.64	291.25	144.160000	167.16	100.0000	1.286396
2006-02-01	7.4900	257.00	59.70	107.130000	106.07	301.50	149.250000	179.84	100.7907	1.315056
2006-03-01	6.9000	257.00	60.93	105.250000	103.66	303.50	142.660000	174.44	100.7567	1.269861
2006-04-01	7.0900	258.00	67.97	107.720000	109.35	302.25	140.820000	180.35	99.4118	1.240099
2006-05-01	6.2000	266.00	68.68	110.650000	113.76	308.00	150.920000	193.17	95.5569	1.198211
...	...	...	...	...	...	...	...	...	...	...
2021-10-01	5.4780	551.95	81.32	239.648763	189.49	401.00	263.599089	354.67	105.6996	2.587783
2021-11-01	5.0176	551.04	79.18	248.719150	189.49	400.00	334.499674	379.45	107.0284	2.789726
2021-12-01	3.7327	554.14	71.53	264.537213	189.49	400.00	327.819675	376.81	108.2017	2.646426
2022-01-01	4.3325	606.22	83.12	276.623189	189.49	427.00	332.059894	374.24	107.6318	2.911201
2022-02-01	4.6577	661.63	91.74	292.622344	189.49	427.00	533.121257	390.50	107.8260	3.051415

Step 2: Inspect dataset and construct dataset for running Prophet Time Series model

Construct dataset for modeling prophet time series model.

```
dataset.info()
dfProphetDataset=dataset.copy()
```

Output:

```
<class 'pandas.core.frame.DataFrame'>
DatetimeIndex: 194 entries, 2006-01-01 to 2022-02-01
Data columns (total 14 columns):
NGAS_US US ($/mmbtu) 194 non-null float64
SOYBEANS ($/mt) 194 non-null float64
CRUDE_PETRO ($/bbl) 194 non-null float64
MAIZE ($/mt) 194 non-null float64
SORGHUM ($/mt) 194 non-null float64
RICE_05 ($/mt) 194 non-null float64
WHEAT_US_SRW ($/mt) 194 non-null float64
WHEAT_US_HRW ($/mt) 194 non-null float64
TWEXAFEGSMTH 194 non-null float64
Cotton, A Index($/kg) 194 non-null float64
Phosphate rock($/mt) 194 non-null float64
DAP($/mt) 194 non-null float64
TSP($/mt) 194 non-null float64
Potassium chloride($/mt) 194 non-null float64
dtypes: float64(14)
memory usage: 22.7 KB
```

```
dfProphetdf = dfProphetDataset.reset_index()[['Date','NGAS_US US ($/mmbtu)', 'SOYBEANS ($/mt)',
'CRUDE_PETRO ($/bbl)',
    'MAIZE ($/mt)', 'SORGHUM ($/mt)', 'RICE_05 ($/mt)',
    'WHEAT_US_SRW ($/mt)', 'WHEAT_US_HRW ($/mt)',
    'TWEXAFEGSMTH', 'Cotton, A Index($/kg)',
    'Phosphate rock($/mt)' , 'DAP($/mt)']].rename({'Date': 'ds','DAP($/mt)':'y'},axis= 'columns')
DfProphetdf
```

Step 3: Split Dataset

Split dataset.

```
from fbprophet import Prophet
## test and train data split
n_obs=30
fbProphet_train, fbProphet_test = dfProphetdf[0:-n_obs],dfProphetdf[-n_obs:]
print(fbProphet_train.shape,fbProphet_test.shape)
```

Output:
(164, 13) (30, 13)

Step 4: Construct Prophet Model & add regressors

Prophet is a procedure for forecasting time series[619] with data based on an additive model where non-linear trends are fit with yearly, weekly, and daily seasonality, plus holiday effects. It works best with time series that have strong seasonal effects and several seasons of historical data. Prophet is robust to missing data and shifts in the trend, and typically handles outliers well. Prophet follows the sklearn model API. We create an instance of the Prophet class and then call its fit and predict methods.

```
multi_model = Prophet(interval_width = 0.95)
```

Add key regressor attributes:

```
Prophet_cols = [ 'NGAS_US US ($/mmbtu)', 'SOYBEANS ($/mt)', 'CRUDE_PETRO ($/bbl)',
   'MAIZE ($/mt)', 'SORGHUM ($/mt)', 'RICE_05 ($/mt)',
   'WHEAT_US_SRW ($/mt)', 'WHEAT_US_HRW ($/mt)', 'TWEXAFEGSMTH',
   'Cotton, A Index($/kg)',
   'Phosphate rock($/mt)']
# adding all columns in add regressor
for col in Prophet_cols:
   multi_model.add_regressor(col)
```

Step 5: Train the Model

Train the model.

```
multi_model.fit(fbProphet_train)
```

```
# filling all columns data in future frame except for y
for col in Prophet_cols:
   future[col] = dfProphetdf[col]
```

```
Future
```

Output:

	ds	NGAS_US US ($/mmbtu)	SOYBEANS ($/mt)	CRUDE_PETRO ($/bbl)	MAIZE ($/mt)	SORGHUM ($/mt)	RICE_05 ($/mt)	WHEAT_US_SRW ($/mt)	WHEAT_US_HRW ($/mt)	TWEXAFEGSMTH
0	2006-01-01	8.6600	257.00	62.46	102.650000	100.64	291.25	144.160000	167.16	100.0000
1	2006-02-01	7.4900	257.00	59.70	107.130000	106.07	301.50	149.250000	179.84	100.7907
2	2006-03-01	6.9000	257.00	60.93	105.250000	103.66	303.50	142.660000	174.44	100.7567
3	2006-04-01	7.0900	258.00	67.97	107.720000	109.35	302.25	140.820000	180.35	99.4118
4	2006-05-01	6.2000	266.00	68.68	110.650000	113.76	308.00	150.920000	193.17	95.5569
...	...	...	...	...	...	...	...	...	...	...
189	2021-10-01	5.4780	551.95	81.32	239.648763	189.49	401.00	263.599089	354.67	105.6996
190	2021-11-01	5.0176	551.04	79.18	248.719150	189.49	400.00	334.499674	379.45	107.0284
191	2021-12-01	3.7327	554.14	71.53	264.537213	189.49	400.00	327.819675	376.81	108.2017
192	2022-01-01	4.3325	606.22	83.12	276.623189	189.49	427.00	332.059894	374.24	107.6318
193	2022-02-01	4.6577	661.63	91.74	292.622344	189.49	427.00	533.121257	390.50	107.8260

Step 6: Prediction

Predict the time series.

```
forecastProphet = multi_model.predict(future)
forecastProphet.tail()
```

Output:

	ds	trend	yhat_lower	yhat_upper	trend_lower	trend_upper	CRUDE_PETRO ($/bbl)	CRUDE_PETRO ($/bbl)_lower	CRUDE_PETRO ($/bbl)_upper	Cotton, A Index($/kg)
189	2021-10-01	475.978730	363.570668	613.308331	475.978720	475.978740	8.048428	8.048428	8.048428	20.274263
190	2021-11-01	476.346511	385.855841	611.098562	476.346501	476.346522	4.815423	4.815423	4.815423	25.800321
191	2021-12-01	476.702429	216.266249	463.226628	476.702418	476.702441	-6.741813	-6.741813	-6.741813	21.878992
192	2022-01-01	477.070211	287.476066	519.868194	477.070199	477.070223	10.767777	10.767777	10.767777	29.124402
193	2022-02-01	477.437992	646.794709	890.286776	477.437980	477.438005	23.790441	23.790441	23.790441	32.961272

5 rows × 52 columns

Plot the time series to visualize:

```
fig1 = multi_model.plot(forecastProphet)
```

Output:

To see the forecast components, Prophet.plot_components[620] is useful. In the figure below, we can see the trend, yearly seasonality, and weekly seasonality of the time series. Black dots are regressor data points used to train the model. The line plot consists of the prediction values of DAP. It is supposed to be a line, but it looks like a weird wide blue area.[621]

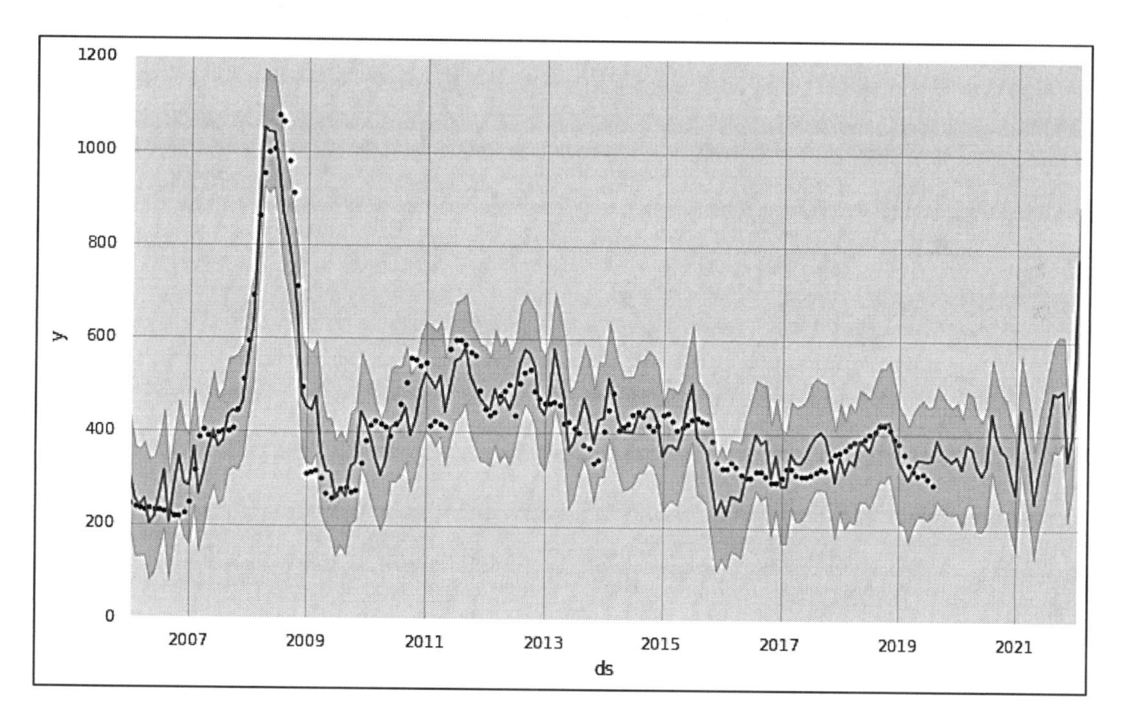

Step 7: Evaluate the Model

Evaluate the model by calculating MAPE and others.

```
# calculate MAE between expected and predicted values for december
y_true = dfProphetdf['y'].values
y_pred = forecastProphet['yhat'].values

print(y_true.shape)
print(y_pred.shape)
```

Output:
(194,)
(194,)

```
from sklearn.metrics import mean_absolute_error
from matplotlib import pyplot
plt.figure(figsize=(18, 9))
mae = mean_absolute_error(y_true, y_pred)
print('MAE: %.3f' % mae)
# plot expected vs actual
pyplot.plot(y_true, label='Actual')
pyplot.plot(y_pred, label='Predicted')
pyplot.legend()
pyplot.show()
```

Output:
MAE: 62.556

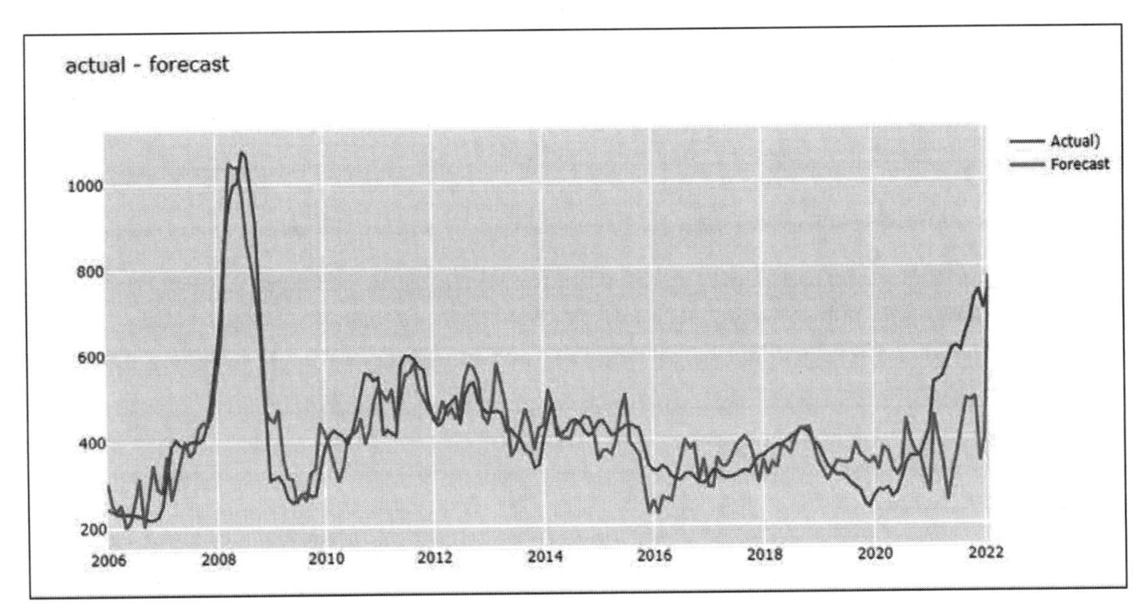

As can be seen, the prophet forecast follows very closely with the actual.

Heat waves could lead to energy and foreign reserves deficit. The endogenous impact is an increase in overall food insecurity for countries that depend on fossil fuels.

After reading this chapter, you should be able to answer questions on the impact of energy prices on macroeconomic drivers. Next, you should be able to construct a ML model with changes in energy prices from the U.S. economic point of view and analyze the impact of energy price fluctuations on the agricultural commodities. Next, you should be able to use agricultural farm inputs, fertilizers, and develop a linkage model. Finally, you should be proficient in applying the food security bell curve model to develop Machine Learning Models for Energy Prices & Fertilizer Costs—Urea, Energy Prices and Fertilizer Costs—Phosphate, and Commodities Demand and Energy Shocks on Phosphate Model.

References

[1] Lutz Kilianm. The Economic Effects of Energy Price Shocks, July 12, 2008, https://www.aeaweb.org/conference/2009/retrieve.php?pdfid=145, Access Date: May 11, 2022.

[2] David Knowles. Record-breaking heat wave gripping India and Pakistan threatens crops, leaves millions sweltering, Tue, April 26, 2022, 1:59 PM, https://www.yahoo.com/news/record-breaking-heat-wave-gripping-india-and-pakistan-threatens-crops-leaves-millions-sweltering-205937185.html, Access Date: May 05, 2022.

[3] Kishore, P., Basha, G., Venkat Ratnam, M. Amir AghaKouchak, Qiaohong Sun, Isabella Velicogna and Ouarda, T.B.J.M. 2022. Anthropogenic influence on the changing risk of heat waves over India. Sci. Rep. 12: 3337. https://doi.org/10.1038/s41598-022-07373-3.

[4] Patrick Galey. Heat wave in India threatens residents and crucial wheat harvest, PUBLISHED TUE, APR 26 20229:26 PM EDTUPDATED TUE, APR 26 202211:01 PM EDT, https://www.cnbc.com/2022/04/27/heat-wave-in-india-threatens-residents-and-crucial-wheat-harvest.html, Access Date: May 13, 2022.

[5] Sara, E. Pratt. 2022. Early Season Heat Waves Strike India, April 27, 2022, https://earthobservatory.nasa.gov/images/149766/early-season-heat-waves-strike-india, Access Date: May 13, 2022.

[6] Bloomberg. India turns to expensive foreign gas to ease its power crisis, Last Updated: May 05, 2022, 05:53 PM IST, https://economictimes.indiatimes.com/industry/energy/oil-gas/india-turns-to-expensive-foreign-gas-to-ease-its-power-crisis/articleshow/91333653.cms, Access Date: May 10, 2022.

[7] Arshad, R. Zargar. Severe heat wave kills dozens in India and Pakistan in a "snapshot" of what's to come from climate change, expert says, MAY 9, 2022/2:53 PM, https://www.cbsnews.com/news/india-heat-wave-pakistan-climate-change-snapshot/,Access Date: May 10, 2022.

[8] Dirk Jan van de Ven and Roger Fouquet, Historical Energy Price Shocks and their Changing Effects on the Economy, April 2014, https://www.lse.ac.uk/granthaminstitute/wp-content/uploads/2014/03/WP153-Historical-energy-price-shocks-and-their-changing-effects-on-the-economy.pdf, Access Data: May 11,2022.

[9] Charlotte Tuttle. Unexpected Hikes in Energy Prices Increase the Likelihood of Food Insecurity, July 13, 2017, https://www.ers.usda.gov/amber-waves/2017/july/unexpected-hikes-in-energy-prices-increase-the-likelihood-of-food-insecurity/, Access Data: May 11, 2022 .

[10] Charlotte Tuttle and Timothy K.M. Beatty. The Effects of Energy Price Shocks on Household Food Security in Low-Income Households, July 2017, https://www.ers.usda.gov/publications/pub-details/?pubid=84240, Access Data: May 11, 2022.

[11] Charles, A. Radin. US Business Cycle Expansions and Contractions, Business cycle data last updated: 07/19/2021,https://www.nber.org/research/data/us-business-cycle-expansions-and-contractions, Access Date: May 12, 2022.

[12] U.S. Energy Information Administration, SHORT-TERM ENERGY OUTLOOK, Release Date: May 10, 2022 | Forecast Completed: May 5, 2022 | Next Release Date: Jun. 7, 2022, https://www.eia.gov/outlooks/steo/realprices/, Access Date: May 12, 2022.

[13] Timothy Puko. Biden Exploring Release of Diesel Fuel Reserves Amid High Prices, May 23, 2022 3:14 pm ET, https://www.wsj.com/articles/biden-exploring-release-of-diesel-fuel-reserves-amid-high-prices-11653333243, Access Date: May 25, 2022.

[14] Federal Reserve Bank of San Francisco. What are the possible causes and consequences of higher oil prices on the overall economy? November 2007, https://www.frbsf.org/education/publications/doctor-econ/2007/november/oil-prices-impact-economy/, Access Date: May 12, 2022.

[15. Paul Page. Rising Diesel Costs Are Straining U.S. Truckers, Shipping Operations,May 12, 2022 1:20 pm ET,https://www.wsj.com/articles/rising-diesel-costs-are-straining-u-s-truckers-shipping-operations-11652376035?mod=djemRTE_h, Access Date: May 13, 2022.

[16] Sill, Keith. 2007. The Macroeconomics of Oil Shocks. FRB Philadelphia Business Review, 2007:Q1,https://www.philadelphiafed.org/-/media/frbp/assets/economy/articles/business-review/2007/q1/br_q1-2007-3_oil-shocks.pdf, Access Date: May 12, 2022.

[17] John Fernald and Bharat Trehan. Why Hasn't the Jump in Oil Prices Led to a Recession? 2005-31 | November 18, 2005, https://www.frbsf.org/economic-research/publications/economic-letter/2005/november/why-the-jump-in-oil-prices-has-not-led-to-a-recession/, Access Date: May 12, 2022.

[18] Steve Hargreaves. Gasoline prices averaged the highest ever in 2012, despite never topping the single day spikes seen in 2008, December 31, 2012: 4:11 PM ET, https://money.cnn.com/2012/12/31/news/economy/gas-prices/, Access Date: May 12, 2022.

[19] Ben, S. Bernanke. Irreversibility, Uncertainty, and Cyclical Investment, The Quarterly Journal of Economics, Volume 98, Issue 1, February 1983, Pages 85–106, https://doi.org/10.2307/1885568.

[20] Robert, S. Pindyck. Irreversibility, Uncertainty, and Investment, September 1991, http://web.mit.edu/rpindyck/www/Papers/IrreverUncertInvestmentJEL1991.pdf, Access Date: May 12, 2022.

[21] Roberto, Mosheim. Fertilizer Use and Price, Last updated: Wednesday, October 30, 2019, https://www.ers.usda.gov/data-products/fertilizer-use-and-price/, Access Date: May 08, 2022.

[22] Sellars, S. and V. Nunes. 2021. Synthetic Nitrogen Fertilizer in the U.S. farmdoc daily (11): 24, Department of Agricultural and Consumer Economics, University of Illinois at Urbana-Champaign, February 17, 2021, Access Date: Ma.

[23] Ben Dummett. Bayer's Troubled Monsanto Megadeal Finally Shows Promise, Updated June 7, 2022, 4:37 pm ET, https://www.wsj.com/articles/russia-ukraine-war-threatens-wheat-supply-jolts-prices-11647115099?mod=article_inline,Access Date: June 10, 2022.

[24] Apodaca, Lori E. 2020a. Nitrogen (Fixed) – Ammonia. Mineral Commodity Summaries. U.S. Geological Survey.

[25] RENEE CHO, Phosphorus: Essential to Life—Are We Running Out? APRIL 1, 2013, https://news.climate.columbia.edu/2013/04/01/phosphorus-essential-to-life-are-we-running-out/, Access Date: May 14, 2022.

[26] Colussi, J., G. Schnitkey and C. Zulauf. War in Ukraine and its Effect on Fertilizer Exports to Brazil and the U.S. farmdoc daily (12): 34, Department of Agricultural and Consumer Economics, University of Illinois at Urbana-Champaign, March 17, 2022., Access Date: May 08, 2022.

[27] Elliot Smith, Fertilizer prices are at record highs. Here's what that means for the global economy, PUBLISHED TUE, MAR 22 20228:28 AM EDTUPDATED TUE, MAR 22 20228:36 AM EDT, https://www.cnbc.com/2022/03/22/fertilizer-prices-are-at-record-highs-heres-what-that-means.html, Access Date: May 08, 2022.

[28] Aczel, M. 2019. What is the Nitrogen Cycle and Why is it Key to Life? Front. Young Minds. 7: 41. doi: 10.3389/frym.2019.00041.

[29] Chris Dawson, Potassium a Nutrient Essential for Life, 2014, https://www.ipipotash.org/uploads/udocs/388-ipi-booklet-potassium-a-nutrient-for-life.pdf, Access Date: May 14, 2022.

[30] USDA Press, USDA Announces Plans for $250 Million Investment to Support Innovative American-made Fertilizer to give US Farmers more choices in the Marketplace, WASHINGTON, March 11, 2022, https://www.usda.gov/media/press-releases/2022/03/11/usda-announces-plans-250-million-investment-support-innovative, Access Date: May 10, 2022.

[31. Kimberly Amadeo and Reviewed by Erika Rasure, U.S. Dollar Index: What It Is and Its Recent History, Updated May 02, 2022, https://www.thebalance.com/u-s-dollar-index-historical-data-3306249, Access Date: May 21, 2022.

[32] Chad, E. Hart. U.S. Agriculture and the Value of the Dollar, Summer 2003, https://www.card.iastate.edu/iowa_ag_review/summer_03/article5.aspx, Access Date: May 21, 2022.

[33] Stephanie Glen. Bimodal Distribution: What is it? From StatisticsHowTo.com: Elementary Statistics for the rest of us! https://www.statisticshowto.com/what-is-a-bimodal-distribution/, Access Date: May 21, 2022.

[34] Economic Times, how global currencies fared in the calendar year 2016, Updated: 23 Dec 2016, 12:37 PM IST, https://economictimes.indiatimes.com/investments-markets/how-global-currencies-fared-in-the-calendar-year-2016/us-dollar/slideshow/56135255.cms, Access Date: May 21, 2022.

[35] Dollar surges to highest level since March 2003 after US economic data, PUBLISHED WED, NOV 23 20162:39 PM ESTUPDATED WED, NOV 23 20162:39 PM EST, https://www.cnbc.com/2016/11/22/dollar-nears-13-12-year-high-buoyed-by-multiple-rate-hike-view.html , Access Date: May 21, 2022.

[36] Chuck Kowalski and Gordon Scott. How the Dollar Impacts Commodity Prices, Updated December 29, 2021, https://www.thebalance.com/how-the-dollar-impacts-commodity-prices-809294, Access Date: May 20, 2022.

[37] Shelby Myers and Veronica Nigh. Too Many to Count: Factors Driving Fertilizer Prices Higher and Higher, December 13, 2021, https://www.fb.org/market-intel/too-many-to-count-factors-driving-fertilizer-prices-higher-and-higher, Access Date: May 08, 2022.

[38] Jeffrey Gillespie. Commodity Costs and Returns, Last updated: Tuesday, May 03, 2022, https://www.ers.usda.gov/data-products/commodity-costs-and-returns/commodity-costs-and-returns/#Historical%20Costs%20and%20Returns:%20Milk, Access Date: May 08, 2022.

[39] Elizabeth Elkin. Fertilizer Prices Climb to New North American Peak, Squeezing Crop Margins, November 5, 2021, 10:06 AM PDT, https://www.bloomberg.com/news/articles/2021-11-05/soaring-fertilizer-prices-climb-to-new-north-american-peak, Access Date: May 21, 2022.

[40] Stephen, M. Jasinski. Mineral Resource of the Month: Phosphate Rock, Wednesday, January 28, 2015, https://www.earthmagazine.org/article/mineral-resource-month-phosphate-rock/, Access Date: May 21, 2022.

[41] Nikolay Khabarov and Michael Obersteiner. Global Phosphorus Fertilizer Market and National Policies: A Case Study Revisiting the 2008 Price Peak, 14 June 2017, https://www.frontiersin.org/articles/10.3389/fnut.2017.00022/full, Access Date: May 21, 2022.

[42] Jason Brownlee. How to Check if Time Series Data is Stationary with Python, December 30, 2016, https://machinelearningmastery.com/time-series-data-stationary-python/, Access Date: May 21, 2022.

[43] Peter Watson. Testing normality including skewness and kurtosis, last edited 2018-08-14 09:28:35, https://imaging.mrc-cbu.cam.ac.uk/statswiki/FAQ/Simon, Access Date: May 21, 2022.

[44] Thomas, J. Gumbley. Price Relationships in the U.S. Nitrogen Fertilizer Industry, April 2018, https://scholarworks.montana.edu/xmlui/bitstream/handle/1/14550/GumbleyT0518.pdf?isAllowed=y&sequence=1, Access Date: May 22, 2022.

[45] Keith Good. Prices of Most Fertilizers Continue Rise, as Wheat Prices Climb on Supply Concerns, May 6, 2022, https://farmpolicynews.illinois.edu/2022/05/prices-of-most-fertilizers-continue-rise-as-wheat-prices-climb-on-supply-concerns/, Access Date: May 22, 2022.
[46] Angelica, Williams and Amy Boline. Fertilizer prices spike in leading U.S. market in late 2021, just ahead of 2022 planting season, Last updated: Wednesday, February 09, 2022, https://www.ers.usda.gov/data-products/chart-gallery/gallery/chart-detail/?chartId=103194, Access Date: May 22, 2022.
[47] Gary, Schnitkey. Current Fertilizer Prices, Natural Gas, and Longer-Run Supply, October 30, 2012, https://farmdocdaily.illinois.edu/2012/10/current-fertilizer-prices-natu.html, Access Date: May 22, 2022.

[537] The Economic Effects of Energy Price Shocks - https://www.aeaweb.org/conference/2009/retrieve.php?pdfid=145

[538] Extreme heat impacting millions across India and Pakistan - https://news.un.org/en/story/2022/04/1117272

[539] Early Season Heat Waves Strike India - https://earthobservatory.nasa.gov/images/149766/early-season-heat-waves-strike-india

[540] Anthropogenic influence on the changing risk of heat waves over India - https://www.nature.com/articles/s41598-022-07373-3#citeas

[541] Heatwave data- https://www.epa.gov/sites/default/files/2021-04/heat-waves_fig-3.csv

[542] Climate Change Indicators: Heat Waves https://www.epa.gov/climate-indicators/climate-change-indicators-heat-waves#ref9 -

[543] India Energy Use - https://www.eia.gov/international/analysis/country/IND

[544] India turns to expensive foreign gas to ease its power crisis - https://economictimes.indiatimes.com/industry/energy/oil-gas/india-turns-to-expensive-foreign-gas-to-ease-its-power-crisis/articleshow/91333653.cms

[545] Severe heat wave kills dozens in India and Pakistan in a "snapshot" of what's to come from climate change, expert says - https://www.cbsnews.com/news/india-heat-wave-pakistan-climate-change-snapshot/

[546] bp Statistical Review of World Energy July 2021 - https://www.bp.com/en/global/corporate/energy-economics/statistical-review-of-world-energy.html

[547] Heat Waves - https://www.who.int/india/heat-waves

[548] Unexpected Hikes in Energy Prices Increase the Likelihood of Food Insecurity - https://www.ers.usda.gov/amber-waves/2017/july/unexpected-hikes-in-energy-prices-increase-the-likelihood-of-food-insecurity/

[549] The Effects of Energy Price Shocks on Household Food Security in Low-Income Households - https://www.ers.usda.gov/publications/pub-details/?pubid=84240

[550] real_prices.xlsx - https://www.eia.gov/outlooks/steo/realprices/real_prices.xlsx

[551] US Business Cycle Expansions and Contractions - https://www.nber.org/research/data/us-business-cycle-expansions-and-contractions

[552] SHORT-TERM ENERGY OUTLOOK (Real Prices Viewer)- https://www.eia.gov/outlooks/steo/realprices/

[553] Biden Exploring Release of Diesel Fuel Reserves Amid High Prices - https://www.wsj.com/articles/biden-exploring-release-of-diesel-fuel-reserves-amid-high-prices-11653333243

[554] Rising Diesel Costs Are Straining U.S. Truckers, Shipping Operations - https://www.wsj.com/articles/rising-diesel-costs-are-straining-u-s-truckers-shipping-operations-11652376035?mod=djemRTE_h

[555] Diesel fuel explained Factors affecting diesel prices- https://www.eia.gov/energyexplained/diesel-fuel/factors-affecting-diesel-prices.php

[556] Record Diesel Prices Pressure European Drivers, U.S. Deliveries - https://www.wsj.com/articles/record-diesel-prices-pressure-european-drivers-u-s-deliveries-11652416873?mod=hp_lead_pos4

[557] Why Hasn't the Jump in Oil Prices Led to a Recession? - https://www.frbsf.org/economic-research/publications/economic-letter/2005/november/why-the-jump-in-oil-prices-has-not-led-to-a-recession/

[558] The Macroeconomics of Oil Shocks- https://www.philadelphiafed.org/-/media/frbp/assets/economy/articles/business-review/2007/q1/br_q1-2007-3_oil-shocks.pdf

[559] eia - https://www.eia.gov/outlooks/steo/realprices/

[560] All real and nominal price series - https://www.eia.gov/outlooks/steo/realprices/real_prices.xlsx

[561] real_prices.xlsx - https://www.eia.gov/outlooks/steo/realprices/real_prices.xlsx

[562] Gasoline prices averaged the highest ever in 2012, despite never topping the single day spikes seen in 2008 - https://money.cnn.com/2012/12/31/news/economy/gas-prices/

[563] fotw#915_web.xlsx - https://www.energy.gov/sites/default/files/2016/03/f30/fotw%23915_web.xlsx

[564] Fertilizer Use and Price - https://www.ers.usda.gov/data-products/fertilizer-use-and-price/

[565] U.S. Fertilizer Use and Price - https://www.ers.usda.gov/data-products/fertilizer-use-and-price/

[566] All Fertilizer Use - https://www.ers.usda.gov/webdocs/DataFiles/50341/fertilizeruse.xls?v=1847.1

[567] Synthetic Nitrogen Fertilizer in the U.S. - https://farmdocdaily.illinois.edu/2021/02/synthetic-nitrogen-fertilizer-in-the-us.html

[568] as food shortages stoked by Russia's invasion of Ukraine drive demand for seeds and pesticides to boost global crop production. - https://www.wsj.com/articles/bayers-agricultural-unit-thrives-as-ukraine-war-threatens-food-supplies-11654588988?page=1

[569] The Fertilizer Institute. 2019. 2019 State of the Fertilizer Industry. Washington, D.C. https://www.fertilizerreport.org/

[570] Top Countries in Nitrogen Fertilizer Production Metric Tons - 2002 to 2019 - https://www.nationmaster.com/nmx/ranking/nitrogen-fertilizer-production

[571] Phosphorus: Essential to Life—Are We Running Out? - https://news.climate.columbia.edu/2013/04/01/phosphorus-essential-to-life-are-we-running-out/

[572] War in Ukraine and its Effect on Fertilizer Exports to Brazil and the U.S - https://farmdocdaily.illinois.edu/2022/03/war-in-ukraine-and-its-effect-on-fertilizer-exports-to-brazil-and-the-us.html

[573] Potassium a Nutrient Essential for Life - https://www.ipipotash.org/uploads/udocs/388-ipi-booklet-potassium-a-nutrient-for-life.pdf

[574] What Is the Nitrogen Cycle and Why Is It Key to Life? - https://kids.frontiersin.org/articles/10.3389/frym.2019.00041

[575] Follow the nitrogen and phosphorus cycles and learn why farmers fertilize fields to keep them productive - https://www.britannica.com/science/nitrogen-cycle

[576] USDA Announces Plans for $250 Million Investment to Support Innovative American-made Fertilizer to give US Farmers more choices in the Marketplace - https://www.usda.gov/media/press-releases/2022/03/11/usda-announces-plans-250-million-investment-support-innovative

[577] U.S. Dollar Index: What It Is and Its Recent History - https://www.thebalance.com/u-s-dollar-index-historical-data-3306249

[578] Nominal Advanced Foreign Economies U.S. Dollar Index (TWEXAFEGSMTH) - https://fred.stlouisfed.org/series/TWEXAFEGSMTH

[579] Bimodal Distribution: What is it? From StatisticsHowTo.com: Elementary Statistics for the rest of us! https://www.statisticshowto.com/what-is-a-bimodal-distribution/

[580] How global currencies fared in the calendar year 2016 - https://economictimes.indiatimes.com/investments-markets/how-global-currencies-fared-in-the-calendar-year-2016/slideshow/56135250.cms

[581] Dollar surges to highest level since March 2003 after US economic data - https://www.cnbc.com/2016/11/22/dollar-nears-13-12-year-high-buoyed-by-multiple-rate-hike-view.html

[582] How the Dollar Impacts Commodity Prices - https://www.thebalance.com/how-the-dollar-impacts-commodity-prices-809294

[583] external-data-indices-only may excel database - https://www.imf.org/-/media/Files/Research/CommodityPrices/Monthly/external-data-indices-onlymay.ashx

[584] FACT SHEET ON 2019 NATIONAL TRADE ESTIMATE: Fighting to Open Foreign Markets to American Agriculture - https://ustr.gov/about-us/policy-offices/press-office/fact-sheets/2019/march/fact-sheet-2019-national-trade-estimat-1

[585] Too Many to Count: Factors Driving Fertilizer Prices Higher and Higher - https://www.fb.org/market-intel/too-many-to-count-factors-driving-fertilizer-prices-higher-and-higher

[586] Commodity Costs and Returns - https://www.ers.usda.gov/data-products/commodity-costs-and-returns/

[587] Corn Cost Return - https://www.ers.usda.gov/webdocs/DataFiles/47913/CornCostReturn.xlsx?v=2880.3

[588] Cotton Cost Return - https://www.ers.usda.gov/webdocs/DataFiles/47913/CottonCostReturn.xlsx?v=2880.3

[589] Wheat Cost Return - https://www.ers.usda.gov/webdocs/DataFiles/47913/WheatCostReturn.xlsx?v=2880.3

[590] Fertilizers by Product - https://www.fao.org/faostat/en/#data/RFB

[591] Fertilizer Prices Climb to New North American Peak, Squeezing Crop Margins - https://www.bloomberg.com/news/articles/2021-11-05/soaring-fertilizer-prices-climb-to-new-north-american-peak

[592] Phosphate Rock - Historical Statistics (Data Series 140) 2018 update - https://www.usgs.gov/media/files/phosphate-rock-historical-statistics-data-series-140-2018-update

[593] Mineral Resource of the Month: Phosphate Rock, Wednesday - https://www.earthmagazine.org/article/mineral-resource-month-phosphate-rock/
[594] Commodities Monthly Prices - https://thedocs.worldbank.org/en/doc/5d903e848db1d1b83e0ec8f744e55570-0350012021/related/CMO-Historical-Data-Monthly.xlsx
[595] Nominal Advanced Foreign Economies U.S. Dollar Index (TWEXAFEGSMTH) - https://fred.stlouisfed.org/series/TWEXAFEGSMTH
[596] World Bank Pink Sheet Data - https://thedocs.worldbank.org/en/doc/5d903e848db1d1b83e0ec8f744e55570-0350012021/related/CMO-Historical-Data-Monthly.xlsx
[597] Nominal Advanced Foreign Economies U.S. Dollar Index- https://fred.stlouisfed.org/series/TWEXAFEGSMTH
[598] Nominal Advanced Foreign Economies U.S. Dollar Index- https://fred.stlouisfed.org/series/TWEXAFEGSMTH
[599] Global Phosphorus Fertilizer Market and National Policies: A Case Study Revisiting the 2008 Price Peak - https://www.frontiersin.org/articles/10.3389/fnut.2017.00022/full
[600] How to Check if Time Series Data is Stationary with Python- https://machinelearningmastery.com/time-series-data-stationary-python/
[601] Testing normality including skewness and kurtosis - https://imaging.mrc-cbu.cam.ac.uk/statswiki/FAQ/Simon
[602] Granger Causality in Time Series - Explained using Chicken and Egg problem - https://www.analyticsvidhya.com/blog/2021/08/granger-causality-in-time-series-explained-using-chicken-and-egg-problem/
[603] AIC - https://towardsdatascience.com/introduction-to-aic-akaike-information-criterion-9c9ba1c96ced
[604] Fertilizer Use and Price - https://www.ers.usda.gov/data-products/fertilizer-use-and-price.aspx
[605] Near Record High Prices - https://blogs.worldbank.org/opendata/soaring-fertilizer-prices-add-inflationary-pressures-and-food-security-concerns
[606] A Perfect Storm in Fertilizer Markets - https://cap.unl.edu/crops/perfect-storm-fertilizer-markets
[607] Climate Change Indicators: Heat Waves - https://www.epa.gov/climate-indicators/climate-change-indicators-heat-waves#ref7
[608] A Perfect Storm in Fertilizer Markets - https://cap.unl.edu/crops/perfect-storm-fertilizer-markets
[609] Prices of Most Fertilizers Continue Rise, as Wheat Prices Climb on Supply Concerns - https://farmpolicynews.illinois.edu/2022/05/prices-of-most-fertilizers-continue-rise-as-wheat-prices-climb-on-supply-concerns/
[610] Fertilizer prices spike in leading U.S. market in late 2021, just ahead of 2022 planting season - https://www.ers.usda.gov/data-products/chart-gallery/gallery/chart-detail/?chartId=103194
[611] High natural gas prices could lead to spike in food costs through fertilizer link- https://www.spglobal.com/commodityinsights/en/market-insights/blogs/agriculture/011922-fertilizer-costs-natural-gas-prices#
[612] Fertilizer prices spike in leading U.S. market in late 2021, just ahead of 2022 planting season - https://www.ers.usda.gov/data-products/chart-gallery/gallery/chart-detail/?chartId=103194
[613] Current Fertilizer Prices, Natural Gas, and Longer-Run Supply - https://farmdocdaily.illinois.edu/2012/10/current-fertilizer-prices-natu.html
[614] PRICE RELATIONSHIPS IN THE U.S. NITROGEN FERTILIZER INDUSTRY - https://scholarworks.montana.edu/xmlui/bitstream/handle/1/14550/GumbleyT0518.pdf?isAllowed=y&sequence=1
[615] Current Fertilizer Prices, Natural Gas, and Longer-Run Supply - https://farmdocdaily.illinois.edu/2012/10/current-fertilizer-prices-natu.html
[616] MAP AS A STARTER FERTILIZER FOR CORN - https://www.cropnutrition.com/resource-library/map-as-a-starter-fertilizer-for-corn
[617] MAP AS A STARTER FERTILIZER FOR CORN - https://www.cropnutrition.com/resource-library/map-as-a-starter-fertilizer-for-corn
[618] A Perfect Storm in Fertilizer Markets - https://cap.unl.edu/crops/perfect-storm-fertilizer-markets
[619] Prophet Model - https://facebook.github.io/prophet/
[620] Prophet - https://facebook.github.io/prophet/docs/quick_start.html#python-api
[621] Prophet plot explained - https://www.mikulskibartosz.name/prophet-plot-explained/

Section IV: Conclusion

CHAPTER 8

Future

The World is at a critical juncture. Compounding the damage from the COVID-19 pandemic, the Russian invasion of Ukraine has magnified the slowdown in the global economy, which is entering what could become a protracted period of feeble growth and elevated inflation. This raises the risk of stagflation, with potentially harmful consequences for middle and low income economies alike. The increased tension and deficit to feed people around the world and continued unprecedented demand for greener fuel & biodiesel, have elevated the need for food security across the world and will remain an important social & economic necessity cum issue over the next several decades. As food-versus-fuel tension becomes more intense, the day will come when more agricultural products will be used for energy than food. Adding to the conundrum, the war in Ukraine and the COVID-19 pandemic have changed the face of the earth in terms of supply chain, resource availability, and human labor and have exposed our vulnerabilities to food security to an even greater extent. In essence, humanity is at a critical juncture and what this unprecedented movement in our lives has thrusted upon us—the practitioners of the agriculture and technologists of the world—is to develop climate smart food security enhanced innovative products to address the multi-pronged food security challenges.

Agricultural innovation is key to overcoming concerns of food security; the infusion of data science, artificial intelligence, sensor technologies, and advanced analytics models with traditional agricultural practices such as soil engineering, fertilizers, and agronomy is one of the best ways to achieve enhanced food security.

Finally, advanced analytics proposed in this book will enable practitioners of economics, data science, public service agencies, risk managers, academics, and thought leaders to address the needs of current and future population food security needs!

Appendices

Appendix A—Food Security & Nutrition

The 17 Sustainable Development Goals (SDGs)

The 2030 Agenda for Sustainable Development, adopted by all United Nations Member States in 2015, provides a shared blueprint for peace and prosperity for people and the planet, now and into the future. At its heart are the 17 Sustainable Development Goals[622] (SDGs), which are an urgent call for action by all countries—developed and developing—in a global partnership. They recognize that ending poverty and other deprivations must go hand-in-hand with strategies that improve health and education, reduce inequality, and spur economic growth—all while tackling climate change and working to preserve our oceans and forests.

Food Security Monitoring System (FSMS)

The Food Security Monitoring System[623] (FSMS), generally, covers 35 key indicators, that are classified along the four dimensions of food security:[624] availability (5 indicators), access (10 indicators), utilization (13 indicators) and stability (7 indicators). Indicators are classified along the four dimensions of food security: availability, access, utilization, and stability.[625]

Number	Type of Indicator	Source
	Availability Indicators	
1	Average dietary energy supply adequacy	FAO
2	Average value of food production	FAO
3	Share of dietary energy supply derived from cereals, roots and tubers	FAO
4	Average vegetarian protein supply	FAO
5	Average supply of protein of animal origin	FAO
	Accessibility Indicators	
6	Gross domestic product per capita (in purchasing power equivalent)	WB
7	Domestic food price level index	WB/ILO/FAO
8	Percent of paved roads over total roads	WB
9	Road density	World Road Statistics
10	Rail lines density	WB
11	Prevalence of undernourishment (PoU)	FAO
12	Number of people undernourished	FAO
13	Share of food expenditure of the poor	FAO
14	Depth of the food deficit	FAO
15	Prevalence of food inadequacy	FAO
	Utilization Indicators	
16	Access to improved water sources	WHO/UNICEF
17	Access to improved sanitation facilities	WHO/UNICEF
18	Percentage of children under 5 years of age affected by wastage	WHO/UNICEF
19	Percentage of stunted children under 5 years of age	WHO/UNICEF
20	Percentage of underweight children under 5 years of age	WHO/UNICEF
21	Percentage of underweight adults	WHO
22	Prevalence of anaemia among pregnant women	WHO/WB
23	Prevalence of anaemia among children under 5 years of age	WHO/WB
24	Prevalence of vitamin A deficiency	WHO
25	Prevalence of iodine deficiency	WHO
26	Prevalence of undernourishment (PoU)	FAO
27	Mortality rate	WHO/WB/UNICEF
28	Food Consumption	
	Stability Indicators	
29	Cereal import dependency ratio	FAO
30	Percent of arable land equipped for irrigation	FAO
31	Value of food imports over total merchandise exports	FAO
32	Political stability and absence of violence/terrorism	WB/WWGI
33	Domestic food price volatility	WB/ILO/FAO
34	Per capita food production variability	FAO
35	Per capita food supply variability	FAO

NHANES 2019–2020 Questionnaire Instruments—Food Security

Questionnaires are administered to NHANES participants both at home and in the MECs. The questionnaires and a brief description of each section follows.[626]

- FSQ.041 In the last 12 months, since last {DISPLAY CURRENT MONTH AND LAST YEAR}, did {you/you or other adults in your household} ever cut the size of your meals or skip meals because there wasn't enough money for food?
- FSQ.061 In the last 12 months, did you ever eat less than you felt you should because there wasn't enough money for food?
- FSQ.071 [In the last 12 months], were you ever hungry but didn't eat because there wasn't enough money for food?
- FSQ.081 [In the last 12 months], did you lose weight because there wasn't enough money for food?
- FSQ.092 [In the last 12 months], did {you/you or other adults in your household} ever not eat for a whole day because there wasn't enough money for food?
- FSQ.111 In the last 12 months, since {DISPLAY CURRENT MONTH AND LAST YEAR} did you ever cut the size of {CHILD'S NAME/any of the children's} meals because there wasn't enough money for food?
- [In the last 12 months], did {CHILD'S NAME/any of the children} ever skip meals because there wasn't enough money for food?

Macroeconomic Signals that Could help Predict Economic Cycles

Signal #	Type	Description
Oil price swings	Leading Indicator	oil price swings[627] appear to be consistent and frequent historical precursors to U.S. recessions.[628] A spike in oil prices has preceded or coincided with 10 out of 12 post-WWII recessions.
Monetary Trends	Lagging Indicator	A period of expansionary monetary policy in the years prior to the recession, sometimes to help fund government war spending or to re-inflate the economy after the previous round of recession. Once the resulting debt bubbles pop or the end of a war leads to cutbacks in monetary expansion, several years' worth of overextended, debt-based investments and malinvestments tend to be wiped out in a process of debt deflation[629,630] in a relatively short period. This spikes unemployment and drags down GDP.
During recession, GDP & Credit Contraction Occur	Lagging Indicator	Country GDP/Credit contraction, in percent[631]

List of World Development Indicators (WDI)

Indicator	Code
Growth and economic structure	
GDP (current US$)	NY.GDP.MKTP.CD
GDP growth (annual %)	NY.GDP.MKTP.KD.ZG
Agriculture, value added (annual % growth)	NV.AGR.TOTL.KD.ZG
Industry, value added (annual % growth)	NV.IND.TOTL.KD.ZG
Manufacturing, value added (annual % growth)	NV.IND.MANF.KD.ZG
Services, value added (annual % growth)	NV.SRV.TOTL.KD.ZG
Final consumption expenditure (annual % growth)	NE.CON.TOTL.KD.ZG
Gross capital formation (annual % growth)	NE.GDI.TOTL.KD.ZG
Exports of goods and services (annual % growth)	NE.EXP.GNFS.KD.ZG
Imports of goods and services (annual % growth)	NE.IMP.GNFS.KD.ZG
Agriculture, value added (% of GDP)	NV.AGR.TOTL.ZS
Industry, value added (% of GDP)	NV.IND.TOTL.ZS
Services, value added (% of GDP)	NV.SRV.TOTL.ZS
Final consumption expenditure (% of GDP)	NE.CON.TOTL.ZS
Gross capital formation (% of GDP)	NE.GDI.TOTL.ZS
Exports of goods and services (% of GDP)	NE.EXP.GNFS.ZS
Imports of goods and services (% of GDP)	NE.IMP.GNFS.ZS
Income and savings	
GNI per capita, Atlas method (current US$)	NY.GNP.PCAP.CD
GNI per capita, PPP (current international $)	NY.GNP.PCAP.PP.CD
Population, total	SP.POP.TOTL
Gross savings (% of GDP)	NY.GNS.ICTR.ZS
Adjusted net savings, including particulate emission damage (% of GNI)	NY.ADJ.SVNG.GN.ZS
Balance of payments	
Export value index (2000 = 100)	TX.VAL.MRCH.XD.WD
Import value index (2000 = 100)	TM.VAL.MRCH.XD.WD
Personal remittances, received (% of GDP)	BX.TRF.PWKR.DT.GD.ZS
Current account balance (% of GDP)	BN.CAB.XOKA.GD.ZS
Foreign direct investment, net inflows (% of GDP)	BX.KLT.DINV.WD.GD.ZS

Wheat Data Disappearance and End Stocks

This data product contains statistics on wheat—including the five classes of wheat: hard red winter, hard red spring, soft red winter, white, and durum and rye. It includes data published in the monthly Wheat Outlook previously the annual Wheat Yearbook. Data is monthly, quarterly, and/or annual, depending upon the data series. Most data is on a marketing-year basis, but some is in calendar years.[632]

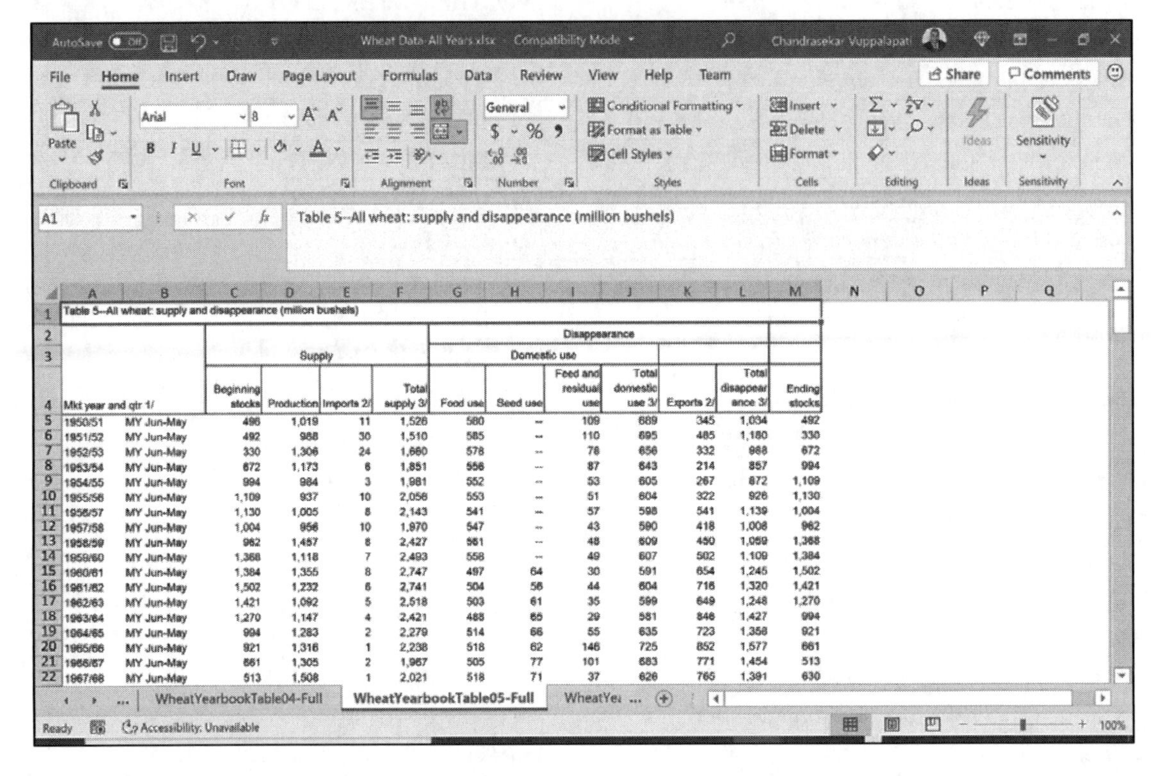

Table 5--All wheat: supply and disappearance (million bushels)

Mkt year and qtr 1/		Supply				Disappearance						
						Domestic use						
		Beginning stocks	Production	Imports 2/	Total supply 3/	Food use	Seed use	Feed and residual use	Total domestic use 3/	Exports 2/	Total disappearance 3/	Ending stocks
1950/51	MY Jun-May	496	1,019	11	1,526	580	--	109	689	345	1,034	492
1951/52	MY Jun-May	492	988	30	1,510	585	--	110	695	485	1,180	330
1952/53	MY Jun-May	330	1,306	24	1,660	578	--	78	656	332	988	672
1953/54	MY Jun-May	672	1,173	6	1,851	556	--	87	643	214	857	994
1954/55	MY Jun-May	994	984	3	1,981	552	--	53	605	267	872	1,109
1955/56	MY Jun-May	1,109	937	10	2,056	553	--	51	604	322	926	1,130
1956/57	MY Jun-May	1,130	1,005	8	2,143	541	--	57	598	541	1,139	1,004
1957/58	MY Jun-May	1,004	956	10	1,970	547	--	43	590	418	1,008	962
1958/59	MY Jun-May	962	1,457	8	2,427	561	--	48	609	450	1,059	1,368
1959/60	MY Jun-May	1,368	1,118	7	2,493	558	--	49	607	502	1,109	1,384
1960/61	MY Jun-May	1,384	1,355	8	2,747	497	64	30	591	654	1,245	1,502
1961/62	MY Jun-May	1,502	1,232	6	2,741	504	56	44	604	716	1,320	1,421
1962/63	MY Jun-May	1,421	1,092	5	2,518	503	61	35	599	649	1,248	1,270
1963/64	MY Jun-May	1,270	1,147	4	2,421	488	65	29	581	846	1,427	994
1964/65	MY Jun-May	994	1,283	2	2,279	514	66	55	635	723	1,358	921
1965/66	MY Jun-May	921	1,316	1	2,238	518	62	146	725	852	1,577	661
1966/67	MY Jun-May	661	1,305	2	1,967	505	77	101	683	771	1,454	513
1967/68	MY Jun-May	513	1,508	1	2,021	518	71	37	626	765	1,391	630

Hard Red Wheat Contracts - https://www.barchart.com/futures/quotes/KEN20/interactive-chart

Food Aids

PL-480 or Food for Peace

The Government of India has received generous PL 480 from the United States during the years1956 to 1968; USAID and PL–480 I in the years 1961to1969.[633] The administrations of John F. Kennedy and Lyndon B. Johnson marked a revitalization of the U.S. foreign assistance program, signifying a growing awareness of the importance of humanitarian aid as a form of diplomacy, and reinforced the belief that American security was linked to the economic progress and stability of other nations.

Johnson with Gandhi, March 28, 1966. (White House Photo Office)

The United Nations—17 Sustainable Development Goals (SDGs)

The 17 Sustainable Development Goals (SDGs), which are an urgent call for action by all countries—developed and developing - in a global partnership. They recognize that ending poverty and other deprivations must go together with strategies that improve health and education, reduce inequality, and spur economic growth—all while tackling climate change and working to preserve our oceans and forests.[634]

The Statistical Distributions of Commodity Prices in Both Real and Nominal Terms

Statistical properties of commodities.[635]

Decade	Variable	Mean	Std Dev	Coeff. Of variation	Skewness	Kurtosis
1970/80	banana-n	216.8	64.9	29.9	0.6	–0.6
	banana-r	543.5	89.0	16.4	0.5	–0.0
	cocoa-n	1777.9	1202.5	67.6	0.7	–1.0
	cocoa-r	3970.1	1765.4	44.5	0.9	–0.2
	coffee-n	2219.8	1412.4	63.6	1.2	0.8
	coffee-r	5159.0	2172.5	42.1	1.9	4.1
	cotton-n	1243.3	438.2	35.2	–0.0	–1.4
	cotton-r	3059.3	666.7	21.8	1.7	2.9
	jute-n	322.7	57.9	17.9	1.0	–0.2
	jute-r	845.0	201.4	23.8	0.3	–1.4
	maize-n	98.6	26.7	27.1	–0.4	–0.8
	maize-r	226.2	51.8	22.9	0.4	–0.7
	palmoil-n	466.0	164.8	35.4	–0.1	–1.2
	palmoil-r	1028.6	247.7	24.1	1.3	1.8
	rapeoil-n	496.7	170.7	34.4	0.0	–0.5
	rapeoil-r	1108.4	299.0	27.0	1.3	1.7
	rapeseed-n	263.6	81.3	30.8	–0.3	–0.6
	rapeseed-r	592.2	148.5	25.1	1.2	0.9
	rice-n	285.4	127.2	44.6	0.6	–0.0
	rice-r	682.2	254.6	37.3	2.0	3.3
	rubber-n	676.0	271.5	40.2	0.5	–0.5
	rubber-r	1641.9	344.7	21.0	1.1	2.1
	sisal-n	476.6	271.9	57.0	0.9	0.0
	sisal-r	1141.2	553.3	48.5	1.6	1.5
	soybeans-n	237.9	69.5	29.2	0.0	–0.1
	soybeans-r	540.2	149.1	27.6	2.4	9.0
	soymeal-n	196.3	68.9	35.1	1.5	5.4
	soymeal-r	451.0	181.5	40.3	3.4	14.5
	sugar-n	243.4	198.8	81.7	2.5	7.4
	sugar-r	595.4	424.3	71.3	2.5	7.1
	sunflmeal-n	153.3	44.8	29.2	0.8	2.4
	sunflmeal-r	353.6	118.6	33.5	2.6	8.4
	tea-n	1532.3	607.3	39.6	1.7	3.7
	tea-r	3802.9	892.6	23.5	2.4	8.4
	wheat-n	123.5	43.8	35.5	–0.1	–1.0
	wheat-r	284.0	87.5	30.8	1.7	2.5

1980/90	banana-n	406.3	83.4	20.5	1.2	2.2
	banana-r	557.4	104.4	18.7	0.7	–0.1
	cocoa-n	2060.1	428.7	20.8	0.3	0.6
	cocoa-r	2882.9	769.7	26.7	–0.1	–0.5
	coffee-n	2892.9	569.4	19.7	0.9	1.2
	coffee-r	4031.8	990.4	24.6	0.2	–0.3
	cotton-n	1625.3	319.7	19.7	–0.4	–0.3
	cotton-r	2254.3	524.6	23.3	–0.3	–1.1
	jute-n	361.1	138.7	38.4	2.5	5.8
	jute-r	507.7	247.0	48.6	2.7	6.4
	maize-n	112.8	22.6	20.0	–0.3	–0.5
	maize-r	154.5	42.2	27.3	0.1	–1.1
	palmoil-n	448.4	157.0	35.0	0.9	0.5
	palmoil-r	622.5	273.7	44.0	0.9	0.2
	rapeoil-n	454.1	124.7	27.5	0.9	0.6
	rapeoil-r	626.1	232.4	37.1	1.0	0.1
	rapeseed-n	267.5	58.7	21.9	0.1	–0.0
	rapeseed-r	369.1	115.3	31.2	0.1	–0.7
	rice-n	294.4	92.7	31.5	1.0	0.2
	rice-r	404.1	132.7	32.8	1.0	0.4
	rubber-n	997.3	203.3	20.4	1.2	0.8
	rubber-r	1379.5	313.5	22.7	0.8	–0.5
1980/90	sisal-n	594.5	85.8	14.4	1.3	1.1
	sisal-r	822.7	145.6	17.7	0.4	–0.1
	soybeans-n	259.6	43.1	16.6	0.6	–0.6
	soybeans-r	351.0	71.6	20.4	0.5	–0.5
	soymeal-n	219.3	41.8	19.1	0.4	–0.4
	soymeal-r	294.4	56.1	19.0	0.5	–0.3
	sugar-n	236.6	168.3	71.1	1.9	3.7
	sugar-r	320.9	224.0	69.8	2.0	3.6
	sunflmeal-n	143.2	38.5	26.9	0.3	–0.5
	sunflmeal-r	193.7	58.6	30.3	0.6	–0.9
	tea-n	2131.0	564.5	26.5	1.8	2.8
	tea-r	2991.0	1035.2	34.6	1.6	2.0
	wheat-n	149.2	23.1	15.5	–0.5	–0.7
	wheat-r	203.2	39.8	19.6	–0.4	–1.2

1990/00	banana-n	480.5	104.3	21.7	0.7	0.5
	banana-r	514.0	115.3	22.4	0.7	0.3
	cocoa-n	1340.0	226.5	16.9	0.0	-0.9
	cocoa-r	1434.4266.9	18.6	0.5	–0.5	
	coffee-n	2095.6	805.6	38.4	0.9	0.1
	coffee-r	2238.4	860.0	38.4	0.9	0.2
	cotton-n	1632.6	307.1	18.8	0.3	0.0
	cotton-r	1741.3	306.7	17.6	–0.1	–0.5
	jute-n	329.7	80.7	24.5	0.6	–0.6
	jute-r	351.1	80.2	22.8	0.5	–0.8
	maize-n	111.2	23.9	21.5	2.0	4.9
	maize-r	118.1	24.1	20.4	1.6	3.7
	palmoil-n	477.6	125.7	26.3	0.3	–1.3
	palmoil-r	509.1	139.1	27.3	0.4	–0.9
	rapeoil-n	507.2	100.2	19.8	0.1	–1.3
	rapeoil-r	541.5	117.1	21.6	0.2	–1.2
	rapeseed-n	248.9	47.4	19.0	–0.0	–1.5
	rapeseed-r	265.7	55.1	20.7	0.0	–1.3
	rice-n	285.5	41.6	14.6	0.1	–0.3
	rice-r	304.1	44.4	14.6	0.2	–0.5
	rubber-n	993.4	66.9	6.7	0.2	–1.0
	rubber-r	1061.2	71.4	6.7	–0.2	–0.5
	sisal-n	700.4	109.6	15.6	–0.2	–0.6
	sisal-r	749.0	121.5	16.2	–0.3	–0.3
	soybeans-n	250.6	33.7	13.5	0.5	–0.2
	soybeans-r	267.1	39.4	14.8	0.2	–0.5
	soymeal-n	205.5	40.0	19.5	0.6	–0.2
	soymeal-r	218.8	43.8	20.0	0.6	–0.2
	sugar-n	234.7	51.3	21.9	–0.1	–0.4
	sugar-r	251.0	55.6	22.2	–0.1	–0.1
	sunflmeal-n	116.8	24.5	21.0	–0.2	–0.7
	sunflmeal-r	124.3	26.3	21.2	–0.3	–0.7
	tea-n	2002.6	334.3	16.7	1.1	1.6
	tea-r	2147.0	409.0	19.0	1.2	2.1
	wheat-n	146.4	32.2	22.0	1.1	1.3
	wheat-r	155.5	32.3	20.8	0.7	0.8

(r-real, n-nominal)

The kurtosis for a normal distribution is 3.

Poverty Thresholds for 2019 by Size of Family and Number of Related Children Under 18 Years

Size of family unit	Weighted average thresholds	Related children under 18 years								
		None	One	Two	Three	Four	Five	Six	Seven	Eight or more
One person (unrelated individual):	13,011									
Under age 65...	13,300	13,300								
Aged 65 and older...	12,261	12,261								
Two people:	16,521									
Householder under age 65...	17,196	17,120	17,622							
Householder aged 65 and older...	15,468	15,453	17,555							
Three people...	20,335	19,998	20,578	20,598						
Four people...	26,172	26,370	26,801	25,926	26,017					
Five people...	31,021	31,800	32,263	31,275	30,510	30,044				
Six people...	35,129	36,576	36,721	35,965	35,239	34,161	33,522			
Seven people...	40,016	42,085	42,348	41,442	40,811	39,635	38,262	36,757		
Eight people...	44,461	47,069	47,485	46,630	45,881	44,818	43,470	42,066	41,709	
Nine people or more...	52,875	56,621	56,895	56,139	55,503	54,460	53,025	51,727	51,406	49,426

Source: U.S. Census Bureau.[636]

Appendix B—Agriculture

Agricultural Data Surveys

USDA NASS

The Unites States Department of Agriculture – National Analytics Statistics Services: Online Survey Form[637]

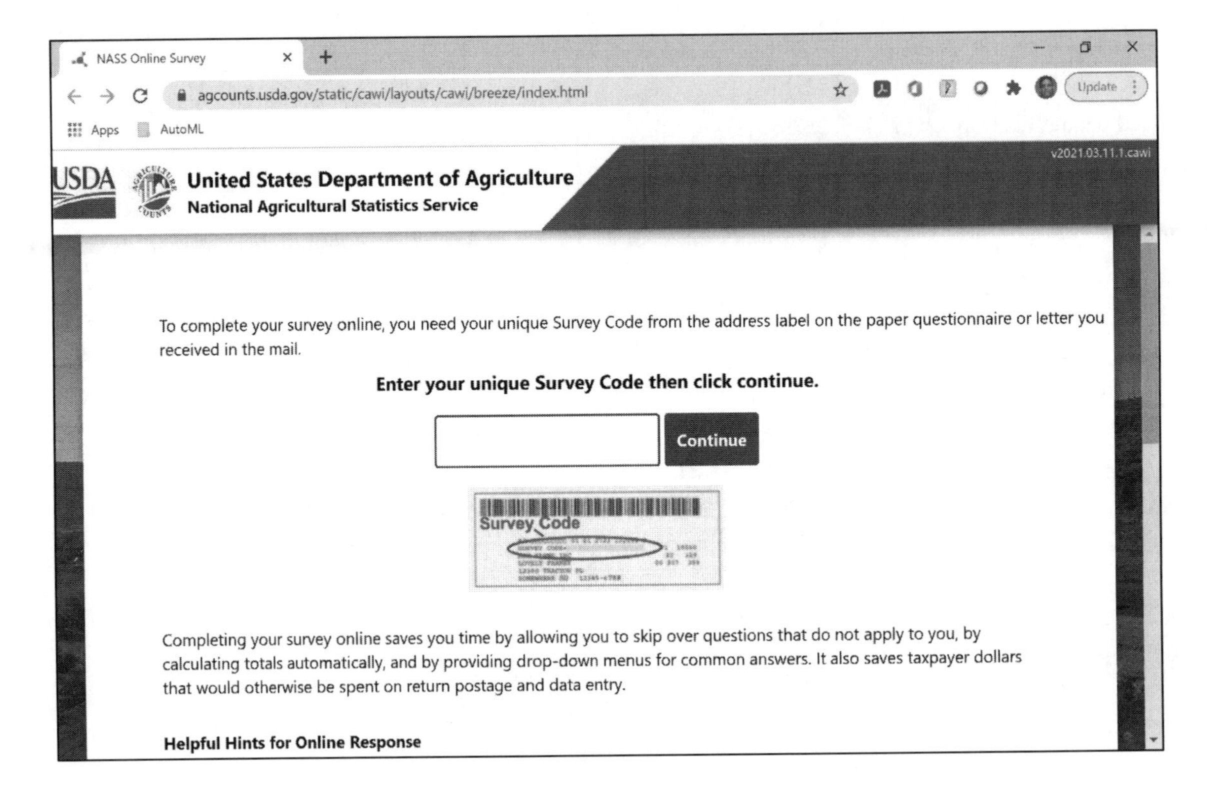

USDA—The Foreign Agricultural Service (FAS) Reports and Databases

World Agricultural Production

Current: https://www.fas.usda.gov/data/world-agricultural-production

Archive: https://usda.library.cornell.edu/concern/publications/5q47rn72z?locale=en

USDA's Foreign Agricultural Service (FAS) publishes a monthly report on crop acreage, yield and production in major countries worldwide. Sources include reports from FAS's worldwide offices, official statistics of foreign governments, and analysis of economic data and satellite imagery. The reports reflect official USDA estimates released in the monthly World Agricultural Supply and Demand Estimates (WASDE).

World Markets and Trade

Current: https://www.fas.usda.gov/data

Archive: https://usda.library.cornell.edu/catalog?f%5Bmember_of_collections_ssim%5D%5B%5D=Foreign+Agricultural+Service&locale=en

USDA's Foreign Agricultural Service (FAS) publishes monthly and quarterly reports which include data on U.S. and global trade, production, consumption and stocks, as well as analysis of developments affecting world trade in oilseeds, grains, cotton, livestock and poultry. The reports reflect official USDA estimates released in the monthly World Agricultural Supply and Demand Estimates (WASDE).

Global Agricultural Information Network (GAIN)

https://www.fas.usda.gov/databases/global-agricultural-information-network-gain

USDA's Foreign Agricultural Service (FAS) provides timely reports on foreign markets through the Global Agriculture Information Network (GAIN) database. An average of 2,000 reports are added each year, with reports going back to 1995. GAIN reports are compiled by FAS' global market intelligence network, which includes FAS foreign service officers and locally engaged staff in over 90 overseas offices world-wide. They provide on-the-ground intelligence, insight, and analysis on nearly 200 countries, delivering information on foreign agricultural markets, crop conditions, and agro-political dynamics of interest to U.S. agriculture. GAIN reports contain assessments of commodity and trade issues made by USDA staff and are not necessarily statements of official U.S. government policy.

Production, Supply and Distribution (PS&D) Online

https://apps.fas.usda.gov/psdonline/app/index.html#/app/home

PSD Online is the public repository for USDA's Official Production, Supply and Distribution forecast data and reports for key agricultural commodities. PSD Online datasets are reviewed and updated monthly by an interagency committee chaired by USDA's World Agricultural Outlook Board (WAOB). The committee consists of representatives from Foreign Agricultural Service (FAS), the Economic Research Service (ERS), the Farm Service Agency (FSA), and the Agricultural Marketing Service (AMS).

USDA and NASA Global Agricultural Monitoring (GLAM)

https://glam1.gsfc.nasa.gov/

The USDA and NASA Global Agricultural Monitoring (GLAM) system provides near real-time and science quality Moderate Resolution Imaging Spectroradiometer (MODIS) Normalized Difference Vegetation Index (NDVI) from the satellites Terra and Aqua. The public can view and retrieve MODIS 8 day composited, global NDVI satellite imagery and time series data. GLAM was developed by NASA's Global Inventory Modeling and Mapping Studies (GIMMS) group for USDA's Foreign Agricultural Service.

Global Agricultural and Disaster Assessment System (GADAS)

https://geo.fas.usda.gov/GADAS/index.html

USDA's Foreign Agricultural Service (FAS) provides the Global Agricultural and Disaster Assessment System (GADAS), a web-based Geographic Information System (GIS) tool which integrates a vast array of highly detailed earth observation data streams, particularly targeted towards agricultural and disaster assessment analysis. GADAS is an interactive website which provides analysts with a wide variety of routine geospatial products (maps, charts, tables) they require for comprehensive situational investigations and recurring assessments. GADAS is an interactive global web analysis system, capable of displaying, comparing, analyzing, and sharing geospatial data.[638]

Export Sales Reporting

https://apps.fas.usda.gov/esrquery/

USDA's Export Sales Reporting Program monitors U.S. agricultural export sales on a daily and weekly basis. Export sales reporting provides a constant stream of up-to-date market information for 40 U.S. agricultural commodities sold abroad. The weekly U.S. Export Sales report is the most currently available source of U.S. exports sales data. The data is used to analyze overall levels of export demand, to determine where markets exit, and to assess the relative position of U.S. commodities in foreign markets

Global Agricultural Trade System (GATS)

https://apps.fas.usda.gov/gats/default.aspx

The Global Agricultural Trade System (GATS) is a searchable database containing monthly U.S. Census Bureau trade data organized by agricultural commodity and agricultural related product groups. Trade data is searchable by partner countries and partner groups. Historical U.S. agricultural trade data is available back to 1967. In addition, U.N. trade statistics (UN Comtrade) may be queried through GATS. UN trade data is available for nearly 200 countries or areas, dating from the inception of the Harmonized System (HS) of trade codes in 1989 to present. The database is continuously updated. U.S. trade data is updated monthly according to the U.S. Census Bureau's reporting system. UN Comtrade datasets are updated in GATS after nationally submitted data to the UN is standardized by the UN Statistical Division and added to the UN Comtrade database.

USDA Data Products

Data Products - https://www.ers.usda.gov/data-products/#!topicid=14829&subtopicid=14847

Commodity Costs and Returns

USDA has estimated annual production costs and returns and published accounts for major field crop and livestock enterprises since 1975. Cost and return estimates are reported for the United States and major production regions for corn, soyabeans, wheat, cotton, grain sorghum, rice, peanuts, oats, barley, milk, hogs, and cow-calf. These cost and return accounts are "historical" accounts based on the actual costs incurred by producers.[639]
05/02/2022

Fertilizer Use and Price

This data summarizes fertilizer consumption in the United States by plant nutrient and major fertilizer products—as well as consumption of mixed fertilizers, secondary nutrients, and micronutrients—for 1960 through the latest year for which statistics are available. The share of planted crop acreage receiving fertilizer, and fertilizer applications per receiving acre (by nutrient), are presented for major producing States for corn, cotton, soyabeans, and wheat (nutrient consumption by crop data starts in 1964). Fertilizer farm prices and indices of wholesale fertilizer prices are also available.[640]
10/30/2019

Data Sources

UN data marts

Source: http://data.un.org/Explorer.aspx?d=FAO

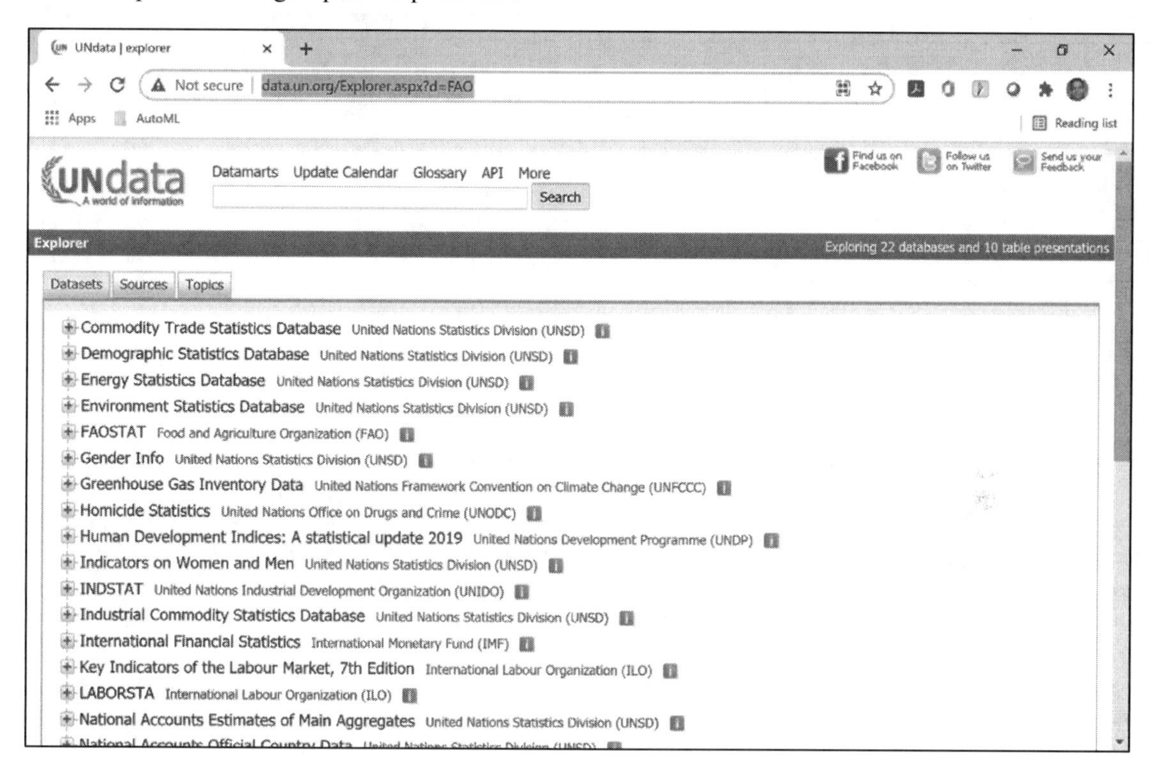

World Development Indicator tool:[641]

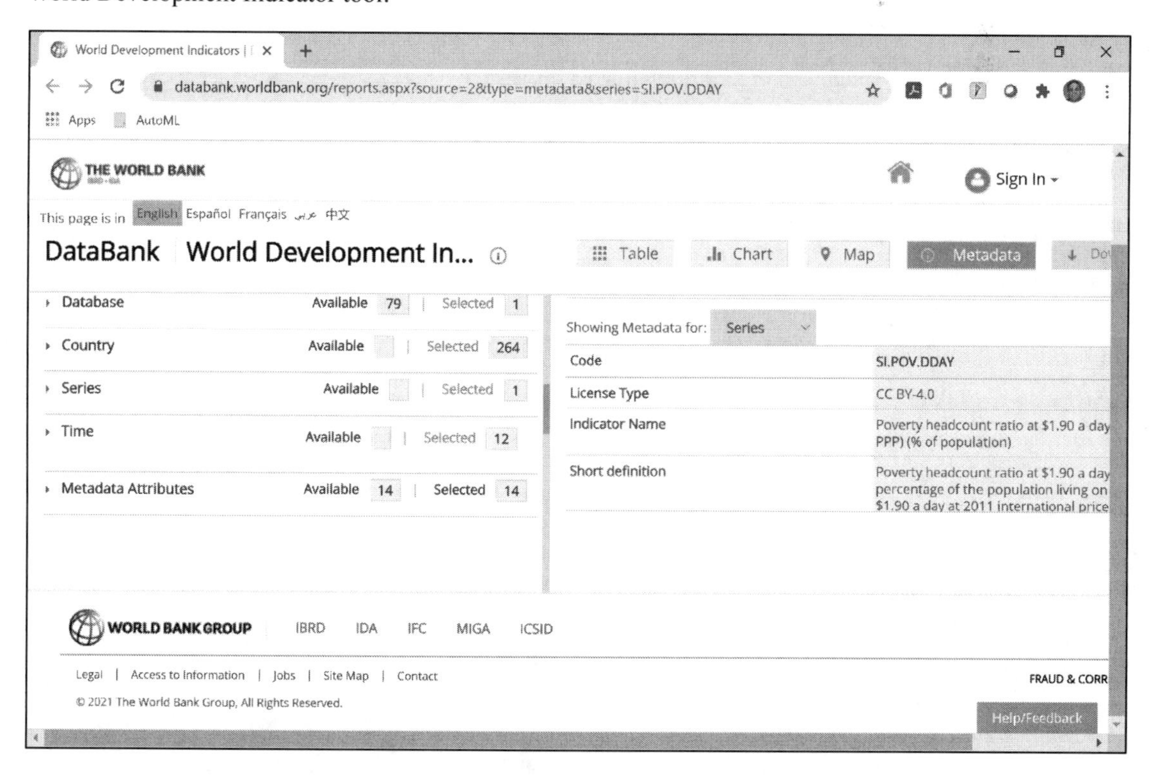

Food and Agriculture Organization of the United Nations

FAOSTAT Data:[642] FAO provides following data domains: Production, Prices, Emission-agriculture, Inputs, Trade, Population, emissions-Land Use, Investment, Macro Economic Data, and others.

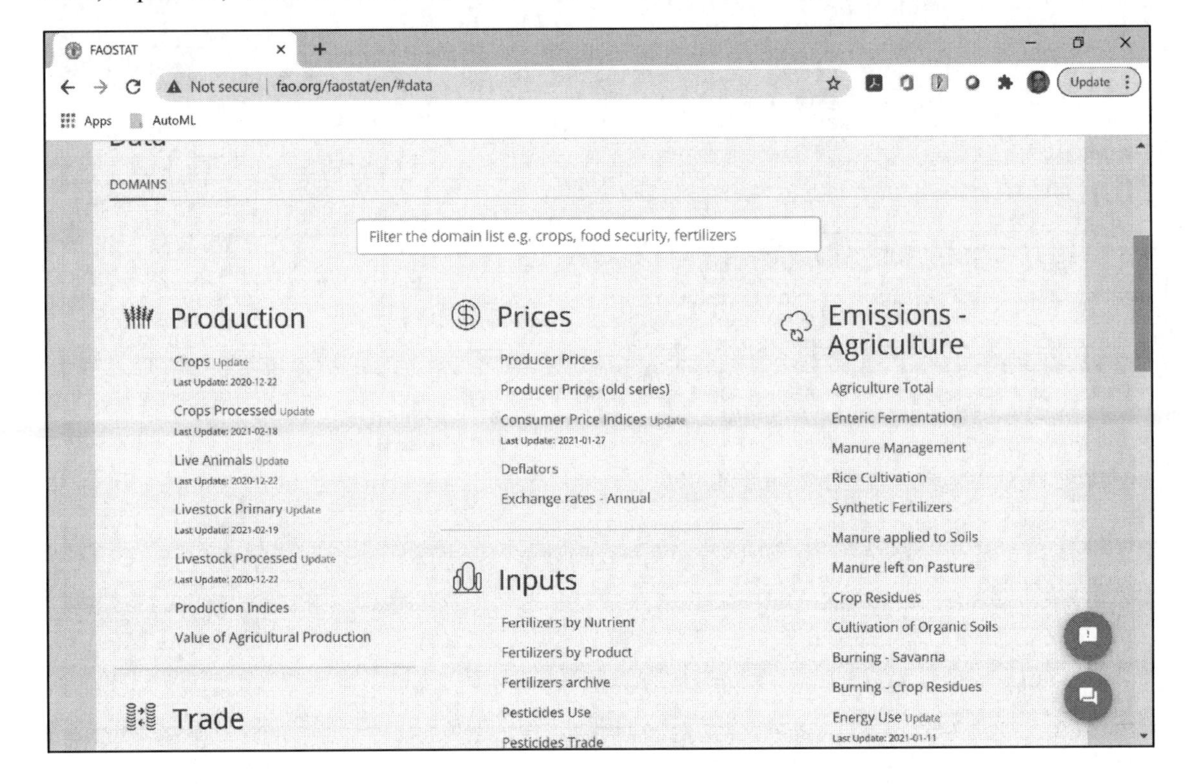

Conversion Factors

Here are a few useful conversion factors:[643]

- 1 bushel of wheat = 60 pounds or 77.2 kilograms/hectoliter
- 1 metric ton = 2,204.622 pounds
- Bushels x 0.0272155 = metric tons
- Metric tons x 36.7437 = bushels
- Price per bushel x 36.7437 = price per metric ton
- Price per metric ton x 0.0272155 = price per bushel
- 1 acre = 0.4047 hectares
- 1 hectare = 2.4710 acres
- Bushels/acre x 0.06725 = metric tons/hectare
- Metric tons/hectare x 14.87 = bushels/acre

National Dairy Development Board (NDDB) India

The National Dairy Development Board's (NDDB)[644] creation is rooted in the conviction that our nation's socio-economic progress lies largely on the development of rural India.

The Dairy Board was created to promote, finance and support producer-owned and controlled organizations. NDDB's programs and activities seek to strengthen farmer owned institutions and support national policies that are favorable to the growth of such institutions. Fundamental to NDDB's efforts are cooperative strategies and principles.

NDDB's efforts transformed India's rural economy by making dairying a viable and profitable economic activity for millions of milk producers while addressing the country's need for self-sufficiency in milk production.

NDDB has been reaching out to dairy farmers by implementing other income generating innovative activities and offering them sustainable livelihoods.

Milk production[645] and per capita availability of milk in India[646]		
Year	**Production (Million Tonnes)**	**Per Capita Availability (gms/day)**
1991–92	55.6	178
1992–93	58.0	182
1993–94	60.6	186
1994–95	63.8	192
1995–96	66.2	195
1996–97	69.1	200
1997–98	72.1	205
1998–99	75.4	210
1999–2000	78.3	214
2000–01	80.6	217
2001–02	84.4	222
2002–03	86.2	224
2003–04	88.1	225
2004–05	92.5	233
2005–06	97.1	241
2006–07	102.6	251
2007–08	107.9	260
2008–09	112.2	266
2009–10	116.4	273
2010–11	121.8	281
2011–12	127.9	290
2012–13	132.4	299
2013–14	137.7	307
2014–15	146.3	322
2015–16	155.5	337
2016–17	165.4	355
2017–18	176.3	375
2018–19	187.7	394

Worldwide—Artificial Intelligence (AI) Readiness

IN TERMS OF READINESS FOR AI, COUNTRIES APPEAR TO FALL INTO FOUR GROUPS[647]—Varying conditions among countries imply different degrees of AI adoption and absorption, and therefore economic impact.

Above threshold[1] Within threshold[1] Below threshold[1]

Group		AI-related			Enablers					
	Readiness areas	AI investment	AI research activities	Productivity boost from automation	Digital absorption	Innovation foundation	Human capital	Connectedness	Labor-market structure	
	Examples of indicators included	VC, PE, M&A, seed,[2] grant	Patents, publications, citations	Automation potential of activities	Technology utilization	R&D investment, business-model creation	PISA score, STEM graduates, GHCI[3]	MGI Connectedness Index	Redundancy costs, indexes on worker-employer collaboration	
	Data sources	Dealogic, S&P, Capital IQ	WIPO, Scimago Journal Rank	MGI	GTCI[4] (INSEAD)	OECD, INSEAD, WIPO	INSEAD, WEF, UNESCO, Eurostat	MGI	World Bank, INSEAD	**Total score**[5]
1	China									
	United States									
2	Australia	n/a								
	Belgium	n/a								
	Canada									
	Estonia	n/a								
	Finland	n/a								
	France									
	Germany									
	Iceland	n/a								
	Israel	n/a								
	Japan									
	Netherlands	n/a								
	New Zealand	n/a								
	Norway	n/a								
	Singapore	n/a								
	South Korea									
	Sweden									
	United Kingdom									

Above threshold[1] Within threshold[1] Below threshold[1]

Group		AI-related			Enablers					
	Readiness areas	AI investment	AI research activities	Productivity boost from automation	Digital absorption	Innovation foundation	Human capital	Connectedness	Labor-market structure	
	Examples of indicators included	VC, PE, M&A, seed,[2] grant	Patents, publications, citations	Automation potential of activities	Technology utilization	R&D investment, business-model creation	PISA score, STEM graduates, GHCI[3]	MGI Connectedness Index	Redundancy costs, indexes on worker-employer collaboration	
	Data sources	Dealogic, S&P, Capital IQ	WIPO, Scimago Journal Rank	MGI	GTCI[4] (INSEAD)	OECD, INSEAD, WIPO	INSEAD, WEF, UNESCO, Eurostat	MGI	World Bank, INSEAD	**Total score**[5]
3	Chile	n/a								
	Costa Rica	n/a								
	Czech Republic	n/a								
	India	n/a								
	Italy	n/a								
	Lithuania	n/a								
	Malaysia	n/a								
	South Africa	n/a								
	Spain									
	Thailand	n/a								
	Turkey	n/a								
4	Brazil	n/a								
	Bulgaria	n/a								
	Cambodia	n/a								
	Colombia	n/a								
	Greece	n/a								
	Indonesia	n/a								
	Pakistan	n/a								
	Peru	n/a								
	Tunisia	n/a								
	Uruguay	n/a								
	Zambia	n/a								

IN TERMS OF READINESS FOR AI, COUNTRIES APPEAR TO FALL INTO FOUR GROUPS[648]

Appendix C—Data World

Global Historical Climatology Network monthly (GHCNm)

The Global Historical Climatology Network monthly (GHCNm)[649] dataset provides monthly climate summaries from thousands of weather stations around the world. The initial version was developed in the early 1990s, and subsequent iterations were released in 1997, 2011, and most recently in 2018. The period of record for each summary varies by station, with the earliest observations dating to the 18th century. Some station records are purely historical and are no longer updated, but many others are still operational and provide short time delay updates that are useful for climate monitoring. The current version (GHCNm v4) consists of mean monthly temperature data, as well as a beta release of monthly precipitation data.

NCEI uses GHCN monthly to monitor long-term trends in temperature and precipitation. It has also been employed in several international climate assessments, including the Intergovernmental Panel on Climate Change 4th Assessment Report, the Arctic Climate Impact Assessment, and the "State of the Climate" report published annually by the Bulletin of the American Meteorological Society.
Select a monthly time series: Historical observations

To retrieve historical data of weather stations, please consider the following:[650]

- Step 1: Select database for Historical observations

In the selection below, minimum temperature was selected GHCN-M (all).

GHCN-M (adjusted)	GHCN-M (all)	other
○precipitation ⓘ	○precipitation ⓘ	○GLOSS sealevel ⓘ
○mean temperature ⓘ	○mean temperature ⓘ	○world river discharge (RivDis)
○minimum temperature ⓘ	◉minimum temperature ⓘ	○USA river discharge (HCDN)
○maximum temperature ⓘ	○maximum temperature ⓘ	○N-America snowcourses (NRCS)
(full lists)	○sealevel pressure	○european SLP (ADVICE)

- Step 2: Select stations

Select a location on the World Map to get the data for the nearest weather station.

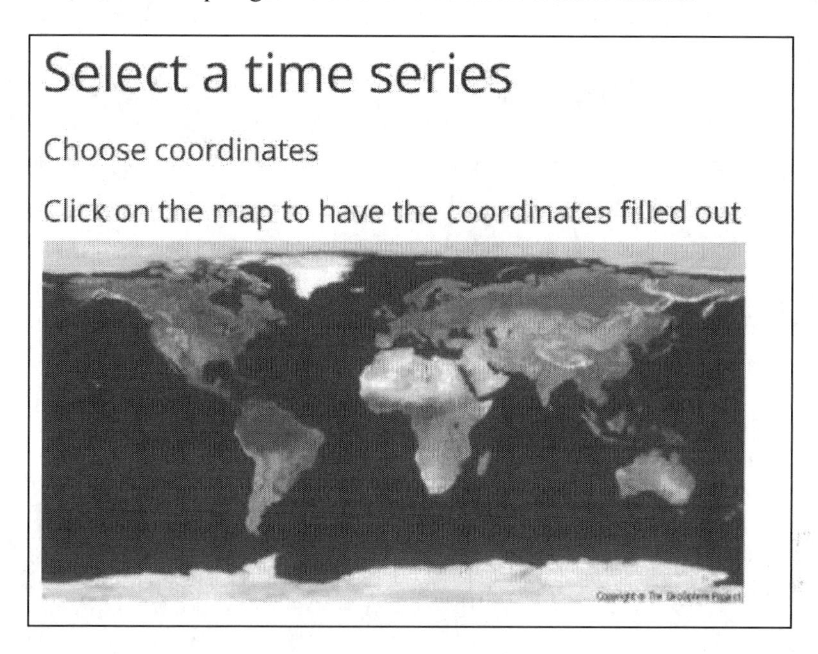

In the coordinates screen, the location in Brazil was selected and the coordinates are populated in stations near fields.

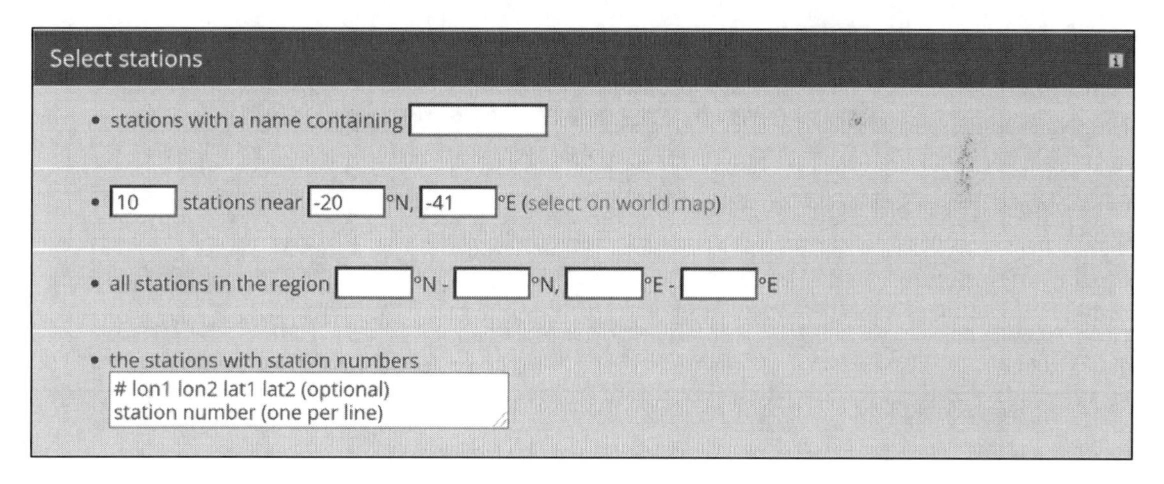

- Step 3: Choose time, distance

Provide stations to be retrieved with the time series data keyed in. The following selection can be interpreted as : "*At least 10 years of data in the monthly season starting in all the months in the year 2000–2021*".

Time, distance

At least 10 years of data in the monthly season starting in all months in years 2000 - 2021

At least ° apart and with m < elevation <

Get stations Clear Form

Select station of the nearest region for which data is to be analyzed. In the above case, I have selected

MONTES CLAROS (BRAZIL)
coordinates: –16.72N, –43.87E, 646.0 m
(prob: 849m)
WMO station code: 83437 (get data)
Associated with urban area (pop. 152000)
Terrain: hilly WARM GRASS/SHRUB
Found 18 years with data in 2002–2019

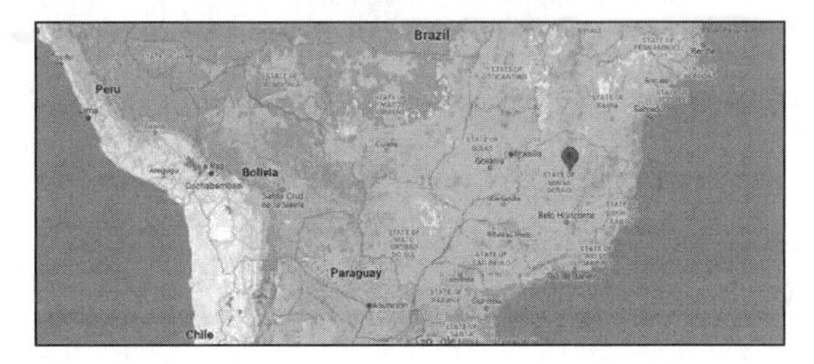

The selected data is as described in the figure below:

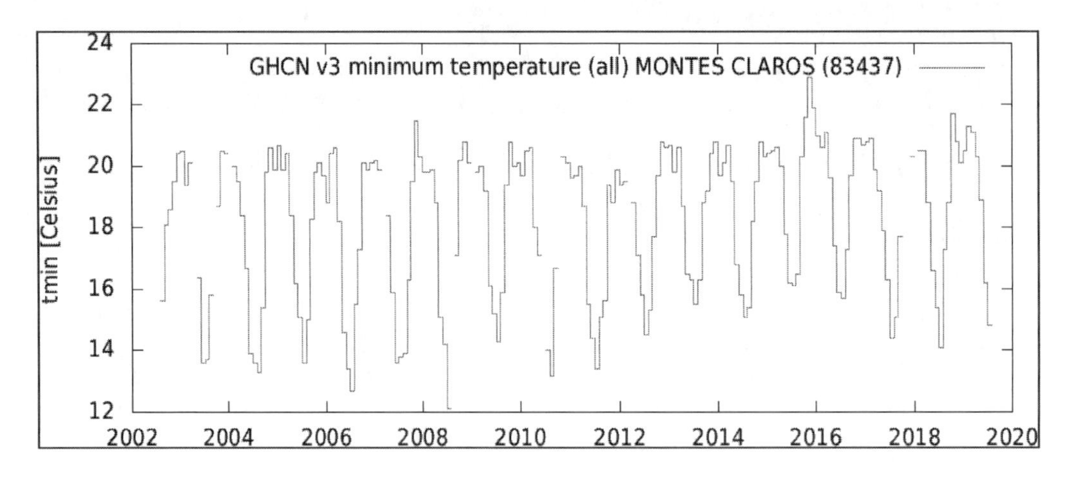

MONTES CLAROS (BRAZIL), coordinates: –16.72N, –43.87E, 646.0m (prob: 849 m), WMO station code: 83437 MONTES CLAROS, tmin [Celsius] daily minimum temperature (unadjusted) from GHCN-M v3.3.0.20190817 (eps, pdf, metadata, raw data, netcdf)

The raw data looks as follows: https://climexp.knmi.nl/data/ma83437.dat

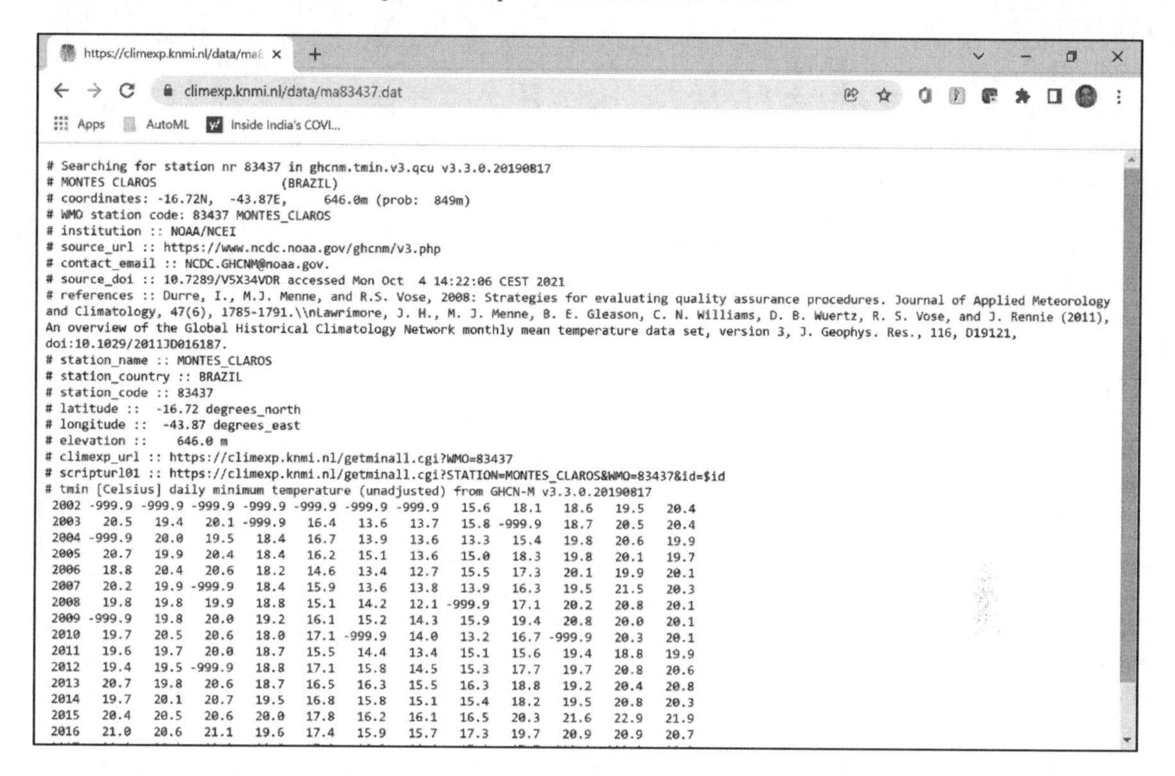

```
# Searching for station nr 83437 in ghcnm.tmin.v3.qcu v3.3.0.20190817
# MONTES CLAROS                 (BRAZIL)
# coordinates: -16.72N,  -43.87E,     646.0m (prob:  849m)
# WMO station code: 83437 MONTES_CLAROS
# institution :: NOAA/NCEI
# source_url :: https://www.ncdc.noaa.gov/ghcnm/v3.php
# contact_email :: NCDC.GHCNM@noaa.gov.
# source_doi :: 10.7289/V5X34VDR accessed Mon Oct  4 14:22:06 CEST 2021
# references :: Durre, I., M.J. Menne, and R.S. Vose, 2008: Strategies for evaluating quality assurance procedures. Journal of Applied Meteorology
and Climatology, 47(6), 1785-1791.\\nLawrimore, J. H., M. J. Menne, B. E. Gleason, C. N. Williams, D. B. Wuertz, R. S. Vose, and J. Rennie (2011),
An overview of the Global Historical Climatology Network monthly mean temperature data set, version 3, J. Geophys. Res., 116, D19121,
doi:10.1029/2011JD016187.
# station_name :: MONTES_CLAROS
# station_country :: BRAZIL
# station_code :: 83437
# latitude ::  -16.72 degrees_north
# longitude ::  -43.87 degrees_east
# elevation ::    646.0 m
# climexp_url :: https://climexp.knmi.nl/getminall.cgi?WMO=83437
# scripturl01 :: https://climexp.knmi.nl/getminall.cgi?STATION=MONTES_CLAROS&WMO=83437&id=$id
# tmin [Celsius] daily minimum temperature (unadjusted) from GHCN-M v3.3.0.20190817
 2002 -999.9 -999.9 -999.9 -999.9 -999.9 -999.9 -999.9   15.6   18.1   18.6   19.5   20.4
 2003   20.5   19.4   20.1 -999.9   16.4   13.6   13.7   15.8 -999.9   18.7   20.5   20.4
 2004 -999.9   20.0   19.5   18.4   16.7   13.9   13.6   13.3   15.4   19.8   20.6   19.9
 2005   20.7   19.9   20.4   18.4   16.2   15.1   13.6   15.0   18.3   19.8   20.1   19.7
 2006   18.8   20.4   20.6   18.2   14.6   13.4   12.7   15.5   17.3   20.1   19.9   20.1
 2007   20.2   19.9 -999.9   18.4   15.9   13.6   13.8   13.9   16.3   19.5   21.5   20.3
 2008   19.8   19.8   19.9   18.8   15.1   14.2   12.1 -999.9   17.1   20.2   20.8   20.1
 2009 -999.9   19.8   20.0   19.2   16.1   15.2   14.3   15.9   19.4   20.8   20.0   20.1
 2010   19.7   20.5   20.6   18.0   17.1 -999.9   14.0   13.2   16.7 -999.9   20.3   20.1
 2011   19.6   19.7   20.0   18.7   15.5   14.4   13.4   15.1   15.6   19.4   18.8   19.9
 2012   19.4   19.5 -999.9   18.8   17.1   15.8   14.5   15.3   17.7   19.7   20.8   20.6
 2013   20.7   19.8   20.6   18.7   16.5   16.3   15.5   16.3   18.8   19.2   20.4   20.8
 2014   19.7   20.1   20.7   19.5   16.8   15.8   15.1   15.4   18.2   19.5   20.8   20.3
 2015   20.4   20.5   20.6   20.0   17.8   16.2   16.1   16.5   20.3   21.6   22.9   21.9
 2016   21.0   20.6   21.1   19.6   17.4   15.9   15.7   17.3   19.7   20.9   20.9   20.7
```

Datasets for Brazil

Dataset Access: INMET : BDMEP (https://bdmep.inmet.gov.br/)

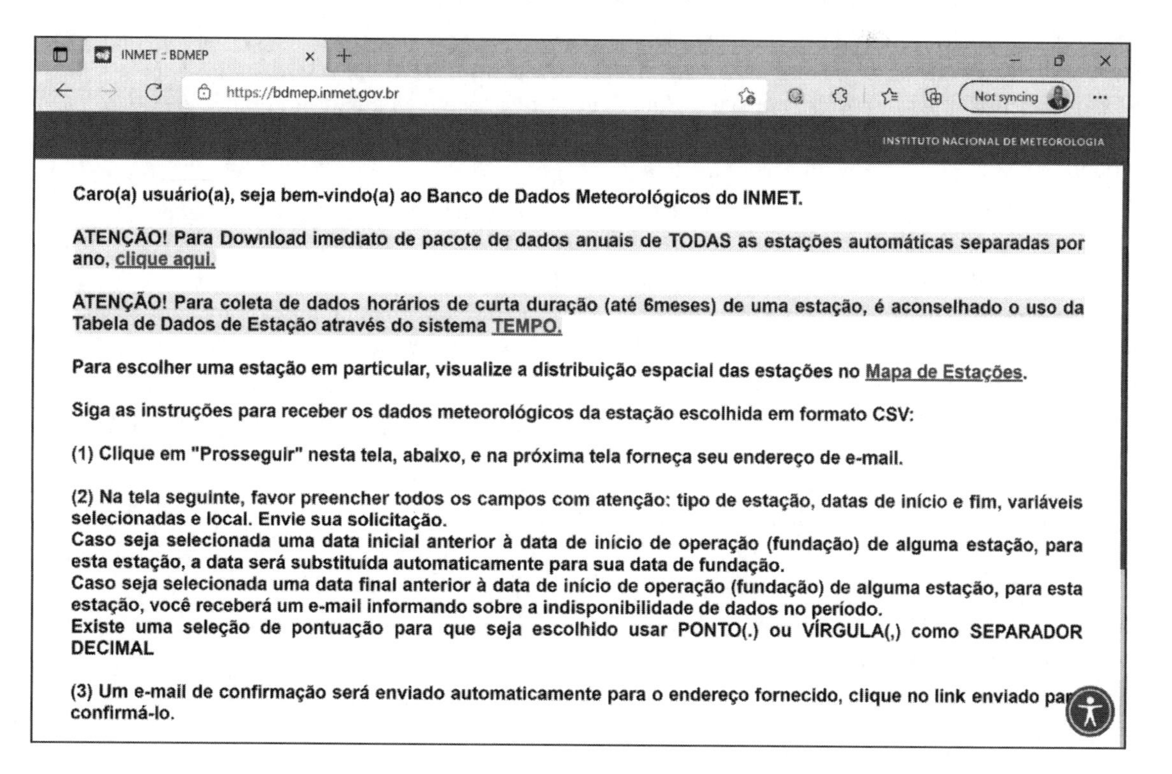

Portuguese	**English**
Caro(a) usuário(a), seja bem-vindo(a) ao Banco de Dados Meteorológicos do INMET.	Dear user, welcome to the INMET Meteorological Database.
ATENÇÃO! Para Download imediato de pacote de dados anuais de TODAS as estações automáticas separadas por ano, clique aqui.	ATTENTION! For Immediate download of annual data package of ALL automatic stations separated by year, click here.
ATENÇÃO! Para coleta de dados horários de curta duração (até 6meses) de uma estação, é aconselhado o uso da Tabela de Dados de Estação através do sistema TEMPO.	ATTENTION! For short-term (up to 6 months) hourly data collection from a station, it is advisable to use the Station Data Table through the TEMPO system.
Para escolher uma estação em particular, visualize a distribuição espacial das estações no Mapa de Estações.	To choose a particular station, view the spatial distribution of stations on the Station Map.
Siga as instruções para receber os dados meteorológicos da estação escolhida em formato CSV:	Follow the instructions to receive the weather data of the chosen station in CSV format:
(1) Clique em "Prosseguir" nesta tela, abaixo, e na próxima tela forneça seu endereço de e-mail.	(1) Click "Continue" on this screen below, and on the next screen then provide your email address.
(2) Na tela seguinte, favor preencher todos os campos com atenção: tipo de estação, datas de início e fim, variáveis selecionadas e local. Envie sua solicitação.	(2) On the next screen, please fill in all the fields carefully: type of station, start and end dates, selected variables and location. Submit your request.
Caso seja selecionada uma data inicial anterior à data de início de operação (fundação) de alguma estação, para esta estação, a data será substituída automaticamente para sua data de fundação.	If an initial date is selected before the start of operation (foundation) date of any station, for this station, the date will be automatically substituted for its foundation date.
Caso seja selecionada uma data final anterior à data de início de operação (fundação) de alguma estação, para esta estação, você receberá um e-mail informando sobre a indisponibilidade de dados no período.	If an end date is selected before the start of operation (foundation) date of any station, for this station, you will receive an email informing you about the unavailability of data in the period.
Existe uma seleção de pontuação para que seja escolhido usar PONTO(.) ou VÍRGULA(,) como SEPARADOR DECIMAL	There is a punctuation selection so that you can choose to use DOT(.) or COMMA(,) as DECIMAL SEPARATOR
(3) Um e-mail de confirmação será enviado automaticamente para o endereço fornecido, clique no link enviado para confirmá-lo.	(3) A confirmation email will be sent automatically to the provided address, click on the sent link to confirm it.
Para efeito de controle, existe uma fila de processamento e quando sua requisição for iniciada você receberá um e-mail alertando.	For control purposes, there is a processing queue and when your request is started you will receive an email alerting you.
Ao término do processo, um e-mail de conclusão será enviado contendo o link de acesso aos dados selecionados; após 48 h esses dados serão apagados.	At the end of the process, a completion email will be sent containing the access link to the selected data; after 48 h these datasets will be deleted.
Para mais de uma requisição, favor aguardar o término do primeiro processamento.	For more than one request, please wait for the first processing to finish.

The National Meteorological Service of Brazil,[651] hereafter INMET, has provided the Brazilian Climate data (Brazilian-CD) online and made it freely available as text CSV to the public. The Brazilian-CD comprises data from all observing weather stations operated and maintained by INMET, by either automatic or conventional stations, with the total number varying according to year, from about 400 in 2000 to 834 stations in 2020. The WMO SMM-CD_NRP was used to assess the temperature, humidity and precipitation data of the Brazilian-CD. The averaged stewardship maturity rating levels for these categories are 2.5 and 1.7 respectively. The assessment was originally made by an expert from INMET, which was moderated by one member of the WMO ET-DDS team. The online publication of the Brazilian CD meets important criteria for publicity, such as data access, portability and preservation, in the operational data management category, but also in this category there is room for improving the documentation and data integrity, due to lack of information, or even application, of such aspects in the web site. Regarding the data stewardship category, it was noticed that no quality assurance or quality control procedures are mentioned in the online documentation of this CD, but a further communication to the assessment point-of-contact (POC) had clarified that the dataset is under a routine procedure of quality control. Furthermore, despite the CD being provided by INMET, no explicit information on governance or POC is given, which lowers the score for this aspect. The Brazilian CD is available online with minimal metadata information, regarding the observing station location, altitude and period of operation. Besides, climate normal parameters can be found as figures and table, by simply consulting the map of the station, though such information is not integrated in the provided CSV downloaded file. Nonetheless, substantial progress has been made recently by INMET for improving the informational content of the Brazilian Climate data to users, as nationally reported and also referred to by relevant centers as NOAA CPC and IRI.

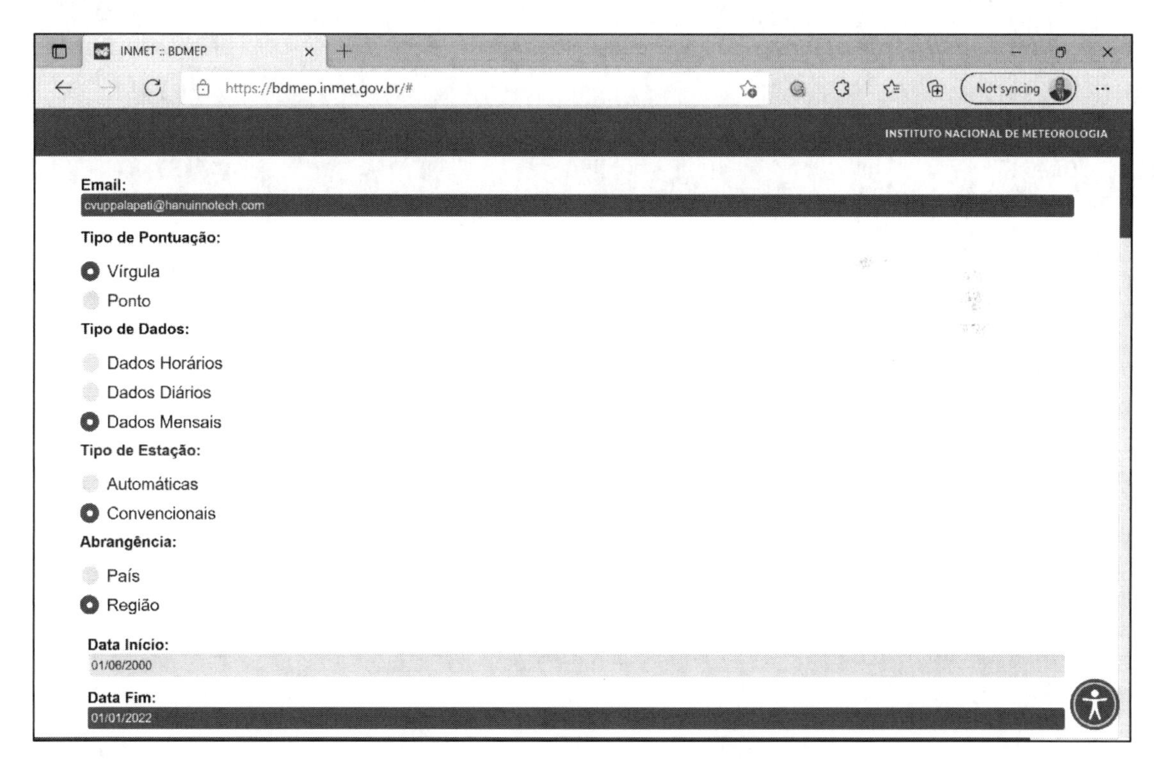

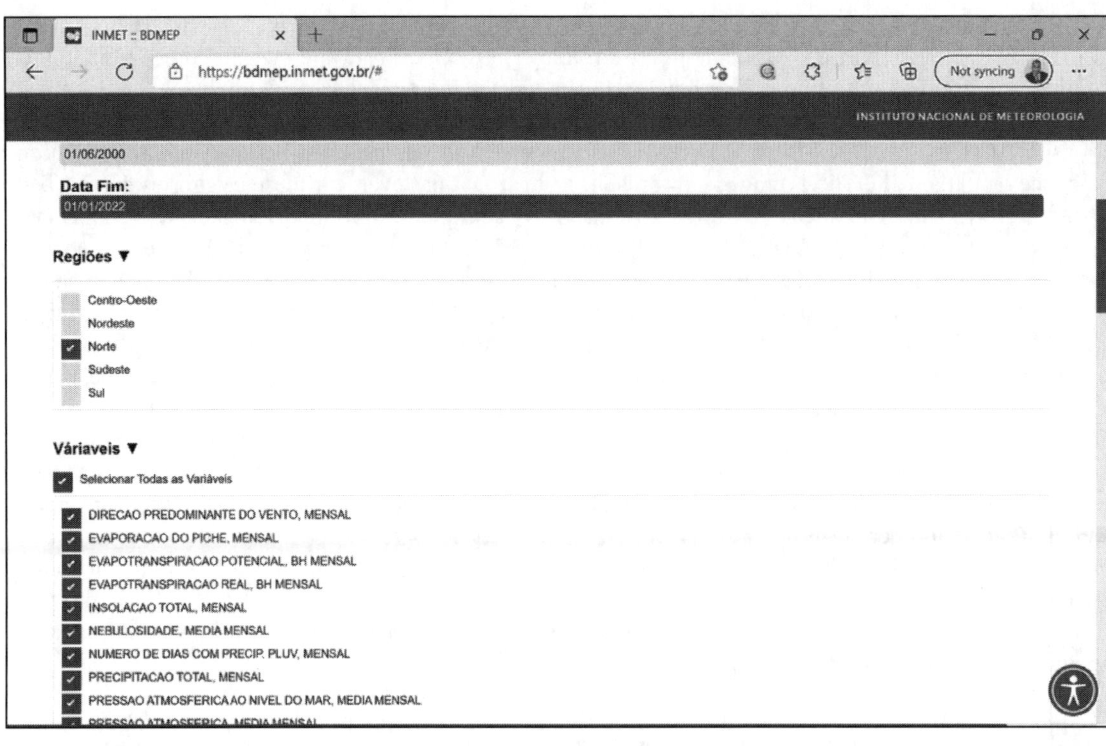
INMET :: BDMEP
https://bdmep.inmet.gov.br/#
Not syncing
INSTITUTO NACIONAL DE METEOROLOGIA
01/06/2000
Data Fim:
01/01/2022
Regiões ▼
Centro-Oeste
Nordeste
Norte
Sudeste
Sul
Váriaveis ▼
Selecionar Todas as Variáveis
DIRECAO PREDOMINANTE DO VENTO, MENSAL
EVAPORACAO DO PICHE, MENSAL
EVAPOTRANSPIRACAO POTENCIAL, BH MENSAL
EVAPOTRANSPIRACAO REAL, BH MENSAL
INSOLACAO TOTAL, MENSAL
NEBULOSIDADE, MEDIA MENSAL
NUMERO DE DIAS COM PRECIP. PLUV, MENSAL
PRECIPITACAO TOTAL, MENSAL
PRESSAO ATMOSFERICA AO NIVEL DO MAR, MEDIA MENSAL

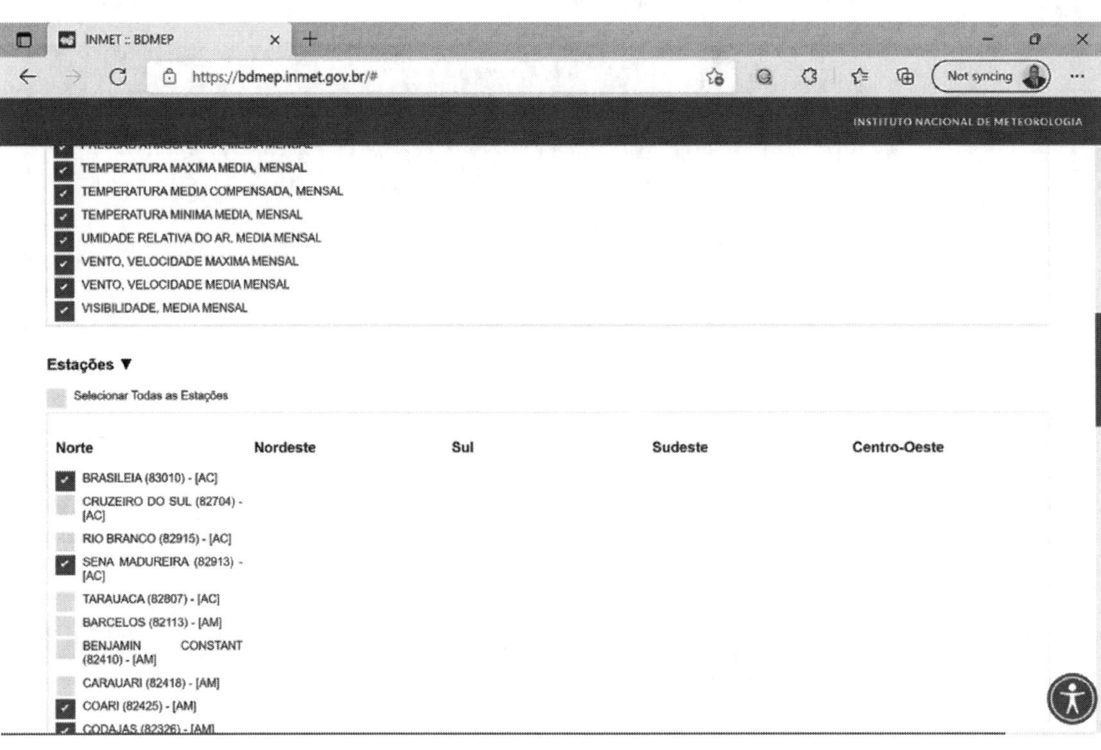
INMET :: BDMEP
https://bdmep.inmet.gov.br/#
Not syncing
INSTITUTO NACIONAL DE METEOROLOGIA
TEMPERATURA MAXIMA MEDIA, MENSAL
TEMPERATURA MEDIA COMPENSADA, MENSAL
TEMPERATURA MINIMA MEDIA, MENSAL
UMIDADE RELATIVA DO AR, MEDIA MENSAL
VENTO, VELOCIDADE MAXIMA MENSAL
VENTO, VELOCIDADE MEDIA MENSAL
VISIBILIDADE, MEDIA MENSAL
Estações ▼
Selecionar Todas as Estações
Norte
Nordeste
Sul
Sudeste
Centro-Oeste
BRASILEIA (83010) - [AC]
CRUZEIRO DO SUL (82704) - [AC]
RIO BRANCO (82915) - [AC]
SENA MADUREIRA (82913) - [AC]
TARAUACA (82807) - [AC]
BARCELOS (82113) - [AM]
BENJAMIN CONSTANT (82410) - [AM]
CARAUARI (82418) - [AM]
COARI (82425) - [AM]

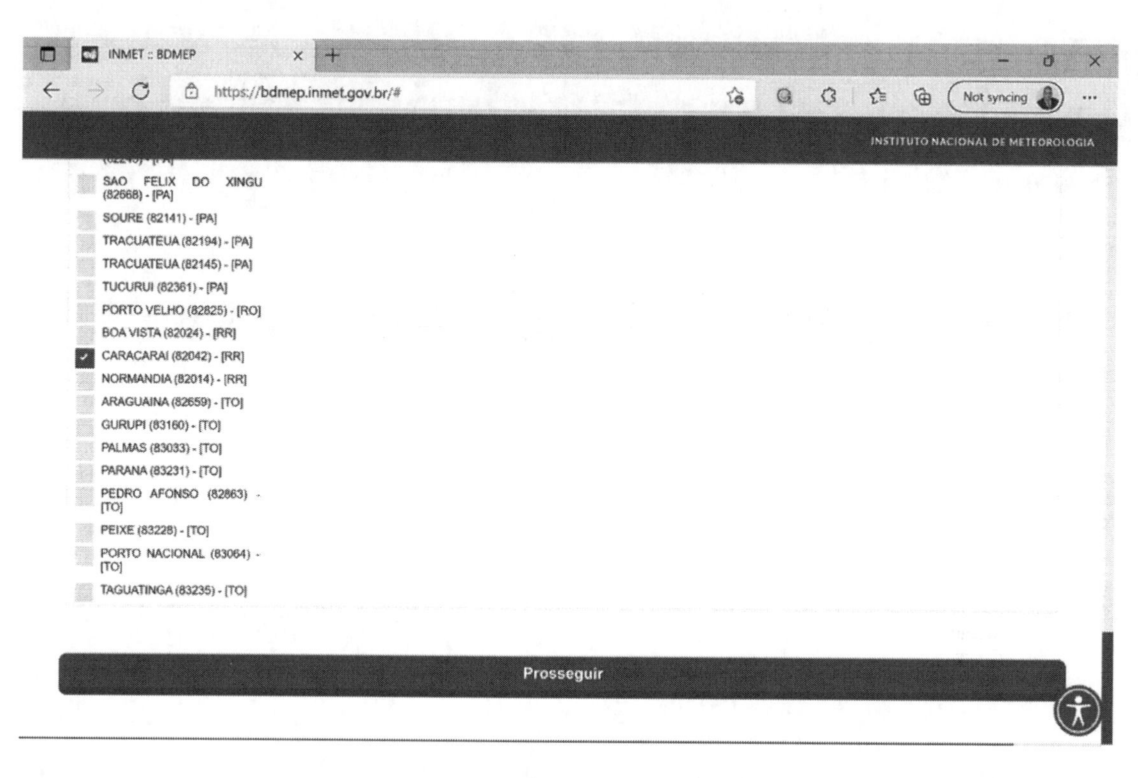
INMET :: BDMEP
https://bdmep.inmet.gov.br/#
Not syncing
INSTITUTO NACIONAL DE METEOROLOGIA
SAO FELIX DO XINGU (82668) - [PA]
SOURE (82141) - [PA]
TRACUATEUA (82194) - [PA]
TRACUATEUA (82145) - [PA]
TUCURUI (82361) - [PA]
PORTO VELHO (82825) - [RO]
BOA VISTA (82024) - [RR]
CARACARAI (82042) - [RR]
NORMANDIA (82014) - [RR]
ARAGUAINA (82659) - [TO]
GURUPI (83160) - [TO]
PALMAS (83033) - [TO]
PARANA (83231) - [TO]
PEDRO AFONSO (82863) - [TO]
PEIXE (83228) - [TO]
PORTO NACIONAL (83064) - [TO]
TAGUATINGA (83235) - [TO]
Prosseguir

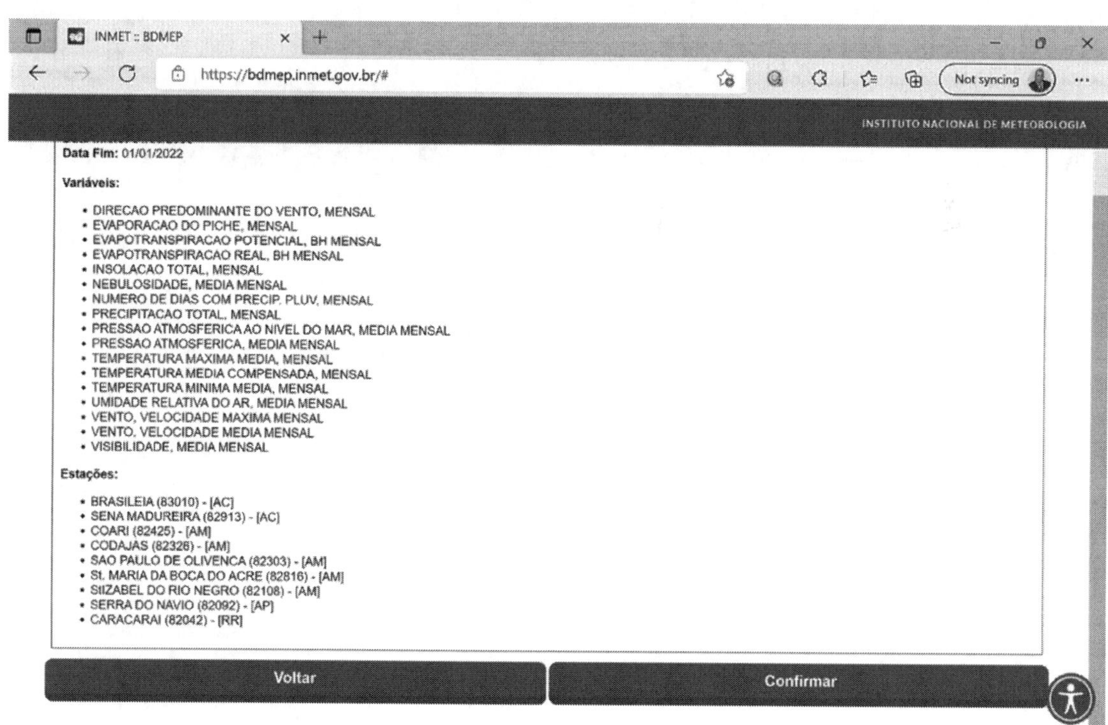
INMET :: BDMEP
https://bdmep.inmet.gov.br/#
Not syncing
INSTITUTO NACIONAL DE METEOROLOGIA
Data Fim: 01/01/2022
Variáveis:
DIRECAO PREDOMINANTE DO VENTO, MENSAL
EVAPORACAO DO PICHE, MENSAL
EVAPOTRANSPIRACAO POTENCIAL, BH MENSAL
EVAPOTRANSPIRACAO REAL, BH MENSAL
INSOLACAO TOTAL, MENSAL
NEBULOSIDADE, MEDIA MENSAL
NUMERO DE DIAS COM PRECIP. PLUV, MENSAL
PRECIPITACAO TOTAL, MENSAL
PRESSAO ATMOSFERICA AO NIVEL DO MAR, MEDIA MENSAL
PRESSAO ATMOSFERICA, MEDIA MENSAL
TEMPERATURA MAXIMA MEDIA, MENSAL
TEMPERATURA MEDIA COMPENSADA, MENSAL
TEMPERATURA MINIMA MEDIA, MENSAL
UMIDADE RELATIVA DO AR, MEDIA MENSAL
VENTO, VELOCIDADE MAXIMA MENSAL
VENTO, VELOCIDADE MEDIA MENSAL
VISIBILIDADE, MEDIA MENSAL
Estações:
BRASILEIA (83010) - [AC]
SENA MADUREIRA (82913) - [AC]
COARI (82425) - [AM]
CODAJAS (82326) - [AM]
SAO PAULO DE OLIVENCA (82303) - [AM]
St. MARIA DA BOCA DO ACRE (82816) - [AM]
StIZABEL DO RIO NEGRO (82108) - [AM]
SERRA DO NAVIO (82092) - [AP]
CARACARAI (82042) - [RR]
Voltar
Confirmar

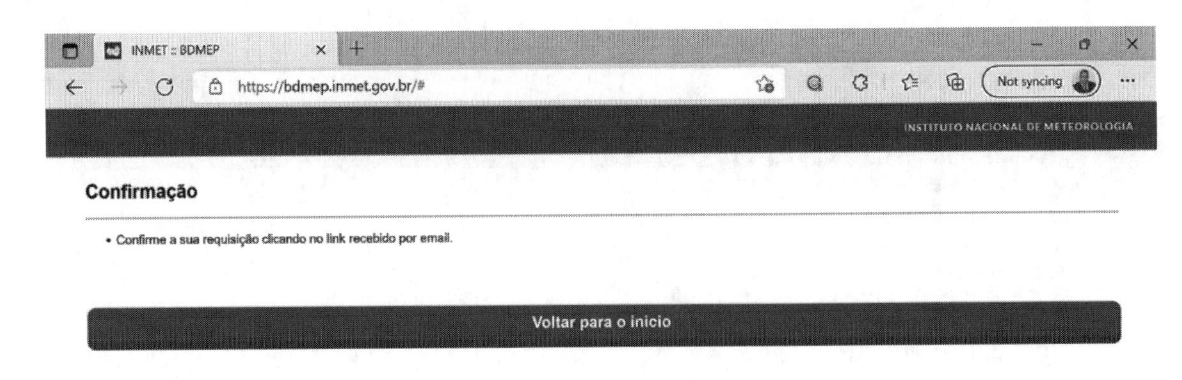
INMET :: BDMEP
https://bdmep.inmet.gov.br/#
Not syncing
INSTITUTO NACIONAL DE METEOROLOGIA
Confirmação
• Confirme a sua requisição clicando no link recebido por email.
Voltar para o inicio

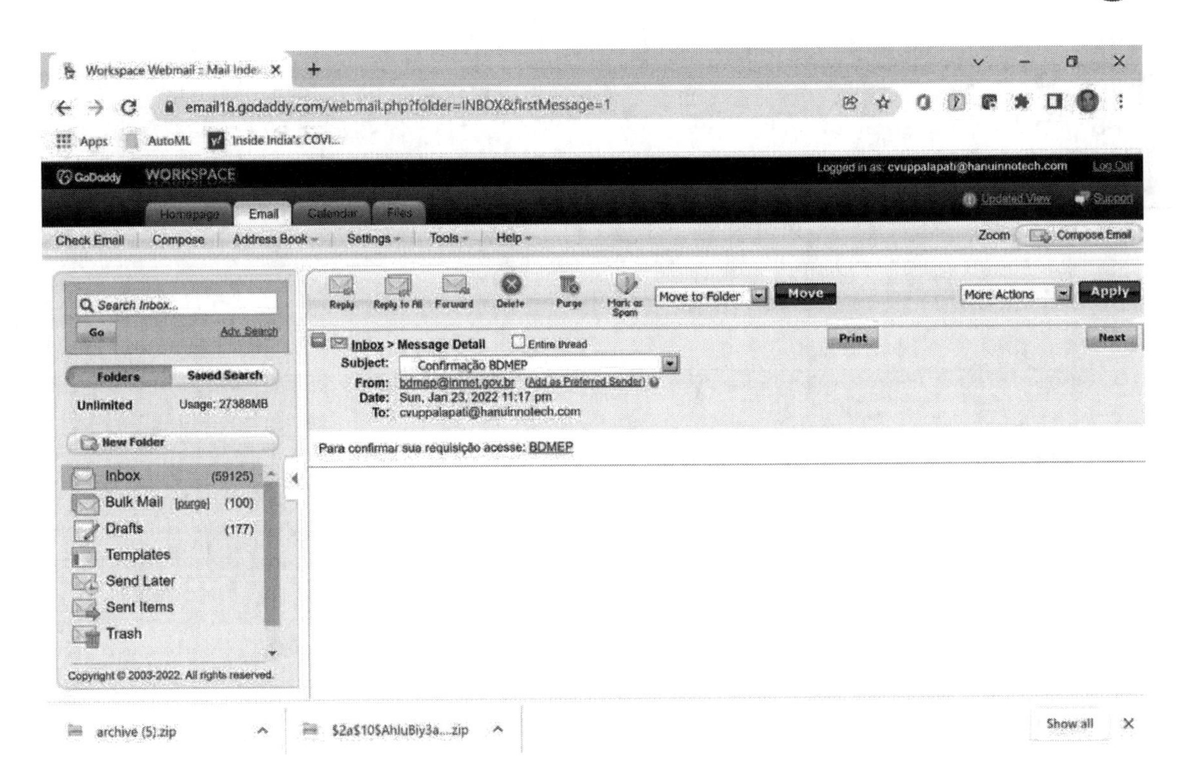
Workspace Webmail :: Mail Inde
email18.godaddy.com/webmail.php?folder=INBOX&firstMessage=1
Apps
AutoML
Inside India's COVI...
GoDaddy WORKSPACE
Logged in as: cvuppalapati@hanuinnotech.com
Log Out
Email
Check Email
Compose
Address Book
Settings
Tools
Help
Zoom
Compose Email
Search Inbox...
Go
Adv. Search
Folders
Saved Search
Unlimited
Usage: 27388MB
New Folder
Inbox (59125)
Bulk Mail [purge] (100)
Drafts (177)
Templates
Send Later
Sent Items
Trash
Copyright © 2003-2022. All rights reserved.
Move to Folder
Move
More Actions
Apply
Inbox > Message Detail
Entire thread
Print
Next
Subject: Confirmação BDMEP
From: bdmep@inmet.gov.br (Add as Preferred Sender)
Date: Sun, Jan 23, 2022 11:17 pm
To: cvuppalapati@hanuinnotech.com
Para confirmar sua requisição acesse: BDMEP
archive (5).zip
Show all

Confirm by clicking the link.

Portages	English
Sua requisição foi confirmada e está na fila para ser processada!	Your request has been confirmed and is queued to be processed!

Portages	English
Sua requisição começou a ser processada. Aguarde novos emails para acessar os dados selecionados. Caso tenha selecionado muitas estações e/ou um período de tempo muito longo, este processamento poderá levar até 24h para terminar. Este é um e-mail automático, não responda.	Your request has started to be processed. Wait for new emails to access the selected data. If you have selected many stations and/or a very long period of time, this processing may take up to 24 hours to complete. This is an automated email, do not reply.

Email confirmation to download the data:

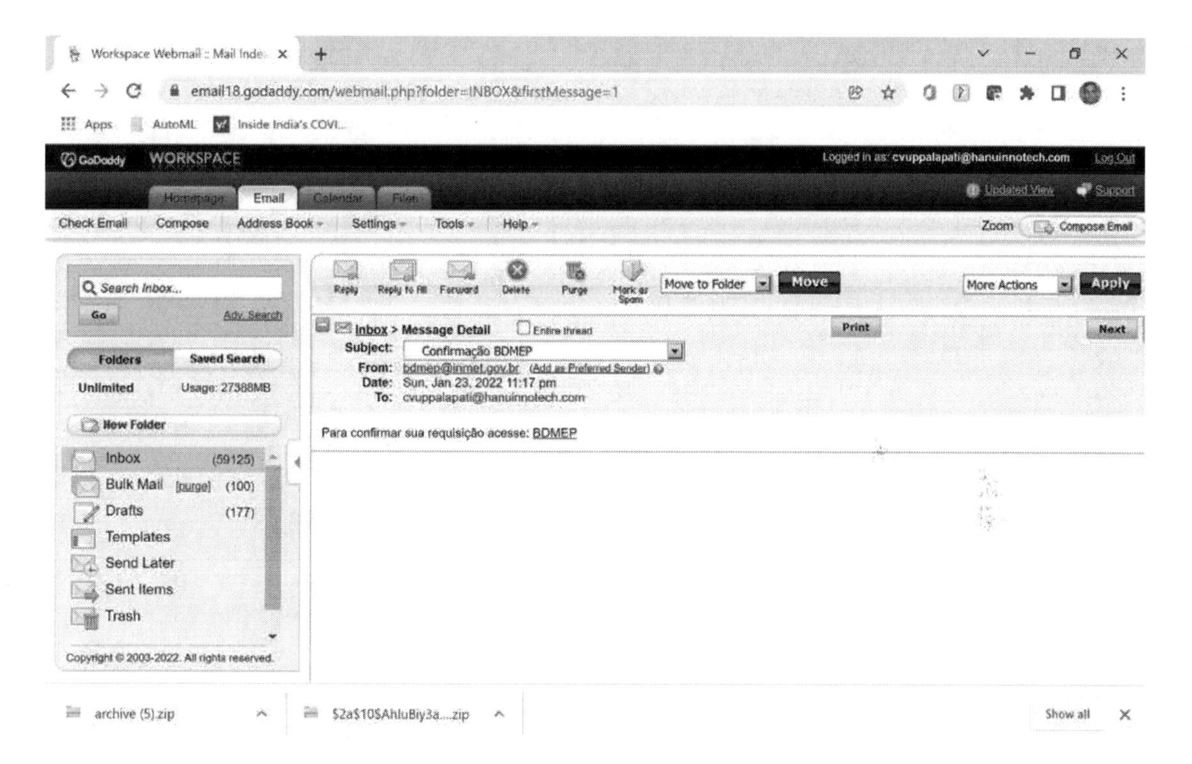

Sua requisição foi concluída. Favor acessar https://bdmep.inmet.gov.br/$2a$10$USgI6pDFd0FOZXyl2UA.seQOFwkz3pixQkiWNCECi1TmWbEoPYAN6.zip para coletar seus dados. Os dados serão apagados em 48horas a contar de 05:21 24-01-2022. Este é um e-mail automático, não responda.	Your request has been completed. Please access https://bdmep.inmet.gov.br/$2a$10$USgI6pDFd0FOZXyl2UA.seQOFwkz3pixQkiWNCECi1TmWbEoPYAN6.zip to collect your data. Data will be deleted within 48 hours from 05:21 2022-01-24. This is an automated email, do not reply.

Source: https://climatedata-catalogue.wmo.int/explore

Labor Force Statistics from the Current Population Survey

Series Id: LNS14000000[652] Seasonally Adjusted Series title: (Seas) Unemployment Rate
Labor force status: Unemployment rate Type of data: Percent or rate Age: 16 years and over

Year	Jan	Feb	Mar	Apr	May	Jun	Jul	Aug	Sep	Oct	Nov	Dec
1948	3.4	3.8	4.0	3.9	3.5	3.6	3.6	3.9	3.8	3.7	3.8	4.0
1949	4.3	4.7	5.0	5.3	6.1	6.2	6.7	6.8	6.6	7.9	6.4	6.6
1950	6.5	6.4	6.3	5.8	5.5	5.4	5.0	4.5	4.4	4.2	4.2	4.3
1951	3.7	3.4	3.4	3.1	3.0	3.2	3.1	3.1	3.3	3.5	3.5	3.1
1952	3.2	3.1	2.9	2.9	3.0	3.0	3.2	3.4	3.1	3.0	2.8	2.7
1953	2.9	2.6	2.6	2.7	2.5	2.5	2.6	2.7	2.9	3.1	3.5	4.5
1954	4.9	5.2	5.7	5.9	5.9	5.6	5.8	6.0	6.1	5.7	5.3	5.0
1955	4.9	4.7	4.6	4.7	4.3	4.2	4.0	4.2	4.1	4.3	4.2	4.2
1956	4.0	3.9	4.2	4.0	4.3	4.3	4.4	4.1	3.9	3.9	4.3	4.2
1957	4.2	3.9	3.7	3.9	4.1	4.3	4.2	4.1	4.4	4.5	5.1	5.2
1958	5.8	6.4	6.7	7.4	7.4	7.3	7.5	7.4	7.1	6.7	6.2	6.2
1959	6.0	5.9	5.6	5.2	5.1	5.0	5.1	5.2	5.5	5.7	5.8	5.3
1960	5.2	4.8	5.4	5.2	5.1	5.4	5.5	5.6	5.5	6.1	6.1	6.6
1961	6.6	6.9	6.9	7.0	7.1	6.9	7.0	6.6	6.7	6.5	6.1	6.0
1962	5.8	5.5	5.6	5.6	5.5	5.5	5.4	5.7	5.6	5.4	5.7	5.5
1963	5.7	5.9	5.7	5.7	5.9	5.6	5.6	5.4	5.5	5.5	5.7	5.5
1964	5.6	5.4	5.4	5.3	5.1	5.2	4.9	5.0	5.1	5.1	4.8	5.0
1965	4.9	5.1	4.7	4.8	4.6	4.6	4.4	4.4	4.3	4.2	4.1	4.0
1966	4.0	3.8	3.8	3.8	3.9	3.8	3.8	3.8	3.7	3.7	3.6	3.8
1967	3.9	3.8	3.8	3.8	3.8	3.9	3.8	3.8	3.8	4.0	3.9	3.8
1968	3.7	3.8	3.7	3.5	3.5	3.7	3.7	3.5	3.4	3.4	3.4	3.4
1969	3.4	3.4	3.4	3.4	3.4	3.5	3.5	3.5	3.7	3.7	3.5	3.5
1970	3.9	4.2	4.4	4.6	4.8	4.9	5.0	5.1	5.4	5.5	5.9	6.1
1971	5.9	5.9	6.0	5.9	5.9	5.9	6.0	6.1	6.0	5.8	6.0	6.0
1972	5.8	5.7	5.8	5.7	5.7	5.7	5.6	5.6	5.5	5.6	5.3	5.2
1973	4.9	5.0	4.9	5.0	4.9	4.9	4.8	4.8	4.8	4.6	4.8	4.9
1974	5.1	5.2	5.1	5.1	5.1	5.4	5.5	5.5	5.9	6.0	6.6	7.2
1975	8.1	8.1	8.6	8.8	9.0	8.8	8.6	8.4	8.4	8.4	8.3	8.2
1976	7.9	7.7	7.6	7.7	7.4	7.6	7.8	7.8	7.6	7.7	7.8	7.8
1977	7.5	7.6	7.4	7.2	7.0	7.2	6.9	7.0	6.8	6.8	6.8	6.4
1978	6.4	6.3	6.3	6.1	6.0	5.9	6.2	5.9	6.0	5.8	5.9	6.0
1979	5.9	5.9	5.8	5.8	5.6	5.7	5.7	6.0	5.9	6.0	5.9	6.0
1980	6.3	6.3	6.3	6.9	7.5	7.6	7.8	7.7	7.5	7.5	7.5	7.2
1981	7.5	7.4	7.4	7.2	7.5	7.5	7.2	7.4	7.6	7.9	8.3	8.5
1982	8.6	8.9	9.0	9.3	9.4	9.6	9.8	9.8	10.1	10.4	10.8	10.8
1983	10.4	10.4	10.3	10.2	10.1	10.1	9.4	9.5	9.2	8.8	8.5	8.3

Year	Jan	Feb	Mar	Apr	May	Jun	Jul	Aug	Sep	Oct	Nov	Dec
1984	8.0	7.8	7.8	7.7	7.4	7.2	7.5	7.5	7.3	7.4	7.2	7.3
1985	7.3	7.2	7.2	7.3	7.2	7.4	7.4	7.1	7.1	7.1	7.0	7.0
1986	6.7	7.2	7.2	7.1	7.2	7.2	7.0	6.9	7.0	7.0	6.9	6.6
1987	6.6	6.6	6.6	6.3	6.3	6.2	6.1	6.0	5.9	6.0	5.8	5.7
1988	5.7	5.7	5.7	5.4	5.6	5.4	5.4	5.6	5.4	5.4	5.3	5.3
1989	5.4	5.2	5.0	5.2	5.2	5.3	5.2	5.2	5.3	5.3	5.4	5.4
1990	5.4	5.3	5.2	5.4	5.4	5.2	5.5	5.7	5.9	5.9	6.2	6.3
1991	6.4	6.6	6.8	6.7	6.9	6.9	6.8	6.9	6.9	7.0	7.0	7.3
1992	7.3	7.4	7.4	7.4	7.6	7.8	7.7	7.6	7.6	7.3	7.4	7.4
1993	7.3	7.1	7.0	7.1	7.1	7.0	6.9	6.8	6.7	6.8	6.6	6.5
1994	6.6	6.6	6.5	6.4	6.1	6.1	6.1	6.0	5.9	5.8	5.6	5.5
1995	5.6	5.4	5.4	5.8	5.6	5.6	5.7	5.7	5.6	5.5	5.6	5.6
1996	5.6	5.5	5.5	5.6	5.6	5.3	5.5	5.1	5.2	5.2	5.4	5.4
1997	5.3	5.2	5.2	5.1	4.9	5.0	4.9	4.8	4.9	4.7	4.6	4.7
1998	4.6	4.6	4.7	4.3	4.4	4.5	4.5	4.5	4.6	4.5	4.4	4.4
1999	4.3	4.4	4.2	4.3	4.2	4.3	4.3	4.2	4.2	4.1	4.1	4.0
2000	4.0	4.1	4.0	3.8	4.0	4.0	4.0	4.1	3.9	3.9	3.9	3.9
2001	4.2	4.2	4.3	4.4	4.3	4.5	4.6	4.9	5.0	5.3	5.5	5.7
2002	5.7	5.7	5.7	5.9	5.8	5.8	5.8	5.7	5.7	5.7	5.9	6.0
2003	5.8	5.9	5.9	6.0	6.1	6.3	6.2	6.1	6.1	6.0	5.8	5.7
2004	5.7	5.6	5.8	5.6	5.6	5.6	5.5	5.4	5.4	5.5	5.4	5.4
2005	5.3	5.4	5.2	5.2	5.1	5.0	5.0	4.9	5.0	5.0	5.0	4.9
2006	4.7	4.8	4.7	4.7	4.6	4.6	4.7	4.7	4.5	4.4	4.5	4.4
2007	4.6	4.5	4.4	4.5	4.4	4.6	4.7	4.6	4.7	4.7	4.7	5.0
2008	5.0	4.9	5.1	5.0	5.4	5.6	5.8	6.1	6.1	6.5	6.8	7.3
2009	7.8	8.3	8.7	9.0	9.4	9.5	9.5	9.6	9.8	10.0	9.9	9.9
2010	9.8	9.8	9.9	9.9	9.6	9.4	9.4	9.5	9.5	9.4	9.8	9.3
2011	9.1	9.0	9.0	9.1	9.0	9.1	9.0	9.0	9.0	8.8	8.6	8.5
2012	8.3	8.3	8.2	8.2	8.2	8.2	8.2	8.1	7.8	7.8	7.7	7.9
2013	8.0	7.7	7.5	7.6	7.5	7.5	7.3	7.2	7.2	7.2	6.9	6.7
2014	6.6	6.7	6.7	6.2	6.3	6.1	6.2	6.1	5.9	5.7	5.8	5.6
2015	5.7	5.5	5.4	5.4	5.6	5.3	5.2	5.1	5.0	5.0	5.1	5.0
2016	4.8	4.9	5.0	5.1	4.8	4.9	4.8	4.9	5.0	4.9	4.7	4.7
2017	4.7	4.6	4.4	4.5	4.4	4.3	4.3	4.4	4.2	4.1	4.2	4.1
2018	4.0	4.1	4.0	4.0	3.8	4.0	3.8	3.8	3.7	3.8	3.8	3.9
2019	4.0	3.8	3.8	3.7	3.7	3.6	3.6	3.7	3.5	3.6	3.6	3.6
2020	3.5	3.5	4.4	14.8	13.3	11.1	10.2	8.4	7.8	6.9	6.7	6.7
2021	6.3	6.2	6.0	6.1	5.8	5.9	5.4	5.2	4.8			

IMF Country Index Weights

Country Indexes and Weights provide macro level consumption and social behavior information of people in terms of purchase preferences such as Food and Non-alcoholic beverages, alcoholic beverages, Clothing, and others.

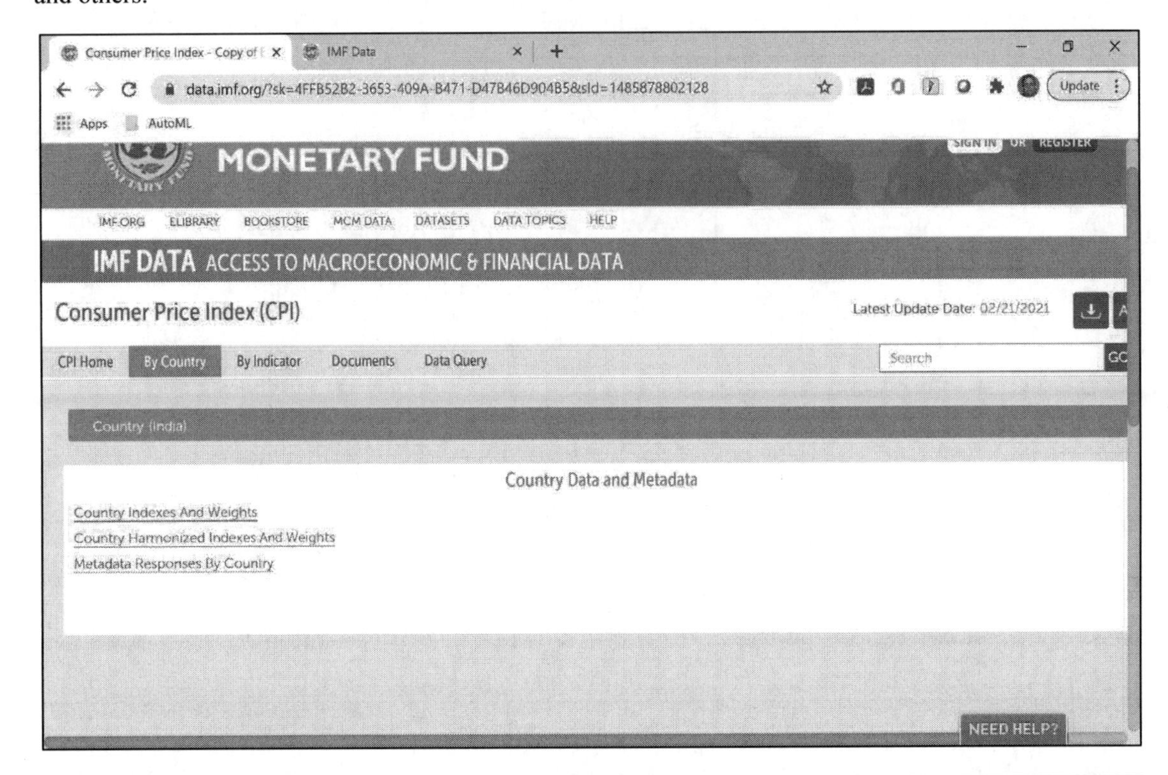

Country: India

Country Indexes And Weight

India

	2012M12	2013M01	2013M02	2013M03	2013M04	2013M05	2013M06	2013M07	2013M08	2013M0
Indexes										
Index Reference Period	2015A	2015A	2015A	2015A	2015A	2015A	2015A	2015A	2015A	2015
Consumer Price Index, All items	83.77	84.54	85.30	85.69	86.45	87.22	88.36	89.89	90.66	91.
Food and non-alcoholic beverages		82.86	83.65	83.73	84.51	85.69	88.29	90.41	92.06	93.
Alcoholic Beverages, Tobacco, and Narcotics		81.20	81.66	82.36	83.05	83.90	84.60	85.22	85.91	86.
Clothing and footwear		84.44	84.91	85.31	85.79	86.27	86.90	87.46	88.01	88.
Housing, Water, Electricity, Gas and Other Fuels		85.37	85.62	85.62	85.79	86.12	89.45	90.29	91.04	91.
Furnishings, household equipment and routine household maintenance		86.56	86.89	87.22	87.80	88.29	88.87	89.37	89.78	90.
Health		87.72	88.14	88.48	88.82	89.32	89.74	90.25	90.67	91.
Transport		93.62	94.61	95.51	94.97	93.80	95.06	97.23	98.13	100.
Communication		90.79	91.33	92.14	92.49	92.94	93.39	93.92	94.46	95.
Recreation and culture		87.54	87.80	87.97	88.39	88.90	89.66	90.51	91.02	91.
Education		83.54	83.70	83.86	84.66	85.15	86.67	88.12	88.77	89.
Restaurants and hotels		82.02	82.78	83.24	83.78	84.54	85.30	86.07	86.68	87.
Miscellaneous goods and services		92.95	92.95	92.77	91.53	90.99	91.44	91.44	93.75	95.
Weights										
Food and non-alcoholic beverages, Weight										

https://data.imf.org/regular.aspx?key=61015892

https://data.imf.org/?sk=388dfa60-1d26-4ade-b505-a05a558d9a42&sId=1479329328660

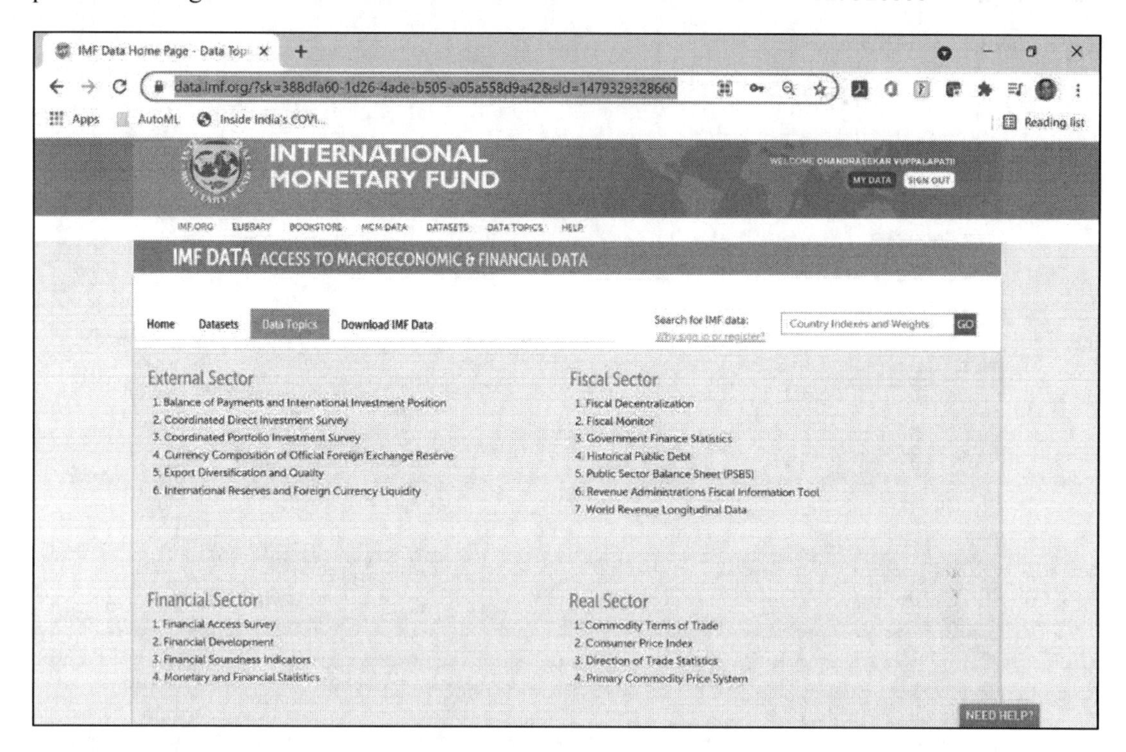

IMF Data Bulk Download

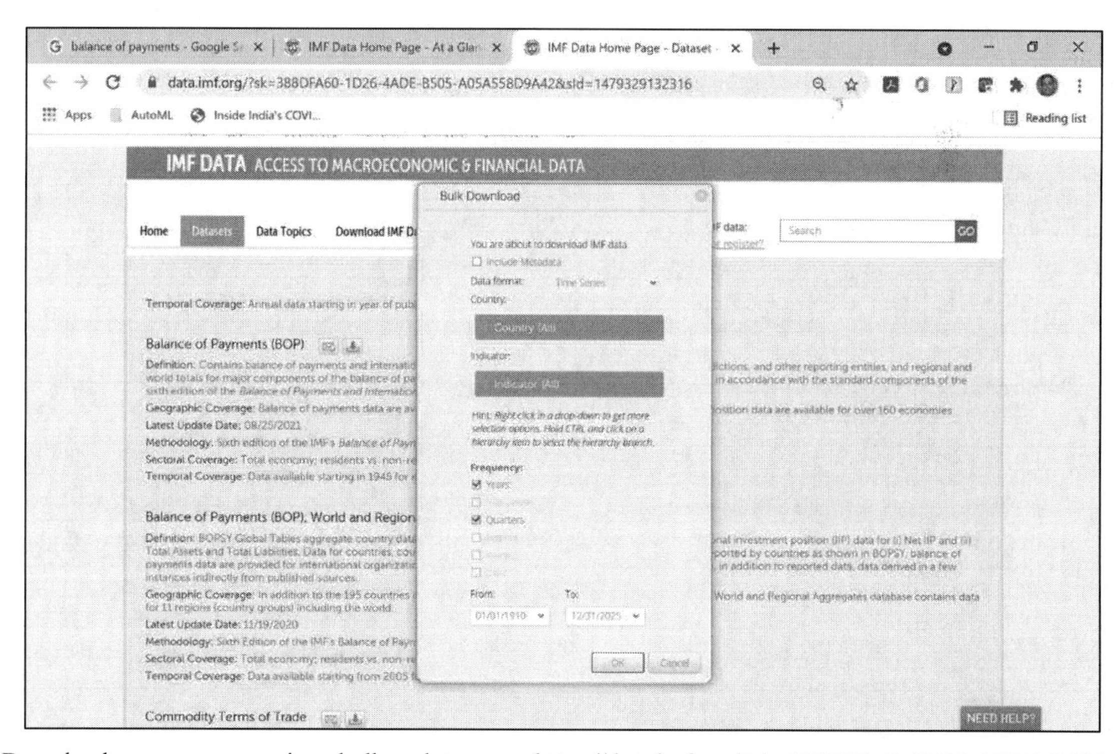

Download macroeconomic bulk data - https://data.imf.org/?sk=388DFA60-1D26-4ADE-B505-A05A558D9A42&sId=1479329132316

https://data.imf.org/?sk=388DFA60-1D26-4ADE-B505-A05A558D9A42&sId=1479329334655

World Bank Data

Macroeconomic world bank data[653]-

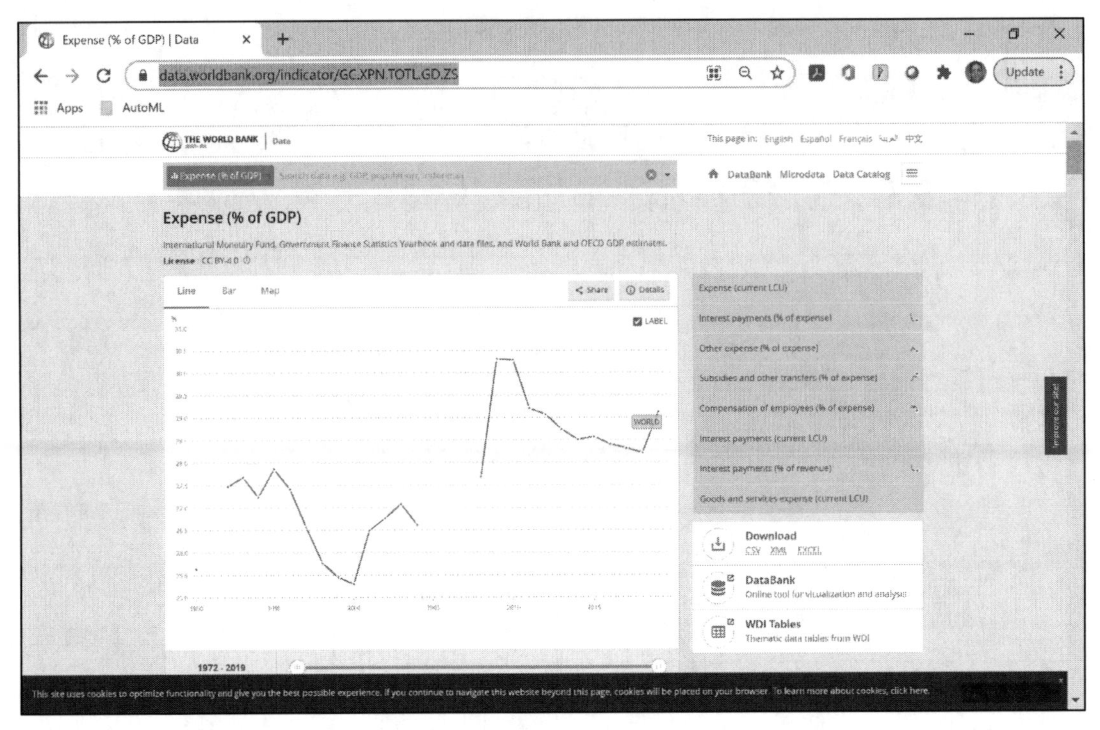

Source: https://data.worldbank.org/indicator/GC.XPN.TOTL.GD.ZS

The World Bank Development Indicators

World Development Indicator tool:[654]

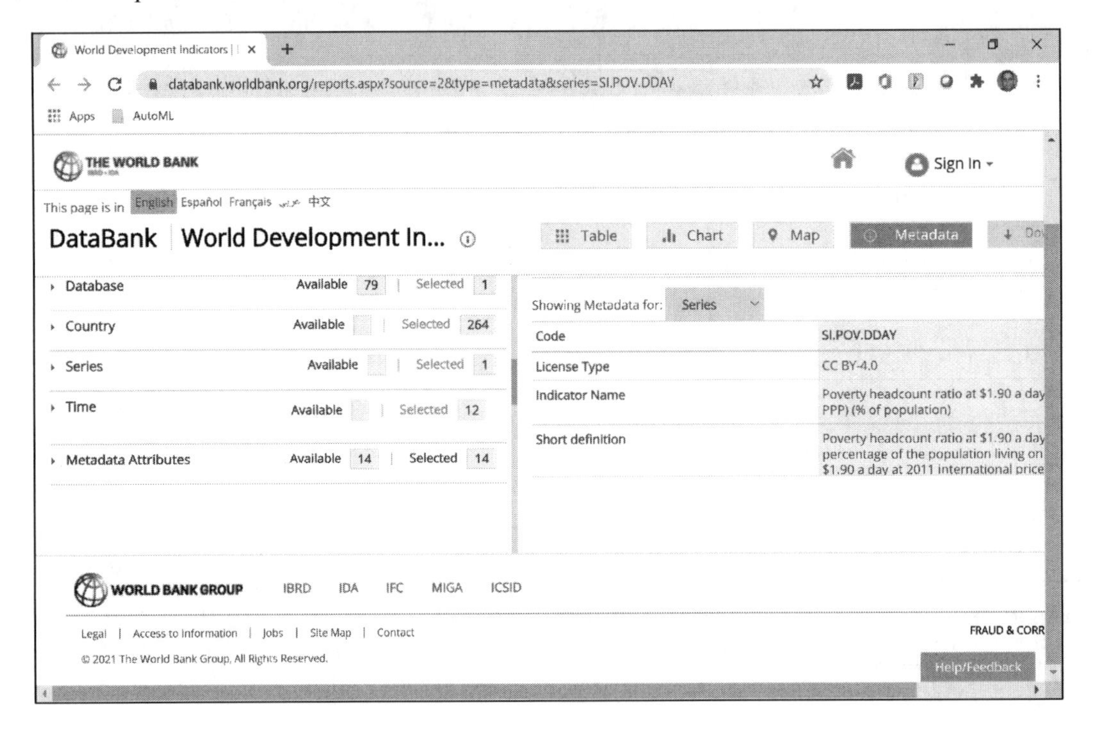

Food and Agriculture Organization of the United Nations

FAOSTAT Data:[655] FAO provides following data domains: Production, Prices, Emission-agriculture, Inputs, Trade, Population, emissions-Land Use, Investment, Macro Economic Data, and others.

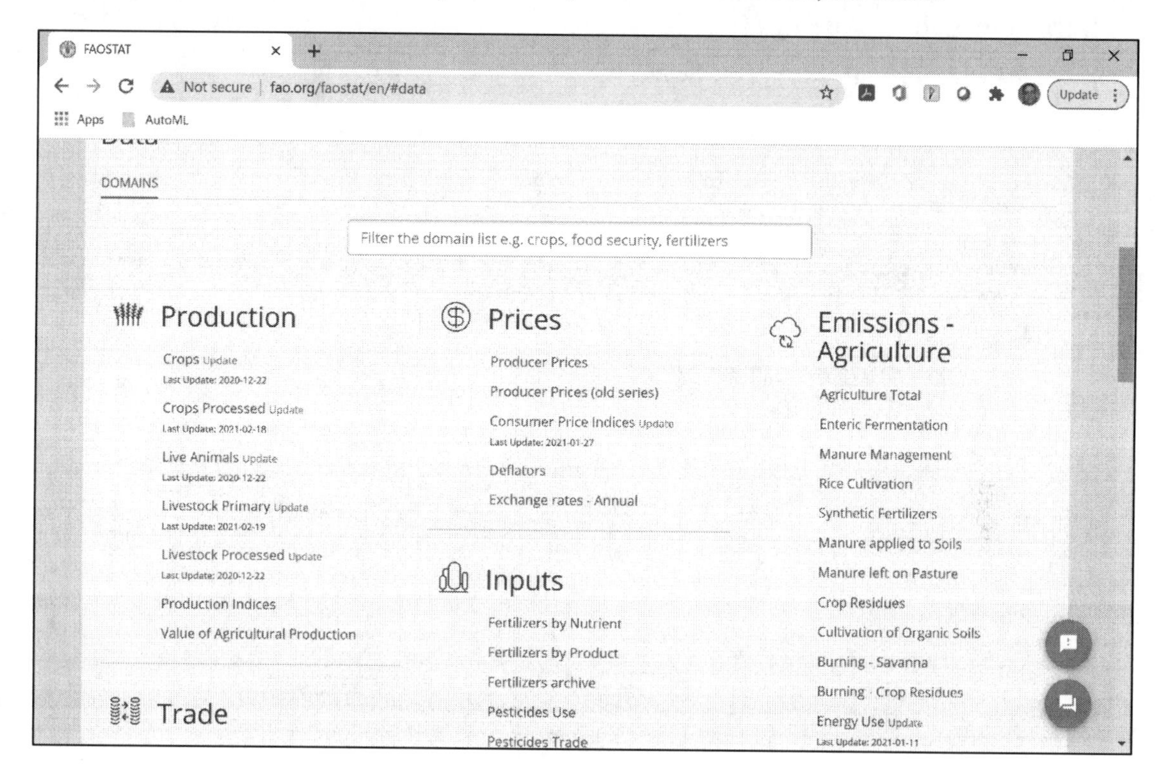

Appendix D—Data U.S.

U.S. Bureau of Labor Statistics

CPI Average Price Data:[656] One-Screen Data Search[657]

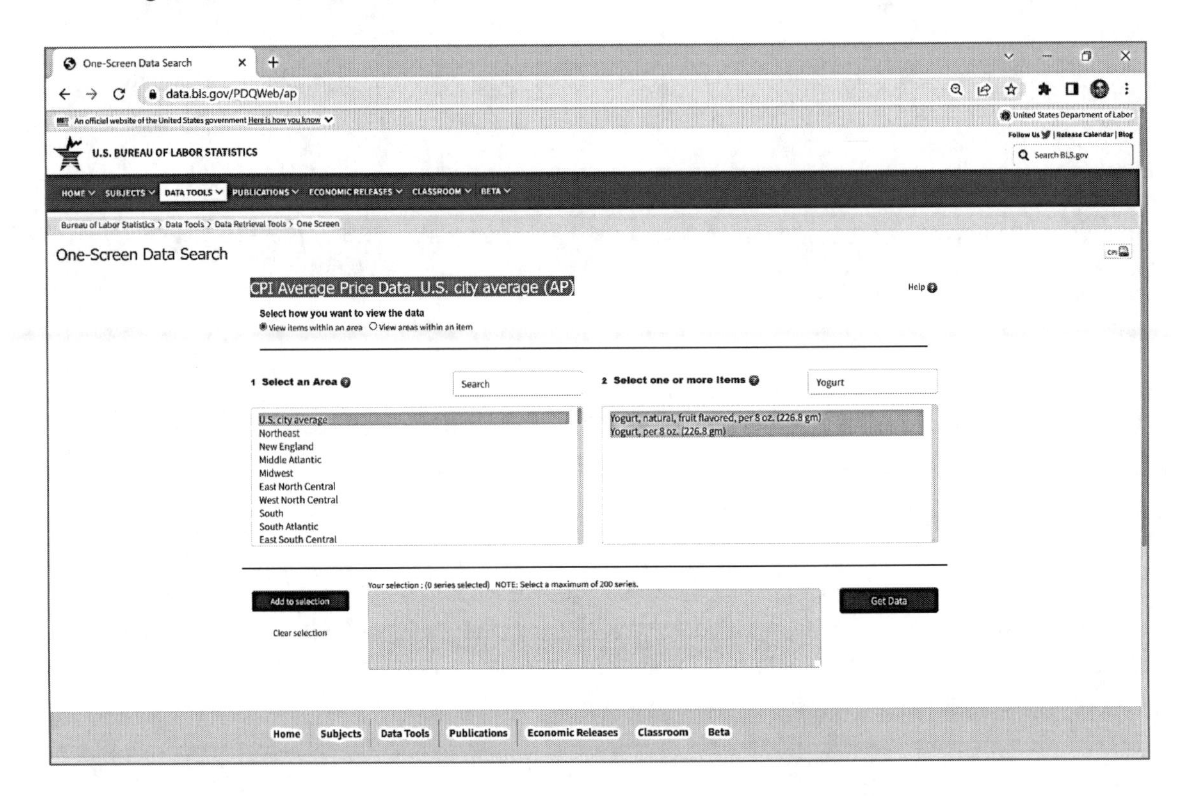

DATA.GOV

Explore data[658] that can help in informed agricultural investments, innovation and policy strategies. If you're interested in agricultural production, food security, rural development, nutrition, natural resources and regional food systems, this page is for you.

Wheat Data - https://www.ers.usda.gov/data-products/wheat-data.aspx

All dairy products (milk-fat milk-equivalent basis): Supply and use

Filename: dymfg.xls[659] (date of retrieval June 6, 2022)

All dairy products (milk-fat milk-equivalent basis): Supply and use[1]

Year	U.S. population, July 1[2]	Supply: Milk Production	Supply: Farm milk fed to calves	Supply: Nonfood use[3]	Supply: For human use	Supply: Imports	Supply: Beginning stocks[4]	Supply: Total supply	Use: Exports[5]	Use: Shipments to U.S. territories	Use: Ending stocks[4]	Food availability, Total: USDA donations	Food availability, Total: Domestic availability, not including USDA donations	Food availability, Total: Total	Food availability, Per capita: USDA donations	Food availability, Per capita: Domestic availability, not including USDA donations
	-- Millions --	Million pounds													Pounds	
1930	123.188	102,984.0	2,986.0	NA	99,998.0	895.0	3,057.0	103,950.0	318.0	135.0	2,612.0	NA	100,885.0	100,885.0	NA	819.0
1931	124.149	105,629.0	2,997.0	NA	102,632.0	664.0	2,806.0	106,102.0	282.0	159.0	1,714.0	NA	103,947.0	103,947.0	NA	837.3
1932	124.949	106,310.0	2,859.0	NA	103,451.0	582.0	1,714.0	105,747.0	180.0	156.0	1,560.0	NA	103,851.0	103,851.0	NA	831.1
1933	125.690	107,162.0	2,878.0	NA	104,284.0	495.0	1,560.0	106,339.0	129.0	161.0	3,807.0	NA	102,242.0	102,242.0	NA	813.4
1934	126.485	104,021.0	2,688.0	NA	101,333.0	407.0	3,807.0	105,547.0	152.0	171.0	2,406.0	NA	102,818.0	102,818.0	NA	812.9
1935	127.362	103,605.0	2,676.0	NA	100,929.0	875.0	2,406.0	104,210.0	133.0	212.0	2,097.0	NA	101,768.0	101,768.0	NA	799.0
1936	128.181	104,710.0	2,755.0	NA	101,955.0	745.0	2,097.0	104,797.0	108.0	195.0	3,070.0	NA	101,424.0	101,424.0	NA	791.3
1937	128.961	104,208.0	2,724.0	NA	101,484.0	758.0	3,070.0	105,312.0	122.0	200.0	2,354.0	NA	102,636.0	102,636.0	NA	795.9
1938	129.969	108,107.0	2,850.0	NA	105,257.0	496.0	2,354.0	108,107.0	155.0	223.0	4,444.0	NA	103,285.0	103,285.0	NA	794.7
1939	131.028	108,992.0	2,967.0	NA	106,025.0	519.0	4,444.0	110,988.0	172.0	249.0	2,723.0	NA	107,844.0	107,844.0	NA	823.1
1940	132.122	111,512.0	2,994.0	NA	108,518.0	290.0	2,723.0	111,531.0	470.0	275.0	2,681.0	NA	108,105.0	108,105.0	NA	818.2
1941	133.402	117,088.0	3,124.0	NA	113,964.0	243.0	2,681.0	116,888.0	2,641.0	352.0	5,629.0	NA	108,266.0	108,266.0	NA	811.6
1942	134.860	120,433.0	3,294.0	NA	117,139.0	623.0	5,629.0	123,391.0	4,349.0	330.0	3,992.0	NA	114,720.0	114,720.0	NA	850.7
1943	136.739	118,517.0	3,276.0	NA	115,241.0	291.0	3,992.0	119,524.0	4,974.0	410.0	6,955.0	NA	107,185.0	107,185.0	NA	783.9
1944	138.397	118,123.0	3,258.0	NA	114,865.0	118.0	6,955.0	121,938.0	6,311.0	500.0	2,789.0	NA	112,338.0	112,338.0	NA	811.7
1945	139.928	120,628.0	3,290.0	NA	117,338.0	156.0	2,789.0	120,283.0	4,498.0	661.0	4,204.0	NA	110,920.0	110,920.0	NA	792.7
1946	141.389	118,697.0	3,228.0	NA	115,469.0	316.0	4,204.0	119,989.0	5,576.0	488.0	2,888.0	NA	111,037.0	111,037.0	NA	785.3
1947	144.126	118,114.0	3,194.0	NA	114,920.0	153.0	2,888.0	117,961.0	4,040.0	363.0	2,636.0	NA	110,922.0	110,922.0	NA	769.6
1948	146.631	113,671.0	3,064.0	NA	110,607.0	195.0	2,636.0	113,438.0	2,762.0	235.0	3,597.0	NA	106,844.0	106,844.0	NA	728.7
1949	149.188	117,003.0	3,163.0	NA	113,840.0	272.0	3,651.0	117,763.0	2,454.0	228.0	5,360.0	260.0	109,461.0	109,721.0	1.7	733.7
1950	151.684	117,302.0	3,286.0	NA	114,016.0	459.0	5,360.0	119,835.0	1,976.0	231.0	4,754.0	1,270.1	111,603.9	112,874.0	8.4	735.8
1951	154.287	115,181.0	3,449.0	NA	111,732.0	525.0	4,754.0	117,011.0	2,248.0	225.0	3,636.0	160.0	110,742.0	110,902.0	1.0	717.8

TableOfContents | AllDairy | AllDairyPcc | TotalCheese | AmCheese | OthCheese | CheesePcc1970-94 | CheesePcc

EIA Short-Term Energy Outlook—Monthly Average Imported Crude Oil Price

SHORT-TERM ENERGY OUTLOOK[660]

All real and nominal price series - https://www.eia.gov/outlooks/steo/realprices/real_prices.xlsx

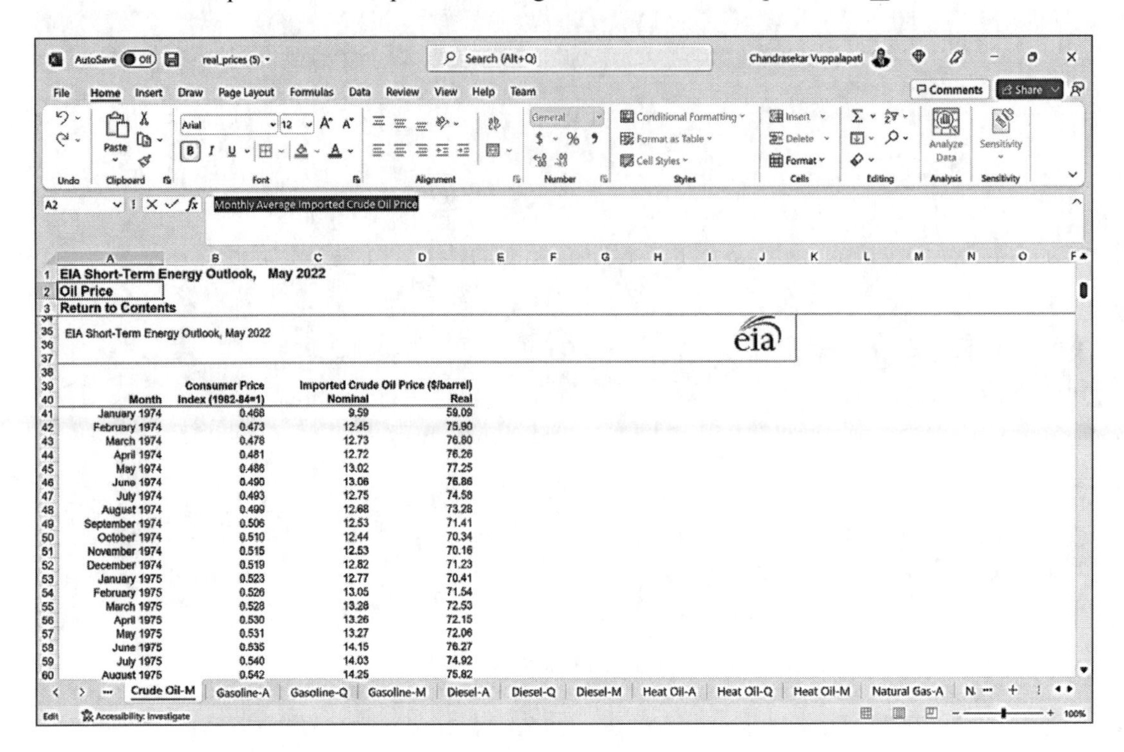

EIA Short-Term Energy Outlook, May 2022

Oil Price

Return to Contents

EIA Short-Term Energy Outlook, May 2022

Month	Consumer Price Index (1982-84=1)	Imported Crude Oil Price ($/barrel) Nominal	Real
January 1974	0.468	9.59	59.09
February 1974	0.473	12.45	75.90
March 1974	0.478	12.73	76.80
April 1974	0.481	12.72	76.26
May 1974	0.486	13.02	77.25
June 1974	0.490	13.06	76.86
July 1974	0.493	12.75	74.58
August 1974	0.499	12.68	73.28
September 1974	0.506	12.53	71.41
October 1974	0.510	12.44	70.34
November 1974	0.515	12.53	70.16
December 1974	0.519	12.82	71.23
January 1975	0.523	12.77	70.41
February 1975	0.526	13.05	71.54
March 1975	0.528	13.28	72.53
April 1975	0.530	13.26	72.15
May 1975	0.531	13.27	72.06
June 1975	0.535	14.15	76.27
July 1975	0.540	14.03	74.92
August 1975	0.542	14.25	75.82

Dollars/Bushel : Dollars/Tonne Converter

Dollars/Bushel Converter:[661]

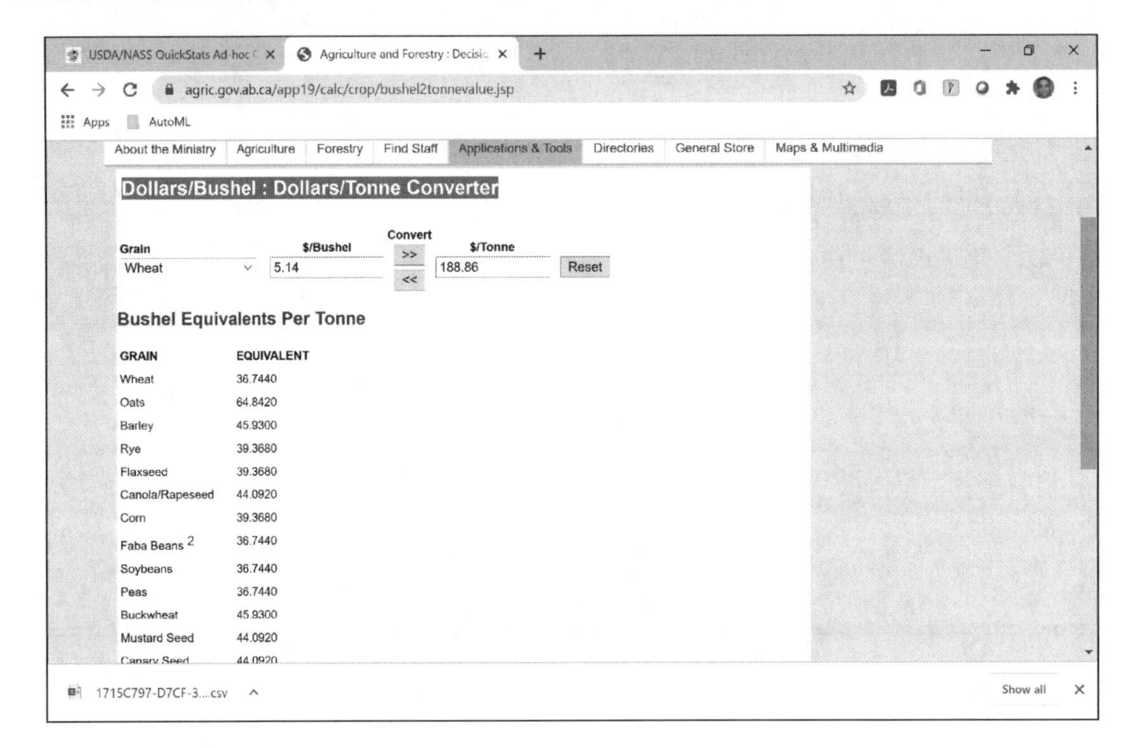

GRAIN	EQUIVALENT
Wheat	36.7440
Oats	64.8420
Barley	45.9300
Rye	39.3680
Flaxseed	39.3680
Canola/Rapeseed	44.0920
Corn	39.3680
Faba Beans 2	36.7440
Soybeans	36.7440
Peas	36.7440
Buckwheat	45.9300
Mustard Seed	44.0920
Canary Seed	44.0920

NOAA - Storm Events Database

Storm Events Database

The Storm Events Database contains the records used to create the official NOAA Storm Data documented publication:[662]

- The occurrence of storms and other significant weather phenomena having sufficient intensity to cause loss of life, injuries, significant property damage, and/or disruption to commerce;
- Rare, unusual, weather phenomena that generate media attention, such as snow flurries in South Florida or the San Diego coastal area; and
- Other significant meteorological events, such as record maximum or minimum temperatures or precipitation that occur in connection with another event.

The database currently contains data from January 1950 to November 2020, as entered by NOAA's National Weather Service (NWS). Due to changes in the data collection and processing procedures over time, there are unique periods of records available depending on the event type. NCEI has performed data reformatting and standardization of event types but has not changed any data values for locations, fatalities, injuries, damage, narratives and any other event specific information. Please refer to the Database Details page for more information.

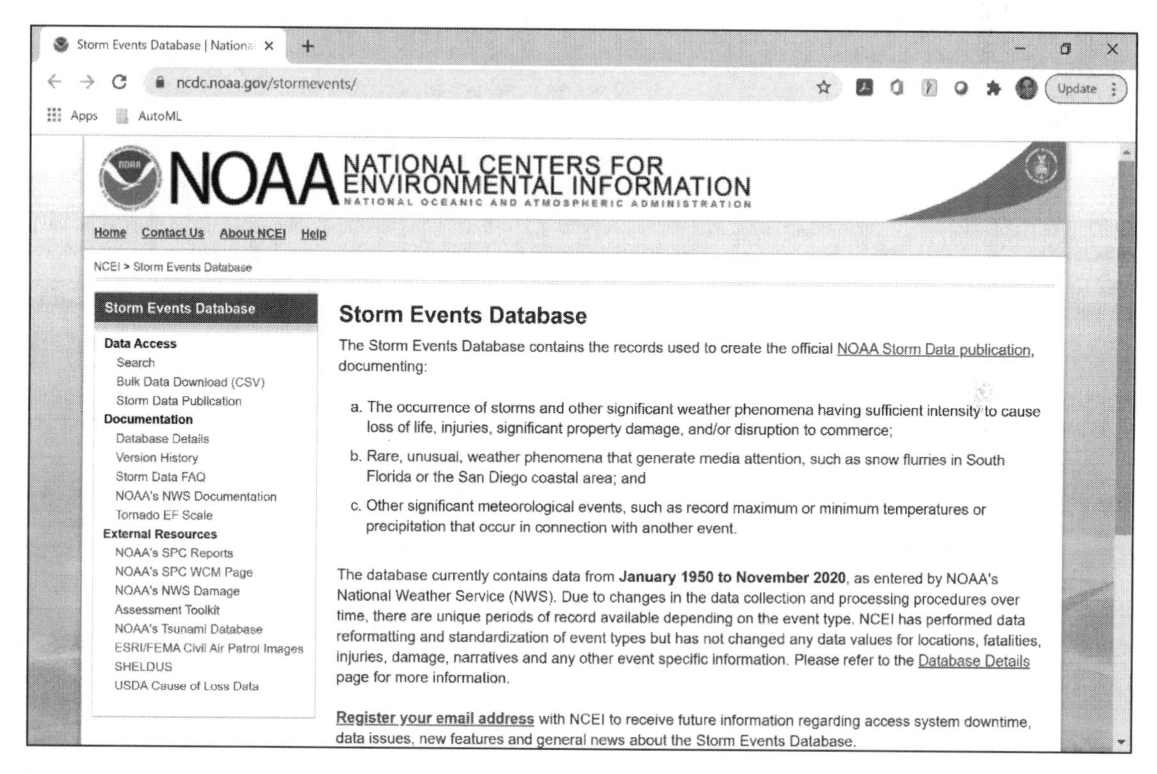

Storm Data Exports Format: describe the storm database meta data.[633]

Storm Data Event Table

The only events permitted in Storm Data are listed in the table below. The chosen event name should be the one that most accurately describes the meteorological event leading to fatalities, injuries and damage, among others. However, significant events, such as tornadoes, having no impact or causing no damage, should also be included in Storm Data.[664]

Event	Designator	Event	Designator
Astronomical Low Tide	Z	Lightning	C
Avalanche	Z	Marine Dense Fog	M
Blizzard	Z	Marine Hail	M
Coastal Flood	Z	Marine Heavy Freezing Spray	M
Cold/Wind Chill	Z	Marine High Wind	M
Debris Flow	C	Marine Hurricane/Typhoon	M
Dense Fog	Z	Marine Lightning	M
Dense Smoke	Z	Marine Strong Wind	M
Drought	Z	Marine Thunderstorm Wind	M
Dust Devil	C	Marine Tropical Depression	M
Dust Storm	Z	Marine Tropical Storm	M
Excessive Heat	Z	Rip Current	Z
Extreme Cold/Wind Chill	Z	Seiche	Z
Flash Flood	C	Sleet	Z
Flood	C	Sneaker Wave	Z
Frost/Freeze	Z	Storm Surge/Tide	Z
Funnel Cloud	C	Strong Wind	Z
Freezing Fog	Z	Thunderstorm Wind	C
Hail	C	Tornado	C
Heat	Z	Tropical Depression	Z
Heavy Rain	C	Tropical Storm	Z
Heavy Snow	Z	Tsunami	Z
High Surf	Z	Volcanic Ash	Z
High Wind	Z	Waterspout	M
Hurricane (Typhoon)	Z	Wildfire	Z
Ice Storm	Z	Winter Storm	Z
Lake-Effect Snow	Z	Winter Weather	Z
Lakeshore Flood	Z		

Legend: There are three designators: C - County/Parish; Z - Zone; and M – Marine Zone.

Consumer Price Index, 1913

Historical data from the era of the modern U.S. consumer price index (CPI).[665]

Year	Annual Average CPI(-U)	Annual Percent Change (rate of inflation)
1913	9.9	
1914	10.0	1.3%
1915	10.1	0.9%
1916	10.9	7.7%
1917	12.8	17.8%
1918	15.0	17.3%
1919	17.3	15.2%
1920	20.0	15.6%

1921	17.9	–10.9%
1922	16.8	–6.2%
1923	17.1	1.8%
1924	17.1	0.4%
1925	17.5	2.4%
1926	17.7	0.9%
1927	17.4	–1.9%
1928	17.2	–1.2%
1929	17.2	0.0%
1930	16.7	–2.7%
1931	15.2	–8.9%
1932	13.6	–10.3%
1933	12.9	–5.2%
1934	13.4	3.5%
1935	13.7	2.6%
1936	13.9	1.0%
1937	14.4	3.7%
1938	14.1	–2.0%
1939	13.9	–1.3%
1940	14.0	0.7%
1941	14.7	5.1%
1942	16.3	10.9%
1943	17.3	6.0%
1944	17.6	1.6%
1945	18.0	2.3%
1946	19.5	8.5%
1947	22.3	14.4%
1948	24.0	7.7%
1949	23.8	–1.0%
1950	24.1	1.1%
1951	26.0	7.9%
1952	26.6	2.3%
1953	26.8	0.8%
1954	26.9	0.3%
1955	26.8	–0.3%
1956	27.2	1.5%
1957	28.1	3.3%
1958	28.9	2.7%
1959	29.2	1.08%
1960	29.6	1.5%
1961	29.9	1.1%
1962	30.3	1.2%

1963	30.6	1.2%
1964	31.0	1.3%
1965	31.5	1.6%
1966	32.5	3.0%
1967	33.4	2.8%
1968	34.8	4.3%
1969	36.7	5.5%
1970	38.8	5.8%
1971	40.5	4.3%
1972	41.8	3.3%
1973	44.4	6.2%
1974	49.3	11.1%
1975	53.8	9.1%
1976	56.9	5.7%
1977	60.6	6.5%
1978	65.2	7.6%
1979	72.6	11.3%
1980	82.4	13.5%
1981	90.9	10.3%
1982	96.5	6.1%
1983	99.6	3.2%
1984	103.9	4.3%
1985	107.6	3.5%
1986	109.6	1.9%
1987	113.6	3.7%
1988	118.3	4.1%
1989	124.0	4.8%
1990	130.7	5.4%
1991	136.2	4.2%
1992	140.3	3.0%
1993	144.5	3.0%
1994	148.2	2.6%
1995	152.4	2.8%
1996	156.9	2.9%
1997	160.5	2.3%
1998	163.0	1.6%
1999	166.6	2.2%
2000	172.2	3.4%
2001	177.1	2.8%
2002	179.9	1.6%
2003	184.0	2.3%

2004	188.9	2.7%
2005	195.3	3.4%
2006	201.6	3.2%
2007	207.3	2.9%
2008	215.3	3.8%
2009	214.5	–0.4%
2010	218.1	1.6%
2011	224.9	3.2%
2012	229.6	2.1%
2013	233.0	1.5%
2014	236.7	1.6%
2015	237.0	0.1%
2016	240.0	1.3%
2017	245.1	2.1%
2018	251.1	2.4%
2019	255.7	1.8%
2020	258.8	1.2%
2021	271.0	4.7%

Inflation Calculator

What's a dollar worth? How far does a past dollar stretch to equal the modern dollar? What would past prices be today?[666]

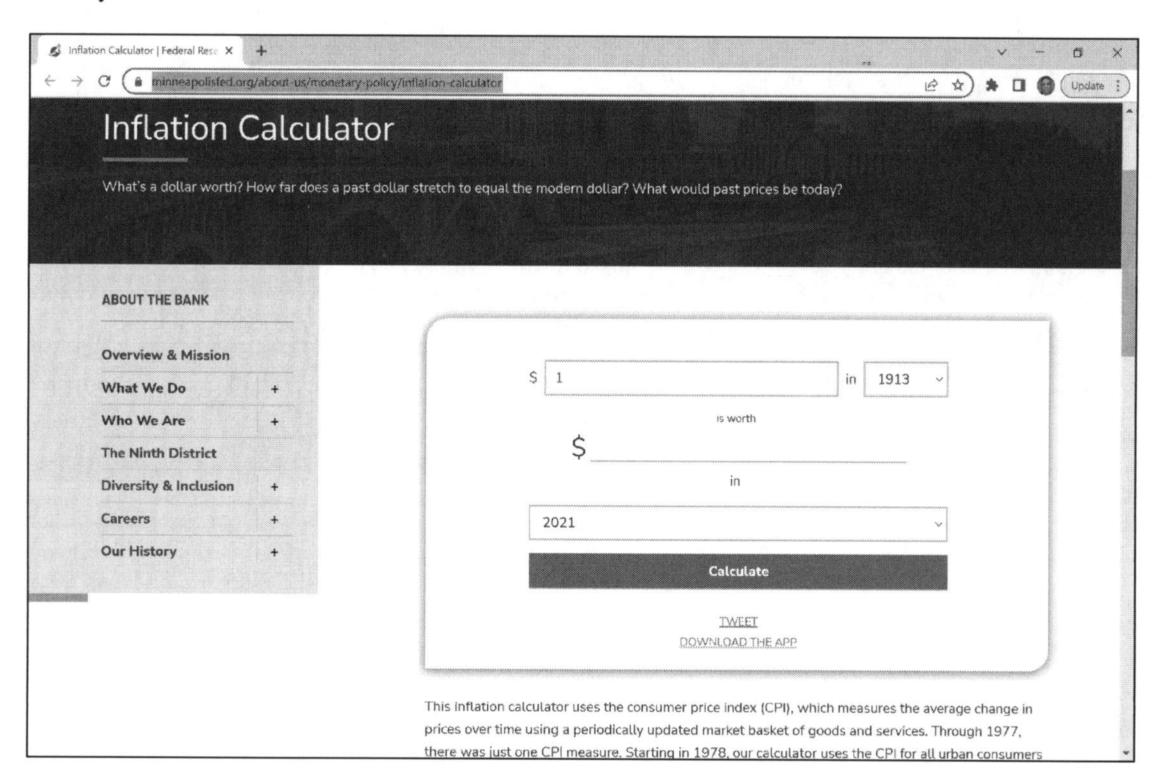

Appendix E—Economic Frameworks & Macroeconomics

Macroeconomic Signals that could help Predict Economic Cycles

Signal #	Type	Description
Oil price swings	Leading Indicator	oil price swings[667] appear to be consistent and frequent historical precursors to U.S. recessions.[668] A spike in oil prices has preceded or coincided with 10 out of 12 post-WWII recessions.
Monetary Trends	Lagging Indicator	A period of expansionary monetary policy in the years prior to the recession, sometimes to help fund government war spending or to re-inflate the economy after the previous round of recession. Once the resulting debt bubbles pop or the end of a war leads to cutbacks in monetary expansion, several years' worth of overextended, debt-based investments and malinvestments tend to be wiped out in a process of debt deflation[669,670] in a relatively short period. This spikes unemployment and drags down GDP.
During recession, GDP & Credit Contraction Occur	Lagging Indicator	Country GDP / Credit contraction, in percent[671]

The U.S. Recessions

The bottom line: So, what do all these very different recessions have in common? In many cases, the most important single factor is a period of expansionary monetary policy in the years prior to the recession, sometimes to help fund government war spending or to re-inflate the economy after the previous round of recession.

Once the resulting debt bubbles pop or the end of a war leads to cutbacks in monetary expansion, several years' worth of overextended, debt-based investments and malinvestments tend to be wiped out in a process of debt deflation in a relatively short period. This spikes unemployment and drags down GDP.

Beyond the underlying monetary trends, real economic shocks often help to trigger the turning point into recession. For one, oil price swings appear to be consistent and frequent historical precursors to U.S. recessions.[672]

A spike in oil prices has preceded or coincided with 10 out of 12 post-WWII recessions.[673] This highlights that while global integration of economies allowing for more effective cooperative efforts between governments has increased over time, the integration itself ties the world economies more closely together, making them more susceptible to problems outside their borders.

Recession	Details
U.S. Recession—January 31, 2020–March 31, 2020[674]	
The Great Recession (December 2007 to June 2009)[675]	The collapse of the housing bubble of the 2000s and resulting record foreclosures and a financial crisis that threw markets worldwide into a tailspin. Summary: Duration: Eighteen months[676] GDP decline: 4.3%[677] Peak unemployment rate: 10.0%[678]
Dot-com recession (March 2001 to November 2001)	The dot-com bubble burst in 2000, when an over-inflated Nasdaq lost more than 75% of its value and wiped out a generation of tech investors. Those losses left the stock market in a vulnerable place that got worse in the fall of 2001, when the devastation of the September 11, 2001,[679] terrorist attacks and a series of major accounting scandals at corporations like Enron and Swissair spurred a stock market crash. The resulting recession was relatively short, at just eight months, and also shallow, as GDP dipped only 0.6% and unemployment reached 5.5%. Summary: Duration: Eight months[680] GDP decline: 0.3% Peak unemployment rate: 5.5%[681]
Gulf War recession (July 1990 to March 1991)	A mild recession kicked off in 1990, as the Federal Reserve had been slowly raising interest rates for over two years to keep inflation in check. Those moves slowed down the economy, which then took a hit when Iraq invaded Kuwait in the summer of 1990 (followed by U.S. involvement and the Gulf War) and caused global oil prices to more than double. The recession lasted just eight months, with GDP declining[682] 1.1% during that period and unemployment reaching roughly 7%. Summary: Duration: Eight months[683] GDP decline: 1.5% Peak unemployment rate: 6.8%[684]
Energy crisis recession (July 1981 to November 1982)[685]	In 1981, having emerged from a recession just a year before (the early ’80s are described as having a “double dip recession because there were two so close together), the Federal Reserve tried to tame rising inflation with a stricter monetary policy that raised interest rates and slowed the economy. Those policies managed to reduce inflation to around 4% by 1983, but the cost was a 16-month recession that saw GDP drop by around 3% and unemployment spiked to 10.8%. Summary: Duration: 16 months GDP decline: 2.9% Peak unemployment rate: 10.8%

1980 Recession (January 1980 to July 1980)	Inflation rates rose throughout the late-1970s, reaching double-digit levels in 1979 and peaked at 22%[686] in 1980. As a result, the Federal Reserve raised interest rates to stop the rising inflation, which slowed down the economy (GDP dropped over 2%)[687] and caused unemployment to spike to 7.8%. The Fed lowered interest rates again in mid-1980, giving the economy a chance to rebound and end a brief, six-month recession. Summary: Duration: Six months GDP decline: 2.2% Peak unemployment rate: 7.8%
Oil embargo recession (November 1973 to March 1975)	In the fall of 1973, the Organization of the Petroleum Exporting Countries, or OPEC, put an embargo on oil imports from multiple countries, including the U.S., over their support of Israel's military. Oil prices roughly quadrupled[688] as a result, putting a major crunch on the economy as gas prices soared for consumers, reducing their spending on other items. The resulting recession lasted for 16 months (it even outlasted the oil embargo itself, which OPEC lifted in 1974) and saw GDP decline by 3.4%[689] while unemployment climbed from 4.8% to nearly 9%. Summary: Duration: 16 months GDP decline: 3% Peak unemployment rate: 8.6%
Recession of 1969–1970 (December 1969 to November 1970)	The "mild recession"[690] that ensued caused unemployment to peak at around 6% while the GDP dropped less than 1% before the Fed eased its monetary policies to restart economic growth in 1970. Summary: Duration: 11 months GDP decline: 0.6% Peak unemployment rate: 5.9%
Recession of 1960–1961 (April 1960 to February 1961)	Even though two previous recessions in the '50s stemmed from tighter monetary policies giving rise to interest rates, the Federal Reserve began slowly raising interest rates following the end of the previous recession in 1958, leading to another short-lived recession at the start of the 1960s. The 10-month recession saw the GDP drop by nearly 2% and unemployment peaked at 6.9%, while President John F. Kennedy spurred a rebound in 1961 with stimulus spending that included tax cuts and expanded unemployment and Social Security benefits. Summary: Duration: 10 months GDP decline: 1.6% Peak unemployment rate: 6.9%

Recession of 1957–1958 (August 1957 to April 1958)	This recession in the late-1950s lasted eight months. GDP fell by 3.7% and unemployment peaked at 7.4% as the government's tighter monetary policy in the mid-1950s raised interest rates in an effort to curb inflation. As a result, consumer prices also continued to rise, which led to a decline in spending. Meanwhile, a global recession (which also happened to coincide with the 1957 Asian flu pandemic that killed 1.1 million people[691] worldwide) further hurt the U.S. economy as the country's exports declined by more than $4 billion.[692] Summary: Duration: Eight months13 GDP decline: 3.7%18 Peak unemployment rate: 7.4%24
Post-Korean War recession (July 1953 to May 1954)	As with previous post-war recessions, this downturn was spurred by a shift in government spending after the end of the Korean War (which lasted from 1950 to 1953). The country's GDP dropped by 2.2% and unemployment peaked at roughly 6%,[693] as the government wound down security spending following the war and the U.S. Federal Reserve tightened monetary policy[694] to curb inflation (which includes increasing interest rates). However, spiking interest rates hurt consumer confidence in the economy and decreased consumer demand. The Fed eased its policies in 1954, allowing the economy to rebound after a 10-month. Summary: Duration: 10 months[695] GDP decline: 2.7%[696] Peak unemployment rate: 5.9%[697]
Post-WWII slump (November 1948 to October 1949)	After the war there was an eight-month recession (see below), but the economic challenges stemming from the end of World War II again caught up with the U.S. economy during the last stretch of the 1940s. However, this 11-month recession—in which the country's GDP dropped by less than 2%—was considered "very mild"[698] by economists, who attribute the downturn in part to consumer demand leveling off after previously spiking when wartime rationing efforts ceased. Economists also point to a decline in fixed investments, while the influx of veterans returning from war and competing[699] for limited civilian jobs helped the unemployment rate climb as high as 7.9%,[700] according to the U.S. Bureau of Labor Statistics. Summary: Duration: 11 months[701] GDP decline: 1.7%1 Peak unemployment rate: 5.7%

Post-World War II recession (February 1945 to October 1945)	This downturn was caused primarily by a significant drop in government spending and GDP (which fell 11%) as the U.S. pivoted from a wartime economy built around manufacturing supplies for the World War II effort to a peacetime economy focused on creating civilian jobs for returning veterans. Summary:[702] Duration: Eight months[703] GDP decline: 10.9%[704] Peak unemployment rate: 5.2%[705]
The Roosevelt recession (May 1937 to June 1938)[706]	This recession was essentially a 13-month pause in the nation's recovery from the Great Depression and modern economists have called the episode a "cautionary tale". In 1937, President Franklin D. Roosevelt cut government spending at a time when the country's economic recovery was still fragile enough to be derailed. As a result, unemployment jumped from roughly 14% to nearly 20% and the real GDP fell by 10%. The following year, Roosevelt signed a $3.75 billion spending bill that restarted the economic recovery. Summary:[707] Duration: 13 months GDP decline: 10% Peak unemployment rate: 20%[708]

Labor Force Statistics from the Current Population Survey

Series Id: LNS14000000[709] Seasonally Adjusted Series title: (Seas) Unemployment Rate
Labor force status: Unemployment rate Type of data: Percent or rate Age: 16 years and over

Year	Jan	Feb	Mar	Apr	May	Jun	Jul	Aug	Sep	Oct	Nov	Dec
1948	3.4	3.8	4.0	3.9	3.5	3.6	3.6	3.9	3.8	3.7	3.8	4.0
1949	4.3	4.7	5.0	5.3	6.1	6.2	6.7	6.8	6.6	7.9	6.4	6.6
1950	6.5	6.4	6.3	5.8	5.5	5.4	5.0	4.5	4.4	4.2	4.2	4.3
1951	3.7	3.4	3.4	3.1	3.0	3.2	3.1	3.1	3.3	3.5	3.5	3.1
1952	3.2	3.1	2.9	2.9	3.0	3.0	3.2	3.4	3.1	3.0	2.8	2.7
1953	2.9	2.6	2.6	2.7	2.5	2.5	2.6	2.7	2.9	3.1	3.5	4.5
1954	4.9	5.2	5.7	5.9	5.9	5.6	5.8	6.0	6.1	5.7	5.3	5.0
1955	4.9	4.7	4.6	4.7	4.3	4.2	4.0	4.2	4.1	4.3	4.2	4.2
1956	4.0	3.9	4.2	4.0	4.3	4.3	4.4	4.1	3.9	3.9	4.3	4.2
1957	4.2	3.9	3.7	3.9	4.1	4.3	4.2	4.1	4.4	4.5	5.1	5.2
1958	5.8	6.4	6.7	7.4	7.4	7.3	7.5	7.4	7.1	6.7	6.2	6.2
1959	6.0	5.9	5.6	5.2	5.1	5.0	5.1	5.2	5.5	5.7	5.8	5.3
1960	5.2	4.8	5.4	5.2	5.1	5.4	5.5	5.6	5.5	6.1	6.1	6.6
1961	6.6	6.9	6.9	7.0	7.1	6.9	7.0	6.6	6.7	6.5	6.1	6.0
1962	5.8	5.5	5.6	5.6	5.5	5.5	5.4	5.7	5.6	5.4	5.7	5.5
1963	5.7	5.9	5.7	5.7	5.9	5.6	5.6	5.4	5.5	5.5	5.7	5.5
1964	5.6	5.4	5.4	5.3	5.1	5.2	4.9	5.0	5.1	5.1	4.8	5.0
1965	4.9	5.1	4.7	4.8	4.6	4.6	4.4	4.4	4.3	4.2	4.1	4.0
1966	4.0	3.8	3.8	3.8	3.9	3.8	3.8	3.8	3.7	3.7	3.6	3.8
1967	3.9	3.8	3.8	3.8	3.8	3.9	3.8	3.8	3.8	4.0	3.9	3.8
1968	3.7	3.8	3.7	3.5	3.5	3.7	3.7	3.5	3.4	3.4	3.4	3.4
1969	3.4	3.4	3.4	3.4	3.4	3.5	3.5	3.5	3.7	3.7	3.5	3.5
1970	3.9	4.2	4.4	4.6	4.8	4.9	5.0	5.1	5.4	5.5	5.9	6.1
1971	5.9	5.9	6.0	5.9	5.9	5.9	6.0	6.1	6.0	5.8	6.0	6.0
1972	5.8	5.7	5.8	5.7	5.7	5.7	5.6	5.6	5.5	5.6	5.3	5.2
1973	4.9	5.0	4.9	5.0	4.9	4.9	4.8	4.8	4.8	4.6	4.8	4.9
1974	5.1	5.2	5.1	5.1	5.1	5.4	5.5	5.5	5.9	6.0	6.6	7.2
1975	8.1	8.1	8.6	8.8	9.0	8.8	8.6	8.4	8.4	8.4	8.3	8.2
1976	7.9	7.7	7.6	7.7	7.4	7.6	7.8	7.8	7.6	7.7	7.8	7.8
1977	7.5	7.6	7.4	7.2	7.0	7.2	6.9	7.0	6.8	6.8	6.8	6.4
1978	6.4	6.3	6.3	6.1	6.0	5.9	6.2	5.9	6.0	5.8	5.9	6.0
1979	5.9	5.9	5.8	5.8	5.6	5.7	5.7	6.0	5.9	6.0	5.9	6.0
1980	6.3	6.3	6.3	6.9	7.5	7.6	7.8	7.7	7.5	7.5	7.5	7.2
1981	7.5	7.4	7.4	7.2	7.5	7.5	7.2	7.4	7.6	7.9	8.3	8.5
1982	8.6	8.9	9.0	9.3	9.4	9.6	9.8	9.8	10.1	10.4	10.8	10.8
1983	10.4	10.4	10.3	10.2	10.1	10.1	9.4	9.5	9.2	8.8	8.5	8.3
1984	8.0	7.8	7.8	7.7	7.4	7.2	7.5	7.5	7.3	7.4	7.2	7.3
1985	7.3	7.2	7.2	7.3	7.2	7.4	7.4	7.1	7.1	7.1	7.0	7.0
1986	6.7	7.2	7.2	7.1	7.2	7.2	7.0	6.9	7.0	7.0	6.9	6.6
1987	6.6	6.6	6.6	6.3	6.3	6.2	6.1	6.0	5.9	6.0	5.8	5.7
1988	5.7	5.7	5.7	5.4	5.6	5.4	5.4	5.6	5.4	5.4	5.3	5.3
1989	5.4	5.2	5.0	5.2	5.2	5.3	5.2	5.2	5.3	5.3	5.4	5.4
1990	5.4	5.3	5.2	5.4	5.4	5.2	5.5	5.7	5.9	5.9	6.2	6.3

Year	Jan	Feb	Mar	Apr	May	Jun	Jul	Aug	Sep	Oct	Nov	Dec
1991	6.4	6.6	6.8	6.7	6.9	6.9	6.8	6.9	6.9	7.0	7.0	7.3
1992	7.3	7.4	7.4	7.4	7.6	7.8	7.7	7.6	7.6	7.3	7.4	7.4
1993	7.3	7.1	7.0	7.1	7.1	7.0	6.9	6.8	6.7	6.8	6.6	6.5
1994	6.6	6.6	6.5	6.4	6.1	6.1	6.1	6.0	5.9	5.8	5.6	5.5
1995	5.6	5.4	5.4	5.8	5.6	5.6	5.7	5.7	5.6	5.5	5.6	5.6
1996	5.6	5.5	5.5	5.6	5.6	5.3	5.5	5.1	5.2	5.2	5.4	5.4
1997	5.3	5.2	5.2	5.1	4.9	5.0	4.9	4.8	4.9	4.7	4.6	4.7
1998	4.6	4.6	4.7	4.3	4.4	4.5	4.5	4.5	4.6	4.5	4.4	4.4
1999	4.3	4.4	4.2	4.3	4.2	4.3	4.3	4.2	4.2	4.1	4.1	4.0
2000	4.0	4.1	4.0	3.8	4.0	4.0	4.0	4.1	3.9	3.9	3.9	3.9
2001	4.2	4.2	4.3	4.4	4.3	4.5	4.6	4.9	5.0	5.3	5.5	5.7
2002	5.7	5.7	5.7	5.9	5.8	5.8	5.8	5.7	5.7	5.7	5.9	6.0
2003	5.8	5.9	5.9	6.0	6.1	6.3	6.2	6.1	6.1	6.0	5.8	5.7
2004	5.7	5.6	5.8	5.6	5.6	5.6	5.5	5.4	5.4	5.5	5.4	5.4
2005	5.3	5.4	5.2	5.2	5.1	5.0	5.0	4.9	5.0	5.0	5.0	4.9
2006	4.7	4.8	4.7	4.7	4.6	4.6	4.7	4.7	4.5	4.4	4.5	4.4
2007	4.6	4.5	4.4	4.5	4.4	4.6	4.7	4.6	4.7	4.7	4.7	5.0
2008	5.0	4.9	5.1	5.0	5.4	5.6	5.8	6.1	6.1	6.5	6.8	7.3
2009	7.8	8.3	8.7	9.0	9.4	9.5	9.5	9.6	9.8	10.0	9.9	9.9
2010	9.8	9.8	9.9	9.9	9.6	9.4	9.4	9.5	9.5	9.4	9.8	9.3
2011	9.1	9.0	9.0	9.1	9.0	9.1	9.0	9.0	9.0	8.8	8.6	8.5
2012	8.3	8.3	8.2	8.2	8.2	8.2	8.2	8.1	7.8	7.8	7.7	7.9
2013	8.0	7.7	7.5	7.6	7.5	7.5	7.3	7.2	7.2	7.2	6.9	6.7
2014	6.6	6.7	6.7	6.2	6.3	6.1	6.2	6.1	5.9	5.7	5.8	5.6
2015	5.7	5.5	5.4	5.4	5.6	5.3	5.2	5.1	5.0	5.0	5.1	5.0
2016	4.8	4.9	5.0	5.1	4.8	4.9	4.8	4.9	5.0	4.9	4.7	4.7
2017	4.7	4.6	4.4	4.5	4.4	4.3	4.3	4.4	4.2	4.1	4.2	4.1
2018	4.0	4.1	4.0	4.0	3.8	4.0	3.8	3.8	3.7	3.8	3.8	3.9
2019	4.0	3.8	3.8	3.7	3.7	3.6	3.6	3.7	3.5	3.6	3.6	3.6
2020	3.5	3.5	4.4	14.8	13.3	11.1	10.2	8.4	7.8	6.9	6.7	6.7
2021	6.3	6.2	6.0	6.1	5.8	5.9	5.4	5.2	4.8			

United Nations Statistics Department (UNSD)

The following table provides the main economic indicators. MEI and Food Security show correlation, especially during economic downturns.

The OCED—Main Economic Indicators (MEI)

National Accounts

1. GDP (Value)
2. GDP (Volume)
3. Implicit Price Value

Production

4. Industry excluding construction
5. Manufacturing
6. – Consumer goods: total
7. – Consumer non-durable goods
8. – consumer durable goods
9. – Investment goods
10. Intermediate goods including energy
11. Intermediate goods excluding energy
12. Energy
13. Construction
14. Services
15. Rate of capacity utilization

Commodity Output

16. Cement
17. Crude steel
18. Crude Petroleum
19. Natural Gas
20. Commercial Vehicles
21. Passenger Cars

Manufacturing —sales (volume)

22. Total
23. – Domestic
24. – Export
25. – Consumer Goods
26. – Consumer non-durable goods
27. – Consumer durable goods
28. Investment goods
29. Intermediate goods including energy

Manufacturing—New Orders (volume)

30. Total
31. – Domestic
32. – Export
33. Consumer Goods
34. – Consumer non-durable goods
35. – Consumer durable goods
36. Investment goods
37. Intermediate goods including energy

Manufacturing – Stocks (volume)

38. Total
39. Finished Goods
40. Work InProgress
41. Intermediate goods

OCED composite leading indicator

42. Trend restored
43. 6-month rate of change (annual rate)

Construction

44. Orders/Permits: total construction
45. Orders/Permits: residuals
46. Work put in place: total construction
47. Work put in place: residential

Business tendency surveys

48. Industrial business climate
49. Industry Production: future tendency
50. Industrial orders inflow: tendency
51. Industrial order books: level
52. Industrial finished goods stocks: level
53. Industrial export order books or demand: level
54. Industrial rate of capacity utilization
55. Industrial employment: future tendency
56. Industrial selling prices: future tendency
57. Construction order inflows: future tendency
58. Construction employment: future tendency
59. Retail/wholesale: present business situation
60. Retail/wholesale business situation: future tendency
61. Retail/wholesale stocks: level
62. Other services: present business situation
63. Other services business situation: future tendency
64. Other services employment: future tendency

Consumer tendency surveys:

65. Consumer confidence indicator
66. Consumer expected inflation
67. Consumer expected economic situation

Retail Sales

68. Total Retail Sales (Value)
69. – Total Retail sales (volume)
70. New Passenger car registration (level)

Labour

78. Employment :total
79. – Employment: agriculture
80. – Employment: industry
81. – Employment: services
82. Total Employees
83. – Part-Time employees
84. – Temporary employees
85. Total unemployment (level)
86. Total unemployment (rate)
87. Unemployment: short-term index
88. Worked hours
89. Job vacancies

Wages

90. Hourly earnings: all activities
91. Hourly earnings : manufacturing
92. United labor costs : manufacturing

Producer Prices

93. Total
94. Manufacturing
95. – Consumer goods
96. – Investment Goods
97. – Intermediate goods including energy
98. – intermediate goods excluding energy
99. – Energy
100. Food
101. Services

Consumer Prices

102. Total
103. Food
104. All Items less food and energy
105. Energy
106. All Services less rent
107. Rent
108. National core inflation

Domestic finance

109. Narrow money
110. Broad Money
111. Domestic Credit to total economy
112. New Capital issues
113. Fiscal balance
114. Public debt

	International Trade 71. Imports c.i.f or f.o.b (value) 72. Exports c.i.f or f.o.b (value) 73. Net trade (value) 74. Imports c.i.f or f.o.b (volume) 75. Exports c.i.f or f.o.b (volume) 76. Import Prices 77. Export Prices	**Balance of Payments** 115. Current account balance 116. – Balance on goods 117. – Balance on services 118. – Balance on income 119. – Balance on current transfer 120. Capital and financial account balance 121. – Reserve assets 122. Net errors and omission
		Interest rates – share price 123. 3-month interest rate 124. Prime interest rate 125. Long-term interest rate 126. All shares price index
		Foreign finances 127. U.S. dollar exchange rate: spot 128. Euro exchange rate: spot 129. Reserve assets excluding gold

Reserve Bank of India—HANDBOOK OF STATISTICS ON INDIAN ECONOMY

Macro-Economic Aggregates (At Current Prices)[710] & Component of GDP[711]

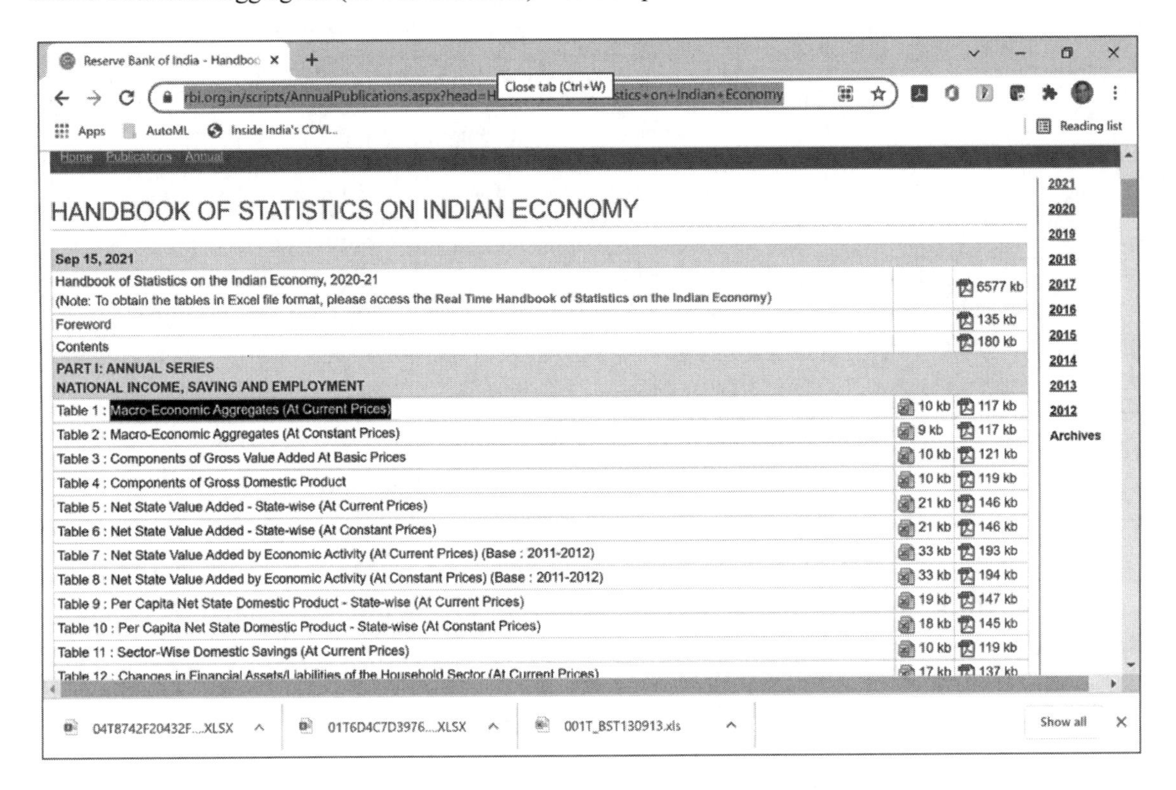

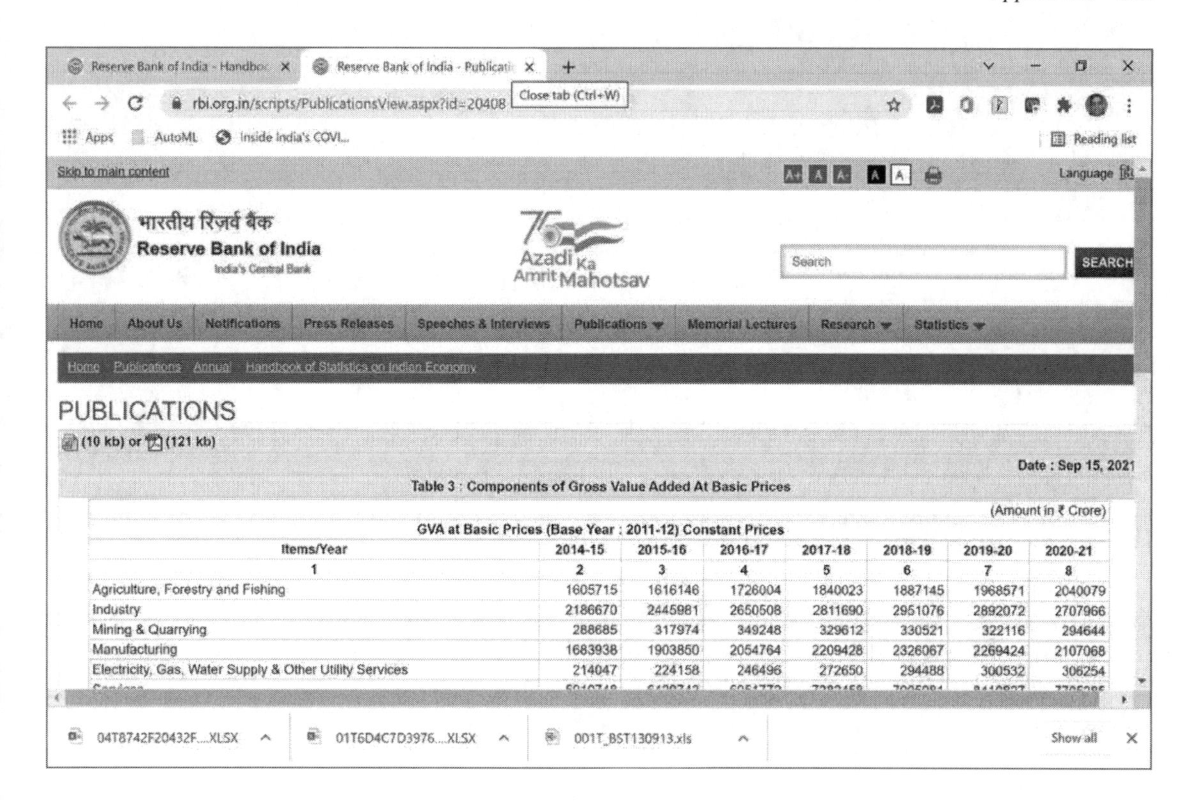

Table 3 : Components of Gross Value Added At Basic Prices

(Amount in ₹ Crore)

GVA at Basic Prices (Base Year : 2011-12) Constant Prices							
Items/Year	2014-15	2015-16	2016-17	2017-18	2018-19	2019-20	2020-21
1	2	3	4	5	6	7	8
Agriculture, Forestry and Fishing	1605715	1616146	1726004	1840023	1887145	1968571	2040079
Industry	2186670	2445981	2650508	2811690	2951076	2892072	2707966
Mining & Quarrying	288685	317974	349248	329612	330521	322116	294644
Manufacturing	1683938	1903850	2054764	2209428	2326067	2269424	2107068
Electricity, Gas, Water Supply & Other Utility Services	214047	224158	246496	272650	294488	300532	306254

Components of Gross Value Added At Basic Prices[712]
https://www.rbi.org.in/scripts/PublicationsView.aspx?id=19736

Department of Commerce

Bureau of Economic Analysis – The United States Department of Commerce

- NIPA Handbook: Concepts and Methods of the U.S. National Income and Product Accounts

  NIPA Handbook: Concepts and Methods of the U.S. National Income and Product Accounts | U.S. Bureau of Economic Analysis (BEA)

United Nations Data Sources

UN DATA MARTS

Source: http://data.un.org/Explorer.aspx?d=FAO

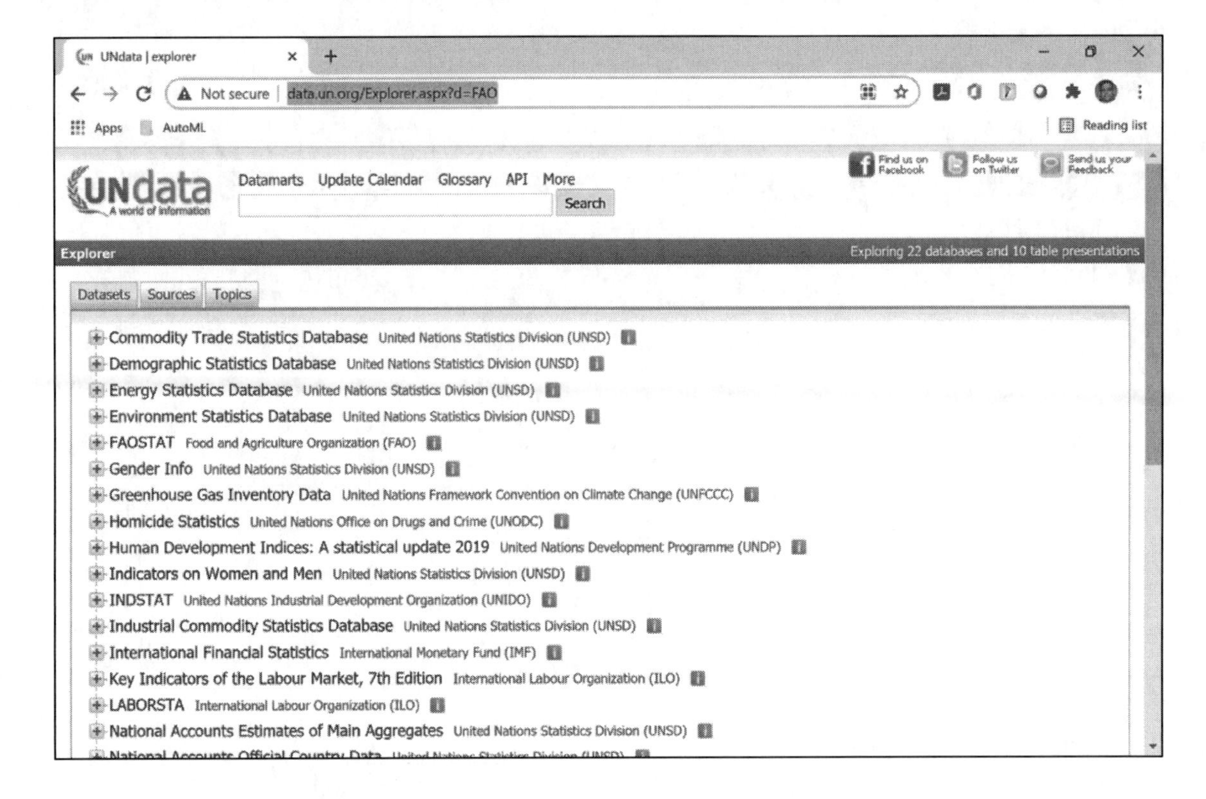

IHS Global Economy Data

IHS[713] provides a global economic database covering 200+ countries. The Data includes: Balance of payments, Cyclical indicators, Finance and financial markets, Government finance, Housing and construction, Output, capacity and capacity utilization, Merchandise trade, National accounts, Labor market, Population, Prices, and Wholesale and retail trade.

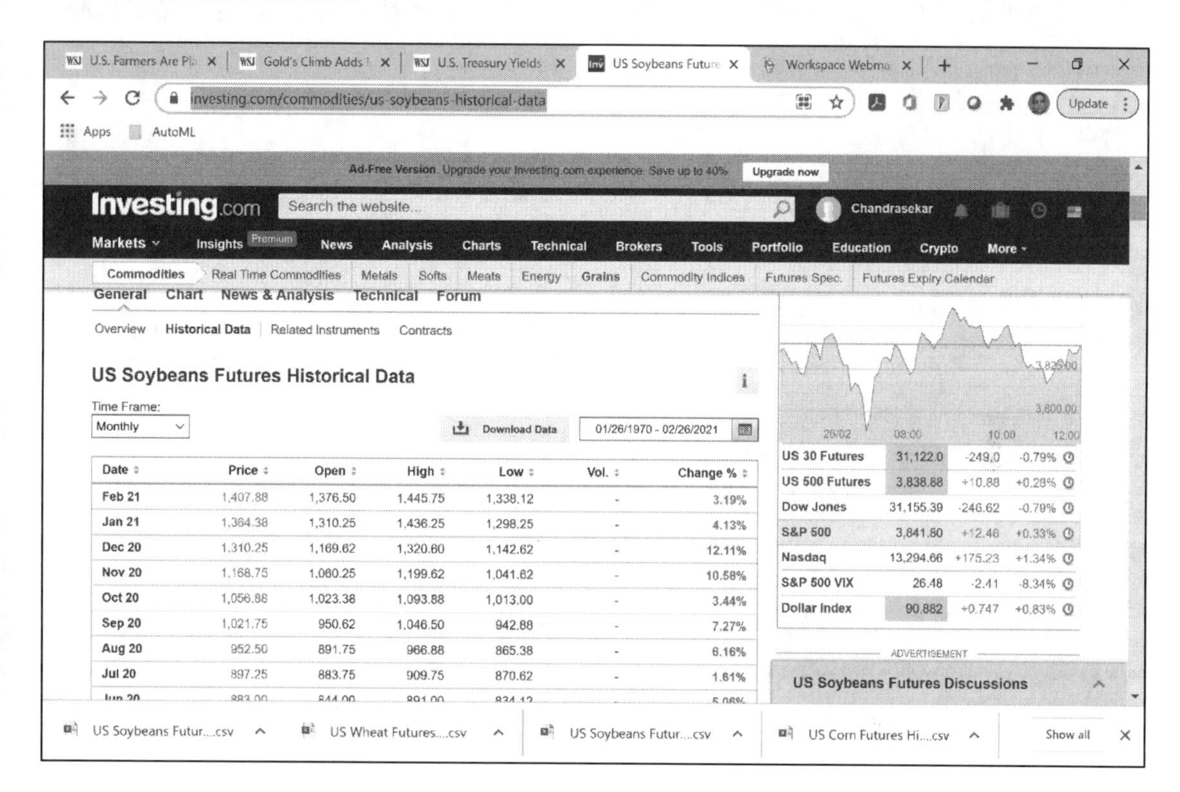

U.S. Commodities Futures Data

Investing.com Future prices[714]

IMF Data Access to Macroeconomic & Financial Data

The International Monetary Fund (IMF)[715] covers the following topics:

- External Sector
- Fiscal Sector
- Financial Sector
- Real Sector
- And Cross Domain.

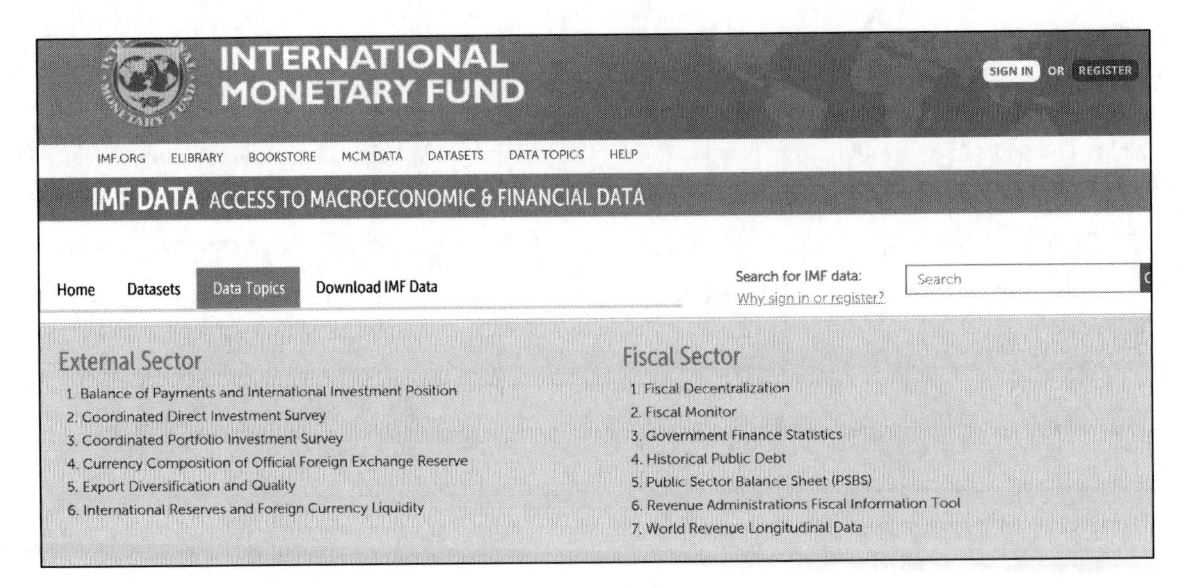

Some of the macro level indicators: Real GDP, percentage change from previous year, CPI, percentage change from previous year, External Trade, and Real Effective exchange rate.

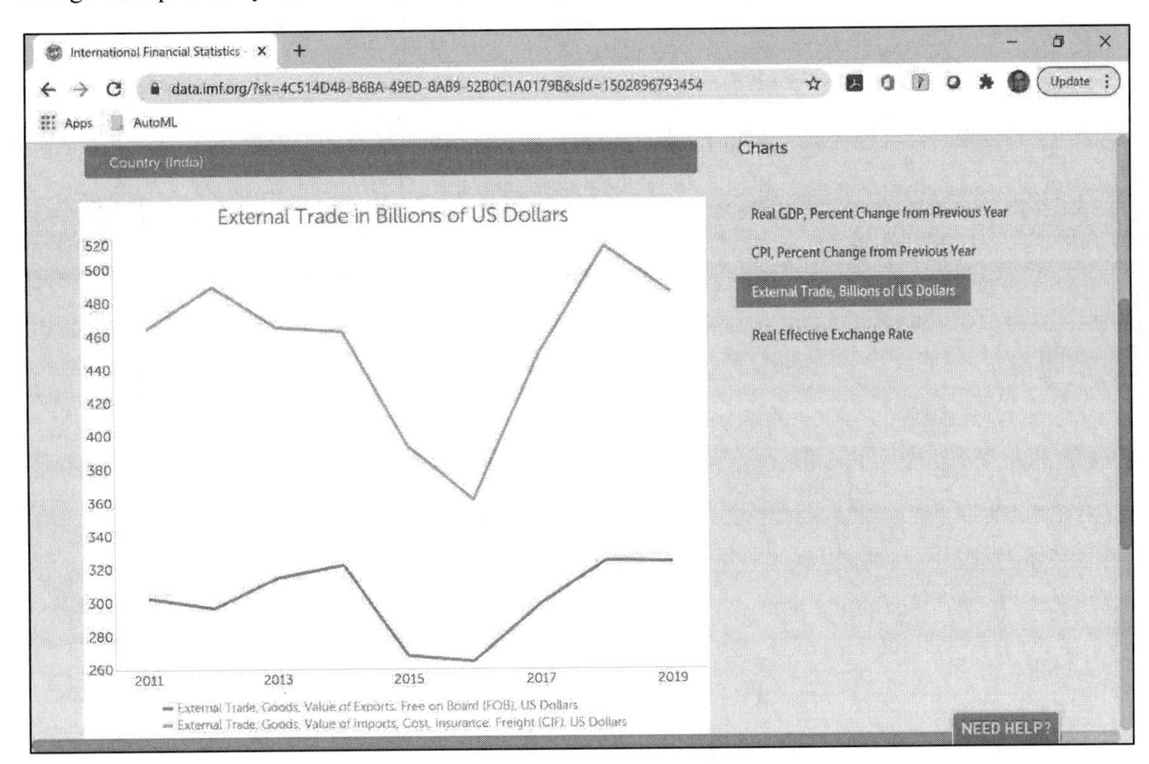

https://data.imf.org/?sk=4C514D48-B6BA-49ED-8AB9-52B0C1A0179B&sId=1502896793454

IMF Country Index Weights

Country Indexes and Weights provide macro level consumption and social behavior of people in term purchase preferences, Food and Non-alcoholic beverages, alcoholic beverages, Clothing, and others.

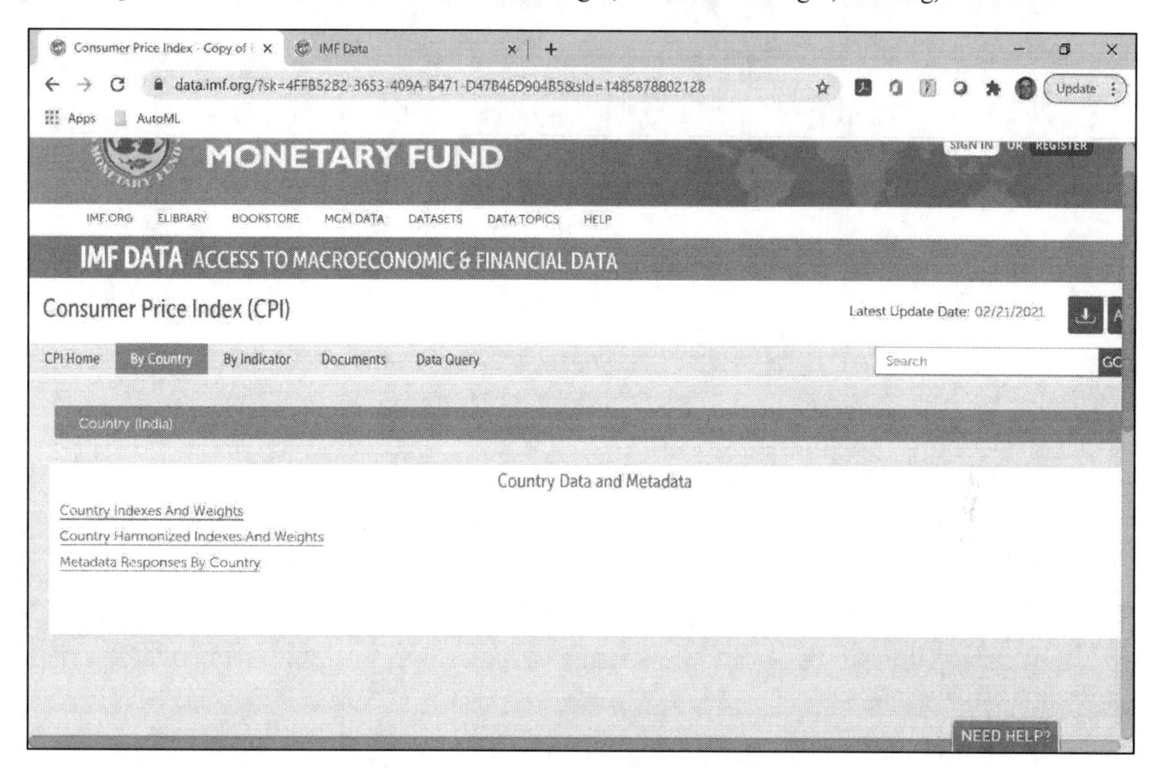

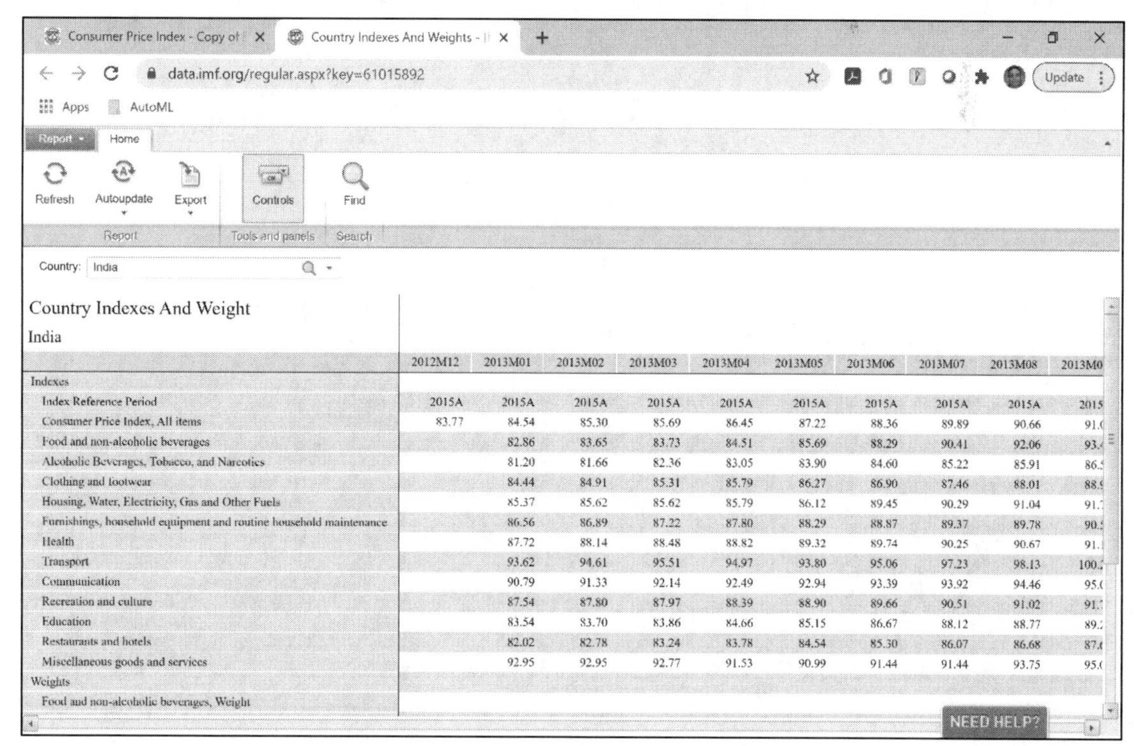

Country Indexes And Weight

India

	2012M12	2013M01	2013M02	2013M03	2013M04	2013M05	2013M06	2013M07	2013M08	2013M0
Indexes										
Index Reference Period	2015A	2015A	2015A	2015A	2015A	2015A	2015A	2015A	2015A	2015
Consumer Price Index, All items	83.77	84.54	85.30	85.69	86.45	87.22	88.36	89.89	90.66	91.
Food and non-alcoholic beverages		82.86	83.65	83.73	84.51	85.69	88.29	90.41	92.06	93.
Alcoholic Beverages, Tobacco, and Narcotics		81.20	81.66	82.36	83.05	83.90	84.60	85.22	85.91	86.
Clothing and footwear		84.44	84.91	85.31	85.79	86.27	86.90	87.46	88.01	88.
Housing, Water, Electricity, Gas and Other Fuels		85.37	85.62	85.62	85.79	86.12	89.45	90.29	91.04	91.
Furnishings, household equipment and routine household maintenance		86.56	86.89	87.22	87.80	88.29	88.87	89.37	89.78	90.
Health		87.72	88.14	88.48	88.82	89.32	89.74	90.25	90.67	91.
Transport		93.62	94.61	95.51	94.97	93.80	95.06	97.23	98.13	100.
Communication		90.79	91.33	92.14	92.49	92.94	93.39	93.92	94.46	95.
Recreation and culture		87.54	87.80	87.97	88.39	88.90	89.66	90.51	91.02	91.
Education		83.54	83.70	83.86	84.66	85.15	86.67	88.12	88.77	89.
Restaurants and hotels		82.02	82.78	83.24	83.78	84.54	85.30	86.07	86.68	87.
Miscellaneous goods and services		92.95	92.95	92.77	91.53	90.99	91.44	91.44	93.75	95.
Weights										
Food and non-alcoholic beverages, Weight										

https://data.imf.org/regular.aspx?key=61015892

https://data.imf.org/?sk=388dfa60-1d26-4ade-b505-a05a558d9a42&sId=1479329328660

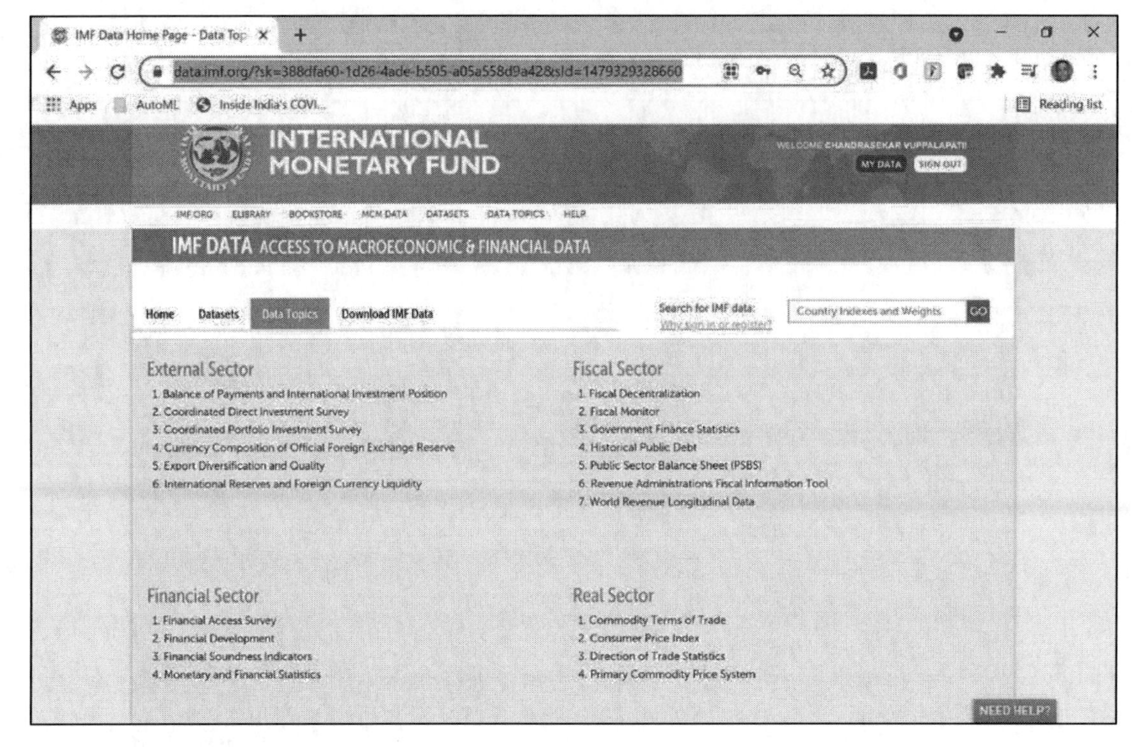

IMF Data Bulk Download

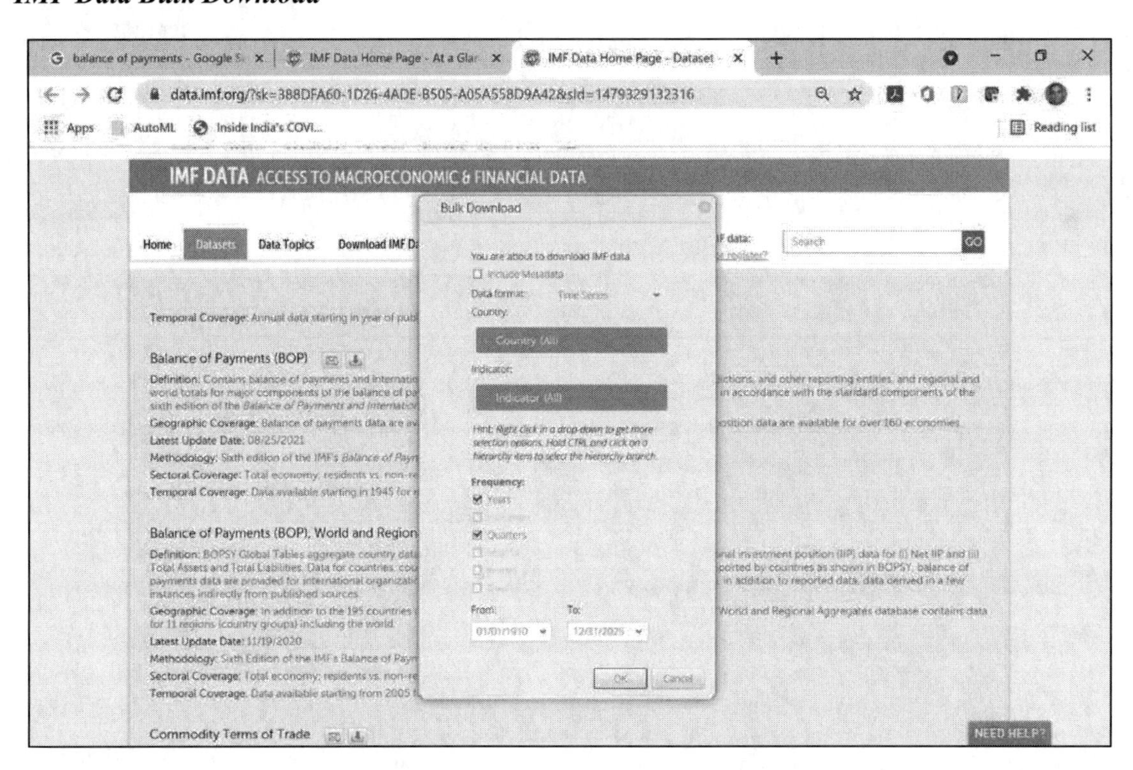

Download macroeconomic bulk data - https://data.imf.org/?sk=388DFA60-1D26-4ADE-B505-A05A558D9A42&sId=1479329132316

https://data.imf.org/?sk=388DFA60-1D26-4ADE-B505-A05A558D9A42&sId=1479329334655

World Bank Data

Macroeconomic world bank data[716] -

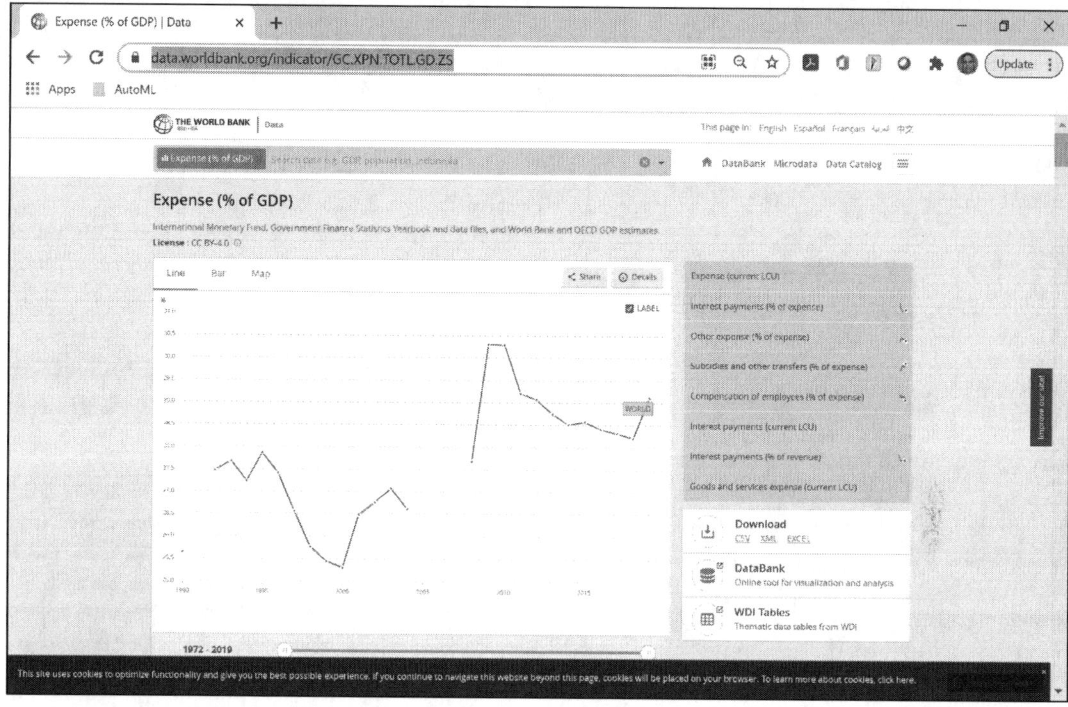

Source: https://data.worldbank.org/indicator/GC.XPN.TOTL.GD.ZS

The World Bank Development Indicators

World Development Indicator tool:[717]

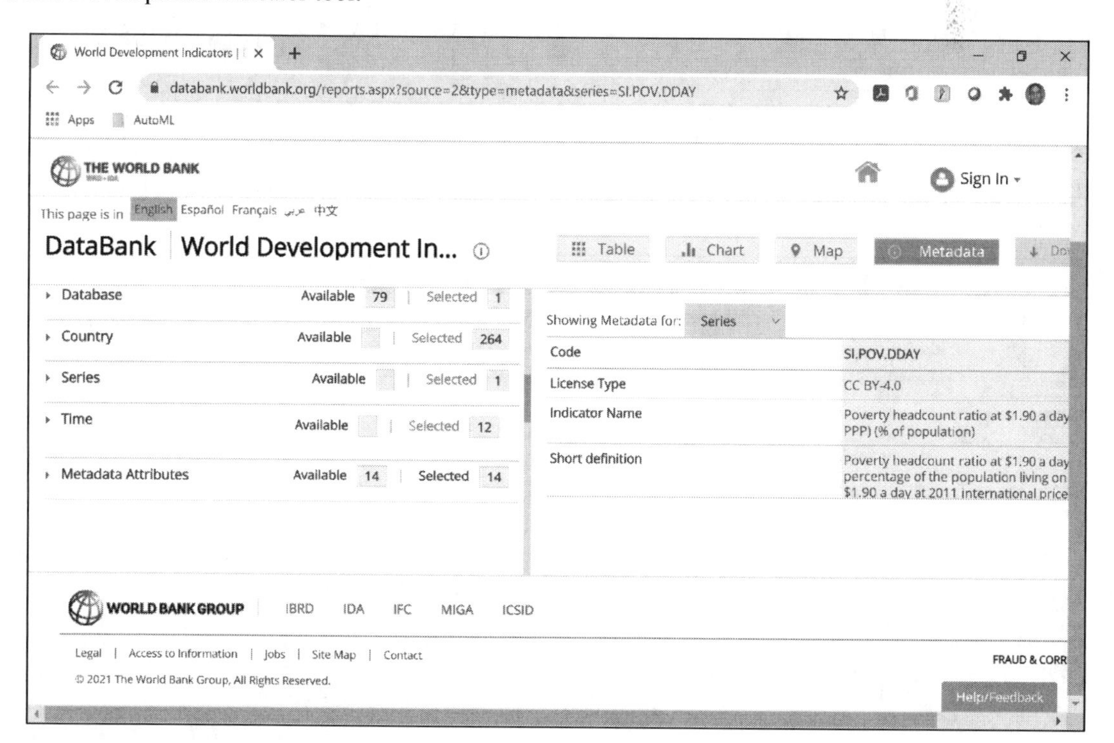

Food and Agriculture Organization of the United Nations

FAOSTAT Data:[718] FAO provides following data domains: Production, Prices, Emission-agriculture, Inputs, Trade, Population, emissions-Land Use, Investment, Macro Economic Data, and others.

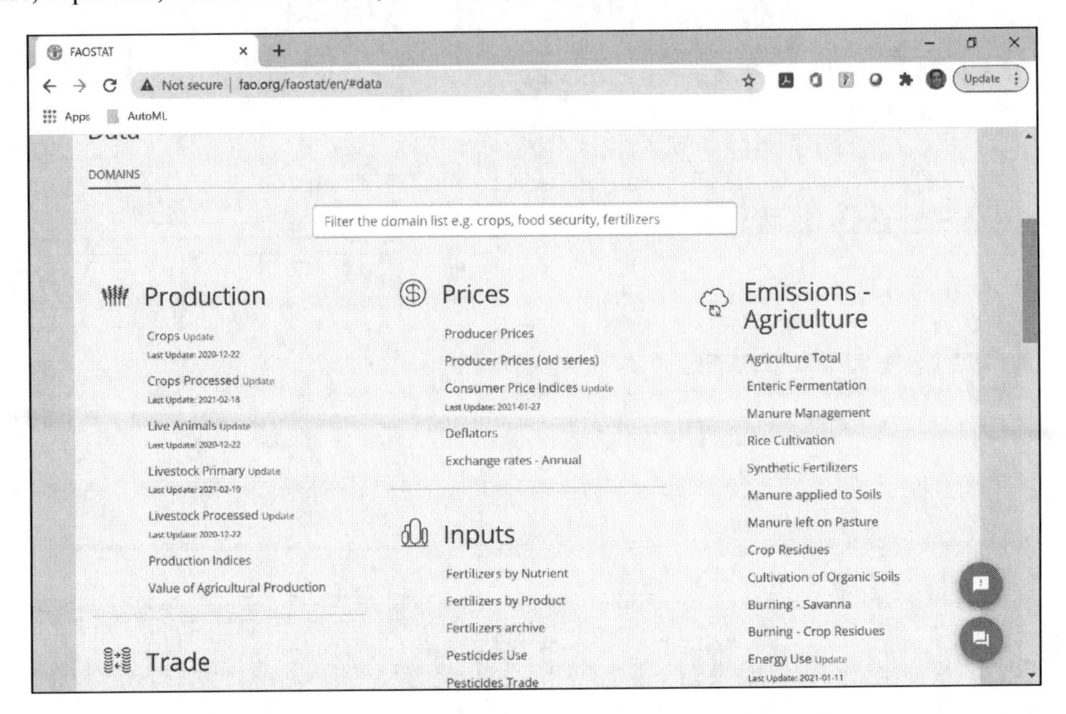

Wheat Data Disappearance and End Stocks

This data product contains statistics on wheat—including the five classes of wheat: hard red winter, hard red spring, soft red winter, white, and durum and rye. Includes data published in the monthly Wheat Outlook

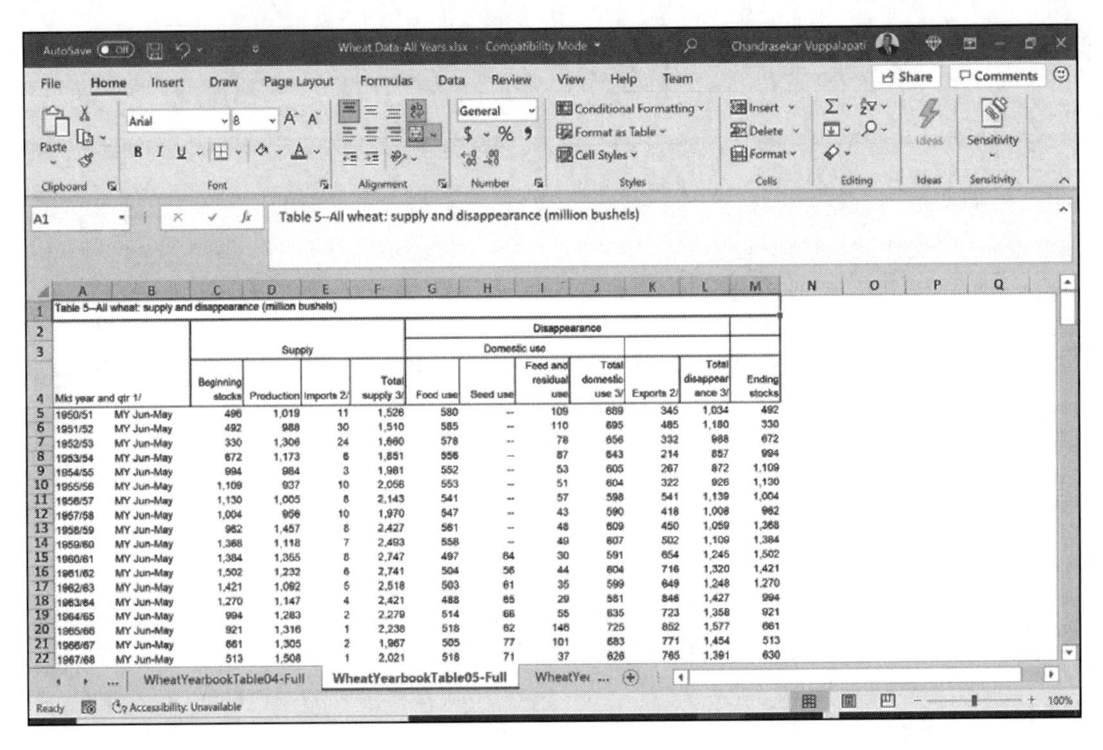

Table 5--All wheat: supply and disappearance (million bushels)

Mkt year and qtr 1/		Supply				Disappearance						
						Domestic use						
		Beginning stocks	Production	Imports 2/	Total supply 3/	Food use	Seed use	Feed and residual use	Total domestic use 3/	Exports 2/	Total disappearance 3/	Ending stocks
1950/51	MY Jun-May	496	1,019	11	1,526	580	--	109	689	345	1,034	492
1951/52	MY Jun-May	492	988	30	1,510	585	--	110	695	485	1,180	330
1952/53	MY Jun-May	330	1,306	24	1,660	578	--	78	656	332	988	672
1953/54	MY Jun-May	672	1,173	6	1,851	556	--	87	643	214	857	994
1954/55	MY Jun-May	994	984	3	1,981	552	--	53	605	267	872	1,109
1955/56	MY Jun-May	1,109	937	10	2,056	553	--	51	604	322	926	1,130
1956/57	MY Jun-May	1,130	1,005	8	2,143	541	--	57	598	541	1,139	1,004
1957/58	MY Jun-May	1,004	956	10	1,970	547	--	43	590	418	1,008	962
1958/59	MY Jun-May	962	1,457	8	2,427	561	--	48	609	450	1,059	1,368
1959/60	MY Jun-May	1,368	1,118	7	2,493	558	--	49	607	502	1,109	1,384
1960/61	MY Jun-May	1,384	1,355	8	2,747	497	64	30	591	654	1,245	1,502
1961/62	MY Jun-May	1,502	1,232	6	2,741	504	56	44	604	716	1,320	1,421
1962/63	MY Jun-May	1,421	1,092	5	2,518	503	61	35	599	649	1,248	1,270
1963/64	MY Jun-May	1,270	1,147	4	2,421	488	65	29	581	846	1,427	994
1964/65	MY Jun-May	994	1,283	2	2,279	514	66	55	635	723	1,358	921
1965/66	MY Jun-May	921	1,316	1	2,238	518	62	146	725	852	1,577	661
1966/67	MY Jun-May	661	1,305	2	1,967	505	77	101	683	771	1,454	513
1967/68	MY Jun-May	513	1,508	1	2,021	518	71	37	626	765	1,391	630

previously the annual Wheat Yearbook. Datasets are monthly, quarterly, and/or annual, depending upon the data series. Most datasets are on a marketing-year basis, but some are calendar year based.[719]

Hard Red Wheat Contracts - https://www.barchart.com/futures/quotes/KEN20/interactive-chart

Poverty Thresholds for 2019 by Size of Family and Number of Related Children Under 18 Years

Size of family unit	Weighted average thresholds	Related children under 18 years								
		None	One	Two	Three	Four	Five	Six	Seven	Eight or more
One person (unrelated individual):	13,011									
Under age 65...	13,300	13,300								
Aged 65 and older ...	12,261	12,261								
Two people:	16,521									
Householder under age 65...	17,196	17,120	17,622							
Householder aged 65 and older..	15,468	15,453	17,555							
Three people...	20,335	19,998	20,578	20,598						
Four people...	26,172	26,370	26,801	25,926	26,017					
Five people...	31,021	31,800	32,263	31,275	30,510	30,044				
Six people...	35,129	36,576	36,721	35,965	35,239	34,161	33,522			
Seven people...	40,016	42,085	42,348	41,442	40,811	39,635	38,262	36,757		
Eight people...	44,461	47,069	47,485	46,630	45,881	44,818	43,470	42,066	41,709	
Nine people or more...	52,875	56,621	56,895	56,139	55,503	54,460	53,025	51,727	51,406	49,426

Source: U.S. Census Bureau.

Rice Production Manual

Rice Production manual provides comprehensive view of rice & agriculture.[720]

[622] The 17 Goals of Sustainable Development - https://sdgs.un.org/goals

[623] Sudan - Food Security Monitoring System, http://fsis.sd/SD/EN/FoodSecurity/Monitoring/

[624] Handbook for Defining and Setting up a Food Security Information and Early Warning System (FSIEWS) - https://www.fao.org/3/X8622E/x8622e00.htm#TopOfPage

[625] Food Security Information and Knowledge Sharing System – Sudan - https://www.fao.org/family-farming/detail/en/c/450039/

[625] Food Security (FSQ) - https://wwwn.cdc.gov/nchs/data/nhanes/2019-2020/questionnaires/FSQ_Family_K.pdf

[627] Oil Price Swings - https://www.macrotrends.net/1369/crude-oil-price-history-chart

[628] Oil futures volatility and the economy - https://www.eurekalert.org/news-releases/843803

[629] Michael Assous (2013) Irving Fisher's debt deflation analysis: From the Purchasing Power of Money (1911) to the Debt-deflation Theory of the Great Depression (1933). The European Journal of the History of Economic Thought, 20:2, 305-322, DOI: 10.1080/09672567.2012.762936

[630] Debt Deflation - https://www.investopedia.com/terms/d/debtdeflation.asp

[631] What Happens During Recessions, Crunches and Busts? - https://www.imf.org/external/pubs/ft/wp/2008/wp08274.pdf
[632] Wheat Data - https://www.ers.usda.gov/data-products/wheat-data/
[633] USAID – PL480 - https://history.state.gov/milestones/1961-1968/pl-480
[634] The 17 Goals - https://sdgs.un.org/goals
[635] The statistical distributions of commodity prices in both real and nominal terms - http://www.fao.org/3/y4344e/y4344e0i.htm
[636] Source: https://www.census.gov/data/tables/time-series/demo/income-poverty/historical-poverty-thresholds.html
[637] NASS Online Survey - https://www.agcounts.usda.gov/static/cawi/layouts/cawi/breeze/index.html
[638] World Agricultural Production - https://apps.fas.usda.gov/psdonline/circulars/production.pdf
[639] Commodity Cost and Returns - https://www.ers.usda.gov/data-products/commodity-costs-and-returns/
[640] Fertilizer Use and Price- https://www.ers.usda.gov/data-products/fertilizer-use-and-price/
[641] The World Bank Development Indicator - https://databank.worldbank.org/reports.aspx?source=2&type=metadata&series=SI.POV.DDAY
[642] FAPSTAT - http://www.fao.org/faostat/en/#data
[643] The USDA Conversion Factors - https://www.ers.usda.gov/data-products/wheat-data/documentation/
[644] About NDDB - https://www.nddb.coop/
[645] Milk Production in India - https://www.nddb.coop/information/stats/milkprodindia
[646] NDDB Daily Milk Data - https://www.nddb.coop/sites/default/files/statistics/Mp%20India-ENG-2019.pdf
[647] AI Readiness - https://www.mckinsey.com/~/media/McKinsey/Featured%20Insights/Artificial%20Intelligence/Notes%20from%20the%20frontier%20Modeling%20the%20impact%20of%20AI%20on%20the%20world%20economy/MGI-Notes-from-the-AI-frontier-Modeling-the-impact-of-AI-on-the-world-economy-September-2018.ashx
[648] AI Readiness - https://www.mckinsey.com/~/media/McKinsey/Featured%20Insights/Artificial%20Intelligence/Notes%20from%20the%20frontier%20Modeling%20the%20impact%20of%20AI%20on%20the%20world%20economy/MGI-Notes-from-the-AI-frontier-Modeling-the-impact-of-AI-on-the-world-economy-September-2018.ashx
[649] GHCNm - https://www.ncei.noaa.gov/products/land-based-station/global-historical-climatology-network-monthly
[650] GHCNm - https://climatedata-catalogue.wmo.int/explore
[651] World Metrological Data - https://climatedata-catalogue.wmo.int/beta/datasets-brazil
[652] U.S. Unemployment data - https://data.bls.gov/timeseries/LNS14000000?years_option=all_years
[653] The World Bank Data - https://www.worldbank.org/en/topic/macroeconomics
[654] The World Bank Development Indicator - https://databank.worldbank.org/reports.aspx?source=2&type=metadata&series=SI.POV.DDAY
[655] FAPSTAT - http://www.fao.org/faostat/en/#data
[656] CPI: Average price data - https://www.bls.gov/cpi/factsheets/average-prices.htm
[657] BLS – One-Screen Search - https://data.bls.gov/PDQWeb/ap
[658] Data.gov - https://www.data.gov/food/
[659] Dairy - https://www.ers.usda.gov/webdocs/DataFiles/50472/dymfg.xls?v=739.7
[660] Short Term Energy Look - https://www.eia.gov/outlooks/steo/realprices/
[661] Dollar Bushel Converter - https://www.agric.gov.ab.ca/app19/calc/crop/bushel2tonnevalue.jsp
[662] Storm Events Database - https://www.ncdc.noaa.gov/stormevents/
[663] Storm Data Export Format - https://www1.ncdc.noaa.gov/pub/data/swdi/stormevents/csvfiles/Storm-Data-Export-Format.pdf
[664] NATIONAL WEATHER SERVICE INSTRUCTION 10-1605 STORM DATA PREPARATION https://www.nws.noaa.gov/directives/sym/pd01016005curr.pdf
[665] Consumer Price Index, 1913- - https://www.minneapolisfed.org/about-us/monetary-policy/inflation-calculator/consumer-price-index-1913
[666] Inflation Calculator - https://www.minneapolisfed.org/about-us/monetary-policy/inflation-calculator
[667] Oil Price Swings - https://www.macrotrends.net/1369/crude-oil-price-history-chart
[668] Oil futures volatility and the economy - https://www.eurekalert.org/news-releases/843803
[669] Michael Assous (2013) Irving Fisher's debt deflation analysis: From the Purchasing Power of Money (1911) to the Debt-deflation Theory of the Great Depression (1933). The European Journal of the History of Economic Thought, 20:2, 305-322, DOI: 10.1080/09672567.2012.762936
[670] Debt Deflation - https://www.investopedia.com/terms/d/debtdeflation.asp
[671] What Happens During Recessions, Crunches and Busts? - https://www.imf.org/external/pubs/ft/wp/2008/wp08274.pdf
[672] Oil futures volatility and the economy - https://www.eurekalert.org/news-releases/843803
[673] Crude Oil Prices - 70 Year Historical Chart - https://www.macrotrends.net/1369/crude-oil-price-history-chart
[674] U.S. Recession - https://fred.stlouisfed.org/series/DTWEXBGS
[675] The Great Recession - https://www.federalreservehistory.org/essays/great-recession-of-200709
[676] Business Cycle Dating - https://www.cepr.net/clearing-up-some-facts-about-the-depression-of-1946/
[677] The Great Recession - https://www.federalreservehistory.org/essays/great-recession-of-200709
[678] The Great Recession - https://www.federalreservehistory.org/essays/great-recession-of-200709
[679] 2001 Recession - https://files.stlouisfed.org/files/htdocs/publications/review/03/09/Kliesen.pdf
[680] Business Cycle Dating - https://www.cepr.net/clearing-up-some-facts-about-the-depression-of-1946/
[681] The Current Economic Recession: How Long, How Deep, and How Different From the Past? - https://web.archive.org/web/20091010043009/http://fpc.state.gov/documents/organization/7962.pdf

[682] Deep Recessions, Fast Recoveries, and Financial Crises: Evidence from the American Record - https://www.nber.org/system/files/working_papers/w18194/w18194.pdf
[683] Business Cycle Dating - https://www.cepr.net/clearing-up-some-facts-about-the-depression-of-1946/
[684] Recession of 1981–82 - https://www.federalreservehistory.org/essays/recession-of-1981-82
[685] Unemployment continued to rise in 1982 as recession deepened - https://www.bls.gov/opub/mlr/1983/02/art1full.pdf
[686] Mysteries of monetary policy - https://www.aei.org/articles/mysteries-monetary-policy/
[687] The 2001 Recession: How Was It Different and What Developments May Have Caused It? - https://files.stlouisfed.org/files/htdocs/publications/review/03/09/Kliesen.pdf
[688] Oil Shock of 1973–74 - https://www.federalreservehistory.org/essays/oil-shock-of-1973-74
[689] Deep Recessions, Fast Recoveries, and Financial Crises: Evidence from the American Record - https://www.nber.org/system/files/working_papers/w18194/w18194.pdf
[690] Deep Recessions, Fast Recoveries, and Financial Crises: Evidence from the American Record - https://www.nber.org/system/files/working_papers/w18194/w18194.pdf
[691] 1957-1958 Pandemic (H2N2 virus) - https://www.cdc.gov/flu/pandemic-resources/1957-1958-pandemic.html
[692] The 1957-58 Recession in World Trade - https://fraser.stlouisfed.org/files/docs/publications/FRB/pages/1955-1959/14330_1955-1959.pdf
[693] Unemployment rate and long-term unemployment rate, January 1948–December 2011, seasonally adjusted (in percent) - https://www.bls.gov/spotlight/2012/recession/data_cps_unemp_1948.htm
[694] Deep Recessions, Fast Recoveries, and Financial Crises: Evidence from the American Record - https://www.nber.org/papers/w18194
[695] Business Cycle Dating - https://www.cepr.net/clearing-up-some-facts-about-the-depression-of-1946/
[696] The Current Economic Recession: How Long, How Deep, and How Different From the Past? - https://web.archive.org/web/20091010043009/http://fpc.state.gov/documents/organization/7962.pdf
[697] Labor Force Statistics from the Current Population Survey 1948-2021 - https://data.bls.gov/timeseries/LNS14000000?years_option=all_years
[698] A Case Study: The 1948-1949 Recessions - https://www.nber.org/system/files/chapters/c2798/c2798.pdf
[699] A Review of Past Recessions - https://www.investopedia.com/articles/economics/08/past-recessions.asp
[700] Unemployment rate and long-term unemployment rate, January 1948–December 2011, seasonally adjusted (in percent) - https://www.bls.gov/spotlight/2012/recession/data_cps_unemp_1948.htm
[701] The Current Economic Recession: How Long, How Deep, and How Different From the Past? - https://web.archive.org/web/20091010043009/http://fpc.state.gov/documents/organization/7962.pdf
[702] A Review of Past Recessions - https://www.investopedia.com/articles/economics/08/past-recessions.asp
[703] Business Cycle Dating - https://www.cepr.net/clearing-up-some-facts-about-the-depression-of-1946/
[704] Clearing Up Some Facts About the Depression of 1946 - https://www.cepr.net/clearing-up-some-facts-about-the-depression-of-1946/
[705] ANNUAL ESTIMATES OF UNEMPLOYMENT IN THE UNITED STATES, 1900-1954 - https://www.nber.org/system/files/chapters/c2644/c2644.pdf
[706] How many recessions you've actually lived through and what happened in every one - https://www.cnbc.com/2020/04/09/what-happened-in-every-us-recession-since-the-great-depression.html
[707] A Review of Past Recessions - https://www.investopedia.com/articles/economics/08/past-recessions.asp
[708] Recession of 1937–38 - https://www.federalreservehistory.org/essays/recession-of-1937-38
[709] U.S. Unemployment data - https://data.bls.gov/timeseries/LNS14000000?years_option=all_years
[710] Macro-economic aggregates - https://www.rbi.org.in/scripts/PublicationsView.aspx?id=20406
[711] Components of GDP - https://www.rbi.org.in/scripts/PublicationsView.aspx?id=20409
[712] Components of Gross Value - https://www.rbi.org.in/scripts/PublicationsView.aspx?id=20408
[713] IHS Global Economy Data - https://ihsmarkit.com/products/global-economic-data.html
[714] Investing.com - https://www.investing.com/commodities/us-soybeans-historical-data
[715] IMF - https://data.imf.org/?sk=388DFA60-1D26-4ADE-B505-A05A558D9A42&sId=1479329328660
[716] The World Bank Data - https://www.worldbank.org/en/topic/macroeconomics
[717] The World Bank Development Indicator - https://databank.worldbank.org/reports.aspx?source=2&type=metadata&series=SI.POV.DDAY
[718] FAPSTAT - http://www.fao.org/faostat/en/#data
[719] Wheat Data - https://www.ers.usda.gov/data-products/wheat-data/
[720] Rice Production Manual - http://rice.ucanr.edu/files/288581.pdf

Index

A

ADASYN 7, 76, 113, 117, 124, 125, 128, 129
Advanced Very High-Resolution Radiometer 324
affordability 77, 147, 148, 177, 454
Afghanistan 141, 158, 207, 210, 213–226, 229, 231, 232, 234, 238–241, 244–246, 257, 263, 264, 266, 267
AIC 22, 30, 51, 56, 57, 192, 201, 296, 298, 431, 436, 439, 446, 449, 469
Anhydrous ammonia 412, 419
APU0000709112 8, 43, 73–75
APU0000710212 42, 44, 75
Arable land 79, 80, 131, 146, 212, 222, 228, 229, 239–241, 243, 249, 250, 255, 257, 266, 267, 275
Artificial Intelligence 76, 307, 328, 329, 473
Augmented Dickey Fuller 19–23, 29–31, 53, 56, 224, 254, 283, 295, 296, 298, 430, 436, 438, 446, 449
autocorrelation 3, 15–19, 24, 26, 51, 52, 59, 72, 74, 293, 300, 434, 435, 440, 445, 451
AVHRR 324–326, 335

B

Brightness Temperature 114, 324, 325, 335, 366, 391
bushels 113, 117, 122–124, 126–128, 331, 332, 334, 346, 368, 459

C

cereals 79, 146, 168, 371
Change in Equilibrium 154, 155
Cheddar Cheese 8–11, 13, 29, 33, 37–39, 42–46, 49, 51, 56–58, 68, 72, 74, 75
Chicago Mercantile Exchange 10
Climate shocks 154
Climate-Smart sensors 15
climatology 6, 325, 335, 409
CMIP 313, 326–331, 370, 378, 379, 397, 399
Composite 4, 324, 325, 335
Composite Index 325
conflicts 6, 138, 139, 156, 202, 212, 214, 220, 278, 306, 313, 314, 330, 402
conundrum 333, 473
COVID-19 76, 129, 130, 136, 139, 154, 161, 207, 213, 214, 224, 246, 253, 263, 272, 273, 313, 402, 407, 415, 473
Cow Necklace 15, 390, 393, 401
CPI 3, 6–8, 15, 36, 37, 41–47, 49, 61, 62, 67, 73, 75, 104, 148, 159, 160, 168, 169, 208, 226, 227, 229, 247, 254, 256, 257, 266, 267, 273, 275, 279, 281, 283, 286, 287, 290–292, 296, 298–304, 406, 441, 452
CTOT 273, 274, 306
cycles 5, 14, 31, 56, 151, 161, 207, 208, 271, 314, 399, 414, 467, 468

D

DAP 419, 423, 425, 426, 428, 429, 431, 433, 444, 449–452, 454, 455, 457–459, 461, 463
Data Enrichment 76, 86, 129
Dekad 325
demographics 329
Dietary 37, 135, 138, 146, 147, 185, 209, 217, 286, 313
disastrous 224, 265
disrupted 126, 156, 214, 220, 313, 332, 361
disruption 75, 138, 139, 160, 202, 210, 212, 214, 217, 263, 275, 282, 306, 321, 325, 413, 415, 417, 423, 455, 458
domino effect 138, 306
Drift 20, 27, 29–31
drought 5, 113, 114, 132, 138, 139, 148, 202, 210, 213, 214, 220, 222, 224, 245, 250, 306, 313, 314, 318, 320–322, 324–327, 332–336, 338, 339, 342, 346, 351, 356, 390, 397, 399, 400, 402

E

EDA 5–7, 103, 266, 267
Education 75, 132, 138, 147, 153, 161, 167, 175, 186, 188, 189, 191–194, 196, 198–201, 214, 265, 272, 275, 306, 311, 330, 331, 399–401
EIA 15, 49, 74, 75, 409, 467, 468
End hunger 207
Endogenous 56, 140, 141, 283, 310, 314, 402, 412, 464
enhance food security 77, 371
Exogenous 5, 7, 21, 37, 56, 102, 140, 141, 214, 313, 412
expensive 10, 77, 98, 159, 163, 276, 334, 365, 403, 406, 467
Exploratory Data Analysis 3, 5, 7, 62, 103, 287, 429
export restrictions 235, 455, 458

F

farmer 8, 9, 15, 175, 224, 322, 332, 360, 399, 400
Fertilizer 80, 86–89, 91, 93, 94, 96, 97, 99–101, 131, 138, 173, 202, 246, 306, 308, 314, 323, 366, 402, 407, 411–414, 417–419, 421–423, 426, 428, 429, 444, 449, 454, 455, 458, 459, 464, 468, 469
Fertilizer Consumption 80, 86, 411
Fertilizer costs 88, 131, 402, 423, 444, 455, 464, 469
FFPI 168, 208
FIES-SM 149–151
Food inflation 74, 168, 185, 209, 210, 246, 265, 271, 275, 279–281, 306, 310, 311, 321, 324, 421, 422, 454

food insecurity 34, 77, 78, 114, 130, 131, 135–141, 143–145, 149–154, 161, 162, 164–169, 171, 173, 175–181, 185, 187–192, 194–200, 202, 206–210, 213, 214, 217, 218, 222, 224, 229, 230, 246, 254, 258, 263, 269, 271–274, 280, 309, 311–313, 335, 365, 379, 390, 398, 405, 411, 421, 464, 467
Food Security Bell Curve 135, 149, 152, 166, 202, 264, 272, 389, 464
food-fuel 6
Foreign Currency Reserves 247
Futures 157, 158, 207, 278, 323, 324, 338, 387, 388, 399–401, 454, 460

G

gasoline 15, 77, 405–407, 409–411, 468
GDP 7, 80, 104, 140, 148, 153, 161, 162, 166, 167, 174, 176, 178–181, 186, 188, 191–194, 196, 200, 201, 208, 212, 216, 226–229, 232, 235, 237, 238, 241, 243–246, 248, 251, 254–257, 259, 261, 263, 266, 267, 271, 274, 275, 403
GDP per capita 80, 140, 153, 162, 176, 178–180, 186, 188, 191–194, 196, 200, 201, 208, 246
germination stage 317
GFSI 135, 148, 152, 153, 170, 177–179, 181–183, 202, 207, 209, 267
Giffen Goods 158
gold 105, 153, 166, 178–181, 186, 188, 191, 194, 196, 200, 208, 212, 226, 229, 232, 241, 243, 244, 254, 257, 259, 261, 266, 274, 415
Grain 74, 79, 86, 113, 156, 157, 173, 207, 213, 224, 271, 275, 279–281, 283, 284, 288, 290, 303, 310, 311, 315, 317, 318, 321, 324, 332, 334, 356, 357, 359, 366, 371, 403, 418
Granger's Causality Test 33, 297, 298, 437, 438, 448, 449

H

Hard Red Winter Wheat 113, 334, 428
Heatwave 313, 319, 358, 361–366, 371, 380, 384, 385, 387, 388, 390, 397, 400, 402, 405, 467
Hot Encoding 78, 85, 86
hunger 78, 130, 135, 137, 149, 170, 171, 197, 199, 207, 209, 213, 214, 216, 231, 258, 266, 388

I

IMF 131, 163, 166, 168, 207, 208, 247, 267, 417, 468
Imputation 5, 7, 44, 76, 87, 88, 91–93, 95, 97, 98, 100, 102, 129, 131, 180, 233, 240, 260, 374, 386
India 77, 141, 157, 165, 167, 174, 176, 190, 198, 210, 222, 265, 319–321, 329–331, 357–367, 369–373, 375, 380, 384, 385, 387, 388, 390, 398–401, 403, 404, 413, 418, 422, 429, 467
IPCC 321, 327, 328, 362, 384, 388, 399

J

JOSEPH R. BIDEN JR 76

K

Kansas Drought 333
KDD 3
KPSS 19, 22–24, 34, 53, 54, 74

L

La Niña 325, 399
LAG 5, 15–19, 21–23, 56, 60, 283, 293, 295, 300, 430, 437, 439–441, 449, 452
Livestock 77, 145, 156, 167, 176, 215, 216, 224, 248, 251, 322, 371, 388
LOCF 44, 88

M

Machine Learning 6, 40, 76, 78, 86, 95, 114, 124, 133, 135, 137, 141, 149, 150, 152, 154, 158, 160, 177, 191, 194, 199, 202, 210, 212, 213, 264, 271, 278, 281, 303, 310, 313, 338–340, 371, 380, 384, 390, 397, 402, 423, 444, 454, 460, 464
maturity 107, 131, 315, 318
METOP 326
MICE 7, 76, 87, 98, 100, 101, 103, 129, 131
Modern Test Theory 150
MODIS 366, 372, 401
multivariate panel regressions 151

N

natural disasters 214, 246, 275
NCDEX 387, 388, 401
NDM 11, 12, 57, 58, 66, 67, 75
NDVI 7, 114, 325, 335
NOCB 88
Nonfat Dry Milk 11, 57, 58
Normalized Difference Vegetation Index 7, 114, 313, 324, 325, 335, 356, 397
Null Hypothesis 21–24, 30, 31, 33, 74, 87, 283, 290, 295–297, 430, 431, 438, 448
Nutrition gap 152, 153, 212, 213, 219

O

outbreak 213

P

PACF 18, 19, 40, 51, 52
Paradox 158
PCPS 158
Phenological Stages 313, 316, 397
plot_acf 17, 40, 51, 283, 293, 434, 445
Population Models 269
Potash 91, 94, 96, 100, 173, 412, 413, 417, 419, 425, 450, 458, 468
PPCA 87, 131
Prevalence 34, 36, 131, 135, 137, 140, 141, 146, 149–153, 161, 162, 164–166, 168, 171, 175–181, 185, 187–192, 194–202, 206, 209–212, 214, 217, 224–226, 228–232, 235, 237, 238, 241–246, 254–259, 261, 263, 264, 266, 269, 271–274, 311, 390, 398, 405
price hikes 324, 459
probabilistic 87
Producer Price Index (PPI) 159, 275, 276
prognosis 263, 319, 336
Proxy 6, 77, 114, 174, 272, 325, 335, 356, 367, 391

R

random walks 5
Rasch Model 150, 207
Representative Concentration Pathway (RCP) 328, 404
resolution 4, 37, 45, 103, 104, 159, 276, 324
Response Theory 149, 150, 207
Rice Futures 400
ripening 317, 318, 321, 356
Russian Federation 142

S

satellite 7, 114, 132, 324, 334, 335, 341, 342, 390, 391, 393, 395, 396, 400
SDG 214, 258
SDR 143, 208
serial correlation 15, 18, 51, 293, 295, 434, 445
Shocks 6, 135, 136, 139, 144, 149, 154, 156, 160, 166, 167, 173, 202, 212, 213, 218, 220, 222, 224, 246, 247, 250, 272, 273, 275, 276, 278, 281, 303, 310, 323, 324, 399, 402–405, 409, 411, 421, 454, 460, 464, 467, 468
SMN 4, 114, 115, 119, 313, 324, 325, 335, 339, 341, 342, 347, 350, 352, 355, 356, 367, 368, 372, 390, 397
SMOTE 7, 76, 113, 114, 117, 121, 125, 128, 129
Spring wheat 315, 316, 318, 351, 356, 357
Sri Lanka 139, 140, 141, 206, 207, 210, 213, 246, 247, 249, 250, 252, 253, 257–259, 263, 264, 266, 267
SSP 313, 328–331, 378, 397
STAR 4, 73, 114, 118–120, 132, 313, 324, 341, 390, 397, 399–401
stationary 3, 5, 17, 19–25, 27–32, 34, 51, 53, 54, 74, 293–296, 416, 423, 430, 431, 435–437, 445–447, 469
Status Quo 152, 153, 212, 213
stochastic 27, 28, 31, 290
Sustainable 77, 137, 139, 142, 150, 202, 207, 211, 214, 247, 266, 267, 271, 329, 331, 396, 414, 424
sustainable agriculture 207
sustainable food 77, 139, 142

T

Temperature Condition index 114, 313, 324, 325, 335, 336, 351, 356, 391, 397
THE WHITE HOUSE 76, 129, 156
the World Food Program (WFP) 163, 208, 209, 213, 266
Total reserves 153, 166, 178–181, 186, 188, 191–194, 196, 200, 201, 208, 212, 217, 218, 226, 232, 241, 243, 244, 254, 257, 259, 261, 266
trade restrictions 218, 263
Transformation 17, 27, 34, 53, 60, 61, 87, 119, 175, 209, 247, 293, 294, 300, 301, 435, 441, 445, 446, 452
transportation 140, 143, 170, 209, 246, 275, 324, 405, 407, 419
trends 5, 15, 28, 48, 56, 67, 72, 74, 77, 88, 131, 158, 160, 171, 303, 329, 343, 407, 461

U

undernourishment 146, 153, 210, 211, 224–226, 228, 229, 231, 232, 235, 237–239, 241–246, 254–257, 259, 261, 263, 264, 266, 269, 390
Unit Root 20–22, 27, 74, 290, 295, 296, 430, 431
USD 105, 106, 142, 163, 164, 170, 208, 216, 217, 246, 404, 427, 428
USDA 3, 8, 20, 24, 29, 38, 39, 48, 49, 50, 67, 73–75, 77, 86, 131, 132, 136, 161, 173, 175, 206–209, 219, 266, 272, 311–313, 331, 371, 398–401, 412, 414, 415, 458, 467–469
Uttar Pradesh 319, 329, 357, 359, 361, 362, 365–368, 371, 379, 389

V

VAR 3, 11, 26, 33, 36, 37, 55–57, 59, 60, 62, 64, 66, 67, 72, 90, 116, 184, 185, 231, 258, 294, 297–300, 303, 343, 426, 437–440, 445, 448, 449, 450, 452, 459
Vegetation Condition index 114, 313, 324, 325, 335, 351, 356, 367, 397

W

wars 6, 135, 139, 156, 202, 276, 278, 314, 402, 413
water stress 250, 3188
weather patterns 319, 366
Wheat Futures 157, 207, 323, 324, 358, 387, 388, 399, 401
Wholesale Price Index 159, 276

Y

Yogurt 7, 9, 11, –13, 33, 37, 39, 41, 43–46, 49, 57, 58, 66–68, 75